The Art of Electronics: The x-Chapters

The Art of Electronics: The x-Chapters expands on topics introduced in the best selling third edition of *The Art of Electronics*, completing the broad discussions begun in the latter. In addition to covering more advanced materials relevant to its companion, *The x-Chapters* also includes extensive treatment of many topics in electronics that are particularly novel, important, or just exotic and intriguing. You'll find here techniques and circuits that are available nowhere else!

Within this abundance of essential material you can find topics as diverse as:

- *Real* components, for example: capacitors (loss, ESR, dielectrics, frequency dependence,...); resistors (transient power, excess noise, nonlinearity, ESL, speedup tricks,...); wire (skin effect, twisted pairs, shielding, ...); inductors (magnetic materials, core loss, gaps, ...); diodes (loss, reverse recovery, step-recovery,...).
- Specialized tables, such as high-speed VFB and CFB op-amps, gate and bridge drivers, high-side switches and current-sense ICs, and more.
- High-voltage low-capacitance current sources; bipolarity current mirrors.
- Precision high-voltage amplifiers.
- High-voltage pulsed energy, HV reversible polarity, three-state HV pulser, SiC.
- JFETs: a guided tour, and a closer look.
- Power MOSFETs as linear elements, with many examples.
- Silicon photomultipliers and fast LED pulsers.
- Transimpedance amplifiers – bandwidth, stability, and a 7-decade *linear* TIA.
- Rail-to-rail op-amps: gremlins are lurking!
- Current-feedback amplifiers demystified.
- Low-noise ultra-isolated power.
- Anatomy of a counterfeit iPhone charger.
- BJT amplifier distortion – a deep exploration.
- Sending power on a beam of light.
- PWM for dc motors – a myth demolished.
- Transient voltage protection and transient thermal response.

The x-Chapters can be thought of as the missing pieces of *The Art of Electronics*, to be used either as its complement, or as a direct route to exploring some of the most exciting and oft-overlooked topics in advanced electronic engineering. This enticing spread of electronics wisdom and expertise will be an invaluable addition to the library of any student, researcher, or practitioner with even a passing interest in the design and analysis of electronic circuits and instruments.

Paul Horowitz is Professor of Physics and of Electrical Engineering, emeritus, at Harvard University, where in 1974 he originated the Laboratory Electronics course from which emerged *The Art of Electronics*. In addition to his work in circuit design and electronic instrumentation, his research interests have included observational astrophysics, X-ray and particle microscopy, and optical interferometry. He is one of the pioneers of the search for intelligent life beyond Earth (SETI). He is the author of some 200 scientific articles and reports, has consulted widely for industry and government, and is the designer of numerous scientific and photographic instruments.

Winfield Hill is by inclination an electronics circuit-design guru. After dropping out of the Chemical Physics graduate program at Harvard University, and obtaining an E.E. degree, he began his engineering career at Harvard's Electronics Design Center. After 7 years of learning electronics at Harvard he founded Sea Data Corporation, where he spent 16 years designing instruments for Physical Oceanography. In 1988 he was recruited by Edwin Land to join the Rowland Institute for Science. The institute subsequently merged with Harvard University in 2003. As director of the institute's Electronics Engineering Lab he has designed some 500 scientific instruments. Recent interests include high-voltage RF (to 15 kV), high-current pulsed electronics (to 1200 A), low-noise amplifiers (to sub-nV and pA), and MOSFET pulse generators.

The Art of Electronics:
The x-Chapters

Paul Horowitz HARVARD UNIVERSITY, MASSACHUSETTS

Winfield Hill ROWLAND INSTITUTE AT HARVARD

Shaftesbury Road, Cambridge CB2 8EA, United Kingdom

One Liberty Plaza, 20th Floor, New York, NY 10006, USA

477 Williamstown Road, Port Melbourne, VIC 3207, Australia

314–321, 3rd Floor, Plot 3, Splendor Forum, Jasola District Centre,
New Delhi – 110025, India

103 Penang Road, #05–06/07, Visioncrest Commercial, Singapore 238467

Cambridge University Press is part of Cambridge University Press & Assessment,
a department of the University of Cambridge.

We share the University's mission to contribute to society through the pursuit of
education, learning and research at the highest international levels of excellence.

www.cambridge.org
Information on this title: www.cambridge.org/9781108499941
DOI: 10.1017/9781108753029

First published 2020
Third printing with corrections and additions 2021 (version 2, April 2024)

Printed in the United Kingdom by TJ Books Limited, Padstow Cornwall, April 2024

A catalogue record for this publication is available from the British Library

ISBN 978-1-108-49994-1 Hardback

To Ava & Vida

and

Hank & Zadie

CONTENTS

LIST OF TABLES

PREFACE

Why the "x-Chapters"? Simple answer: they are the e*X*tra Chapters that couldn't be shoehorned into the 1250-page third edition of *The Art of Electronics*; they are the completion of the broad discussions begun in the latter.

But that's too simple. First some background: We're often asked "is your book *The Art of Electronics* (AoE) a textbook, or is it a reference book?" to which we answer "yes." Although it originated as a handwritten samizdat-style set of course notes[1] for a newly-born circuit-design course in laboratory electronics at Harvard in 1974, it has acquired a majority readership among circuit designers, both professional and, increasingly, among the happily growing ranks in the self-taught maker community.

There's an inherent tension between the text and reference book structures: a textbook should lay out its subject in a logical progression, whereas a reference book should treat topics as capsules. For example, in AoE we visit and revisit the subject of current sources successively as new tools come to hand: in Chapter 1 we use a voltage source in series with a resistor; in the next chapter we use BJTs with emitter resistors (simple version, then V_{BE} compensated, then the Wilson mirror or cascode to defeat Early effect); in Chapter 3 it's no surprise that we make current sources with FETs and with hybrid BJT–FET configurations; Chapter 4 provides the powerful op-amp tool to create precision currents sources, a topic that continues in the following chapter on precision design. It doesn't stop there – in Chapter 7 we see more current sources (this time in connection with sawtooth oscillators); in the power chapter (9) they're back (hijacked 3–terminal regulators); and of course they can't resist a cameo appearance in the conversion chapter (13), as digitally-programmable precision sources, nanoamp sources, high-voltage floating sources, coil-driving sources, and more.

That structure, good for class- and self-study, is suboptimal for someone with plenty of electronics background who needs to design a current source and needs to balance the engineering tradeoffs. For a challenge like that, a better choice would be a (reference) book with a section called *Current Source Design*. Our initial plans for the third edition of AoE included a set of chapter annexes, the x-Chapters,[2] which would capture advanced topics and applications in a less linear style. But including such annexes seemed awkward, and anyway by the time the book had reached 1200 pages (without the x-chapters) it was clear we needed to peel off the latter as a supplementary volume. We did keep a handful of advanced topics in AoE3, largely in Chapters 5 and 8 (Precision Circuits, Low-Noise Techniques), but this new annex includes all the advanced material that we had targeted for Chapters 1, 2, 3, 4, and 9 (with corresponding numbering).

Now for *The x-Chapters*: Freed from the constraints of linear textbook-style organization, we've written these like a set of short stories, on topics that are, variously, advanced, important, novel, or just plain fun. We think of them as little gems, a collection of "pearls of electronics" (our editor's suggested title). As such, they are different in character from the topics of the main volume, AoE3.

Here are some examples, from the nearly 100 major topic headings:

- wiring: skin effect, coupling, shielding, PCB impedance
- resistors and capacitors: nonlinearity, tempco, parasitics, pulse endurance, and more
- inductors demystified (at last!); cores, gapping, and all that...
- simplified treatment of poles, zeros, and the *s*-plane
- switches and relays: contact degradation, dry switching, coil driving, and more
- diodes: leakage, reverse recovery, step-recovery, tunnel diodes
- bipolarity current mirrors
- what is the *actual* leakage current of BJTs and JFETs?
- BJT amplifier distortion: a SPICE exploration
- parasitic oscillation in the emitter follower
- why the emitter follower output looks inductive

---

[1] See for yourself: https://artofelectronics.net/prehistory/.

[2] But also, in a nod to Mulder and Scully, a suggestion that "the truth is in there."

- designs by the masters:

  (a) ±20 V, 5 ns, 50 Ω amplifier;
  (b) bulletproof input protection;
  (c) the Monticelli ouput stage;
  (d) wide-range (7-decade) linear transimpedance amplifier

- JFETs: a guided tour, and a closer look
- MOSFETs as linear transistors; CMOS linear amplifiers
- depletion-mode MOSFET current sources
- high-voltage, low-capacitance current sources
- pulse energy in power MOSFETs
- MOSFET gate drivers
- high-voltage pulsers
- high-voltage probe with $>1\text{G}\Omega$ input impedance
- IGBTs and other power semiconductors
- low-voltage switching: MOS vs BJT
- op-amp history: from Philbrick to SMT
- feedback stability and phase margin
- bias-current cancellation in BJT-input op-amps
- transimpedance amplifiers – bandwidth and stability
- precision high-voltage amplifier
- ratio (normalizing) transimpedance amplifier
- unity-gain buffers
- current-feedback amplifiers demystified
- the gotchas of rail-to-rail op-amps
- silicon photomultipliers
- analog function circuits – the Lorenz attractor
- high-voltage bipolarity current source
- ripple reduction in PWM
- reverse-polarity protection
- transformer + rectifier + capacitor = giant spikes!
- low-voltage clamp/crowbar
- high-efficiency ("green") power switching power supplies
- power-factor correction
- high-side current sensing
- the "dc transformer" (bus converter)
- beware counterfeit chargers (or, don't bite into *that* Apple!)
- low-noise ultra-isolated power
- negative-input switching converters
- precision negative bias supply for silicon photomultipliers
- high-voltage negative regulator
- precision low-noise laboratory power supply
- PWM for dc motors – a myth demolished
- sending power on a beam of light
- fast LED pulsers
- "it's too hot" redux – many methods of thermometry

- transient voltage protection and transient thermal response
- low-capacitance MOSFET-gate protection
- charge-dispensing piezo positioner
- precision 1500 V 1 microsecond ramp
- fast shutoff of high-energy magnetic field

We hope these will appeal to a varied audience. We enjoyed experimenting with their circuits, and finding a way to set down on paper the essential and digestible takeaway. *Bon appetít!*

Acknowledgments

Once again topping the list of people to whom we owe a giant debt of gratitude is David Tranah, our indefatigable editor: he is our linchpin, our helpful LaTeXpert, our wise advisor of all things bookish, and (would you believe?) our *compositor*! He put up with a pair of fussy authors, and he hosted us in a fine visit to the Cambridge University Press mother-ship.

We are grateful to Jim Macarthur, circuit designer extraordinaire, for his careful reading of chapter drafts, and invariably helpful suggestions for improvement; we adopted every one. Our colleague Peter Lu taught us the delights of Adobe Illustrator, and appeared at a moment's notice when we went off the rails; the book's figures are testament to the quality of his tutoring. We are indebted to Mike Burns for his careful reading, and careful computation of thorny functions.

For their many helpful contributions we thank Steve Cerwin, Tom Hayes, Phil Hobbs, Peter Horowitz, John Larkin, Maggie McFee, Ali Mehmed, and Jim Thompson. We thank also others whom (we're sure) we've here overlooked, with apologies for the omission. Additional contributors to the book's content (circuits, inspired web-based tools, unusual measurements, etc., from the likes of Keith Billings, Kent Lundberg, and Steve Woodward) are referenced throughout the book in the relevant text.

In the production chain we are indebted to our copy editor Jon Billam for his light but precise touch, and a cast of unnamed graphic artists who converted our pencil circuit sketches into beautiful vector graphics.

And finally, we are forever indebted to our loving, supportive, and ever-tolerant spouses Vida and Ava, who suffered through decades of abandonment as we obsessed over every detail of our *third* encore.

Paul Horowitz
Winfield Hill
August 2019
Cambridge, Massachusetts

Legal notice

In this book we have attempted to teach the techniques of electronic design, using circuit examples and data that we believe to be accurate. However, the examples, data, and other information are intended solely as teaching aids and should not be used in any particular application without independent testing and verification by the person making the application. Independent testing and verification are especially important in any application in which incorrect functioning could result in personal injury or damage to property.

For these reasons, we make no warranties, express or implied, that the examples, data, or other information in this volume are free of error, that they are consistent with industry standards, or that they will meet the requirements for any particular application. THE AUTHORS AND PUBLISHER EXPRESSLY DISCLAIM THE IMPLIED WARRANTIES OF MERCHANTABILITY AND OF FITNESS FOR ANY PARTICULAR PURPOSE, even if the authors have been advised of a particular purpose, and even if a particular purpose is indicated in the book. The authors and publisher also disclaim all liability for direct, indirect, incidental, or consequential damages that result from any use of the examples, data, or other information in this book.

We also make no representation regarding whether use of the examples, data, or other information in this volume might infringe others' intellectual property rights, including US and foreign patents. It is the reader's sole responsibility to ensure that he or she is not infringing any intellectual property rights, even for use which is considered to be experimental in nature. By using any of the examples, data, or other information in this volume, the reader has agreed to assume all liability for any damages arising from or relating to such use, regardless of whether such liability is based on intellectual property or any other cause of action, and regardless of whether the damages are direct, indirect, incidental, consequential, or any other type of damage. The authors and publisher disclaim any such liability.

REAL-WORLD PASSIVE COMPONENTS

CHAPTER **1x**

In the introductory first chapter of AoE3 we saw the basics of passive components – resistors, capacitors, inductors (and transformers), and diodes – and we've treated them as idealized components. In reality things are more complicated: for example, resistors have some "parasitic" capacitance and inductance; their resistance varies with temperature ("tempco"), with applied voltage ("voltage coefficient"), and with the passage of time ("aging"). Even plain old *wire* isn't a simple thing: wire has resistance and inductance, both dependent upon frequency; and it comes in a bewildering variety of sizes, configurations, and varieties of insulation.

Most of the time you can ignore these real-world deviations from the ideal. But a good circuit designer must know about them, and particularly which ones do matter in the design of any particular circuit. In this first "x-chapter" we peel away the ideal (and boring) façade of basic components, revealing their rich interior.

> **Review of Chapter 1 of AoE3**

To bring the reader up to speed, we start this chapter with the end-of-chapter review from the main volume's Chapter 1:

¶A. Voltage and Current.

Electronic circuits consist of components connected together with wires. *Current* (I) is the rate of flow of charge through some point in these connections; it's measured in amperes (or milliamps, microamps, etc.). *Voltage* (V) between two points in a circuit can be viewed as an applied driving "force" that causes currents to flow between them; voltage is measured in volts (or kilovolts, millivolts, etc.); see §1.2.1. Voltages and currents can be steady (dc), or varying. The latter may be as simple as the sinusoidal alternating voltage (ac) from the wallplug, or as complex as a high-frequency modulated communications waveform, in which case it's usually called a *signal* (see ¶B below). The algebraic sum of currents at a point in a circuit (a *node*) is zero (Kirchhoff's current law, KCL, a consequence of conservation of charge), and the sum of voltage drops going around a closed loop in a circuit is zero (Kirchhoff's voltage law, KVL, a consequence of the conservative nature of the electrostatic field).

¶B. Signal Types and Amplitude.

See §1.3. In digital electronics we deal with *pulses*, which are signals that bounce around between two voltages (e.g., +5 V and ground); in the analog world it's *sinewaves* that win the popularity contest. In either case, a periodic signal is characterized by its frequency f (units of Hz, MHz, etc.) or, equivalently, period T (units of ms, μs, etc.). For sinewaves it's often more convenient to use *angular* frequency (radians/s), given by $\omega = 2\pi f$.

Digital amplitudes are specified simply by the HIGH and LOW voltage levels. With sinewaves the situation is more complicated: the amplitude of a signal $V(t) = V_0 \sin \omega t$ can be given as (a) *peak* amplitude (or just "amplitude") V_0, (b) *root-mean-square* (rms) amplitude $V_{rms} = V_0/\sqrt{2}$, or (c) peak-to-peak amplitude $V_{pp} = 2V_0$. If unstated, a sinewave amplitude is usually understood to be V_{rms}. A signal of rms amplitude V_{rms} delivers power $P = V_{rms}^2/R_{load}$ to a resistive load (regardless of the signal's waveform), which accounts for the popularity of rms amplitude measure.

Ratios of signal amplitude (or power) are commonly expressed in *decibels* (dB), defined as dB $= 10\log_{10}(P_2/P_1)$ or $20\log_{10}(V_2/V_1)$; see §1.3.2. An amplitude ratio of 10 (or power ratio of 100) is 20 dB; 3 dB is a doubling of power; 6 dB is a doubling of amplitude (or quadrupling of power). Decibel measure is also used to specify amplitude (or power) directly, by giving a reference level: for example, -30 dBm (dB relative to 1 mW) is 1 microwatt; +3 dBVrms is a signal of 1.4 V rms amplitude (2 Vpeak, 4 Vpp).

Other important waveforms are square waves, triangle waves, ramps, noise, and a host of *modulation* schemes by which a simple "carrier" wave is varied in order to convey information; some examples are AM and FM for analog communication, and PPM (pulse-position modulation) or QAM (quadrature-amplitude modulation) for digital communication.

¶C. The Relationship Between Current and Voltage.

Chapter 1 concentrated on the fundamental, essential, and ubiquitous *two-terminal linear devices*: resistors, capacitors, and inductors. (Subsequent chapters dealt with *transistors* – three-terminal devices in which a signal applied to one terminal controls the current flow through the other pair – and their many interesting applications. These include amplification, filtering, power conversion, switching, and the like.) The simplest linear device is the *resistor*, for which $I = V/R$ (Ohm's Law, see §1.2.2A). The term "linear" means that the response (e.g., current) to a combined sum of inputs (i.e., voltages) is equal to the sum of the responses that each input would produce: $I(V_1 + V_2) = I(V_1) + I(V_2)$.

¶D. Resistors, Capacitors, and Inductors.

The resistor is clearly linear. But it is not the only linear two-terminal component, because linearity does not require $I \propto V$. The other two linear components are *capacitors* (§1.4.1) and *inductors* (§1.5.1), for which there is a time-dependent relationship between voltage and current: $I = C\, dV/dt$ and $V = L\, dI/dt$, respectively. These are the *time domain* descriptions. Thinking instead in the *frequency domain*, these components are described by their *impedances*, the ratio of voltage to current (as a function of frequency) when driven with a sinewave (§1.7). A linear device, when driven with a sinusoid, responds with a sinusoid of the same frequency, but with changed amplitude and phase. Impedances are therefore complex, with the real part representing the amplitude of the response that is in-phase, and the imaginary part representing the amplitude of the response that is in quadrature (90° out of phase). Alternatively, in the polar representation of complex impedance ($Z=|Z|e^{i\theta}$), the magnitude $|Z|$ is the ratio of magnitudes ($|Z|=|V|/|I|$) and the quantity θ is the phase shift between V and I. The impedances of the three linear 2-terminal components are $Z_R=R$, $Z_C=-j/\omega C$, and $Z_L=j\omega L$, where (as always) $\omega=2\pi f$; see §1.7.5. Sinewave current through a resistor is in phase with voltage, whereas for a capacitor it leads by 90°, and for an inductor it lags by 90°.

¶E. Series and Parallel.
The impedance of components connected in series is the sum of their impedances; thus $R_{\text{series}}=R_1+R_2+\cdots$, $L_{\text{series}}=L_1+L_2+\cdots$, and $1/C_{\text{series}}=1/C_1+1/C_2+\cdots$. When connected in parallel, on the other hand, it's the *admittances* (inverse of impedance) that add. Thus the formula for capacitors in parallel looks like the formula for resistors in series, $C_{\text{parallel}}=C_1+C_2+\cdots$; and vice versa for resistors and inductors, thus $1/R_{\text{parallel}}=1/R_1+1/R_2+\cdots$. For a pair of resistors in parallel this reduces to $R_{\text{parallel}}=(R_1R_2)/(R_1+R_2)$. For example, two resistors of value R have resistance $R/2$ when connected in parallel, or resistance $2R$ in series.

The power dissipated in a resistor R is $P=I^2R=V^2/R$. There is no dissipation in an ideal capacitor or inductor, because the voltage and current are 90° out of phase. See §1.7.6.

¶F. Basic Circuits with *R*, *L*, and *C*.
Resistors are everywhere. They can be used to set an operating current, as for example when powering an LED or biasing a zener diode (Fig. 1.16); in such applications the current is simply $I=(V_{\text{supply}}-V_{\text{load}})/R$. In other applications (e.g., as a transistor's load resistor in an amplifier, Fig. 3.29) it is the *current* that is known, and a resistor is used to convert it to a voltage. An important circuit fragment is the *voltage divider* (§1.2.3), whose unloaded output voltage (across R_2) is $V_{\text{out}}=V_{\text{in}}R_2/(R_1+R_2)$.

If one of the resistors in a voltage divider is replaced with a capacitor, you get a simple *filter*: lowpass if the lower leg is a capacitor, highpass if the upper leg is a capacitor (§§1.7.1 and 1.7.7). In either case the -3 dB transition frequency is at $f_{3dB}=1/2\pi RC$. The ultimate rolloff rate of such a "single-pole" lowpass filter is -6 dB/octave, or -20 dB/decade; i.e., the signal amplitude falls as $1/f$ well beyond f_{3dB}. More complex filters can be created by combining inductors with capacitors, described in Chapter 6. A capacitor in parallel with an inductor forms a *resonant circuit*; its impedance (for ideal components) goes to infinity

at the resonant frequency $f=1/(2\pi\sqrt{LC})$. The impedance of a *series LC* goes to zero at that same resonant frequency. See §1.7.14.

Other important capacitor applications in Chapter 1 (§1.7.16) include (a) *bypassing*, in which a capacitor's low impedance at signal frequencies suppresses unwanted signals, e.g., on a dc supply rail; (b) *blocking* (§1.7.1C), in which a highpass filter blocks dc, but passes all frequencies of interest (i.e., the breakpoint is chosen below all signal frequencies); (c) *timing* (§1.4.2D), in which an *RC* circuit (or a constant current into a capacitor) generates a sloping waveform used to create an oscillation or a timing interval; and (d) *energy storage* (§1.7.16B), in which a capacitor's stored charge $Q=CV$ smooths out the ripples in a dc power supply.

Additional applications of capacitors include: (e) *peak detection* and *sample-and-hold* (§§4.5.1 and 4.5.2), which capture the voltage peak or transient value of a waveform, and (f) the *integrator* (§4.2.6), which performs a mathematical integration of an input signal.

¶G. Loading; Thévenin Equivalent Circuit.
Connecting a load (e.g., a resistor) to the output of a circuit (a "signal source") causes the unloaded output voltage to drop; the amount of such *loading* depends on the load resistance, and the signal source's ability to drive it. The latter is usually expressed as the *equivalent source impedance* (or *Thévenin impedance*) of the signal. That is, the signal source is modeled as a perfect voltage source V_{sig} in series with a resistor R_{sig}. The output of the resistive voltage divider driven from an input voltage V_{in}, for example, is modeled as a voltage source $V_{\text{sig}}=V_{\text{in}}R_2/(R_1+R_2)$ in series with a resistance $R_{\text{sig}}=R_1R_2/(R_1+R_2)$ (which is just $R_1\|R_2$). So the output of a 1kΩ–1kΩ voltage divider driven by a 10 V battery looks like 5 V in series with 500 Ω.

Any combination of voltage sources, current sources, and resistors can be modeled perfectly by a single voltage source in series with a single resistor (its "Thévenin equivalent circuit"), or by a single current source in parallel with a single resistor (its "Norton equivalent circuit"); see Appendix D. The Thévenin equivalent source and resistance values are found from the open-circuit voltage and short-circuit current as $V_{\text{Th}}=V_{\text{oc}}$, $R_{\text{Th}}=V_{\text{oc}}/I_{\text{sc}}$; and for the Norton equivalent they are $I_N=I_{\text{sc}}$, $R_N=V_{\text{oc}}/I_{\text{sc}}$.

Because a load impedance forms a voltage divider with the signal's source impedance, it's usually desirable for the latter to be small compared with any anticipated load impedance (§1.2.5A). However, there are two exceptions: (a) a *current source* has a high source impedance (ideally infinite), and should drive a load of much lower impedance; and (b) signals of *high frequency* (or fast risetime), traveling through a length of cable, suffer reflections unless the load impedance equals the so-called "characteristic impedance" Z_0 of the cable (commonly $50\,\Omega$), see Appendix H.

¶H. The Diode, a Nonlinear Component.

There are important two-terminal devices that are not linear, notably the *diode* (or *rectifier*), see §1.6. The ideal diode conducts in one direction only; it is a "one-way valve." The onset of conduction in real diodes is roughly at $0.5\,\text{V}$ in the "forward" direction, and there is some small leakage current in the "reverse" direction, see Figure 1.55. Useful diode circuits include power-supply *rectification* (conversion of ac to dc, §1.6.2), signal rectification (§1.6.6A), *clamping* (signal limiting, §1.6.6C), and *gating* (§1.6.6B). Diodes are commonly used to prevent polarity reversal, as in Figure 1.84; and their exponential current versus applied voltage can be used to fashion circuits with logarithmic response (§1.6.6E).

Diodes specify a maximum safe reverse voltage, beyond which avalanche breakdown (an abrupt rise of current) occurs. You don't go there! But you can (and should) with a *zener diode* (§1.2.6A), for which a reverse breakdown voltage (in steps, going from about $3.3\,\text{V}$ to $100\,\text{V}$ or more) is specified. Zeners are used to establish a voltage within a circuit (Fig. 1.16), or to limit a signal's swing.

1x.1 Wire and Connectors

1x.1.1 Wire gauge: resistance, heating, and current-carrying capacity

Table 1x.1 shows wire sizes, going to extremes at both ends. An easy way to remember it all (if you've left your copy of this book at home) is to note that (a) each wire size is 1 dB (in cross-sectional area, or resistance), and (b) #10 wire is 1 mΩ/ft.[1] At high frequencies the *skin effect* (see §1x.1.6B, below) causes an effective increase in resistance (as does the *proximity effect*, for closely packed multiple wires).

We like #26 Kynar-insulated solid wire for point-to-point wiring on circuit boards, and #22–26 stranded wire (with irradiated PVC insulation) for other internal instrument wiring (as well as for multiwire signal cables), where currents are small (<1 A, say). For larger currents, choose wire sizes according to how much voltage drop and heating you can tolerate. When winding inductors or transformers (see below), the wire size is constrained by power dissipation, quality factor ("Q"), and available core dimensions. For power transformers a typical wire size guideline is 1000 "circular mils" (the square of the wire diameter in mils) per amp, e.g., #20 enamel insulated magnet wire for a 1 A (rms) load current. Figure 1x.1 is a rough guide to the current-carrying capability of a given wire size, based on temperature rise above ambient. But other factors – such as the enclosure or conduit, and the thermal path for heat removal – affect the real-world maximum current.

1x.1.2 Stranding, insulation, and tinning

Stranding
Stranded wire is more flexible and supple than solid wire, and is preferred for cables and for wiring that undergoes motion (e.g., power cords, mouse, keyboard, network patch cables, oscilloscope and voltmeter probes, and so on). But solid wire is often better when wiring between fixed points (such as on a circuit board, or for house wiring) because you don't have to worry about persuading every strand to behave.

---

[1] You can really impress your friends by knowing also that #10 wire is 0.1″ in diameter, or 5 mm² in area.

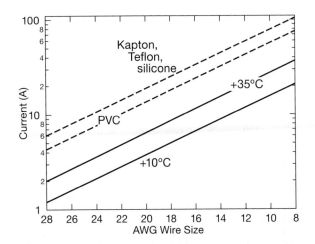

Figure 1x.1. Approximate current-carrying capability versus wire gauge, for 10°C and 35°C rise above ambient (solid lines), and for maximum insulation temperature at 30°C ambient for two insulation families (dashed lines). Derate these values for multiwire cables, by a factor of 0.8 (2–5 conductors), 0.7 (6–15 conductors), and 0.5 (16–30 conductors).

Stranded wire comes with standard numbers of finer strands, often 7 or 19 (the number of coins that fit nicely on a flat surface), with finer stranding providing more flexibility. For *really* supple wiring you want "ropelay" stranding, in which a group of fine stranded wires are themselves stranded into a larger wire. For example, you can get #24 stranded wire as 7 strands of #32 ("7/32"), or 19 strands of #36; as a ropelay you can get 7 groups of 15/44, for a total of 105 strands. In very thick cables the numbers get really large: we bought some #0 "extra-flexible" ropelay cable, stranded as 7x7x86/36 (4214 strands of #36!); the stuff was as supple as clothesline.

There are some exotic forms of stranded wire; two in particular are called *bunched conductor*, and *litz wire*. A bunched conductor consists of a twisted bundle of insulated strands that are stripped and connected together at each end; litz wire also consists of a set of insulated strands, woven however in such a manner that each strand visits the inner and outer portions of the wire as it runs along the length. These unusual forms of stranded wire are used to circumvent skin effect and proximity effect, discussed below in §1x.1.6B.

Insulation
The most common insulation is **PVC** (polyvinylchloride), which has respectable thermal and electrical characteristics. **Polypropylene** and **polyethylene** are also used, the latter particularly in coaxial cable. We favor **irradiated**

Table 1x.1: Copper Wire Table[a]

| AWG | Diameter[b] | | Resistance[c] | | Mass | | Area[e] |
	mils[d]	mm	mΩ/ft	mΩ/m	lb/kft	kg/km	mm²
0	325	8.26	0.098	0.32	320	476	53.5
2	258	6.55	0.156	0.51	201	299	33.6
4	204	5.18	0.249	0.82	126	187	21.1
6	162	4.11	0.395	1.30	79.5	118	13.3
8	129	3.28	0.628	2.06	50.0	74.4	8.36
10	102	2.59	0.999	3.28	31.4	46.7	5.26
12	80.8	2.05	1.59	5.22	19.8	29.5	3.31
14	64.1	1.63	2.53	8.30	12.4	18.5	2.08
16	50.8	1.29	4.02	13.2	7.82	11.6	1.31
18	40.3	1.02	6.39	21.0	4.92	7.32	0.82
20	32.0	0.813	10.2	33.5	3.09	4.60	0.52
22	25.4	0.645	16.1	52.8	1.95	2.90	0.33
24	20.1	0.511	25.7	84.3	1.22	1.82	0.20
26	15.9	0.404	40.8	134	0.769	1.14	0.13
28	12.6	0.320	64.9	213	0.484	0.720	0.084
30	10.0	0.254	103	338	0.304	0.452	0.053
32	7.95	0.202	164	538	0.191	0.284	0.034
34	6.31	0.160	261	856	0.120	0.179	0.021
36	5.00	0.127	415	1361	0.076	0.113	0.013
38	3.97	0.101	660	2164	0.048	0.071	0.0084
40	3.15	0.080	1050	3442	0.030	0.045	0.0053

(a) values at 25°C. (b) for solid conductor; stranded conductors are larger. (c) tempco = +0.4%/°C. (d) 1 mil = 0.001 inch = 0.0254 mm. (e) the area in "circular mils" is the square of the diameter in mils.

From very thick to very thin in the American Wire Gauge (AWG) sizes of copper wire. The resistance values have a positive temperature coefficient of 0.4%/°C. Sizes from #20–26 are typically used in signal cables and in instrument wiring; house wiring uses #14 and #12 for 15 A and 20 A circuits, respectively.

PVC for wiring within instruments, because it is considerably less susceptible to "melt-back" during soldering; it is also tougher, while retaining PVC's flexibility.[2] **Teflon**® insulation is an expensive alternative, with superior thermal, electrical, and chemical properties: it is unaffected by soldering temperatures (no melt-back), it is chemically inert (unaffected by acids, alkalis, hydrocarbons, solvents, ozone, water, oil, and gasoline), and it retains its flexibility at low temperatures; it is rated for operation from −70°C to 200°C (or 260°C for TFE-type Teflon). However, owing to the absence of molecular cross-linking, Teflon is susceptible to "cold-flow" (also known as *creep*, or *compression set*): the Teflon insulation of a wire pulled tightly around a corner tends to cold-flow away from the con-

tact zone.[3] "Magnet wire" consists of a bare solid copper conductor with a tough conformal insulating layer (or layers) (sometimes loosely called **enamel**), intended for inductors and transformers, with correspondingly high temperature ratings; for example, Belden's Beldsol® (nylon over polyurethane) is rated for operation to 270°C. Some other high-temperature insulation types (with excellent low-temperature properties as well) are **silicone** (which excels in flexibility), **Tefzel**®, and **Halar**®; these are rated to 150°C. For more detail see the literature from wire manufacturers such as Alpha and Belden.

Tinning

Nearly all electrical wiring is copper, which is often metal plated ("tinned"), both for compatibility with the insulating material, and for enhanced solderability (compared with bare copper). Tin (or tin alloy) plating is common for most plastic insulations (e.g., PVC), whereas Kynar- and Teflon-insulated wires are usually plated with silver (or nickel); enamel-insulated magnet wire (and litz wire) is ordinarily untinned, as are heavy wiring used in power distribution (e.g., "Romex" house wiring).

1x.1.3 Printed circuit wiring

Most circuit wiring that is not "on-chip" takes the form of PCB traces.[4] Except for the simplest circuits, contemporary PCBs are fabricated as multiple layers with plated-through holes, on a fiberglass-epoxy substrate known as FR-4 (formerly G-10). The standard thickness totals 0.062″ (1.6 mm), with a tough insulating *soldermask*[5] covering all but the exposed *pads*[6] that are to be soldered (to prevent solder bridges and also to protect the surface traces). An informational *silkscreen* legend is applied over the finished board (soldermask and all), indicating parts values and designations, and other generally useful stuff. Components are usually soldered on both sides, using robotic pick-and-place machinery to lay down the parts onto the pads (to which solder paste[7] has been applied), followed

---

[2] We like the 19-strand (versus 7-strand) hookup wire, for greater flexibility; the Alpha part numbers are 7058/19–7054/19 (AWG even-numbered gauges 16 through 24). You can buy equivalents from wire dealers (such as Anixter), spooled from their bulk supplies in the lengths you want, generally at a considerable discount.

[3] This can happen also in a bundle of Teflon-insulated wiring tightly wrapped with cable ties or lacing. NASA has cautioned its spacecraft designers on this point, and suggests Tefzel and Kynar as alternative insulation materials.

[4] A misnomer, because the traces are not *printed*; rather, they are the remnants left after the unwanted copper has been chemically removed.

[5] Usually liquid photo-imageable solder mask over bare copper, "LPI-SMOBC."

[6] The connection points to the electrical components.

[7] An emulsion of solder particles and heat-activated flux.

by a scorching journey through a reflow oven during which the surface-mount components are soldered.[8]

1x.1.4 PCB traces

The wiring traces are copper, with the thickness specified in *ounces*, most commonly "1/2 ounce" or "1 ounce." These strange units refer to the weight of copper per square foot! You can figure it out from first principles, if you like, but the conversion factor is 1 ounce $\leftrightarrow$ 0.00137″ (1.37 mils) $\leftrightarrow$ 35 μm.[9] The *sheet resistance* of 1 oz copper is 0.5 mΩ/square (another strange unit, sometimes written $\Omega/\square$), and varies inversely with copper thickness. When used as a heatsink, a square inch of PCB copper is (very roughly) 50°C/W.

A. Resistance and current-carrying capacity

PCB traces are resistive, which causes a dc voltage drop IR proportional to current, and power dissipation I^2R proportional to the square of current. The dc resistance of a 1 ounce trace is $R = 0.47/w$ ohms/inch (or $0.19/w$ ohms/cm), where w is the trace width in mils (these values scale inversely in copper thickness). Because typical trace widths are in the range of 5–10 mils, the effects of their resistance (~ 0.05 to $0.1\,\Omega$/inch) is generally insignificant, in terms of signal degradation, compared with the effects of capacitance and inductance (see below). However, trace resistance limits current carrying capacity, owing to I^2R heating; see Figure 1x.2.

B. Capacitance and inductance

PCB traces, like all conductors, have capacitance and inductance. To a good approximation, these are proportional to trace length, and are a function of trace width, height above a conductive plane (power/ground planes, usually), and (for capacitance) the dielectric constant of

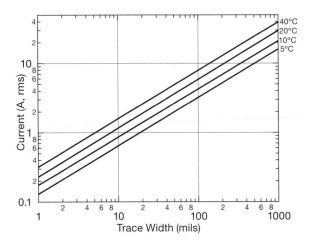

Figure 1x.2. Approximate PCB trace current limits, for 1 ounce (1.37 mil, or 35 μm) copper, as determined by I^2R heating, for the indicated values of temperature rise. For other copper thicknesses scale the x-axis values proportionally.

the PCB substrate and soldermask. PCB layout software sometimes includes routines to calculate capacitance, inductance, transmission line impedance, and even propagation delays. It's useful to know the approximate range of values, though, which we've calculated and plotted in Figure 1x.3, for three values of substrate thickness: 10 mils, 30 mils, and 60 mils (corresponding to a typical multilayer board, a 0.032″ two-layer board, and a standard 0.062″ two-layer board). Note that the values depend primarily on the ratio of trace width to substrate thickness (w/h), and only weakly on the substrate thickness itself. Order of magnitude, traces have about a pF/cm of capacitance, and 5 nH/cm of inductance.

C. Transmission-line impedance and attenuation

Because a PCB trace of given width and spacing has some capacitance per unit length and some inductance per unit length (both dependant on geometry), it behaves like a *transmission line*, a subject treated in some detail in Appendix H. For now, the important facts are (a) for "fast" signals (those that change significantly in the time it takes a signal to travel to the end of a wire and back) you cannot treat a wire as a simple conductor with some single voltage on it – instead, signals travel along it as a wave, and in general will reflect off the far end and return to haunt you; (b) a transmission line has a *characteristic impedance* Z_0, which is the ratio of voltage to current in a traveling wave; (c) if you connect a resistor equal to the characteristic impedance (which is approximately a resistance, $R_0 \approx |Z_0|$, usually 50 Ω) across the far end, there will be no reflections (it will

---

[8] If *through-hole* components are used, the board undergoes *wave soldering*, a terrifying process during which a fountain of molten solder is squirted forcefully against its underside. Lots of smoke! (And any surface-mount components on the bottom side will have been secured with a dot of adhesive, otherwise they will disappear in a cascade of molten solder.)

[9] The copper thickness you specify refers to the *finished* thickness – the process begins with PCB material with thinner cladding, which is then electroplated with copper (required to create plated-through holes and vias) to the final thickness. For example, a PCB with 1 ounce copper would begin life as 1/2 ounce clad board (or inner *core* layers, the thinner sheets that are stacked and laminated to make multilayer boards), plated up to 1 ounce copper thickness. In some PCB processes this might be covered with a thin tin plating.

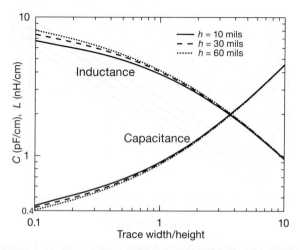

Figure 1x.3. Capacitance and inductance of printed circuit traces, as a function of the ratio of trace width to substrate thickness. These calculated values assume $\varepsilon=4.5$, typical of fiberglass-epoxy FR-4 PCB material.

swallow all signals); and (d) a transmission line so "terminated" at the far end looks purely resistive, with input impedance R_0 – its capacitance and inductance disappear entirely!

As seen in Appendix H, common transmission line configurations are "single-ended" coaxial cable (the familiar black RG-58 with BNC connectors that litter all laboratories), or "differential" twisted pair (the ubiquitous "cat-5" Ethernet used for computer networks). These are the standard forms of cables used to transport fast signals. But you sometimes have fast signals on printed circuit boards, for which you need to make a transmission line from PCB traces. There are two basic forms – *microstrip* (a top-layer trace or pair over an underlying ground plane, with a variant called *coplanar waveguide*), and *stripline* (an internal trace or pair, sandwiched between ground planes) – and either can be single-ended or differential; see Figure 1x.4.

The impedance of single-ended microstrip (a trace above a ground plane) is mostly a function of the ratio of trace width to height, and dielectric constant, and depends only weakly on the trace width itself (Fig. 1x.5); for the usual 50 Ω impedance on standard FR-4 circuit board, you want a trace width approximately 1.7 times the underlying insulation thickness. For a symmetrical stripline sandwich the trace is thinner, roughly 0.7 of the insulation thickness it sees on both sides.

For *differential* signals, which are popular for very fast signals, you use a pair of traces, either side-by-side (for microstrip, or "edge-coupled" stripline), or one above the other ("broadside-coupled" stripline). The impedance now

depends on width, height, and conductor spacing. For differential transmission line the standard impedance is 100 Ω; some suitable PCB trace dimensions are given in Table 1x.2.

Contemporary digital electronics deals in fast signals (risetimes of a nanosecond or less) and wide bandwidths

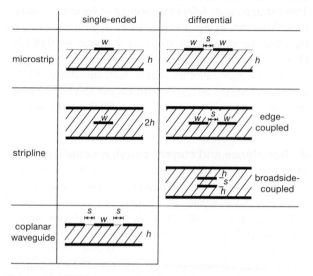

Figure 1x.4. PCB transmission line geometries, for both single-ended and differential signals. Microstrip traces are on an outer layer, whereas Stripline traces are buried.

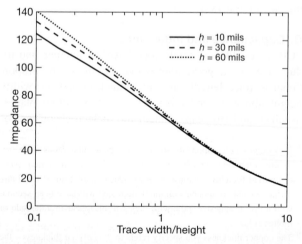

Figure 1x.5. Transmission line impedance of single-ended PCB microstrip, as a function of the ratio of trace width to substrate thickness. These calculated values assume $\varepsilon=4.5$, typical of fiberglass-epoxy FR-4 PCB material. It's usually best to ask the board house which width to use for a 50 Ω (or 100 Ω pair, etc.) line, because FR-4 varies in dielectric constant, and they know what brand they're using.

Table 1x.2: Selected PCB Transmission Lines[a]

	h mils	w mils	s mils
50Ω single-ended			
microstrip	6	10	-
	8	14	-
	10	18	-
	30	55	-
stripline	5	3.5	-
	10	8	-
	15	13	-
100Ω differential			
microstrip	5	8	30
	5	7	11
	5	6	7
	5	5	5
stripline: edge-coupled	7	5	13
	10	6	8.5
	15	6	6
stripline: over/under	5	3.5	10
	7	5	15
	10	9	25

(a) FR-4, $\varepsilon = 4.3$

Dimensions for microstrip and stripline transmission lines, for the standard impedances of 50 Ω (single-ended) and 100 Ω (differential). See Fig. 1x.4 for symbols and geometries.

(to a gigahertz or more), for which even a short length of conductor on a circuit board must be treated as a transmission line, thus requiring proper termination (discussed extensively in Appendix H). It is also frequently the case that timing *skew* (the difference in propagation times) of multiple signals must be kept below very tight limits, as little as 25 picoseconds. Owing to the underlying dielectric, signals propagate along PCB traces at roughly half the speed of light, that is, 60 ps to 70 ps per centimeter. To keep skew less than 25 ps, then, the PCB trace lengths must be equalized to better than 4 mm. Even tighter constraints apply to the two traces of a clocking differential pair – for example, "DDR" memory recommends that the traces be matched in length to 0.5 mm![10] Figure 1x.6 shows an example: a set of single-ended data lines, plus a differential clock line pair, driving a memory chip.[11]

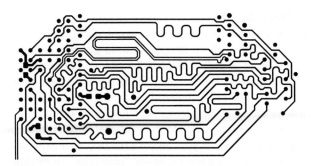

Figure 1x.6. The meandering traces on this portion of a circuit board from our lab are needed to equalize the propagation times of data (single traces) and clock (differential pair near top) going between an FPGA (left) and a DRAM (right).

1x.1.5 Cable configurations

If there's more than one wire, you call it a *cable*. There are many choices here: **Coaxial** cable ("coax") has an inner (usually stranded) conductor, with an outer shield; these are designed as transmission lines, with controlled impedance (see Appendix H on Transmission Lines), and usually designated with an "RG" type number.[12] At low frequencies you can think of coax simply as a shielded cable, with approximately 30 pF/ft (1 pF/cm) of capacitance. **Multiwire** cable (a variety of **instrumentation cable**) is a bundle of strands, color-coded so you can untangle them at the other end; they may be individual wires, **multipair** (multiple pairs), or multiple triads. A multiwire cable may be **unshielded**, or **overall shielded** (with a foil shield, a braided shield, or both). Multipair cables are also available with the pairs **individually shielded**. Multiwire cables generally are not intended as transmission lines; however, certain multipair cables are designed for high-speed data transmission, with controlled impedance. The most common is "cat-5" (or "cat-5e" or "cat-6") 100 Ω (differential) impedance cable, with 4 twisted pairs, used universally for local-area networks (LANs); it's often called "UTP" (unshielded twisted pair) network cable (and shielded twisted pair cable is called "STP").[13]

---

cause the effective "index of refraction" is $\sqrt{(1+\varepsilon)/2}$ ($\sim$1.68), versus $\sqrt{\varepsilon}$ ($\sim$2.1).

[12] For example, RG-58/U is the ubiquitous 50 Ω cable used for BNC patch cables, and RG-174 and RG-316 are skinnier variants, the latter with teflon insulation. Although 50 Ω is the by far the dominant cable impedance, the *video* world has chosen to be different, and uses 75 Ω cable, usually RG-59 or RG-6.

[13] Other popular high-speed data cables are SATA and Firewire (2 individually shielded twisted pairs, 100 Ω and 110 Ω differential impedance, respectively), and USB 2.0 (shielded pair, 90 Ω).

---

[10] See for example "Hardware Tips for Point-to-Point System Design: Termination, Layout, and Routing," Technical Note TN-46-14, Micron Technology, Inc.

[11] PCB propagation is faster on surface traces than on inside layers, be-

Flat cable (also called **ribbon cable**) comes in the familiar unshielded ribbon with 50 mil pitch; variants include ribbon with an integral flexible **ground plane** on one side, or as **fully shielded** flat ribbon. For better signal integrity with differential signals you can get a ribbon of twisted pairs, flattened for 5 cm every 50 cm to allow for the same mass-termination insulation-displacement ("IDC") connectors used for conventional ribbon cable. For single-ended signals you can get a flat ribbon of individual coaxial cables, or a less expensive "near-coax" approximation in which the continuous outer shield does not completely surround the individual strands.

1x.1.6 Inductance and skin effect

A. Inductance
In addition to resistance, *any length of wire has inductance.* That must be so because a current creates a surrounding magnetic field, with its associated field energy; equating that energy with $\frac{1}{2}LI^2$ is often the easiest way to calculate wire inductance. The result is that the inductance is approximately proportional to wire length, and further depends somewhat on spacing to nearby conductors, on wire diameter, and on frequency (via skin effect). For an isolated straight wire of diameter d and length l, the low-frequency inductance is given approximately by

$$L/l \approx 2 \left(\log_e \frac{4l}{d} - 0.75 + \frac{d}{l} \right) \quad \text{nH/cm}$$

So, for example, a 10 cm length of 20-gauge wire ($d = 0.081$ cm) has an inductance of approximately 10.9 nH/cm, when distant from other conductors. The inductance is less for a conductor in proximity to a conductive surface (a "ground plane"), or to other wires; the relevant expressions are

$$L/l \approx 2 \left(\log_e \frac{4h}{d} \right) \quad \text{nH/cm} \quad \text{(wire over plane)}$$

$$L/l \approx 2 \left(\log_e \frac{2\pi h}{w} \right) \quad \text{nH/cm} \quad \text{(flat strip over plane)}$$

$$L/l \approx 4 \left(\log_e \frac{2D}{d} - \frac{D}{l} \right) \quad \text{nH/cm} \quad \text{(wire pair)}$$

where h is the height of the conductor's center above the conducting plane (reasonably accurate for $h > 1.5d$ or $h > w$), D is the center-to-center wire spacing, and w is the width of a flat conductor (such as a PCB trace).[14]

---

[14] There are excellent references on wire inductance and its effects, notably the "Black Magic" volumes by Johnson and Graham, Grover's *Inductance Calculations*, Terman's *Radio Engineers' Handbook*, and Ott's *Noise Reduction Techniques in Electronic Systems.*

A twisted wire pair has slightly less inductance than a parallel pair[15] (and it has considerably less susceptibility to signal pickup via magnetic coupling); you can get a much larger reduction in inductance by creating a twisted pair from a twisted four-wire cable, with diagonal wires tied together.[16] Figure 1x.7 shows inductances for several configurations.

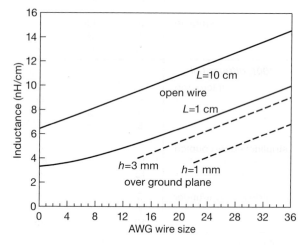

Figure 1x.7. Inductance per centimeter versus wire size for a wire over a ground plane (1 mm and 3 mm heights), and for a wire in isolation (1 cm and 10 cm lengths).

B. Skin effect
When alternating current flows through a conductor, the current is not uniform throughout the bulk – it is concentrated in an outer layer[17] (called the *skin depth*) of thickness $\delta \approx (\pi \mu \sigma f)^{-\frac{1}{2}}$, where σ is the conductivity, μ is the relative magnetic permeability (=1 for non-magnetic materials), and f is the frequency.[18] From a circuit point of view, skin effect causes an increase in the effective resistance (and loss), and a decrease of inductance (because the

---

[15] The D/l term is omitted in the equation above.

[16] We used this trick in the "stopped-light" experiments, where we had to limit the inductance of an 875 A cable to quickly kill the current, allowing trapped sodium ions to fall under gravity, watching the spread to measure their temperature.

[17] To be precise, the current density decreases exponentially, falling to $1/e$ (37%) of its surface value at a depth equal to δ.

[18] If you are interested in the physical origin of skin effect, you can think of it in either of two ways: (i) as the penetration depth of an electromagnetic wave incident on the finite-conductivity metal; or (ii) as resulting from the outward force on the moving charges caused by their motion through the alternating magnetic field they produce.

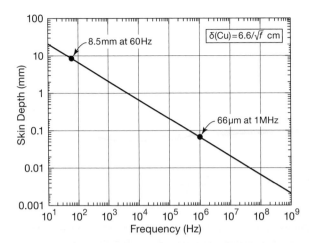

Figure 1x.8. Skin depth in copper as a function of frequency.

Figure 1x.9. We persuaded the guys with the big backhoe to give us a sample of their *impossible* power cable: the diameter of its stranded conductor is six times the skin depth of the 60 Hz ac power. The 50 mm-diameter core is rated to carry 1000 A at 138 kV.

surface currents produce no interior magnetic field). Figure 1x.8 plots the skin depth in copper conductors from 10 Hz to 10 GHz.

Because skin depth decreases with increasing frequency (as $1/\sqrt{f}$), it becomes particularly important at high frequencies, where it is largely responsible for losses in transmission lines (see the Appendix on Transmission Lines). At radio frequencies the skin depth is so shallow (e.g., $\delta = 10\,\mu\text{m}$ at 40 MHz) that you can make low-loss connections, inductors, and so on, by silver plating a poor conductor. A common technique for shielding lightweight instruments and computers at radio frequencies is to apply a thin metallic plating to a plastic enclosure. But it's important to realize that skin depth effects are significant even at low frequencies – for instance, current at the powerline frequency of 60 Hz in a copper conductor is confined to a surface layer of roughly 1 cm. (Copper's skin depth at room temperature is given by $\delta(\text{Cu}) = 6.6/\sqrt{f}$ cm.) For wire much thicker than that, power transmission losses decrease only proportional to the diameter of the wire, rather than to its cross-sectional area (i.e., diameter *squared*).

Fortified with this knowledge, you would not expect to see 60 Hz power transmission cables thicker than about 1 cm; or if they are larger in diameter, they are probably hollow. Imagine our surprise, then, when we encountered the 1000 A copper power cable pictured in Figure 1x.9, with its impressive 50 mm core. There's an elegant trick in operation here: to equalize the currents flowing in each wire strand, the cable is configured so that the individual strands weave in and out along the length of the cable, so on average each strand visits the cable's interior uniformly. Since every strand does this, their currents must be the same: voilà – the cable carries uniform current throughout

its cross-sectional profile, thus violating that high-falutin' skin-depth theory you learned in school![19]

For comparison, we calculated[20] the distribution of 60 Hz ac current[21] that would flow in a *solid* copper core of the same 50 mm diameter, graphed in Figure 1x.11. The

---

[19] There's a bit more to it, as usual in the real world: the strands have to be mildly insulated from each other, implemented in practice with a thin insulating coating. As the inventor Humphreys Milliken put it, in his 1933 patent (#1,904,162, "Electrical Cable"), "The insulation between adjacent strands need be only sufficient to offer a contact resistance between adjacent strands, many times as great as the resistance of one strand for a length of one turn in its progress around the cable, which is in the order of a few thousandths of an ohm." There are other tricks, also: for example, to make the cable manufacturable Milliken teaches how to build it in sectors (five are used in the cable in Fig. 1x.9), each of which is made from forming a twisted bundle of strands into radial wedges. The patent makes great reading – it's a model of clarity, mercifully short (3 pages of text, and just one claim), and with fine illustrations, one of which is reproduced here as Fig. 1x.10.

[20] It involves Bessel functions, as you might guess from the cylindrical geometry; see Ramo, Whinnery, and van Duzer's excellent reference *Fields and Waves in Communication Electronics*, Wiley (1965). We thank Mike Burns for doing the *Mathematica* calculations.

[21] This problem does not afflict *dc*, and large dc power cables are made with conventional stranded construction. DC is increasingly being used for megawatt-scale power transmission, often with a single conductor

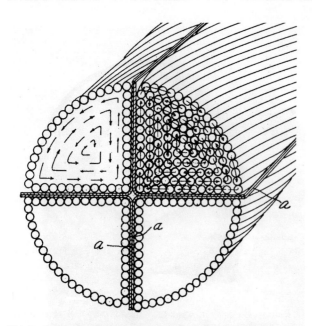

Figure 1x.10. The method invented nearly a century ago to do an end-run around the limitations of skin depth in large ac power cables. The little arrows show the twist pattern of copper strands in each sector of a very thick ($D \gg \delta$) cable. (Fig. 3 of US Patent 1,904,162.)

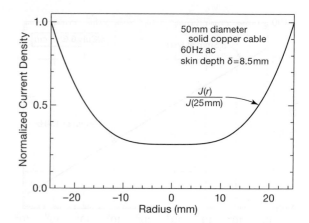

Figure 1x.11. Current density profile for a solid copper conductor of 50 mm diameter carrying 60 Hz ac. The effective ac resistance is approximately 1.7× that of the same-diameter Milliken-wound cable of Fig. 1x.9.

loss in such a cable, owing to reduced current density in the cable's core, is about 1.7 times that of the (uniform current density) Milliken cable. The latter has a resistance of 9 mΩ/km, thus an I^2R loss of 9 W/m at 1000 A ac.

A closely related scheme is used in radiofrequency coils, there known as *litz wire*,[22] in which the individually insulated fine strands are woven in a pattern that equalizes their participation throughout the cross-section of the wire. Litz wire inductors are seen in applications from tens of kilohertz to a few megahertz. Historically, litz wire was in use well before Milliken's 1930 patent application (there are cites going back to 1923), and undoubtedly informed his adaptation of the concept to high-power ac transmission cables.

1x.1.7 Capacitive and magnetic coupling

Signals go where you don't want them to. Often this is caused by capacitive (via the electric field) coupling from signals on adjacent wiring; in that case the coupling is greater into a high-impedance circuit, and more so at higher

frequencies (or faster edge-rates). Perhaps less recognized, signals can couple inductively (via the magnetic field) into circuit loops; in that case the coupled currents are greater between low-impedance circuits.

Which dominates? It depends. We rigged up a test circuit, consisting of adjacent conductors of a coarse-pitch FPC (flexible printed circuit) ribbon (Fig. 1x.12), then we drove one line with a 1 Vpp 1.25 MHz square wave and observed the signal at both ends of the adjacent line. Let's call the driven line the *aggressor*, and the passive line the *victim*. With high-impedance terminations (R_L=10 kΩ at all three ends) the coupling is capacitive, with signals of the same polarity observed at both ends of the victim line.[23] The coupled waveforms (Fig. 1x.13) exhibit the usual exponential decay of a square wave passing through an *RC* highpass filter.

The situation is quite different with the lines terminated in a low impedance (R_L=50 Ω, Fig. 1x.14). The capacitively coupled current is the same as before, but it generates negligible drop across the 25 Ω-terminated victim. But the larger current (40 mApp) in the aggressor creates a magnetic field that induces a current in the (low-impedance) victim loop. Magnetic coupling induces an EMF proportional to dB/dt (the time rate-of-change of magnetic field), so we see a differentiated waveform; furthermore, because the induced current flows in a loop, through both 50 Ω terminating resistors, the polarity is reversed at the opposite ends, as seen in the lower trace pair of Figure 1x.14.

(with ground as the return path); see the products of companies such as ABB Power Grids (owned by Hitachi) for some stunning examples.

[22] From the German *Litzendraht*, or woven wire.

[23] The 20 cm line is short compared with the 240 m signal wavelength of the 1.25 MHz square wave.

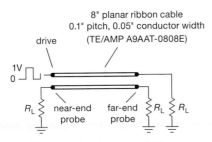

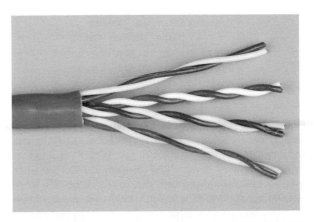

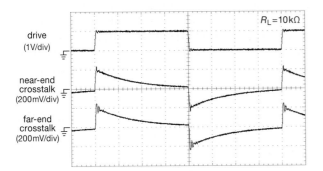

Figure 1x.12. Test circuit to explore coupling from adjacent parallel conductor. Depending on the termination resistance R_L, the coupling can be dominated by capacitive or inductive effects (see Figs. 1x.13 and 1x.14).

Figure 1x.15. Ethernet cable consists of four twisted pairs. This sample is "category-5e" unshielded twisted-pair (UTP), with each pair twisted with a different pitch to reduce signal cross-talk.

Figure 1x.13. Coupling of a 1.25 MHz square wave to an adjacent undriven ribbon conductor. With 10 kΩ termination, capacitive (electric field) coupling dominates, with the same coupled polarity at both ends. Horizontal: 100 ns/div.

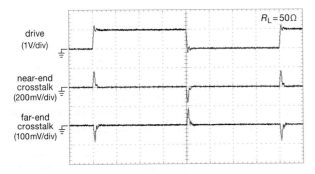

Figure 1x.14. Coupling of a 1.25 MHz square wave to an adjacent undriven ribbon conductor. With 50 Ω termination, inductive (magnetic field) coupling dominates, with opposite induced polarity at the ends. Horizontal: 100 ns/div.

1x.1.8 Mitigation of coupled signals

What to do? Capacitive coupling is less of a problem in low-impedance circuits, and it's suppressed entirely by a conductive ground shield (e.g., coaxial cable, or shielded twisted pair). In general it's best to minimize proximity of unshielded signals, and particularly those carrying signals of high frequency or fast edge rates. On a PCB you can arrange ground planes, and microstrip or stripline geometries (Fig. 1x.4), to reduce fringing fields. Other signal-carrying PCB traces can be routed so that they don't run parallel.

But magnetic fields, particularly at low frequencies, penetrate non-ferrous conductors.[24] You *can* construct magnetic shielding, using layered ferromagnetic materials: low magnetic permeability for the outer shield (to prevent saturation) surrounding high-mu inner layers (for effective shielding). But for most situations you can accomplish the task by avoiding circuit loops whose area permits magnetic induction. For example, twisted wire pairs do a pretty good job, as do coaxial conductors. Ethernet cables consist of four twisted pairs (Fig. 1x.15), with a different winding pitch on each pair to reduce signal "crosstalk" still further (Fig. 1x.16). The twisted-pair construction is effective enough that most Ethernet cable is unshielded ("UTP" – unshielded twisted-pair), but shielded pairs are used for the most demanding applications (Fig. 1x.17E).

1x.1.9 Shielded enclosures

In situations where you need a space free from electromagnetic signals, you can construct (or buy) a *Faraday cage* conductive enclosure. For example, look at the offerings from ETS-Lindgren,[25] where you can choose from tabletop size to full walk-in (or *drive*-in!) habitations. These enclosures are fully shielded, including shielded vents and utility penetrations, and conductive finger-stock strips at movable

---

[24] More on this in §1x.1.9.

[25] www.ets-lindgren.com/testenclosures.

Crosstalk (dB, min) at 100MHz			
	Cat5e	Cat6/6a	Cat7
NEXT	−30.1	−39.9	−62.9
PSNEXT	−27.1	−37.1	−59.9
ACRF	−17.4	−23.3	−44.4
PSACRF	−14.4	−20.3	−41.4

Figure 1x.16. Ethernet cable crosstalk specifications (100 m length), from the 2009 standard (ANSI/TIA-568-C.2). NEXT is near-end crosstalk ratio from an adjacent pair; PSNEXT is similar, but with all other pairs driven. ACRF (formerly ELFEXT) is far-end crosstalk ratio from an adjacent pair, compared with the far-end received signal; PSACRF is similar, but with all other pairs driven.

door openings (Fig. 1x.18); their illustrated catalog shows why it may be better not to attempt to build your own, but to leave the construction of shielded enclosures to the professionals.

There are various specifications for shielding effectiveness of shielded enclosures. One often quoted is NSA 65-6 "Specification for Shielded Enclosures" and its follow-ons (NSA 73-2A, 89-02, 94-106, etc.); the minimum attenuation versus frequency from that document (for both electric and magnetic fields) is graphed in Figure 1x.19.

Of note is the poorer attenuation of low-frequency magnetic fields, for which effective shielding often takes the form of layers of ferromagnetic material (iron, mu-metal, etc.). That is because the skin depth in materials of finite conductivity increases as the inverse square root of frequency (and inversely proportional to the conductivity), as seen for example in Figure 1x.8, where even an excellent conductor (copper) requires nearly a centimeter of thickness to produce even a modest attenuation of an applied 60 Hz magnetic field.

This would appear to render non-magnetic materials pretty much useless for low-frequency magnetic shielding. However, the situation is not so bleak: What is not generally recognized[26] is that a fully-closed conductive shell acts effectively as a far thicker conductor (for the purposes of skin effect) than its actual wall thickness. In fact it can be shown that a closed box of dimension D, with conductive walls of thickness x, attenuates low-frequency uniform magnetic fields as if its walls had a thickness of approximately the geometric mean of those dimensions, i.e., $x' \approx \sqrt{xD}$. Intuitively this is plausible, because the closed path provides a large-area loop to capture magnetic fields and produce a compensating circulating current, evidently

far more effective than the eddy currents circulating locally within the thickness of the conducting shell.

To verify this remarkable claim, we soldered together a 15 cm cube from "1 oz" copper-clad PCB (Fig. 1x.20), for which its 36 μm thick copper layer corresponds to a skin depth at 3.4 MHz; at frequencies well below that the copper-clad material is transparent to oscillating magnetic fields. The "geometric mean" rule, in contrast, predicts that the 0.0036 cm walls of the closed box should attenuate like a 0.23 cm thick copper plate, which is a skin depth in copper at the much lower frequency of 820 Hz. So the test: our box should let through magnetic fields below a kilohertz, but largely attenuate magnetic fields above, say, 10 kHz. To test this we placed the box in a Helmholtz coil pair (Fig. 1x.21), series-resonated with a film capacitor at each of four frequencies (1 kHz, 3 kHz, 10 kHz, and 30 kHz), and driven with an ac current of 2 App; then we measured the interior field with a Hall-effect magnetic probe, comparing the observed field with the applied field.[27] Figure 1x.22 shows the setup.

Did the closed box attenuate as predicted? Indeed! Here are the measured attenuation values (ratio of the interior ac magnetic field to the field in the absence of the box), which we've plotted on log-log axes in Figure 1x.23.

Frequency	Attenuation
1 kHz	1.2 dB
3 kHz	5.7 dB
10 kHz	14.2 dB
30 kHz	21.1 dB

Of importance: for this purpose a "closed box" requires conductivity across all seams comparable to that of the sheet material itself; screwed-together plates are ineffective. We measured the shielding attenuation of our box (before soldering the seams) with all sides joined with short lengths of wire, and found only fractional dB attenuation at 3 kHz and below; in fact that configuration behaved no

---

[26] We are indebted to Richard Garwin for educating the authors on this remarkable fact.

[27] Some details: The Helmholtz coil pair's diameter was 75 cm, with a parallel inductance of 4.7 mH. We used resonating capacitances of 5 μF, 0.6 μF, 55 nF, and 6 nF at the four frequencies, with a 5 Ω non-inductive series current-sensing resistor. The magnetic field was sensed with a DRV5053VA Hall-effect device (9 mV/gauss), and measured with a SRS530 lock-in amplifier. An interesting bump-in-the-road: at first the Hall probe gave a plausible result, but (implausibly) independent of orientation! Turns out we had not yet turned on its dc supply, and what we were seeing was the ac electric field, capacitively coupled to the exposed sensor (the resonant voltage swing at 10 kHz was ±300 V!) So we wrapped the sensor in a magnetic-field-transparent electrostatic shield.

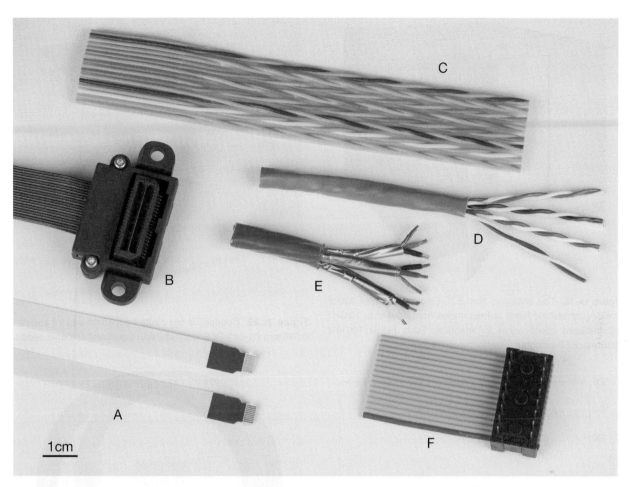

Figure 1x.17. A selection of cables for high-speed signals. A. FFC (flexible flat cable), sometimes called FPC (flexible printed circuit); this sample has 10 conductors on 0.5 mm pitch. B. 40-coax 50 Ω ribbon, terminated in Samtec QSE-series 0.8 mm pitch high-speed ground-plane connectors. C. Amphenol Twist 'N' Flat 8-pair ribbon. D. Cat-5e unshielded twisted-pair (see Fig. 1x.15). E. Individually shielded twisted-pair (STP) cable. F. Unshielded flat ribbon, 0.050″ pitch, terminated in DIP-16 plug. You can get FFC/FPC cables with 1 mm pitch, which are more forgiving of crooked insertion. Jumpers made by rolling wire flat can carry lots of current; but you can also get plated Kapton cables, which are excellent for connecting to cold or hot plates.

differently than a box of six isolated sides with no connections at all.

A Parting Shot

There are many excellent references[28] on the thorny topics of shielding and grounding, which is far more complicated that this short treatment might suggest.

---

[28] For example, (1) Analog Devices MT-095 Tutorial "EMI, RFI, and Shielding Concepts"; (2) H.W. Johnson and M. Graham, *High-Speed Digital Design, A Handbook of Black Magic*, Prentice-Hall (1993); (3) R. Morrison, *Grounding and Shielding Techniques in Instrumentation*, Wiley (1986).

1x.1.10 Connectors

On our bench we used to keep the current DigiKey catalog (the single most useful source of electronic components – get on their mailing list!); the last one that was published as a printed volume had a product index that is four columns wide and seven pages long: roughly 3000 entries in all. The largest single category was "Connectors" ("Integrated Circuits/Semiconductors" is a close second), with 288 major sub-categories, and with the actual entries spanning 435 pages of fine print.[29]

---

[29] Connector catalogs are full of things that they don't actually stock but would be happy to make if you order 100k pieces. Always check for

Figure 1x.18. This enclosure from ETS-Lindgren provides 120 dB shielding of electric fields at frequencies from 10 kHz to 10 GHz, with reduced effectiveness for magnetic fields below 100 kHz. (courtesy ETS-Lindgren, used with permission)

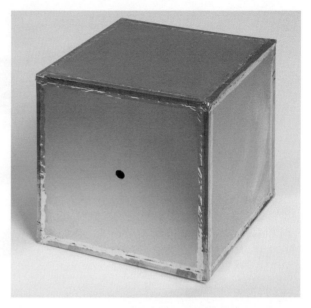

Figure 1x.20. Conductive box made from 15 cm square sheets of 0.036 mm ("1-oz") PCB laminate, with taped and soldered seams.

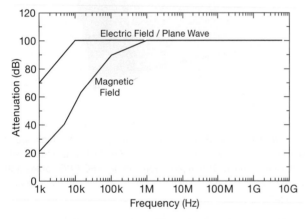

Figure 1x.19. Electromagnetic Attenuation Requirements as illustrated in Figure 1 of NSA 94-106 (24 October 1994, superseding specification NSA 65-6, 30 October 1964).

The point is that connectors are serious business! And not to be taken lightly. As we remarked in earlier book editions,[30] "The most unreliable components in any electronic system will be the following (worst first): **1.** Connectors and cables. **2.** Switches. **3.** Potentiometers and trimmers."

The photographs in Chapter 1 of AoE3 give some idea

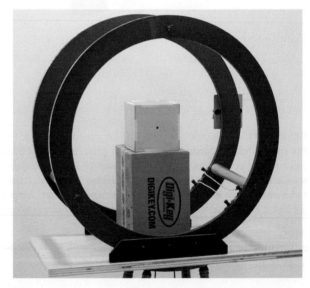

Figure 1x.21. We resonated this Helmholtz coil with series capacitors, and drove it at several frequencies in the kilohertz range to measure the magnetic-field attenuation of the metal-foil box (sitting atop an always-available DigiKey empty box).

of the variety, there categorized as "rectangular," "circular," and "RF and shielded." There are additional ways to slice it up, for example "mass termination" connectors (for ribbon cable[31]) versus individual wire terminations; or panel-

---

ample distributor stock (octopart.com is helpful) before designing in any part, but connectors *especially*.

[30] *The Art of Electronics*, 2nd edition, 1989; page 858.

[31] Also known as IDC (*insulation displacement connectors*). When IDC

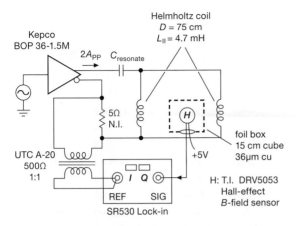

Figure 1x.22. Circuit for measuring low-frequency shielding effectiveness of closed conductive-foil box. Resonating film capacitor C_{res} creates a low impedance at resonance, easily driven to several amps with a few volts sinewave drive. The audio transformer provides reference-signal galvanic isolation to prevent ground loops.

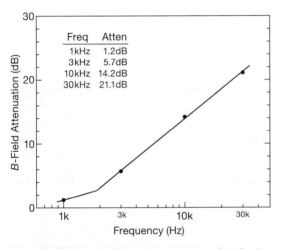

Freq	Atten
1kHz	1.2dB
3kHz	5.7dB
10kHz	14.2dB
30kHz	21.1dB

Figure 1x.23. Measured attenuation of magnetic field for the configuration shown in Fig. 1x.21.

mount versus internal versus cable-joining connectors; or connectors for low-level signals (audio, RF, instrumentation) versus high voltage versus high current; or multiwire versus single-circuit connectors; and so on. As a general statement, you want to choose reliable connectors that are appropriate for the signals involved, and that provide suitable mechanical coupling.

Without going into detail, we'll reveal our biases; we have our favorites, and we have some connectors that we

is done on a connector with a set of wires in one operation, it's called MTA, for *mass termination assembly*.

would *never* use (except possibly at gunpoint). For **multiwire panel-mount** connectors we like the D-subminiature (rectangular) connector series, and particularly the high-quality "screw-machined" type; we also like the RM-series of 12 mm (circular) video connectors made by Hirose. For **higher currents** we like the "Winchester-type" MRA/MRAC shrouded and jackscrew-locking multipin connectors. For **multiwire PCB-mount** connectors we have been happy with the AMPMODU MT series of single-row and double-row socket connectors (for individual wires), and with the 3M series of mass-termination IDC connectors (for ribbon cable). For **50 Ω RF** connectors we like the coaxial SMA and SMB, which can also be used as simple shielded connectors for **low-level** signals. For the latter the standard has been (and doubtless will continue to be) the classic BNC connector; but we have experienced problems with inadequate ground integrity, particularly as these connectors wear from repeated use, and we prefer the less popular TNC. For **high voltage** we like the SHV connector.

Now for the connector types to avoid (reasons withheld – just *trust us!*): Stay away from the "RCA audio connector" (used universally on consumer audio equipment), the "Jones" type multipin connector, the "UHF" and "F" type RF connectors (the latter used universally in consumer-grade television, cable, and satellite), the "blue hexagon" connector, and the 1/8″ and 1/4″ headphone connectors. At a lesser level of dislike, we're not particularly wild about the circular DIN connectors (both regular and mini), the 5 mm (and smaller) coaxial "dc power entry" connectors, and the multipin threaded "microphone connectors" (but the professional XLR series is fine).

1x.1.11 Connectors for RF and high-speed signals

As detailed in Appendix H in AoE3, when dealing with signals at high frequencies (or signals with fast transitions) it is essential to terminate signal-carrying lines with an impedance equal to the cable's characteristic impedance Z_0, in order to prevent reflections from impedance mismatches. The assumption here is that the cable itself has a well-defined characteristic impedance, constant along the length of the cable; if not, reflections will be created at locations of discontinuity in the cable's impedance.

The same goes for a transition at a connector, which ideally should maintain the cable's characteristic impedance across the joint. This becomes more critical, and more difficult, at high frequencies; specialized connectors are needed

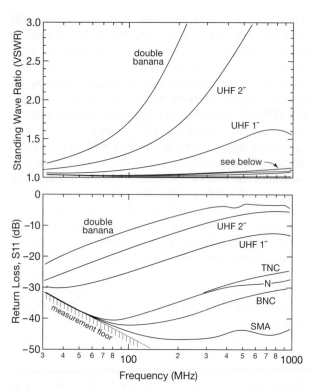

Figure 1x.24. Measured reflection coefficients versus frequency for a selection of connectors. For the worst offenders the corresponding VSWR is also shown. Data courtesy of John S. Huggins, see www.hamradio.me.

once you're operating above, say, 10 GHz.[32] The measured data in Figure 1x.24 illustrates this point, here at more modest frequencies. Among other things, it nicely confirms our personal bias against the euphemistically named "UHF connector."

Table 1x.3 lists most RF and microwave connector types and their maximum operating frequencies, where you can see that the highest frequencies correlate with the smallest diameter (to avoid multimode propagation). A drawback of small connectors is their lower power-handling capability, as seen in Figure 1x.25.

1x.1.12 High-density connectors

In the good ol' days of vacuum tubes, filament transformers, and big chassis with black crinkle finish, we were

---

Table 1x.3: RF Connectors, Approximate f_{max}

Type	f_{max} (GHz)	Type	f_{max} (GHz)
UHF	0.3	SSMA	12.4-36[b]
BNC	4	superSMA	27
SMB	4	3.5mm[a]	34
MCX	6	2.92mm[a,c]	40
MMCX	6	2.40mm	50
SMC	10	APC-7	50
TNC	18	1.85mm	65
SMA[a]	12-26[b]	0.9mm	67
N	18	1.0mm	110

Notes: (a) SMA, 2.92mm, and 3.5mm can mate interchangeably. (b) Variants available with different f_{max}. (c) Also known as a "K" connector.

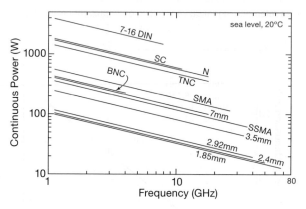

Figure 1x.25. Continuous power-handling capability of a selection of RF and microwave connectors.

happy with connectors that could handle, oh, ten conductors, and weren't much larger than an apricot. We considered the "D-subminiature" connectors really dense, as evidenced by the name.

Things have changed, and contemporary electronics deals in small-footprint ICs with 1000+ "pins," and passive components smaller than a grain of rice. Connectors have to keep up! These come in many flavors – mainly panel-mount, board-to-board, and board-to-cable. The photograph in Figure 1x.26 shows a representative sample from our lab, alongside that D-subminiature of yesteryear (but still a good choice for a reliable panel-mount connector for 9 to 50 pins).

When choosing a high-density connector (assuming you're not constrained by existing components or cables), be sure to consider factors like contact resistance, current-carrying capacity, number of mate–unmate life cycles, and

---

[32] For example, to maintain constant impedance you may have to taper the outer shell and dielectric (if any) to accommodate the changing diameter of the inner conductor. A smaller overall diameter is necessary at the highest frequencies to suppress additional "modes" of wave propagation.

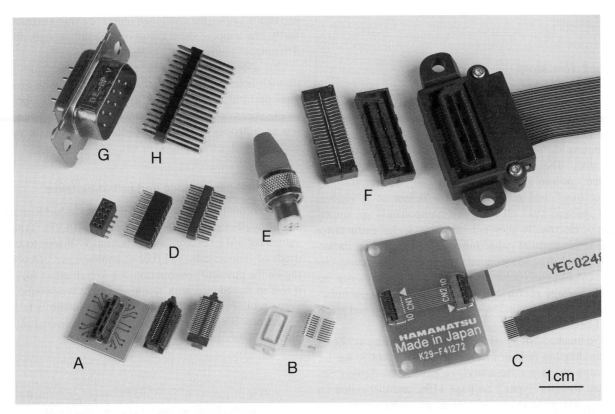

Figure 1x.26. High-density connectors. A. Samtec SS4-20-3.50-L-D-K-TR 0.4 mm-pitch board-to-board connector, good to >15 Gbps, shown here with mating Hamamatsu 4×4 silicon photomultiplier array. B. Hirose DF17(3.0)-20DS-0.5V(57) 0.5 mm-pitch board-to-board connector. C. FFC (flexible flat cable) and mating connector, 0.5 mm pitch. D. Dual-row 0.050″ pin header and socket strips. E. Microtec xx-7 connector for 7-wire shielded cable. F. Samtec QSE-series 40-pin 0.8 mm-pitch high-speed ground-plane connectors and mating cable-end connector with 40 true-coax 50 Ω cables. G and H. DE-9P D-subminiature male connector and 0.1″ pitch dual pin header, for size comparison; in this photo they are giants – hardly "subminiature."

robustness (e.g., surface-mount versus through-hole, presence of mounting flange, shrouded mating area, and so on).

1x.1.13 Connector miscellany

(1) For good reliability, avoid surface-mount connectors where repeated insertions, vibration, or other aggravating forces may detach the delicate traces. Better to use press-fit (which creates a cold weld) or soldered through-hole connectors.

(2) Vibration or repeated thermal expansion and contraction can cause *fretting* of contact material, in which tiny relative movement of contacting metals creates granular wear and consequent corrosion.[33] This does not occur with gold contacts, until there's sufficient wear to expose the under-lying metal. To prevent fretting, prevent the relative motion! (Allow the mating connector to "float," or mechanically anchor the mating parts, etc.) There are dry lubricants that suppress fretting when some relative movement is inevitable.

(3) Gold-plated contacts are desirable, but only when mated to gold. A gold–tin connection is no better than tin–tin.

(4) Insulation-displacement connectors have received criticism for their tendency to fail in the long term. There are many varieties available, and some are better than others. We're told that the inexpensive Tyco 640440-3 (3-pin, low density) is reliable. And we've had good reliability with 3M's IDC ribbon connectors.

---

[33] There's a good discussion at
suddendocs.samtec.com/productspecs/fretting-corrosion.pdf.

1x.2 Resistors

Figure 1.2 in AoE3 shows the range of resistor types, from tiny surface-mount chips to giant wirewound power resistors. The most important characteristics are **resistance**, **power rating**, **tolerance** (accuracy), **stability** (over time), and **temperature coefficient** of resistance. But resistors (like all electronic components) are imperfect – electrically they exhibit some **series inductance** and some **parallel capacitance**,[34] which become important in high-frequency circuits and in power-switching circuits. Additional departures from ideal performance include **voltage coefficient** of resistance and **excess noise**; these are important in low-distortion, low-noise, and precision circuits.

We touched briefly on these less-than-sterling attributes of the humble resistor in several places in AoE3; see for example the Box ("Resistors") on page 5, the Table ("Selected Resistor Types") on page 1106, and discussion on pages 300, 476, and 697–98. Here we elaborate on some of these neglected aspects of a component often taken for granted.

1x.2.1 Temperature coefficient

The ubiquitous thick-film SMT chip resistor (e.g., Vishay CRCW-series) typically has a specified tempco of ±200 or ±100 ppm/°C (designated in the manufacturer's part number). But if you need better, you can get low tempco SMT resistors, for example the inexpensive Panasonic ERJ-xRBD or -xRHD series (±50 ppm/°C), which cost about $0.07 in full-reel quantities (compared with $0.003 for the commodity CRCW types). Still better are some thin-film SMT parts, for example the Panasonic ERA-xAR series or Yageo RTxxxxxRB series (±10 ppm/°C), which cost about $0.18 in full-reel quantities, or the Vishay TNPU-Z series (±5 ppm/°C, $1 in full-reel qty).

For the absolute lowest tempco you can get metal-foil ("Z-foil") SMT resistors from Vishay (VSMP-series, ±0.2 ppm/°C), which exploit a clever thermal compensation trick by bonding the metal foil element to a carefully chosen ceramic substrate whose mechanical coefficient of

expansion causes the combined object to exhibit extraordinarily low tempco; these things cost plenty, though, about $10 apiece.

The above are SMT types; you can, of course, get through-hole (axial or radial lead) resistors with analogous performance. Additional types are available, for example wirewound resistors, which come with tempcos as low as ±20 ppm/°C (though typically they are in the ordinary range of ±100 ppm/°C or so).

1x.2.2 Self-capacitance and self-inductance

Real resistors have some equivalent series inductance and some distributed shunt capacitance (Fig. 1x.27). Typical values for SMT resistors are in the range of tens to hundreds of femtofarads, and 0.01–2 nanohenrys.[35] Depending on the physical construction, the overall effect may be a rise in impedance with frequency (e.g., wirewound resistors), or, if the parallel capacitance dominates, a falling impedance. Both trends can be seen in the measured $|Z(f)|$ plots of Figure 1x.28.

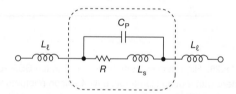

Figure 1x.27. Simplified resistor model, showing parasitic inductances and capacitance. The external inductances L_l represent the inductive contributions of the leads to the impedance of the resistor body (enclosed in dashes).

To explore this further, we measured the impedance of a set of wirewound resistors of the same construction (the classic Ohmite Brown Devil®), with the results plotted in Figure 1x.29. Evidently the inductive contribution dominates at megahertz frequencies, more so for the lower resistor values. For applications at high frequencies, noninductive wirewound resistor types largely eliminate the problem.

What about the parallel capacitance C_p of the model of Figure 1x.27? At some frequency it should form a (damped) parallel resonant circuit, an effect that can be seen in the measured data of Figure 1x.31, where we've extended the frequencies out to 300 MHz for three of the wirewound resistors of Figure 1x.29. You can use a trick to largely compensate for this unseemly behavior, namely

---

[34] Which may be more complicated than a single series L and parallel C, because they are distributed throughout the resistor.

[35] See, for example, Vishay Technical Note 60107, "Frequency Response of Thin Film Chip Resistors."

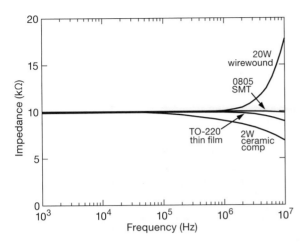

Figure 1x.28. Measured impedance (magnitude) versus frequency for four resistor types. At high frequencies the effects of parasitic inductance and capacitance cause a deviation: upward for wirewound resistors (inductance dominates), downward for non-inductive construction (capacitance dominates).

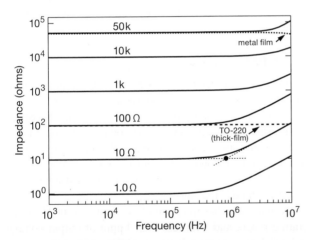

Figure 1x.29. The inductance of wirewound resistors causes the total impedance to rise at high frequencies, as seen in measured data of 20 W "brown devil" types. If this matters in your application, you can get non-inductive wirewound resistors that use bifilar (two windings, connected together at one end), or Ayrton–Perry (two counterwound windings in parallel) winding geometries; the simpler bifilar winding suffers from much higher parasitic capacitance. A "breakpoint" frequency (black dot) fully characterizes a resistor with series inductance (see the wirewound resistor zoo in Fig. 1x.30).

a series *RC* attached across the offending resistor, with *R* equal to the nominal resistance, and *C* selected to flatten the impedance curve.

Once we had the measurement rig set up, we couldn't resist (pun) running a bunch of resistors (of various resis-

tances, and various construction) through it. They all exhibit curves similar to those in Figure 1x.29; to keep the figure uncluttered we plotted just the breakpoints (intersection of nominal resistance with the extrapolated upward slope, see the example in Fig. 1x.29). Figure 1x.30 shows the resulting scatterplot.

The best performers (breakpoints at the highest frequencies) are the carbon composition (RC07 type), the non-inductive Ohmite WN-type (Ayrton–Perry zigzag winding), and the surface-mount small wirewound type. By contrast, the losers are the traditional large-geometry wirewound power resistors. However, some of the latter are available in non-inductive versions: for the Vishay/Dale RS, RH, and LVR types you can get NS, NH, and NI as Ayrton–Perry non-inductive variants.

1x.2.3 Nonlinearity (voltage coefficient)

An ideal resistor maintains $I=V/R$ over time, temperature, frequency, and applied voltage. In the real world resistors exhibit deviations from perfection. A not-insignificant effect is *nonlinearity* – an effective change of resistance with applied voltage.

You can find worst-case specifications in some datasheets: for example, although the commodity Vishay CRCW-style thick-film SMT resistors do not specify a voltage coefficient, their PCAN-series thin-film resistors specify a worst-case resistance change of 0.1 ppm/V, the same as the best-in-class metal foil or metal film types such as the Vishay VSMP and Z-foil series resistors.

Out of curiosity we measured the resistance change versus voltage for a selection of resistor types. You often use a high-resistance voltage divider to monitor a high-voltage dc supply, so we tested high-resistance parts at voltages to 1000 V. Figure 1x.32 shows the results, plotted as log–log and log–linear. The thick-film resistors (curves C–H) are better by some two to three orders of magnitude, compared with the traditional carbon composition type.

Carbon-composition resistors are largely a relic of the past (though they excel in peak power endurance, see §1x.2.6). Sticking with the thick-film types, we explored the nonlinearity versus resistance (for a fixed size), and nonlinearity versus size (for a fixed resistance). Figures 1x.33 and 1x.34 plot the measured results, showing that the nonlinearity increases dramatically with increasing resistance and with decreasing physical size.

When does nonlinearity matter? For low-distortion amplifiers and oscillators, certainly. Also for precision low-voltage monitoring and control of a high voltage source. Note, however, that for the latter what you care about is a

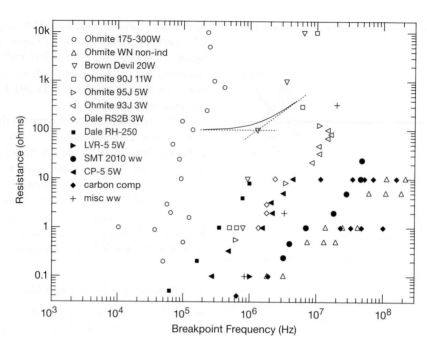

Figure 1x.30. Measured *RL* breakpoints (see Fig. 1x.29) for a sampling of power resistor types. All are through-hole wirewound, except as indicated.

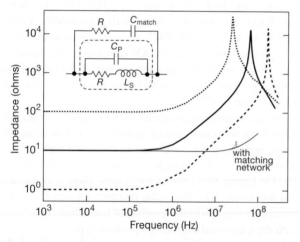

Figure 1x.31. Extending the measurements to 300 MHz reveals *LC* parallel resonances. The high-frequency artifacts are largely suppressed by attaching a matching network of 10 Ω in series with 16.5 nF in parallel with the 10 Ω wirewound resistor. Watch out for the amount of power dissipated in the matching resistor.

precise resistor *ratio*, for which you should probably be using a resistive divider that is designed to maintain a stable ratio as the applied voltage varies – see §1x.2.7, below.

1x.2.4 Excess noise

In our extensive discussion of noise in Chapter 8 we introduced the business of *excess noise* in resistors (§8.1.3); this effect is essentially a fluctuation in resistance, which manifests itself as an added noise voltage (i.e., in addition to Johnson noise, which depends only on the resistance) when current is flowing through the resistor. We've been visited by this phenomenon in some recent instruments we designed; here is the story:

We built a high-voltage amplifier with ±1200 V of operating range, and with less than 1 ppm of output voltage noise.[36] For some experiments at CERN, 160 of these were machine assembled. About 10% of them failed to meet the low-noise goals, with an excess noise level that increased with output voltage. The amplifiers used an Ohmite 150M 1.25 W SM103 high-voltage resistor in the feedback loop. This "Slim-Mox" thick-film-on-Alumina precision planar resistor is 15 mm long and is rated at 7.5 kV. After replacing the offenders we wound up with a small collection of noisy 7.5 kV resistors. Evidently the resistive material has domains that change under the influence of electric fields. (Perhaps the problem would have been avoided if we had made the feedback resistor from fifteen standard 10.0M re-

---

[36] To learn more about the amplifier, ask about the AMP-37 UberElvis project.

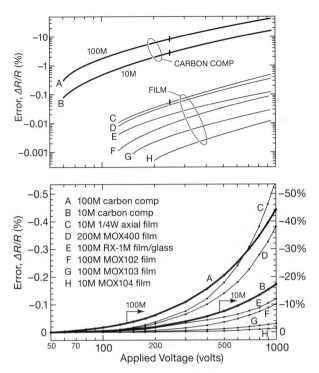

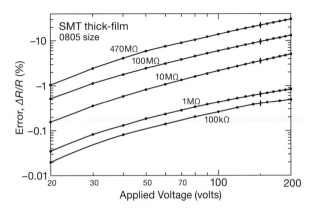

Figure 1x.33. Measured change of resistance for 0805-size thick-film surface-mount resistors, officially rated to 150 V maximum. The nonlinearity increases with resistor value.

Figure 1x.32. Measured change of resistance versus applied voltage for eight resistor types. The 1/4-watt carbon composition (RC07, plots A and B, scale on right) and film (plot C) resistors are rated only to 250 V (marked with vertical stroke), a limit we ignored in our enthusiasm.

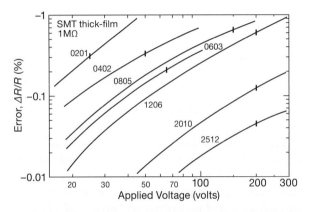

Figure 1x.34. Measured change of resistance for 1 MΩ thick-film surface-mount resistors of different sizes; the rated voltage for each is indicated by vertical strokes. The nonlinearity decreases with increasing physical size.

sistors in series, each limited to less than 80 V in operation.)

1x.2.5 Current-sense resistors and Kelvin connection

We discussed the business of 4-wire sensing ("Kelvin connection") in many places in AoE3; see, for example, pp. 277–78, 294, 350, 365–67, 898, and 1070–71. The basic idea is to eliminate the error in a current measurement by sensing the voltage drop across the current-sensing resistor (often of very low resistance, less than an ohm) with a separate pair of wires (Fig. 1x.35). In that figure, for example, you would suffer a +20% error in the measured current, if you had (foolishly) used the voltage drop between the pair of terminals themselves. Measuring instead the drop between the sense terminals eliminates this error. The assumption, of course, is that the sensing circuit (here the difference amplifier) draws negligible current; this is easily satisfied, especially in high-current circuits where

the sense resistor is of low resistance (and thus prone to error in a 2-wire configuration).

Current-sensing resistors come in an enormous range of current capabilities and physical sizes – see the montage in Figure 1x.36, photoshopped by the authors from datasheets of a half dozen manufacturers, where the scaling varies wildly among the specimens (the little guys labeled "R 010" are 40 times smaller than the big one on the top row, second from right).

1x.2.6 Power-handling capability and transient power

In circuits with pulse waveforms you frequently have situations where components (resistors, diodes, transistors)

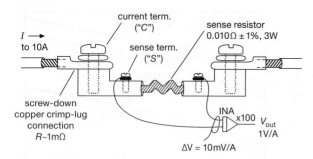

Figure 1x.35. A 4-wire (Kelvin) current-sensing resistor eliminates errors caused by imperfect connections. Here, for example, 4-wire sensing eliminates a 20% error that would be caused by just a milliohm of bolt-on lug resistance.

are subjected to peak power (during the pulse) that is well above the steady-state power rating. That's OK as long as the thermal pulse does not cause the component's temperature to exceed allowable limits. We discuss this further in §9x.25.8 in the context of semiconductor devices (MOSFETs, TVSs), where allowable pulse power is described by the *transient thermal resistance* as a function of pulse duration, $R_{\Theta JC}(\tau)$.

Here we are interested in the humble resistor, where the same effect applies: the peak power during the pulse can be absorbed by the heat capacity of the resistor's mass, as long as the average power does not exceed the part's power rating. Some resistors are designed and specified for such "pulse-withstanding" service. This is usually specified with a graph of peak power (or "pulse load") $\hat{P}_{max}$ versus pulse duration. Figure 1x.37, shows such curves for resistors from seven datasheets, mostly of similar size (1206 SMT for all but curves A and F).

You can see some interesting trends in these plots. Curve B is a Vishay "pulse-proof" resistor, which does considerably better for short pulses than its more conventional curve B′ sibling. Resistor C exploits the thermal conductivity of aluminum nitride to permit high steady-state power (2 W), but with no special attention to short-pulse endurance. Resistor F uses a solid resistive carbon slug (rather than a resistive film), whose mass is able to absorb prodigious peak power (35 kW!) for up to a microsecond – not bad for a quarter-watt resistor. Yet, in spite of its larger size (roughly triple the footprint of the other resistors), it falls below the rest of the pack for pulse durations greater than 10 ms.

In Figure 1x.38 we adapted datasheet plots for some larger pulse-rated resistor types. Here you can see the impressive performance of ceramic composition resistors (the worthy successor to the once-ubiquitous carbon comp),

plots B1–B3; Tyco's CCR-series (plotted) are similar to Ohmite's OX (1 W) and OY (2 W), though the latter do not provide $\hat{P}_{max}$ versus t_i plots.

A word of caution: It is our belief that one should not place complete reliance on the kind of curves provided by manufacturers (Figs. 1x.37 and 1x.38); in part our skepticism is based on their qualitatively different shapes and slopes. For example, in Figure 1x.38 curves B–D have $P_{max} \propto 1/t_i$, whereas curves A, E, and F have $P_{max} \propto 1/\sqrt{t_i}$. If you intend to push these parts close to their limits, you may need to subject sample parts to your own testing (which, conveniently, we discuss next). Generally, though, it's better not to "twist the dragon's tail"; our advice is to derate resistors by 50%. Resistors of higher power rating can cost considerably more – in that case you can use the trick of connecting several low-power (and inexpensive) resistors in series or in parallel (the choice depending on whether you would be happier with "fail open" or "fail short.")

A. Do-it-yourself testing

The ever-creative John Larkin needed to find out the best choice for a 0.33 Ω resistor for a pulse-stress application. Not wanting to rely completely on datasheets, he built the apparatus of Figure 1x.39, which switches a bank of charged energy-storage capacitors across the victim resistor for a known duration, at a known repetition rate.[37] Being a friendly chap, he loaned us his gadget (Fig. 1x.40), which we promised to take good care of. But, um, the very first resistor we blasted emitted a fiery arc downward, causing a cascade that etched away the foil adjacent to the insulating gap (Fig. 1x.41). So much for taking care of valuable (and unique) borrowed equipment (sorry John). To make amends, we built a set of expendable bolt-on "mezzanine" daughterboards; as luck would have it, we have not experienced another foil-destroying explosion.

We tested to destruction many resistors with Larkin's toy, recording their final moments (with a 'scope), and their resting places (with a camera). Read on...

B. Overload to failure

At Larkin's suggestion, we modified the test jig to capture the current-vs-time waveform during extreme overload of a resistor; in this case we applied a 32 V dc step across several types of small 50 Ω resistors (thus 20 W dissipation in resistors rated from 0.25 W to 2 W). All but one were

---

[37] Based upon a set of measurements, he settled on the Vishay/Draloric type AC05 axial-lead 5 W cemented wirewound resistors.

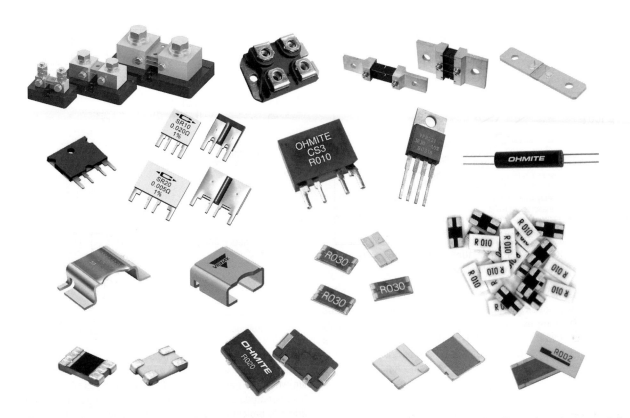

Figure 1x.36. Current-sensing 4-wire resistors range from tiny SMT parts to giant bolt-down 1000 A units. Shown here (not to scale!) are representative parts from six manufacturers (Bourns, Caddock, Ohmite, Riedon, Vishay Intertechnology, and VPG Foil Resistors; photographs used with written permission of the respective manufacturers).

surface-mount types, of size 1206 or similar, soldered to a pair of copper pads each $1''$ square ($6.5\,\mathrm{cm}^2$).

The tests were, uh, both noisy and smelly, in most cases becoming incandescent while belching noxious fumes. Figure 1x.42 includes some carcasses, and Figure 1x.43 shows the death throes of five victims. In the latter figure the initial current of 640 mA corresponds to 3.2 vertical divisions.

Trace A is a "commodity thick-film" chip resistor ($0.005 each in full-reel quantity!), rated at 0.25 W, with no pulse-load specifications; it failed quickly (to an open circuit) at this $80\times$ overload. Trace B is a 0.75 W "pulse-proof" variant ($0.05 each in full-reel quantity) from the same supplier (Vishay), which fares considerably better (lasting about 200 ms before partial opening, followed by arcing (the current spike around 1 s), some sputtering, and final failure. Its specified pulse endurance is curve B in Figure 1x.37.

Trace C is Vishay's 0.4 W CMA "high pulse-load" carbon-film resistor ($0.10 each in full-reel quantity), in

a MELF (cylindrical SMT) package that is only slightly larger than the 1206-size rectangular SMT packages of traces A, B, and D. It does not fare any better than the CRCW-HP (trace B), but, interestingly, it fails through a low-resistance phase (off-scale current spike) before finally opening up. Its specified pulse endurance is curve A in Figure 1x.37.

Trace D is Vishay's 2 W PCAN "high-power aluminum-nitride" SMT resistor ($0.87 each in full-reel quantity), with enlarged terminations to carry heat to the mounting foils; here we're punishing it with a $10\times$ overload, causing some intermittent current dips, but not enough to cause it to fail (even after 10 s). Its specified pulse endurance is curve C in Figure 1x.37.

Finally, trace E is the humble 0.25 W RC07-style (Ohmite OD-series) carbon-composition axial-lead (through-hole) resistor of yesteryear, with claimed "high surge capability" (curve F of Fig. 1x.37). These things used to be inexpensive, but nowadays you'll pay about $0.30 in quantity (50 times as much as a commodity SMT

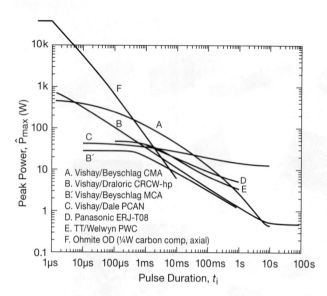

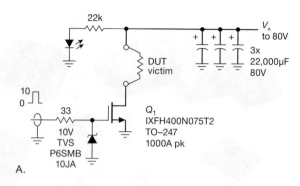

Figure 1x.37. Datasheet plots of single-pulse peak power versus pulse duration, for several pulse-rated SMT resistors. All are 1206 size (3.2×1.6 mm) except the CMA (plot A), which is a similarly sized MELF "0204" (cylindrical, 1.4 mm dia × 3.6 mm long) and the OD (plot F), which is an axial-lead carbon-composition resistor of approximate size "2510" (2.4 mm dia × 6.3 mm long).

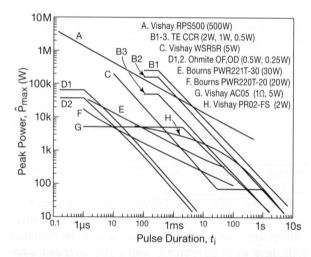

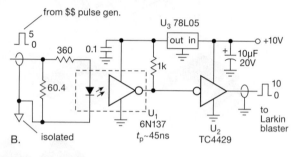

Figure 1x.38. Datasheet plots of single-pulse peak power versus pulse duration, for some larger pulse-rated resistors. Note that the "composition" types (B – ceramic comp, D – carbon comp) battle mightily their larger brethren, in spite of their diminutive size.

resistor). Its failure mode, like the carbon-film CMA, includes a low-resistance current surge and a fail-short endpoint. Its specified pulse endurance is curve F in Figure 1x.37.

Figure 1x.39. John Larkin's pulse-power torture machine. A. A power MOSFET with plenty of muscle switches V_+ across the victim when the gate is pulsed. B. With hundreds of amps flowing (and hundreds of joules of stored energy), we added this isolated driver to protect our expensive pulse generator in the event of a catastrophic fault. (Larkin accused us of excessive caution – but he didn't offer to buy a new pulse generator.)

1x.2.7 Resistor dividers

In §1x.2.3, above, we pointed out that run-of-the-mill film resistors have voltage coefficients of order 10–100 ppm/V (see Figs. 1x.33 and 1x.34), making them unsuitable for precision voltage dividers, especially in high-voltage applications; the best (and pricey!) metal foil types do considerably better, with voltage coefficients in the 0.1 ppm/V range.

But if what you want is a precise voltage *ratio* that does not vary with applied voltage (and, what the heck, stable with temperature as well), you can do no better than a pre-built precision voltage divider. Manufacturers like Caddock are happy to oblige: their 1776-C68 series of precision decade voltage dividers, which cost about $10–15 in unit quantities, have a ratio voltage coefficient of 0.04 ppm/V max (100 to 1200 V), and a ratio tempco of 5 ppm/°C max. These parts have several taps, for example a 9 MΩ part has taps at 900k, 90k, 9k, and 1k (thus voltage ratios of 10:1, 100:1, 1000:1, and 10000:1). And their USVD2 and HVD "ultra-precision voltage dividers"

Figure 1x.40. Photo of the Larkin-blaster. An input pulse train switches on the hefty MOSFET (good for 1000 A peak current), putting the bank of charged capacitors across the resistor.

offer even better linearity: 0.02 ppm/V ratio linearity, and 2 ppm/°C ratio tempco. These go up to 5000 V ratings, with a single ratio (choice of 100:1 or 1000:1); they cost about $40 in unit quantities.

The above parts are intended for high-voltage dividers. At the other end of the spectrum, there are lots of precision dividers intended for low-voltage applications, such as single-ended and difference amplifiers, low-voltage dc supplies, and bridge circuits. You can get these from some semiconductor manufacturers (LTC/ADI, Maxim) and, of course, from the all-encompassing Vishay group. Some examples of the former are Maxim's MAX5490 series, and LTC's LT5400 series, both in surface-mount packages. The MAX5490 is a 100k SOT23-3 surface-mount part with standard ratios (set by the part number) of 1:1, 2:1, 5:1, 10:1, and 25:1, ratio tempco of 2–4 ppm/°C max, and ratio voltage coefficient of 0.1 ppm/V typ. LTC/ADI's part has four separate resistors (two matched pairs) in an MSOP-8 package, with ratios of 1:1, 4:1, 5:1, and 9:1, ra-

tio tempco of 1 ppm/°C max, and ratio voltage coefficient of 0.1 ppm/V typ. These parts cost about $4 (Maxim) and $7–$20 or more, depending on grade (LTC/ADI).

Vishay aims at ultimate performance, with its metal foil (Z-foil) products (but you have to pay the price: $15 to $40 or more, depending on grade). Their wide selection comes in surface-mount (DSMZ, VFCD1505 series) and through-hole varieties (VFD244Z, VSH144Z, 300144Z, 300145Z, 300190Z–300212Z), with ratio tempcos from 0.1 ppm/°C to 4 ppm/°C typ (depending on resistance, ratio, and grade), and ratio voltage coefficients of 0.1 ppm/V max. Because these are constructed from metal foil elements, the resistance range is limited to a maximum of 100 kΩ. By contrast, the Caddock high-voltage parts go to 10 MΩ (1776 and USVD2 series) and 50 MΩ (HVD5 series).

Figure 1x.41. Unexpected result of a 1 Ω 3 W resistor's final fire-belching act – the plasma erupted downward, causing a cascading arc that devoured the foil on facing sides of the gap (compare with Fig. 1x.40).

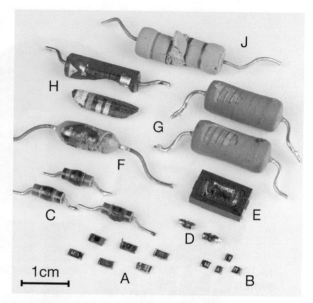

Figure 1x.42. Resistor graveyard: here are some residues of our resistor-torture experiments. A. SMT 1206-size parts that burned but were not consumed. B. SMT 1206 parts that blew into pieces. C. RC07-style 0.25 W carbon-composition resistors developed mid-section bulge. D. MELF carbon-film SMT parts also bulged. E. 5 W SMT 4527-size "power metal strip" resistor blew its top off. F. 3 W wirewound resistor belched fire. G. 5 W wirewound resistors exfoliated some of their skins. H. 2 W wirewound resistor exploded with a bang. J. 2 W ceramic composition was hard to kill, but 250 W for a half minute did the trick.

1x.2.8 "Digital" Resistors

The traditional mechanical potentiometer type of control has largely been superseded by the so-called digital potentiometer – an integrated series-connected array of fixed resistors, with the "wiper" replaced by an array of CMOS switches (Fig. 1x.44).[38] This has many advantages, some of which are (a) elimination of wiper noise, aging effects, and susceptibility to vibration, (b) electronic (digital) control, (c) cold switching, thus elimination of susceptible signal-carrying wiring, (d) accurate tracking of multi-ganged sections, and (e) small size. There are a few drawbacks, also – switch ON-resistance, distributed capacitance, digital signal coupling, limited voltage range, limited choice of resistance – but these are generally minor annoyances, certainly compared with the flaws of the traditional mechanically adjusted panel pot. Some manufacturers of digipots include Analog Devices, Intersil/Renesas, Maxim, and Microchip.

To give some perspective, old-timers will remember (and not with fondness) the early attempts to implement remote control of volume in amplifiers and TVs: a *motor-driven* panel pot! These graybeards will remember, also, the scratchy noise that accompanied rotation of volume controls in audio equipment (hence the admonition "never adjust a volume control while making a professional recording"). Digital potentiometers ("digipots") make it

easy to control the setting with logic signals (SPI, I²C, or UP/DOWN inputs); and they are free of scratchy-wiper noise (though they do exhibit switch-transition clicks). But don't confine your attention to panel-mounted controls – in fact, most digipots are used in trimming applications: sensor calibration, offset trim, current-source setpoint, voltage regulators, bias setpoint, laser drive current, and the like.

A. The digipot zoo

Digital pots come in a bewildering variety of styles and parameters, which we'll attempt to untangle in this section. Some choices include (a) total resistance, (b) number of steps, (c) volatile or non-volatile memory, (d) number of sections ("gangs"), (e) operating voltage range, (f) bandwidth, (g) linear or log steps, (h) digital interface, (i) ratio tolerance, resistance tolerance, and gang matching, (j) tempco, (k) wiper resistance, and (l) distortion. You can get an idea of the digipot family tree from Figure 1x.45, adapted from a figure found on Intersil/Renesas's 2017 product brochure.

---

[38] There are variations on this single-string theme, notably ADI's "segmented architecture," which reduces greatly the number of CMOS switches needed; see *Analog Dialog*, **45-08**, August 2011.

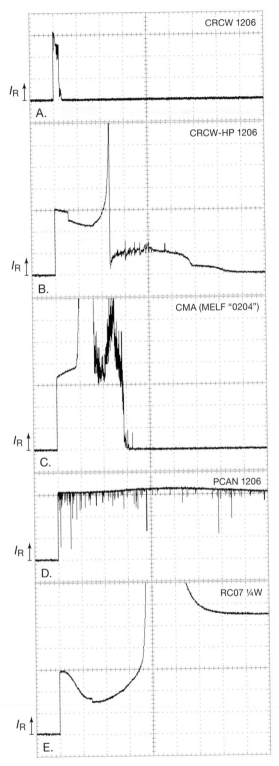

A.   CRCW 1206

I_R

B.   CRCW-HP 1206

I_R

C.   CMA (MELF "0204")

I_R

D.   PCAN 1206

I_R

E.   RC07 ¼W

I_R

Figure 1x.43. We applied 32 V across a selection of 50 Ω resistors and watched their final moments. Vertical: 200 mA/div; Horizontal: 400 ms/div.

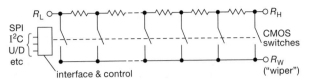

Figure 1x.44. A digital potentiometer IC consists of a series-connected string of resistors, with an array of digitally controlled CMOS analog switches to select the tap. The resistors may be of equal value ("linear taper") or configured to create steps of equal decibels ("log taper" or "audio taper").

Total resistance

Digipots are available in just a few total (end-to-end) resistance choices, ranging from 1 kΩ to 200 kΩ (and, rarely, 1 MΩ). Because many digipots are intended for voltage-divider use, they can have very loose tolerances of total resistance, up to ±25% or more; see below.

Number of steps

Most digipots have 64 to 256 taps, but you can get some with 1024 taps.

Volatile/non-volatile tap register memory

A mechanical pot remembers where you set it, and for many applications that's essential; hence the non-volatile (NV) digipot. Digipots lacking NV memory usually power-up to midscale or bottom.

Number of sections

For applications such as a stereo volume control, or tuning an analog active filter or Wien-bridge sinewave oscillator, you need at least two ganged pot sections. You can get mechanical multisection pots, and digipots offer the same option, usually duals or quads (or, rarely, six sections, e.g., the AD5206). Because of their simple digital control, you can always have as many sections as you want, by commanding multiple digipots to the same tap setting.

Voltage range

Digipots are CMOS devices, usually restricted to rather low total voltage ranges; typical are 5 V total (0 to 5 V, or ±2.5 V, with some "low-voltage" types specified down to 1.7 V total supply), but there are digipots available (particularly from ADI) that run up to 16 V total supply, and even to 33 V (e.g., AD5290, AD5293, AD7376).

Bandwidth

Digipots have internal capacitance, which creates an inherent *RC* roll-off that can be at surprisingly low frequencies. See, for example, Figure 1x.46, which plots the attenuation

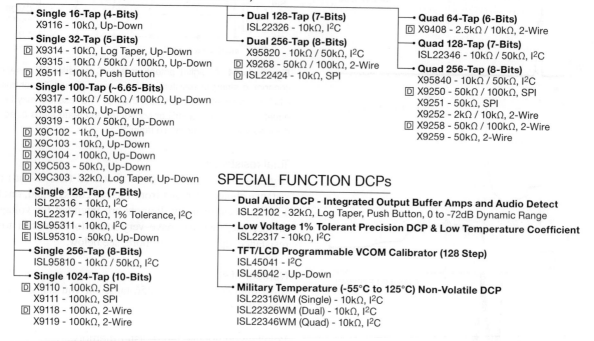

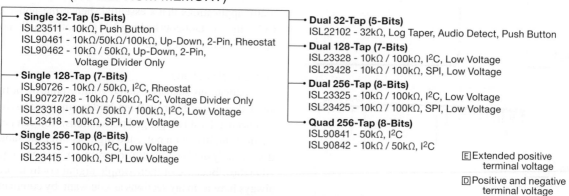

Figure 1x.45. This family tree of Intersil digital pots, adapted from their 2017 product brochure, illustrates a nice selection of parameters: resistance, resolution, interface, linear vs log taper, and volatility. Analog Devices offers an even larger selection (74 devices, each available with two to four resistance values as of this writing), but no handsome genealogical graphic.

versus frequency of a typical digipot, when set to attenuate 6 dB (code 40_H of a 128-tap pot). Digipots of lower resistance roll-off at higher frequencies, as expected, so you might feel comfortable choosing a 100 k digipot for an audio application. But beware – for a given digipot, the frequency at which its attenuation departs from what you expect depends on the tap setting, as seen in Figure 1x.47.

Taper

Most digipots (like most mechanical pots) have a linear "taper," that is, the steps are of equal resistance increments. However, for audio applications you want a logarithmic (or "audio") taper, corresponding to equal ratios, i.e., equal numbers of decibels. There are a few digipots with log tapers, as can be seen in Figure 1x.45. You may wonder why there are not more, given the common use of digipots in audio/video equipment. Good question! There

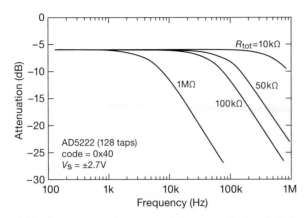

Figure 1x.46. Frequency response of the 128-step AD5222 digipot, in its four available total resistances, when set at half of full-scale.

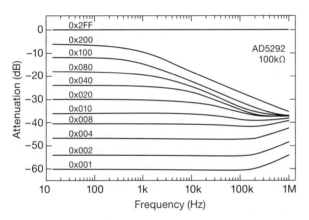

Figure 1x.47. The high-frequency behavior of digipots depends on wiper position, as seen in these curves from the AD5292 1024-tap digipot. This is reminiscent of analogous behavior in MDACs (§13.2.4).

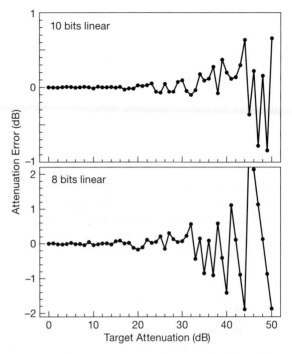

Figure 1x.48. You can get reasonably accurate log (i.e., dB) steps of attenuation from a linear decapot of 256 or 1024 taps, as seen in these plots of error. Note change of vertical scale.

Digital interface

Digipots are controlled digitally, with either the usual SPI or I²C serial bus protocols, or with UP/DOWN control. The latter comes in two forms: one has a CLK input and a U/D′ input, and clocks up or down accordingly; the other kind (sometime called a "push-button" interface) has an UP input and a DOWN input, and connects easily to a pair of buttons (which may have to be debounced, check the specs).

Ratio tolerance, resistance tolerance, and gang matching

The majority of digipots are aimed at potentiometer use, that is, as adjustable voltage dividers; so they tend to have mediocre resistance tolerances, often ±25% or worse (if you want to use a digipot as a 2-wire adjustable resistor, you can get parts with far better resistance tolerance, 1% or so). But digipots do quite well in delivering accurate resistance *ratios*, usually specified in terms of differential and integral nonlinearity (DNL, INL), with typical values around ±0.1 LSB (and rarely worse than 0.5 LSB). Multi-gang digipots specify matching tolerances, typically 0.5 LSB.

are several reasons: (a) It's far easier to make a linear-step digipot, and that's often good enough for log steps – see Figure 1x.48, where we plot the error from perfect dB steps when using nearest-integer steps of a linear taper digipot. (b) For many audio/video applications, the signals are digital along the way, so an attenuator is just a digital multiplier, at zero cost! (c) Also, digipots exhibit unpleasant switch-transition spikes (see below); so even for analog audio gain stages, circuit designers sometimes use impulse-free variable-gain amplifier stages (such as the THAT2181) or digital programmable-gain amplifiers, microcontroller-adjusted from the setting of a (linear) digipot.

Temperature coefficient

You'll usually see two values specified, one for "resistance tempco" (or "rheostat-mode" tempco) and one for "voltage-divider tempco" (or "ratiometric-mode" tempco). When used as a variable resistor, the tempco will be somewhere around either 20–40 ppm/°C or 500–800 ppm/°C (thin-film or polysilicon resistors, respectively), so be sure to check the specs carefully if you want a stable 2-terminal adjustable resistor. As you might expect, when used as a voltage divider these things are better, with typical ratio tempcos of ±5–20 ppm/°C. There are some standouts, for example the AD5291–92 (256 or 1024 steps), which specifies a ratio tempco of ±1.5 ppm/°C (typ), or the MCP42xxx-series from Microchip, with its ±1 ppm/°C (typ); the latter is particularly impressive, given its poor resistance tempco of 800 ppm/°C (typ).

Wiper resistance

The tap ("wiper") is a CMOS switch, with all the benefits and gremlins that accrue thereby. So it's got some series resistance (usually in the range of 10–100 Ω), which (as with all CMOS analog switches) depends on the rail-to-rail supply voltage (discussed extensively in §3.4.2B in AoE3). This rarely matters when you're using a digipot as a voltage divider with a high-impedance load, but it can be serious in 2-terminal (rheostat) mode, particularly for low-voltage digipots. Figure 1x.49 shows a nice example of the increasing R_W with decreasing supply voltage, and also the characteristic peak near the middle of the signal voltage range (when both nMOS and pMOS transistors find themselves with a V_{GS} of only half the supply voltage).

Distortion

For audio or precision applications you care about device linearity (i.e., change of resistance with applied voltage). Distortion specs are often omitted from digipot datasheets; Analog Devices, breaking this code of silence, is pleasantly forthcoming, with typical THD (total harmonic distortion) figures in the neighborhood of 0.01% for many of their parts. If you need better, try their AD5293, with THD of 0.0005% typ (1 Vrms and 1 kHz, a sweet spot for lowest THD); it boasts some other nice features, also, such as operation from 9–33 V total supply, ±1% resistor tolerance max, and 100 kHz bandwidth (flat, all codes).

B. Digipot cautions

Digipots are not perfect, as described above. Here we summarize some additional cautions – things to think about when considering using them (rather than a mechanical trimpot or panel pot) in a circuit.

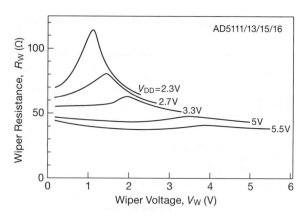

Figure 1x.49. Wiper resistance versus signal voltage, for several supply voltages, for the AD5111/13/15/16-series decapots.

Voltage, current, and power ratings

Most digipots are limited to swings between the supply rails, typically the CMOS value of 5.5 V total supply, whereas mechanical pots can handle far more, up to hundreds of volts (generally limited by power dissipation). Likewise, digipots are good only to a few milliamps and a few tens of milliwatts. Stick with mechanical pots if you need more.

Bandwidth–distortion tradeoff

High-resistance digipots roll off at distressingly low frequencies, but low-resistance digipots suffer from nonlinearity (owing to varying R_{on} of the CMOS switches). Both effects are largely absent in mechanical pots.

Zipper noise

Digipots exhibit switch-transition spikes, typically of microsecond-scale and in the neighborhood of tens of millivolts (but as much as a volt for some higher resistance parts). This is not good in an application such as a volume control. There are tricks to circumvent this problem, such as delaying the switching until the signal's next zero crossing. Mechanical pots are immune from zipper noise, but they get scratchy with time.[39]

C. Wrapup

Digital potentiometers really hit the spot when you want digital control of trims (amplifier or converter offsets, sensor calibration, bias or current setpoints) and where those

---

[39] We find that spinning them back and forth a few dozen times (or much more) can clean them up, especially if you can get some DeoxIT® Gold G-series contact cleaner into the guts. But more often they just refuse to return to their once silky-quiet state.

settings are within low-voltage and low-current circuitry. They are also excellent in applications where you need to have matching (multi-gang) adjustments. The least expensive digipots are hardly more expensive than analogous mechanical trimpots ($0.50 versus $0.30, in unit quantities), and they are compact and reliable. But there are plenty of applications where a simple mechanical trimpot works just fine, requires no interfacing or programming, is available in a wider range of resistance, and can handle much wider ranges of voltage, current, and power.

1x.3 Capacitors

The photograph in Figure 1.30 of AoE3 gives a sense of the capacitive ecosystem's variety and richness. The most important characteristics are **capacitance**, **maximum voltage rating**, **tolerance** (accuracy), **stability** (over time), and **temperature coefficient** of capacitance. But capacitors (like all electronic components) are imperfect – electrically they exhibit some equivalent series resistance R_s and series inductance L_s (commonly called **ESR** and **ESL**, Figure 1x.50), which can be of major importance in high-frequency circuits, and in power-switching circuits.[40] In addition, capacitors don't hold their charge forever, so the model includes some high-value **parallel resistance** R_p, typically so large that the self-discharge time constant of the capacitor ($\tau = R_p C$) is many thousands of seconds (and can be *years*, for capacitors with high-quality film dielectrics). Additional departures from ideal performance include **dissipation factor** (a frequency-dependent loss), **voltage coefficient** of capacitance (nonlinearity), **frequency dependence** of capacitance, **dielectric absorption**, and electromechanical **self-resonances**. We'll describe these briefly, followed by some advice on choosing capacitors for typical applications.

Figure 1x.50. Real capacitors exhibit some small series resistance (R_s) and inductance (L_s), and a large parallel resistance (R_p).

The "family tree" of capacitors in Figure 1x.51 and the contours of available capacitor types in Figure 1x.52 may be helpful in sorting out the many complexities of capacitors. Capacitors are unique among electronic components in spanning more than 16 orders of magnitude in available values (from 0.1 pF to 5000 F, see Figure 1x.52).

1x.3.1 Temperature coefficient

For some applications – oscillators, filters, and timing circuits – you want stability over temperature; for others you may not care. The best capacitors in this regard are "NP0" (also called C0G, a member of EIA Class I) ceramic capacitors, carefully chosen film capacitors (notably those with PPS[41] or polycarbonate dielectric), or mica capacitors. C0G types have tempcos of order $\pm30\,\mathrm{ppm}/°\mathrm{C}$; and PPS and mica capacitors (and some polypropylene types) have tempcos of ~50–$100\,\mathrm{ppm}/°\mathrm{C}$. C0G types excel, furthermore, in constancy of capacitance with frequency, applied voltage, and the ravishes of time; the price you pay is poor volumetric efficiency (i.e., they are physically large relative to their capacitance, and they are available with capacitances topping out at $0.47\,\mu\mathrm{F}$ or so). Other capacitor types are generally inferior, with capacitance changes over typical operating temperatures of a few percent (electrolytic and polyester film types), to $\sim10\%$ (ceramic "Class II" types, e.g., X7R), to an astonishing 50% or more (ceramic "Class III" types, e.g., Y5V[42]) – don't use these for oscillators or timers! Figure 1x.53 shows typical capacitance variation with temperature for several capacitor types. Table 1x.5 lists the official EIA codes corresponding to the maximum capacitance shift over the various operating temperature ranges for class II and III high-κ ceramic capacitors.

Table 1x.4: EIA Temperature Coefficient Codes for Class I Ceramic Dielectric Capacitors

1st character ppm/°C		*	2nd character multiplier		3rd character tempco tolerance[a]	
C	0.0		0	−1	G	±30ppm/°C
B	0.3		1	−10	H	±60ppm/°C
A	0.9		2	−100	J	±120ppm/°C
M	1.0		3	−1000	K	±250ppm/°C
P	1.5		4	−10000	L	±500ppm/°C
R	2.2		5	+1	M	±1000ppm/°C
S	3.3		6	+10	N	±2500ppm/°C
T	4.7		7	+100		
U	7.5		8	+1000	e.g., C0G:	
			9	+10000	tempco = 0±30ppm/°C	

(a) measured as a straight line between values at +25°C and +85°C

The exemplary C0G is a member of the "stable" class I ceramic capacitors, whose other members include

---

[40] But even a mundane circuit like a linear voltage regulator (Chapter 9) may oscillate if the output capacitor's ESR is too high, or too low!

[41] Polyphenylene sulfide.

[42] These capacitors are astonishing, also, in their outrageous *voltage* coefficient of capacitance, see below. It is not clear to the authors why these capacitors even exist.

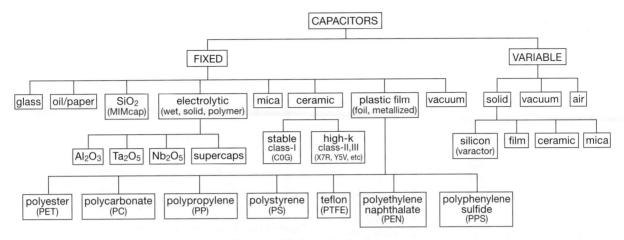

Figure 1x.51. Capacitor family tree.

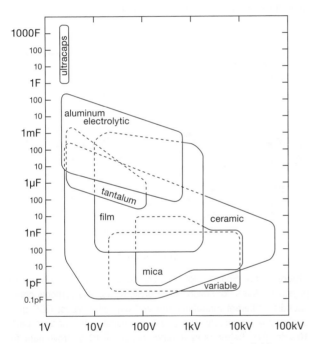

Figure 1x.52. Capacitance and voltage ranges of available capacitor types.

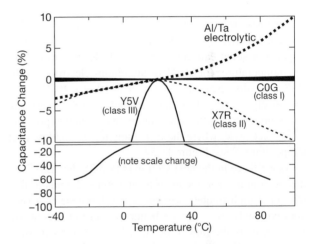

Figure 1x.53. Typical temperature variation of capacitance for three types of ceramic capacitors, and for electrolytics (aluminum or tantalum). Plastic film capacitors (not shown) typically exhibit capacitance variations of $\pm 0.5\%$ to $\pm 2\%$ over the same temperature range.

those with specified non-zero temperature coefficients. Table 1x.4 lists the EIA codes specifying their temperature coefficients. In practice you'll see alternative designations, echoing an earlier era (when "NP0" meant zero tempco, i.e., EIA designation C0G), with an N- or P- prefixing the tempco in units of ppm/°C. For example, an EIA type U2J ($-750\,\text{ppm}/°\text{C}$) is also called an N750. Although you could concoct EIA codes for 81 possible tempcos from the entries in Table 1x.5, in practice you'll find only a half dozen or so

to be available, almost all of them having negative tempcos (NTC).[43]

1x.3.2 ESR

The equivalent series resistance is frequency dependent, generally reaching a minimum around $100\,\text{kHz} - 1\,\text{MHz}$. It is smaller for larger capacitance values, and (for a given capacitance) for a capacitor of higher voltage rating. Typical

---

[43] In addition to the zero tempco C0G, we've found various quantities of P90, N150, N220, N330, N470, and N750 listed at distributors' websites.

Table 1x.5: EIA Temperature Characteristic Codes for Class II & III Ceramic Dielectric Capacitors

1st character T_{min}		2nd character T_{max}		3rd character ΔC, maximum	
X	−55°C	2	+45°C	A	±1.0%
Y	−30°C	4	+65°C	B	±1.5%
Z	+10°C	5	+85°C	C	±2.2%
		6	+105°C	D	±3.3%
		7	+125°C	E	±4.7%
		8	+150°C	F	±7.5%
		9	+200°C	P	±10%
				R	±15%
				S	±22%
				T	+22%/−33%
				U	+22%/−56%
				V	+22%/−82%

e.g., X7R: ±15% maximum change
of capacitance over −55°C to +125°C

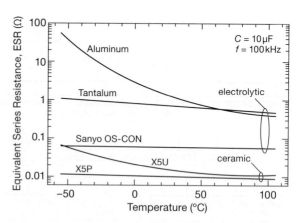

Figure 1x.54. Equivalent series resistance (ESR) versus temperature, of several representative $10\,\mu F$ electrolytic and ceramic capacitors suitable for power supply bypassing. (Adapted from Panasonic Technical Data document AAB8000PE24.)

ESR values are in the range of $1\,m\Omega$ to a few tenths of an ohm, at 100 kHz; at 100 Hz the ESR is typically ten times larger. For some capacitor types the ESR rises rapidly at low temperature: for standard aluminum electrolytics the ESR typically rises by a factor of 5–10 at −40°C, a defect nicely addressed in Panasonic's OS-CON[44] series of aluminum electrolytic capacitors with conductive polymer solid electrolyte, where the ESR remains flat and low over temperature;[45] see Figure 1x.54.

As we'll see in the next section (§1x.3.3), the "ESL" deduced from measurements of a capacitor's Z tends to flatten out at high frequencies, but at low frequencies it rises approximately as $1/f$. So a better measure of loss at low frequencies is the ratio ESR/X_C (the capacitor's *dissipation factor*, DF); this is more honest, also, from a physical point of view, because the low-frequency loss mechanism is dielectric loss, not metallic resistivity.

Within the class of aluminum electrolytic capacitors, we found an interesting correlation between ESR and the capacitor's physical size: the scatterplot in Figure 1x.55 shows that bigger is better, with ESR generally decreasing in inverse proportion to the capacitor's physical volume.

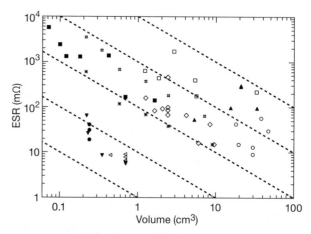

Figure 1x.55. Scatter plot of specified ESR (at 100 kHz) versus physical volume, for several electrolytic types. The symbols correspond to different manufacturers and capacitor types; ESR generally varies inversely with physical volume, with modern parts 10× to 30× better (and another factor of 10 improvement for solid-electrolyte types). The dashed lines are contours of constant figure-of-merit $ESR \cdot V$ (1, 10, 100, 1000 mΩ-cm³). Another trend: volume is proportional to charge $Q=CV$, so, for a given capacitance, volume is proportional to voltage rating.

1x.3.3 ESL

Any length of wire has inductance, and a leaded capacitor is no exception. For bypass and decoupling applications you want to minimize inductance; also, the series inductance forms a series LC circuit, so the overall impedance has a minimum at the resonant frequency $f = 1/2\pi\sqrt{LC}$, where its value equals the ESR. For higher frequencies the

---

[44] Organic semiconductor electrolyte. The OS-CON series was originated by Sanyo Electronics, now manufactured by Panasonic.

[45] This latter type is particularly popular in computer motherboards, switching power supplies, and other applications with substantial ripple currents. Al-poly caps are available with a wide range of ESRs for the same capacitance, voltage rating, and size. This is helpful for use with LDOs and other voltage regulators whose stability depends on particular ranges of ESR. Alternatively you can use a pulse-rated resistor in series with a very low ESR cap.

reactance rises again (like an inductor). See Figures 1x.56 and 1x.57. In Figure 1x.56 we've plotted the effective series resistance with black squares, by which we mean the capacitor model's value of R at each frequency; and this is how such curves are often displayed in datasheets. But a better way to think about it, in the frequency regime below the model's LC series resonance (the minimum impedance), is in terms of the ratio of X_C to R_{eff}, that is, the capacitor's quality-factor Q (the inverse of its dissipation factor, DF), which we've marked on the plot. That is because the "ESR" and X_C track each other at low frequencies – thus a roughly constant Q (but a highly frequency-variable "ESR").

At high frequencies the capacitor's impedance is dominated by inductance. For circuits with fast signals (that's pretty much *every* digital circuit), you must use ceramic surface-mount "chip" capacitors to minimize inductance, which brings the ESL down from several nH to $\sim$1 nH. Because the ESL is dominated by lead inductance, further reductions are possible only with clever geometries, such as multi-leaded "interdigitated" capacitors, or chip arrays with multiple solder-bump connections (Fig. 1x.58).

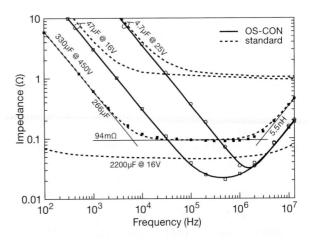

Figure 1x.57. Impedance versus frequency of several electrolytic types, comparing measured and modeled Z, showing the importance of low ESR. The Sanyo/Panasonic OS-CON have solid polymer electrolyte, compared with the liquid electrolyte of standard electrolytic capacitors.

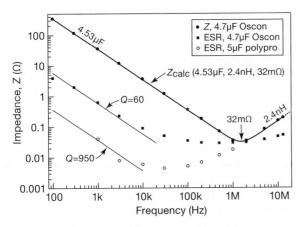

Figure 1x.56. Measured impedance (with HP 4192A *LCR* meter) of a real-world low-ESR capacitor (4.7µF/25V solid-electrolyte aluminum, Sanyo Oscon 25SC4R7M): At low frequencies Z_{total} falls as $1/2\pi fC$, reaching a minimum equal to its ESR (R_s, $\sim$32 mΩ) at its series resonant frequency ($1/2\pi\sqrt{C \cdot ESL}$), then rising again at high frequencies as $2\pi f \cdot ESL$. At low frequencies the ratio of X_C to ESR (i.e., Q) is a better measure. There the polypropylene film capacitor excels – but its large physical size (and high price) disqualifies it for most applications.

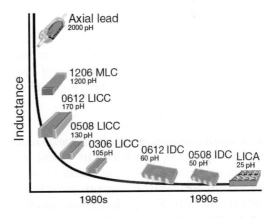

Figure 1x.58. Use ceramic chip (rather than axial-leaded) capacitors for lower parasitic inductance (ESL). Even lower inductance is achieved with multileaded and ball-gridded geometries. (Reprint Courtesy of AVX Corporation, copyright LICENSOR 201X.)

1x.3.4 Dissipation factor

A convenient way to recast capacitor ESR is to express the ratio of ESR to capacitive reactance; the ratio is called *dissipation factor* (DF), and is simply DF=ESR/X_C, or ESR$\times 2\pi fC$. Dissipation factor expresses the ratio of dissipated power to stored (reactive) power, or, equivalently, the ratio of energy lost per radian (as heat in the ESR) to energy stored (in the capacitor's $CV^2/2$). If this sounds familiar, you should not be surprised: it is just the reciprocal of the quality factor Q: DF=1/Q. (§1.7.14). Thinking in terms of phasors, the dissipation factor can be written as DF=tan δ, where the angle δ is called the *loss tangent*. For small DF the loss tangent (in radians) is approximately equal to DF: $\delta \approx \text{DF}$ ($\delta \ll 1$).

Dissipation factor depends on frequency, and at first glance you'd expect it to rise proportional to f. But things are more interesting, because at low frequencies the ESR is dominated by dielectric losses, which decline roughly as $1/f$. You can see this in Figure 1x.56, and the result is that in many cases DF rises only rather slowly with rising frequency, until the effects of metallic losses (resistance and skin-effect) become comparable. Here's some data[46] from a 22 pF ceramic capacitor that illustrates that point:

f (MHz)	$\text{ESR}_{\text{dielec}}$ (mΩ)	R_{metal} (mΩ)	ESR (mΩ)	X_C (Ω)	DF
1	145	7	152	7.2k	2.1×10^{-5}
3	48.2	7.8	56	2.4k	2.3×10^{-5}
30	4.8	9.2	14	241	5.8×10^{-5}
300	0.48	28.5	29	24	1.2×10^{-3}

1x.3.5 Voltage coefficient of capacitance

Capacitance depends on the electrode area and spacing (the "geometry"), and on the dielectric constant of the insulating layers. For example, in the particular case of a parallel plate capacitor, the capacitance C (in pF) is given by

$$C = \kappa \frac{\varepsilon_0 A}{d}$$

where the dimensional units (plate area A and spacing d) are in SI units (meters), $\varepsilon_0 = 8.85 \times 10^{-12}$, and κ is the dielectric constant. For air $\kappa = 1$; for plastic films $\kappa \sim 2$–5; and for ceramic materials κ ranges from $\sim$100 (class-I) to $\sim$10,000 (class-III). To get a feel for magnitudes, note that a pair of plates 1 cm on a side (1 cm^2), separated by an air gap of 1 mm, has a mutual capacitance of approximately 1 pF.[47]

So far, so good. The problem is that nearly all practical capacitors use a dielectric insulating layer; and for those with the highest dielectric constant ("high-κ" ceramic) the dielectric properties depend to a significant degree on the electric field, i.e., on the applied voltage. The most serious offender here is once again the notorious class-III ceramic (whose outrageous temperature characteristics required an expanded scale in Fig. 1x.53!) – see Figure 1x.59, which

---

[46] From "ESR Losses in Ceramic Capacitors," ATC 001-923 rev D, 4/2007, American Technical Ceramics.

[47] You would need a thousand times that area, with separation a thousand times smaller, to get 1 μF; imagine the innards of a (readily available) one *Farad* capacitor!

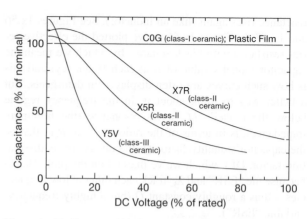

Figure 1x.59. Beware of bigtime voltage effect for certain ceramic capacitors. For most other capacitor dielectrics the effect is negligible: the film capacitors we measured showed ~100.04% of rated capacitance at full rated voltage.

shows that class-II "general-purpose" ceramic capacitors suffer also from this effect, which reduces the capacitance (sometimes dramatically) with applied voltage.[48] For most other capacitor types the effect is far smaller, and can generally be ignored: we measured capacitance changes ranging from a maximum of +500 ppm (+0.05%) for some polypropylene film capacitor samples (at 70% of their rated voltage), down to <10 ppm (0.001%) for glass and C0G ceramic types. To give a sense of their behavior, we've plotted in Figure 1x.60 the capacitance variation versus voltage for these on an expanded scale (compared with Fig. 1x.59).[49]

For many circuit applications (e.g., blocking or bypassing) you may not care about this effect. But it can make a big difference when a capacitor is used for precision wave-

---

[48] These plots were generated with Murata's free "SimSurfing" software (which can be run online, or downloaded as a standalone program). We specified 16 V 0603-size 1 μF ceramic capacitors (except for the C0G type); the values plotted are calculated for an ambient temperature of 25°C and 1 Vrms applied signal at 1 kHz.

[49] The dielectric material by itself is not a complete guide to a capacitor's voltage coefficient: One of our samples of polycarbonate capacitor (Vishay) changed capacitance by +300 ppm at 70% of rated voltage, while that of a different manufacturer (Paktron) showed no variation at our measurement limit (50 ppm), even at full rated voltage. Note particularly that *the dielectric designations for ceramic capacitors specify their temperature coefficients only; they do not speak to the voltage coefficient.* As Mark Fortunato explained in his article ("Temp and voltage variation of ceramic caps, or why your 4.7-uF part becomes 0.33 uF") in the 26Nov12 issue of EDN magazine, "there are many materials that qualify as "X7R." In fact, any material that allows a device to meet or exceed the X7R temperature characteristics, $\pm$15% over a temperature range of -55°C to +125°C, can be called X7R. ... there are no voltage coefficient specifications for X7R or any other types."

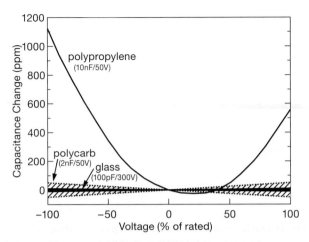

Figure 1x.60. Glass, C0G (but *not* X7R or Y5V) ceramic, and certain film capacitors exhibit low voltage coefficient of capacitance; the polycarbonate and glass capacitors showed no variation, down to the indicated measurement limits. Note expanded scale, compared with Fig. 1x.59.

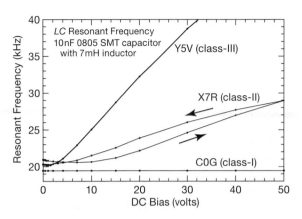

Figure 1x.61. Measured *LC* resonant frequency versus applied dc bias, for three dielectric types of 10 nF ceramic capacitor. For the X7R we ran the bias up to the rated 50 V and back to zero, tracing out the hysteretic curve shown.

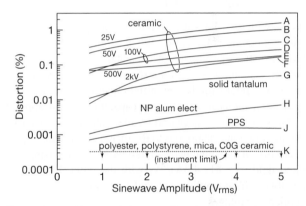

Figure 1x.62. Measured sinewave distortion of an *RC* lowpass filter at its 3 dB frequency, as a function of output amplitude. In all cases there was no significant differences between through-hole and SMT parts. The takeaway: use film, mica, or C0G ceramic capacitors for applications where linearity is important. A, B: Y5V. C, D, E: X7R. F: cera-mite disc. G: CS13B. H: Panasonic ECE 50V. J: PPS 50V. K: C0G radial and SMT; Mylar (polyester); polystyrene; CM07 mica.

form generation. For example, the sinewave output of the ultra-low distortion Wien Bridge oscillator illustrated in §7.1.5B (Fig. 7.22), measured at 0.0002% (!) with a pair of polyester capacitors in the critical frequency-setting circuitry, was degraded a thousandfold (to 0.22% distortion) when X7R (class-II) ceramic capacitors were substituted.

As a measure of this effect, we made bench measurements of the resonant frequency f_{LC} of a parallel *LC* tuned circuit, with each of several 10 nF ceramic capacitors connected across a high-quality toroidal 7 mH inductor. Figure 1x.61 shows the variation of f_{LC} with applied dc bias. The X7R's increase of resonant frequency at 50 V corresponds to a factor-of-2 reduction of capacitance (recall $f_{LC}=1/2\pi\sqrt{LC}$); for the Y5V the capacitance has already fallen by a factor of four at 30 V. By contrast, for the impressive C0G dielectric the resonant frequency was unchanged to within the measurement accuracy of ±10 Hz over the full range of applied bias; for scale, that's about one-tenth the thickness of the plotted line.

We made another set of measurements, this time using the capacitor under test as the lower leg of an *RC* lowpass filter at its 3 dB point (i.e., with $R=1/2\pi fC$). We drove it with a low distortion sinewave (Stanford Research Systems DS360) and measured the distortion of the output (Shibusoku 725B). We used a wirewound resistor, to make sure it was not contributing nonlinearity.[50] Figure 1x.62 shows

the measured distortion, for output amplitudes from 0.7–5 Vrms. As before, the ceramic types disappoint, with the sterling exception of the remarkable C0G, whose performance matches that of the film and mica capacitors (with distortion below the instrument limit of our setup).

TDK's "CeraLink" ceramic capacitors are an interesting exception to this unfortunate trend: their capacitance *increases* with increasing dc voltage. They also have low leakage at high temperature, and they tolerate high ripple current (increasing with frequency).

---

[50] Replacing the capacitor with another wirewound resistor showed that the residual distortion of the setup was around 0.0003%.

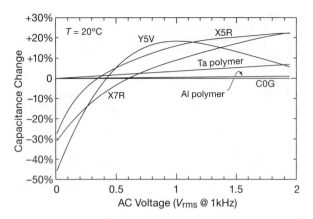

Figure 1x.63. High-κ ceramic capacitors are unsuitable for timing or filtering applications, owing to their large change in capacitance with applied signal. Use film or stable (C0G) ceramic capacitors instead.

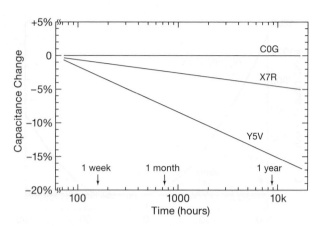

Figure 1x.64. Ceramic capacitors with high-κ dielectrics (X7R and Y5V) exhibit a loss of capacitance over time, a property (among others) in which the stable class I C0G excels.

1x.3.6 AC voltage coefficient

Along with their less-than-stellar constancy of capacitance with applied dc voltage, the high-κ ceramic capacitors exhibit a distressing change of capacitance with applied *ac* voltage, as can be seen in the representative plots of Figure 1x.63. Interestingly, the initial effect (at small ac voltage) is an *increase* in effective capacitance with applied signal voltage. This is caused by the realignment of ferroelectric domains, analogous to the increase in effective permeability of ferromagnetic materials with increasing applied field (see §1x.4.3A). As with the latter, high-κ capacitors exhibit nonlinearity and saturation at very large applied ac voltages.

1x.3.7 Aging

While we're dissing high-κ ceramic capacitors, we'd be remiss in not pointing out another unseemly effect that is not widely appreciated, namely their loss of initial capacitance over time ("aging"). According to a publication by capacitor manufacturer TDK,[51] it is "a decrease in capacitance over time in EIA Class II capacitors. It is a natural and unavoidable phenomenon that occurs in all ferroelectric formulations used in the dielectric material." Figure 1x.64 shows representative curves for three ceramic capacitor types: the barium titanate class III Y5V and class II X7R, and the stable class I C0G that is formulated with titanium oxide or calcium zirconate. Lore has it that aging can be reversed by baking the capacitor for two days at 60°C,

and that soldering produces a similar effect (we wish this were also true for humans); see Kemet App Note V032614.

1x.3.8 Frequency dependence of capacitance

Several effects conspire to produce a variation of capacitance with frequency. Among these are the insulating material's native variation of dielectric constant (ε) with frequency, and the effects of dielectric absorption (which decouples some of the low-frequency capacitance at increasing frequency, see §1x.3.10). As an example of the former, dielectric materials that exploit the orientation (by the applied field) of permanent dipoles pre-existing within the material[52] show a significant reduction of capacitance with increasing frequency, in contrast to the behavior of dielectrics that do not exploit this effect.[53] Figure 1x.65 shows this behavior, with measured data for several capacitor types.

1x.3.9 Electromechanical self-resonance and microphonics

The dielectric in high-κ ceramic capacitors is *piezoelectric*, which means that it deforms mechanically under an applied electric field (and, conversely, when deformed me-

---

[51] Steve Malloy, "What is the Capacitance of this Capacitor?" (June, 1999).

[52] The *ferroelectric* effect, by analogy with the action of permanent magnetic dipoles in ferro*magnetism*.

[53] A well-known example of frequency-dependent dielectric constant is water: at low frequencies $\varepsilon = 80$, dropping to half that value at 20 GHz, and to $\varepsilon = 1.8$ at optical frequencies (10^{15} Hz). TDK's unusual "CeraLink" capacitors buck the trend – they exhibit decreasing ESR with both frequency and temperature.

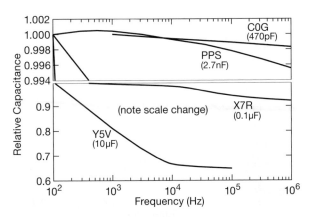

Figure 1x.65. "High-κ" ceramic capacitors (e.g., Y5V) exploit the ferroelectric effect to achieve high capacitance; along with that comes a decrease of capacitance with frequency. By contrast, the effect in film capacitors and C0G ceramics is a thousand times smaller (note change of scale).

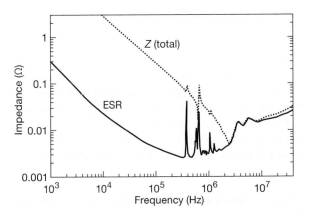

Figure 1x.66. "High-κ" ceramic capacitors are *piezoelectric* – an applied voltage causes the ceramic element to move – resulting in mechanical resonances, and corresponding bumps in their electrical impedance. This graph (adapted from Sanyo technical literature) shows total impedance and ESR for a 10 μF 25 V ceramic multilayer chip capacitor, with 20 Vdc applied bias.

chanically it produces an electric field). It's a tiny loudspeaker (and microphone!) – you can sometimes *hear* these things, when there's an audiofrequency signal across them. The ceramic structure has natural resonant frequencies, typically in the range of tens of kilohertz to a megahertz region, so the capacitor's impedance shows sharp peaks (Fig. 1x.66). This is not a good feature in a bypass capacitor! Because these capacitors convert mechanical strain to voltage, they are *microphonic*, and should be avoided in audio circuit applications where a microphonic voltage would enter the signal path.

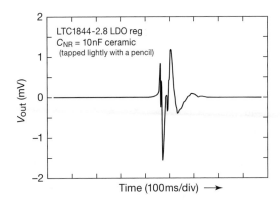

Figure 1x.67. Tapping lightly with a pencil on the ceramic capacitor connected to the "noise bypass" pin of a voltage regulator produced this output transient. The effect is due to the capacitor's microphonic (vibration-induced) sensitivity. (Adapted from the LTC1844 datasheet.)

Another application that's sensitive to this effect is the bypassing (noise filtering) of a precision voltage reference or regulator (a topic discussed in §9.10.6). The LTC1844 datasheet warns of this, and includes an analog 'scope photograph showing "noise resulting from tapping on a ceramic capacitor," which we've redrawn as Figure 1x.67.[54] They don't say what sort of capacitor they used. But that inspired us to try an analogous experiment, namely measuring the voltage produced by an SMT chip capacitor soldered to a PC board, when that board is vigorously flexed by hand. We used a 10 nF capacitor of general-purpose X7R dielectric, and we measured the output signal (filtered to 30 Hz bandwidth) from a bare capacitor (lower trace) and from the same capacitor when biased to +5 V (as it might be in a voltage reference filter). Figure 1x.68 shows the results – a rather dramatic swing of ±2 mV and ±40 mV respectively. Just for fun we tried the inverse experiment – we drove the dc-biased capacitor with an audio sinewave, and we could *hear* the results.[55]

Note that these effects are absent in class-I ceramics (C0G/NP0), and in most other capacitor dielectrics. Although there's an electric field between the plates of any capacitor (proportional to the instantaneous voltage), and a mechanical force (proportional to the voltage squared), microphonic and resonant effects are generally negligible, and only exaggerated in certain materials.

---

[54] And the datasheet suggests that "Similar vibration-induced behavior can masquerade as increased output voltage noise."

[55] Which suggests a subtle "seismic" mechanism of signal coupling between circuits on a PC board.

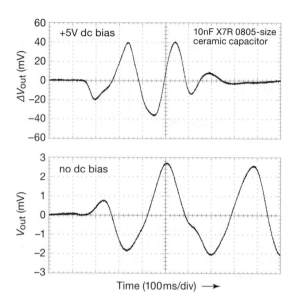

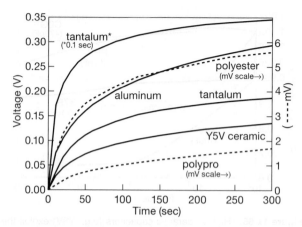

Figure 1x.69. Measured self-recharge of several capacitor samples after momentary discharge from 10 V. Electrolytic capacitors have the memory of an elephant! See the expanded collection in Figure 1x.70.

Figure 1x.68. Stress-induced voltage waveforms produced by vigorous flexing of the supporting PCB. The zero-bias signal is caused by the ceramic's piezoelectric effect, whereas the 20× larger swing in the biased arrangement is due to stress-induced capacitance variation (note change of scale).

1x.3.10 Dielectric absorption

Another important – and underappreciated – effect is *dielectric absorption*. To give a sense of this effect, we selected samples of most of the available capacitor types (and some rarities, such as teflon), and charged them to +10 V. We held them at that voltage for a day or more, then we discharged them through a small resistor, holding them at zero volts for 10 seconds; then we removed the resistor. Figures 1x.69 and 1x.70 show what happened next: they don't just sit there – they remember where they came from, and charge themselves up again!

The origins of dielectric absorption (or dielectric *soakage*, dielectric *memory*) are not entirely understood, but the phenomenon is believed to be related to remnant polarization trapped on dielectric interfaces; mica, for example, with its layered structure, is particularly poor in this regard. From a circuit point of view, this extra polarization behaves like a set of additional series *RC*s across the capacitor (Fig. 1x.71A), with time constants generally in the range of ≈100 µs to several tens of seconds. Dielectrics vary widely in their susceptibility to dielectric absorption; Figure 1x.71B shows data for several high-quality dielectrics, plotted as voltage memory versus time after a 10 volt step of 100 µs duration.

Dielectric absorption can cause significant errors in in-

tegrators (§4.2.6) and other analog circuits that rely on the ideal characteristics of capacitors. In the case of a sample/hold (§4.5.2) followed by precision analog-to-digital conversion (§13.5), the effect can be devastating. In such situations the best approach is to choose your capacitors carefully (Teflon and polystyrene dielectrics seem to be best), retaining a healthy skepticism until proven wrong. In extreme cases you may have to resort to tricks such as compensation networks that use carefully trimmed *RC*s to electrically cancel the capacitor's internal dielectric absorption.

1x.3.11 Capacitor choices for typical applications

With the details of the previous section as background, we offer advice on capacitor selection, by application.

A. Bypass and decoupling

Use *ceramic* bypass capacitors (inexpensive, low ESR/ESL) liberally throughout the circuit, typically 0.1 µF or thereabouts; try to avoid the class-III dielectrics (e.g., Y5V), whose effective capacitance is greatly reduced by voltage and temperature effects – use the physically larger class-II (e.g., X7R) if space permits. Use *at least* one capacitor per active element, and more for fast-switching circuits like microprocessors, video processors, and support chips (where you may see a dozen bypass capacitors clustered tightly around, and under, the IC). Surface-mount capacitors are best, with their low ESL. It's a good idea to include some larger-value *electrolytic* bypass capacitors, for energy storage and damping of resonances; here the "solid dielectric" types (e.g., Sanyo OS-CON) have the

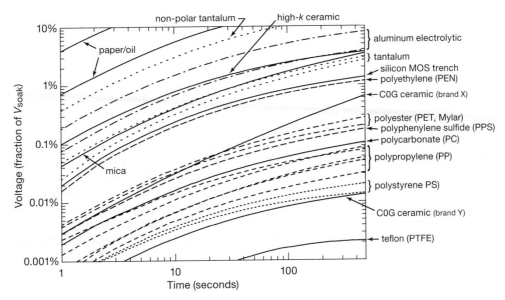

Figure 1x.70. Dielectric absorption causes a "memory effect" in capacitors. Measured recharge curves for several capacitor types, first held at +10 V for a day or more, then at zero volts for 10 seconds.

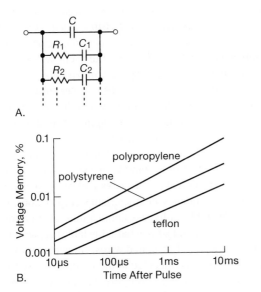

Figure 1x.71. Dielectric absorption in capacitors. A. model; B. values for several high quality film capacitors.

best characteristics (low and stable ESR; long life). Many people believe that **tantalum** bypass capacitors are prone to failure, and should be avoided; our experience here is mixed – we've seen failures with radial dipped tantalum

types, but none with axial hermetically sealed types (e.g., CS13B type).[56]

B. Oscillators, filters, and timing

Don't even *think* about using **high-κ ceramics** (class-II or class-III, e.g., X7R or Y5V)! The best choices here are **mica**, **C0G ceramic**, the somewhat rarer **glass**, and polymer ("plastic") **film** dielectric capacitors. Film capacitors cover the 1 nF–10 µF range; the other types cover the range from several nanofarads down to a fraction of a picofarad. For surface-mount applications, where the part must endure high temperature during fabrication, the choices of film dielectrics are limited, the favorites being **PPS** (polyphenylene sulfide) and **PEN** (polyethylene naphthalate).

C. High frequency

The favorites are **mica** and **C0G ceramic**, where a fixed capacitance is needed; and **air-spaced**, **ceramic**, or **piston**-style capacitors, where a variable or trimmable capacitance is needed. Radiofrequency transmitters often involve high RF voltages, for which you use **"doorknob"**

---

[56] According to John Larkin, MnO$_2$ tantalum capacitors explode (literally detonate) from excessive current (i.e., dV/dt), so they should not be used on input power rails. An s.e.d. participant added that he has an old pair of glasses with a chip right in the middle of one glass where it was hit by an exploding tantalum cap on a board he was testing.

style mica or glass dielectric capacitors, or **vacuum** capacitors, for fixed (non-adjustable) capacitance. For variable (tuning) capacitors you use high-voltage **air-variable** or **vacuum-variable** capacitors.

D. Energy storage

For the high capacitance values needed for energy storage applications (e.g., switchmode power converters, power-supply "filter" capacitors, photoflash capacitors) the usual choice is **aluminum electrolytic** capacitors. The "solid electrolyte" types (popularized by Sanyo as the OS-CON series) have superior electrical characteristics (notably low and stable ESR over temperature). Solid **tantalum electrolytics**, though more expensive, are sometimes preferable. For very high voltages (>1 kV) **oil** capacitors are the favorite.

E. AC line filtering

Capacitors that bridge the ac power line (used, for example, as part of a noise filter at the power input) have to contend with occasional large voltage spikes (sometimes amounting to many kilovolts), which can causing internal arcing. If that causes a permanent short circuit, you've got a problem! Certain capacitor types (in particular, **metallized film** capacitors, in which a thin deposit of aluminum on the dielectric serves as the electrode material) are rated for this service, and are designed to be "self-healing": if the film is punctured by a voltage spike, the transient current causes the nearby metal film to evaporate, clearing the short circuit (see discussion of X1 and X2 capacitors in §9.5.1E). Capacitors rated for across-the-line service will say so, citing the relevant safety ratings (for example UL 1414, and analogous international standards). Some **oil** capacitors are self-healing, and are used for ac line applications.

F. High voltage

High-voltage capacitors (>1 kV, say) are constructed using **vacuum**, **ceramic**, **glass**, **mica**, **plastic film**, or **oil-impregnated** dielectrics (roughly in order, going from lower to higher capacitances). This is a specialty area, not for the timid of heart.

1x.3.12 Capacitor miscellany

Electrolytic capacitors (aluminum, tantalum) are *polarized*, with either + or − markings on the package to indicate correct dc polarity. Electrolytic capacitors do not tolerate reversed polarity of more than a few tenths of a volt; that is because the dielectric consists of a thin oxide coating on the metal anode (positive) electrode, which is destroyed

(chemically "reduced") by significant reverse voltage. Unlike film or ceramic capacitors (which can be operated near or at their rated voltage), electrolytic capacitors should be handled differently. Our experience suggests really conservative derating: in a circuit where an electrolytic capacitor runs at 25 V, say, choose a capacitor rated at 50 V or 63 V (a curious but common rating). Electrolytic capacitors with liquid electrolyte (the usual construction) may explode if you apply overvoltage, significant reverse voltage or excessively rapid charge or discharge; that's why they've got pressure-relief perforations and/or rubber plugs – look for them! However, there do exist **non-polarized electrolytics**, made by anodizing both of the electrodes. **Film** capacitors sometimes have what might be mistaken for a polarity marking (such as a band at one end). This does not indicate polarity (film capacitors are not polarized), but rather designates the *outside foil*, which effectively shields the other electrode, and which you would preferentially choose for the low impedance point, in circuits where pickup matters. **Polystyrene** capacitors (and pretty much anything else made of polystyrene) have a low melting temperature, and they dissolve in some common organic solvents; use care! Ceramic **surface-mount chip** capacitors are vulnerable to damage from flexure of the circuit board; we find ourselves cringing and grinding our teeth when forcing a CPU cooling fan, or memory card, into a sagging computer motherboard as instructed.[57] **High-κ ceramic** capacitors are piezoelectric – they vibrate when driven electrically; sometimes you can hear them singing! The reciprocal effect – the "microphonic" production of a signal voltage when mechanically strained – can be a serious problem as well. The electromechanical properties of ceramic capacitors are responsible for peaks and resonances in their ESR versus frequency characteristics (Fig. 1x.66), particularly undesirable in bypass applications. A nice trick for making a small (and somewhat adjustable) capacitor is the **gimmick** – a pair of insulated wires, connected into the circuit at one end, unconnected at the far end, and twisted together; the capacitance is roughly 0.5–1 pF/cm, depending on the insulation thickness and how tightly the wires are

---

[57] The same problem can occur in production, when an SMT ceramic cap is near a V-cut where the PCB is to be separated. The stress can create micro-cracks in the ceramic, causing delayed failure. Some automotive capacitors include internal stress-reducing flexibility to alleviate this problem. See
https://www.electronicdesign.com/boards/pcb-designers-need-know-these-panelization-guidelines and
https://ec.kemet.com/knowledge/flexible-termination-reliability-in-harsh-environments.

twisted.[58] Among the rarefied types of capacitors, monolithic **glass** capacitors (made by AVX) have the interesting property (among their many outstanding features[59]) of being highly resistant to ionizing radiation: they will withstand 100 megarads with hardly a complaint; that's two hundred thousand times the lethal dose to a human. **Teflon** capacitors, another hard to find species, excel in absence of dielectric absorption. **Ultracapacitors** (or "supercapacitors") exploit molecular-size dielectric separation at a liquid–conductor interface, along with techniques that produce greatly enhanced electrode surface area, to achieve capacitances ranging from farads to kilofarads (but with a maximum voltage rating of 2.5 V). In the smaller sizes (0.1–1 farad) they are routinely used for short-term memory backup; and they're serious candidates for energy storage in high peak-power applications such as electric transportation.

---

[58] We measured 1.0 pF/cm, using kynar-insulated #26 solid wire with one to two twists per centimeter, dropping to ~0.8 pF/cm when more loosely twisted. At a more sophisticated level, you can think of the twisted pair as a short *transmission line* (Appendix H in AoE3), for which the capacitance per unit length works out to $pF/cm = 33/v_p Z_0$, where Z_0 is the characteristic impedance (e.g., 50 Ω for coax, or 100 Ω for twisted-pair network cable), and v_p is the "velocity factor" (the ratio of signal speed to the speed of light, e.g., 0.66 for solid polyethylene, or 1.0 for air. The latter can be written in terms of the effective dielectric constant ε, namely $v_p = 1/\sqrt{\varepsilon}$). You can make a "PCB gimmick" by putting a few extra pads, to be jumpered at assembly time as needed.

[59] Zero voltage coefficient, absence of microphonics, absence of aging, operation to 125°C (to 200°C for some types), exceptional stability, inertness, and reliability.

1x.4 Inductors

Inductors and transformers – sometimes grouped as "magnetics" – can present the most complex problems in a circuit design: the possible parameter "space" is so large (Fig. 1x.72 suggests some of the variety of the inhabitants of that space) that you often have to select a core material[60] and configuration,[61] iterate the actual design, then wind the inductor (or transformer) and validate its performance.

In this section we'll explore inductors in a progression of steps, departing somewhat from the usual text- or reference-book treatment. Here is our plan:

```
Inductors
  air-core
    solenoidal
      end effects
      approximations
      adding a magnetic core
    toroidal
  magnetic materials
    types
    permeability
    B-H curve, saturation, loss
    gapping -- effective mu, stability,
      energy storage
  basic inductor design
    ungapped (maximum L, dc saturation)
    gapped (adjustable L, stability, energy,
      saturation)
```

1x.4.1 The basics

Inductance (more properly, *self* inductance) is the circuit property caused by an electrical circuit linking some of the magnetic field that it produces when a current flows through it.[62] According to Faraday's Law of Induction, a *changing* current then induces a voltage source within the circuit itself (a *"back EMF"*), given by $V = L\,dI/dt$; in the

---

Table 1x.6: Conversion Factors Between SI and CGS

	Gaussian CGS → SI	SI → Gaussian CGS
B	1 gauss = 10^{-4} tesla	1 tesla = 10^4 gauss
H	1 oersted = $\dfrac{10^3}{4\pi}$ amp/m	1 amp/m = $4\pi\times10^{-3}$ Oe
	≈ 80 amp/m	≈ 12.6 mOe
V	1 statvolt = 300 volts	1 volt = $\dfrac{1}{300}$ statvolt
I	1 esu/s = $\dfrac{1}{3}$ nanoamp	1 amp = 3×10^9 esu/s

Note: all factors with 3's are the speed-of-light "3", i.e., 2.998

Conversion factors between the two orthodoxies.

frequency domain this is the familiar reactive impedance $Z_L = j\omega L$.[63] An inductor stores energy in its magnetic field (provided by the circuit's voltage as the current builds up),[64] given by $U_{\mathrm{mag}} = \frac{1}{2}LI^2$.

1x.4.2 Air-core inductors

The simplest inductor is a coil of wire.[65] If the inductor is wound without a magnetic core material, it is (somewhat loosely) called an "air coil," or an "air-cored coil," even though it will usually be wound on a rigid non-conducting plastic or ceramic coil form. At one extreme it may be a single circular loop (or multiple turns bunched into a single circular loop); at the other extreme it may be a long single-layer (or multilayer) coil (a "solenoid"). Quite often the coil is wound on a core of magnetic material, whose magnetic permeability provides a path for the magnetic field; somewhat more accurately, the core's magnetic properties cause the magnetic field B (which is the field responsible for inducing voltages) to be enhanced by a factor of μ_r (the "relative permeability") for the same coil currents and geometry, compared with a non-magnetic core material.[66]

Note: when we use the term "inductance" here, we are

---

[60] From among the many varieties of ferrites, powdered iron, or laminated (or tape-wound) ferrometals.

[61] E.g., C-cores, E-cores, pot cores, or toroids.

[62] The precise definition is $L = \Phi/I$, where Φ is the magnetic flux linked by the circuit: $\Phi = \int_S B \cdot dS$.

[63] The *reactance* is the magnitude of the (imaginary) impedance: $X_L = \omega L = 2\pi f L$.

[64] From a physical point of view there's an energy associated with the magnetic field itself, to the tune of an energy *density* $B^2/2\mu_0$ (joules per cubic meter); you integrate this expression over volume to get the stored energy, which is returned to the circuit when the current is removed.

[65] Though we remind the reader that a length of wire, coiled or not, is inductive, as described above in §1x.1.6.

[66] An accurate statement for an inductor with a magnetic core is that the "conduction currents" in the coil conductors determine the so-called H field, from which the B field (enhanced by the "bound currents" of the ferromagnetic material) is given by $B = \mu_r\mu_0 H$; here μ_0 is the

Figure 1x.72. A collection of inductor types, plucked from our various parts bins. The selection includes both cored and air-core types, slug-tuned (adjustable) types, and itty-bitty chip inductors. You can get helpful "designer's kits," loaded with sample inductors, from companies like Coilcraft.

referring to the coil's *self* inductance, that is, the effect of the coil's current on itself (as in $V = L\,dI/dt$). Later we'll deal with *mutual* inductance, the effect of coil A's currents on coil B, and vice-versa. Mutual inductance is at the heart of the transformer, and is not to be ignored!

Scaling
For any coil size and shape (loop, solenoid, etc.), the inductance scales as N^2, where N is the number of turns fit into the same winding envelope. That is because the magnetic field (for a given current) scales as N, and the field (whatever its configuration) is linked N times; so $\Phi_{\text{linked}}/I \propto N^2$.

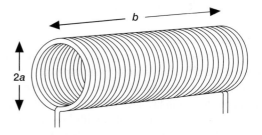

Figure 1x.73. An air-core solenoid inductor.

Note that this scaling assumes that the greater number of turns is fit into the same geometry, e.g., by going to a smaller wire gauge, or by winding several layers compactly onto the same form.

A. Solenoid – approximate
For a solenoid the exact inductance calculation is complicated, owing to end effects. However, if we're willing to ig-

---

defined constant $4\pi \times 10^{-7}$, and μ_r is the core material's "relative permeability," often simply called "mu." Non-magnetic materials have $\mu_r = 1$, whereas (ferro)magnetic materials have mu values generally in the range of 10^2–10^4. Though accurate, this description ignores many important issues; a full narrative here would have to deal with the complexities of nonlinearity, inhomogeneity, and anisotropy of the core material, as well as boundary conditions of the B and H fields.

nore such annoyances, it's easy enough to calculate the approximate inductance of a "long" solenoid (i.e., one whose length b is much greater than its diameter $2a$ (Fig. 1x.73): From the free-space Maxwell equation $\int B \cdot ds = \mu_0 NI$ we get

$$B \approx \mu_0 NI/b = 4\pi \times 10^{-7} NI/b, \qquad (1x.1)$$

under the reasonable assumption that the field is approximately uniform within the solenoid. Note the units here, which are in the SI (Système International d'unités) units:[67] I (amperes), b (meters), and B (tesla); the factor μ_0 (the "permeability of free space") is defined as $4\pi \times 10^{-7}$ (with units of kg·m·(amp·s)$^{-2}$, if you really want to know). To figure the inductance, we calculate the flux per turn $\Phi = B \cdot \pi a^2$ and take the ratio $L = \Phi_{\text{linked}}/I$ to get:

$$L \approx \mu_0 \pi a^2 N^2/b = 4\pi^2 10^{-7} a^2 N^2/b \quad \text{henrys.} \qquad (1x.2)$$

Given the kinds of air-core inductors they use, circuit designers aren't fond of measuring dimensions in meters, nor results in henrys. So you'll see eq'n 1x.2 written in mongrel units as

$$L \approx 4\pi^2 a^2 N^2/b \quad \text{nH,} \qquad (1x.3)$$

where the dimensions a and b are now in handy centimeters.

An example. A coil of 100 turns has radius $a = 0.5$ cm and length $b = 3$ cm. If we stick our head in the sand and ignore the fact that the length of the coil is not much greater than its diameter, we find from eq'n 1x.3 that $L \approx 32.9\,\mu$H. We use the "approximately equal to" symbol ($\approx$) here, because we know our answer is optimistically high. Read on . . .

B. Solenoid – exact

What about those pesky fringing fields? Many fine minds of centuries past[68] (undistracted by smartphones, the Internet, social media, or even television) have worked on the problem, producing inductance formulas framed as terrifying power-law expansions riddled with logs, elliptic integrals, and other nightmares. Happily, these fine folks also went to the trouble of calculating tables that list the correction factor K to go in front of eq'n 1x.3; these give values of K as a function of the coil's *aspect ratio* $b/2a$ (i.e., the ratio of length to diameter). An often-referenced

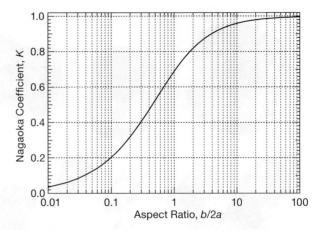

Figure 1x.74. Nagaoka's coefficient K for eq'ns 1x.2 and 1x.3, to account for "fringing field" correction to inductance of a finite-length air-core solenoid.

set is due to Nagaoka,[69] which can be found in many references (for example, Tables 37 and 38 in *Inductance Calculations*, F.W. Grover, Dover, 2004). Nagaoka's table (graphed in Fig. 1x.74), applied to this example, gives a factor $K = 0.872234$ for $b/2a = 3$, thus a corrected value of inductance $L = 28.70\,\mu$H. Having ignored the weaker fields near the coil ends, our earlier approximate result was high by 15%.

If you're disinclined to consult tables, you can instead exploit some charming semi-empirical inductance approximations, of which the most popular is probably the remarkable Wheeler formula:[70]

$$L = \frac{a^2 N^2}{9a + 10b} \quad \mu\text{H.} \qquad (1x.4)$$

Here the coil radius a and length b are, somewhat weirdly, in *inches*.[71] This simple formula is claimed to be accurate to within 1% for $l/r > 0.8$, i.e., for a coil that is not too stubby. Let's give it a try on our 100-turn example coil, where $a = 0.1969$ inches and $b = 1.181$ inches; Wheeler's formula then gives $L = 28.54\,\mu$H – in error by 0.6%, not bad! And if we are willing to use the "unrounded" coefficients ($9 \rightarrow 9.12$ and $10 \rightarrow 9.92$) we get

---

67 See Table 1x.6 for unit conversions.

68 Would you believe: Kelvin, Kirchhoff, Lorenz, Maxwell, Rayleigh, and Wien? *It's true!* – see the 237-page tome, "Formulas and tables for the calculation of mutual and self-inductance," *Bulletin of the Bureau of Standards*, **8**, 1 (1912).

69 "Formulas and tables for the inductance of solenoids," H. Nagaoka, *J. Coll. Sci. Tokyo*, **27**, 6, 18–33 (1909).

70 H.A. Wheeler, "Simple inductance formulas for radio coils," *Proc. I.R.E.*, **16**, 1398 (1928). A variant with unrounded values of 9.12 and 9.92 is claimed to be accurate to 0.1% for $1.3 \leq b/a \leq 10$, according to the 1942 edition of Radiotron Designer's Handbook.

71 The reader can convert the formula to metric, but it will no longer look neat and clean.

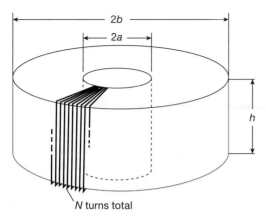

Figure 1x.75. In an ideal toroidal inductor the magnetic field is non-zero only inside the enclosed volume, eliminating the "end effects" of a solenoid.

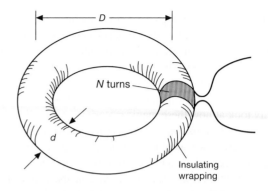

Figure 1x.76. An N-turn air-core loop.

$L=28.69\,\mu\text{H}$ – wow, almost perfect agreement (error of 0.035%) with Nagaoka's elaborate 6-digit tables!

Soon enough we'll see how to enhance the inductance of a simple coil by inserting a ferromagnetic core. Before that, though, let's look at a common coil geometry that eliminates end effects – by eliminating the "ends."

C. Toroid

Imagine a long solenoidal coil, bent around in a circle so its ends meet. The B-field chases its tail, and (in the ideal case of uniform sheets of current) there are no end effects. Figure 1x.75 shows the idea, here drawn with a rectangular cross section, both to reflect typical toroids and also to simplify calculation. It's straightforward to calculate its inductance from first principles. The B-field goes around in circles, with magnitude (from Maxwell's equation) of

$$B(r) = \mu_0 \frac{NI}{2\pi r} \quad \text{tesla} \qquad (1x.5)$$

at radius r meters; and the inductance is the N-turn linked flux divided by the current, or

$$L = \frac{1}{I} N \int_a^b B(r)\,h\,dr = \frac{\mu_o}{2\pi} N^2 h \log_e \frac{b}{a} \quad \text{henrys.} \qquad (1x.6)$$

As we did with eq'n 1x.2, we can present this in convenient units of centimeters and nanohenrys:

$$L = 2N^2 h \log_e \frac{b}{a} \quad \text{nH.} \qquad (1x.7)$$

We'll be revisiting the solenoid and the toroid presently, when we add in the "secret sauce" of ferromagnetic cores.

D. Loop

For a circular loop of diameter D cm, consisting of N turns of wire bunched into a bundle of thickness d cm (Fig. 1x.76), the inductance is approximately

$$L = 16N^2 D \left(\log_e \frac{8D}{d} - 2 \right) \quad \text{nH.} \qquad (1x.8)$$

A similar equation applies for the inductance of a single-turn loop: in that case $N=1$ and d is the wire diameter.

Here's an amusing challenge, first tackled by the legendary Maxwell: If you have a given length of wire and want to maximize the inductance of an N-turn loop, what should the dimensions be? He found $D \approx 3.7d$, but it turns out that the answer is almost exactly $D \approx 3d$, a configuration proposed in 1931 by B. Brooks (and hence a "Brooks coil").

1x.4.3 Magnetic-core inductors

Air-core inductors (and transformers) have their place in circuit design, primarily for radiofrequency applications where there's little energy involved, and where small values of inductance (in the nH to μH) suffice. But far more common are applications where serious amounts of power are moving around, the primary example being the ubiquitous switchmode power supply (SMPS). Cored inductors and transformers are necessary, also, when dealing with low frequencies, for example inductors for audiofrequency filters, or transformers for the ac powerline. Let's take a look at this often-confusing business of (ferro)magnetic materials and their application to inductors and transformers.

A. Ferromagnetic materials

The "air-core" inductors we've been describing are typically wound on some rigid form, so it's not really *air*; but it might as well be, because materials like plastic or ceramic are, to a good approximation, non-magnetic.[72] By contrast, the class of *ferromagnetic* materials (named after iron, but characteristic of numerous elements, alloys, and ceramic-like compounds) exhibit a mighty effect – a typical high-mu ferrite[73] might enhance the air-core B-field (and thus the inductance) by a factor of 1000–5000 or more. Let's briefly look at how these magic materials accomplish this.

If we could look with microscopic eyes at a ferromagnetic material, we would see a hodgepodge of small regions, in each of which the billions of atomic magnets (due to electron spin) are all aligned in the same direction. They do this for mysterious quantum-mechanical reasons that energetically favor alignment within each of these magnetic "domains." For most ferromagnetic materials the magnetization of the various domains points in different directions, averaging out in an unmagnetized sample; this, too, is a consequence of minimizing the energy of the system, which would be higher if all domains were aligned with each other. A magnetic field applied to such a sample, however, makes it energetically favorable for more domains to partially align.

And that's how the field enhancement happens: the weak applied field coaxes the built-in atomic magnets into partial alignment, contributing their (larger) field to the applied field. The enhancement factor is the relative permeability, μ_r, which can range up to 10^4 or so in a high-μ material.

Free and bound currents: *B* and *H*

It's useful to view the magnetic field contributed by the atomic magnets (the magnetized domains) as if they are produced by tiny "bound" currents.[74] These are to be distinguished from the "free" currents (charges flowing in conductors), which we control. It is the field produced by the latter (the "independent variable") that causes the bound currents to respond (the "dependent variable"); and it is the combination of the fields due to free currents and bound currents that comprises the total magnetic B-field.

Viewed this way, it's helpful to separate the magnetic fields caused by these two kinds of currents – this is the reason for dealing with the so-called H-field:

$$\int H \cdot ds = I_{\text{free}} \qquad (1x.9)$$

$$\int B \cdot ds = \mu_0(I_{\text{free}} + I_{\text{bound}})$$

$$= \mu_0 I_{\text{total}}, \qquad (1x.10)$$

Since it's I_{free} that we control, and it's B that we want to know (because it creates induced voltages, deflects charged particles, etc.), we write, finally,

$$B = \mu_r \mu_0 H, \qquad (1x.11)$$

which displays the magnetic material's response, in the form of its relative permeability μ_r, to the free-current-induced H-field. Following Purcell,[75] and ignoring precedent to the contrary, we call B the "magnetic field," and we call H simply the "H-field."

B–H curves

At a simplistic level, a given magnetic material responds to the H-field, created by the free currents we control, producing a B-field that is larger by a factor μ_r. It's (much) more complicated, though, because μ_r is not a good approximation to a constant; rather, it depends on the applied H-field, and it exhibits hysteresis.

Figure 1x.77 shows schematically what happens in a ferromagnet as the applied amp-turns (therefore the magnitude of H) is increased. At a small applied field the domains begin to align (they do not physically move; instead the domain "walls," shown as dashed lines, move[76] to allow favorably aligned domains to grow, at the expense of their neighbors). When the domains are nearly aligned, the material has reached saturation.

You might think the curve would retrace its steps when H is reduced, but in fact the material has some memory – it's "sticky" – and the B–H curve heads off to the second quadrant. We ran some B–H curves on sample ferrite cores, with the results shown in Figures 1x.78–1x.80. Figure 1x.78 is a $MnZnFe_2O_4$ soft ferrite, intended for power inductors and transformers; the term "soft" means that it exhibits a smooth B–H curve that returns nearly to zero

---

[72] Physics-savvy readers will complain that even "non-magnetic" materials have *some* response to an applied external magnetic field. That is indeed correct: *diamagnetic* materials weakly repel an applied B-field, and *paramagnetic* materials exhibit a (usually somewhat greater) tendency to concentrate an applied field. However, these effects are *tiny* – they alter an applied magnetic field by roughly 1 part in 10^5, and are completely ignorable for anything having to do with inductors, transformers, or circuits of any kind. Take that, you fussy physicists!

[73] A term we'll use, loosely, to mean any ferromagnetic core material unless stated otherwise.

[74] Indeed, Ampère hypothesized that the behavior of magnetic materials was caused by permanently circulating current loops inside, a view consistent with contemporary scientific thinking.

[75] E.M. Purcell and D.J. Morin, *Electricity and Magnetism, 3rd edn.*, Cambridge University Press (2013), §11.10.

[76] The movement is not smoothly continuous, but rather occurs with little jumps; if you connect an amplifier to a coil wound around an iron bar, you can hear the rasping sound as a permanent magnet is moved toward the bar. This is called Barkhausen noise – check out a nifty demonstration at www.youtube.com/watch?v=YLycGnOCqLc.

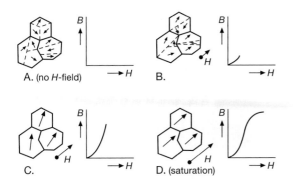

Figure 1x.77. Alignment of magnetic domains in a ferromagnetic material, as the applied H-field is increased (going from A to D). The hexagonal tiles represent the material grains, within each of which are several magnetic domains (separated by dashed lines).

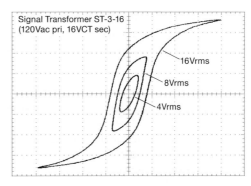

Figure 1x.79. Measured B–H curve set of a small ac powerline transformer, driven on the 16 Vrms secondary at 60 Hz. Note near saturation at rated voltage.

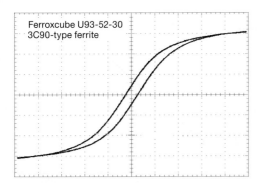

Figure 1x.78. Measured B–H curve of a "soft ferrite" U-core, showing a small degree of hysteresis. Horizontal: 40 A/m per div.; Vertical: 150 mT/div.

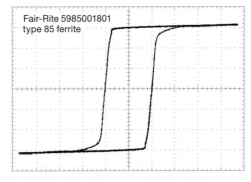

Figure 1x.80. This *hard* ferrite ("square-loop") core saturates abruptly, and maintains its saturated B-field when excitation is removed. Horizontal: 40 A/m per div.; Vertical: 150 mT/div.

magnetization, as contrasted with the "square-loop" characteristic of the type 85 ferrite shown in Figure 1x.80, which retains nearly saturated magnetization ("residual flux density," B_r) at $H=0$. For Figure 1x.79 we used an ac powerline transformer, whose core consists of ferromagnetic metal laminations; its B–H curves enclose more area (thus greater loss, see below) than the ferrite material. Interestingly, this transformer is normally operated over the outermost curve, running the core nearly to saturation.[77]

Relation of core loss and B–H curve

It's straightforward to demonstrate mathematically that the area inside the BH curve is equal to the energy dissipated per cycle, by simply figuring out the product of the driving current and the induced voltage in the excitation winding, and integrating over a cycle. To keep things simple we'll ignore niceties like the variation of B field across the core's cross-section.

Here's how it goes. If we have a winding of N turns around a toroidal core (or other closed magnetic geometry linked by the winding) of length l and cross-sectional area A, and the winding carries an oscillatory current $i(t)$, the magnitude of the H-field[78] is $H(t)=Ni(t)/l$.

Owing to the core material's magnetic properties, there results a B-field, which depends on the instantaneous value of H as well as its past history (as described by the core's B–H curve). The changing B-field induces an oscillatory

---

[77] The near saturation at rated voltage is a manufacturer's design decision, to minimize the amount of iron and copper in the transformer. This results in a fairly high no-load current, e.g., high enough to prevent the transformer from having an "Energy-Star" rating. However, ac-line transformers can be made as efficient as one desires; we used 230-volt Signal Transformer parts at 120 Vac (half the rated input voltage) and observed very low no-load currents.

[78] From the Maxwell equation $\nabla \times \vec{H} = \vec{J} + \partial \vec{D}/dt$.

voltage in the winding,[79] to the tune of $v(t) = -N d\Phi/dt = NA\, dB/dt$. So, we've got the (changing) voltage and current, from which we can find the energy lost per cycle (call it U_c):

$$U_c = \int_t^{t+T} v(t)i(t)\, dt = lA \int_t^{t+T} H(t)\frac{dB(t)}{dt}\, dt.$$

The core's volume is $V = lA$, so the energy lost per cycle per unit volume is the quantity in the integral; and so, multiplying by the number of cycles per second (the frequency, f) the *power* dissipated is given by

$$P = fV \oint H\, dB$$

where the integral around the loop is just the area enclosed by the $B-H$ curve. "Soft" ferrites exhibit small enclosed area, and are favored for low-loss inductors and transformers. Let's see how to use these handy materials.

B. Ferrite-core solenoid

Back in §1x.4.2 we calculated the inductance of an air-core solenoid. What happens if instead we wrap the coil around a ferromagnetic core (Fig. 1x.81) that has a $B-H$ curve akin to that of Figure 1x.78? Intuitively we expect that the coil's H-field (produced by the conduction currents in the winding) will coax the atomic magnets into partial alignment, adding their magnetic field contributions; and we might hope for an enhancement approaching the relative permeability μ_r (of order 10^3), and thus an increase by that same factor in the coil's inductance.

Indeed, the inductance is significantly greater than that of the air-core inductor – but by a disappointingly smaller factor. A way to think about it is that the coil does not live in a space fully occupied by ferrite material; rather, much of the magnetic field is outside the core. We will make this quantitative shortly, when we consider the toroidal core inductor, which (just as with the air-core toroid versus solenoid) turns out to be the simpler case.

Continuing with the solenoid-over-ferrite, the effective permeability approaches that of the material when the cylindrical core is long compared with its diameter. Figure 1x.82 plots μ_{eff} versus the core's aspect ratio (length/diameter) for a fully-wound cylindrical ferromagnetic core, based on a semi-empirical approximation. Materials of higher μ require a greater aspect ratio to obtain that advantage; for example, the core would have to be 50 times as long as its diameter for a ferrite with $\mu = 1000$ to produce an effective mu of 500. Although there are some tricks you can play to tease a higher μ_{eff} from a short

---

[79] From the Maxwell equation $\nabla \times \vec{E} = -\partial\vec{B}/dt$.

Figure 1x.81. A ferrite core increases the inductance of a solenoid inductor.

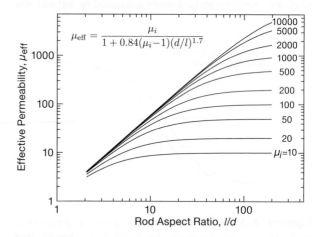

Figure 1x.82. The effective permeability of a ferrite rod approaches that of the bulk material only when its length is substantially longer than its diameter (and especially so for materials of high μ). (Semi-empirical formula, V.I. Khomich, "Ferrite Antennas," 1989.)

core (e.g., putting most of the winding near the center, see Fig. 1x.83), a far better approach is to wrap the core back around itself (analogous to the air-core solenoid) so the B-field stays (almost) entirely within the high-μ material.

C. Ferrite-core toroid

Now let's revisit the toroidal inductor of Figure 1x.75, imagining that it is instead wound on a ferrite core. Because the free current in the winding creates the interior H-field, eq'n 1x.5 is replaced by

$$H(r) = \frac{NI}{2\pi r} \quad \text{tesla}, \tag{1x.12}$$

and, assuming an ideal ferrite (linear, with $B = \mu_r \mu_0 H$) the inductance formulas (eq'ns 1x.6 and 1x.7) are simply multiplied by the factor μ_r:

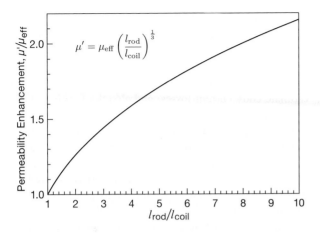

Figure 1x.83. A coil wound in the middle of a ferrite rod sees a somewhat larger permeability (μ') than predicted by the graphs of Fig. 1x.82 for a fully-wound rod. (Cube-root law, from W.J. Polydoroff, "High Frequency Magnetic Materials," 1960.)

$$L = \frac{1}{I} N \int_a^b B(r)\, h\, dr = \mu_r \frac{\mu_o}{2\pi} N^2 h \log_e \frac{b}{a} \quad \text{henrys,} \quad (1x.13)$$

$$L = 2\mu_r N^2 h \log_e \frac{b}{a} \quad \text{nH,} \qquad (1x.14)$$

where the first equation's dimensions are meters, and the second centimeters.

Shortcuts

Life doesn't have to be this complicated – you shouldn't have to integrate over core cross-sections, or fiddle with logs, etc. Circuit designers like shortcuts, and manufacturers are happy to oblige. In that spirit, they provide, for any given core geometry and size, the following parameters:

l_e : The core's effective magnetic path length.

A_e : The core's effective cross-sectional area.

A_{min} : The core's minimum cross-sectional area.

$\Sigma(l/A)$: The "core factor" (sometimes called C_1), mathematically equivalent to $\int dl/A$ around the magnetic path.

V_e : The core's effective magnetic volume.

In addition, for such a core made from a particular magnetic material, they provide also:

A_L : The core's inductance, in nanohenrys per turn-squared.

μ_e : The core's effective (relative-) permeability, sometimes called μ_r.

These make your life easy. Let's see how.

Calculating L

A_L is your friend, here, you can't go wrong:

$$L = N^2 A_L \quad \text{nH.} \qquad (1x.15)$$

So, for instance, a TDK B65813JR87 core (25 mm by 13 mm "RM10"-style core, type N87 ferrite) specifies $A_L = 5200\,\text{nH}/t^2$. If you want an inductance of 1 mH, for example, just wind 14 turns, and off you go.

Presently we'll see that it's possible to adjust the A_L value by introducing a gap in the otherwise-closed magnetic path. A gap reduces A_L, which might at first seem like a bad idea – who wants *less* permeability? – but in fact it offers important advantages in some applications (energy storage, tunability, freedom from saturation).

Calculating B

It's essential to make sure you aren't bringing the core close to saturation; that is because, as we saw in the B–H curves of §1x.4.3A, magnetic materials are not linear. A given core material will saturate at some value of B-field, typically in the range of 200–600 mT (2000–6000 gauss, in Gaussian cgs units that are still commonly used in magnetics) for ferrites.[80] You can saturate a core by providing a dc current

$$I_{dc} \geq \frac{B_{sat} l_e}{\mu_r \mu_0 N} \qquad (1x.16)$$

where l_e is the length of the magnetic path through the core (for example, $l_e \approx 2\pi r$ for a toroid of mean radius r). If you must support a dc amp-turn product that would saturate a given core, choose a core material with lower μ_r, or a core whose material has a higher B_{sat}, or a core that has an air gap; the latter reduces the core's effective mu, as we'll see in the next subsection.

If you know the current, you can rearrange eq'n 1x.16 to find

$$B = \mu_r \mu_0 \frac{NI}{l_e}, \qquad (1x.17)$$

where we've written I without the subscript. That's because a core can be saturated by a purely ac excitation (or by a combination of ac and dc), if it asserts a sufficient "volt-time product." It works like this: a cored inductor with some inductance L will see its current increase as $dI/dt = V/L$, thus producing a current $I = Vt/L$ after time t (if the voltage is changing, it's a "volt–time integral" you calculate). So it will saturate after a volt–time product $Vt = LI_{sat}$, where I_{sat} is the winding current that produces

---

[80] Iron powder cores have somewhat higher values of B_{sat}, in the range of 1000–1600 mT, but considerably lower permeability.

saturation (eq'n 1x.16). And of course the same thing can happen with simple sinewave excitation, since a drive voltage $V_0 \sin 2\pi f t$ applied to an inductance L produces a current $I_0 = V/\omega L = V_0/2\pi f L$, from which you can get B_{peak} from eq'n 1x.17. As we'll see, for interesting reasons a gapped core has reduced susceptibility to dc saturation, but not to ac saturation.

Now for the **shortcuts** for figuring out the peak B-field, to ensure you are not saturating a core:

It's not hard to show[81] that $A_L = \mu_r \mu_0/\Sigma(l/A)$, from which you can readily find[82] that

$$B_{\text{peak}} = \frac{V_{\text{rms}}}{\pi\sqrt{2}fNA_e} \quad \text{tesla,} \qquad (1x.18)$$

where A_e is in square meters. In the common case where the applied voltage is a square wave, the peak B-field is instead

$$B_{\text{peak}} = \frac{V_{\text{pp}}(\text{square})}{8fNA_e} \quad \text{tesla.} \qquad (1x.19)$$

If you prefer to measure in centimeters, and you like your B-fields in gauss, use these equivalent formulas:

$$B_{\text{peak}} = \frac{10^8 V_{\text{rms}}}{4.44fNA_e(\text{cm}^2)} \quad \text{gauss.} \qquad (1x.20)$$

$$B_{\text{peak}} = \frac{10^8 V_{\text{pp}}(\text{square})}{8.89fNA_e(\text{cm}^2)} \quad \text{gauss.} \qquad (1x.21)$$

Finally, if you've got a non-zero dc current in addition to the sinewave, the peak B-field becomes

$$B_{\text{peak}} = \frac{V_{\text{rms}}}{\pi\sqrt{2}fNA_e} + NI_{\text{dc}}\frac{A_L}{A_e} \quad \text{tesla,} \qquad (1x.22)$$

or for a square wave with dc current:

$$B_{\text{peak}} = \frac{V_{\text{pp}}(\text{square})}{8.89fNA_e} + NI_{\text{dc}}\frac{A_L}{A_e} \quad \text{tesla,} \qquad (1x.23)$$

with all units in SI (volts, amps, meters, and henrys per turn-squared).

Calculating core loss

Cores are not perfect – they exhibit some power absorption, both from eddy currents in the partially conductive material, and, as we saw earlier, from hysteresis in the B–H curve. Eddy current losses are sometimes characterized by specifying a complex permeability, where μ' is the real (non-lossy) part, and μ'' is the imaginary (lossy) part. The usual model is an equivalent resistance R_s in series with the intended inductance L_s, where the "s" subscript refers

to the series RL model.[83] For the series model the ratio of impedances $R_s/\omega L_s$ equals μ_s''/μ_s' (the subscripts refer to the series model), and can be represented by a *loss tangent* $\tan\delta = R_s/\omega L_s$. Ferrite datasheets provide graphs of μ_s' and μ_s'' as a function of frequency, from which you can estimate eddy current losses, and also the limiting Q of a tuned circuit: $Q = \omega L/R_s = 1/\tan\delta$.

But there are also hysteresis losses, which depend roughly linearly on frequency but, unhappily, non-linearly on the B–H curve excursion. How do you figure out the total core loss, then? Convenient for this purpose are families of curves of total core loss versus peak B-field, for several frequencies – see Figure 1x.84, where we've replotted data from Ferroxcube's "Soft Ferrites and Accessories" databook.

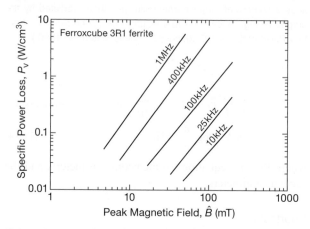

Figure 1x.84. Specific core loss (power per unit volume) versus peak B-field and frequency for a particular Ferroxcube ferrite blend. Gaussian readers note: 1 mT equals 10 gauss.

Copper loss

When a core is used as a *transformer*, the winding inductance is non-critical (as long as it is significantly larger than the impedances caused by the load), so you have some freedom in setting the number of turns in the primary and secondary, as long as you preserve the desired turns ratio. Fewer turns reduces I^2R heating in the wire (doubly so, because you can use a larger-gauge wire), but increases the B-field excursion, therefore the core heating. A commonly observed rule-of-thumb in magnetics design is to adjust the number of turns so that the wire ("copper") loss is roughly equal to the core loss. So you can estimate the total dissipation (core plus copper) for a given design. You don't want

---

[81] Try it!
[82] Try this one, too.

[83] A parallel model is sometimes preferred when dealing with transformers.

things to get too hot, so it's helpful to have a rough idea of the thermal resistance of a wound core. Figure 1x.85 shows *approximate* values of R_θ versus core volume V_e, for cores of various shapes, including the heating produced by the windings (assuming $P_{\text{wire}} \approx P_{\text{core}}$).

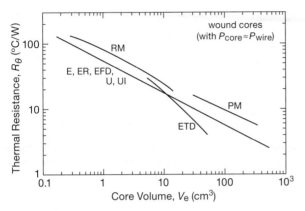

Figure 1x.85. Approximate thermal resistance R_θ versus core volume V_e for ferrite transformer cores (with windings) of several common shapes, assuming comparable core and copper power loss. (Replotted from Epcos/TDK Ferrites Databook, 2013.)

Winding capacity

When designing an inductor or transformer, you've got to make sure the windings will fit on the core. Manufacturers provide handy curves for this purpose, for example like the one in Figure 1x.86.

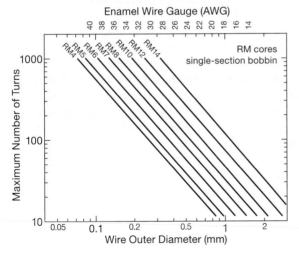

Figure 1x.86. Wire capacity of RM-style cores, wound on single-section bobbins ("coil formers"). Two-section bobbins accommodate approximately 10% fewer turns. (Reformatted from Epcos/TDK Ferrites Databook, 2013.)

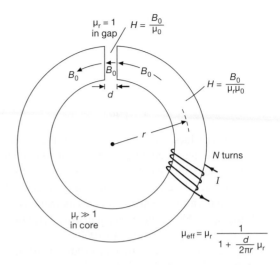

Figure 1x.87. Introducing even a small air gap in a ferrite core significantly reduces its effective mu (μ_{eff}), see text. The perpendicular component of the B-field is continuous across an interface, a consequence of the absence of "magnetic charge"; so the H-field is larger in the gap by a factor of μ_r, the permeability of the ferrite material.

D. Gapped core

Now to the business of core gapping. Look at Figure 1x.87, where we've drawn a ferrite toroid with material of relative mu (μ_r), assumed much larger than unity; that's fair, because most ferrites have μ_r values in the range of 1000–10,000. The core has a small gap, length d, which is much shorter than the mean magnetic path length $l_e = 2\pi r$. Some current I is flowing through N turns. We want to know the effective mu (μ_{eff}), that is, the equivalent relative permeability of an ungapped core of similar dimensions.

This turns out to be quite straightforward. Readers with a background in electromagnetism will recognize the "boundary conditions" on the B and H fields, specifically that the normal component of B ($B_\perp$) is continuous across an interface, and likewise for the tangential component of H ($H_\parallel$). Of course, everywhere we have $B = \mu H = \mu_r \mu_0 H$. So the magnitude of B is the same throughout the core (call it B_0), and H takes on values of B_0/μ_0 (in the gap) and $B_0/\mu_r\mu_0$ (in the core); H is *much* larger (by the factor μ_r, of order 10^3–10^4) in the gap than in the core.

That's all[84] we need: applying eq'n 1x.9 we have

$$\frac{B_0}{\mu_0}\left(\frac{l_e}{\mu_r} + d\right) = NI$$

---

[84] More than all: we don't need the boundary condition on H.

which can be written in the form

$$\frac{B_0 l_e}{\mu_{\text{eff}}\mu_o} = NI$$

with effective mu given by

$$\mu_{\text{eff}} = \frac{\mu_r}{1 + \dfrac{d}{l_e}\mu_r}. \qquad (1\text{x}.24)$$

Look at some values: with zero gap we get the expected $\mu_{\text{eff}} = \mu_r$ (i.e., the ungapped core exhibits its native relative permeability); but with a gap that's μ_r times smaller than l_e (here equal to $2\pi r$) the effective permeability is reduced to half ($\mu_r/2$). That's a small gap: a 5 cm (2″) diameter core of permeability $\mu_r = 2000$ would need a gap of $157\,\mu$m (about 0.006″) to reduce its permeability to 1000.

So, gapping a core is a way to adjust (downward) its effective permeability. OK, fine. But here's an *amazing fact* – it also *stabilizes* the permeability. Consider this example: take a core with $\mu_r = 4000$, and add a gap to reduce it to $\mu_{\text{eff}} = 500$. The required gap is $d = 2\pi r/571$. Now imagine that the core's permeability drifts downward by 10% (from a change of temperature, or by operating over a larger portion of the B–H curve, or by mechanical stress, etc.); now the effective permeability of the same gapped core is $\mu_{\text{eff}} = 493$. Hah, that's a change of just -1.4%!

Think about it – by throwing away some permeability, we gain stability. Remind you of something? Recall negative feedback, where the closed-loop gain is

$$G_{\text{CL}} = \frac{G_{\text{OL}}}{1 + BG_{\text{OL}}},$$

which has the same form as eq'n 1x.24! So, gapping trades off permeability to improve its constancy, just the way feedback trades off open-loop gain to improve linearity (i.e., constancy of gain). To make this point graphically we plotted in Figure 1x.88 the effective mu (μ_{eff}) versus the ungapped core mu (μ_r), for several values of gap length, assuming a core with a magnetic path length $l_e = 100$ mm.

Analogous to the "desensitivity" factor in traditional feedback (§2.5.3, eq'n 2.17), you can calculate the stabilizing effect of core gapping by taking the derivative $d\mu_{\text{eff}}/d\mu_r$ in eq'n 1x.24, to find

$$\frac{d\mu_{\text{eff}}}{\mu_{\text{eff}}} = \frac{d\mu_r}{\mu_r} \cdot \frac{\mu_{\text{eff}}}{\mu_r}, \qquad (1\text{x}.25)$$

i.e., that gapping a core reduces the effects of a fractional change in core permeability by a factor of μ_{eff}/μ_r. The latter (the factor by which the core permeability is reduced by gapping) is called the "dilution ratio."

Quite apart from the stability benefits, there are additional reasons for using a gapped core: the lower μ_{eff} al-

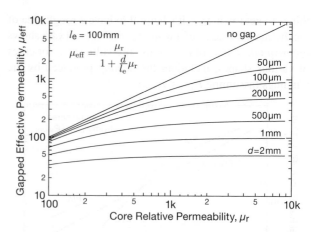

Figure 1x.88. A gapped core exhibits better constancy of effective permeability μ_{eff} over variations in the core's native permeability μ_r.

lows increased dc excitation before the core saturates, as seen in Figure 1x.89. To elaborate on this important point: you might think that gapping would offer no improvement in the inductance for a given allowable dc current (a useful figure-of-merit), because any advantage in dc current-carrying capability would be offset by the lowered inductance (from the lowered μ_{eff}). But you'd be wrong. To see why, imagine we've started an inductor design, choosing an ungapped core with some μ_r (characteristic of the ferrite) and some corresponding A_L (characteristic of both the ferrite and the core geometry), and that we've calculated the number of turns N to get the inductance L_0 that we want (i.e., $N = \sqrt{L_{\text{(nH)}}A_L}$). But then we look at a graph like Figure 1x.89 (or, equivalently, the tabulated values of B_{sat}) and discover that the core will saturate. So we choose instead a gapped core that reduces μ_{eff} (and therefore A_L) by a factor of 100; now to get the same L_0 we increase N by a factor of 10. Result: we've got the inductance we want, but with one-tenth of the B-field (10 times as many amp-turns – thus ten times as high an H-field – but 1/100th the effective permeability).[85]

As we'll see presently, introducing a gap also increases greatly the energy stored in the core's magnetization; this is essential in many varieties of switching power converters in which the inductor (or transformer) is used to store and release energy; these include boost, buck, invert, and flyback topologies.[86]

---

[85] At this point we have to worry about the losses caused by the smaller wire gauge needed to fit, perhaps choosing a larger core size. There are some good graphical methods that help you choose core and gap size with less iteration; the so-called *Hanna curve* is a nice example.

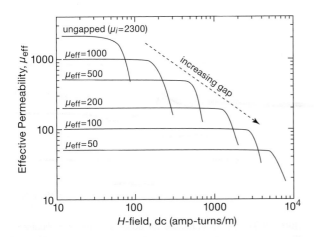

Figure 1x.89. A gapped core exhibits reduced effective permeability μ_{eff}. So a given dc excitation causes a smaller B-field, according to $B = \mu_{\text{eff}} \mu_0 H$, and therefore a gapped core can withstand a larger amp-turn product. (Ferroxcube 3C94 ferrite, reformatted from Soft Ferrites and Accessories Data Handbook, 2013.)

A complication

We've made some assumptions and approximations in our calculation of the effective permeability μ_{eff} of a gapped toroid. In particular, we've assumed a uniform cross-sectional area of both the toroid and its gap. For a toroidal geometry that's perfectly reasonable; but it's not terribly accurate in the case of other core geometries (pot cores, RM cores, E cores, and the like), and it's these latter geometries that are by far the most commonly used to make gapped cores. So an accurate calculation has to take into account the variation of cross-sectional area as it follows flux lines around the core; to do this correctly you have to integrate the (variable) quantity $B/A_e \mu_0 \mu_r$ (instead of the constant H in eq'n 1x.9) around the magnetic path. This is not an easy calculation, because the flux can take a winding path as it navigates the shape of the ferrite. Fortunately, it's not your job to calculate this – the core manufacturers specify all the data you need.

For example, Ferroxcube specifies for their RM12 core geometry to have an effective length $l_e = 56.6$ mm, and an effective area $A_e = 146$ mm^2. They tabulate several factory-gapped options (each with its own part number) made from type 3C90 or 3C94 ferrites; those ferrites have a relative permeability of $\mu_{\text{eff}} = 1730$. If, as we've done above, we make the dubious assumption that the core has uniform cross-section,[87] we can use eq'n 1x.24 to esti-

mate the effective gapped mu (μ_{eff}); and from that we can estimate the A_L value (nH per turn–squared), using[88] $L/N^2 = \mu_e \mu_0 A_e / l_e$. The table in Figure 1x.90 lists calculated results, using as input their datasheet values of "air gap" length d:

	calculated		datasheet	
gap (μm)	μeff	A_ℓ (nH/t²)	μeff	A_ℓ (nH/t²)
0	1730	5570	1730	5600 ±25%
300	170	547	194	630 ±5%
510	104	335	123	400 ±5%
680	79	254	97	315 ±5%
900	61	196	77	250 ±3%
1570	35	113	49	160 ±3%

Ferroxcube RM12[a] Gapped Core[b]

(a) $A_\ell = 146$mm2, $\ell_e = 56.6$mm. (b) type 3C90 or 3C94 ferrite.

Figure 1x.90. Calculated effective mu and A_L values, compared with datasheet values, for a selection of core gap lengths.

The calculated values are consistently low, by some 10%–30%. Our assumption of uniform cross-section is no doubt partly to blame. But the datasheet gap dimensions are likely only approximate, as evidenced, for example, by the fact that the data for the same RM12 core with type 3F3 ferrite shows identical values for the gap length, μ_{eff}, and A_L; yet that ferrite has a lower mu ($\mu_r = 1560$, versus the $\mu_r = 1730$ for the 3C90 and 3C94 ferrites).

Tunable inductors

Factory-gapped cores are made with accurately controlled gaps, typically less than a millimeter, and with well-controlled values of A_L, as evidenced in the tabular data for the Ferroxcube core above. There are situations, though, in which you want to have some adjustability of the finished inductor, for example in an LC filter or resonant circuit. In such cases you can get ferrite cores in certain geometries (RM-type cores and P-type pot cores) that have a hole going through the gap, into which you can screw a ferrite tuning plug to alter the effective gap; typically you get 10%–30% tunability.

You get similar tunability with what might be termed an "open magnetic" style inductors: these are small coils with a threaded core that accommodates a threaded ferrite piece (second row in Fig. 1x.72); they're called "slug-tuned," and you can get them in nominal inductances going from tens of nH to a few mH, with tunability in the range of ±5%

---

[86] Put another way, only the so-called *forward* converters are exempt.

[87] Manufacturers provide a helpful quantity, the "core factor" C_1, which is simply the line integral $\int ds/A_e$ around the flux path.

[88] In two easy steps: $B = \mu_e \mu_0 NI/l_e$, from which $L = \mu_e \mu_0 N^2 A_e / l_e$.

to ±30% or so. They are available unshielded or with an enclosing metal shield. Look at offerings from Coilcraft, Murata, and Toko.

E. Noise and spike suppression

A favorite technique to suppress noise, and to tame annoying high-frequency oscillations, is the use of lossy inductors. These are commonly available as surface-mount "chip" inductors, but, unlike ordinary inductors, they are designed to be lossy at high frequencies. They are usually specified in terms of their impedance at a standard frequency (often 100 MHz), and their dc resistance and maximum rated current. Murata's BLM-series is typical, with 100 MHz impedances in the range of $10\,\Omega$ to $1k\Omega$, and current ratings from a few hundred milliamps to several amps; they come in 0603 SMT size, and cost about $0.05 in 100pc quantities.

A recent addition to the quiver of spike suppression armaments are amorphous square-loop "Amobeads" from Toshiba. These come as bare beads or as 2-terminal gull-wing SMT components; they work by presenting a high inductance for a short duration event (a spike), but saturating to a low impedance for steady-state signals. We played around with these things, driving a low-impedance load ($10\,\Omega$) through 4 turns wound around an Amobead (Toshiba SS10X7X4.5W) with a bipolarity pulse pair (bottom trace of Fig. 1x.91), comparing the output (top trace) with that of a similar arrangement using a soft ferrite (FB73-110).

The soft ferrite, acting like a linear inductor, produces a rising waveform (from $V=LdI/dt$), but the square-loop Amobead suppresses the output for $10\,\mu s$, after which it passes the full amplitude through. So spikes less than $5\,\mu s$ would be completely suppressed – very good!

1x.4.4 Inductors and transformers for power converters

There's lots in play when designing a transformer for a switching power supply; whole books (large, and many) are devoted to this subject. There are some helpful guidelines available, for example "reference designs" that include core and winding specifications (or, simpler, manufacturer and part numbers for commercially available transformers). If you're designing your own, though, a useful starting point is Keith Billings' core-size nomogram, a modified version of which is shown in Figure 1x.92. It can

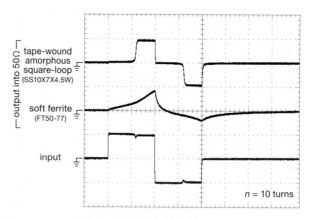

Figure 1x.91. Series inductors for spike suppression, illustrated with a bipolarity rectangular wave input and $50\,\Omega$ load. The soft ferrite behaves like a roughly constant inductance, whereas the square-loop inductor exhibits very large inductance until its abrupt saturation. Vertical: 5 V/div; Horizontal: $10\,\mu s$/div.

be shown[89] that the power rating of a ferrite transformer in a switching power supply is a function of the core's "area product" (the product of effective core area A_e and winding area A_w). Figure 1x.92's nomogram evaluates this relationship for a typical ferrite (Epcos/TDK type N27). To use it, you start with the desired power rating, then follow the dashed line to its intersection with the desired switching frequency. The vertical coordinate tells you the approximate core size (as A_eA_w, with some standard cores listed on the right side), and the horizontal coordinate tells you the optimum B-field excursion.

We will resist the temptation to delve into this world, instead pointing out (the heading of the next subsection) that you can often finesse the whole problem by just *buying* the inductor you need.

1x.4.5 Why *build* it, when you can *buy* it?

If you're in the business of highly optimized power converters, or you're pushing the envelope in a custom design, you may be forced to design your own magnetics. But, happily, there are inductor and transformer manufacturers waiting in the wings to bring you a finished product. Some companies to check are Bourns, Coilcraft, Coiltronics, Hammond, Murata, Panasonic, Pulse Engineering, Renco, Schott, TDK, Toko, and Wurth.

Better still, the semiconductor manufacturers of switch-

---

[89] See, for example, his "Derivation of Area Product Equations for Transformer Design," in *Switchmode Power Supply Handbook*, Appendix 4.A.

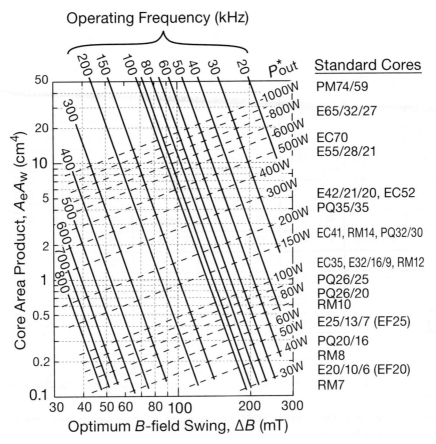

Figure 1x.92. Nomogram for choosing a ferrite core for a full- or half-bridge power converter, given a required output power and operating frequency, as taught by Keith Billings. The core types listed correspond to the core area product at left. (* reduce the power rating by 35% for single-ended converters.)

ing converters helpfully provide tables of inductor values and ordering part numbers for the example circuits and reference designs in their datasheets and application notes (see Fig. 1x.93 for an example). Even if you go with a custom inductor or transformer, these designs give you a good starting point.

1x.4.6 Inductor examples

A. Radiofrequency "chokes" and bias-T's

Most of the time when you think of inductors, you're interested in them as a component of a resonant circuit or of a filter, or as an energy-storage component in a switching converter. In these applications you care primarily about its inductance, and about parameters like its ESR, its Q, its saturation current, and the like.

In this section we are interested instead in the use of in-

ductors as ac-blocking and bias-coupling components;[90] such an inductor is called an *RF choke*, often packaged with a capacitor in a device called a *bias-T*. In these applications the particular inductance is non-critical; what's important is that the choke pass dc while presenting a high impedance (compared to the signal impedance) over the full range of signal frequencies. As we'll see, having sufficient inductance (thus high impedance) is the least of your problems – the trick is minimizing parasitic winding capacitance, which causes the impedance to plummet at high frequencies.

---

[90] The reverse of a blocking capacitor (see §1.7.1C, and a hundred examples scattered throughout the book), an analogous application where the component value (i.e., capacitance) is non-critical.

Table 2. Inductor Selection Guide

Inductance	Current	Schott		Renco		Pulse Engineering		Coilcraft
(µH)	(A)	THT	SMT	THT	SMT	THT	SMT	SMT
68	0.32	67143940	67144310	RL–1284–68–43	RL1500–68	PE–53804	PE–53804–S	DO1608–68
	0.58	67143990	67144360	RL–5470–6	RL1500–68	PE–53812	PE–53812–S	DO3308–683
	0.99	67144070	67144450	RL–5471–5	RL1500–68	PE–53821	PE–53821–S	DO3316–683
	1.78	67144140	67144520	RL–5471–5	–	PE–53830	PE–53830–S	DO5022P–683
100	0.48	67143980	67144350	RL–5470–5	RL1500–100	PE–53811	PE–53811–S	DO3308–104
	0.82	67144060	67144440	RL–5471–4	RL1500–100	PE–53820	PE–53820–S	DO3316–104
	1.47	67144130	67144510	RL–5471–4	–	PE–53829	PE–53829–S	DO5022P–104
150	0.39	–	67144340	RL–5470–4	RL1500–150	PE–53810	PE–53810–S	DO3308–154
	0.66	67144050	67144430	RL–5471–3	RL1500–150	PE–53819	PE–53819–S	DO3316–154
	1.20	67144120	67144500	RL–5471–3	–	PE–53828	PE–53828–S	DO5022P–154
220	0.32	67143960	67144330	RL–5470–3	RL1500–220	PE–53809	PE–53809–S	DO3308–224
	0.55	67144040	67144420	RL–5471–2	RL1500–220	PE–53818	PE–53818–S	DO3316–224
	1.00	67144110	67144490	RL–5471–2	–	PE–53827	PE–53827–S	DO5022P–224
330	0.42	67144030	67144410	RL–5471–1	RL1500–330	PE–53817	PE–53817–S	DO3316–334
	0.80	67144100	67144480	RL–5471–1	–	PE–53826	PE–53826–S	DO5022P–334

NOTE: Table 1 and Table 2 of this Indicator Selection Guide shows some examples of different manufacturer products suitable for design with the LM2575.

Figure 1x.93. This example from ON Semiconductor's LM2575 buck converter datasheet provides part numbers and manufacturers of suitable inductors. Reproduced with permission.

RF chokes

Figure 1x.94 shows several typical circuits with RF chokes ("RFC"). The broadband amplifier circuits (A and B) exploit the choke's high impedance at signal frequencies to supply operating current while simultaneously allowing plenty of gain (compared with the alternative of a low-value resistor). And, as with a resonant parallel *LC* collector load (e.g., Fig. 13.23 in AoE's second edition) the inductive pull-up permits signal peak-to-peak signal swings up to twice the supply voltage (e.g., the ADL5530 in circuit B specifies swings to 21.8 dBm, or 7.8 Vpp into a 50 Ω load, at the so-called "1 dB compression point.") Figure 1x.94C shows the use of RF chokes in a tuned (not broadband) RF power amplifier, where you want to keep the pi-network's tuning components (C_T – tuning; C_L – loading) at ground potential. RFC_1 biases the amplifier (as in circuits A and B), and RFC_2 collaborates with C_{block} to enforce dc ground on the chassis-mounted tuning capacitors.

Figure 1x.95 shows some RF chokes intended for operation in the so-called "HF band" (roughly 3–30 MHz).

They're charming, but they don't look like ordinary inductors (e.g., Fig. 1.51) – what's going on? *Here's* what's going on: in an inductor wound the ordinary way ("layer-wound," like the bottom right inductor in the picture), the parasitic capacitance wreaks havoc – it produces a parallel resonance (in this case at 300 kHz), and above that frequency it looks capacitive, with reactance falling as $1/f$, see the measured data in Figure 1x.97, where it is compared with the RF choke specimen (of the same inductance and wire resistance) pictured just to its left.

Evidently the folks who make RF chokes know what they're doing. They have several tricks: (a) the winding is divided into several series-connected sections, so the windings don't overlap the way they do when wound in layers; and (b) within each section the winding zigzags back and forth (variously called *basket-weave, lattice-wound, honeycomb,* or *duolateral*), further reducing the effective capacitance. The net result is impressive: a reduction of the effective shunt capacitance from 112 pF (resonance at 300 kHz)

Figure 1x.94. An RF choke (see Fig. 1x.95) provides a dc bias path while presenting a high impedance over some range of radiofrequencies. A. collector pullup for an *npn* amplifier stage. B. external pullup provides operating current for a tiny GaAs integrated wideband amplifier (a "MMIC," here providing fixed 16 dB gain from dc to 1 GHz). C. high-power (many kilowatts) vacuum-tube amplifier, with RF choke in the plate circuit so the tunable "pi-network" coupler components operate at dc ground (enforced by the output choke).

to a mere 1 pF (broad resonance at ∼4 MHz, look ahead to Fig. 1x.98).[91]

Variations on these winding techniques can be used to minimize the parasitic capacitance in inductors, even when they are not wound in a multi-section form. One technique is "bank-winding," in which a layered inductor is wound as shown in Figure 1x.96.[92] Another technique (popular in bias-T construction, as we'll see) is to connect in series a set of dissimilar chokes, such that the overall device ex-

hibits high impedance at all frequencies of interest. This is seen nicely in the elegant R-175A choke, whose lumpy impedance plot is high at the frequencies corresponding to the HF amateur radio bands; the choke's svelte figure (lengthy single-layer) hints at its high RF voltage rating (to ∼10 kV, with currents to 800 mA), and its low effective capacitance (0.8 pF).

"Peaking" coil

A favorite trick in wideband amplifier design is to add a small inductor in series with the collector (or drain) resistor; such a *peaking coil* extends the bandwidth by raising the load impedance at the high frequency end of the bandpass.

Bias-T's

A *bias-T* is, most simply, an RF choke and a blocking capacitor (Fig. 1x.99), usually seen in its "connectored" form, that is, with the components in a shielded box with RF connectors for the signal path, and either a connector or a feedthrough capacitor for the bias. Several of these are pictured in Figure 1x.102. Bias-T's are used to separate (or combine) signals and dc bias (power) on a single line (usually coax). A common example is a residential satellite receive dish, with its front-end amplifiers, oscillators, and mixers (the "low-noise block," or LNB) at the dish focus, powered and controlled through the same 75 Ω coax that brings the converted signal back to the satellite set-top box.[93] Other bias-T applications include biasing laser diodes, photodetectors, active antennas, or amplifier stages, or powering remote relays or semiconductor RF switches; the "dc bias" terminal has enough bandwidth (typically 10 kHz) to carry low-frequency control signals along with the actual dc.

decades of bandwidth; a typical RF bias-T covers the range from <1 MHz to a few gigahertz, and costs $50–100. Wider bandwidths are possible, for example the PSPL5542 bias-T from Picosecond Pulse Labs (now Tektronix) covers 10 kHz to 50 GHz (Fig. 1x.100), and includes pulse specifications such as risetime (7 ps) and delay time (140 ps). Other manufacturers of very wideband bias-T's include Marki Microwave, API/Inmet, and Anritsu, with units going to 65 GHz. For such millimeter-wavelength bias-T's, the high-frequency end is limited primarily by the connector style; SMA connectors are good to 26 GHz, whereas the 2.92 mm K-connector is rated to 40 GHz, and the 2.4 mm V-connector goes to 65 GHz.

---

[91] For those who like to think in terms of fields and their energy content, the RF choke construction minimizes the voltage difference between adjacent windings (thus the $\propto E^2$ field energy, which constitutes the $\frac{1}{2}CV^2$ capacitive term), while preserving the linked magnetic field (the inductive term). The broad resonance is due in part to the pole-spreading effect of the several sections of coupled same-frequency resonators.

[92] Some useful references, for those interested in this arcane subject, are H. P. Miller, "Multi-band r-f choke coil design," *Electronics*, **8**, 254–55 (1935), and H. A. Wheeler, "The design of radio-frequency choke coils," *Proc. I.R.E.* **24**, 6, 850–58 (1936). You can find some material, also, in Terman's legendary *Radio Engineers' Handbook*, McGraw-Hill (1943) on pp. 87–89. Evidently this was a hot topic in electrical engineering in the 1930s.

---

[93] Does *anyone* actually put the STB on top of their "television set"?

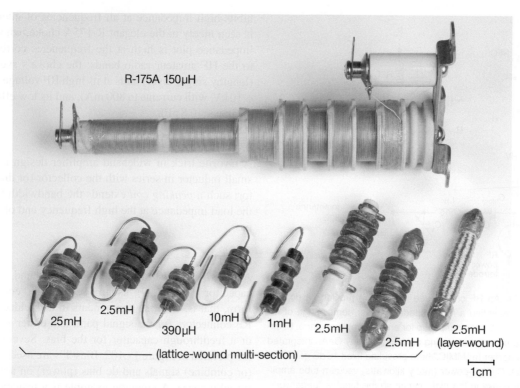

R-175A 150µH

25mH   2.5mH   390µH   10mH   1mH   2.5mH   2.5mH   2.5mH (layer-wound)

(lattice-wound multi-section)

1cm

Figure 1x.95. A sampling of some RF chokes. The impressive beast at the top was manufactured by the National Co., intended for high-power transmitters in the amateur radio bands. The row below are wideband lattice-wound multi-section chokes, whose performance is compared in Figs. 1x.97 and 1x.98 with that of a simple layer-wound inductor of the same inductance and resistance (lower right).

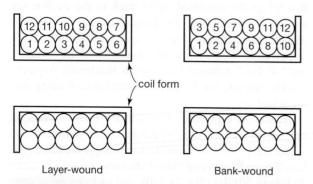

coil form

Layer-wound   Bank-wound

Figure 1x.96. Minimizing self-capacitance by *bank-winding*, which reduces the voltage difference between adjacent windings (and thus the energy stored in the electric field).

Achieving this kind of performance requires exquisite attention to detail, particularly in the design of the RF choke. The latter is implemented as a series string of chokes, with the smallest closest to the signal line; the latter is often implemented as a tapered conical inductor (Fig. 1x.101). Additional passive components (damping resistors, equalizing capacitors) are added to the recipe, as needed, to ensure flat response and good transient behavior.

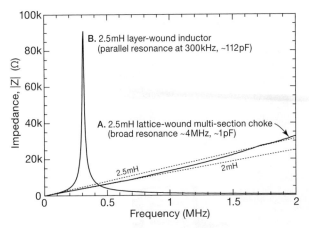

Figure 1x.97. Impedance versus frequency for a 2.5 mH RF choke, and for a layer-wound inductor of the same inductance and resistance, the pair shown in the lower right of Fig. 1x.95. The layer-wound inductor exhibits a sharp low-frequency resonance and low impedance (capacitive) above resonance, whereas the multi-section RF choke behaves nearly like an ideal inductance.

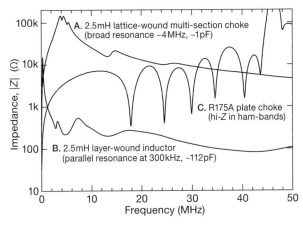

Figure 1x.98. When extended to higher frequencies, there is seen broad high-Z resonance in the multi-section choke, whose low ($\sim$1 pF) effective shunt capacitance keeps the overall impedance high throughout the frequency range, in sharp contrast to the pathetic performance of the layer-wound inductor. Also shown is the impedance of the unusual R175A choke (top, Fig. 1x.95), whose sections are tuned to produce high impedance on the several amateur-radio bands.

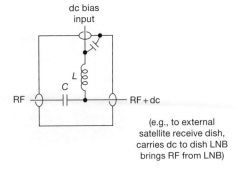

(e.g., to external satellite receive dish, carries dc to dish LNB brings RF from LNB)

Figure 1x.99. A "bias-T" couples an RF or pulse signal through (in both directions) while applying a dc bias to the device connected to one (RF+dc) port only. The inductor L is a wideband RF choke, often implemented as several series inductors (for wide bandwidth).

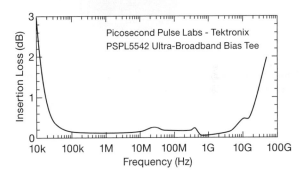

Figure 1x.100. An example of an extraordinarily broadband bias-T, effective over seven decades of frequency. (But if you have to ask the price, you can't afford it – $1,300.)

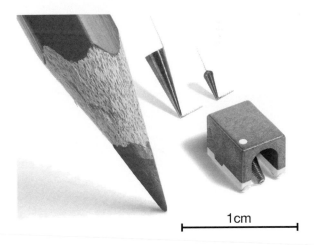

Figure 1x.101. Tapering an inductor suppresses resonant LC behavior, and forms an effective choke for a gigahertz bias-T. (Illustration used with permission of Coilcraft, Inc.)

63

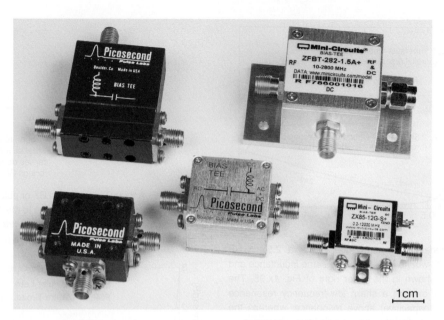

Figure 1x.102. Radiofrequency bias-T's with SMA connectors for the RF path., The Mini-Circuits unit at lower right has a bandwidth of 0.2 MHz to 12 GHz, and costs $100 in unit quantities. Bias-T's from Picosecond Pulse Labs (now Tektronix) push the limits of performance, with ultra-wideband units such as the PSPL5542 (10 kHz to 50 GHz, see Fig. 1x.100).

1x.5 Poles and Zeros, and the "s-Plane"

In the frequency domain it's certainly OK to think about a simple *RC* circuit (for example a single-section *RC* lowpass filter) basically as a circuit with a characteristic 3 dB frequency ($f_{3dB} = 1/2\pi RC$), and an asymptotic rolloff (6 dB/octave), with corresponding phase shift ($0°$ for $f \ll f_{3dB}$, going smoothly through $-45°$ at f_{3dB}, and on to $-90°$ for $f \gg f_{3dB}$). This is how we treated it in Chapter 1 of AoE3 (see for example §1.7.8 and Figure 1.104), where we dealt also with the overall scaling of resistor values so that its input impedance did not heavily load the input signal source, and its output impedance was able to drive the load attached to its output. For a more complicated circuit you can write down complex impedances and nodal equations (Kirchhoff's voltage and current laws) to figure out the response; or you can punt and run a SPICE simulation (see Appendix J).

But there's another approach, favored by folks who spend their days designing filters, or worrying about stability of feedback and control circuits.[94] As one of them recently proclaimed, when asked about the utility of the complex frequency plane, "I *live* in the *s*-plane!" Where he lives (with lots of company) is in a plane that is an extension of real frequency (ω, or $2\pi f$) into the complex plane. Just to make things confusing, the real axis corresponds to imaginary frequency: $s = j\omega$ (you can blame this on the Laplace transform). So real frequencies lie along the *y*-axis, and imaginary frequencies (i.e., exponentials) lie along the *x*-axis.[95]

The utility of the *s*-plane comes about because the transfer function (usually written $H(s)$, for the complex ratio of V_{out} to V_{in}) of any linear circuit you can concoct from *R*'s, *L*'s, and *C*'s (or their active circuit analogs) can be written as a fraction that looks like

$$H(s) = a\frac{(s-Z_1)(s-Z_2)\cdots(s-Z_m)}{(s-P_1)(s-P_2)\cdots(s-P_n)} \tag{1x.26}$$

where the *Z*'s ("zeros") are points in the complex plane

where the transfer function has value zero (i.e., a null), the *P*'s ("poles") are points where its value is divergent (i.e., infinite), and *a* is an overall multiplicative gain factor. The response of such a circuit is fully described by the location of its poles and zeros (along with the multiplying factor *a*). You'll commonly find engineers talking about the response of some system entirely in terms of the locations of its poles and zeros,[96] evidence of the intuition that can be acquired from such a technique. Let's see how this goes, with a couple of simple examples.

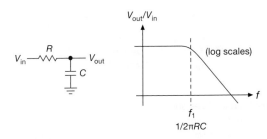

Figure 1x.103. Single-section *RC* lowpass filter circuit, and its frequency response (magnitude of V_{out}/V_{in}) plotted on logarithmic axes.

Example: Single-section *RC* lowpass

Figure 1x.103 shows the familiar single-section *RC* lowpass filter, along with the magnitude of its response V_{out}/V_{in} versus frequency (i.e., $|H(f)|$), plotted on logarithmic axes. As we saw in §1.7.9, it's simply a frequency-dependent voltage divider, whose (complex) response is found by taking ratios of (complex) impedances:

$$\frac{V_{out}}{V_{in}} = \frac{Z_C}{Z_R + Z_C} = \frac{-j/\omega C}{R - j/\omega C}$$
$$= \frac{1}{1 + j\omega RC} = \frac{1}{1 + sRC} \tag{1x.27}$$

where in the last step we've substituted $s = j\omega$. This has the form of eq'n 1x.26, with unit gain and with a pole lying on the *x*-axis at location $s = -1/RC$ (Fig. 1x.104A). The filter's response would be infinite at a (complex) frequency corresponding to such a pole; but *real* frequencies correspond to points on the *y*-axis, at which the response here is happily finite, and falling with increasing frequency, i.e., with increasing distance from the fearsome pole. We'll see some more intuition presently (look at Fig. 1x.106 if you cannot wait); for now you can deduce, from eq'n 1x.27, that the magnitude of the lowpass filter's response, at any real frequency *f* on the *y*-axis, is inversely proportional to

---

[94] Topics discussed in Chapters 2, 4, 4x and 6, and Appendix E.

[95] Physicists seem content with the concept of frequency as a complex number. But we've noticed that engineers sometimes exhibit discomfort; they prefer to say that the *s*-plane represents frequency along the *y*-axis, and another quantity, σ, along the *x*-axis. That is, a point in the *s*-plane has coordinates $\mathbf{s} = \sigma + j\omega$.

[96] And of their trajectories in the complex plane as various parameters are adjusted; this is known as "root locus."

the distance between that point and the pole, scaled by unit gain at dc. And the corresponding phase shift is just the angle (relative to the *x*-axis) of the line joining those points.

Exercise 1x.1. Take the challenge: show that we tell the truth, starting with the last form of eq'n 1x.27, and confining your thinking to the *s*-plane.

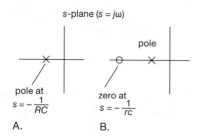

Figure 1x.104. Poles and zeros, plotted in the *s*-plane, of the (complex) frequency response $H(s)$ of simple *RC* circuits. A. Single-section lowpass filter (Fig. 1x.103). B. "Pole-zero" lowpass filter with added resistor *r* (Fig. 1x.105).

More complex circuits will have additional poles (and perhaps zeros, as we'll see next). For example, a filter consisting of a pair of cascaded *RC* lowpass sections will have a pair of poles on the negative *x*-axis; its amplitude response is the product (or sum of dB's) of the gain magnitudes for each pole (i.e., inverse of distances from each pole), and its phase shift is the sum of the phase shifts (i.e., angles from each pole). It follows that the ultimate rolloff of a cascade of *n* lowpass sections goes as $1/f^n$ (i.e., $6n$ dB/octave), with a phase shift of $-90n°$. But you knew that already, right?

Note that a pole on the *y*-axis corresponds to infinite response at the corresponding real frequency; a stable system has all its poles in the left half plane. And you can guess what happens if there's a pole in the right half plane.[97]

Example: Pole–zero network

When you design amplifiers with negative feedback (the meat of Chapter 4), you often find yourself in a situation where the amplifier itself (without feedback: "open-loop") has a low-frequency pole, and thus open-loop gain dropping at 6 dB/octave with 90° lagging phase shift at higher frequencies. In that case feedback is stable. But there may be a second pole at a higher frequency, causing the open-loop phase shift to approach 180° at a frequency where

feedback can produce an oscillation.[98] An elegant solution is the use of a so-called "pole-zero" network, in which an *s*-plane zero is used to cancel the offending second pole.

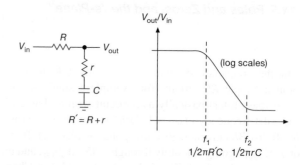

Figure 1x.105. Lowpass filter circuit with added resistor *r*, and its "pole-zero" frequency response. Its rolloff flattens to an asymptotic attenuation $V_{out}/V_{in} \rightarrow r/(r+R)$ above the *rC* characteristic frequency $f_2 = 1/2\pi rC$.

The addition of a resistor (Fig. 1x.105) converts the simple *RC* lowpass into a pole-zero network. The response is easy to figure:

$$H(s) = \frac{V_{out}}{V_{in}} = \frac{r+1/sC}{R+r+1/sC} = \frac{1+srC}{1+s(R+r)C} \qquad (1x.28)$$

where we've sidestepped the *j* and *ω* stuff by going directly to the form $Z_C = 1/sC$. For this circuit we'll assume that $r \ll R$, and by inspection we see that the pole has moved to a slightly lower frequency $s = -1/(R+r)C$, and there is now a zero at a higher frequency $s = -1/rC$; these are indicated conventionally by the symbols × and o, as in Figure 1x.104B. The corresponding phase shift (not shown) goes from 0° at low frequencies, to nearly −90° well above f_1, and back down to 0° well above f_2.

Another place you encounter pole–zero response is in the output capacitor of a voltage regulator: the regulator's output resistance forms a pole with the capacitance, and the capacitor's ESR forms a zero. Low-dropout regulators are fussy about their output bypassing, frequently requiring the electrolytic capacitor's ESR to fall in some range of minimum and maximum – see, for example, the discussion in §9.3.7.

Visualizing the response

The transfer function $H(s)$, with its amusing poles and zeros, is defined over the whole complex *s*-plane. But if you want to know the network's response to signals of (real)

---

[97] "Do you see what happens, Larry?! Do you see what happens?!"

[98] As we'll see in §4.9.2, the condition is that the phase shift reaches 180° – negative feedback becomes positive feedback – at a frequency where the loop gain is unity.

frequency f, all that matters is the value (amplitude and phase) of $H(s)$ along the *s*-plane's *y*-axis. You can visualize a scalar function of two variables as a contour map (a "fishnet"), as in Figure 1x.106 where the magnitude of the transfer function (i.e., $|H(s)|$, the ratio of output amplitude to input amplitude; the circuit's "gain") is shown[99] along slices parallel to the axes. The slice along the imaginary axis reveals the response magnitude, here seen as quite flat out to unit angular frequency (the filter's -3 dB point).

Some folks tell us that they find helpful the image of a rubber sheet, draped over the poles (way up at infinity!), and nailed down to the *x–y* plane at the zeros and at the far horizons. Hey, whatever works for *them* . . .

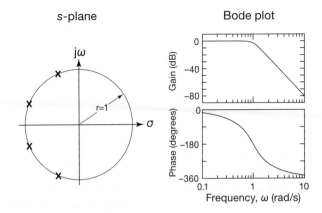

Figure 1x.107. The poles of a Butterworth lowpass filter lie on a circle of radius equal to its 3 dB bandwidth, seen here for a 4th-order (4-pole) implementation. In the stopband its gain falls off at 24 dB/octave, with a lagging phase shift approaching $-360°$.

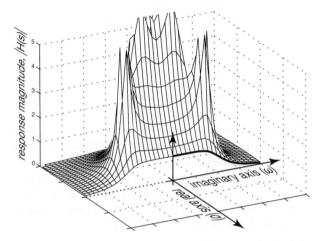

Figure 1x.106. Contour map ("fishnet") of the magnitude of the transfer function V_{out}/V_{in} for a 4-pole Butterworth lowpass filter. The slice along the $j\omega$-axis reveals the response to real frequencies ω. Filters with performance like this cannot be implemented as a passive *RC* circuit; to create off-axis poles you must include inductors, or (equivalently) use an active-filter implementation, see Chapter 6.

A few loose ends

A single-section *RC highpass* filter has a zero at the origin, and a pole at $s = -1/RC$. The zero corresponds to no response (a null) at dc, rising 6 dB/octave (and with a leading $+90°$ phase shift) at increasing frequency. That would continue forever, were it not for the pole that stops the rise at the characteristic 3 dB frequency $f_{3dB} = 1/2\pi RC$; it also cancels the leading phase shift, so the phase shift at

frequencies well above f_{3dB} is zero. But, once again, you knew that already.

In a real *RLC* circuit there are at least as many poles as zeros. Poles and zeros can be real (as in the preceding examples) or complex ("off-axis"). The latter always occur as conjugate pairs, i.e., as a pole (or zero) pair located symmetrically above and below the *s*-plane's *x*-axis, as for example in the 4-pole Butterworth lowpass filter of Figures 1x.106 and 1x.107. For a network with both poles and zeros, the magnitude of the gain at any frequency (represented by a point along the *y*-axis) is proportional to the product of the distances to each zero, divided by the product of the distances to each pole (replace multiplications and divisions by sums and differences, when figuring gain as dB's). The phase shift is the sum of the angles from the zeros, minus the sum of the angles from the poles.

Exercise 1x.2. Try the recipe for yourself: (a) As you run your finger up the *y*-axis (frequency) in Figure 1x.104B, draw a sketch of the approximate phase shift of the pole-zero network (Fig. 1x.105) that it models. Does it agree with the statement made in the paragraph that includes eq'n 1x.28? (b) Now puzzle over the fact that, in the limit of high frequency, the ratio of distances to the zero and to the pole approaches unity, which would appear to say that the response goes to unity, a result we know to be wrong. Perhaps eq'n 1x.28 needs to be rearranged into the form of eq'n 1x.26, to bring the overall gain factor in front? Take it from there.

---

[99] Calculated, with our thanks, by Kent Lundberg, using a combination of his Matlab script and his keen intuition of network behavior.

1x.6 Mechanical Switches and Relays

Some years ago we bought a rather expensive stereo receiver/amplifier, which had relays to connect the speaker outputs after a time delay (as nearly all audio amplifiers do), to prevent a loud thump at turn-on. The amplifier worked fine, but the relays didn't: after a few years the sound would be distorted, or absent, in one or both channels; you could restore health, for a while, by forcing the relays to open and close a few dozen times (preferably while playing something Really Loud).

This is a nice example of "dry switching," and the non-intuitive fact that a mechanical switch or relay rated to switch moderate or high currents will generally perform poorly with a low current load.[100] The lesson here is that these simplest of components have their quirks and variations, including (in addition to this problem of contact ratings) such things as make-before-break (or vice-versa), asymmetrical load ratings, magnetized armatures and polarized coils, contact bounce, and so on. In this section we summarize some of the more important things you'll want to know.

1x.6.1 Why use *mechanical* switches or relays?

In general you can switch circuits and loads with mechanical switches and relays (see examples in Figs. 1x.108, 1x.109, and 1x.110), or with solid-state devices (transistors, SCRs, Triacs). There are advantages to each. To put it most simply, mechanical switching excels in robustness, electrical isolation, low cost, and the ability to switch the widest variety of signals; it's also a natural for things like front panel buttons.[101]

---

[100] The normal load causes a small arc that cleans corrosion from the contacts.

[101] By comparison, electronic switching (which relies on transistors or other silicon switching devices to control a switched circuit) excels in speed, ease of electronic drive, absence of wearout and contact bounce, and immunity to vibration and shock. See discussions in Chapters 3, 9, and 12.

1x.6.2 So what's the problem?

A. Relay and switch contact life

Switch suppliers present you with *too many* choices – see Figure 1x.111 for the impressive offerings from C&K's 7000-series miniature toggle switches alone! Lots of the choices have to do with mechanical configurations (the "actuator" style, the length of the threaded bushing, the size and shape of the terminals) and with the circuit configuration (how many poles, whether spring-return, etc.). In this sea of choices it's easy to underestimate the importance of the contact material.

The likely problem in the stereo amplifier was sulfidization (or oxidation) of a silver-alloy contact material intended for power circuit applications, where some arcing (from circuit currents) is required to clean the contacts. Such materials form a film, which can be resistive, or (worse) they can present a nonlinear "interface voltage" of order several hundred millivolts.

"Dry" (low-level) switching

For low-level switching (less than 10 volts and less than 0.2 amp) you want gold – usually clad (pressure welded) or plated onto a base metal like nickel. It doesn't form an oxide film, and it effectively switches circuits in which the voltages and currents are negligible. But don't use gold contacts for power switching; the arc will quickly burn off the gold, exposing the underlying metal.[102] In the C&K chart of Figure 1x.111 you would select contact material option "B" (or possibly "G," see below).

Power switching

Gold is good – for low-level switching – but it is burned away by the arc that forms when you open a circuit of more than about 10 volts carrying more than about 0.5 A.[103] For such power switching the usual material is a silver alloy (silver cadmium oxide, silver nickel, silver tin oxide, silver copper nickel, silver tungsten, or silver palladium), the particular choice involving trade-offs of resistance to arcing and welding, metal transfer between contacts, ability to handle peak inrush currents, required contact pressure, electrical resistance, and mechanical life. In Figure 1x.111 you would choose contact material option "Q" (or possibly "G," see below). These silver alloys suffer from sulfidization and oxidation, however, and only work properly when cleaned by some degree of arcing. So, in addition to

---

[102] But note the comments on gold-over-silver in the next paragraphs.

[103] The threshold for arc formation depends somewhat on the contact material; see the excellent Tyco application note "Relay Contact Life," 13C3236, for more details.

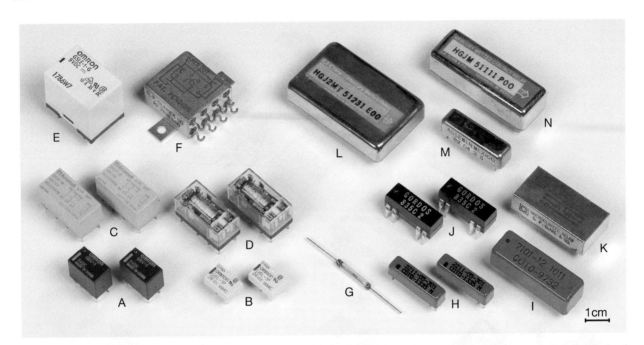

Figure 1x.108. A selection of small relays. A–D: PCB-mounting "telecom-style" relays, with contact ratings to ~1 A. E: a larger plug-in relay, with 10 A contacts. F: hermetically sealed relay. G: magnetically operated reed switch. H–K: reed relays (reed switch with integral actuating coil). L–N: mercury-wetted reed relays, long life ($>10^9$ operations), stable contact resistance, and no contact bounce.

their maximum voltage and current ratings, power relays or switches[104] usually specify *minimums*; for example, the datasheet for the Tyco T92-series power relays specifies a minimum load rating of 500 mA @ 12 Vac/Vdc, and the smaller Omron G5CA series specifies a "minimum permissible load" of 5 Vdc, 100 mA.

"Universal" contacts?

Figure 1x.111 offers "G" contacts, "gold over silver," which sounds like the universal stuff. It is, in the sense that you can use it for either low-level or power switching. The datasheet even gives you permission, saying "ratings: 0.4 VA max @ 20 V ac or dc max, or 5 A @ 120 vac or 28 vdc; 2 A @ 250 vac." But *beware* – once used to switch a 2 amp load, say, the gold is burned off, exposing the silver (power switching) material; and now the switch cannot be used reliably for low-level switching.

Overtravel (wiping)

The mechanical design of some switches and relays (particularly those with silver or palladium contacts) causes the contacts deliberately to slide across each other by a small amount during closure; this wiping action helps clean the contact surfaces.

---
[104] Also called "general purpose" relays or switches.

AC vs DC

Note the much lower voltage rating for dc: 28 V (dc) versus 120 V (ac). That's because it's easier to sustain a damaging dc arc (the ac arc is quenched when the current passes through zero). Arc welders know all about this.

One way switch designers deal with the problem of arcing is by tailoring the speed of contact closure: switches intended for ac service employ a "slow-make, slow-break" design, so the contacts separate slowly enough to allow the ac current waveform to go through zero (extinguishing the arc) while the contacts are still in close proximity. By contrast, switches intended for dc service are "quick-make, quick-break," to open the contact gap rapidly during the inevitable arc.

B. Contact protection

A small amount of arcing is beneficial in power switching, to keep the contacts clean. But it's easy to have too much of a good thing: excessive arcing causes erosion, pitting, and material transfer (Fig. 1x.112). For this reason you need to provide contact protection, in the form of a series *RC*, diode, zener, or varistor (see §9x.25.3B).

Arcing can occur when relay (or switch) contacts either open or close, in a circuit which subjects the contacts to voltages greater than roughly 10 V and currents greater

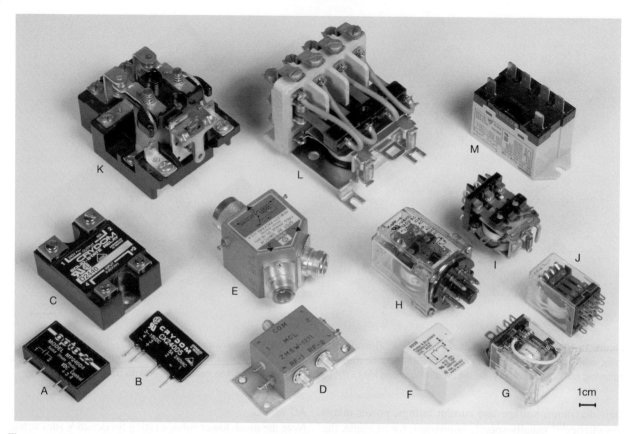

Figure 1x.109. A selection of larger relays. A–C: solid-state (thyristor) relays. D–E: RF relays. F–J: relays for moderate currents (10–20 A), with various plug-in and solder-lug connection options. K–M: muscular multi-pole relays for switching substantial currents (to 30 A or more).

than roughly 0.5 A. Ten volts may seem too small to cause electrical breakdown, but it corresponds to very large local electric fields at moments when the contact spacing is sufficiently small.[105] The resulting current flow, concentrated near local peaks in the surface roughness, causes localized heating. An "arc" involves actual melting of the contact material, forming a metal-vapor conductive arc, with transfer of metal from cathode to anode.

Depending upon how much energy is involved, arcing can lead to long-term degradation of the contacts, or it can destroy them quickly; it can also cause the contacts to stick shut. Relay life expectancy ratings take into account the effects of normal arcing, and that is why the "mechanical life" (no load, or low-level switching) is greater than the "electrical life" (full power load). For example,

the datasheet for the Omron G5CA series of 10 A PCB-mounting relays specifies a mechanical life of 20 million operations (minimum), but an electrical life of 300,000 operations (minimum) with a resistive load of 10 A at 250 Vac.[106]

Contact protection circuits aim to reduce or eliminate arcing, and can extend the electrical life well beyond the (unprotected) ratings; and protective networks are especially important when switching inductive loads.[107] Similar methods of contact protection are used for both induc-

---

[105] For example, 10 V across a gap of 4 microinches (0.1 μm) generates an electric field of a megavolt per centimeter, sufficient to cause electrons to be emitted ("field emission") and accelerate across the gap.

[106] Interestingly, that figure drops to 100,000 operations for a 10 A *dc* resistive load (and that at the reduced voltage of 30 Vdc). An ac load is less destructive for two reasons: the arc is quenched at each current reversal, and the alternating current flow tends to minimize one-way material transfer.

[107] See Ott's fine reference, *Noise Reduction Techniques in Electronic Systems*, for more detail; also, see the relevant publications from relay manufacturers, e.g., Tyco App Notes 13C3202, 13C3311, and 13C3264, "Relay Technical Information" from Aromat/NAIS, "Relay

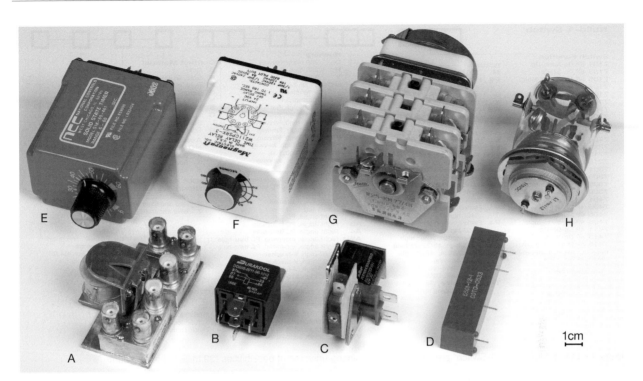

Figure 1x.110. A few more relays we found in our parts drawers. A: DPDT coax relay. B: automotive relay, perhaps the most abundant variant on Earth. C: solenoid-actuated microswitch. D: high-voltage (10 kV) encapsulated reed relay. E–F: time-delay relays. G: we're not sure what this is . . . a clock motor drives a DPDT switch, creating a periodic reversing switch (readers – chime in, please!). H. high-voltage (20 kV) two-pole vacuum relay.

tive and resistive loads; the following subsections illustrate some commonly used techniques.

C. Relay coil suppression

Whatever drives an electromechanical relay has to deal with the inductance of the coil, which (like any inductive load) can generate impressive spikes at turn-off (recall §1.6.7). For example, the popular and diminutive G5V PCB-mounted relay generated 440 V positive spikes when we energized it from its rated 5 Vdc and then switched it off without spike-limiting circuitry; and a somewhat larger relay (P&B KRP11DG), energized from 24 Vdc, spiked to 1100 V (limited by MOSFET breakdown – it wanted to go even higher).

Figure 1x.113 shows three ways to limit the turn-off voltage transient, for a relay actuated with dc. Simplest is a diode across the coil, with polarity as shown (circuit A); its minimum voltage and current ratings are the supply voltage and the coil current, respectively, both easily

satisfied. The disadvantage of this method is the slow decay of inductor current; ignoring the unimportant forward diode drop, the current decays as a simple exponential, of time constant L_{coil}/R_{coil}. If we assume the coil inductance is constant (i.e., ignoring the effects of armature movement and core saturation), the decay is just the classic RL circuit (model A on the right-hand side of Fig. 1x.113):

$$L\frac{dI}{dt} + IR_{coil} = 0, \qquad (1x.29)$$

with solution

$$I(t) = I_0 e^{-t/\tau}, \qquad (1x.30)$$

where $\tau = L/R_{coil}$ and $I_0 = V_{coil}/R_{coil}$.

The time constant τ is generally in the range of a few milliseconds (for small relays) to tens of milliseconds (for larger relays). For some applications you don't care about a few extra milliseconds, but the *relay* may care – that's because a slow decay of coil current (and corresponding magnetic field) causes the moving contacts to separate slowly, preventing an abrupt termination of switched current. Contacts are prone to damage from arcing, so you want to

---

User's Guide" (Z54-E1-1) and "Relay Technical Information" (X301-E3-1b) from Omron.

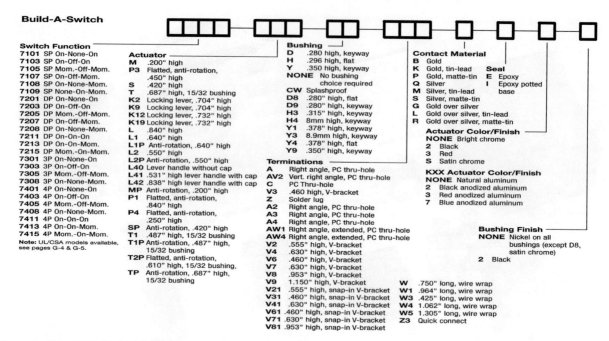

Build-A-Switch

Switch Function
7101	SP On-None-On
7103	SP On-Off-On
7105	SP Mom.-Off-Mom.
7107	SP On-Off-Mom.
7108	SP On-None-Mom.
7109	SP None-On-Mom.
7201	DP On-None-On
7203	DP On-Off-On
7205	DP Mom.-Off-Mom.
7207	DP On-Off-Mom.
7208	DP On-None-Mom.
7211	DP On-On-On
7213	DP On-On-Mom.
7215	DP Mom.-On-Mom.
7301	3P On-None-On
7303	3P On-Off-On
7305	3P Mom.-Off-Mom.
7308	3P On-None-Mom.
7401	4P On-None-On
7403	4P On-Off-On
7405	4P Mom.-Off-Mom.
7408	4P On-None-Mom.
7411	4P On-On-On
7413	4P On-On-Mom.
7415	4P Mom.-On-Mom.

Note: UL/CSA models available, see pages G-4 & G-5.

Actuator
M	.200" high
P3	Flatted, anti-rotation, .450" high
S	.420" high
T	.687" high, 15/32 bushing
K2	Locking lever, .704" high
K9	Locking lever, .704" high
K12	Locking lever, .732" high
K19	Locking lever, .732" high
L	.840" high
L1	.640" high
L1P	Anti-rotation, .640" high
L2	.550" high
L2P	Anti-rotation, .550" high
L40	Lever handle without cap
L41	.531" high lever handle with cap
L42	.838" high lever handle with cap
MP	Anti-rotation, .200" high
P1	Flatted, anti-rotation, .840" high
P4	Flatted, anti-rotation, .250" high
SP	Anti-rotation, .420" high
T1	.487" high, 15/32 bushing
T1P	Anti-rotation, .487" high, 15/32 bushing
T2P	Flatted, anti-rotation, .610" high, 15/32 bushing
TP	Anti-rotation, .687" high, 15/32 bushing

Bushing
D	.280 high, keyway
H	.296 high, flat
Y	.350 high, keyway
NONE	No bushing choice required
CW	Splashproof
D8	.280" high, keyway
D9	.280" high, keyway
H3	.315" high, keyway
H4	8mm high, keyway
Y1	.378" high, keyway
Y3	8.9mm high, keyway
Y4	.378" high, flat
Y9	.350" high, keyway

Terminations
A	Right angle, PC thru-hole
AV2	Vert. right angle, PC thru-hole
C	PC Thru-hole
V3	.460 high, V-bracket
Z	Solder lug
A2	Right angle, PC thru-hole
A3	Right angle, PC thru-hole
A4	Right angle, PC thru-hole
AW1	Right angle, extended, PC thru-hole
AW4	Right angle, extended, PC thru-hole
V2	.555" high, V-bracket
V4	.630" high, V-bracket
V6	.460" high, V-bracket
V7	.630" high, V-bracket
V8	.953" high, V-bracket
V9	1.150" high, V-bracket
V21	.555" high, snap-in V-bracket
V31	.460" high, snap-in V-bracket
V41	.630" high, snap-in V-bracket
V61	.460" high, snap-in V-bracket
V71	.630" high, snap-in V-bracket
V81	.953" high, snap-in V-bracket

W	.750" long, wire wrap
W1	.964" long, wire wrap
W3	.425" long, wire wrap
W4	1.062" long, wire wrap
W5	1.305" long, wire wrap
Z3	Quick connect

Contact Material
B	Gold
K	Gold, tin-lead
P	Gold, matte-tin
Q	Silver
M	Silver, tin-lead
S	Silver, matte-tin
G	Gold over silver
L	Gold over silver, tin-lead
R	Gold over silver, matte-tin

Seal
E	Epoxy
I	Epoxy potted base

Actuator Color/Finish
NONE	Bright chrome
2	Black
3	Red
S	Satin chrome

KXX Actuator Color/Finish
NONE	Natural aluminum
2	Black anodized aluminum
3	Red anodized aluminum
7	Blue anodized aluminum

Bushing Finish
NONE	Nickel on all bushings (except D8, satin chrome)
2	Black

Figure 1x.111. C&K's "Build-A-Switch" presents an overwhelming universe of possibilities (39 million, by our count!).

Figure 1x.112. Material transfer caused by excessive arcing results in a "spike and crater." From Tyco App Note 13C3203: "Contact Arc Phenomenon," used with permission of TE Connectivity.

open them quickly, particularly when switching circuits with significant voltages, currents, or inductive loads.

One way to speed things up is shown in circuit B (a resistor of value R across the coil). Here the decay of coil current is again a simple exponential, but with the time constant τ now shortened to $L/(R_{coil}+R)$. We rigged up a test setup and measured some enlightening waveforms. Curve B of Figure 1x.114, where the external 392 Ω resistance is added to the coil resistance of 167 Ω, shows the faster decay of current and correspondingly shorter release time, when compared with the simple diode-clamped coil (curve A).[108]

Circuit C, though requiring an extra component, pro-

---

[108] You sometimes see a variation with a capacitor in series with a resistor.

Figure 1x.113. Relay coil suppression circuits for dc relays (left), redrawn to show situation when drive is removed. A diode is the simplest, but it extends the release time compared with a resistor or a zener-plus-diode.

duces the fastest decay, therefore the shortest release time. You can think of it as clamping the persistent coil current to the zener voltage V_Z (plus a negligible diode drop), until the current reaches zero. At first you might imagine that this would produce a *linear* decay, according to $V = L\,dI/dt$; but, surprise, the current decay waveform is still an exponential, but with an offset zero and a larger overall multiplier. With only a passing familiarity with differential equa-

---

This eliminates the additional ON-state static current. Note, however, that a capacitor alone is a *terrible* idea, because it creates an enormous inrush current; the latter causes rapid degradation of contact material, and may cause the contacts to stick shut.

tions, it's easy enough to see how it goes: The addition of the V_Z term adds a constant term to eq'n 1x.29, converting it to the "inhomogeneous" equation

$$L\frac{dI}{dt} + IR_{coil} = -V_Z, \qquad (1x.31)$$

with solution, after satisfying the initial condition $I(t)=0$ at $t=0$,

$$I(t) = \left(I_0 + \frac{V_Z}{R_{coil}}\right)e^{-t/\tau} - \frac{V_Z}{R_{coil}}, \qquad (1x.32)$$

and with the same exponential time constant as with a diode clamp, $\tau = L/R_{coil}$. We can cast this into a nice form by replacing I_0 with its value V_{coil}/R_{coil}, so that the coefficient of the exponential in eq'n 1x.32 becomes $(V_{coil}+V_Z)/R_{coil}$. Put another way, compared with the simple diode clamp (eq'n 1x.29), the coefficient of the exponential is increased by a factor $1+V_Z/V_{coil}$, with a corresponding negative offset of V_Z/R_{coil}. So, although the "time constant" τ is the same as with a simple diode clamp, the waveform gets to zero much faster.

This behavior can be seen in the traces of Figure 1x.114, where the diode-clamped decay (trace A) is longest, followed by the exponential of shorter time constant (series resistor shunt, trace B), and the steepest (and truncated at zero current) current decay with a zener clamp (trace C). For the latter the contact release delay is roughly shortened by the estimated factor $1+V_Z/V_{coil}=3$; but real life is invariably more complicated, and even humble relays are no exception. The wiggles in the decay waveforms are caused by the changing inductance as the armature opens; and the contact release delays, though showing the correct trends in traces A→B→C, do not conform accurately to predictions. As an example, with *no* suppression components, the observed peak voltage was 440 V (impressive, for a small 5 V relay), yet the release delay was 0.52 ms, i.e., not much faster than the 0.68 ms with the 10 V zener clamp. The reason is the mechanics – finite spring force on the armature mass creates a minimum contact release time, even if the current (and thus the magnetic field) falls to zero instantaneously. We've collected additional measured values for both relays in Table 1x.7 on the following page.

Component ratings
The diode clamp (circuit A) carries the rated coil current at turn-off, and holds off the rated coil voltage while the relay is energized. Easily satisfied – for example, a 1N4148 signal diode (100 V, 300 mA, \$0.04 qty 100) is fine for the G5V-1-DC5 (5 V, 30 mA) or KRP11DG (24 V, 50 mA) we tested. In general, you'll have no problem finding diodes with adequate ratings.

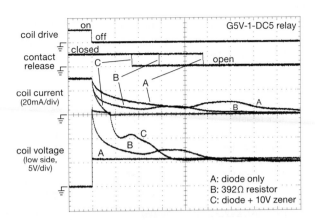

Figure 1x.114. Measured voltage and current waveforms, and contact timing for an Omron G5V small PCB-mount relay. The labels correspond to the coil-suppression circuits in Fig. 1x.113. Horizontal: 400 μs/div.

The resistor (circuit B) suffers more, because it sees the same coil current, but with a larger voltage drop. Our 392 Ω resistor dissipated 350 mW peak, dropping after 200 μs to 35 mW. This sounds bad, for something like a 0.125 W (0805 size) chip resistor. But, not to worry, we're saved by the component's transient thermal impedance (see §§1x.2.6 and 9x.25.8), which allows considerably greater pulsed power dissipation: according to the Vishay CRCW0805 thick-film chip-resistor datasheet, you can dissipate 4 W for 1 ms, or 1.5 W for 10 ms (these are single-pulse ratings; for multiple pulses the corresponding pulse power ratings are 1.6 W and 0.5 W, as long as the average power stays within its 0.125 W limit). So, we're fine, in this example. You can do a bit better with *thin*-film resistors, where the corresponding single-pulse ratings are 10 W and 4 W, dropping to 6 W and 2.2 W for multiple pulses.

The zener clamp (circuit C) dissipates a peak power of $I_{coil}V_Z$, which is the same as a resistor that is sized to produce the same peak voltage. And, similarly, zeners can handle considerable peak power for short times. For example, the 1N5221-5281 series (2.4 V–200 V) of "500 mW" zeners have a transient thermal impedance $Z_{\Theta JC}$ of 15°C/W for a 1 ms pulse, and 27°C/W for a 10 ms pulse. So, for the 10 V zener (1N5240) we used in Figure 1x.113 the peak power (30 mA × 10 V, or 300 mW) caused the junction temperature to rise less than 5°C. If we had used a 1N5271 100 V zener, the corresponding thermal step would be less than 30°C, even though the peak power (3 W) is well above the 0.5 W average power rating.

These are puny zeners, and still they do the job for small relays. You can get considerably heftier zeners, intended for transient voltage clamping, for example the P6KEvvvA

Table 1x.7: Relay Coil Suppression Measurements

Suppression Method	Release Time (ms)	Peak Voltage (volts)
G5V-1-DC5: 5V 167Ω coil (30mA)		
no suppression	0.52	440
30V zener + diode	0.54	35
1.2k resistor	0.77	35
10V zener + diode	0.68	16
392Ω resistor	1.15	16
4.3V zener + diode	0.94	9.5
130Ω resistor	1.37	9.5
diode only	1.91	5.6
KRP11DG: 24V 470Ω coil (50mA)		
no suppression	1.9	1100*
200V zener + diode	2.1	220
4k resistor	3.5	220
47V zener + diode	3.3	72
24V zener + diode	5.0	47
diode only	14.0	25

* flat-topped waveform, clamped by avalanche breakdown of the
1kV MOSFET switch (IXTA05N100).

zener TVS series are rated at 600 W pulse power (1 ms) or
250 W (10 ms); they are available as back-to-back bidirectional zener clamps, see next paragraph. The P6K series are
the small guys; next up the ladder is the 1.5KE series, rated
at 1500 W (1 ms) or 1000 W (3 ms). These will take care of
your big-relay coil-suppression problems.

Relays with ac coils

We've been dealing with dc relay coils only; for relay coils
energized by *ac* you can't use a diode clamp (nor even a
pair of back-to-back series diodes). Instead you can use a
series pair of zeners,[109] a varistor, or a series[110] *RC*, see
Figure 1x.115. As with inductive dc loads, a voltage clamp
(circuits B or C) produces the fastest decay. Switch and relay manufacturers generally advise placing the suppression
network across the relay coil for lower voltage operation
(up to 48 V, say), but for relays or other loads operating
at powerline voltages they suggest putting the suppression
network across the contacts of the switch.

D. Improving relay switching speed

As seen in Figure 1x.114 (and discussed in §1x.6.2C), the
best way to speed up a relay's contact *release* is to allow
the inductive voltage to rise, either with a zener clamp or

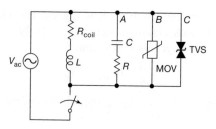

Figure 1x.115. Relay coil suppression for ac relays.

resistive network,[111] to produce a faster decay of current
(and thus magnetic field) according to $dI/dt=V/L$. But,
how to speed up contact *closure*? Just do the analogous
thing, namely overdrive the initial coil voltage.

Figure 1x.116 shows several ways to do it. If you have
a dc rail higher than the relay's coil rating, you can use
Circuit A, where the dropping resistor R_1 is chosen to produce the rated coil voltage in the steady state. If the relay
has been off for at least a few tens of milliseconds, C_1 is
charged to 12 V, and jolts the relay into fast closure; the
12 V rail is a good place to clamp the coil decay, also.
Figure 1x.117 shows the release-time improvement from
clamping to 12 V (trace B versus A), and the closure-time
improvement from the full "kickstart" (trace C versus A).
Trace C shows an ×3 speedup in contact closure, and an
×2 speedup in contact release.

You can eliminate the large capacitor C_1 if you're willing
to use a second MOSFET (Fig. 1x.116B); Q_2 provides the
12 V kick, with Q_1 holding the fort thereafter. Note that
the clamp diode is connected from the bottom of R_2 (rather
than bridging the relay) for faster release.

If you don't have a dc rail at an elevated voltage, you
can use the circuit of Figure 1x.116C: here the actuation
of Q_1 pulls the low side of the relay coil to ground, and
switches on Q_2 to bring the high side of C_1 to $2V_+$ (i.e.,
+24 V). Diode D_1 maintains rated steady-state coil voltage after the startup transient. Figure 1x.118 shows contact
dynamics under three conditions: normal 12 V drive and
diode clamp (trace A), 12 V drive with 15 V zener clamp
to the rail (trace B), and the full kickstart circuit (trace C,
in which the bidirectional gate-protection zener acts also as
the discharge clamp). Trace C shows an ×2 improvement
in switching speed.

---

[109] Available as bidirectional transient suppressors, e.g., the P6KEvvvCA
series, where the vvv designators specify the voltage, in the range of
6.8 V to 400 V.

[110] Or just a resistor alone, as in Fig. 1x.113B.

[111] If you drive the coil with a MOSFET, you can often omit a suppression network altogether, exploiting the MOSFET's allowable avalanche
(zener-like) breakdown.

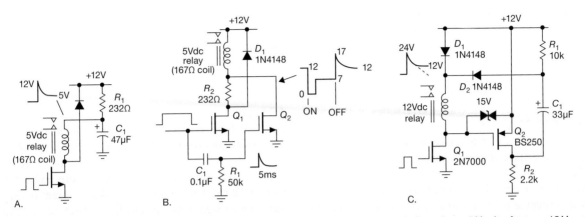

Figure 1x.116. Overdriving a relay coil with a "kickstart" circuit shortens the turn-on delay. A. Powering a 5 V relay from a +12 V supply, with R_1 sized to drop 7 volts after C_1 has kicked the relay closed. B. A second MOSFET lets you get rid of the large-value electrolytic. C. For a 12 V relay you can double the initial coil voltage as shown. The bidirectional zener does double duty: it protects the pMOS gate and speeds up the contact release; see the measured waveforms in Figs. 1x.117 and 1x.118.

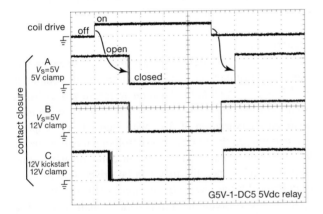

Figure 1x.117. Measured dynamics of a 5 Vdc relay: A. 5 V supply, diode clamp to 5 V. B. Faster contact release with a diode clamp to 12 V. C. Faster closure with the kickstart circuit of Fig. 1x.116A (C). Horizontal: 2 ms/div.

Figure 1x.118. Measured dynamics of a 12 Vdc relay, showing improved switching speed with a zener clamp (B) and with the kickstart circuit of Fig. 1x.116C (C), compared with the sluggish performance of a simple diode clamp (A). Horizontal: 4 ms/div.

1x.6.3 Other switch and relay parameters

A. Switches: Function, actuator, bushing, terminals

The number of poles is self-explanatory, but note in Figure 1x.111 the options for a center position, as well as a spring-loaded momentary: e.g., "On-None-Mom" has two positions, with spring return from momentary, whereas "On-Off-Mom" returns to the middle of three positions, and "Mom-off-Mom" springs back to the middle from either end.

The "actuator" is the lever; we prefer the "flatted" type to the standard bat-handle. If you don't want the switch bumped accidentally, get the locking lever type (you pull it out to change positions; it springs back and locks).

You'll want "solder lug" terminals, unless the panel components solder onto a PCB behind the panel (in which case you want "PC Thru-hole"). "Wire wrap" might be chosen by wire-wrappers; but you know who you are!

B. Relays: Moving-armature, reed, and solid-state
Moving-armature relays

The classic *moving-armature* relay consists of a contact assembly (attached to some magnetic material) that is actuated by the magnetic field from a nearby coil. These things go *click* (or maybe *clack*, for the big ones) when you provide operating current; the latter may be dc or ac, with common ratings of 5, 12, or 24 Vdc, or 24 or 120 Vac. As described earlier, they come in various contact ratings, and they generally are designed with wiping contacts (the contacts slide slightly during closing) that helps clean the contacts and also discourage sticking; you can pretty much count on contact bounce during closure (but contact *release* is usually bounce-free).

For switching power loads you can find plenty of relays ("power relays") rated to 30 A or 50 A, for example the TE/P&B T92 series or the Omron G7L series, with contact ratings of 20–30 A at 250 Vac; these are ∼60 cm^3 rectangular plastic modules with quick-connect terminals, and with various coil voltage choices, requiring about 2 W coil excitation.

At the other end of the power spectrum, the smallest moving-armature relays ("signal relays") have contact ratings up to a few amps, they are physically quite small, and some can be driven from logic outputs (our favorite Omron G5V series are 1 cm^3, switch up to 1 A loads, and require 150 mW coil excitation; the 1 A TE/P&B V23026 series are 0.7 cm^3 and require only half that, thus 14 mA at 5 V or 22 mA at 3 V). These small plastic relays are PCB-mounting, either with through-holes or surface-mount.

Reed relays

If you want even smaller relays for switching signals, or those requiring less coil power, *reed* relays are a good choice. Some popular manufacturers include Littelfuse (Hamlin), Coto, TE/P&B, and Pickering. There consist of a pair of thin ferromagnetic vane contacts sealed in a glass tube, surrounded with an energizing coil. Although you can get SPDT or SPST-NC, the most common configuration is normally-open SPST. These things are small and low power – the Hamlin HE3600 or Coto 9007 series come in a single-inline (SIL) 5 mm-wide package of 0.7 cm^3 volume, requiring a logic-friendly 10 mA at 5 V coil excitation – and they cost about $1 in small quantities. Reed relays are intended for low-level switching, with maximum current ratings of 1 A or less; they switch faster than moving-armature types, and have greater lifetimes (e.g., 10^8 operations while switching 10 mA and 1 V for the inexpensive Coto type above, 10^9 for the better-grade Coto 7000 series). They come in shielded varieties, and there are a few with non-bouncing mercury-wetted contacts.

Reed relays are excellent for reliable signal switching in low-level analog circuits (e.g., for switching bandwidth-setting resistors and capacitors in analog filters, or setting gain in 'scope front-ends), because their high OFF-resistance (typically 10^{10}–$10^{12}\,\Omega$), low ON-resistance (typically $\lesssim 0.2\,\Omega$), and very low capacitance (typically $\lesssim 1$ pF) are gentle to signals being switched.

Solid-state relays

These are discussed in detail in Chapters 3 (§§3.4.1–3.4.3, FET analog switches) and 12 (§12.7.6, triac SSRs) of AoE3, and we remark here only a few points of comparison with mechanical relays:

AC loads

For switching *powerline loads* from logic control signals, SSRs are excellent: logic-level isolated control, zero-voltage turn-on, zero-current turn-off, no wearout, no bounce, and load ratings to 125 A or more and 280 V or more. One caution, though: the leakage current in the OFF state can be significant, as much as 10 mA, so light loads may need a shunt resistor to suppress OFF-state output voltage (as we found in §15.4). Note also that SSRs are more expensive than mechanical relays of comparable ratings; for example, the Crydom D2425 (240 Vac, 25 A) costs $45 in small quantities, compared with $10 for the comparably rated Omron G7L-1A.

Analog switching

MOSFET analog switches are fast, free of bounce, and inexpensive. But they have numerous limitations, detailed in §3.4.2, among them *limited voltage range*, *lack of galvanic isolation from the control signal*, *non-zero ON-resistance*, *nonlinearity* (variation of R_{ON} with signal level), *capacitive coupling* in the OFF-state, control-signal *charge transfer*, and *limited bandwidth*. And, in common with semiconductor ICs generally, they can be driven into *latchup*, or destroyed by an *overvoltage transient*. There are many applications involving high impedance, small signals, low currents, low distortion, or wide bandwidths where a reed relay will beat the pants off any MOSFET analog switch ... don't forget about these little gems!

1x.7 Diodes

The ideal diode has zero forward drop and series inductance, zero reverse current and shunt capacitance, instantaneous onset of forward current when forward biased, and zero delay when transitioning from forward conduction to reverse blocking.

Real diodes are different. In this section we visit the ways in which they can disappoint us, and what we can do about it. We also look at some interesting diode variants – Schottky diodes, fast-recovery and soft-turnoff diodes, step-recovery ("snapoff") diodes, varactors ("tuning diodes"), PIN diodes, and tunnel (Esaki) diodes.

1x.7.1 Diode characteristics

A. The family tree

Diodes and rectifiers span a wide range of voltage and current ratings, within which they are distinguished by additional characteristics (forward drop, recovery time, etc.). To get oriented, take a look at Figure 1x.119, where we've indicated the favored technology choices in regions of maximum V and I. Diodes with current ratings above an amp or so are often called *rectifiers*, contrasted with low-current *signal diodes*. We'll illustrate the range of characteristics graphically, with a set of curves from both datasheets and bench measurements. We trust the reader will prefer such visual stimulation to the alternative of yak yak yak...

B. Reverse (leakage) current

With the exception of zeners,[112] diodes aren't supposed to conduct in the reverse direction. Real diodes exhibit leakage current, which can range from picoamps (for signal diodes) to many milliamps (for high-current power rectifiers). Leakage currents tend to be a weak function of reverse voltage, but a strong function of temperature (with a typical doubling for every $\sim$10°C rise). This behavior is seen in Figures 1x.120–1x.123.

C. Forward voltage drop

Ideal diodes would conduct forward currents with zero voltage drop, whereas real diodes exhibit a soft onset of

---

[112] And step-recovery diodes, see §1x.7.2F.

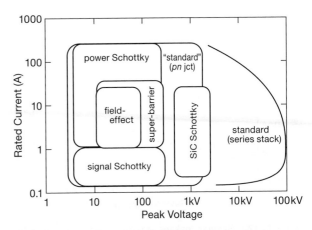

Figure 1x.119. These are the diode and rectifier technologies currently favored for the voltage and current regions shown.

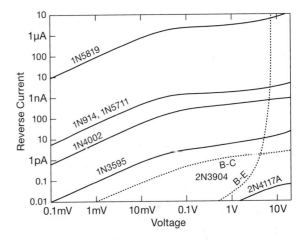

Figure 1x.120. Measured diode reverse leakage currents, at room temperature; expect a doubling for every 10°C rise. The 2N3904's base-emitter junction is excellent at low voltage, but exhibits a zener-like characteristic near its rated BV_{BE} of $\sim$6 V.

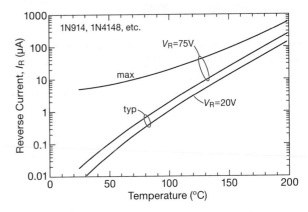

Figure 1x.121. Reverse leakage versus temperature for the popular 1N4148-type signal diodes (adapted from an NXP datasheet).

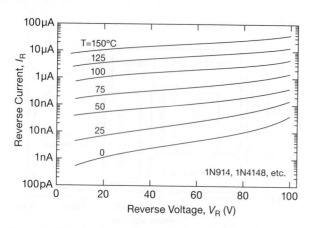

Figure 1x.122. Reverse leakage versus voltage and temperature for the popular 1N4148-type signal diodes (adapted from a Toshiba datasheet).

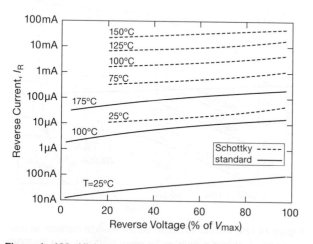

Figure 1x.123. High-current (power) rectifiers exhibit considerably higher leakage currents, as seen in these datasheet plots of two 20A parts (Schottky: Fairchild FSV20100V, V_{max}=100V; standard: Motorola MUR2020R, V_{max}=200V).

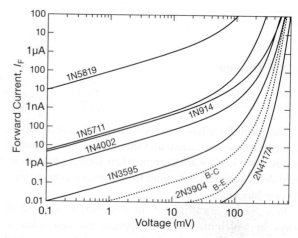

Figure 1x.124. Measured forward current versus applied voltage, for a selection of diodes. The 2N3904 is a small-signal *npn* BJT; the 2N4117A curve is the gate-channel junction of a small-geometry JFET (I_{DSS}=90 µA max).

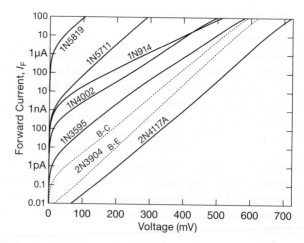

Figure 1x.125. The same data as Fig. 1x.124, here plotted on log–linear axes.

forward current, as we saw in many places in AoE3 (beginning at §1.6.1); Figures 1x.124 and 1x.125 show the onset of forward current for a selection of diodes, a graph we used in §5.2 in connection with op-amp input protection.

It's sometimes instructive to flip the axes on plots like this. Figure 1x.126 shows nicely the lower forward voltage drop of Schottky diodes, compared with their standard silicon *pn*-junction brethren. And, conversely, diodes made from semiconductors with larger bandgaps (gallium arsenide, an example of a "III–V" semiconductor) turn on at higher forward voltages, as seen in Figure 1x.127. Such semiconductors are used in optoelectronics, and also in high-speed or high-temperature applications (less leak-

age at high temperature because of their larger bandgap; GaAs's higher mobility allows higher speeds).

Tempco of forward voltage

Similar to the base-emitter junction, the diode's forward drop decreases with increasing temperature, typically −2.1 mV/°C, but less at higher currents because of the effect of series resistance. Figure 1x.128 shows a datasheet plot of the 1N4148's tempco versus forward current. Such a plot is exceedingly rare – we found this one in a vintage GE Semiconductor databook; evidently the novelty of small-signal silicon diodes back in those heady 1970s mo-

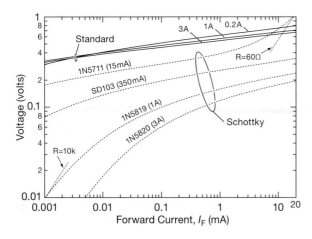

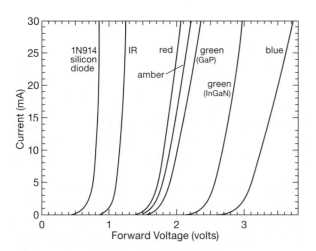

Figure 1x.126. Here we reversed the axes to illustrate the lower forward drop of Schottky-barrier diodes.

Figure 1x.127. Compound semiconductors, with their larger bandgaps, begin forward conduction at higher voltages than silicon, as seen in this plot of various LED formulations.

tivated the engineers at GE to measure and plot an unusually thorough set of parametric data.

D. Dynamic impedance

Diodes are nonlinear, so they don't have a "resistance"; but you can talk about a dynamic resistance, that is, the ratio of change of voltage to change of current at some operating current.

Forward impedance

The forward current I_f through a diode is approximately exponential in applied voltage V_f ("approximately" because

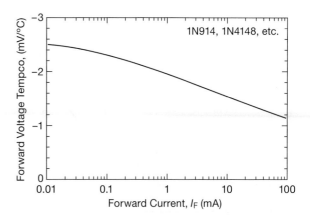

Figure 1x.128. Temperature coefficient (typ) of forward drop for the omnipresent 1N4148. (from GE datasheet)

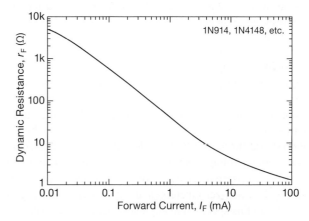

Figure 1x.129. Plot of dynamic forward resistance for the ever-popular 1N4148-type signal diode, from a General Semiconductor datasheet.

of non-zero internal resistance), as given by a modified Ebers–Moll equation

$$I_f = I_0 \exp\left(\frac{V_f}{nV_T}\right), \tag{1x.33}$$

where $V_T = kT/q \approx 25$ mV and n is somewhere in the range 1–2. From this we can get the incremental resistance

$$R_{incr} = nV_T/I_f, \tag{1x.34}$$

or roughly 25 to 50 ohms divided by the current in milliamps. Figure 1x.129 shows a datasheet plot of this behavior. It follows well an inverse relationship with current over much of the range (with $n \approx 2$), flattening at the low-current end (owing to non-zero conductivity at zero volts, see next subsection) and at the high-current end (owing to internal series resistance).

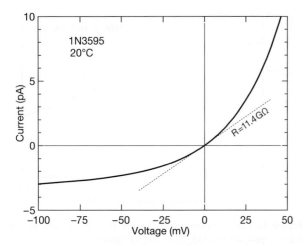

Figure 1x.130. This sample of a 1N3595 low-leakage diode has an incremental resistance of ~11 GΩ in the forward–reverse crossover region at zero applied voltage, as seen in these measured data.

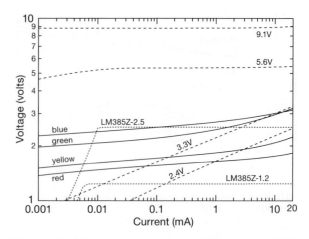

Figure 1x.131. Measured *I–V* curves of some zeners (dashed lines), IC shunt references (dotted lines), and LEDs of various colors (solid lines). One might question the judgment of folks who make (or use!) zeners of 3.3 V or below.

Zero-voltage impedance

Equation 1x.34 predicts infinite incremental resistance at zero applied voltage; but real life is a bit more complicated. Look at Figure 1x.130, where we've plotted the measured current of the 1N3595 low-leakage signal diode in the region straddling zero volts.[113] Indeed there is zero current at zero applied voltage; but the zero-voltage point is trapped between non-zero reverse and forward leakage currents, resulting in a non-zero incremental conductivity. Sadly, you won't find this specified in any manufacturer's datasheet.

Reverse impedance

Figure 1x.130 shows an increasing incremental resistance for greater reverse voltages. This can't go on forever, of course – at higher reverse voltages there is onset of zener-like avalanche breakdown, and the like. Zener diodes, in particular, are intended for operation in the breakdown region, where their incremental resistance can be just a few ohms. This was discussed in AoE's §9.10.1, where Figure 9.88 plots zener impedance Z_Z versus zener voltage, showing best performance for zeners around 7 volts. Zeners of much lower voltage perform miserably, as seen in the measured data of Figure 1x.131. Happily, there are excellent alternatives: a 2-terminal (shunt) IC voltage reference (like the pair of LM385s plotted in the figure), or even (if you're feeling adventuresome) an LED of appropriate bandgap, operated in the forward direction.

E. Peak current

As with any power device, diodes and rectifiers can withstand a short pulse of peak power considerably greater than their average power rating. This property, sometimes characterized as a reduced thermal resistance, is simply caused by the pulse energy being absorbed by the heat capacity of the junction (or other masses involved, for example in a power resistor or a MOSFET, see §§1x.2.6, 9.4.2B, and 9x.25.8).

Figures 1x.132 and 1x.133 plot peak current versus pulse duration for a 1N4148-style silicon signal diode and for a high-voltage silicon-carbide rectifier, respectively. For the latter, the datasheet shows the same characteristic plotted in terms of its equivalent transient thermal resistance $R_{\Theta JC}$ versus pulse duration. In both cases you can see that the effect decreases for repetitive pulses, asymptoting to the steady-state limit as the duty cycle t_{on}/T approaches unity.

F. Reverse capacitance

Reverse biasing a *pn* junction creates an insulating *depletion region*, whose thickness increases with increasing reverse bias, thus causing reduced junction capacitance. All diodes participate in this effect, which is exploited (for example for electrical tuning) in the *varactor* diodes we saw back in §7.1.5D and Figure 7.27. Generally, though, parasitic capacitance is undesirable – the less the better. This is particularly the case where diodes are in a signal path (active rectifier, clamp, etc.), where the diode's capacitance couples fast signals.

Figures 1x.134 and 1x.135 plot typical diode capaci-

---

[113] We wanted to know this for the design of the precision high-impedance millivoltmeter in Fig. 5.1 (§5.2).

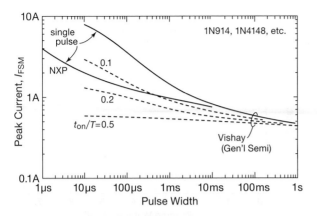

Figure 1x.132. Maximum peak forward current, single-pulse and repetitive, $T_J=25°C$ prior to pulse. (adapted from respective datasheets)

tance versus reverse voltage. As expected, small signal diodes (with their small junction areas) have less capacitance than their larger rectifier brethren. In rectifier applications (particularly in switchmode power converters), another effect – stored charge and reverse recovery – are of greater importance; they are discussed in §1x.7.2.

G. Zener capacitance

Zener diodes are often used to protect sensitive components from destructive overvoltage. A classic example is the MOSFET gate–source terminal pair, where even a brief excursion past 20 V or so can destroy the device. See, for example, Figure 3x.57 in the section on high-voltage current-source circuits (§3x.6). For large MOSFETs the added capacitance of the zener is usually insignificant compared with the MOSFET's gate–source capacitance C_{iss} (see Table 3.4). But in a high-speed circuit with small MOSFETs the protective-zener capacitance it can become bandwidth-limiting dominant effect. We discuss this in §4x.23.9, where the capacitance of an ordinary 250 mW zener (55 pF) swamps the combined 20 pF capacitance of the other (carefully chosen) components. In that circuit we used instead the D1213A low-capacitance protective device (intended for protecting signal lines) with its pleasantly diminutive $C \sim 1$ pF.

There are other choices. Figure 1x.136 shows the innards of some good zener-type components, and Figure 1x.137 shows their capacitance compared with ordinary zeners and with the base–emitter junction of a small BJT. You don't get something for nothing, however: the low-capacitance parts are generally less robust in the presence of large transient currents; for example, 5 A peak for the D1213A versus 20 A for a 1N5240, for an 8/20 µs

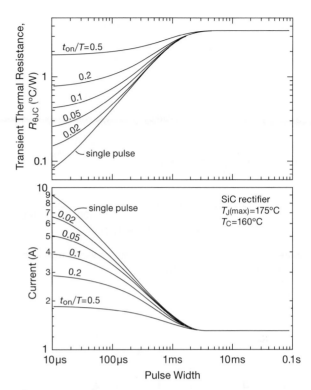

Figure 1x.133. Maximum peak forward current and transient thermal resistance for a 650 V silicon-carbide Schottky rectifier in a small (3.6×4.3 mm) surface-mount package (GeneSiC type GB01SLT06-214, DO-214AA pkg), adapted from the datasheet.

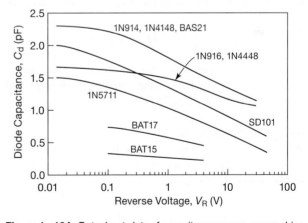

Figure 1x.134. Datasheet plots of capacitance versus reverse bias of signal diodes. The BAT15, BAT17, SD101, and 1N5711 are Schottky types, with maximum forward current of 100 mA, 30 mA, 30 mA, and 15 mA, respectively; the other diodes are rated at 200–300 mA.

81

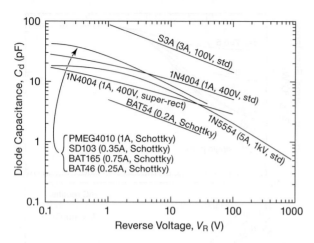

Figure 1x.135. Datasheet plots of capacitance versus reverse bias for selected medium-current diodes.

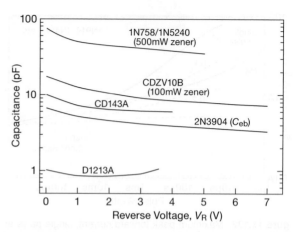

Figure 1x.137. Capacitance versus reverse bias for zener-type protective components.

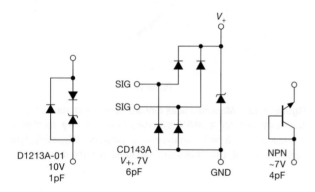

Figure 1x.136. Protective zener-like devices with low capacitance.

pulse (see §9x.25.7A). The CD143A seems to be an exception – according to the datasheet[114] it can withstand 40 A for diode-to-rail conduction. We haven't tried that, yet.[115] There's more discussion of zener clamps (both large and small) in Chapter 9x (§9x.25.3C).

BJT base–emitter "zener"

We've used the ∼7 V base–emitter reverse breakdown of small-signal BJTs (like the ubiquitous 2N3904) as a protective device for signal inputs (e.g., Figure 3x.57); it has

pleasantly low capacitance (less than 7 pF) – see our measured curve in Figure 1x.137 – and the reverse avalanche breakdown is very sharp, with no conducted current below 7 V. Once it starts conducting it has a low dynamic resistance, with the breakdown voltage increasing only 5 mV per decade up to 1 mA, where its dynamic resistance is about 5 Ω. The voltage increases by a volt at 100 mA (where R_{dyn} is about 5 Ω) and by another volt by 0.5 A, where the resistance is still lower. Above 50 mA the breakdown voltage will be a little higher if the collector pin is left open: you need it tied to the base to benefit from its assistance with inverted-mode gain;[116] see Figure 1x.138 for measured data. The 2N3904 can withstand currents up to 4 A for pulse durations under 20 μs, suitable for human-body discharge (see §9x.25.7). It's worth noting that the $r_{bb'}$ base-resistance we measured and reported in Table 8.1a (110 Ω for the 2N3904), is not relevant to the effective zener resistance during breakdown.

Two-stage protection

As nice as these zener-clamp devices are, there's an inherent problem when trying to prevent damage to an input that expects signals of, say, +3.3 V (e.g., a logic input to a complex microcontroller), where there may be a high-voltage incoming transient that would force 10 A through the clamping device. The worry is that such high currents result in a destructive overvoltage, owing to the clamp's finite clamping impedance. Look at Figure 1x.139, where we've clamped a pair of I²C bus signals with the 3.3 V version of the CD143A. According to the datasheet, the device

---

[114] The datasheet states that it "provides ESD protection for the external ports of portable electronic devices such as cell phones, handheld electronics, and personal computers.... withstands a minimum ±8 kV Contact/±15 kV Air Discharge per the ESD test method specified in IEC 61000-4-2.... will meet IEC 61000-4-2 (ESD) to 30 kV, IEC 61000-4-4 (EFT) to 40 A, and IEC 61000-4-5 (Surge) to 12 A."

[115] We measure 9 V at low currents, rising to 14 V at 5 A. It was perfectly happy with 15 A 20 μs pulses.

[116] Above modest current levels we measured most of the breakdown current going through the collector pin, rather than the base.

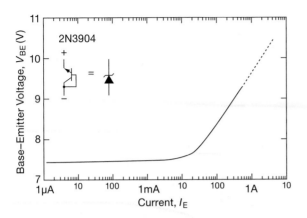

Figure 1x.138. Measured clamping voltage of a 2N3904's base–emitter junction in reverse breakdown.

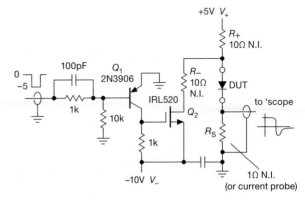

Figure 1x.140. Simple test circuit for displaying diode "reverse recovery." The forward current is set by V_+ and R_+; V_- is then adjusted to set the reverse current.

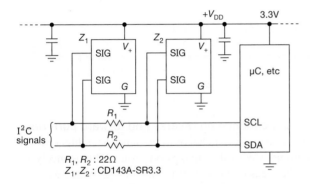

Figure 1x.139. In a situation like this, where vulnerable I²C signals may be going off-board, a two-stage zener clamp provides robust protection against oversized incoming transients.

will clamp a 10 A transient to +8.2 V, maximum. That's OK – but not nearly good enough! So in our circuit the 22 Ω series resistors (intentionally small to prevent slowing the bus signals) limit the second-stage current to about 200 mA, a current at which the second clamp is fully able to prevent destructive overvoltage. (Another set of 22 Ω resistors and another pair of CD143A's would be required at the far end of the off-board cable).

1x.7.2 Stored charge and reverse recovery

One non-ideal property of real diodes is related to the business of *charge storage*, which affects the transition from conducting to non-conducting states. It's something you usually don't want, especially in rectifier applications[117]

---

[117] We saw some undesirable effects of reverse recovery in ac line-powered rectifier circuits in §9.5.3C, where the "snap-off" at the end of reverse

or switching converters,[118] although the effect can be exploited[119] (in a suitably optimized *step-recovery* diode) to generate short pulses (down to sub-nanoseconds), or in a so-called PIN diode to switch or attenuate radiofrequency signals. Let's look at the effect first, then some description of what's going on.

A. Reverse recovery test circuit

Figure 1x.140 is a simple test circuit, configured to abruptly switch a device-under-test (DUT) from a steady dc forward current to a reverse (non-conducting) state, under control of a negative-going input pulse (Q_1 brings MOSFET Q_2 into heavy conduction, pulling the top side of the DUT negative). The positive and negative supplies, together with resistors R_+ and R_-, set the corresponding current levels. If you try it on a plain ol' 1N4001 rectifier (rated at 1 A and 50 V), you'll see something like the traces in Figure 1x.141. Here we've adjusted things for both forward and peak reverse currents of 0.5 A.[120]

When the external circuit attempts to stop current flow by reversing the applied voltage, what we see is a continuation of conduction, now in the reverse direction and with approximately constant current, during which the voltage drop across the diode continues to be in the "forward" direction for a *storage time* duration t_s. After that the volt-

---

conduction produced nasty spikes. These can be tamed with filtering (as in Fig. 9.49), or eliminated entirely with Schottky rectifiers. See the detailed discussion, with scope traces, in §9x.6.

[118] See especially §9.8.3E.

[119] Paraphrasing Elbert Hubbard, "When life gives you lemons, make lemonade."

[120] A common set of datasheet test conditions for rectifiers like this are 0.5 A forward and 1 A peak reverse currents.

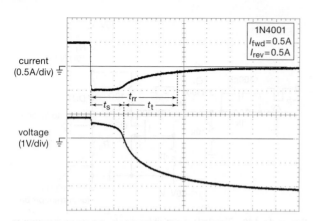

Figure 1x.141. Reverse recovery of a rectifier (1N4001: 50 V and 1 A rating), as measured with the circuit of Figure 1x.140. During *storage time* t_s the minority carriers are being swept from the junction, after which the diode voltage reverses sign; the *reverse recovery time* t_{rr} is the sum of t_s and the transition time t_t, the latter usually taken to end after the reverse current has decayed to a fraction (here 25%) of its peak value. Horizontal: 1 μs/div.

age across the diode reverses, with current (still in the reverse direction) trailing off toward zero during a *transition time* t_t. The total duration of reverse current is the sum, and is called the *reverse recovery time*: $t_{rr} = t_s + t_t$. For ordinary diodes the termination of reverse current is gradual (as here), so datasheets typically state a criterion for the end of t_{rr} such as decay to 25% of peak reverse current; the 1N4001 datasheet uses that percentage, and specifies a t_{rr} of 2 μs for 0.5 A forward current and 1 A peak reverse current (with those parameters we measured a reverse recovery time of about 1.5 μs, see Figure 1x.142).

The phenomenon of reverse recovery originates in the physics of the semiconductor *pn* junction: during forward conduction there is a large concentration of excess *minority carriers*[121] in the region of the junction. To bring the diode into the non-conducting state you've got to sweep these excess carriers back across the junction (by the externally applied reverse current), a process that occupies a *storage time* t_s; during that time interval the voltage across the diode is still in the "forward" direction (see Fig. 1x.141). After that the voltage across the diode changes sign, relaxing to the asymptotic reverse bias with a characteristic *transition time* t_t; depending on the diode construction (doping profile) the current waveform during this phase may be gradual or abrupt.[122]

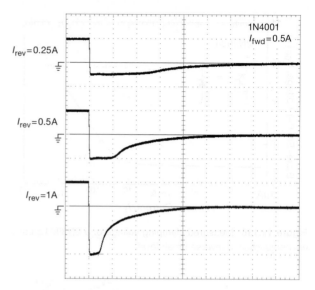

Figure 1x.142. Reverse recovery of a 1N4001 rectifier (50 V and 1 A rating), from a dc forward current of 0.5 A, for three values of reverse current. The recovery time is inverse in the reverse current, corresponding to a fixed stored charge of $\sim$0.8 μC. Horizontal: 1 μs/div; Vertical: 0.5 A/div.

B. Dependence on reverse and forward currents

For a given stored charge in the junction, a higher reverse current sweeps the charges out faster. This is nicely seen in the measured traces of Figure 1x.142, where we've followed a forward dc current of 0.5 A with three values of peak reverse current (0.25 A, 0.5 A, and 1 A). The reverse recovery time is approximately inverse in the reverse current.

Likewise, the stored charge increases with increasing forward current; so the recovery time, for a given reverse current, is approximately proportional to the dc current that precedes the current reversal, as can be seen in the measured traces of Figure 1x.143. Here we've used the same peak reverse current (0.5 A) following three values of forward dc current (0.25 A, 0.5 A, and 1 A). The observed reverse recovery time is approximately proportional to the pre-step forward dc current.

---

[121] Electrons in the *p*-type semiconductor, holes in the *n*-type semiconductor, in each case originating from the other side of the junction, where they are plentiful (thus "majority" carriers).

[122] During this phase the residual current can be thought of as due to junc-

tion capacitance in combination with the external circuit's drive. However, the diode's capacitance is far from constant – just as in BJTs or FETs, the capacitance versus reverse bias is highly nonlinear (see for example Fig. 3.100). In the most extreme case (a "hyper-abrupt" varactor diode, or step-recovery diode) the precipitous drop of capacitance with increasing reverse voltage causes an abrupt (and sometimes useful) termination of reverse current, see §1x.7.2F.

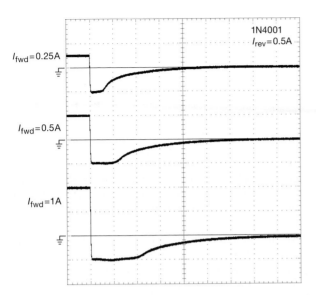

Figure 1x.143. Similar to Figure 1x.142, but here shown with fixed 0.5 A reverse current and three values of dc forward current. The stored charge is proportional to the forward current. Horizontal: 1 μs/div; Vertical: 0.5 A/div.

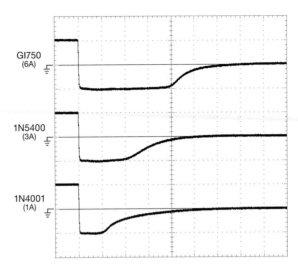

Figure 1x.144. Comparison of reverse recovery for three 50 V rectifiers of different current ratings, all at 0.5 A forward and reverse currents. The recovery time (thus stored charge) scales linearly with diode size. Horizontal: 1 μs/div; Vertical: 0.5 A/div.

C. Dependence on diode size

For a given forward current, the stored charge increases with increasing junction area (for diodes of similar construction). We dug out three ordinary rectifier diodes, all rated at 50 V peak reverse voltage, but with forward current ratings of 1 A, 3 A, and 6 A. Figure 1x.144 shows their respective recovery times, with 0.5 A peak reverse current following an applied forward dc current of 0.5 A.

D. Schottky and fast-recovery diodes

For applications where diodes must rectify high frequencies (such as switching power supplies, or fast signal rectification) you want a diode with little or no stored charge, i.e., with very fast reverse-recovery time. Schottky diodes (metal–semiconductor junction) to the rescue – they have no stored charge, and, as a bonus, they have lower forward voltage drop than conventional (*pn*-junction) diodes. Schottky diodes come in small-signal types (e.g., SD101, SD103, 1N5711), and they come in high-current versions. Generally they are limited to relatively low breakdown voltages (100 V or less), and, because their leakage rises rapidly with temperature, they do have a tendency toward thermal runaway.

A recent entry into the rectifier arena is the "field-effect rectifier diode" (FERD), which claims to compete favorably with Schottky rectifiers, owing to its good V_f/I_R ratio, and lower tempco.

Fast-recovery diodes do have stored charge, but typically a factor of five or so less than conventional diodes of comparable voltage and current ratings. We show some measured recovery waveforms for these diode types in Figure 1x.145. See also §1x.7.2E, where we discuss diodes that combine fast (also "ultra-fast" and "hyper-fast"[123]) recovery with "soft recovery."

E. Soft-recovery diodes

For most diode and rectifier applications[124] you'd rather have *no* stored charge; but if the application does not easily accommodate Schottky diodes (which have no stored charge), say because it runs at high-voltage, you can still benefit from a diode that limits its dI/dt slope during recovery. These are called *soft-recovery* diodes or rectifiers, and the diode construction (doping profile, etc.) is engineered to create a smooth reverse-current waveform; this is often beneficial, because it's abrupt changes in current that generate large voltage transients in circuits with inductance ($V = L\,dI/dt$).

Soft-recovery diodes will be labeled as such, and sometimes will be characterized by a "softness factor" SF=t_t/t_s (see Fig. 1x.141).[125] An example of the very antithesis of

---

[123] Curb your adjectival enthusiasm!

[124] With the exception of *step*-recovery diodes, see §1x.7.2F.

[125] Though a better measure is the ratio of *peak dI/dt* during t_t to that during t_s, because it is the peak slew rate of current that generates inductive transients.

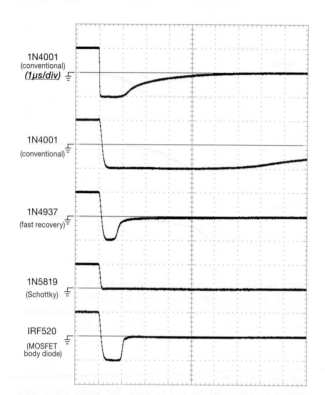

Figure 1x.145. Not all diodes are created equal, as seen here with three 1 A diodes: the 1N4937 "fast-recovery" diode has tenfold less stored charge than the conventional 1N4001; and the 1N5819 Schottky diode has no detectable stored charge at this scale. The body diode of a typical power MOSFET is comparable to a fast-recovery diode. Vertical: 0.5 A/div; Horizontal: 200 ns/div, except top trace.

soft recovery is seen in Figure 1x.147, a diode designed to have a nearly zero softness factor.

It's possible to combine favorable attributes: for example, in switch-mode power supplies you want both fast and soft recovery. You'll find products that deliver both, for example ONsemi's line of "Soft Ultrafast Recovery Power Rectifiers" and "Hyperfast Soft Recovery Rectifiers" which extend to parts with voltage ratings of 1200 V and current ratings of 75 A (e.g., RHRG75120), areas where Schottkys dare not tread.

F. Step-recovery diodes

In the 1960's the designers at Hewlett-Packard exploited the phenomenon of charge storage in *pn* junctions by creating the so-called *step-recovery* diode.[126] This device be-

---

[126] See, for example, HP Application Note AN-918 ("Pulse and waveform generation with step recovery diodes"), AN-920 ("Harmonic generation with step recovery diodes"), and "Microwave harmonic genera-

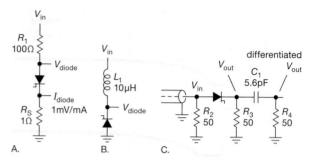

Figure 1x.146. Test circuits used for the waveforms in Figures 1x.147–1x.152. The unusual diode symbol with the little step-shaped stub signifies step-recovery.

haves in most respects like an ordinary 2-terminal diode; but its doping profiles have been crafted to enhance the charge storage (in one variant it is actually a PIN diode, with an undoped "intrinsic" region between the *p*-type and *n*-type semiconductors on either side). Its essential characteristic is an abrupt transition from conducting to nonconducting state at the end of the storage time t_s: referring to Figure 1x.141, although the storage time t_s is unexceptional, the transition time t_t is extremely short, typically a fraction of a nanosecond. This abrupt termination of reverse conduction can be used to generate sharp edges or pulse trains, with the production of copious harmonics, as we shall see presently.

We tested a few samples of vintage HP axial-lead step-recovery diodes (types 5082-0112 and 5082-0180), using the circuits in Figure 1x.146. We're dealing with currents in the milliamps, here, so we can use simple signal-level circuits. The first circuit (Fig. 1x.146A) is a simplified version of Figure 1x.140; an input signal (a symmetrical sinewave, say) drives the diode alternately through forward and reverse polarity, with the current set by R_1. With a 1 MHz sinewave you get the amusing 'scope traces of Figure 1x.147. Here the storage time (from reversal of diode current to reversal of diode voltage) is about 120 ns, followed by an unresolved transition time (decay of reversed current) at the coarse timescale of 100 ns/div. Note that the voltage across the diode remains in the forward direction,

---

tion and nanosecond pulse generation with the step-recovery diode," *Hewlett-Packard Journal* **16** 4 (Dec. 1964). The effect was first published in the 1960 IEEE ISSCC conference proceedings, by Boff, Moll, and Shen: "A new high-speed effect in solid-state devices." The introductory sentence reads "The recovery characteristics of certain types of pn-junction diodes exhibit a discontinuity which may be used to advantage for the generation of harmonics or for the production of millimicrosecond [nanosecond] pulses." They go on to relate the effect to an abrupt drop of capacitance when reverse biased.

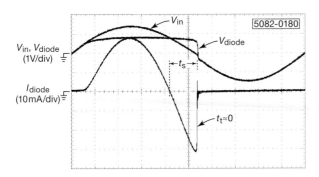

Figure 1x.147. Waveforms for a 5082-0180 step-recovery diode, when driven with a 1 MHz sinewave of ∼1 V amplitude in the circuit of Fig. 1x.146A. Horizontal: 100 ns/div.

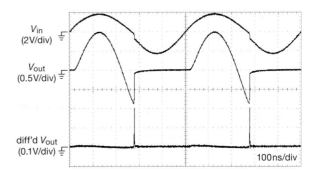

Figure 1x.149. Waveforms for a 5082-0112 step-recovery diode, when driven with a 2 MHz sinewave in the pulse-generating circuit of Fig. 1x.146C. Horizontal: 100 ns/div.

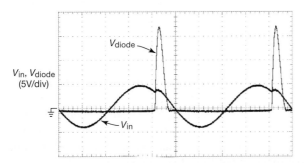

Figure 1x.148. Waveforms for a 5082-0180 step-recovery diode, when driven through a 10 μH series inductor with a 2 MHz sinewave in the circuit of Fig. 1x.146B. Horizontal: 100 ns/div.

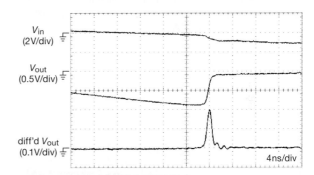

Figure 1x.150. Same as Fig. 1x.149, but zoomed in to resolve pulse width. Horizontal: 4 ns/div.

even as the current has reversed, during the interval t_s when stored charge is being removed; this was seen earlier, in Figure 1x.141, where however the transition time t_t was the very antithesis of a "step-recovery."

The abrupt cessation of current suggests a way to exploit this effect to produce pulses of relatively large amplitude: simply drive the diode with a series inductor, whose craving for continuity of current will produce voltage spikes (Fig. 1x.146B, where we've reversed the diode polarity to produce positive pulses). This is seen in the waveforms in Figure 1x.148. Similar (but unwanted) behavior is seen in power-supply rectifier circuits, even with ordinary rectifier diodes; see the discussion in §9x.6.

We reconfigured the test setup as shown in Figure 1x.146C, where the diode's current generates a corresponding voltage waveform across R_3, differentiated by $C_1 R_4$ to create a short pulse. We also substituted a faster diode (5082-0112). Figure 1x.149 shows the speedy pulses, unresolved on the same 100 ns/div timescale as used for Figure 1x.148. Crank up the horizontal sweep speed to zoom in, and you can see the pulse width

(Fig. 1x.150), a bit less than 1 ns on this 400 MHz scope. That's close to the scope's bandwidth limit;[127] viewing it on a 1 GHz scope (Tek MSO4104) revealed its full-width-half-maximum (FWHM) to be 0.65 ns.

Physicists, engineers, and mathematicians are fond of the properties of the *delta function*, a unit-area step of zero width. A single delta function has a flat frequency spectrum, out to infinity, whereas the spectrum of an imperfect approximation to a delta function (a "spike," of width δ, and amplitude of $1/\delta$ to preserve unit area) extends to a high-frequency cutoff of roughly $f_{max} \approx 0.3/\delta$. And a periodic train of spikes produces a spectrum consisting of spectral lines at the fundamental frequency and its integer multiples (harmonics), again extended roughly out to a maximum frequency $0.3/\delta$.

Can it be that easy to make a frequency "comb," we wondered. Imagine our excitement as we pulled the spectrum analyzer down off the shelf, hooked it to the periodic train of spikes shown in the bottom trace of Figure 1x.149, and set it to run a power spectrum.

---

[127] Rule-of-thumb: risetime is approximately $t_r \approx 0.35/BW$.

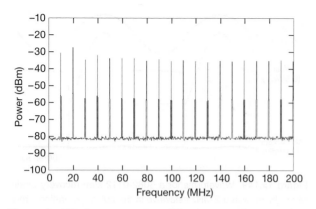

Figure 1x.151. Power spectrum of the pulse output ("differentiated V_{out}") of the circuit of Fig. 1x.146C, with a type 5082-0112 step-recovery diode, when driven with a 10 MHz sinewave. The frequency "comb" is flat to 200 MHz.

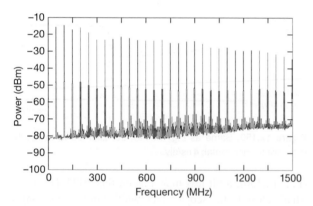

Figure 1x.152. Same as Fig. 1x.151, but with 50 MHz sinewave input and with analyzer set to 1.5 GHz span.

And, as commonly happens with a complex instrument that you haven't used for a while, it *isn't* that easy! – you have to relearn which buttons to push. An RF spectrum analyzer has a lot of parameters to set (resolution bandwidth, video bandwidth, averaging, measurement type, scale factors, input reference and gain, and so on). After some fiddling with our Rigol DSA815TG we got it marching, making the spectra shown in Figures 1x.151 and 1x.152. The former shows a comb of harmonics of a 10 MHz sinewave input, reasonably flat in amplitude out to 200 MHz. For the latter we drove the circuit with a 50 MHz sine, and set the spectrum analyzer to its full range of 1.5 GHz, revealing a falloff of some 15 dB for the highest frequencies. Step-recovery diodes are useful for generating copious high-order harmonics from a reference sinewave, as seen here. They see application also in high-speed trigger circuits, with some diode types capable of risetimes

of 10 picoseconds or less. Step-recovery diodes intended for really high frequencies come in microwave-style "pill" packages, to minimize parasitic inductance. With them it's possible to generate frequency combs to 100 GHz and beyond.

G. A far-out step-recovery application: Larkin's 40-amp kilovolt pulser

Here's an impressive bit of pulse power engineering: the ever-creative John Larkin[128] exploited power-diode reverse-recovery to generate kilovolt-amplitude 20 ampere pulses (into a 50 Ω load) of several nanosecond duration – that's 20 kW instantaneous power – at rates up to 200 kHz or so. As he describes it, he uses power MOSFETs (a) to switch +48 V across the diode in the forward direction for about 80 ns until the current ramps up to 100 A, then (b) he drives it in the reverse direction (from a −400V supply) through a small inductor to about −80 A until it snaps off, generating a 1 kV negative spike into a 50 Ω coax (paralleled by an internal 50 Ω termination that sets the source impedance). Seen at the coax output, you get a 3.5 ns pulse of −1 kV amplitude into a 50 Ω load, or double that into a high-impedance load.[129] By thus mistreating the diode to the max, he can run the thing up to 500 kHz, but not without some serious forced-water cooling.

Figure 1x.153 shows the beast's output pulse, delivered to a 2.5 pF load through ten feet of RG142 coaxial cable. The slower sweep reveals a non-negligible ±100 V reflection, caused by the mismatched load.[130]

H. What about *forward* recovery?

There is an analogous effect that causes a delay in the onset of current flow in a diode, after voltage is applied in the forward direction. This is called *forward recovery*, somewhat of a misnomer since there's no history of flowing current from which to "recover." But anyway.

This effect is usually of little importance, with typical timescales of just a couple of nanoseconds. But it can matter in a clamp circuit subjected to transients of fast rise-

---

[128] Check out some of his creations at highlandtechnology.com.

[129] As usual in the real world, there is more to say, and to do: Larkin first tried a high-voltage PIN diode (a pair of FR802's in series), but he found that the collector-base junction of an *npn* CRT-driver (Fairchild FJAF6810: 1500 V, 10 A) worked even better. His final circuit includes some peaking inductors, a blocking capacitor, and some silicon-carbide Schottky diodes to keep the output unipolar.

[130] A "local-electrode atom probe," in which a giant voltage pulse rips ions from a nanometer-radius metal tip. As a load it is both capacitive and highly nonlinear.

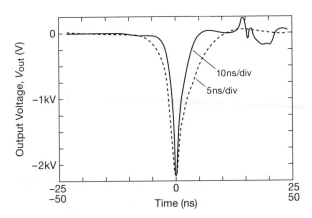

Figure 1x.153. Highland Technology's T220 high-voltage pulser generates impressive 2 kV (open-circuit) fast pulses, seen in this scope tracing (redrawn to show two sweep rates). It can drive a kilovolt across a 50 Ω load, at rates to 200 kHz. Vertical: 250 V/div.

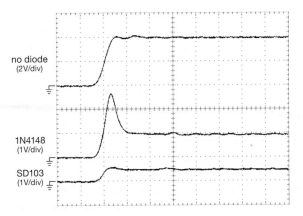

Figure 1x.154. "Forward-recovery" overshoot in a 1N4148 signal diode, wired across a 50 Ω line as a diode-drop voltage clamp, and driven with a 4 V (unclamped) step (thus 120 mA forward current). By comparison, the SD103 Schottky diode exhibits none. Horizontal: 4 ns/div.

time. Out of curiosity[131] we coaxed a 1N4148 small-signal diode (100 V, 200 mA) into revealing this unseemly behavior, by shunting it directly across the 50 Ω BNC scope input and driving it with a large-amplitude step.[132] Figure 1x.154 shows what the scope saw, when driven with a 4 V step on the 50 Ω line. There's overshoot to about +2.6 V, then the expected clamping, which at the 120 mA forward current (3 V drop across the 25 Ω double termination) is at ∼ 0.95 V. In the same test configuration an SD103A small-signal Schottky diode (40 V, 350 mA) driven with a 140 mA step shows negligible overshoot. (Sharp-eyed readers may imagine they see a small bump in the latter. They would be correct, though a simple calculation shows it to be consistent with estimated inductive overshoot: the current is rising through 70 mA in 2 ns, and the DO-35 diode package has an inductance of ∼ 3 nH, so

$V = L\,dI/dt$ predicts an overshoot of 0.1 V, approximately what is seen in the bottom trace of Fig. 1x.154.)

1x.7.3 The tunnel diode

Back when the authors were geeky teenagers, there was a lot of excitement about a weird and wonderful new 2-terminal device, the *tunnel diode* (also known as an Esaki diode, after its inventor[133]). To us it was, well, *exotic* – it had just two terminals (like any other diode), but over a portion of its I–V curve it exhibited negative incremental resistance; that is, an increasing applied voltage produced a decreasing current.

What good is that? Well, as we mentioned in §1.2.3, you can make an amplifier by using a negative resistance as one component of a humble resistive voltage divider. And, if you can make an amplifier, you can make an oscillator. And digital circuits like gates and flip-flops

Perhaps even more thrilling to us youngsters was the mechanism of this circuit behavior – the curious fact that the Esaki diode exploited quantum-mechanical tunneling, whereby an electron can surmount an energy barrier that, by classical physics, is guaranteed to prevent its passage.[134] *Quantum mechanics! classically forbidden tunneling!* What wonderful stuff – even a whiff of

---

[131] Stimulated by a brief discussion in Tom Lee's terrific *Planar Microwave Engineering* (and don't miss his other hit, *The Design of CMOS Radio-Frequency Integrated Circuits*), and also by some scope traces from John Larkin.

[132] To do this effectively, you must suppress series inductance to negligible levels, otherwise what you'll see is voltage overshoot as the inductive current ramps up. Our method, after a few tries, was simply to drill a small hole in the side of a BNC elbow, into which we jammed the leadless diode until it contacted the center conductor; then we soldered the other terminal directly to the metal shell. We put this "elbow-clamp" on the 50 Ω scope input, driven through a short length of coax from a pulse generator. To judge the effect of residual series inductance we repeated the tests, with the diode replaced by an equivalent piece of its lead-wire; that "zero-voltage clamp" was perfect – we could not see any voltage step, right down to the thickness of the scope trace. The traces in Fig. 1x.154 are authentic.

[133] Leo Esaki, who happened onto it somewhat by accident, and who happily was rewarded with a Nobel prize in 1973. His discovery paper is barely over one page: "New phenomenon in narrow germanium p–n junctions," *Phys. Rev.*, **109**, 603–04 (1958).

[134] For a goofy analogy, imagine inmates milling about in a prison yard, occasionally brushing up against the concrete walls. Then imagine one

Einstein's spooky action-at-a-distance ("spukhafte Fern-wirkung")... what's not to love about these things?

We weren't the only ones thrilled by these little diodes. The tunnel diode is very fast – with response times considerably less than a nanosecond (mμs, "millimicrosecond" in the parlance of that time), thus the promise of wideband amplifiers, gigahertz oscillators (kmc, "kilomegacycle" in those days), and fast switching logic (gates, flip-flops, and the like). You can sense some of this optimism in the Forward to GE's 1961 Tunnel Diode Manual: "Tunnel diodes, together with other semiconductor devices,[135] will make possible the practical design of equipment now either impossible or impractical." The Forward goes on to mention just one quantitative parameter (magnificently displayed with all its commas and leading zeros intact), namely that their newest tunnel diodes, intended for microwave use, have "an inductance of less than [0].000,000,000,4 henries."

Despite these rosy predictions, tunnel diodes pretty much disappeared, largely overtaken by greatly improved silicon transistors. The latter are far more flexible, both in voltage range (tunnel diodes must operate at less than a volt), and in the isolation between input and output that comes naturally in a 3-terminal device. And, at least in our experience, tunnel diodes are fussy devices, prone to parasitic oscillation. On a more positive note, a variant constructed with two (or more) barriers, known as a "resonant tunneling diode" (RTD), has recently shown itself capable of extraordinarily high frequency oscillation. At time of writing these devices hold the record, having reached a terahertz (1000 GHz).[136] We could, of course, stop here... but the tunnel diode is a remarkable device in its own right (it's not its fault that transistors sprinted ahead), and perhaps we can be forgiven for indulging in a bit of teenage nostalgia.

A. Current versus voltage: Region of negative resistance

Ignoring the details of device physics (as is our habit), and going straight to its circuit properties, we show in Fig-

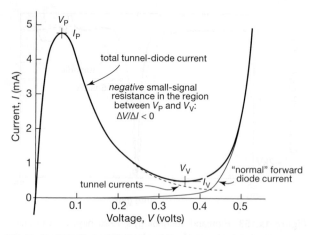

Figure 1x.155. Tunnel diode *IV* characteristic, in the quadrant corresponding to "forward" direction of ordinary diode conduction. The peak at low forward voltage is tunneling current, which extends into the "reverse" direction.

ure 1x.155 a plot of tunnel-diode current versus applied voltage, not unlike that shown in Esaki's original paper. This is the forward conduction direction of a germanium tunnel diode, with "normal" conduction rising rapidly at a forward voltage of about 0.5 V. The tunneling phenomenon adds a current at low voltages (with zero offset – it goes right through the origin, and in fact continues down into the reverse direction); the tunnel current drops off rapidly by a few tenths of a volt, producing the overall curve shown.[137]

We've labeled the dc device parameters: Peak current (I_P) and corresponding forward voltage (V_P), and "valley current" (I_V) with its forward voltage (V_V). Typical tunnel diodes have peak currents of a few mA, at voltages around 50–90 mV (for germanium) or double that voltage for gallium arsenide. The region between peak and valley exhibits a negative small-signal (or *differential*, or *dynamic*) resistance

$$R_{dyn} = dV/dI, \qquad (1x.35)$$

with maximum negative resistance (at the inflection point of the downward-sloping portion) typically in the range of $-10\,\Omega$ to $-100\,\Omega$.

B. Measuring the tunnel diode characteristic curve

This should be easy – after all, it's a simple 2-terminal device. Not so fast! If you put a tunnel diode in a general-

---

of them just disappearing, poof, and emerging intact on the outside; nice!

[135] The author of this manual was, no doubt, mindful of the fact that GE was heavily invested in the manufacture of many "other semiconductor devices."

[136] See for example Sugiyama et al., "Room-temperature resonant-tunneling-diode terahertz oscillator based on precisely controlled semiconductor epitaxial growth technology," *NTT Tech. Rev.*, **9** (Oct. 2011). The authors' enthusiasm is evident right up front, where in the Abstract they proclaim that their 1.04 THz oscillator achieves "the highest oscillation frequency ever reported in single solid-state electron devices."

[137] The ideal tunnel current drops even more rapidly, but additional currents (usually called *excess current*, produced by impurity tunneling phenomena) combine to produce what we've loosely labeled "tunnel currents."

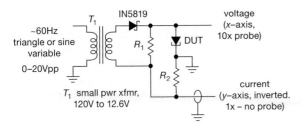

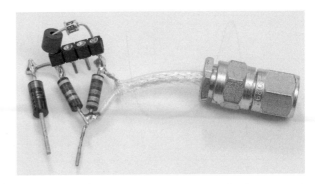

Figure 1x.156. Simple curve-tracer circuit. R_1 and R_2 must be in close proximity to the device under test, to minimize stray inductance, and R_1+R_2 must be less than the tunnel diode's minimum negative resistance. Increase the drive voltage until the full curve is displayed.

Figure 1x.157. Test setup for tunnel-diode curve-tracer. The ferrite bead achieved modest (but incomplete) suppression of oscillations. For scale, the SIP socket is 0.1″ pitch, and the connector is an SMA male.

purpose curve tracer, you don't get anything like Figure 1x.155; instead you get an awful tangled fuzzy mess. The problem is the negative resistance, which loves to combine with parasitic inductances and capacitances to make an oscillator.

To do this right you have to minimize all stray inductances, easily enough done with a simple (but carefully configured) custom setup. Figure 1x.156 shows the circuit we used to do the job, and Figure 1x.157 shows the wiring. But you're not done yet – to get the full $I-V$ curve you have to keep the series resistance seen by the tunnel diode less than the magnitude of its minimum negative resistance: if you don't, the load line (see Appendix F) will not be steep enough to trace out the full curve; instead the operating point will snap across from near the peak, and return from a point near the valley. This behavior is seen nicely in Figure 1x.158, a 'scope capture made with a 4.7 mA tunnel diode in the test circuit of Figure 1x.156 with $R_1{=}6.8\,\Omega$ and $R_2{=}51\,\Omega$; the $60\,\Omega$ series resistance is approximately double that of the tunnel diode ($-30\,\Omega$ at steepest point). Interestingly, you can see bits of the $I-V$ trajectory in this averaged ($n{=}32$) trace.

OK, lower the resistor values, keeping the total resistance less than $20\,\Omega$. But, be careful – the trace will now carry us through the region of negative resistance, where "there be dragons." And dragons there are, see Figure 1x.159, where an oscillation is evident. In fact, the averaged trace conceals the unpleasant reality that the oscillation in fact swings over nearly the full vertical amplitude, even with the addition of the oscillation-damping ferrite bead seen in Figure 1x.157.

The problem is that there's too much circuit inductance, even with the tightest arrangement of resistors we could manage. And the solution is to exploit the compact geometry of surface-mount resistors, to reduce circuit inductance to an absolute minimum. We finally achieved

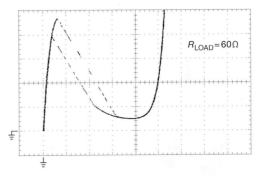

Figure 1x.158. First try at plotting I versus V for a 4.7 mA germanium tunnel diode, with the circuit of Fig. 1x.156, as shown in Fig. 1x.157. The 'scope is displaying current (vertical, 1 mA/div) versus voltage (horizontal, 0.1 V/div), in "XY mode," with zero levels as marked with ground symbols. The bistable behavior is caused by a load resistance ($60\,\Omega$ here) larger than the device's maximum negative resistance ($-30\,\Omega$, see Fig. 1x.160).

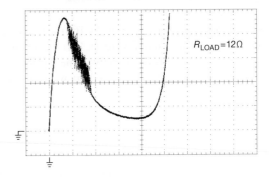

Figure 1x.159. Second try at plotting I versus V for a 4.7 mA germanium tunnel diode, with the circuit of Fig. 1x.156, as shown in Fig. 1x.157. Horizontal: 0.1 V/div; Vertical: 1 mA/div.

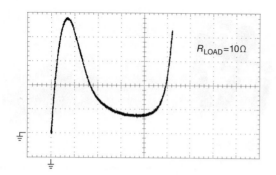

Figure 1x.160. Oscillations-be-gone! A clean *I–V* curve, obtained by squeezing down the inductance (and everything else), as seen in Fig. 1x.161; the 'scope was set at full bandwidth (400 MHz). Same scales as Fig. 1x.159.

Figure 1x.161. Minimizing inductance in the curve tracer: the axial-lead resistors of Fig. 1x.157 are replaced by 0805-size SMT resistors, soldered tightly to the socket's topside. All oscillations were eliminated, when this setup was wrapped with copper foil, reducing still further the residual circuit inductance.

an oscillation-free *I–V* curve (Fig. 1x.160) with the compact configuration shown in Figure 1x.161, where we soldered the SMT chip resistors directly to the topside of the lead sockets (and wrapped the whole thing with conductive tape).

C. Tunnel diode trigger circuit

Tunnel diodes were used in Tektronix oscilloscopes in the 1960s and 1970s for precise triggering. In that switching application (as in Fig. 1x.158, where we used a Tektronix replacement germanium tunnel diode, STD704, Tek p/n 152-0125) they are well behaved, because they are not operated in the linear negative-resistance region. To see how fast these things can go, we rigged up a simple trigger cir-

cuit (Fig. 1x.162), this time using a Russian GaAs tunnel diode ($I_P = 10$ mA, type 1I305A, labeled "1И305A"). Figure 1x.163 shows the abrupt output waveform, when gently nudged by a lazily rising 1 MHz sinewave; the measured transition time is <1 ns (in fact, 0.8 ns when viewed on our snazzy new 16 GHz 100 Gsps 'scope) – not bad.

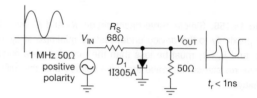

Figure 1x.162. Tunnel-diode trigger circuit; series resistor R_S was chosen for a trigger threshold of 1.0 V. We bought twenty of the Russian GaAs parts on eBay; they arrived in a handsomely decorated packet all the way from Bulgaria.

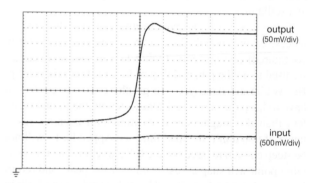

Figure 1x.163. Measured waveforms of the trigger circuit of Figure 1x.162, showing a nanosecond output transition time. The slope of the 1 MHz input waveform is barely visible in this screenshot, where the horizontal scale is 2 ns/div.

Not bad at all... but high-speed BJTs caught up, and rendered tunnel diodes pretty much obsolete (apart from the GHz–THz oscillator applications mentioned earlier). One can look back, with some amusement (and perhaps nostalgia), at the many application circuits found in GE's 1964 *Transistor Manual*, whose Chapter 14 ("Tunnel Diode Circuits") shows 23 pages of circuits (with parts values). These include oscillators (temperature sensing, voltage controlled, variable amplitude, microwave, delay line, sinewave with plug-in coils, quartz crystal, citizens band, fire department, high-accuracy chronometer, remote control, self modulated, FM wireless microphone), frequency converters (civil air patrol, citizens band, community TV up-converter, L-band, FM with AFC, and an amusing AM-to-FM downconverter that uses one tunnel diode as the oscillator and another as a "back-diode" RF detector), and a collection of other circuits (radiation detector, light detector, amplifiers, one-shots, flip-flops, dividers, logic gates, and peak-sensing and sampling circuits).

1x.8 Miscellaneous Circuits with Capacitors and Inductors

Here we collect some amusing, and sometimes amazing, circuits that exploit properties of the passive trio: R, L, and C.

1x.8.1 Improved leading-edge detector

From Richard Pickvance comes this friendly amendment to our brief discussion in §1.4.3 of a "leading-edge" detector, shown there in Figure 1.43. That circuit will do the job, as advertised – but it subjects the output buffer to a negative spike when the input waveform returns to the LOW state. That forces the input protection diodes into conduction, which is unhealthy (at minimum), and possibly destructive for a large value coupling capacitor; it also puts current transients onto the supply rail.

A better circuit is shown in Figure 1x.164, where the NAND gate's inputs stay happily within the supply rails. The only way the input protection diodes can be driven into conduction is if the logic power is abruptly shut down while the input is in the HIGH state. If that worries you, add a small resistor ($\sim100\,\Omega$) from point X to the gate's input terminal (which, if similarly situated in Fig. 1.43, would alleviate any anxiety).

Note added in later printings: After we wrote this section, we moved happily on to the next topic (capacitance multipliers), blissfully unaware that, in fact, this circuit *does not work!*. Sharp-eyed reader Jake Thomas pointed out that the termination of the output pulse causes point X to ramp back up: in his words, "it's a gated oscillator, and not a very good one." The moral, if there is one lurking here, is that (as with human relationships) it's surprisingly easy to overlook a fatal flaw in something that looks attractively dazzling at first sight.

Figure 1x.165 shows the right way to do it, though it costs you another gate. A small price to pay for a totally reliable circuit.

1x.8.2 Capacitance multipliers

Back in §8.12.3 (Fig. 8.92) of AoE3 we showed the use of a "capacitance multiplier" to reduce ripple and noise on a

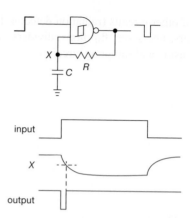

Figure 1x.164. "Improved" (but fatally flawed, see text) leading-edge detector; a Schmitt-trigger gate prevents multiple transitions.

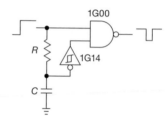

Figure 1x.165. This improved leading-edge detector actually works; we've been using it for over 50 years. For short output pulse widths the Schmitt trigger is unnecessary; instead use the other half of an 'LVC2G00, for example (more than a million in factory stock!).

sensitive supply rail, and in §9x.20 we've extended the discussion of this trick for suppressing power-rail ripple. But the concept of using active gain to increase the effective capacitance (or, generally, to transform the impedance) of a passive component is widely applicable,[138] as for example in active filter circuits, or the generalized impedance converter (§6.2.4D).

Here (Fig. 1x.166) are some example circuits[139] in which op-amps are used to increase the capacitance seen at a node. The top circuit (A) is easy to understand: it's simply a "Miller machine," in which the far side of capacitor C is driven with an inverting signal that is R_2/R_1 times larger than the input signal, thus boosting the effective capacitance by the factor shown (in practice you should add a fixed resistor on the R_1 side of the variable resistor, to limit the maximum gain).

The second circuit (B) is a minor variation on the negative-impedance converter (§6.2.4B, Fig. 6.12), here

---

[138] We thank Richard Pickvance for urging their inclusion.

[139] From the vintage "Op Amp Circuit Collection," National Semiconductor Application Note 31, 1978.

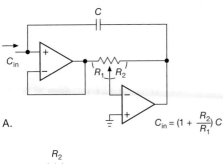

A. $C_{in} = (1 + \frac{R_2}{R_1})\,C$

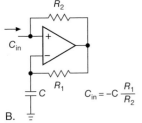

B. $C_{in} = -C\,\frac{R_1}{R_2}$

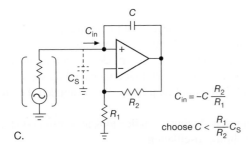

C. $C_{in} = -C\,\frac{R_2}{R_1}$

choose $C < \frac{R_1}{R_2}C_S$

Figure 1x.166. Capacitance multiplier circuits. A. Variable capacitance multiplier. B. Negative capacitance multiplier. C. Canceling parasitic input capacitance.

generalized to allow a multiplicative factor R_1/R_2 (rather than the fixed unity factor of Fig. 6.12). The bottom circuit (C) exploits the negative capacitance produced by feedback capacitor C to (mostly) cancel the bandwidth-limiting effect of parasitic capacitance at the amplifier's input. We've seen other ways to accomplish the same goal (for example, with a bootstrapping "guard shield" driven by a buffered replica of the input, see Figs. 5.78 and 8.138), but this method does not require an input buffer. Because you're introducing positive feedback here, be careful to limit its amplitude – don't be greedy, be content with canceling, say, $\sim 90\%$ of C_S.

ADVANCED BJT TOPICS

In the second chapter of AoE3 we introduced the bipolar transistor (BJT). We began with a simple current-amplifier model, good enough to understand some basic circuits: transistor switch, emitter follower, current source, and common-emitter amplifier. We then looked more closely, with the Ebers–Moll transconductance model, which allowed us to understand deficiencies in these circuits: nonlinearity, bias stability, and the like. With that understanding we were able to deal with more complex circuits, for example current mirrors and differential amplifiers. We concluded the chapter with a look at the essential tool of negative feedback.

There's yet more to learn, and in this chapter we explore advanced topics, including a detailed look at amplifier distortion, Early effect, the Sziklai configuration, BJT bandwidth, and emitter-follower oscillation.

Review of Chapter 2 of AoE3

To bring the reader up to speed, we start this chapter with the end-of-chapter review from AoE3's Chapter 2:

¶A. Pin-Labeling Conventions.

The introduction (§2.1) described some transistor and circuit-labeling conventions. For example, V_B (with a single subscript) indicates the voltage at the base terminal, and similarly I_B indicates current flowing into the base terminal. V_{BE} (two subscripts) indicates base-to-emitter voltage. Symbols like V_{CC} and V_{EE} (repeated subscripts) indicate the positive and negative supply voltages.

¶B. Transistor Types and Polarities.

Transistors are three-terminal devices capable of amplifying signals. They come in two broad classes, *bipolar junction transistors* (BJTs, the subject of this chapter), and *field-effect transistors* (FETs, the subject of Chapter 3). BJTs have a control terminal called the *base*, and a pair of output terminals, called the *collector* and the *emitter* (the corresponding terminals in a FET are *gate*, *drain*, and *source*). A signal applied to the base controls the current flowing from collector to emitter. There are two BJT polarities available, *npn* and *pnp*; for *npn* devices the collector is more positive than the emitter, and the opposite is true for *pnp*. Figure 2.2 illustrated this and identifies intrinsic diodes that are part of the transistor structure, see ¶D and ¶F below. The figure also showed that the collector current and the (much smaller) base current combine to form the emitter current.

Operating modes. Transistors can operate as *switches* – turned ON or OFF – or they can be used as *linear* devices, for example as amplifiers, with an output current proportional to an input signal. Put another way, a transistor can be in one of three states: *cutoff* (non-zero V_{CE} but zero I_C), *saturated* (non-zero I_C but near-zero V_{CE}), or in the *linear region* (non-zero V_{CE} and I_C). If you prefer prose (and using "voltage" as shorthand for collector-to-emitter voltage V_{CE}, and "current" as shorthand for collector current I_C), the cutoff state has voltage but no current, the saturated state has current but near-zero voltage, and the linear region has both voltage and current.

¶C. Transistor Man and Current Gain.

In the simplest analysis, §2.1.1, the transistor is simply a current amplifier, with a *current gain* called *beta* (symbol β, or sometimes h_{FE}). A current into the base causes a current β times larger to flow from collector to emitter, $I_C = \beta I_B$, if the external circuit allows it. When currents are flowing, the base-emitter diode is conducting, so the base is ~0.65 V more positive (for *npn*) than the emitter. The transistor doesn't *create* the collector current out of thin air; it simply throttles current from an available supply voltage. This important point is emphasized by our "transistor man" creation (Figure 2.7), a little homunculus whose job is to continuously examine the base current and attempt to adjust the collector's current to be a factor of β (or h_{FE}) times larger. For a typical BJT the beta might be around 150, but beta is only loosely specified, and a particular transistor type may have a 3:1 spread (or more) in specified beta at some collector current (and further 3:1 spreads of β versus I_C and β versus temperature, see for example Figure 2.76).

¶D. Switches and Saturation.

When operated as a switch, §2.2.1, a current must be injected into the base to keep the transistor "ON." This current must be substantially more than $I_B = I_C/\beta$. In practice a value of 1/10th of the maximum expected collector current is common, but you could use less, depending on the manufacturer's recommendations. Under this condition the transistor is in *saturation*, with 25–200 mV across the terminals. At such low collector-to-emitter voltages the base-to-collector diode in Figure 2.2 is conducting, and it robs some of the base-current drive. This creates an equilibrium at the saturation voltage. In ¶K below we look at some circuit examples. See also the discussion of transistor saturation in this chapter (Chapter 2x).

¶E. The BJT is a Transconductance Device.

As we point out in §2.1.1, "A circuit that depends on a particular value for beta is a bad circuit." That's because β can vary by factors of 2 to 3 from the manufacturer's nominal datasheet value. A more reliable design approach is to use other highly-predictable BJT parameters that take into account that it is a *transconductance* device. In keeping with the definition of transconductance (an output current proportional to an input voltage), a BJT's collector current, I_C, is controlled by its base-to-emitter voltage, V_{BE}, see §2.3. (We can then rely on $I_B = I_C/\beta$ to estimate the base current, the other way around from the simple approach in ¶C.) The transconductance view of BJTs is helpful in many circumstances (estimating gain, distortion, tempco), and it is essential in understanding and designing circuits such as differential amplifiers and current mirrors. However, in many situations you can circumvent the beta-uncertainty problem with circuit design tricks such as dc feedback or emitter degeneration, without explicitly invoking Ebers–Moll (¶F). Note also that, just as it would be a bad idea to bias a BJT by applying a base current calculated from I_C/β (from an assumed β), it would be even worse to attempt to bias a BJT by applying a calculated V_{BE} (from an assumed I_s, see ¶F); more on this in ¶Q, below. We might paraphrase this by saying "a circuit that depends on a particular value for I_s, or for operation at a precise ambient temperature, is a bad circuit."

¶F. Ebers–Moll.

Figure 2.41 showed a typical *Gummel plot*, with V_{BE} dictating I_C, and thus an approximate I_B. Equations (2.8) and (2.9) displayed the exponential (or logarithmic) nature of this relationship. A simple form of the equation, $I_C=I_s\exp(V_{BE}/V_T)$ and its inverse, $V_{BE}=V_T\log_e(I_C/I_s)$, where the constant $V_T=25\,\text{mV}$ at $25°C$, reveals that collector current is determined by V_{BE} and a parameter I_s, the latter related to the transistor die size and its current density. I_s is a very small current, typically some 10^{11} times smaller than I_C. The Ebers–Moll formula accurately holds for the entire range of silicon BJT types, for example those listed in Table 8.1. The integrated-circuit (IC) industry relies on Ebers–Moll for the design of their highly-successful BJT linear circuits.

¶G. Collector Current versus Base Voltage: Rules of Thumb.

See §2.3.2. It's useful to remember a few rules of thumb, which we can derive from Ebers–Moll: I_C increases by a factor of ten for a $\approx 60\,\text{mV}$ increase in V_{BE}; it doubles for an $\approx 18\,\text{mV}$ V_{BE} increase, and it increases 4% for a 1 mV V_{BE} increase.

¶H. Small Signals, Transconductance and r_e.

See §2.3.2B. It's convenient to assume operation at fixed I_C, and look for the effect of small changes ("small signals"). First, thinking about the rules of thumb above, we can calculate (eq'n 2.13,) the transconductance, $g_m=\partial I_C/\partial V_{BE}=I_C/V_T$. This evaluates to $g_m=40\,\text{mS}$ at 1 mA, with g_m proportional to current. To put it another way, we can assign an effective internal resistance r_e in series with the emitter, $r_e=1/g_m=V_T/I_C$, see eq'n 2.12. (The small r indicates *small signal*.) A useful fact to memorize: r_e is about $25\,\Omega$ at a collector current of 1 mA, and it scales inversely with current.

¶I. Dependence on Temperature.

See §2.3.2C. In ¶F we said $V_T=25\,\text{mV}$ at $25°C$, which suggests it's not exactly a constant, but changes with temperature. Because $V_T=kT/q$ (§2.3.1), you might guess that V_{BE} is proportional to absolute temperature, thus a temperature coefficient of about $+2\text{mV}/°C$ (because $V_{BE}\approx 600\,\text{mV}$ at $T=300\text{K}$). But the scaling parameter I_s has a large opposite tempco, producing an overall tempco of about $-2.1\,\text{mV}/°C$. Memorize this fact also! Because V_T is proportional to absolute temperature, the tempco of transconductance at fixed collector current is inversely proportional to absolute temperature (recall $g_m=I_C/V_T$), and thus drops by about $0.34\%/°C$ at $25°C$.

¶J. Early Effect.

See §2.3.2D. In our simple understanding so far, base voltages (or currents) "program" a BJT's collector current, independent of collector voltage. But in reality I_C increases slightly with increasing V_{CE}. This is called the *Early effect*, see eq'n 2.14 and Figure 2.59, which can be characterized by an *Early voltage* V_A, a parameter independent of operating current; see eq'n 2.15. If the Early voltage is low (a common drawback of *pnp* transistors) the effect can be quite large. For example, a *pnp* 2N5087 with $V_A=55\,\text{V}$ has $\eta=4\times 10^{-4}$, and would experience a 4 mV shift of V_{BE} with a 10 V change of V_{CE}; if instead the base voltage were held constant, a 10 V increase of collector voltage would cause a 17% increase of collector current. We hasten to point out there are circuit configurations, such as *degeneration*, or the *cascode*, that alleviate the Early effect. For more detail see the discussion in this chapter (Chapter 2x).

Circuit Examples

With this summary of basic BJT theory, we circle back and review some circuit examples from Chapter 2. One way to review the circuits is to flip through the chapter looking at the pictures (and reading the captions), and refer to the associated text wherever you are uncertain of the underlying principles.

¶K. Transistor Switches.

BJT switches were discussed in §2.2.1, and circuit examples appear in Figures 2.9 (driving an LED), 2.10 (high-side switching, including level shifting), and 2.16 (with an emitter-follower driver). Simply put, you arrange to drive a current into the base to put the transistor into solid saturation for the anticipated collector load current (i.e., $I_B \gg I_C/\beta$), bringing its collector within tens of millivolts of the emitter. More like this appears in Chapter 12 (Logic Interfacing). Looking forward to Chapter 3 in AoE3 (and Chapter 3x here), the use of *MOSFET* switches often provides a superior switching solution (§§3.4.4 and 3.5); their control terminal (the gate) conveniently requires *no* static gate current, though you may have to provide significant transient currents to charge its gate capacitance during rapid switching.

¶L. Transistor Pulsers.

Basic timer and pulse generator circuits were shown in Figures 2.11 (pulse from a step) and 2.12 (pulse from a pulse). These are simple, but not terribly accurate or stable; better to use a dedicated timer or pulse generator IC, see §7.2.

¶M. Schmitt Trigger.

A *Schmitt trigger* is a threshold level-detecting circuit (Figure 2.13) with hysteresis to prevent multiple transitions when noisy input signals go though the threshold(s). Although you can make a Schmitt trigger circuit with discrete transistors, good design practice favors the use of dedicated *comparator* ICs, see §§4.3.2 and 12.3.

¶N. Emitter Follower.

The emitter follower is a linear amplifier with an ideal voltage gain of unity, see §2.2.3. The beta of the transistor increases the follower's input impedance and reduces its output impedance, see §2.2.3B and eq'n 2.2. There's more detail in §2.3.3 and Figure 2.43, where the effect of the intrinsic emitter resistance r_e is taken into account. In simplified form $R_{out} = r_e + R_s/\beta$, where R_s is the signal source resistance seen at the base. The dc output voltage is offset from the dc input by V_{BE}, about 0.6 V to 0.7 V, unless a cancelling circuit is used, see §2.2.3D and Figure 2.29. Emitter followers are also used as voltage regulators, see §2.2.4 and Figures 2.21 and 2.22. A precision alternative is the *op-amp follower*, see §4.2.3 in Chapter 4.

¶O. Current Source (or Current Sink).

In contrast to the familiar *voltage source* (which delivers a constant voltage regardless of load current, think of a battery), a *current source* delivers a constant current regardless of the load's voltage drop, see §2.2.6 and Figure 2.31; there's no everyday "battery equivalent." Transconductance devices like BJTs, with their relatively constant collector output currents, are natural candidates for making current sources. For the simplest current source, the base is biased with a voltage, say V_b, with respect to a reference point (often ground), and the emitter is connected through a resistor to the same reference. For an *npn* transistor with ground reference the output (sinking) current will be $I_C = (V_b - V_{BE})/R_E$, see Figure 2.32. For better stability and predictability the V_{BE} term can be cancelled, see Figure 2.33. The operating voltage range of a current source is called its *compliance range*, set on the low end by collector saturation, and on the high end by the transistor's breakdown voltage or by power-dissipation issues. Current sources are frequently created using current-mirror circuits, see ¶P below. Precise and stable current sources can be made with op-amps (§4.2.5); there are also dedicated current-source integrated circuits (§9.3.14).

¶P. Current Mirrors.

A current mirror (§2.3.7) is a three-terminal current-source circuit that generates an output current proportional to an input "programming" current. In a typical configuration (Figures 2.55 and 2.58) the mirror attaches to a dc rail (or to ground), reflecting the programming current, the latter perhaps set by a resistor. The circuit often omits any emitter resistors, thus achieving compliance to within a fraction of a volt of the rail. Ordinarily you wouldn't attempt to apply exactly the right V_{BE} to generate a prescribed I_C (à la Ebers–Moll); but that's exactly what you're doing here. The trick is that one transistor (Q_1) of the matched pair inverts Ebers–Moll, creating from the programming current I_P exactly the right V_{BE} to re-create the same current in the output transistor Q_2. Cute!

These circuits assume matched transistors, such as you would find inside an IC (recall from ¶G that even a 1 mV difference of V_{BE} produces a 4% change of current). Figure 2.62 graphs base-emitter voltage difference versus collector current ratio, $\Delta V_{BE} = V_T \log_e(I_{C2}/I_{C1})$. You can exploit this effect to generate a "ratio mirror," as discussed in this chapter (Chapter 2x).

As nice as it looks, the basic current mirror of Figure 2.55 suffers from Early-effect change of output current when the output voltage changes. The effect is particularly serious with *pnp* transistors: in the example of a 2N5087 in ¶J above, the 4 mV change of V_{BE} (for a 10 V output change) would cause a 17% current error. One solution (Figure 2.60) is to add emitter degeneration resistors, at the expense both of compliance near the reference rail and of dynamic range. A more elegant solution is the Wilson mirror (Figure 2.61), which defeats Early effect by exploiting the ever-useful *cascode* configuration (Figure 2.84B). Cascode transistor Q_3 passes output transistor Q_2's collector current to the load, while Q_2 operates with a fixed V_{CE} of one diode drop (its own V_{BE}). The Wilson mirror's ingenious configuration also cancels base-current errors (an ordinary mirror with BJTs having $\beta = 100$ has a current error of 2%). Degeneration resistors can be added, as shown in circuit B, for additional suppression of Early effect, but they would be omitted in a "pure Wilson mirror." Linear ICs are full of Wilson mirrors. See further discussion of *bipolarity* current mirrors in this chapter (Chapter 2x).

¶Q. Common-Emitter Amplifiers.

See §§2.2.7 and 2.3.4, and Figures 2.35, 2.48 and 2.50. The simplest form of BJT amplifier has a grounded emitter, a load resistor R_L from the collector to a supply V_+, and a dc bias plus a small signal voltage applied to the base. The gain is $G_V = -R_L/r_e$. If the base bias is carefully set so that the collector current pulls the collector halfway to ground, then $I_C = V_s/2R_L$, $r_e = V_T/I_C = 2R_L V_T/V_s$, and so the voltage gain (recall $V_T \approx 25$ mV) is $G_V = -20V_s$, where V_s is in units of volts. For $V_s = 20$ V, for example, the voltage gain is -400.

That's a lot of gain! Unless the signals are small, however, there's a serious problem: the gain is inverse in r_e, thus proportional to I_C. But the latter changes as the output voltage swings up and down, producing first-order changes in gain, with resulting severe distortion (Figure 2.46). This can be alleviated (at the expense of gain) by adding *emitter degeneration* in the form of an emitter resistor R_E. The gain is then $G_V = -R_L/(R_E+r_e)$, with greatly reduced effect of varying r_e; see Figure 2.47, where emitter degeneration was added to reduce the gain by a factor of ten ($R_E = 9r_e$). This is also a form of negative feedback, see §2.3.4B and ¶W below. You can think of this circuit as a classic current source (¶O) driving a resistor as load; the voltage gain is the current source's transconductance multiplied by the load resistance, $G_V = g_m R_L$, where $g_m = -1/r_e$.

We've sidestepped the important issue of setting the base bias voltage to produce the desired quiescent collector current. But we don't know the appropriate voltage V_{BE}, and a small change has a big effect, see ¶G above (e.g., a 60 mV uncertainty in V_{BE}, which is about what you might encounter from different batches of a given transistor, produces a 10× error in I_C!). There are many circuit solutions, see §2.3.5, but the simplest involves adding emitter degeneration at dc, bypassed as necessary to produce higher gain at signal frequencies (Figure 2.50 and 2.51). Another approach is to use a matching transistor to set the bias, analogous to the current mirror (Figure 2.52); this method is inherent in the widely-used *differential amplifier* (Figure 2.65). A third approach is to exploit feedback to set the bias (Figures 2.53 and 2.54), a method that figures centrally in op-amp circuits (Chapter 4).

¶R. Differential Amplifiers.

The differential amplifier (§2.3.8) is a symmetrical configuration of two matched transistors, used to amplify the difference of two input signals. It may include emitter degeneration (Figure 2.64), but need not (Figure 2.65). For best performance the emitter pulldown resistor is replaced by a current source, and (for highest gain) the resistive collector load is replaced by a current mirror (Figure 2.67). Differential amplifiers should reject strongly any common-mode input signal, achieving a good common-mode rejection ratio (CMRR, the ratio $G_{\text{diff}}/G_{\text{CM}}$). Differential amplifiers can be used to amplify single-ended input signals (ground the other input), where the inherent cancellation of V_{BE} offsets allows accurate dc performance (§2.3.8B). Ordinarily you use only one output from a differential amplifier; that is, it is used to to convert a balanced input to a single-ended output. But you can use both outputs (a "fully-differential amplifier," §5.17) to drive a balanced load, or to create a pair of signals 180° out of phase (a *phase splitter*). See also the sections on the emitter-input differential amplifier and on BJT amplifier distortion in this chapter (Chapter 2x), and §5.13–§5.16 (precision differential and instrumentation amplifiers).

¶S. Comparators.

A differential amplifier with lots of gain G_{diff} is driven into differential saturation with a small differential input (§2.3.8E). For example, just a few millivolts of input difference is adequate to saturate the output if $G_{\text{diff}} = 1000$ (easily accomplished with a current-mirror collector load). When operated in this way, the differential amplifier is a voltage *comparator*, a circuit used widely to sense thresholds or compare signal levels; it's the basis of analog-to-digital conversion, and figures importantly in Chapter 12 (see §12.3 and Tables 12.1 and 12.2).

¶T. Push–Pull Amplifiers.

A single transistor conducts in one direction only (e.g., an *npn* transistor can only sink current from its collector, and source current from its emitter). That makes it awkward to drive a heavy load with alternating polarity (e.g., a loudspeaker, servomotor, etc.), although it can be done, wastefully, with a single-ended stage ("class-A") with high quiescent current, see Figure 2.68. The push-pull configuration uses a pair of transistors connected to opposite supply rails (§2.4.1), an arrangement that can supply large output currents of either polarity with little or no quiescent current. Figure 2.69 shows a push-pull follower with complementary polarities, and with zero quiescent current ("class-B"); this produces some crossover distortion, which can be eliminated by biasing the pair into quiescent conduction ("class-AB," Figure 2.71),. The output transistors can be beta-boosting configurations like the Darlington or Sziklai (¶U), see for example Figure 2.78. The push-pull configuration is widely used in logic circuits (see Figure 10.25), gate driver ICs (see Figure 3.97), and in combination with op-amps to deliver greater output currents (see Figure 4.26).

¶U. The Darlington and Sziklai Connections.

These simple combinations of two transistors create a 3-terminal equivalent transistor with $\beta=\beta_1\beta_2$. The Darlington (Figures 2.74 and 2.75) cascades two transistors of the same polarity and has a base-emitter drop of $2V_{BE}$; the Sziklai (Figure 2.77) pairs opposite polarities, and has a single base-emitter drop (which is only weakly dependent on output current, thanks to R_B). For either configuration a resistor R_B should be connected across the output transistor's base-emitter terminals. For more about this subject see the discussion in this chapter (Chapter 2x).

¶V. Miller Effect.

Like all electronic components, transistors have inter-terminal capacitances, designated (by terminal pairs) C_{be}, C_{ce}, and C_{cb}.[1]

While C_{be} and C_{ce} slow the input and output waveforms by creating lowpass filters with the source and load resistances, the effect of the feedback capacitance C_{cb} is more insidious: it creates an additional input capacitance to ground equal to C_{cb} multiplied by the stage's inverting voltage gain, thus its effective input capacitance becomes $C_{eff}=(G_V+1)C_{cb}$. This is the infamous *Miller effect* (§2.4.5B), whose impact can be devastating in high-speed and wideband amplifiers. Some circuit solutions include the grounded-base amplifier, the differential amplifier, and the cascode configuration (see the discussion of cascode in this chapter).

¶W. Negative Feedback.

If there were a Nobel prize for grand-concepts-in-circuit-design, it would surely go to Harold Black for his elegant elucidation of *negative feedback*. In its simplest form, it consists of subtracting, from the input signal, a fraction B of an amplifier's output signal V_{out} (Figure 2.85). If the amplifier's open-loop gain is A, then the closed-loop gain becomes (eq'n 2.16) $G_{cl}=A/(1+AB)$. The quantity AB, which generally is large compared with unity, is called the *loop gain*, and it (more precisely the quantity $1+AB$) is the multiplier by which negative feedback improves the amplifier's performance: improved linearity and constancy of gain, and (in this *series feedback* circuit configuration) raised input impedance and lowered output impedance; see §2.5.3.

Feedback is the essence of linear design, and it is woven deeply into the DNA of op-amp circuits (the subject of Chapter 4), and power circuits (Chapter 9). With negative feedback you can make amplifiers with 0.0001% distortion, voltage sources with 0.001 Ω output impedance, and many other wonders too magnificent here to relate.

---

[1] These have many aliases (a common set uses initials for "in" and "out" instead of "base" and "collector," thus C_{ie}, C_{oe}, and C_{ob}, respectively), see the section on BJT bandwidth in this chapter (Chapter 2x).

2x.1 What's the Actual *Leakage Current of BJTs and JFETs?*

In implementing a tricky circuit to measure extreme ranges of input currents (the "Starlight-to-Sunlight Linear Photometer," §4x.3.8), we encountered a problem that required real measurements of real transistors: the *actual* leakage current you can expect in samples of real-world BJTs and JFETs. We needed values down in the picoamps with $V_{BE}=0\,V$ and $V_{CE}=10\,V$, but the datasheets were spectacularly, uh, *unhelpful*: for example, the workhorse 2N3904 specifies no more than 50,000 pA (50 nA), and that with the base–emitter helpfully reverse-biased to $-3\,V$, but with $V_{CE}=30\,V$. The situation with JFETs was not better: for our MMBF5460 the brave manufacturers were willing to specify only a gate leakage current of 5,000 pA (5 nA), and they offered no usable limits on the off-state channel current.[2]

So, we took the challenge: we warmed up our Keithley 6514 "Programmable[3] Electrometer," and set out to measure what these puppies actually do. The results are comforting: for small-signal BJTs (2N3904/06 200 mA-class) you can expect room-temperature zero-bias leakage currents to be at most 1 nA, and often one to three orders of magnitude less (Fig. 2x.1). The measurements plotted in that figure include some 20 different '3904 samples, including parts from 11 manufacturers spread over 17 different date codes. Compared with BJTs, the small JFETs fare considerably better (Fig. 2x.2), with leakage currents rarely more than 1 pA (spread only over one order of magnitude, and correlated roughly with current rating).

This was good news for our Chapter 4x photometer. But we offer a *caution*: you cannot depend on unspecified parameters. Even though you may suspect that the manufacturers are low-balling the capabilities of their offerings (or just being lazy in their testing), you should be not unprepared for an unpleasant surprise, someday. So, if you want to exploit components beyond their official specifications, you must be willing to perform tests on incoming parts, to ensure they do what you need them to do. Don't forget,

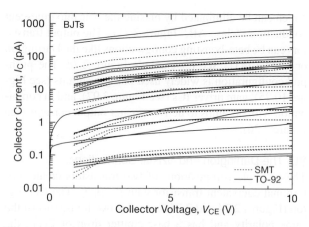

Figure 2x.1. Collector leakage current for a selection of small-signal transistors (2N3904, 2N3906, 2N4401, MPSA42, and SMT equivalents), measured at room temperature with $V_{BE}=0$.

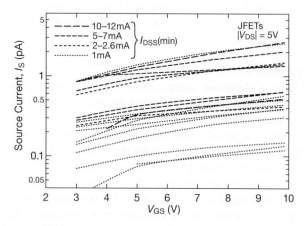

Figure 2x.2. Channel (source-terminal) current versus gate–source voltage for a selection of JFETs (2N5457, 2N5460, and SMT equivalents; and J309, J175, 2SJ74BL, LSK170B, LSK389B, BF861A, BF861B, BF862, and PMBF4395), measured at room temperature with $V_{DS}=5\,V$.

also, that leakage currents rise exponentially with temperature, typically doubling for every 10°C increase in temperature.

---

[2] For which the only information was the threshold drain current that defined "cutoff," namely 1,000,000 pA (1 μA)

[3] But, consistent with our old-school instincts, we ran it manually. Takes too long to learn the magic commands.

2x.2 Current-Source Problems and Fixes

We introduced the BJT current source and current mirror in AoE3 (§§2.2.6 and 2.3.7, respectively), with a promise of more to come later. That later is *now*. Current sources can be built with BJTs, or FETs, or with op-amps (alone, or in combination with discrete transistors). An op-amp current source provides better stability (and control), but there are situations where you might choose a simple BJT current source, owing to its simplicity, speed, and lack of auxiliary supply rails. Here we look at a few ways to improve BJT current-source performance.

There are several effects conspiring to degrade the performance of the simple BJT current source of Figure 2.32 (repeated here as Fig. 2x.3):

Early effect. At any given collector current the base–emitter voltage V_{BE} varies slightly with changes in collector-to-emitter *voltage* V_{CE}. This is the Early effect, which made a cameo appearance in §2.3.2, and which is discussed further in this chapter in §2x.5. The changes in V_{BE} produced by voltage swings across the load cause the output current to change: even with a fixed applied base voltage, the emitter voltage (and therefore the emitter current) changes.

Current gain (beta). Changes in beta with varying V_{CE} produce small changes in output (collector) current even with fixed emitter current, because $I_C = I_E - I_B$; in addition, there are small changes in applied base voltage produced by the variable loading of the nonzero bias source impedance as beta (and therefore the base current) changes.

Temperature effects. V_{BE} and also β depend on temperature. This causes drifts in output current with changes in ambient temperature. In addition, the transistor junction temperature varies as the load voltage is changed (because of variation in transistor dissipation), resulting in departure from ideal current source behavior. The change of V_{BE} with ambient temperature can be compensated with a circuit like that shown in Figure 2x.4, in which Q_2's base–emitter drop is compensated by the drop in emitter follower Q_1, with similar temperature dependence. Here R_3 is a pull-up resistor for Q_1, needed because Q_2's base sinks current, which Q_1's emitter is powerless to source.

These effects can be tamed, for example by biasing the base so that at least a volt or two appears across the emitter resistor (greatly reducing the effect of a small variation in V_{BE}), and biasing the base with a voltage source of low enough impedance that changes in beta are irrelevant. To put some numbers on this, the current from the circuit in Figure 2x.3A varied about 0.5% in actual measurements with a 2N5088 transistor. In particular, for load voltages varying from zero to 8 volts, the Early effect contributed 0.5%, and transistor heating effects contributed 0.2%. In addition, variations in beta contributed 0.05% (note the stiff divider). Thus these variations result in a less-than-perfect current source: The output current depends slightly on voltage and therefore has less than infinite impedance. We'll see methods that get around this difficulty.

2x.2.1 Improving current-source performance

In general, the effects of variability in V_{BE}, whether caused by temperature dependence (approximately $-2\,\mathrm{mV/°C}$) or by dependence on V_{CE} (the Early effect), can be minimized by choosing the emitter voltage to be large enough (at least 1 V, say) so that changes in V_{BE} of tens of millivolts will not result in large fractional changes in the voltage across the emitter resistor (remember that the *base* voltage is what is held constant by your circuit). For instance, choosing $V_E = 0.1$ volt (i.e., applying about 0.7 V to the base) would cause 10% variations in output current for 10 mV changes in V_{BE}, whereas the choice $V_E = 1.0$ volt would result in 1% current variations for the same V_{BE} changes. Don't get carried away, though. Remember that the lower limit of output compliance is set by the emitter voltage. Using a 5 volt emitter voltage for a current source running from a $+10$ volt supply limits the output compliance to slightly less than 5 volts (the collector can go from about $V_E + 0.2\,\mathrm{V}$ to V_{CC}, i.e., from 5.2 V to 10 V).

Wilson mirror. An elegant solution, if you're willing to squander a third transistor, and don't need compliance to the supply rail, is the Wilson current mirror, discussed in §2.3.7B (Figure 2.61). It is popular within linear integrated circuits (and used to be available as a discrete component, see below). Like the cascode we'll see next, it does its magic by clamping the collector voltages of both transistors in the mirror pair, thus defeating Early effect.

Cascode. Figure 2x.5 shows another circuit configuration that improves current-source performance significantly. Current source Q_1 functions as before, but with collector voltage held fixed by Q_2's emitter. The load sees the same current as before, since Q_2's collector and emitter currents are nearly equal (large beta). But with this circuit the V_{CE} of Q_1 doesn't change with load voltage, thus

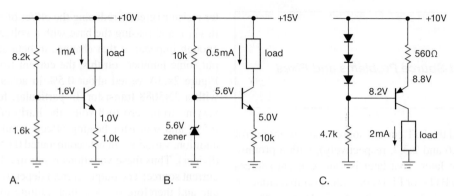

Figure 2x.3. Simple BJT current sources.

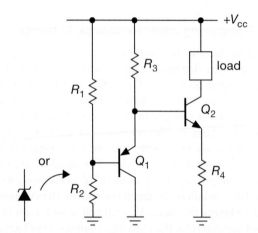

Figure 2x.4. One method of temperature-compensating a current source.

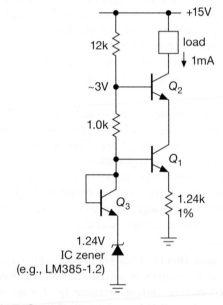

Figure 2x.5. Cascode current source for improved current stability with load voltage variations.

eliminating the small changes in V_{BE} from Early effect and dissipation-induced temperature changes. Measurements with 2N5088s gave 0.1% current variation for load voltages from 0 to 8 volts; to obtain performance of this accuracy it is important to use stable 1% resistors, as shown. This circuit connection is known as the "cascode"; its property of insulating Q_1 from load voltage variations finds use in many applications, for example in high-frequency amplifiers where a fixed collector voltage prevents bandwidth-spoiling via capacitive feedback (the Miller effect). Of course, the use of an op-amp nicely circumvents the problem of V_{BE} variation altogether, as we saw for example in §§4.2.5, 5.16.9, and 13.3.3.

The effects of variability of beta can be minimized by choosing transistors with large beta (or a Darlington), so that the base current contribution to the emitter current is relatively small. Base current can be eliminated altogether if a FET (MOSFET or JFET) replaces the BJT; as nice as that sounds, the spread of gate characteristics (see §3.1.5)

makes it impractical to construct a current source with discrete FETs.[4] However, with an op-amp controlling the discrete device (e.g., Fig. 4.12A) a FET works well, and eliminates all base-current error.

As good as FETs are (owing to their pleasant lack of dc gate current), it is sometimes preferable to use a BJT in a current source, because of the significantly lower capacitance. See §4x.17 for an elegant scheme that eliminates base-current errors.

Figure 2x.6 shows simple current source/sinks that let

---

[4] It's common to see MOSFET current mirrors within a monolithic integrated circuit, however, where you have the ability to lay down well-matched transistors.

Q_1's base–emitter drop set the current in Q_2's emitter: $I_{out}=V_{BE}/R_2$. In circuit A, for instance, pull-down resistor R_1 biases Q_2, and it happily holds Q_1's collector at two diode drops below V_{CC}, eliminating Early effect as in the previous circuit. These circuits are not temperature-compensated; the voltage across R_2 decreases approximately 2.1 mV/°C, causing the output current to decrease approximately 0.3%/°C. But these are simple current sources that don't need a reference voltage, and whose current is relatively stable with supply voltage.

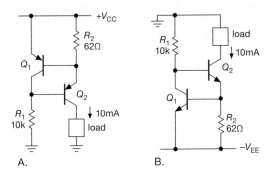

Figure 2x.6. Transistor V_{BE}-referenced current source. A. current source; B. current sink.

2x.2.2 Current mirrors: multiple outputs and current ratios

Current mirrors can be expanded to source (or sink, with *npn* transistors) current to several loads. Figure 2x.7 shows the idea. Note that if one of the current source transistors saturates (e.g., if its load is disconnected), its base robs current from the shared base reference line, reducing the other output currents. The situation is rescued by adding another transistor (Fig. 2x.8).

Figure 2x.9 shows two variations on the multiple-mirror idea. These circuits mirror twice (or half) the control current. In the design of integrated circuits, current mirrors

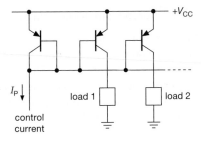

Figure 2x.7. Current mirror with multiple outputs. This circuit is commonly used to obtain multiple programmable current sources.

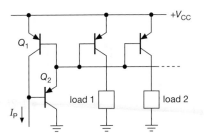

Figure 2x.8. Buffered base drive in the multiple current mirror.

with any desired current ratio can be made by adjusting the size of the emitter junctions appropriately.

Texas Instruments used to offer complete monolithic Wilson current mirrors in convenient TO-92 transistor packages (the TL011 series, with 1:1, 1:2, 1:4, and 2:1 ratios, with output compliance from 1.2 V to 40 V). The Wilson configuration gave good current source performance – at constant programming current the output current increased by only 0.05% per volt – and they were inexpensive. Unfortunately, they've gone the way of the hula hoop and the dial telephone. You *can* get 1:1 monolithic current mirrors without the Wilson configuration (i.e., two matched transistors), for example the BCM61B (*npn*) or BCM62B (*pnp*). Their output impedance (i.e., variation of current with voltage) is only mediocre; but you can "Wilsonize" them by adding an outboard transistor to get reasonable performance.

2x.2.3 Widlar logarithmic current mirror

Another way to generate an output current that is a fraction of the programming current is to insert a resistor in the emitter circuit of the output transistor (2x.10). This circuit was devised by Widlar,[5] to solve the problem of generating microampere current sources on-chip, where it is difficult to integrate high-value resistors. In any circuit where the transistors are operating at different current densities, the Ebers-Moll equation predicts that the difference in V_{BE} depends only on the ratio of the current densities. For matched transistors, the ratio of collector currents equals the ratio of current densities. The graph in Figure 2x.11 is handy for determining the difference in base-emitter drops in such a situation. This makes it easy to design a "ratio mirror."

The Widlar ratio mirror can be "Wilsonized," as shown

---

[5] See for example "Some circuit design techniques for linear integrated circuits," *IEEE Trans. Circuit Theory*, **CT-12**, 4, 586–590 (1965); and US Patent 3,320,439, "Low-Value Current Sources for Integrated Circuits," issued 16 May 1967.

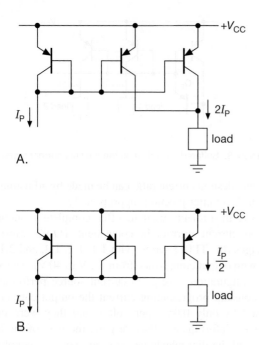

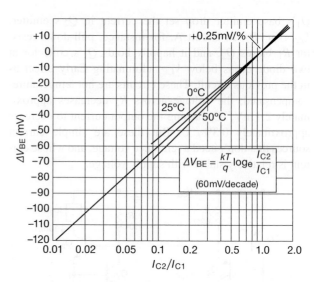

Figure 2x.11. Collector current ratios for matched transistors as determined by the difference in applied base–emitter voltages. See Table 8.1b for low-noise matched BJTs.

Figure 2x.9. Current mirrors with current ratios other than 1:1, as implemented with discrete matched transistors. In an integrated circuit the current ratio is set by the physical ratio of emitter areas.

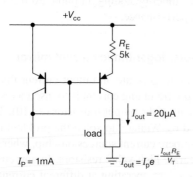

Figure 2x.10. Widlar's ratio current-mirror configuration, with values disclosed in his 1967 patent. Note that the output current is no longer a simple multiple of the programming current.

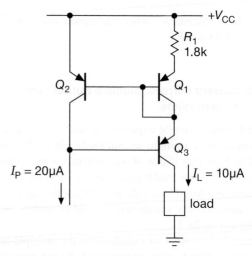

Figure 2x.12. Wilson mirror configuration applied to the logarithmic ratio mirror of Figure 2x.10.

2x.2.4 Current source from Widlar mirror

The Widlar current mirror (along with some conventional current mirrors) is pressed into service in the LM334 current-reference IC, see Figure 2x.13. The reference current is roughly $60\,\mathrm{mV}/R_{\mathrm{set}}$ over a very wide current range; and, as expected, its tempco is proportional to absolute temperature (PTAT). See §3x.6.4, where this IC is used in a high-voltage current source.

in Figure 2x.12, to greatly improve constancy of current over the compliance range (i.e., greatly increased output resistance) by suppressing Early effect in the output transistor.

Exercise 2x.1. Show that the ratio mirror in Figure 2x.12 works as advertised.

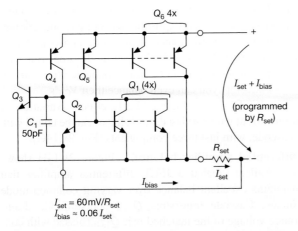

$I_{set} = 60\,mV/R_{set}$
$I_{bias} \approx 0.06\,I_{set}$

Figure 2x.13. The LM334 is a 2-terminal current source that is programmed by a single external resistor R_{set}. We've modified the ratios (16:1 in reality) for ease of understanding; it's still, uh, *challenging*.

2x.3 The Cascode Configuration

The so-called *cascode connection* may well be the "aspirin of electronics": it cures many diseases.[6] We've seen most of these in AoE3 (by actual count, the word "cascode" appears 109 times, not counting the index or table of contents). At times we could hardly contain our admiration, with purple prose like this snippet: "Thunderous applause, yet again, for the remarkable cascode." Many circuit folks associate the cascode exclusively with high-frequency amplifiers, where it is an effective antidote to the bandwidth-killing Miller effect. But its blessings spread over many other design problems, so we think it helpful to our readers to pull together a collection of the many applications of the cascode.

The basic cascode: Miller killer.   Figure 2x.14A is the simplest cascode (here shown with BJTs, but implementable with JFETs, MOSFETs, or hybrid combinations[7]). Ignoring for the moment transistor Q_2, this would be a grounded-emitter amplifier with Q_1 driving load resistor R_L. Transistor Q_2 (with fixed dc base bias) forms the cascode, passing through Q_1's collector current while clamping its collector (at $V_{bias}-V_{BE}$).[8] So the amplifier works as before (though the output cannot swing as low), but by holding Q_1's collector nearly fixed, Q_2 prevents signal swing across C_{cb}, eliminating Miller effect. Figures 2x.14B,C, and D show JFET/BJT implementations of the cascode.

Early effect.   But collector clamping has other benefits: by eliminating variations in V_{CE} you eliminate its effect on V_{BE} (Early effect), thus creating a far better current mirror

(Fig. 2x.14E). The Wilson mirror (Fig. 2x.14F) saves a few components while accomplishing the same result. Though each transistor of the matched pair (Q_{1a} and Q_{1b}) run at fixed V_{CE}, their V_{CE}'s differ by one diode drop. But with one additional component (a diode-connected transistor in Q_{1a}'s collector) you can even balance their V_{CE}'s.

For JFETs the analogous "Early effect" (variation of drain current with varying drain voltage) can be tamed with a cascode, with just three components (Fig. 2x.14G).

Differential amplifier balance.   Figure 2x.14H shows how Agilent created a JFET differential amplifier that maintains excellent balance over varying common-mode voltage. Cascode transistors $Q_{2a,b}$ bootstrap the drain-source voltage of the matched pair $Q_{1a,b}$ (biased with constant current), so the latter have no idea that anything is happening as the common-mode input voltage changes. Figure 2x.14I shows a less-precise variation on this theme, exploiting as cascodes a JFET pair of higher I_{DSS}.

TIA cascode.   The cascode sees effective application in a transimpedance amplifier, where the bandwidth-lowering capacitance of the input signal (from a photodiode, say) is isolated by the action of the cascode (Fig. 2x.14J). Several problems with this simple circuit are nicely circumvented in the popular *regulated cascode* of Figure 2x.14K (usually implemented with MOSFETs).

High-voltage circuits.   A string of cascode transistors lets you distribute the high voltage, keeping the parts within their ratings; see Figure 2x.14L. The same trick with MOSFETs requires a few additional components, to prevent gate breakdown (Fig. 2x.14M).

References in AoE3.   The cascode and its relatives are discussed on pages 102, 114–15, 146, 148ff, 345, 369, 377, 498, 502, 534, 537, 544, 548–52, 693, and 697.

---

[6] Aspirin: reducing fever, pain, swelling, and risk of stroke or heart attack. Cascode: curing bandwidth reduction from Miller effect, curing current-source variation from Early effect, protecting low-voltage components in high-voltage circuits, reducing JFET-amplifier input current, maintaining precision over common-mode variations in differential amplifiers, and more.

[7] A historical note: the cascode was named and invented in 1939, well before the invention of the transistor. See Fig. 12b and footnote 14 in "On electronic voltage stabilizers," by Hunt and Hickman, *Rev. Sci. Ins.*, **10**, 6–19 (1939).

[8] This is a simple form of the so-called *current conveyor*, see "The *current conveyor*: A new circuit building block," by Smith and Sedra, *Proc. IEEE*, **56**, 8, 1368–1369 (1968).

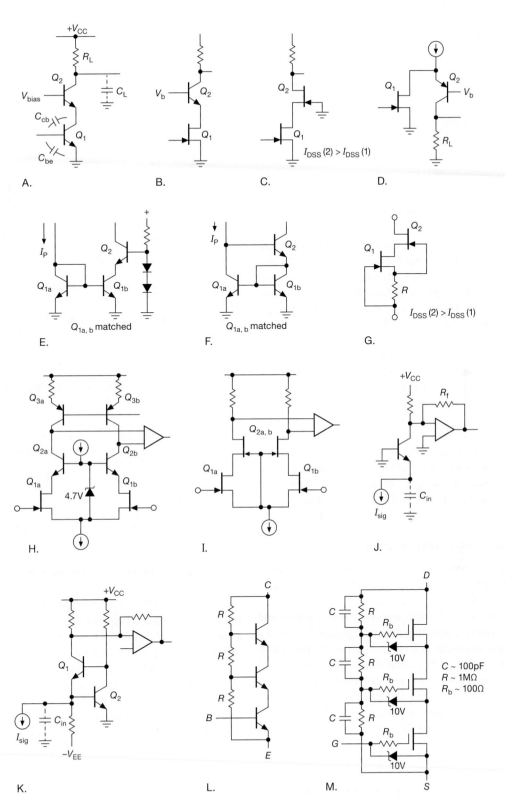

Figure 2x.14. The many faces of the remarkable cascode. (A) Conventional BJT cascode, to defeat Miller effect. (B) Hybrid JFET/BJT cascode. (C) Self-biased JFET cascode. (D) Folded cascode (or "inverted cascode"). (E) Current mirror with cascode. (F) Wilson mirror. (G) JFET 2-terminal self-biased current source. (H,I) JFET differential amplifier with bootstrapped cascode. (J) Cascode transimpedance amplifier. (K) Regulated cascode (RGC). (L,M) Cascode stack to extend transistor operating voltage.

109

2x.4 BJT Amplifier Distortion: a SPICE Exploration

Bipolar transistors are remarkably useful when used as signal amplifiers, but when used in simple circuits they do have serious imperfections, such as distortion. We know that feedback around multiple stages of a high-gain amplifier works well to reduce distortion, but it's useful to explore the BJT-amplifier distortion scene without the use of feedback. Transistor operation is well explained by semiconductor theories such as the Ebers–Moll base-voltage collector-current equations, the Early effect, and the Gummel–Poon charge equations. Formulas from these theories have been incorporated into the SPICE circuit analysis engine developed at UC Berkeley. SPICE can be used for accurate analysis of transistor-circuit performance, provided accurate component models are used. In this section we'll do some "lab experiments," but without the lab bench. Instead we'll use the free demo version of IntuSoft's SPICE program, which they call ICAP/4; download it and follow along with us! Readers unfamiliar with SPICE are advised to read the brief SPICE Primer in Appendix J.

2x.4.1 Grounded-emitter amplifier

Here we take up in more detail the subject first explored in §2.3.4 and Figure 2.46, and especially the circuit we introduced in §2.3.5B and Figure 2.52A.

We've chosen the 2N5088 transistor (the 2N5088 and 2N5089 are popular parts in good audio preamplifiers) in a simple common-emitter (CE) amplifier circuit. We biased it at $I_c=1$ mA and $V_c=10$ V (half the supply voltage), and evaluate the amplifier's gain and distortion. Figure 2x.15 shows the circuit entered into the SpiceNet schematic capture and SPICE management program. As shown, we assigned values of AC=1 and DC=20 to the V1 and V2 voltage generators.

First let's check the frequency response of the amplifier. We click the SIMULATION SETUP button (pencil over wavy line; or ACTIONS | SIMULATION SETUP | EDIT) and select AC ANALYSIS. We'll use 20 points per octave, from 1 Hz to 10 MHz, and we'll SAVE the project at this point. NOTE: If you enter "10M" for the ENDING FREQUENCY, you will get an error telling you that F_ending must be greater

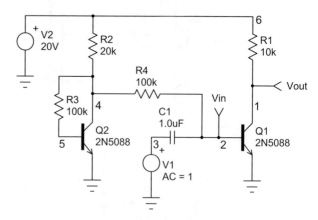

Figure 2x.15. First circuit: grounded-emitter amplifier.

than F_starting; that's because SPICE (which doesn't distinguish upper and lower case) interprets "M" as milli: you must enter "10meg."

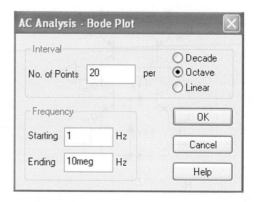

Figure 2x.16. Frequency response setup screen.

Clicking the RUN SIMULATION button (running person button) starts the IsSpice4 SPICE engine, and launches a new window (ISSPICE4) and associated subwindows (SIMULATION STATUS, SIMULATION CONTROL, OUTPUT, and ERRORS AND STATUS); it may also launch the IntuScope display window, with associated subwindows ADD WAVEFORM and SCALING. Next we'll press EDIT TEXT (SpiceNet window: pencil over paper, or ACTIONS | TEXT EDIT), which launches a new window (ISED), and select the OUT file (use the WINDOW tab), so we can check the DC OPERATING CONDITIONS that SPICE calculated for our circuit before doing the AC ANALYSIS. Part way down we see

```
       Node      Voltage
***
   V( 1 )        9.564778e+000
   V( 2 )        4.817968e-001
```

(The node numbers may differ, depending on the order in which you placed the parts.) This tells us the collector is sitting at 9.56 V, close enough to our 10 V goal, and the collector current must be close to 1 mA. The base-emitter voltage is biased at 482 mV.

Further down in the OUT file we see hundreds of frequency response datapoints, but it's more convenient to plot the data than to examine these numbers, so we click the SCOPE button (crosshairs with sinewave, or ACTIONS | SCOPE) to bring up the IntuScope display program. Once there we click ADD WAVEFORM (if the dialog hasn't started automatically), check the box labeled TEST PTS ONLY, select the VOUT signal, and click ADD. We adjust the plot by clicking OPTIONS and selecting LOG-LOG under GRAPH TYPES. Expand the outer window enough so you can resize the graph itself by dragging the corner. Rename the vertical axis (double click on the default label, then uncheck AUTO GENERATE LEGEND). You should get something like Figure 2x.17.

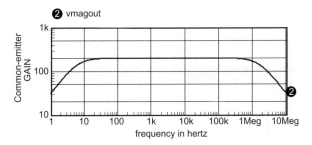

Figure 2x.17. Frequency response, 1 Hz–10 MHz, of the circuit of Figure 2x.15.

The frequency response is flat from about 20 Hz to 500 kHz. The grounded-emitter amplifier gain formula is $G=R_L/r_e$ (where the dynamic resistance $r_e=1/g_m=V_T/I_c$, with $V_T=kT/q=25$ mV at room temperature). Since $r_e=25\,\Omega$ at 1 mA (a handy number we should all have memorized), we expect the gain to be about 400, rather than 200 on the plot above. That's a factor-of-two error; something must be seriously wrong!

2x.4.2 Getting the model right

SPICE can provide accurate answers, but only if it is given accurate component models. The SPICE engine has an accurate built-in processing model for BJTs, which it runs based on parameter values it gets from a component library. Let's use EDIT TEXT to examine the transistor model for the 2N5088 in our SPICE library. The part we used is iden-

tified by IntuSoft as a generic "amplifier" part, labeled QN5088. The model looks like this:

.MODEL QN5088 NPN BF=780 BR=4 CJC=7.83P
+ CJE=11.8P IKF=30M IKR=45M IS=21.0P ISE=41.8P
+ NE=2 NF=1 NR=1 RB=92.6 RC=9.26 RE=23.1
+ TF=3.18N TR=127F VAF=98.5 VAR=18 XTB=1.5

The terms RB, RC, and RE are external resistances that SPICE adds to the three transistor pins in its internal model. Do these values make sense? Looking at a 2N5088 datasheet, we see that the transistor works well to 30 mA or more. We know that the dynamic emitter resistance r_e should be less than an ohm at 30 mA, so how are we likely to fare if an additional 23 Ω is added by the model? It's going to be a disaster!

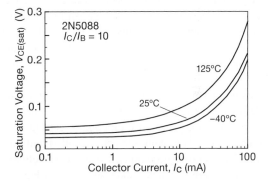

Figure 2x.18. Checking the model: datasheet's $V_{CE(sat)}$.

What value should RE have? We can get an idea from a datasheet plot of the 2N5088's saturation voltage, $V_{CE(sat)}$ (Figure 2x.18). We see that V_{CE} is about 210 mV at 100 mA, and perhaps 75 mV of that is due to simple transistor action. So the sum of RC and RE should be no more than $R=135$ mV/100mA, or 1.35 Ω. Compare that to 32.4 Ω in the model, whew! We could modify the values for the QN5088 part, but rather than trust the rest of the model, let's see if we can use another part in the library. Under the Fairchild heading in IntuSoft's library, we find both a 2N5088 and a 2N5089. Here's what you see if you drop those parts into the schematic, then look at the *.out file:

.MODEL 2N5088F NPN BF=1.122K BR=1.271
+ CJC=4.017p CJE=4.973p EG=1.11 FC=.5
+ IKF=14.92m IKR=0 IS=5.911f ISC=0
+ ISE=5.911f ITF=.35 MJC=.3174
+ MJE=.4146 NC=2 NE=1.394 RB=10 RC=1.61
+ TF=821.7p TR=4.673n VAF=62.37 VJC=.75
+ VJE=.75 VTF=4 XTB=1.5 XTF=7 XTI=3
.MODEL 2N5089 NPN BF=1.434K BR=1.262
+ CJC=4.017p CJE=4.973p EG=1.11 FC=.5

+ IKF=15.4m IKR=0 IS=5.911f ISC=0 ISE=5.911f
+ ITF=.35 MJC=.3174 MJE=.4146 NC=2 NE=1.421
+ RB=10 RC=1.61 TF=822.3p TR=4.671n VAF=62.37
+ VJC=.75 VJE=.75 VTF=4 XTB=1.5 XTF=7 XTI=3

The first thing we see is that there's no RE at all in these models (SPICE assumes RE=0), and RC is 1.61 Ω, so that looks pretty good. The only difference we see between the 2N5088 and 2N5089 models are BF=1.122K and BF=1.434K, the transistor's beta values. Both of these parts are on the same datasheet, which shows h_{FE} current gain (or beta) values of 350 and 450 respectively. Only minimum values are shown at 1 mA (although the typical values shown for 0.1 mA are 3× higher than the minimum values). Many manufacturers provide SPICE models for their parts, and they often choose the worst-case values. Here we are given optimistic values, but since beta isn't critical in our circuit, we'll accept that and substitute Fairchild's 2N5088 model in our circuit. (If the value of beta mattered much in our circuit, we would edit the SPICE model to match reality.)

Using the new transistor model (highlight old transistor, then DELETE key or EDIT | CLEAR; then place new part), we try the frequency response plot again, getting the plot of Figure 2x.19. Aha! Now we get a small-signal gain of 340, much closer to expectations. We also have better high-frequency response, which makes sense given Fairchild's lower capacitance values in their model.

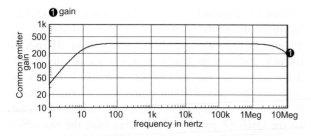

Figure 2x.19. Frequency response with better transistor model.

Rechecking the DC operating conditions, we see

```
V( 1 )     1.045805e+001
V( 2 )     6.690726e-001
```

We have 10.45 V on the collector, OK, that's fine. But it's interesting that now we have V_{BE}=669 mV, compared with 482 mV earlier. This is a more sensible value, and is due to a more realistic value I_S=5.911 fA, compared with 21 pA in the first (generic) model. Fairchild's datasheet plot (Figure 2x.20) shows about 630 mV at 1 mA, but we're happy enough with this model.

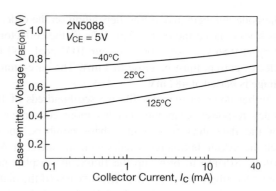

Figure 2x.20. Checking the model: datasheet's $V_{BE(on)}$.

2x.4.3 Exploring the linearity

Now we're ready to continue and evaluate the nonlinearity of the common-emitter amplifier. Going back to our schematic in SpiceNet, we double-click the V1 voltage source. In the dialog box, under TRAN GENERATORS, click on the button showing "none," and select PWL (Figure 2x.21). We'll create a triangle test waveform by entering a few time–voltage datapoints; a triangle wave input makes it easy to spot deviations from linearity.

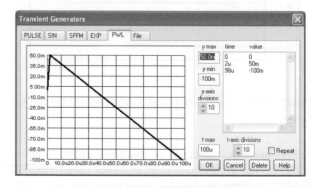

Figure 2x.21. Voltage sweep setup screen.

The gain of a common-emitter transistor stage is $G= - g_m R_L$, where $g_m = V_T / I_c$. Our concern with this equation is that a changing output requires a changing collector current, and the equation (see §2.3.4A and Figures 2.45 and 2.46) tells us to expect a substantial gain variation from this changing collector current, thereby causing high distortion. We're looking to see if the amplifier produces a rounded response to the linear triangle wave input, an indication of changing gain over the waveform, with corresponding distortion.

We have to decide what our test-voltage range should be. With a gain of 350, a +50 mV input should easily drive the collector voltage down to zero volts before we start the ramp; and we end the ramp at -100 mV to drive the output close to the positive supply rail. We select $2\,\mu$s for the initial step to 50 mV, and $98\,\mu$s to complete the ramp. This is fast enough to avoid droop from the ac coupling capacitor, and slow enough to avoid errors from the high-frequency roll-off. We enter the values with commas between the number pairs, and a CR after each pair. We select scaling factors (Y MAX, T MAX, etc.) for the plot, as shown, so the waveform will display nicely.

Next we select TRANSIENT ANALYSIS under SIMULATION SETUP, and use a $100\,\mu$s TOTAL ANALYSIS TIME and a $0.1\,\mu$s DATA-STEP TIME (1000 points) to capture our waveform.

We run the SPICE engine again, and then go to IntuScope and select FILE | NEW GRAPH. Using the ADD WAVEFORM dialog's TRAN1 mode, we select vout and click ADD. If the amplifier were linear, we'd see a linear output ramp – an inverted, amplified version of the input. Instead we see a curving, squashed output waveform (Figure 2x.22).

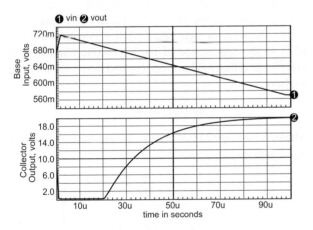

Figure 2x.22. Output response to an input ramp for the circuit of Figure 2x.15.

First, we see that a $+50$ mV signal (669 mV(quiescent) $+ 50$ mV $= 719$ mV) was more than adequate to drive the output to zero volts; in fact the output doesn't start to rise until the base voltage falls to about 690 mV, or $+21$ mV input signal. This is further confirmation that the gain increases with I_c.

Looking at the shape of the output curve, we see that the first part looks reasonably linear, but the gain falls dramatically at high collector voltages (i.e., at low collector currents).

A. Input–output transfer function

Ideally we'd like to have a plot of output voltage directly versus input voltage, rather than each plotted separately versus time. We can do this on a new graph, by changing the X-AXIS choice in the ADD WAVEFORM dialog, from default–time, to the input signal V1 (we added a test point V1 on the input). Figure 2x.23 shows the result.

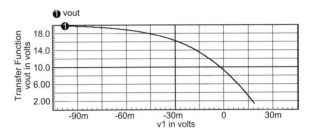

Figure 2x.23. Transfer function of the circuit of Figure 2x.15. The output goes (nonlinearly) from 20 V down to 1 V for inputs going from -100 mV to $+25$ mV.

At this point we also made a few more changes: we set the INITIAL CONDITION (in the SpiceNet schematic) of coupling cap C1 to 669 mV (double-click the cap and enter IC=-669m), and we checked USE INITIAL CONDITIONS (UIC) in the TRANSIENT ANALYSIS dialog. We set V1's initial value to $+21$ mV (in the transient generator PWL setup). We set the MAXIMUM TIME STEP to $0.02\,\mu$s, to reduce noise in the output data. Finally, we set the TIME TO START RECORDING DATA to $0.3\,\mu$s to skip past the startup transient. We refined the plot with the SCALING dialog to adjust the graph's x-axis ("x scale" $= 15$m, "x-offset" $= -30$m).

B. Gain versus input

IntuScope lets us do some valuable post-processing analysis. By differentiating the V_{out}-vs-V_{in} plot with CALCULATOR | CALCULUS, we get a plot of the amplifier's gain versus output voltage, shown in Figure 2x.24. (We adjusted the graph's x- and y-axis for appearance. If you want to follow along, highlight the gain trace, then go to the SCALING window: we used 15m for the x-scale with -30m offset, and 75 for the y-scale with -300 offset.)

Line #1 is the output voltage (from 20 V down to 0.38 V) versus input voltage, and line #2 is the amplifier's gain versus input voltage. The gain varies from nearly 0 to more than -500; the gain at the quiescent point (v1 $= 0$ V) is -341.6 (drag the cursor until x=0 in the display at the bottom of IntuScope; or just enter 0 there), as we saw in the small-signal frequency response plot. To take these measurements, we used IntuScope's CURSOR tool, which

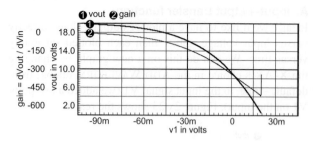

Figure 2x.24. Gain (trace 2) calculated from the transfer function (trace 1) of Figure 2x.23; the gain goes from zero to −500, ouch!

shows us X and Y values for the selected trace, as shown in Figure 2x.25. A gain varying from 0 to 500 isn't very good, but we know how to do much better.

2x.4.4 Degenerated common-emitter amplifier

Clearly a simple grounded-emitter amplifier has a great deal of distortion, but we can trade off higher gain for lower distortion by adding an emitter resistor (recall §2.3.4B and Figure 2.52B), Figure 2x.26. Let's aim for a gain of ten.

Some like to think of emitter degeneration as negative feedback. But no explicit external feedback path is involved, so purists who worry about transient response issues, etc., with ordinary feedback, needn't be concerned. Furthermore, time delays, phase shifts, and feedback stabilization issues at high frequencies are avoided with emitter degeneration.

For the simulation we set v1's PWL signal for a +1000 mV to −1000 mV range, and plotted transfer function and gain versus v1 as before. In terms of percentage gain variation, the result (Figure 2x.27) is much better (note expanded scale), but it's still a mediocre amplifier, with the gain (plot 2) changing from about $G=9.8$ at an output voltage of 2 V, down to 9.68 at 10 V (about 1.2% lower), and then continuing down to 8.81 at 18 V (another 9% lower).

Furthermore, the drop in gain is not symmetrical about the quiescent point, so the amplifier produces second-harmonic distortion, insulting the ears of the audiophile. If we could balance the positive and negative gain losses, the even harmonic distortion products would be eliminated. That can be done by going to a balanced (differential) circuit configuration.

2x.4.5 Differential amplifier

Realizing that our single-ended common-emitter amplifier circuit already requires two transistors (one to bias the other), we are motivated to ask why we shouldn't instead use these same two transistors in a differential amplifier. In essence, one will still be biasing the other, but in a more useful manner. If the long-tail pair (see §2.3.8) is biased with a 2 mA current, each side carries 1 mA, so the output load resistor will be biased at half the positive supply voltage (+10 V) as before. This allows for a symmetrical output swing up to ±8 V or so. Let's test[9] the circuit (Figure 2x.28) by driving it with a ±100 mV dc-coupled ramp of 100 μs duration, via the v1 PWL dialog.

The output transfer function exhibits a nice symmetry about the quiescent point, reflected in the plot of calculated gain; the symmetry implies absence of distortion at even harmonics. But there remains a large variation in gain: more than a factor of two over the 2 V to 18 V output range. That is, the amplifier continues to have substantial distortion, in fact more than that of the amplifier it replaces. We should not be surprised, though, because this circuit lacks the linearizing emitter degeneration of the previous circuit (Figure 2x.26); we'll fix that, presently.

A. Estimating the distortion

We can evaluate the amplifier's performance analytically. The derivation of the gain of a differential amplifier involves the ratio of two Ebers–Moll exponentials, which leads to the appearance of the hyperbolic tangent function.[10] For differential current I_{out} and long-tail current I_E we have

$$I_{out}/I_E = 2\tanh\frac{V_{in}}{2V_T},$$

where $V_T = kT/q = 25$ mV at room temperature. A series expansion gives us

$$I_{out}/I_E = \frac{V_{in}}{V_T} - \frac{2}{3}\left(\frac{V_{in}}{2V_T}\right)^3 + \cdots$$
$$= \frac{V_{in}}{V_T}\left[1 - \frac{2}{3}\frac{V_T}{V_{in}}\left(\frac{V_{in}}{2V_T}\right)^3 + \cdots\right] \quad (2x.1)$$

The first term is $I_{out}/I_E = V_{in}/V_T$, equivalent to the familiar $G = R_L/r_e$. The second term shows us how this drops off for inputs greater than 25 mV or so (i.e., V_T), and should be enough to evaluate the nonlinear gain dropoff.

A single-ended differential amplifier like ours has half the gain of a full differential output, or $G = R_L/2r_e$. At the

---

[9] Here we've lazily adopted the parlance of our time, referring to a pure numerical simulation as a "*test*"! It gets worse – you'll hear people say something like "I built this circuit, and measured . . . ," when in fact they built nothing, and measured nothing. The starry-eyed delusions of the SPICE-obsessed designer.

[10] Written tanh x," and pronounced "tansh." The tanh function goes from 0 to 1 as its argument goes from zero to infinity, and $\tanh(x)\approx x$ for $x \ll 1$).

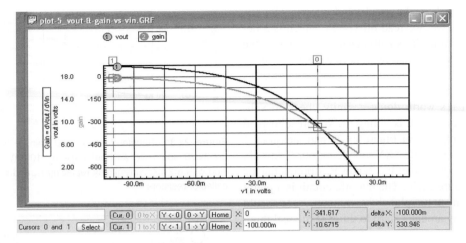

Figure 2x.25. Reading gain values from the plot with a cursor.

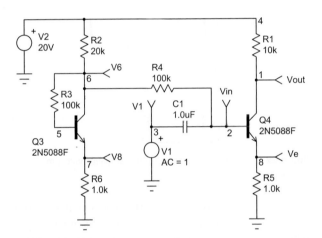

Figure 2x.26. Second circuit: degenerated common-emitter amplifier.

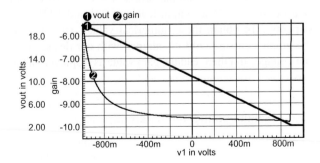

Figure 2x.27. Transfer function and gain of the circuit of Figure 2x.26. For this circuit the gain changes from −5.5 to −9.8, about 80%.

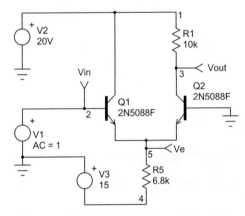

Figure 2x.28. Third circuit: differential amplifier.

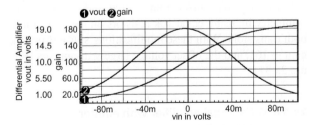

Figure 2x.29. Transfer function and gain of the circuit of Figure 2x.28. The gain error involves the hyperbolic tangent, see text.

default SPICE temperature of 27°C (where V_T=25.8 mV) this predicts our amplifier should have a gain of 10k/51.7, or G=193, in reasonable agreement with the SPICE result of G=180 for input signals less than 10 mV. Let's try a larger input, say 50 mV. Equation 2x.1 predicts the gain should decrease by a factor of $2/3 \times 25.8/50 \times (50/51.7)^3$, or 31%. Going back to the graph, and allowing for the −4 mV offset voltage, we read off $G \approx 96$, which is a decrease of −47% from the small-signal gain. Though the formula gives us a good idea of what to expect, it's likely that SPICE, with its Early-effect corrections, etc., gives us a more accurate answer.

At this point it's worth doing a sanity check to validate the stories that SPICE is telling us. We breadboarded the differential amplifier of Figure 2x.28 and measured[11] its transfer function (using "XY" mode on our Tektronix lunchbox-style 'scope to plot V_{out} versus V_{in}), producing the screenshot of Figure 2x.30. The real-life circuit is a pretty good replica of the SPICE plot above (hmmm..., or should it be the other way around? Hard to think of the real thing as a "replica"!)

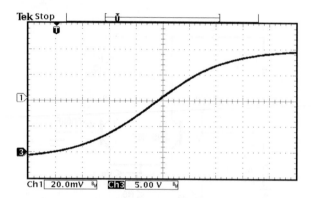

Figure 2x.30. Measured transfer function of the circuit of Figure 2x.28.

With renewed confidence in SPICE, let's try some variations. An obvious improvement is to replace the emitter tail resistor with a 2 mA current source, and to improve the symmetry by using equal 10k resistors on the collector of both transistors. With these modifications SPICE drops the gain to 161, and the offset voltage largely disappears (Figure 2x.31).

2x.4.6 Differential amplifier with emitter degeneration

As with the simple single-ended common-emitter amplifier, we can improve differential amplifier performance with emitter degeneration, as in Figure 2x.32. In the simplified gain formula $G = R_1/(R_2 + R_3 + r_{e1} + r_{e2})$ it's primarily the current-dependent r_e terms that degrade the linearity.

Figure 2x.33 shows the greatly improved linearity, with

[11] For real: on the bench this time, with wires and stuff!

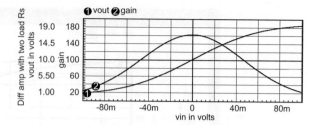

Figure 2x.31. Transfer function and gain of the circuit of Figure 2x.28, again over an input range of −100 mV to +100 mV, modified with a 2 mA current sink replacing R_5 and with matching 10k collector resistors.

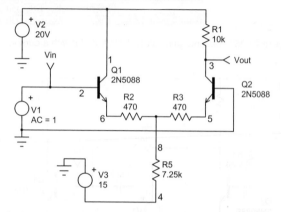

Figure 2x.32. Fourth circuit: differential amplifier with emitter degeneration.

$G = 9.654$, close to our goal of $10.0 \times$. The output waveform looks nice and straight, and the gain plot has the nice symmetry that indicates an absence of second-harmonic distortion. Nonetheless, the distortion at large amplitudes is still high (by audiophile standards), with a gain reduction of 1.2% (9.537/9.654) at output levels of 4 V and 16 V (and soaring to −12.5% at output levels of 2 V and 18 V, near clipping). To obtain these numerical values, slide one of the graph's cursors until it's aligned with our desired output voltage, and then read the gain value as displayed in the box.

2x.4.7 Sziklai-connected differential amplifier

The circuit in Figure 2x.34 improves the linearity, and it is popular among microphone preamp designers. The basic idea is to maintain a constant current (and hence constant g_m) for the matched *npn* input transistors by making them into Sziklai pairs.[12] In this way the complementary

[12] See §§2.4.2A and 2x.6 for a discussion of the Sziklai connection.

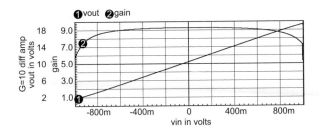

Figure 2x.33. Transfer function and gain of the circuit of Figure 2x.32. The input voltage range is −1 V to +1 V for this circuit of reduced gain.

pnp transistor of each Sziklai pair will do the work of servicing the changing output current. The new gain formula is $G=R_1/(R_2+R_3)$, without any bothersome r_e terms. Figure 2x.35 shows the resulting transfer function and gain. Now we're beginning to see some seriously good low-distortion performance! The gain is 9.988 near zero volts.

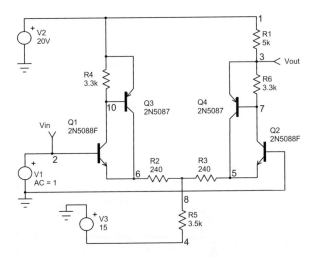

Figure 2x.34. Fifth circuit: Sziklai-connected differential amplifier with emitter degeneration.

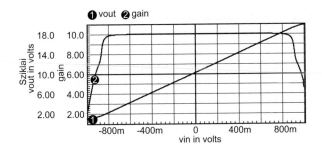

Figure 2x.35. Transfer function and gain of the circuit of Figure 2x.34.

The same data is shown in Figure 2x.36, expanding the gain axis to show only the top 5%. Compared with the unadorned diff amp with degeneration (Figure 2x.32), we see a smaller −0.4% gain variation over a 4 V to 16 V output or nearly 10× better, and less than −2% at 2 V to 18 V outputs. We could play with the values of R_4 and R_6 to improve on that 2% value (at the low currents necessary for large positive output swing, a low value of R_6 combined with a relatively large current in Q_2 leaves insufficient current to operate the Sziklai-transistor Q_4).

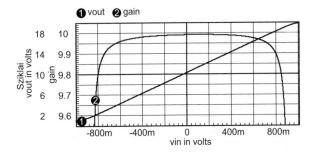

Figure 2x.36. Transfer function and gain, expanded scale.

2x.4.8 Sziklai-connected differential amplifier with current source

We can further improve the circuit by replacing R_5 with a current source, as in Figure 2x.37. Now we're enjoying an even lower −0.25% gain-reduction distortion over the 4 V to 16 V output range (Figure 2x.38). While it's true that many folks don't consider 0.25% to be "low distortion," it's also true some prefer soft limiting to the hard limiting that one experiences with conventional feedback circuits. That's one argument made for vacuum-tube amplifiers. This circuit has only 0.1% peak gain distortion over a 6 V to 14 V output swing (half of full output range). Note that what we've called "distortion" is the *peak-to-peak* gain deviation; the more usual measure is *rms* distortion, which is smaller typically by a factor of five, as seen in the measured harmonic distortion plots of Figure 2x.49.

If we increase the gain of this circuit, say to about $G\sim50$ by substituting $50\,\Omega$ emitter resistors, we might expect the distortion to increase. As Figure 2x.39 shows, we see gain reductions of 1% over a 4 V to 16 V swing, and 0.4% over 6 V to 14 V.

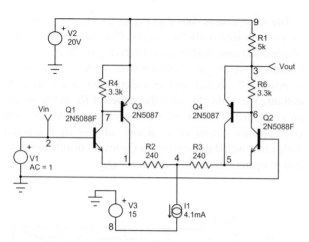

Figure 2x.37. Sixth circuit: Sziklai differential amplifier with emitter current source.

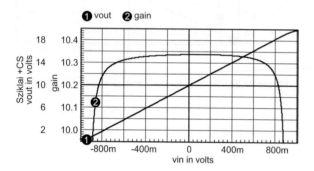

Figure 2x.38. Transfer function and gain of the circuit of Figure 2x.37.

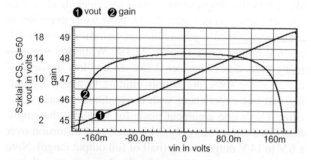

Figure 2x.39. Transfer function and gain, with reduced emitter resistors.

2x.4.9 Sziklai-connected differential amplifier with cascode

We've dealt with several sources of distortion while improving our circuit, but we haven't yet dealt with the Early effect, which describes the variation of V_{BE} with changing

V_{CE} (see §§2.3.2 and 2x.5). We can try to reduce this effect with a cascode circuit, to eliminate the changes in collector voltage across the differential pair during large output swings. Figures 2x.40 and 2x.41 show the trial circuit and simulation results.

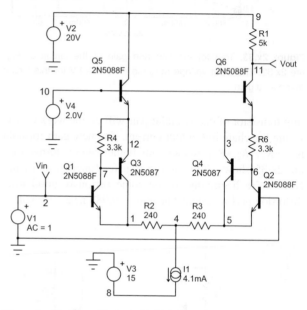

Figure 2x.40. Seventh circuit: Sziklai differential amplifier with cascode.

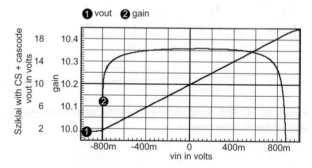

Figure 2x.41. Transfer function and gain of the circuit of Figure 2x.40.

Well, that doesn't seem to have made much improvement – still 0.1% for half scale. It may be we've suffered an offsetting degradation from operation at such a low $V_{CE}=1.4$ V. Another way to combat the Early effect, at least for small signals, is to balance the load resistances, so we tried a 5k resistor in the Q_1Q_3 collector, in place of the cascode. Aha, a bit better, we get 0.07% gain falloff at half of full swing (not shown).

2x.4.10 Caprio's quad differential amplifier, with cascode

Caprio's quad[13] is a unique configuration (Figure 2x.42) in which normal changes in V_{BE} are canceled. Here's how to understand this diabolically clever circuit: the voltage drop from Q_1's base to the right-hand side of R_3 is the sum of two base–emitter drops, one corresponding to the left-hand collector current, and the other corresponding to the right-hand collector current. But exactly the same statement goes for the voltage drop from Q_2's base to the left-hand side of R_3. So, even when the collector currents become unbalanced (from an input signal excursion), the input signal differential is faithfully conveyed across the gain-setting resistor R_3. In other words, the exact input signal voltage appears across R_3, without distortion caused by changing transistor base–emitter voltages. Cute!

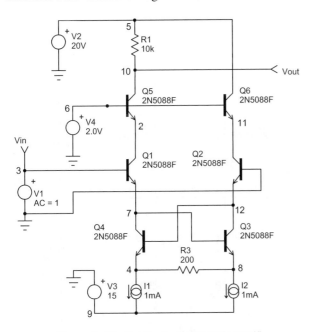

Figure 2x.42. Eighth circuit: "Caprio's quad."

Caprio's quad is limited to operation with small input signals, say under 400 mV, to avoid saturating transistor Q_3 or Q_4. In our circuit the gain has been set to 50, so that ± 160 mV can drive the output over a swing of ± 8 V. Figure 2x.43 shows the results.

The gain reduction is 0.36% at 4 V and 16 V outputs, and 0.12% at 6 V and 14 V. This is three times better than the 0.4% we observed for the Sziklai-connected amplifier

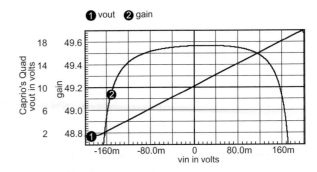

Figure 2x.43. Transfer function and gain of the circuit of Figure 2x.42, with inputs from -200 mV to $+200$ mV.

at the same gain. Barrie Gilbert points out[14] that Caprio's quad has another issue to worry about, namely a negative input resistance, leading to instabilities with slightly reactive input sources.

With lower gain and high input signals, the circuit exhibits a phase inversion when the input is overdriven. Figure 2x.44 shows what happens when you drive a Caprio quad designed for a gain of 25 with a 1 Vpp input triangle. It has a nice linear range, as we saw, but gets into big trouble if overdriven.

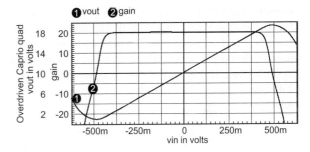

Figure 2x.44. Overdriving Caprio's quad.

2x.4.11 Caprio's quad with folded cascode – I

The Caprio quad with cascode has admirably low distortion (this is open-loop, mind you!), owing to the quad's cancellation of V_{BE}s in the input stage, and the cascode's suppression of Early effect. But we can do even better: by "folding" the cascode we can take advantage of the full rail-to-rail supply voltage range. Our first try is a gain-of-100 circuit (Figure 2x.45), in which the *pnp* output stage

---

[13] R. Caprio, "Precision differential voltage–current convertor," *Electron. Lett.*, **9**, 147–148 (1973).

[14] Toumazou, ed., *Analogue IC Design: The Current-Mode Approach*, Peregrinus Ltd. (1990), page 72.

is biased to mid-supply when Q_1's quiescent collector current of 1 mA leaves 0.5 mA of emitter current for the output cascode Q_3.

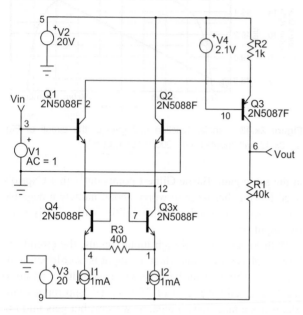

Figure 2x.45. Ninth circuit: Caprio's quad with folded cascode.

Running the SPICE engine, we get a plot of the nearly rail-to-rail output swing, but with disappointing linearity (Figure 2x.46). The gain drops off markedly at the negative portion of the output swing; it's down about 20% from the peak gain when V_{out} is -18 V (the diagonal trace is V_{out}, and the curvy trace is the gain. (Trace 7, at top, is a preview of what comes next.)

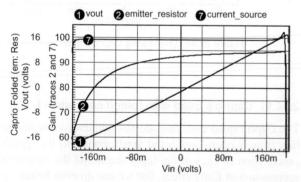

Figure 2x.46. Transfer function and gain of the circuit of Figure 2x.45.

The reason is this: although Q_1 sinks a current that is accurately linear with input signal, the residual current provided by R_2 is *not* constant (because the V_{BE} of Q_3 varies with collector current, approximately -60 mV/decade).

So, for example, when the output is close to the negative rail, Q_3's V_{BE} is reduced, which increases R_2's residual current and therefore the signal V_{out}. The folded cascode has greatly degraded the distortion of Caprio's quad! Something needs to be done.

2x.4.12 Caprio's quad with folded cascode – II

Not to fret – there's an easy fix. Just replace R_2 with a current source, chosen to bias V_{out} at mid-supply after allowing for Q_1's quiescent current. Figure 2x.47 shows the circuit (where we've been lazy and used a current source symbol; in reality you'd use a BJT current source, or current mirror with small emitter resistors; we've also replaced the pair of redundant current sinks with a functionally identical single 2 mA sink).

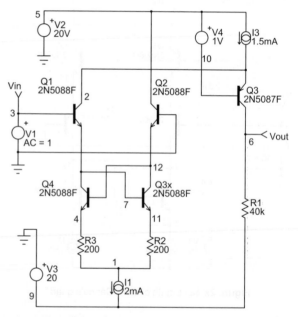

Figure 2x.47. Tenth circuit: Caprio's quad with folded cascode and emitter current source.

This circuit change produces excellent output linearity (trace 7 plots the gain versus input voltage), while preserving the nearly rail-to-rail output swing. Look at the gain plot in Figure 2x.48, where the vertical scale has been expanded to show the residual variation of gain with output swing – the peak-to-peak gain variation is less than 0.4% over the output range of -19 V to $+17$ V.

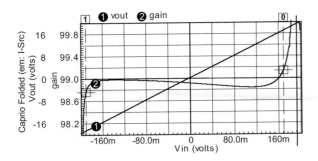

Figure 2x.48. Transfer function and gain of the circuit of Figure 2x.47.

2x.4.13 Measured distortion

As we remarked in §2x.4.8, the peak "gain-reduction distortion" figures represent the worst deviation from perfect linearity over the swing, and so they're quite a bit larger than the usual measure of audio amplifier distortion – rms total harmonic distortion (THD).

Knowing the transfer function, it's possible to calculate the distortion numerically. But it's more fun to breadboard several of these circuits and *measure* their THD on the bench. It's also a chance to escape the lure of computer modeling, and get a grip on reality.

We did that, with the results shown in Figure 2x.49, where the logarithmic vertical axis tends to downplay the substantial range of measured distortions. Look closely, and you'll see that emitter degeneration produces reasonable distortion values in the differential amplifier, but the Caprio quad with cascode does an order of magnitude better; and in the current-source-fed folded cascode it reigns triumphant, with an admirably low distortion of just 0.01% (especially for an amplifier without linearizing negative feedback) even at 25 Vpp output swing. Reality nicely follows theory!

2x.4.14 Wrapup: amplifier modeling with SPICE

Our tour of distortion in BJT amplifiers has been fun, and easy. It let us explore the properties of various configurations, making changes with both little effort and great reward.

But, a *warning*: SPICE takes its models literally, and therefore goes seriously off the rails when the models are inaccurate. We saw this here, with the poor generic QN5088 model we tried first. Similar problems are found with MOSFET models, many of which fail completely in modeling the important "subthreshold region" (§3.1.4A).

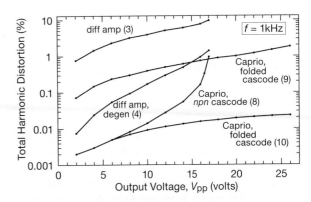

Figure 2x.49. Measured harmonic distortion at 1 kHz of the open-loop amplifier circuits of Figures 2x.28, 2x.32, 2x.42, 2x.45, and 2x.47. The numbers in parentheses refer to the *n*-th circuit iteration).

It's essential to validate your SPICE models with real-world behavior before placing trust in simulations.

Readers interested in delving deeply into audio amplifier design will find inspiration in the excellent *Audio Power Amplifier Design, 6th edition* by Douglas Self (Focal Press, 2013).

2x.5 Early Effect and Early Voltage

In §2.3, where we supplemented our first simple transistor model (current amplifier) with the Ebers–Moll transistor model (voltage-to-current amplifier, or *transconductance amplifier*), we briefly introduced the Early effect. We did that because it is an important deviation from the ideal transconductance amplifier: collector current is primarily programmed by applied base voltage (V_{BE}), but it is not entirely independent of collector voltage (V_{CE}). This has important consequences for circuits such as current sources (where it limits the output impedance) and common-emitter amplifiers with active loads (where, for the same reason, it limits the maximum voltage gain).

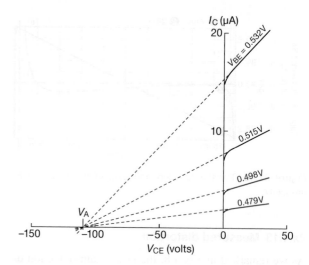

Figure 2x.51. The collector current curves, extrapolated backward, meet the voltage axis at a common point V_A (the Early voltage). I_{C0} signifies the extrapolated collector current at zero collector voltage. These are measured curves for a 2N5963 *npn* "superbeta" small-signal transistor (higher beta version of the 2N5962).

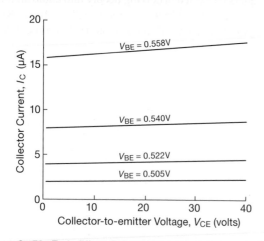

Figure 2x.50. Early Effect: Collector current, programmed primarily by applied base–emitter voltage, varies slightly with collector–emitter voltage. These are measured curves for a 2N3904 *npn* small-signal transistor..

2x.5.1 Measuring Early effect

Figure 2x.50 shows measured curves of collector current versus voltage (I_C vs V_{CE}) for several values of applied base–emitter voltage (V_{BE}). The curves show the Ebers–Moll exponential increase of I_C with V_{BE}, easily seen as the approximate doubling of collector current with each 20 mV step in applied base voltage. Of course, I_C depends also upon temperature; but that is not shown in these plots, for which the temperature was carefully held constant.

The curves show additionally that collector current depends somewhat on collector *voltage* – the curves slant slightly upward. This is the Early effect, caused by a modulation of effective base width by the collector potential. To the eye, it appears that the degree of slant increases as we go up the family of curves. In fact, if you were to extrapolate the curves to the left, they would all intersect the horizontal axis at a common point, V_A, known as the *Early voltage* (the minus sign is ignored); see Figure 2x.51. Typically V_A ranges from a few tens of volts to a few hundred volts. Low-voltage and high-beta transistors tend to have larger Early effect (lower Early voltage), as can be seen in Figure 2x.52, where we compare measured curves of I_C versus V_{CE} for three *npn* transistors spanning a wide range of current gain and maximum rated voltage. Early effect is also quite a bit more prominent in *pnp* transistors, compared with their *npn* brethren, as seen in the comparison of the 2N3904 (*npn*) and 2N3906 (*pnp*) complementary pair (Figure 2x.53).

The Early voltage, although very useful, is not usually specified on transistor datasheets; sometimes you'll get a set of characteristic curves like Figure 2x.50 or 2x.51, from

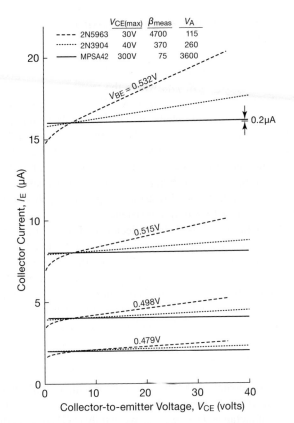

Figure 2x.52. Low-voltage and high-beta transistors exhibit greater Early effect, as seen in these measured data for three exemplary *npn* small-signal transistors. Approximate doubling of collector current corresponds to an $\sim$18 mV increase in base–emitter voltage (shown for the superbeta 2N5963, but similar for the other two transistors).

which you can estimate V_A.[15] Another source of information is the SPICE-model parameter "VA."[16]

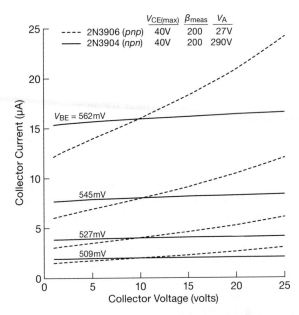

Figure 2x.53. Early effect is far greater in *pnp* than in the analogous *npn* transistor, as seen in this comparison of one sample each of the 2N3904 and 2N3906. The base voltages (shown only for the 2N3904) were chosen to produce the same set of currents (2 μA, 4 μA, 8 μA, 16 μA) in both transistors at $V_{CE}=10$ V. Voltages and currents are negative for the *pnp* 2N3906.

2x.5.2 Some Early effect formulas

The Early effect's linear growth of I_C with V_{CE} is most simply expressed as a modified Ebers–Moll equation

$$I_C = I_S \left(e^{V_{BE}/V_T} \right) \left(1 + \frac{V_{CE}}{V_A} \right)$$

which extrapolates to zero collector current at $V_{CE} = -V_A$, as it should; note that we have omitted the insignificant -1 term from the Ebers–Moll equation, assuming operation at currents significantly above the reverse leakage current I_S. As before, $V_T = kT/q = 25.3$ mV at room temperature.

We used an alternative formulation of the Early effect when we introduced it in §2.3.2; there we spoke instead of the change in V_{BE} required to maintain constant collector current as the collector voltage is changed. It's a simple exercise in differentiation to show that this "feedback factor" η (as Early termed it, and for which he used the symbol μ_{ec}) is given by

$$\eta \equiv \frac{\partial V_{BE}}{\partial V_{CE}} = \frac{V_T}{V_A} \frac{I_{C0}}{I_C} = \frac{V_T}{V_A + V_{CE}},$$

where I_{C0} is the "zero-voltage" collector current (see Fig. 2x.51), and therefore that the change in V_{BE} required to maintain constant collector current with a change in col-

---

[15] We've measured Early voltages for a collection of useful transistors, listed in Table 8.1 in AoE3. From the table values it is evident that Early voltages for *npn* transistors are typically in the range of 70 to 400 volts, while for *pnp* transistors they are in the range of 25 to 200 volts. The Early voltage for high-voltage parts like the MPSA42 and MPSA92 (*npn* and *pnp* respectively, both 300 V) are much higher (1.2kV and 350 V), but you pay the price in very much lower current gain ($\beta \approx 50$ to 75).

[16] But a caution: we have found some published models to be excessively conservative, for example a BC847 model (a high-beta *pnp*) assigning $V_A=10$ V when a more realistic value is 30 V; the latter is already hard enough to deal with in a design.

lector voltage is

$$\Delta V_{BE} = -\eta \Delta V_{CE}$$

as in §2.3.2. For collector voltages that are small compared with the Early voltage, this reduces to $\eta \approx V_T/V_A$. With an Early voltage of ~100 V (typical for an *npn* transistor), the coefficient η (the ratio of V_T to $V_A + V_{CE}$) is approximately $\eta \approx 2.5 \times 10^{-4}$ at small collector voltages; a four volt increase of collector voltage requires a corresponding 1 mV drop in V_{BE} to maintain constant collector current. Equivalently, at constant V_{BE} that same 4 V increase in V_{CE} would result in roughly a 4% increase in collector current. A *pnp* transistor with characteristically lower Early voltage, say 25 V, would show a larger effect, a 16% increase of I_C over a 4 V collector swing, or 80% for a 20 V swing; this would degrade greatly the performance of a simple current mirror (Figure 2.58), and argues forcefully for the use of voltage-clamping configurations like the Wilson mirror (Figure 2.61) or the cascode.

Exercise 2x.2. Derive the above expression for η. Hint: implicit differentiation is helpful.

2x.5.3 Consequences of Early effect: Output resistance

An easy way to understand the circuit consequences of Early effect is to consider the transistor's collector output resistance: a collector current that increases linearly with collector voltage is equivalent to a resistance in parallel with an ideal transistor (i.e., one that has constant output current independent of voltage, or infinite output resistance). That parallel resistance r_o is different for each curve in Figure 2x.51, of course, since it is equal to the inverse of the slope at the operating point (I_{C1} at V_{CE1}): $r_o = V_A/I_{C0} = (V_A + V_{CE1})/I_{C1}$. For example, a transistor with V_A=100 V, running at 1 mA, has an output resistance of $r_o \approx 100$ k; for a collector current of 0.1 mA the corresponding figure is 1 MΩ;[17] that is, the output resistance r_o is inversely proportional to collector current. You sometimes see things expressed in terms of output *conductance* (g_o), the inverse of resistance: $g_o = 1/r_o$; the latter is proportional to the collector current operating point I_{C1}. Thinking of Early effect in terms of collector output resistance, it's easy to derive two useful results.

A. Maximum single-stage voltage gain

We saw earlier in the chapter that the maximum voltage gain G_V of a common-emitter amplifier was achieved with grounded emitter (no external R_E), and was given by $G_V = g_m R_C = R_C/r_e$, where the transconductance g_m is given by $g_m = 1/r_e = I_C(mA)/25\Omega$. For a grounded-emitter stage biased to half the positive supply voltage, for example, the gain is $G_V = 20V_{CC}(volts)$, as shown in Exercise 2.15 in AoE3. For this calculation we had assumed that the transistor was an ideal transconductance amplifier, i.e., that there was no Early effect; this is reasonable, because the collector load resistor is typically much smaller than the transistor's output resistance r_o.

But what happens if the collector load is not a resistor (a "passive load"), but a current source (an "active load"), with its very high impedance?[18] Now the unloaded gain is given by $G_V = g_m r_C$, where r_C represents the parallel combination of the amplifier's output resistance r_o and the current source resistance r_{cs}. If we assume a perfect current source ($r_{cs} = \infty$), then the single-stage voltage gain is

$$G_V = g_m r_o = \frac{I_C}{V_T} \frac{V_A + V_{CE}}{I_C} = \frac{V_A + V_{CE}}{V_T},$$

or approximately 4,000 for a transistor with an Early voltage of 100 V.[19] Writing this expression in terms of output conductance, $G_V = g_m/g_o$. We call this "GMAX" in the context of JFETs; values are listed in Tables 3.1 and 3x.1, with discussion in §§3.3.2 and 3x.4.

Other things being equal, a transistor with a higher Early voltage is a better approximation to an ideal transconductance amplifier. It permits higher single-stage voltage gain with an active (current-source) load; it exhibits better power-supply rejection; and, when connected as a current source, it is closer to the ideal of infinite output impedance (i.e., no change of output current with swings of output voltage). You might consider Early voltage by itself a good figure of merit. However, transistors optimized for high current gain (beta) generally have lower Early voltage, owing to their thinner base regions (which suffer a larger relative variation with changing V_{CE}).[20] A figure of merit that

---

[17] Although V_{CE} appears in the equation for r_o, the output resistance shows little change with collector voltage for a given bias condition, because the collector current moves up and down along a nearly constant slope.

[18] Such an arrangement is not bias-stable without feedback; but current-source active loads are common in operational amplifiers.

[19] You sometimes see this result approximated as $G_V \approx V_A/V_T$, though a more accurate calculation requires a multiplying factor of I_C/I_{C0}, which is equivalent to the result above.

[20] High beta also correlates with lower breakdown voltages, particularly evident in the so-called "superbeta" transistors; for example, the input-stage superbeta transistor pair in the OP-97 amplifier is operated at a V_{CE} equal to one diode drop!

is sometimes used is the product of Early voltage and current gain: FOM=βV_A.

B. Current-source output impedance

As we saw in the last section, the impedance looking into the collector of a transistor with fixed applied V_{BE} is just

$$r_0 = \frac{V_A}{I_{C0}} = \frac{V_A + V_{CE}}{I_C}.$$

This is the output impedance of a simple current mirror (Fig 2.55), since the "programming" transistor serves only to apply the appropriate (and temperature-compensated) V_{BE} to its identical companion, the current-sourcing transistor. With typical Early voltages of $\sim 100\,$V, you can therefore expect to get output impedances of order 100k for a 1 mA current mirror. Ideally, of course, you'd like it to be much higher, since this figure corresponds to a variation of 1% per volt of output voltage change (this is, not coincidentally, the ratio of 1 volt to the Early voltage).

One solution is to use a pair of emitter degeneration resistors (Fig. 2.60). This makes the output current less sensitive to variations in V_{BE}, for the same reason it linearizes and makes stable biasing possible with a common-emitter amplifier. As an example, if we chose emitter resistor values such that there is one diode drop across them at the operating current (e.g., 600 Ω for a 1 mA current mirror), then it's easy to estimate that a 1 volt change at the output now has a much smaller effect: the Early effect change in V_{BE} (at constant current) is

$$\Delta V_{BE} = \eta \Delta V_{CE} = \frac{V_T}{V_A + V_{CE}} \Delta V_{CE},$$

about 0.25 mV per volt variation of collector voltage for V_A=100 V; that causes a current change of about 0.4 μA per volt, or 0.04% per volt of output voltage change.[21] This is 25 times better than the bare current mirror without emitter resistors, though one price you pay is the inability to bring the collector load all the way down to the emitter rail.

This raises an interesting question: Is there an "effective Early voltage" that characterizes an emitter-degenerated transistor? There is, and its value is

$$V_A(\text{effective}) = V_A \left(1 + \frac{I_E R_E}{V_T} \right),$$

which works out to about 2.5 kV in the above example; this corresponds to the output coefficient of current change

of 0.04%/V we found above. The factor of improvement is approximately the ratio of the voltage drop across the emitter resistor to 25.3 mV (V_T at room temperature).

A second solution to Early-effect reduction of collector output impedance is to use a transistor configuration that prevents changes of collector voltage – the cascode (§2x.3). Also see Figure 2.84 in the main book, and (for JFETs) §3.2.3C and Figure 3.27.

---

[21] This small current change justifies the approximation of constant current that we used in its derivation; those inclined to excessive precision may iterate, or solve a transcendental equation, arriving ultimately at a result that differs in no significant way.

2x.6 The Sziklai Configuration

In AoE's Chapter 2 we introduced the two basic beta-boosting configurations: the Darlington (Figure 2.74) and the Sziklai (Figure 2.77, sometimes called "complementary Darlington"). When configured without any base resistors, they share the property of exhibiting a combined current gain β equal to the product of the respective betas at each transistor's operating current: $\beta = \beta_1 \beta_2$; they also share the less desirable property of a saturation voltage no smaller than one diode drop.

In practice a resistor is usually added across the base–emitter terminals of the second transistor in either configuration, as in Figures 2.75 and 2x.54A, to improve the turn-off speed, reduce leakage current, and stabilize the base–emitter drop for currents above that at which Q_2 is brought into conduction. These advantages come at the cost of reduced beta, particularly at low currents (where the drop across R_1 does not bring Q_2 into conduction), but also at higher currents (where R_1 robs current that could have been amplified by Q_2). Assume for the following discussion that the Sziklai (or Darlington) includes the base resistor(s); the resistor value is normally chosen to bring Q_2 into conduction at a small fraction of the full operating current, in order to get the benefit of the composite configuration over most of the operating range.

2x.6.1 Two-transistor "standard" Sziklai

Figure 2x.54 shows base current, current gain, and base–emitter drop, on a logarithmic current axis spanning collector current over four orders of magnitude (0.1 mA to 1 A). For the moment look only at the traces marked "A," which correspond to the standard 2-transistor Sziklai of Figure 2x.54A. With its base resistor of 1 kΩ, the onset of conduction in Q_2 begins, as expected, at about 0.6 mA (i.e., when the drop across R_1 is about 0.6 V). The plots clearly reveal the flattening of V_{BE} with higher current; the Sziklai's relative constancy of V_{BE} is due to Q_1's nearly constant operating current I_{C1} of 0.6 mA (the current through R_1), which holds up until an output current I_{C2} for which Q_2's limited beta requires a base current comparable to I_{C1}. The 2N4403 used in this circuit has a typical beta of 200, so we expect an upturn in V_{BE} of 18 mV (from a doubling

of I_{C1}) at about $I_{C2} = 100$ mA, in good agreement with the plotted SPICE simulation.

The analogous Darlington circuit, with the same transistor types and resistor values, does not do as well, as can be seen in the "A" plots of Figure 2x.55. Not only is the base–emitter drop roughly double that of the Sziklai; it also increases with at least the usual 60 mV/decade Ebers–Moll characteristic, because the overall V_{BE} is the sum of that of Q_1 and Q_2, and the latter's base–emitter drop corresponds to nearly the full output current. Put another way, the Sziklai sidesteps this pitfall by exposing only Q_1's base–emitter drop to the outside world. The smaller and more constant V_{BE} of the Sziklai (compared with Darlington) was exploited to bring about stable biasing in the audio amplifier push–pull power output stage of Figure 2.78 in AoE3. Another favorite application of the Sziklai exploits the complementary polarity of Q_2 – so you can rig up a push–pull output stage using only *npn* power transistors (Darlington on top, Sziklai on the bottom, as illustrated in Figure 2.63 of the second edition).

Some additional points:

Transitions The Sziklai with its Q_2 base resistor exhibits a changeover in characteristics, from $g_m = g_{m1}$ and $\beta = \beta_1$ at low currents, to $g_m = g_{m1}g_{m2}$ and $\beta = \beta_1\beta_2$ at high currents.[22] This behavior is no different from that of the Darlington with analogous base resistor.

Emitter output resistance The Sziklai's output resistance seen at the emitter, r_e (i.e., as an emitter follower) is far lower than that of a single transistor (or Darlington). This can be seen from the greatly reduced change in V_{BE} versus collector current at higher currents. To put this in equations, the output resistance of a single transistor $r_e = 1/g_m = 25\Omega/I_{C(mA)}$, and somewhat higher for a Darlington, whereas for the Sziklai it tends toward some factor "x" times smaller, where x is the ratio of the I_{C2} output current to the relatively constant current $I_{C1} = V_{BE}/R_1$. For example, we get $r_e = 0.5\,\Omega$ (instead of $2.5\,\Omega$) at 10 mA.

2x.6.2 Three-transistor "enhanced" Sziklai

If you're willing to squander another transistor (Figure 2x.54B), you can do even better. At high currents this circuit benefits from the current gain of the third transistor; better still, we get far better constancy of V_{BE}, as seen

---

22 As long as $R_1 \gtrsim 0.7\beta_2/I_{C2}$. For example, let's say we have an amplifier with 1 A full output current, and we want to maximize the Sziklai's beta for output currents above 100 mA – then we would choose $R_1 = 680\,\Omega$ or thereabouts.

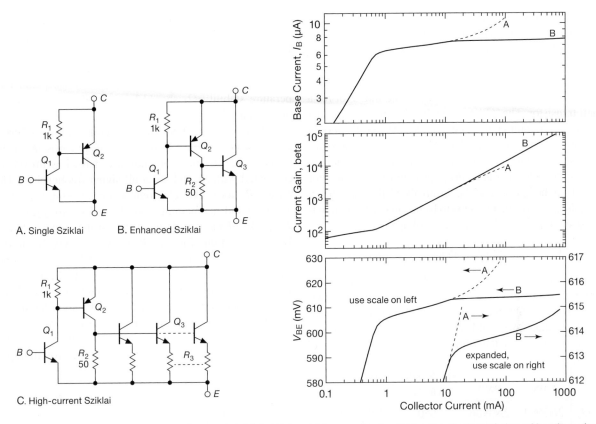

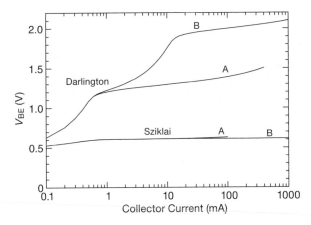

Figure 2x.54. The Sziklai configuration of cascaded BJTs (A) has high current gain, β (like the Darlington), but with only a single-transistor base–emitter drop V_{BE}. Adding a third transistor (enhanced Sziklai, B) stabilizes the collector current of Q_2, thus stabilizing V_{BE} over current. The plots were generated with SPICE, with 2N3904, 2N4403, and ZTX618 transistors for Q_1–Q_3, respectively.

Figure 2x.55. Base–emitter voltage drops for Darlington pairs and triples, analogous to the Sziklai circuits A and B in Figure 2x.54, and with the same resistor values. The base–emitter drop of a Darlington pair (A) reaches two diode drops once the second transistor begins conducting; for the Darlington-connected triple (B) you get two such transitions, reaching three diode drops.

in the corresponding "B" plots. Here's how to understand the magic: First note that, for output currents large enough to bring Q_3 into conduction (about 10 mA), R_2 shields Q_2 from having large variations in collector current (just as R_1 did, in the 2-transistor Sziklai). So Q_2's base–emitter drop hardly varies, thus keeping R_1's contribution to Q_1's collector current constant. Next note that R_2 is sized to put Q_2's relatively constant collector current around 10 mA, thus a base current of roughly 0.05 mA ($\beta \sim 200$), negligible compared with R_1's 0.6 mA. Aha, that does it, because that the thing we call the "V_{BE}" of the Sziklai is really the base–emitter drop of the input transistor Q_1 alone; and Q_1 lives in a world of delightfully constant collector current.

As with the 2-transistor Sziklai, changes in the composite V_{BE} are much smaller than that of a single transistor operating at the same current; that is, the output resistance of a Sziklai emitter follower compares favorably with that of an analogous 3-transistor Darlington. A single transistor has an output resistance falling inversely with current, down to $0.025\,\Omega$ for an ideal BJT at 1 A, driven from a low-impedance signal source, whereas in Figure 2x.54B the 3-transistor Sziklai composite has $r_0=0.0023\,\Omega$ at 1 A, thus $10\times$ better.

The curves in Figure 2x.54 extend all the way to a collector current of an amp, which may sound like a lot for a small transistor. The ZTX618 can handle an amp OK (it's rated to 3.5 A continuous collector current), but it's of only modest size (it comes in a TO-92 small-signal transistor package), and there's some small amount of internal resistance that adds a voltage drop (proportional to collector current) to the V_{BE} seen at the terminals. Calling that internal resistance $\mathcal{R}_E$ (there seems to be no generally acknowledged symbol[23]), the base–emitter voltage is

$$V_{BE} = V_{BE}(@1mA) + V_T \log_{10}\frac{I_C}{1mA} + I_C\mathcal{R}_E$$

where the last term represents an ohmic high-current correction to the ideal Ebers–Moll law given by the first two terms. You can infer the value of $\mathcal{R}_E$ from the high-current end of datasheet curves of $V_{CE(sat)}$ versus I_C. For the plots in Figure 2x.54 we used the following values for $V_{BE}(@1mA)$ and $\mathcal{R}_E$:

	NPN	PNP	V_{BE} @ 1mA (mV)	$\mathcal{R}_E$ (Ω)
Q_1	2N3904	2N3906	650	1.0
Q_2	2N4401	2N4403	620	0.3
Q_3	ZTX618	ZTX718	550	0.04

It's interesting to see that the 3-transistor Sziklai has an emitter output resistance r_0 about $17\times$ lower than the output transistor's $\mathcal{R}_E$. Feedback works!

2x.6.3 Push–pull output stage: a Sziklai application

Fig. 2x.56 shows an example of a push–pull BJT audio output stage that uses only *npn* power transistors, taking advantage of the Sziklai configuration.[24] The Sziklai pair Q_5Q_6, though constructed with *npn* power transistor Q_6, acts like a *pnp* power transistor; this is a good thing, given the generally better performance (and availability) of *npn* devices.

Transistor Q_2 biases the bases of the push–pull stage into

class-AB operation (some quiescent current at crossover), with a voltage difference $\Delta V=V_{BE}(1+R_3/R_4)$; this is adjusted to be greater than the push–pull pair's $3V_{be}$ combined base–emitter drops (two for the Darlington, one for the Sziklai) by an amount $I_Q(R_7+R_8)$. For acceptable temperature stability Q_2 should be thermally coupled to the output-stage heatsink.

In this circuit, C_2 bootstraps the collector load of driver transistor Q_1, keeping a constant voltage across R_2 as the signal varies (C_2 must be chosen to have low impedance compared with R_1 and R_2 at all signal frequencies). That makes R_2 look like a current source, raising Q_1's voltage gain and maintaining good base drive to Q_3, even at the peaks of the signal swing. When the signal gets near V_{CC}, the junction of R_1 and R_2 rises above V_{CC} because of the stored charge in C_2. In this case, if $R_1=R_2$ (not a bad choice) the junction between them rises to 1.5 times V_{CC} when the output reaches V_{CC}. This circuit has enjoyed considerable popularity in commercial audio amplifier design, although a simple current source in place of the bootstrap is superior, since it maintains the improvement at low frequencies and eliminates the undesirable electrolytic capacitor.

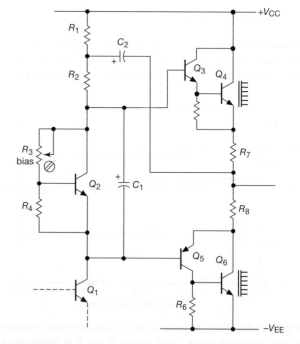

Figure 2x.56. "Quasi-complementary" push–pull output stage with only *npn* power transistors.

---

[23] SPICE uses "RE," but we've chosen the calligraphic $\mathcal{R}_E$ to distinguish from R_E, a symbol used often to designate an external resistor in the emitter circuit.

[24] Devised by H.C. Lin in 1956. Interestingly, it was the unavailability of *npn* power transistors that motivated his design, in which he used germanium *pnp* power transistors for Q_4 and Q_6 (with corresponding polarity changes).

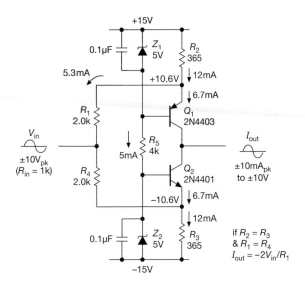

Figure 2x.57. Fast bipolarity current source. See §4x.24 for a high-voltage elaboration of this basic circuit, aided and abetted by the magic of op-amps.

2x.7 Bipolarity Current Mirrors

Here we continue the discussion of current sources and current mirrors, begun in AoE3's §§2.2.6 and 2.3.7, respectively, with an example that exploits the Sziklai's property of stable base–emitter voltage drop.

2x.7.1 A simple high-speed bipolarity current source

To get warmed up, look first at the simpler high-speed bipolarity current source circuit in Figure 2x.57. The circuit is a push–pull configuration, with both transistors in conduction ("class-A"), in which the output current is controlled by increasing the current in one transistor while decreasing the current in the other. Complementary transistors Q_1 and Q_2 are in a symmetrical "folded cascode" configuration, biased at equal quiescent currents. Assuming the emitter voltages do not change, a voltage V_{in} applied to the input unbalances the currents by $\Delta V_{in}=V_{in}/R_1+V_{in}/R_4$, i.e., by $V_{in}/R_\parallel$, here $V_{in}/1\mathrm{k}\Omega$.[25] That unbalance appears as the output current I_{out}. The circuit is fast, because the output capacitance is low (about 9 pF), and the common-base (cascode) is not bedeviled by Miller effect. With the values shown, a step command to $I_{out}=10\,\mathrm{mA}$ would cause the circuit's output to slew at $dV/dt=I/C\approx1000\,\mathrm{V}/\mu\mathrm{s}$ into an open circuit, if the speed were limited only by the circuit's self-capacitance (i.e., if no other mechanisms act to limit the speed).

Of course, when estimating the ultimate output step response you need to worry about the additional capacitance of the load itself. And for large current steps (where the calculated slew rate approaches numbers like 1 V/ns) you need to consider the transistor's limited bandwidth, f_T. For inputs other than a step, it may be the input signal that sets the output slew rate. A good way to think about output speed, then, is this: (a) the output response is limited first by the input waveform, (b) next by the transistor's f_T, and (c) finally by the capacitance at the output node. The latter is a slew rate $S=I/C$ in the high resistance load limit, or

---

[25] The quiescent bias current of the output transistors must be at least half the maximum output current, to ensure current flow at all output currents.

---

an RC time constant for a finite resistive load shunted by circuit and load capacitances.

The measured traces in Figure 2x.58 shows these regimes nicely. We rigged up the circuit of Figure 2x.57, attached various loads, and drove it with a pulse that created an output current step going from a negative current to a positive current of equal magnitude. The top trace shows what happens with a 10 kΩ load resistor, probed with a 1 pF FET 'scope probe. Here the slew rate is limited by the current available to charge the $\sim10\,\mathrm{pF}$ total capacitance; as the waveform passes zero volts the slew rate is $S=I/C=50\mu\mathrm{A}/10\mathrm{pF}$, or 5 V/$\mu$s. The overall waveform is the familiar RC charging curve, because the resistor robs charging current from the parallel capacitance as the waveform rises (as it does in the simplest *series RC* subjected to a *voltage* step).

The middle trace shows the situation with a highly capacitive load (1000 pF), here connected in parallel with a base–emitter transistor junction that acts as a $\sim7\,\mathrm{V}$ zener. The slew rate is a constant $S=I/C$, or 1 V/μs, clamped at $-0.8\,\mathrm{V}$ by base–emitter diode conduction, and at $+7\,\mathrm{V}$ by base–emitter reverse breakdown. The bottom trace shows what happens with a large current step into a low-impedance load (doubly terminated 50 Ω coax); the waveform slews at 50 V/μs, far less than the ideal slew rate of 1000 V/μs estimated earlier. That is because the input step comes from a pulse generator of 5 ns risetime, echoed nicely in the output current waveform; with a much faster pulse generator we would expect to see the ultimate slew

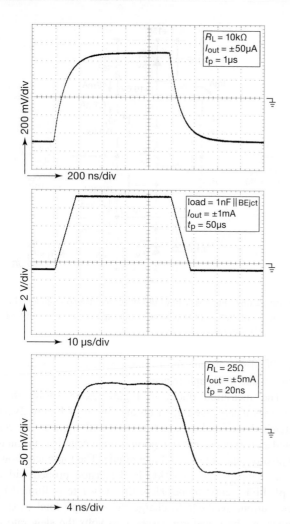

Figure 2x.58. Measured output waveforms for the circuit of Figure 2x.57, when driven by inputs that command a symmetrical output current step as indicated. Note different loads, current step amplitudes, and *xy* scale factors.

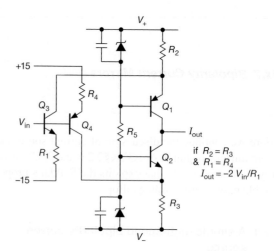

Figure 2x.59. Bipolarity current source with reduced drive current, capable of operation at high voltages.

A. Reducing input current

The first issue is nicely addressed with the circuit of Figure 2x.59, where a pair of input transistors, configured as current source/sinks, convert the input programming voltage to a pair of currents to drive the folded cascode as before; this reduces the input current by the beta of the input transistors. In Chapter 4x (§4x.24) we'll see how to use op-amps to eliminate input current entirely.

B. Operating at higher voltages

Note that the supply voltages V_+ and V_- in this circuit can be quite a bit larger than ± 15 V, as long as Q_1–Q_4 voltage and power ratings are respected, to create a high-voltage bipolarity current source.

Figure 2x.60 shows several variations on this high-voltage theme. Circuits A and B are elaborations of the input circuit, whose output currents connect to the folded cascode output circuit of Figure 2x.59: adding a cascode (A) keeps power dissipation low on the critical current-setting transistors Q_3Q_4 (which can be low-voltage high-β types, unlike Q_5 and Q_6), and insulates them entirely from Early effect; as before, $I_1 = V_{in}/R_1$. A problem with this circuit is excessive dissipation, solved by replacing each resistor R_1 with a divider pair, thus permitting separate gain and class-A quiescent output-stage bias (in the gain equation R_1 is replaced by the parallel resistance of the divider).

Circuit C is a variation on the folded cascode output stage (only the positive-rail half is shown). The current mirror configuration (Q_8Q_9) produces a current gain of $I_{out}^+/I_1 = R_2/R_3$ (compared with unity for the simple folded cascode), thus (when combined with circuits A or B) an

rate limited by the f_T of the transistors, but there is no hint of that in these measured waveforms.

Now for some drawbacks of this circuit: (a) The input signal has to supply the full output current; and (b) the circuit is imprecise (even if trimmed by adjusting R_2, say), because the base–emitter voltage drop of the transistors *does* change with output current (according to Ebers–Moll), thus creating a nonlinearity; put another way, the voltage across R_2 or R_3 is not constant (as we assumed), thus their contributions to the output current vary with input voltage.

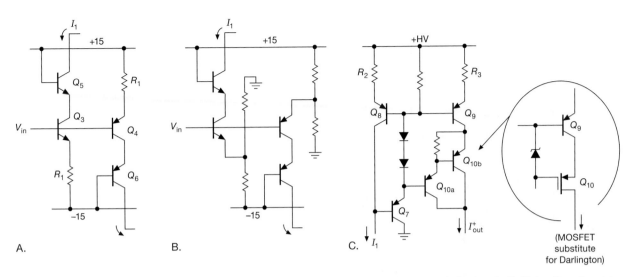

Figure 2x.60. Variations on the high-voltage bipolarity current source of Figure 2x.59. A: cascode front-end with high-voltage transistors Q_5 and Q_6. B. separate resistors allow independent programming of gain and quiescent current. C. positive-rail output stage with current gain R_2/R_3, with MOSFET alternative to Darlington BJT.

output current $I_{out}^+=(V_{in}/R_1)(R_2/R_3)$. In this circuit the mirror's base current derives from *pnp* follower Q_7, so it doesn't rob significant programming current; Q_7 clamps the input current node, suppressing Miller effect at the collector of Q_5. The cascode output Q_{10} and *pnp*-boosted input forces the mirror pair to operate at low (and essentially constant) V_{CE}, suppressing Early effect; so Q_8 and Q_9 can be low-voltage high-beta parts. The output transistor $Q_{10a,b}$ can be a power Darlington, helpful in dealing with characteristically low *pnp* beta. The inset shows a power MOSFET substitute for the Darlington, biased with a zener to accommodate the MOSFET's larger V_{GS} (compared with the Darlington's $2 \times V_{BE}$); power MOSFETs are nice, but the large input capacitance C_{iss} may cause extended turn-off time. See Chapters 3 and 3x for extensive discussion of MOSFETs.

We'll have more to say on the subject of high-voltage bipolarity current sources in §4x.24.

2x.7.2 Precision bipolarity current source with folded cascode

Figure 2x.61 shows a bipolarity current source with improved accuracy and linearity compared with that of the simple circuit of Figure 2x.57. As before, the output current is programmed by V_{in} and R_{in} alone. Similarly numbered Q_1 and Q_2 are situated as before, but assisted by Q_7 and Q_8 to form Sziklai pairs. We've also replaced the emitter resistors with current sources (Q_5 and Q_6). The stable

base–emitter drop of Q_1 and Q_2 largely eliminates the nonlinearity of the simpler circuit (Figure 2x.57), and the use of stable current sources instead of emitter resistors puts a final nail into the coffin of nonlinearity.

Because this circuit aspires to precision, we added a trimmer to Q_6's emitter resistor, to balance the quiescent currents of Q_5 and Q_6; when balanced to, say, 1% of the trimmer's range, the residual current offset is down to $\pm2.5\,\mu$A, or 0.01% of full-scale. In this circuit R_7 and R_8 improve the stability (the Sziklai's internal feedback can lead to some ringing and overshoot). With 2N5087/8 drivers and 2N4401/3 output transistors, the output capacitance is about 25 pF. This circuit isn't as fast as Figure 2x.57, but it's quite a bit more precise.

When driving some device with a programmed current, you often want to know the corresponding voltage drop across the device under test ("DUT"). Figure 2x.62 shows a simple way to add such *source-monitor* capability to the bipolarity current source of Figure 2x.61.[26] The fast (LT1220: 250 V/µs, 75 ns settle to 0.01%) follower A_2 buffers the voltage seen across the device, with a 50 Ω series resistor for driving a length of coax line without instability. We've added input inverter A_1 to undo the inversion that's inherent in the current source itself (recall the latter's gain is -2 mA/V).

---

[26] A commercial *source measure unit* (SMU, see §13.3.3A) does this job, among its other tricks (for example, sourcing a *voltage* while measuring the current).

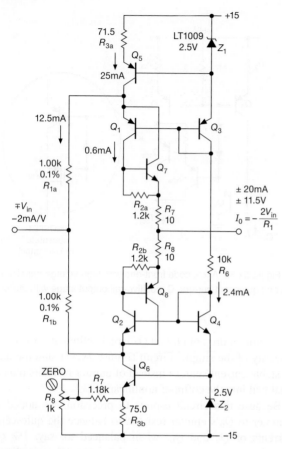

Figure 2x.61. Precision bipolarity current source. The Sziklai configuration of Q_1Q_7 and Q_2Q_8 maintains nearly constant base–emitter drop in Q_1 and Q_2.

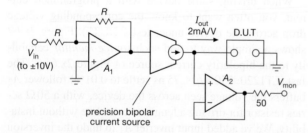

Figure 2x.62. Source-monitor for use with the current source circuit of Figure 2x.61.

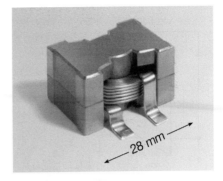

2x.8 The Emitter-Input Differential Amplifier

The most common differential amplifier configuration is the long-tailed pair (§2.3.8 in AoE3) and its elaborations (e.g., current-source emitter pull-down, current-mirror collector load, collector cascode, and their various combinations), along with analogous configurations constructed with FETs (§3.2.4). Those circuits exploit the relatively high input impedance at the control inputs (base, or gate), a consequence of the BJT's current gain (β), or the FET's very low input leakage current (I_{GSS}).

But you can reconfigure the differential amplifier such that the inputs are at the *emitters*, effectively forming a common-base amplifier pair. You don't get the desirable high input impedance, but in exchange you get improved speed (no Miller effect), and (depending on configuration) operation all the way to the supply rail. Let's look at an example.

2x.8.1 An application: High-current, high-ratio current mirror

Here's a real-world application of the emitter-input differential amplifier, namely a circuit to measure the inductance versus current of a power inductor of the kind found in switch-mode power supplies (the topic of SMPS is discussed in §9.6). People like to put as much electronics as possible onto the PC board, and component manufacturers have responded with a vast array of surface-mount components. Figure 2x.63 shows an example of a surface-mount power inductor, along with the manufacturer's curves of inductance versus current for this particular inductor family.

An elegant way to measure such curves directly is to measure the voltage across the inductor while applying a linear ramp of current. The applied current has constant dI/dt, to which the inductor responds with a voltage across it equal to $V_L = L\,dI/dt$. Thus the measured voltage is proportional to the inductance at any moment; that is, the graph of V versus time is a plot of inductance versus current, see Figure 2x.64. So, we've got a "curve tracer" for inductor core saturation. Excellent!

For this application we want a current source that (a) can go to the high currents at which power inductors saturate, (b) with good linearity (any distortion appears as an error in

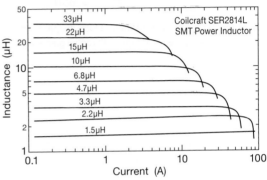

Figure 2x.63. Inductors with magnetic core material exhibit *saturation*, seen as a loss of inductance at high currents. (Adapted from the Coilcraft SER2800-series datasheet.)

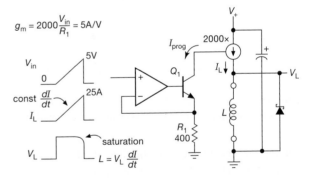

Figure 2x.64. Current-mirror application: identifying inductor saturation via measurement of L versus I.

measured inductance), (c) with plenty of slew rate (so the measurement effectively takes place at frequencies characteristic of switching supplies), (d) with programming via an easily generated low-level signal, and (e) with compliance nearly to the positive rail (to minimize power dissipation).

Figure 2x.65 is a block diagram for the full circuit of Figure 2x.66. It is configured as a current mirror with an accurate 2000:1 current transformation ratio, so a 0–10 mA sinking programming current generates a 0–20 A

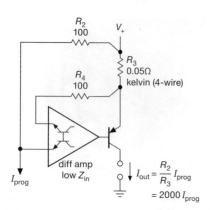

Figure 2x.65. High-ratio current mirror block diagram. The amplifier symbol indicates an emitter-input differential gain stage.

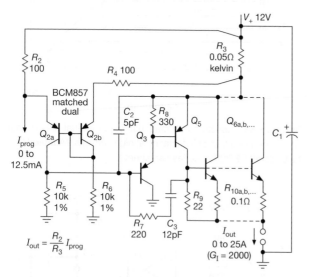

Figure 2x.66. High-current (25 A) , high-ratio (2000:1) current mirror, with emitter-input differential amplifier and Sziklai output stage.

sourcing current output. The amplifier symbol represents a differential amplifier of moderate gain (<100 in our circuit), with low-impedance inputs at the emitters of a differential pair. Feedback forces a current through sense resistor R_3 to make its voltage drop balance the drop through R_2 produced by the programming current. That is, $I_{out}R_3/R_2=I_{prog}$, thus a current gain of R_2/R_3. For the resistor values shown the current gain is $G_I=2000$.

The full circuit (Figure 2x.66) looks rather more complicated, a general fact of life in circuit design. In any case, it's instructive (hey, no pain no gain). Let's walk through it.

The BCM857 is an accurate *pnp* monolithic matched pair with unusually high current gain – you can see its specifications in Tables 8.1a (the BCM860 single version) and

8.1b, and a plot of measured beta as trace 17 in Figure 8.39. Its 50 V rating lets you run this circuit over a wide supply range; the circuit is arranged so that transistors are both operated at approximately one diode drop below the emitters.[27]

The power stage consists of Sziklai pair Q_5Q_6 (the latter drawn as a parallel set of emitter-ballasted small transistors), assisted by Darlington-connected Q_3. The latter adds two diode drops on the way up to the current-sense resistor, setting the V_{CE} of Q_{2a} at about one diode drop; it also provides the additional current gain needed for output currents to 25 A, given the $\sim 1\,mA$ quiescent current of the differential input stage.

While we're discussing high currents, what about those little TO-92 output transistors Q_6? Well, they are small (TO-92 "E-line" pkg), but they are *mighty*: they're rated at 5 A continuous collector current (20 A peak), so a set of eight easily handles the maximum current. But the big gorilla here is *power*: into a short circuit an output current of 25 A corresponds to 300 W! You can't do that with these puppies (even a gang of eight), at least not for long. But you can exploit the thermal mass of the transistor, with infrequent (i.e., low duty cycle) short pulsed ramps. The E-line "transient thermal resistance"[28] $R_{\Theta JA}$ is about 2°C/W for pulses no longer than 2 ms; so a team of eight transistors are allowed to dissipate approximately $P=8\times100°C/R_{\Theta JA}$, or about 400 W. So the circuit is OK with a 2 ms linear ramp to twice that power, thus allowing a ramp to 25 A with a power supply as high as 30 V.

There's no reason, though, not to use output transistors of more substantial power rating. An obvious choice is an *n*-channel MOSFET, which substitutes nicely for Q_6 (Figure 2x.67).

This circuit is pretty fast: with a BC857 for Q_3, a ZTX951 for Q_5, and eight ZTX851s for Q_6, a simulation shows an f_{3dB} of 10 MHz (at 10 A), a current slew rate of 150 A/μs, and an output voltage slew rate of 30 V/μs.

2x.8.2 Improving the emitter-input differential amplifier

The differential amplifier of Figure 2x.66 works well in its intended application, where (a) we want to operate all the way to the positive rail, (b) we have the luxury of low

---

[27] You might worry about Early effect, which tends to loom large in *pnp* transistors generally; we measured a mediocre Early voltage $V_A=30\,V$, which however reduces only slightly the open-loop voltage gain from the ideal $G=R_5/R_2=100$, see §2x.5.

[28] See §9x.25.8 for further discussion of transient thermal resistance.

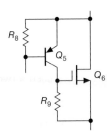

Figure 2x.67. MOSFET substitute for paralleled BJTs Q_6Q_7 in Figure 2x.66.

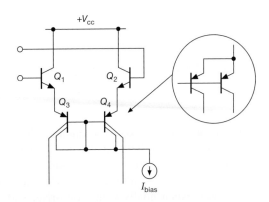

Figure 2x.69. Biasing the operating current via a second collector permits operation to the positive rail (and a bit beyond).

One drawback of this scheme is that the input signal cannot go all the way to the positive rail, because Q_8 consumes a precious diode drop. There is a way to fix this, though, by controlling the collector current with a second collector on each of Q_3 and Q_4 (which you can think of as a matched transistor sharing base and emitter connections), see Figure 2x.69.[30] This kind of circuitry is well suited to integrated circuit design, which innately provides good matching of transistor characteristics, and easy integration of multiple collectors (or drains, for FETs).

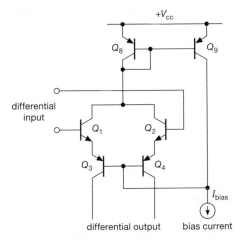

Figure 2x.68. The emitter-input differential amplifier, with input followers (for higher input impedance) and fully differential outputs.

source impedance, and (c) we need only a single-ended output. But there are applications where you would like a higher input impedance, and a fully differential output.

Figure 2x.68 shows how to do this. It is based on the circuit used in the legendary μA741 op-amp,[29] introduced in 1968 and in vigorous production still (they cost about \$0.10 in quantity).

The input emitter followers Q_1 and Q_2 raise the input impedance of the emitter-input differential pair Q_3Q_4 by a factor of their current gain. This circuit (in contrast to Figure 2x.66) also has a fully differential output. That involves more than simply exposing both collectors, however, because you need to set the collector currents without touching the collectors. In this circuit that is done by sensing and mirroring the operating current with current mirror Q_8Q_9, the output of which adjusts the base voltage of the differential pair to set the quiescent current equal to the programmed I_{bias}.

---

[29] See Figure 4.28 in the second edition of this book, which shows the internal circuit of the μA741.

[30] This circuit is used in Bob Widlar's LM301, see "Design techniques for monolithic operational amplifiers," *IEEE J. Solid-State Ckts.*, **SC-4**, 4 (Aug 1969), which includes a nice discussion of the design details.

2x.9 Transistor Beta versus Collector Current

Semiconductor engineers like to use Gummel plots (Figure 2x.70) rather than h_{FE} beta-versus-current plots. The Gummel plot embodies Ebers–Moll thinking, with the base-emitter voltage V_{BE} along the x-axis. Collector current I_C increases exponentially with base drive, with its current level set by the transistor's size. The I_C trace displays the transistor's performance. In the central region the base current plot shows constant beta, tracking the collector current, with beta given by the current ratios. Recombination current losses reduce the beta at low currents, and at high current densities "current crowding" and other effects reduce the beta. The measurements and plot can be made quickly, and IC engineers learn to interpret the information they see at a glance.

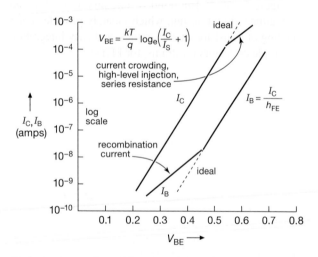

Figure 2x.70. A "Gummel plot," showing BJT base and collector currents over the operating range, shows why transistor beta falls at low currents (recombination current) and at high currents (high-level injection, current crowding, internal resistance).

When making a Gummel plot of transistor measurements, you may find that the higher current region fails to follow Ebers–Moll, with the rate of current increase declining as V_{BE} is increased. The rate of base current increase itself may fall in a similar way, especially if the transistor's beta has not yet significantly decreased. The decrease at high currents is due to high-level injection in the base, sometimes called *current-crowding*. This is dealt

with in the Gummel–Poon model (which is used in SPICE) to decrease I_s at high V_{BE}, causing bending in the I_C curve. All transistors encounter this effect at sufficiently high currents.

To explore this further we used a two-channel SMU to take measurements on an ZTX618, a small 20 V transistor with low saturation voltage, meant to compete with power MOSFETs working to 10 A. The data showed a healthy beta all the way to 10 A, but the Gummel plot showed the slope of I_C beginning to tail off above 50 mA. The I_C/I_B ratio appeared to be degrading above 100 mA, and the I_B plot showed recombination losses starting below 1 mA. These two observations match up with a peak in beta between 50 mA and 200 mA, where the beta gets as high as 450. There's almost no classic flat h_{FE} region in the Gummel plot. We've noticed that the Zetex low-saturation transistors are unusual in many ways, all good. They took the top low-noise transistor honors in Table 8.1a, and they were able to make the world-record 70 pV/$\sqrt{\text{Hz}}$ amplifier in §8.5.9!

In Figure 2x.71 we've measured beta versus collector current, going from 1 μA to 50 mA, for some sixty small-package BJTs (including superbeta types). Most are also small-signal types, but some can work well at currents up to 10 A. However, the maximum measuring current in the plots was 50 mA, which was appropriate for evaluating parts for use in low-noise circuits (this plot was imported from Chapter 8). About a dozen transistors in the two graphs have rising beta all the way to the 50 mA cut-off; these have their maximum beta peak at higher currents, off-scale to the right. Many discrete BJTs show a humped beta plot, thus deviating from the classic (level beta curve) Gummel plot. That's likely one of the things semiconductor designers look at. It's evident that they don't always seek to optimize that aspect, in whatever trade-offs they are making. For example, a 2N4401 (trace 31) conforms well to the ideal of constant beta, whereas a 2N5963 (trace 43) does not.

We were curious to see how the Gummel plots of these two transistors compared (curve 43, with its highly peaked beta, versus curve 31, with nearly constant beta). So we rummaged in our history bins, where we found the actual samples of those transistors that we used, those many years ago, to make Figure 2x.71; these we Gummeled with a source-measure unit (SMU), producing the plots of Figures 2x.72 and 2x.73. They confirm nicely the graphed trends in current gain versus collector current.[31]

---

[31] We also discovered an amusing effect at very low currents: when we ran them first at a modest collector voltage (5 V), the plots of base and

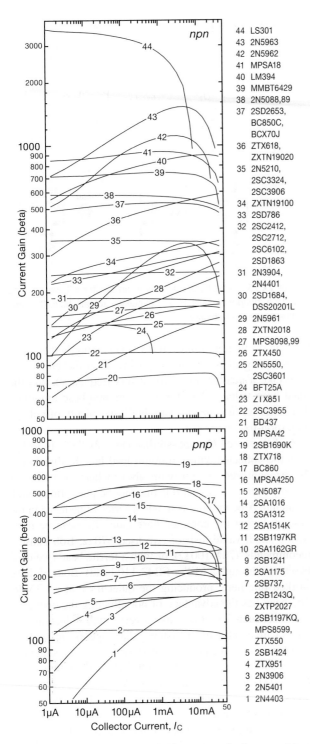

44	LS301
43	2N5963
42	2N5962
41	MPSA18
40	LM394
39	MMBT6429
38	2N5088,89
37	2SD2653, BC850C, BCX70J
36	ZTX618, ZXTN19020
35	2N5210, 2SC3324, 2SC3906
34	ZXTN19100
33	2SD786
32	2SC2412, 2SC2712, 2SC6102, 2SD1863
31	2N3904, 2N4401
30	2SD1684, DSS20201L
29	2N5961
28	ZXTN2018
27	MPS8098,99
26	ZTX450
25	2N5550, 2SC3601
24	BFT25A
23	ZTX851
22	2SC3955
21	BD437
20	MPSA42
19	2SB1690K
18	ZTX718
17	BC860
16	MPSA4250
15	2N5087
14	2SA1016
13	2SA1312
12	2SA1514K
11	2SB1197KR
10	2SA1162GR
9	2SB1241
8	2SA1175
7	2SB737, 2SB1243Q, ZXTP2027
6	2SB1197KQ, MPS8599, ZTX550
5	2SB1424
4	ZTX951
3	2N3906
2	2N5401
1	2N4403

Figure 2x.71. Measured current gain versus collector current for a selection of BJTs. See §8.5.8, from which this plot was taken, for discussion of low-noise design with BJTs.

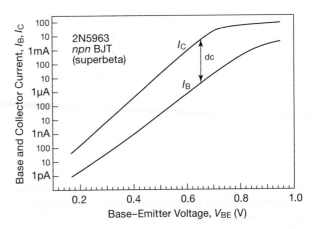

Figure 2x.72. This measured Gummel plot of a 2N5963 superbeta *npn* transistor exhibits a large falloff of beta at both high and low currents, seen also in curve 43 of Fig. 2x.71.

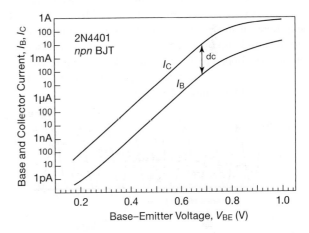

Figure 2x.73. By contrast, the measured Gummel plot of a 2N4401 (curve 31 of Fig. 2x.71) shows near constancy of beta (which however is considerably lower than that of the superbeta transistor of Fig. 2x.72).

collector currents took a dip at low currents, with I_B even going slightly *negative* for collector currents around a few picoamps. This was caused by collector-to-base leakage current – that is why a BJT has a higher collector leakage current (and a lower breakdown voltage) when the base is left open, versus tied to the emitter. To get honest Gummel plots we reduced the collector voltage to 1 V for the lower-current portion of the plots.

2x.10 Parasitic Oscillations in the Emitter Follower

The emitter follower has a voltage gain slightly less than unity,[32] and so there would appear to be no opportunity for it to become an oscillator. And many experienced circuit designers have been fooled by this simple reasoning. Let's take a closer look.

The fine sinewave in Figure 2x.74 is what we measured at the output (emitter) when we hooked up a 2N3904 as a follower, with +15 Vdc on the collector, +5 Vdc on the base, and a 1k emitter load resistor (the circuit of Fig. 2x.75A). It's screaming along at 90 MHz, with 3 Vpp amplitude! It successfully jammed the FM reception in our lab. It showed up on a 'scope ten feet away, even when the probe was clipped to ground. The only good thing you can say about it is that the bottom tips of the wave are at the expected output voltage of ~4.3 V.

Figure 2x.74. Feed it 5 V dc, and the emitter follower sings like a bird! Output waveform of the circuit of Fig. 2x.75A. Vertical: 1 V/div. Horizontal: 10 ns/div.

Figure 2x.75B shows what's happening: the inductance of the long collector wire resonates with the collector capacitance (and stray wiring capacitance) to form a parallel resonant circuit as collector load. In our case the wire was a 10 cm loop ($L \approx 0.3\,\mu$H), the collector capacitance $C_{CB} \approx 4$ pF, and $C_{stray} \approx 4$ pF (because we rigged this up on a "solderless breadboard"). That works out to a resonant frequency of $f_{LC} \approx 100$ MHz.

That's fine, but how do you get voltage gain, needed to

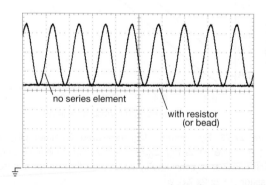

Figure 2x.75. Parasitic oscillations in an emitter follower. A. The circuit we *thought* we built. B. The circuit we actually built. C. Cures for this unseemly behavior.

keep an oscillation going? What's happening is that some collector signal is coupled capacitively to the emitter (it only take a picofarad or two, at this frequency), and the fixed base voltage then forms a common-base amplifier. The latter has no *current* gain, but it has plenty of voltage gain (non-inverting).

And the cure? One possibility (Fig. 2x.75C) is to bypass the collector with minimum lead length; that forces the resonant frequency way up, above where the transistor has any current gain; but this may not be powerful enough medicine. A better cure is to add a series impedance in the base, reducing the common-base gain. This can be done with a small-value resistor (we used 470 Ω; Bob Pease[33] recommends 47 Ω–100 Ω), or with a ferrite bead (we used two turns through a Miller FB43-110 bead). Both of these worked fine here, producing the quiet trace at +4.3 Vdc. The bead has the advantage of not affecting the operation at low frequencies: It acts like a very small inductance, in series with an effective resistance that appears only at high frequencies (the bead we used specifies an impedance of 3 Ω or less at frequencies below 1 MHz, and 50 Ω or more above 100 MHz, for a single turn). A third trick that sometimes works is to add a small series resistor between the emitter and the load.

You can find some rather sophisticated analysis of this problem, if you look in the literature.[34] These generally involve a transformation of a reactive load impedance into

---

[32] To be precise, $G_V = R_{load}/(r_e + R_{load})$.

[33] *Troubleshooting Analog Circuits*, ISBN 978-0750694995 (1991).

[34] See, for example, "Prevent emitter-follower oscillation" by Chess-

a negative input impedance, which supports an oscillation; seen in this light, a (positive resistance) base resistor of equal or greater magnitude then eliminates the oscillation.

This is discussed also in §7.1.5E (*intentional* oscillators). See also the discussion in §2x.13.

man and Sokal (*Electronic Design*, **24**, 13 (1976)), or "Stability of capacitively-loaded emitter followers: a simplified approach" by Glen DeBella (*HP Journal*, **17**, 8 (1966)).

2x.11 BJT Bandwidth and f_T

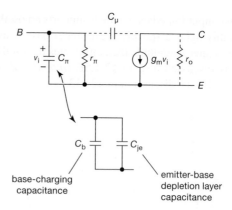

Figure 2x.76. Hybrid-π model of the BJT, showing explicitly the transconductance g_m, and the feedback (Miller) and input capacitances C_μ and C_π. The latter includes a current-dependent component C_b.

As we discussed in §2.4.5, an amplifier's ultimate bandwidth depends critically on the effects of various capacitances (in the external circuit and load, and in the transistors themselves). At a frequency of 100 MHz, for example, the reactance of 15 pF is a very unignorable 100 Ω. Techniques for dealing with this were discussed in AoE3 (e.g., in §§3.2.3 and 8.11.7). Here we provide an introduction to BJT amplifiers at high frequencies, and to the ultimate bandwidth of the transistor itself – the "transition frequency" f_T at which its current gain β has fallen to unity. A transistor with no current gain isn't good for much.

2x.11.1 Transistor amplifiers at high frequencies: first look

The simple common-emitter amplifier with resistive collector load shows a roll-off of gain with increasing signal frequency, mostly owing to the effects of load capacitance and junction capacitance. Looking back at Figure 2.83, for example, C_L forms a lowpass filter of time constant $R_L C_L$ in combination with the amplifier's collector load resistance R_L. Remember that at signal frequencies V_+ is the same as ground; hence the equivalent circuit shown. C_L includes collector-to-emitter and collector-to-base capacitances, as well as load capacitance. At frequencies approaching $f \approx 1/R_L C_L$ the amplifier's gain begins dropping rapidly.

A. Reducing the effect of load capacitance

The simplest therapy consists of measures to reduce the product $R_L C_L$. For example:

1. Choose a transistor (or FET) with low interelectrode (junction and lead) capacitance; these are usually designated as RF or switching transistors.
2. Isolate the load with an emitter follower, thus reducing the capacitive load seen at the collector.
3. Reduce R_L. If you keep I_C constant, the gain drops, owing to reduced $g_m R_L$. Remember that for a BJT $g_m = 1/r_e = I_C/V_T$, or $I_C(\text{mA})/25$ for an amplifier with bypassed emitter. To keep the gain constant with decreasing R_L you have to raise the collector current by

keeping V_+ constant. Thus,

$$f_{max} \approx 1/R_L C_L \propto I_C/C_L$$

which accounts for the rather high currents often used in high-frequency circuits.

2x.11.2 High-frequency amplifiers: the ac model

Load capacitance is not the only effect reducing amplifier gain at high frequencies. As we mentioned in the discussion of Miller effect (§2.4.5), the feedback capacitance (C_{cb}) from output to input can dominate the high-frequency roll-off, especially if the input signal source impedance is not low. In order to determine where an amplifier will roll off, and what to do about it, it is necessary to introduce a relatively simple ac model of the transistor. We will do that now.

A. ac model

The "hybrid-π" small-signal transistor model of Figure 2x.76 is just about the simplest possible, yet it is reasonably useful in estimating the performance of high-speed circuits. The base–emitter input signal voltage v_i programs the output current source $I_C = g_m v_i$ (Ebers–Moll), with an input impedance r_π (that we know as βr_e), a shunt output resistance r_o that represents the change of collector current with collector voltage (Early effect, §2x.5), and two capacitances C_μ and C_π that represent the feedback (Miller) and input capacitances.[35] As we'll see, the latter determines the transistor's bandwidth f_T.

---

[35] These capacitances have multiple names: some aliases for C_π are C_{ib}, C_{ie} or C_{be}; for C_μ you'll see C_{cb} or C_{ob} (the latter includes C_{ce}, a small

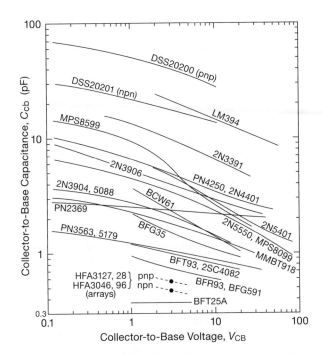

Figure 2x.77. Collector-to-base capacitance versus voltage for selected bipolar transistors, from manufacturers' datasheets. Some interesting transistors are clear off the bottom, see §2x.11.2F on page 144.

B. Effects of collector voltage and current on transistor capacitances

The collector-to-base feedback capacitance C_μ consists of a combination of the capacitance of the die and transistor leads (of order $\sim 0.5\,\mathrm{pF}$) and of the semiconductor junction itself. The latter dominates, and behaves like a reverse-biased diode, with a capacitance that decreases gradually with increasing back-bias, as shown in Figure 2x.77 (this effect is exploited in the voltage variable capacitors known as "varactors"). The capacitance varies with voltage approximately as $C = k(V - V_\mathrm{d})^n$, where n is in the range of $-1/2$ to $-1/3$ for transistors and V_d is a "built-in" voltage of about 0.6 V.

The base-to-emitter input capacitance C_π is different, because you're dealing with a *forward*-biased junction. In this case the effective capacitance rises dramatically with increasing base current (V is near V_d), and it would make little sense to specify a value for C_π on a transistor data sheet. However, it turns out that the effective C_b (the current-dependent component of C_π) increases with increasing I_E (and therefore decreasing r_e) in such a way that the RC

increment to C_cb). Sometimes an additional output capacitance C_o is included, with alternative names C_oe or C_ce.

product $(r_\mathrm{b} C_\pi)$ remains roughly constant. As a result, the transistor's gain at a particular frequency depends primarily on the ratio between current lost into C_π and current that actually "drives the base," and is not strongly dependent on collector current. Therefore, instead of attempting to specify C_π, the transistor manufacturer usually specifies f_T, the frequency at which the current gain β has dropped to unity.[36]

It goes like this:[37] the total input capacitance C_π consists of a term C_je that is relatively constant[38] (it depends on the roughly constant V_BE), combined with a term C_b (the "base charging capacitance") that depends approximately linearly on I_C. The latter goes as $C_\mathrm{b} = \tau_F g_m = \tau_F I_\mathrm{C}/V_\mathrm{T}$, where τ_F is the "forward transit time," approximately constant for a given transistor design.[39]

The transistor's current gain is down 3 dB at a frequency where the magnitude of signal current through C_π (the current robbed from the base) is equal to that through r_π (the current that drives the base). Remembering that $r_\pi = \beta r_\mathrm{e} = \beta/g_m = \beta V_\mathrm{T}/I_\mathrm{C}$, we can write expressions for f_3dB and f_T as

$$f_\mathrm{3dB} = \frac{1}{2\pi r_\pi C_\pi} = \frac{1}{2\pi \beta\, r_\mathrm{e} C_\pi} = \frac{I_\mathrm{C}}{2\pi \beta\, C_\pi V_\mathrm{T}} \qquad (2\mathrm{x}.2)$$

and (because f_T corresponds to $\beta = 1$)

$$f_\mathrm{T} = \frac{1}{2\pi r_\mathrm{e} C_\pi} = \frac{I_\mathrm{C}}{2\pi (C_\mathrm{b} + C_\mathrm{je}) V_\mathrm{T}} = \frac{1}{2\pi \left(C_\mathrm{je} \frac{V_\mathrm{T}}{I_\mathrm{C}} + \tau_F\right)}. \qquad (2\mathrm{x}.3)$$

At medium to high currents the rising C_b dominates over the constant C_je, so the expressions for f_3db and f_T reduce to

$$f_\mathrm{3dB} = \frac{1}{2\pi r_\pi C_\mathrm{b}} = \frac{1}{2\pi \beta\, \tau_F} \qquad (2\mathrm{x}.4)$$

and

$$f_\mathrm{T} = \frac{I_\mathrm{C}}{2\pi C_\mathrm{b} V_\mathrm{T}} = \frac{1}{2\pi\, \tau_F}. \qquad (2\mathrm{x}.5)$$

For small-signal BJTs, τ_F is typically in the range of 10–500 ps, thus corresponding f_T values of 300 MHz to 15 GHz. Transistors intended for RF applications will

---

[36] A transistor's f_T depends also on collector voltage, V_CE; though commonly specified at, say, $V_\mathrm{CE} = 5$ V, some datasheets helpfully show curves of f_T versus I_C for several values of V_CE. See §2x.11.2D for some discussion of the voltage effect.

[37] For the estimation of f_T the feedback capacitance C_μ can be ignored, as can the output resistance r_o.

[38] Lead and stray capacitances, though usually negligible by comparison, can be considered part of C_je.

[39] Physically, τ_F represents the mean time for minority carriers to cross between the emitter and collector.

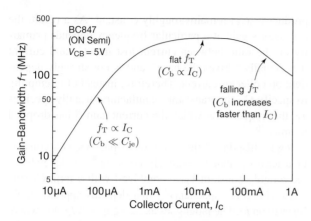

Figure 2x.78. Transistor gain–bandwidth product f_T versus collector current I_C, as calculated from the manufacturer's SPICE parameters for the BC847 *npn* transistor.

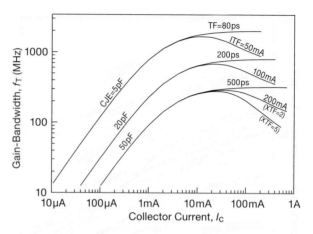

Figure 2x.79. SPICE simulation of f_T versus I_C, using typical small-signal parameters and equations 2x.3 and 2x.6; see §2x.11.2D.

have low values of τ_F, high f_T, and (generally) low maximum voltage ratings. For example, the wideband (5 GHz) BFT25A near the top of Figure 2x.80 is rated at $V_{CE(max)}=5$ V, whereas general-purpose transistors like the 2N4401 and 2N3904 (300 MHz) are rated at 40 V. The high-voltage MPSA42 (300 V) has an f_T of 50 MHz.

C. Low- and high-current regions

As we've just seen, when the collector current is reduced the C_b contribution falls linearly, so that at low currents the fixed capacitances C_{je} (plus any additional wiring capacitance) ultimately dominate. They consume an increasing share of the input current (because r_π is rising as $1/I_C$), so f_T falls proportional to collector current, as indicated in Figure 2x.78, and seen in the SPICE simulation of Figure 2x.79 and datasheet curves of Figure 2x.80.

Other effects[40] take over at the high current end, causing an increase of τ_F and therefore a reduction of f_T (because C_b increases faster than I_C), see Figures 2x.78–2x.80.

D. SPICE parameters

The SPICE circuit simulator incorporates a set of parameters (as many as 50, when various correction terms are included) describing bipolar transistors; these are generally provided by the transistor manufacturer (though you are free to modify them based upon measured behavior). These include dc parameters such as I_S, β and its dependence on temperature and current,

V_A, and the like. Of relevance here are the following, whose names mirror the variables involved in f_T:

BF	ideal maximum beta, β
CJE	base–emitter zero-bias capacitance, C_{jbe}
TF	ideal forward transit time, τ_F
ITF	f_T degrades for I_C above ITF
VTF	f_T drops at low V_{CE} and high I_C
XTF	multiplier for ITF and VTF parameters

The last three parameters (based upon the 1970 Gummel–Poon model[41] of BJTs) describe the variation of τ_F with both collector current and collector voltage. It's interesting to see how SPICE computes τ_F from TF and its correction coefficients, namely[42]

$$\tau_F = \text{TF}\left\{ 1 + \text{XTF}\left(\frac{I_C}{I_C + \text{ITF}}\right)^2 \exp\left(\frac{V_{BC}}{1.44 \cdot \text{VTF}}\right)\right\}. \quad (2x.6)$$

Typical values for a 2N4401 might be CJE=50 pF, TF=500 ps, ITF=0.2 A, and XTF=1.5. By way of eq'ns 2x.3 and 2x.6 these SPICE parameters predict a rising f_T with I_C, passing 100 MHz at 1 mA, flattening above 10 mA (where τ_F is dominant), and falling off at currents over 50 mA (owing to an increase in τ_F); this behavior is in reasonable agreement with the manufacturer's plot of f_T shown in Figure 2x.80.

Table 2x.1 lists f_T and C_{cb} for a selection of bipolar transistors, as specified on the respective datasheets.

E. Comparing SPICE models with measured f_T

These commonly used formulas are convenient for modeling f_T, but we have found them to be inaccurate in several

---

[40] Known as "high-level injection," "base push-out," or base widening (the Kirk effect: C.T. Kirk, "A theory of transistor cutoff frequency (f_T) falloff at high current densities," *IRE Trans. Electron Dev.*, **ED-9**, 2, 164–174 (1962).

[41] H.K. Gummel and H.C. Poon, "An integral charge control model of bipolar transistors," *Bell Sys. Tech. J.*, **49**, 5, 827–852 (1970).

[42] Note that V_{BC} in this equation is a negative quantity.

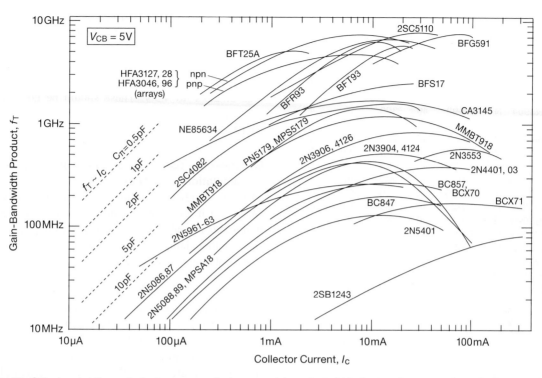

Figure 2x.80. Gain–bandwidth product, f_T, versus collector current for selected bipolar transistors, as shown in the manufacturers' datasheets. Some wideband transistors are clear off the top, see §2x.11.2F.

ways: they predict too little reduction of f_T at high currents (the formula predicts eventual flattening after initial falloff); they predict too little reduction of f_T at low voltages and low-to-medium currents; and they overestimate f_T at high voltages and currents. In addition, we have found that manufacturers' SPICE parameters relevant to modeling f_T are sometimes seriously flawed.

To explore this further, we gathered eight SPICE models for the popular BC846–50 series of *npn* transistors, as provided by different manufacturers. The calculated f_T versus I_C is plotted in Figure 2x.81, along with datasheet graphical plots where available (in this case we found only three). As seen here, the manufacturer's SPICE models and the datasheet f_T plots may not be in agreement at low currents; and you may see as much as a factor of three variation of f_T for the same part type, as modeled by different manufacturers. Note also that SPICE models that omit the ITF term are not to be trusted if your design runs at the higher end of the current range. In spite of these shortcomings, you can always adjust the manufacturer's SPICE parameters in these equations, based upon trustworthy or measured f_T's, to get good estimates of bandwidth in the important low-to-medium current regions, as shown for example by our ad-

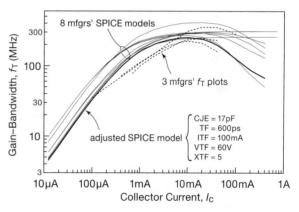

Figure 2x.81. The f_T values calculated from manufacturers' SPICE models for the BC846–50 at $V_{CB}=5$ V differ by factors of two or more, and several of them omit the ITF term, predicting flat f_T at high I_C. We measured samples from four manufacturers, and found $f_T=5$–10 MHz at 10 µA. The bold curve represents a conservative SPICE model.

justed "conservative f_T" SPICE model (bold curve) in Figure 2x.81. This lets you get reasonable estimates of transistor gain–bandwidth over the full range of collector current, information that is often missing from datasheets.

**Table 2x.1: Bandwidth and Capacitance
of Selected Discrete BJTs[a]**

Part #	Type[b]	Pkg[c]	V_{CEO} (V)	f_T typ (MHz)	@ I_C (mA)	C_{cb} max (pF)
2N5088	N	TO-92	30	50[m]	0.5	4
2N5087	P	TO-92	50	40[m]	0.5	4
2N3904	N	TO-92	40	300	10	4
2N3906	P	TO-92	40	250	10	4.5
2N4401	N	TO-92	40	250	20	6.5
2N4403	P	TO-92	40	200	20	8.5
BCX70	N	SOT23	45	250	10	2.5[t]
BCX71	P	SOT23	45	200	10	4.5[t]
BC847	N	SOT23	45	300	10	6
BC857	P	SOT23	45	150	10	6
MPS8099	N	TO-92	80	150	10	6
MPS8599	P	TO-92	80	150	10	8
2N5550	N	TO-92	150	200	10	6
2N5401	P	TO-92	150	200	10	6
DSS20201L	N	SOT-23	20	300	100	45
DSS20200L	P	SOT-23	20	200	100	100
2SD1760	N	DDPAK	50	80	500	40[t]
2SB1184	P	DDPAK	50	80	500	50[t]
MMBT918	N	SOT-23	15	600[m]	4	1.7
MPS5179	N	TO-92	12	900[m]	5	1.0
MMBTH81	P	SOT-23	20	600	5	0.85
BFS17	N	SOT-23	25	1000	25	1.5
2SC4082	N	SC-70	20	1500	10	1.5
BFG35	N	SOT-223	18	4000	100	2[t]
BFR93	N	SOT-23	15	5000	30	0.8[t]
BFT93	P	SOT-23	12	5000	30	1.0[t]
BFG591	N	SOT-223	15	7000	70	0.7[t]
BFT25A	N	SOT-23	5	5000	1	0.45
HFA3046	5P	S-16, Q-16	8	8000	10	0.5[t3]
HFA3128	5N	S-16, Q-16	8	5500	10	0.6[t3]

(a) f_T and C_{cb} at 5V. (b) N=npn, P=pnp. (c) S-16=SOIC-16,
Q-16=QFN-16. (m) minimum. (t) typical. (t3) typical at 3V.

Relevant data for some of the transistors shown in Figures 2x.77 and 2x.80.

F. Wideband micropower BJTs

With the proliferation of handheld wireless devices, wideband micropower circuits are all the rage. As the curves in Figure 2x.80 suggest, you do better by operating at currents above 1 mA, say. But there are some attractive BJT choices, even if you need to starve the thing of collector current. One trick is the use of emitter contacts between base and collector, which shields the collector–base capacitance (reducing Miller effect) while simultaneously reducing emitter lead inductance.

Three low-cost parts that use this technique are completely off the bottom of the capacitance plot of Figure 2x.77, and they peak well over the top of the gain–bandwidth plot of Figure 2x.80. They are the

gallium-arsenide BFP650 (0.26 pF, 37 GHz), the silicon-germanium BFP620 (0.12 pF, 65 GHz!), and the plain ol' silicon BFG410[43] (0.04 pF!, 22 GHz). These may be good candidates for obtaining high f_T at micropower currents, provided you can live with their low maximum V_{CEO} ratings of 2.3 V to 4.5 V. These should be good for $f_T >$ 10 MHz at 1 μA (as well as might the commonplace 5 V BFT25A).

G. Collector–base time constant and maximum oscillation frequency

The gain–bandwidth product f_T is a useful index of BJT speed, but it does not take into account some bandwidth limitations in real-world circuits. For example, f_T is measured with a clamped collector, thus blissfully ignores the speed-robbing Miller effect (§2.4.5), which looms large in high-speed circuits. So it's important to fold in some knowledge of a transistor's feedback capacitance C_{cb}.

From the simple Miller effect model we considered in Figure 2.83 you would conclude that the only quantities that matter are C_{cb} and the signal's source resistance R_s, and that by simply reducing the latter to zero you could banish Miller effect altogether. That's not quite right, however: the BJT has some built-in "base spreading resistance" $r_{bb'}$, in the range of a few ohms to a few hundred ohms for typical small-signal transistors (see Table 8.1a for some $r_{bb'}$ values, obtained from noise measurements[44]), so even when driven with a zero-impedance signal source there is a Miller effect roll-off, characterized by a time constant $r_{bb'}C_{cb}$.

Manufacturers rarely specify $r_{bb'}$ on datasheets, but they may provide a tabulated value, or even a plot, of the "collector–base time constant" $r_{bb'}C_{cb}$. For example, Toshiba specifies $r_{bb'}C_{cb}$=30 ps (max) for their 2SC2714 high-frequency amplifier; and for the transistors in Figure 2x.82 the manufacturer provides curves of typical $r_{bb'}C_{cb}$ versus collector current.

The product $r_{bb'}C_{cb}$ figures into a quantity known as the "maximum oscillation frequency" (f_{max}); that is the maximum possible frequency of self-sustaining sinusoidal oscillation in a circuit where the BJT is the only active component. It is the frequency at which the *power* gain has

---

[43] We are sad to report that the BFG410 is on life support; microwave circuit designers are eschewing classic silicon.

[44] The base spreading resistance figures prominently in low-noise design, because the thermal Johnson noise voltage generated across $r_{bb'}$ sets a lower limit on an amplifier's voltage noise; see, for example, Figure 8.12.

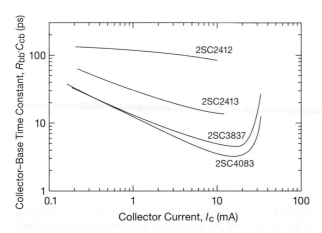

Figure 2x.82. Collector–base time constant $r_{bb'}C_{cb}$ versus collector current, as shown in datasheets from Rohm Semiconductor.

fallen to unity, and is given by

$$f_{max} = \sqrt{f_T/8\pi r_{bb'}C_{bc}}$$

or half the geometric mean of f_T and the characteristic roll-off frequency ($1/2\pi r_{bb'}C_{cb}$) set by the base–collector time constant. This equation can be derived rigorously, and demonstrates that a high f_T alone is not the whole story; you've got to minimize the base–collector time constant as well. The maximum oscillation frequency can be greater or less than f_T.

2x.11.3 A high-frequency calculation example

In the second edition of this book we showed an example where this simple transistor model is applied to the design of a high-frequency broadband amplifier. You can find it there in §13.05 on page 869. (We plead lack of space here – this edition is crammed full of new material; *too* full, some might say.)

2x.12 Two-terminal Negative Resistance Circuit

Here's a circuit that can come in handy: a 2-terminal device that exhibits negative incremental resistance over some useful range of voltages. We saw such behavior in the *tunnel diode* (§1x.7.3), but it operates over a small range of voltage (a few tenths of a volt), is not adjustable, and is highly prone to oscillation. With op-amps you can make a more useful negative resistor (§6.2.4B), but it is not a 2-terminal device, because it requires external power. And neither of these can handle high voltages, or high power dissipation.

There have been collections of circuits aimed at this goal,[45] and you'll find references to the "Lambda diode," which uses a pair of complementary JFETs, or a JFET–BJT combination.[46] But a superior circuit that brought us to attention was one that Erich Wagner casually mentioned[47] on the electronics newsgroup "s.e.d."; in his words "easy with two BJTs."

Erich provided his circuit details in a SPICE listing, which we reverse-engineered to get the circuit shown in Figure 2x.83B. You can think of it as a conventional V_{BE}-biased current source (Figure 2x.83A), to which a foldback current-limit has been added (similar to the improved foldback circuit shown in §9x.4, Figure 9x.10B). Erich told us it was popular with amplifier designers in the 1970s, to protect class-AB output stages.

With the values shown, it has a large negative-resistance region, as seen in the SPICE simulation of Figure 2x.84. You can adjust the component values to fit your application: maximum current $I_{max} \approx 0.7 V_{BE}/R_1$, resistance in the negative region $R \approx -1.1 R_1 R_3/R_4$, and maximum negative-resistance operating voltage $V_{max} \approx V_{BE} R_3/R_4$ (above V_{max} the current through R_2 and R_3 creates a compensating positive resistance).

[45] For example L. Chua et al., "Bipolar–JFET–MOSFET negative resistance devices," *IEEE Trans. Circuits*, **32**, 1, 46–61 (1985).

[46] See, for example, T. Hiromitsu and K. Gota, "Complementary JFET negative-resistance devices," *IEEE Jour. Solid-state Cir.*, **SC-10**, 6, 509–515 (1975), and "Simulation of a novel bipolar-FET type-S negative resistance circuit," *Active and passive Elec. Comp.*, **26**, 129-132 (2003).

[47] Usenet forum, sci.electronics.design, July 6, 2019.

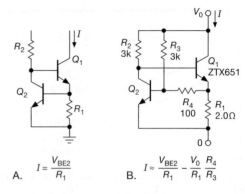

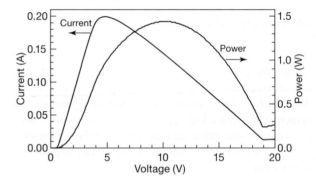

Figure 2x.83. If a foldback path is added to the conventional V_{BE}-biased current sink (A), the result is a two-terminal floating current source with a large negative-resistance region (B).

Figure 2x.84. SPICE simulation of the current and total power dissipation for the circuit of Fig. 2x.83B, with the component values shown.

In the circuit Q_2's V_{BE}, running at currents below 6 mA, was about 0.55 V. You can derive more accurate formulas if you like, but we're already far better off than we would be with most other simple 2-terminal negative-resistance circuits.

The foldback concept can be very useful in limiting power dissipation, to a peak of about 1.4 W at 10 V in this 20 V example.

With proper choice of components this circuit can be run at considerably higher voltages. We ran simulations with the 400 V ZTX458, with $R_1 = 90.9\,\Omega$ for a maximum current of 5 mA, and for R_2, R_3, and R_4 chosen to be 1M, 3.9M, and 4.99k respectively; the results are shown in Figure 2x.85. These values maintain a 2 mA operating current at 350 V (700 mW), with a maximum dissipation of 770 mW at 260 V.

In this context it's interesting to compare the improvement in peak power dissipation that this method achieves, compared with a 2-terminal constant-current source. For

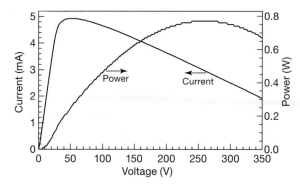

Figure 2x.85. SPICE simulation of the current and total power dissipation for the circuit of Fig. 2x.83B, with $R_1 = 90.9\,\Omega$, $R_2 = 1M$, $R_3 = 3.9M$, and $R_4 = 4.99k$.

example, in §3x.6.5 we designed high-voltage current sources for use in high-voltage amplifiers. But we struggled mightily with power dissipation constraints, which would have been greatly alleviated had we added some foldback current reduction (albeit at the expense of constancy of pullup current).

You can make a low-voltage version of the circuit that reaches a peak current at 650 mV (this may be about the best that can be done, given the two-V_{BE} nature of the circuit). For an *npn* non-floating version, use 2N5089 BJTs, and try $R_1 = 3.3k$, $R_3 = 83k$, and $R_4 = 22k$, for a negative-resistance region peaking at 150 μA, with $R = -12.05k$. Use $R_2 = 150k$, and connect it to +2.5 V, etc., so it's not part of the output node. The node isn't very fast (e.g., R_3 and Q_2's $C_{cb} = 3pF$ create a 0.25 μs time constant for a 650 kHz pole, and an apparent node capacitance of 20 pF), but you could speed it up by scaling the resistors to run at higher currents.

2x.13 If It Quacks Like an Inducktor...

...then you might as well call it an inductor.[48] More specifically (and in the context of this chapter), an emitter follower's output can look *inductive* at high frequencies, even though there's no inductor in sight. And so it can form a resonant circuit, complete with ringing and overshoot, when asked to drive a capacitive load.

This was brought home to us recently, when we faced the problem of driving a substantial capacitive load with fast pulses. More specifically, we had a thousand channels, each carrying amplified pulses from a photomultiplier; and each of these needed to drive 8 inputs of a complex integrated circuit (an FPGA,[49] see Chapter 11), totaling about 64 pF of load capacitance. That may not sound like much; but in fact it presents a tough load for these signals, with their characteristic rise times of ~1 ns. You can think of it this way: the reactance of that capacitive load at 300 MHz (characteristic of 1 ns rise and fall times) is about 8 Ω, which severely attenuates the signal that is coming from a 50 Ω source (Thévenin) impedance. Equivalently, you can notice that the time constant of 50 Ω and 68 pF is 3.4 ns, quite a bit longer than the signal's rise and fall times.

What to do? A reasonable solution here was to add an emitter follower, to reduce the signal's source impedance. And, because the signal (consisting of ~100 mV negative-going pulses) was riding on a +4 V quiescent bias, the simplest configuration was a *pnp* follower (Figure 2x.87). These are fast signals, so we chose a fast transistor: an MMBTH81, whose f_T (roughly speaking,[50] the frequency at which the current gain has dropped to 1) is around 600 MHz. We arranged things for 10 mA quiescent current

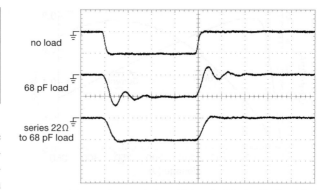

Figure 2x.86. Waveforms from an MMBTH81 *pnp* emitter follower driving a 68 pF load. The 100 mV negative-going input pulse has 50 Ω source resistance. The ringing can be simply understood as arising from the follower's inductor-like rising output impedance, resonating with the load capacitance. A series resistor critically damps the series *RLC*. Vertical: 100 mV/div; Horizontal: 10 ns/div.

(these small-geometry transistors have a relatively large $V_{BE} \approx 0.8$ V at I_C=10 mA).

Now, here's where the duck comes in: the follower's output impedance is r_e (the usual $25\Omega/I_C$ [mA]) in series with the beta-reduced source resistance. But beta is falling as $1/f$ at high frequencies, so the output impedance begins rising proportional to frequency – and that's just what an inductor does! So the output looks inductive at these frequencies, with $L = R_s/2\pi f_T$, and it forms a series *LC* circuit. The resonance is somewhat damped by r_e, which however is too small to prevent ringing; that's because the impedance of *L* (or of *C*) is about 14 Ω at resonance, which is the value of damping resistor needed to suppress ringing.

Indeed, ringing was observed, and cured with an added series resistor at the emitter output. Figure 2x.86 shows this ringing behavior, and the fix. Here we used a pulse generator to create a clean 100 mV negative pulse of 40 ns duration. The observed ringing was at ~120 MHz, somewhat below our simple estimate of 175 MHz;[51] the 22 Ω resistor value was then chosen approximately equal to the capacitor's reactance at that frequency.[52]

---

[48] Paraphrasing a saying that is often attributed to James Whitcomb Riley: "when I see a bird that walks like a duck and swims like a duck and quacks like a duck, I call that bird a duck."

[49] "It's a complicated case, Maude. Lotta ins. Lotta outs. And a lotta strands to keep in my head, man. Lotta strands in old Duder's..."

[50] More precisely, the frequency at which the extrapolation of the curve of current gain (β) versus frequency intercepts the frequency axis. This is usually specified as the "current-gain × bandwidth product," typically specified at a frequency at which the current gain β has dropped to somewhere in the range of 5–10. For the MMBTH81, for example, f_T is specified as 600 MHz (minimum), measured at 100 MHz.

---

[51] Transistor behavior at these frequencies is more complicated than the simple $\beta = f/f_T$ first-order estimate.

[52] Which however slowed the transition times (yet happily retaining the full pulse amplitude), as seen in the bottom trace.

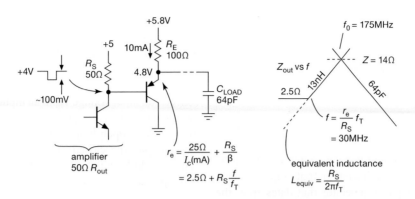

Figure 2x.87. A *pnp* emitter follower buffers a fast pulse coming from an amplifier with $50\,\Omega$ output impedance. But the falling beta at high frequencies produces a rising output impedance, mimicking an inductor. This combines with the load capacitance to produce a resonant circuit ... and trouble!

2x.14 "Designs by the Masters": ±20 V, 5 ns, 50 Ω Amplifier

You can learn a lot by studying the circuit designs of the masters. Here we look at a discrete output amplifier stage found in a Tektronix pulse generator that dates from the 1970s. Their PG508 was a member of the TM500 modular instrument family, and it featured fast clean transitions (adjustable risetimes, 5 ns to 50 ms, no overshoot or ringing) with plenty of drive (20 Vpp open-circuit, 10 Vpp into 50 Ω) and dc offset control so that the output could reach ±20 V into an open circuit (±10 V into 50 Ω). The output circuit is a linear amplifier, needed to reproduce the variable rise- and fall-time ramps; it includes a bipolarity folded cascode whose basic structure is not unlike that of our bipolarity current source in Figure 2x.59, but decorated with numerous real-world refinements. As we'll see, the overall circuit, though incorporating the same folded-cascode topology, converts an input *current* signal into an output *voltage* signal, quite the opposite of Figure 2x.59's voltage-to-current behavior! Let's take a tour, beginning with the block diagram of Figure 2x.88; it may be helpful to refer also to the full circuit diagram in Figure 2x.89.

2x.14.1 Output stage block diagram

The 6-transistor gain block Q_1–Q_6 is a bipolarity transconductance amplifier, with gain g_m set by a pair of emitter resistors. The output node (which we earlier exploited as a current source) drives a complementary push–pull emitter follower Q_7–Q_{10}, thus generating a voltage output. This may sound strange – an unloaded current source acting as a voltage source – but it is quite OK as long as negative feedback is wrapped back around to the input. (In fact, a high-gain transconductance stage driving a push–pull follower is the basic structure of the operational amplifier, see for example Figure 4.43.) Because the transconductance stage sees primarily a small capacitive load, the open-loop voltage gain at low frequencies is very large: $G_{V_{OL}} = g_m X_C$, about 160 at 10 MHz for the component values in Figure 2x.89, with a $1/f$ dependence on frequency.

The feedback resistor R_f (R_1 in the full circuit diagram) closes the loop, creating a low impedance "summing junction" input node, exactly as in Figure 2.89A. With the feed-

back resistor, the overall configuration is now a transresistance amplifier: an applied input current generates an output voltage $V_{out} = -I_{in}R_f$. That is the configuration used in the PG508. If you add an input resistor R_g, the circuit becomes a voltage amplifier (as in Figure 2.89B), because R_g converts an input voltage V_{in} to an input current $I_{in} = V_{in}/R_g$, thus generating an output voltage $V_{out} = -V_{in}R_f/R_g$ (in the limit of large open-loop voltage gain). For $R_f = R_g = 2\,k\Omega$, for example, the closed-loop voltage gain is $G_{V_{CL}} = -1$. (This should be *terra cognita* for readers already familiar with op-amps.)

2x.14.2 Output stage: the full enchilada

In this amplifier (Figure 2x.89), transistors Q_3–Q_6 comprise the voltage-amplifying stage, or VAS. Transistors Q_5 and Q_6 form a complementary folded cascode, biased to a substantial quiescent current of 30 mA by followers Q_{11} and Q_{12} (with $10\,\Omega$ isolating resistors to prevent oscillation when loaded with 10 nF bypass capacitors at the cascode bases). The cascode's output (node X) is buffered by push–pull follower Q_7–Q_{10}, biased into class-A conduction by the dc drop across $D_1D_2R_4$, and conveyed to the output terminal by a network of resistors that sets the output impedance to 50 Ω (and with a bit of capacitive peaking for best transient response). The folded cascode is a good topology, because it is fast (no Miller effect), and it permits nearly rail-to-rail swings (because it is "folded").

Turning our attention to the input, complementary current sources Q_3 and Q_4 convert an input voltage signal to a push–pull current to drive the cascode. The current sources are driven by followers Q_1 and Q_2, each with a dc offset to establish Q_3Q_4's quiescent bias of 30 mA.

Ignoring for the moment the feedback resistor R_1, we can estimate the open-loop voltage gain of the circuit from input to node X as $G = g_m X_L$, where g_m is the transconductance from input to the collectors of Q_5 and Q_6, and X_L is the load impedance seen by the collectors. The latter is the capacitive reactance of the combined collector-to-base capacitances C_{cb} of Q_5–Q_{10}, about $C_L = 20$ pF total (including wiring capacitance) for these high-frequency transistors; and the transconductance driving that load is $g_m = 2/R_3$ (twice the transconductance of each current source, $1/R_{3a}$, etc., because the push–pull currents are combined). So the magnitude of the voltage gain is $G = 2X_C/R_3 = 1/10\pi fC_L$, or roughly $G \approx 80$ at 20 MHz. The estimated slew rate, for a full-scale input of ± 0.3 V (thus driving ∓ 60 mA into ΣC_{cb}) is impressive, $S = I/C = 3000$ V/μs. The measured output swing (Figure 2x.90) with the controls set for minimum rise time (5 ns) shows an actual slew rate somewhat

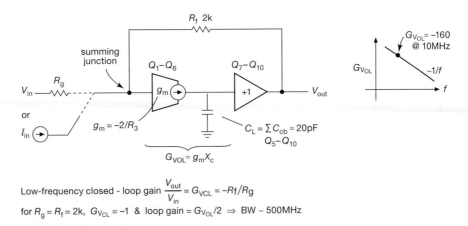

Low-frequency closed - loop gain $\frac{V_{out}}{V_{in}} = G_{VCL} = -R_f/R_g$

for $R_g = R_f = 2k$, $G_{VCL} = -1$ & loop gain = $G_{VOL}/2 \Rightarrow$ BW ~ 500MHz

Figure 2x.88. Tektronix PG508 pulse generator output stage block diagram.

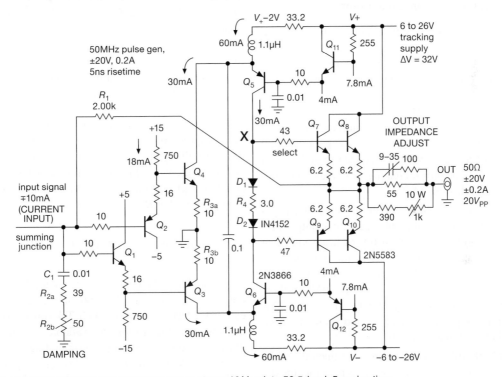

Figure 2x.89. Tektronix PG508 pulse generator output stage: 10 Vpp into 50 Ω load, 5 ns rise time.

smaller, about 2500 V/μs, the discrepancy likely due to the finite input signal speed from the "variable transition time generator" circuitry (which tops out at 5 ns).

Closing the loop with R_1 creates a low-frequency transresistance gain of $-2k\Omega$, with signal and dc offset currents combined at the input summing junction. In high-speed circuits you commonly see signals conveyed as *currents*, circumventing Miller effect while simultaneously allowing large dynamic range. Here there's another advan-

tage: the V_+ and V_- supply rails are maintained at 32 V difference, the pair moving up or down as needed to accommodate the settings of output pulse amplitude and offset. This is necessary because the high-speed transistors in the output stage are limited to 30 V. An input signal in the form of a voltage would have to track the peregrinations of the output supplies, whereas an input *current* signal is unaffected by a dc offset of the entire output stage.

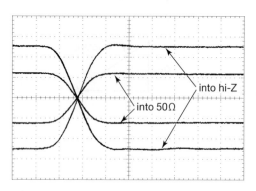

Figure 2x.90. Measured step response of a PG508 pulse generator at maximum bipolarity amplitude, both open-circuit and into a 50 Ω load. Horizontal: 4 ns/div; Vertical: 4 V/div.

2x.14.3 Output stage: some fine points

Biasing. The voltage between the emitter terminals of input followers Q_1 and Q_2 is $2V_{BE}$, so the 16 Ω resistors provide the needed additional 600 mV drop to bias current sources Q_3 and Q_4 at a quiescent current of 30 mA. The folded cascode stage gets its bias from the drop across the base resistors of Q_{11} and Q_{12}, imposed across the emitter resistors of Q_5 and Q_6, thus a quiescent current of $I_Q=7.8\text{mA} \times 255/33.2=60$ mA. And the output push–pull follower transistors see a voltage between their bases of 90 mV plus two diode drops, creating an output stage quiescent bias of roughly 40 mA.[53]

Supply rails. The regulators for the positive and negative rails are controlled by the pulse generator output amplitude settings, to provide at least 6 V of headroom; that is, the positive supply V_+ is 6 V above the pulse's HIGH level, and the negative supply V_- is 6 V below the pulse's LOW level. For maximum pulse amplitude (20 Vpp), then, the supply rails are separated by $\Delta V=V_+-V_-=32$ V. This scheme minimizes power dissipation, and keeps the transistors within their voltage ratings.

Peaking and stability. To maintain good linearity while extending the frequency response as far as possible, the designers included some "peaking" components, for example the $1.1\,\mu$H inductors in the cascode emitters, and the 9–35 pF capacitor in the output network.

A more interesting adjustment is the curious "damping" control at the input. Its job is to ensure stability, and provide best transient response, by limiting the high-frequency

loop gain. To understand this, imagine the circuit with C_1 and R_2 absent. The loop gain would then be equal to the circuit's open-loop voltage gain (recall that the input signal is a *current*, i.e., a high impedance), shown in the right-hand side of Figure 2x.88. The $1/f$ roll-off shown there would be OK, in terms of stability. The problem is that additional roll-offs are inevitable in a multistage amplifier like this, and they create transient ringing, and even oscillation, if they combine to bring the phase shift to 180° before the loop gain has dropped to unity.

As hinted in §2.5.4B, and as discussed in detail in §4.9.2, the solution is to arrange things so that the loop gain falls to unity well before the phase shift is approaching 180°. One way to do this is to reduce the loop gain by attenuating the feedback signal, see for example Figure 4.102. This is the function of R_2 here: at frequencies above about 0.25 MHz (where the reactance of C_1 is small compared with R_2) the feedback voltage (and therefore the loop gain) is reduced by the ratio $R_2/(R_1+R_2)$, about a factor of 30 for a nominal R_2 of 64 Ω (R_{2b} trimmer at middle position).[54]

Figure 2x.91 shows the effect of this "damping" network on a 5 V output step into a 50 Ω load, with selected values spanning the full range of 50 Ω trimmer R_{2b}. Optimum pulse response is obtained near the middle of the range (the designers likely chose the fixed and trimmer resistor values by trial), with sluggish response at lower loop gains (smaller R_{2b}), and substantial 20% overshoot at the highest loop gain (higher R_{2b}).

2x.14.4 Epilogue: 120 V, 5 A, dc-10 MHz Laboratory Amplifier

We would have liked to build our own replica of the PG508 output stage, but the 2N2866 and 2N5583 RF metal-can TO-5 transistors, on which Tektronix mounted giant heatsink fins, are no longer available; and the newer replacements don't have the power-handling capability of the older parts. What's more, we had several applications for a high-speed power amplifier working at higher voltages and

---

[53] You might initially estimate 15 mA (30 mA×3 Ω puts 90 mV across 6.2 Ω), but the drop across a signal diode like a 1N4154 at 30 mA is somewhat greater than the V_{BE} of a transistor like Q_7–Q_{10} running at these collector currents. A better approach to biasing is to use a (thermally coupled) transistor in place of $D_1D_2R_4$, as in Figure 2x.56.

[54] Restated in language that will resonate with readers whose familiarity includes op-amp circuits, it goes like this: We've got a multistage wideband amplifier with a single feedback resistor R_f, and with no compensation capacitors. So, with a signal source of high impedance (i.e., a *current* signal) it's operating at unity closed-loop gain, a sure recipe for instability. Thinking of it as a decompensated amplifier, then, we can ask what is its minimum stable closed-loop voltage gain. And, based on the circuit values and the observed step response, the answer for this amplifier is $G_{CL}=50$ (with moderate overshoot), or $G_{CL}=30$ (no overshoot). The damping network ensures this minimum gain at the higher end of its bandwidth where oscillation is possible.

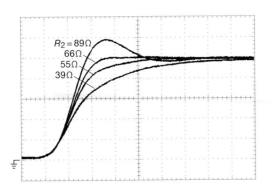

Figure 2x.91. Step response of a PG508 pulse generator for several settings of "damping" resistor R_2. Horizontal: 4 ns/div; Vertical: 1 V/div.

with more bandwidth than previous designs. But the techniques the Tektronix engineers had developed looked like a good place to start.

Circuit designs like this are very much limited by the availability of specialized high-performance components. We searched for candidate parts and found some nice video output-stage transistors, originally created to drive the electrodes of high-resolution large-screen CRT computer monitors, popular in the late 1990s. Somehow semiconductor designers managed to combine low C_{cb} capacitance with plenty of power dissipation capability, a combination that had not appeared before (and has not reappeared since!).

Below is a mini-table of some useful parts you might consider using. It includes an unusual Figure-of-Merit, FoM, for power/capacitance, i.e., watts per picofarad (W/pF). You likely haven't seen that one before, but it's very useful![55]

The final five parts may not be candidates for a 10 MHz amplifier; they're included for their power capability. We chose the 180-volt TO-220F parts, and, employing the concepts in the PG508, created an amplifier with up to 160 V of output range (rather than 30 V) and with 4 A or 5A of output current capability (rather than 0.2 A). Our 100 V version has a 10 MHz bandwidth and slews at 1000 V/μs; its

high voltage and high current capabilities have been very useful.[56]

			P_{diss}	V_{BR}	C_{ob}	f_T	FOM
npn	*pnp*	Pkg	(W)	(V)	(pF)	(MHz)	(W/pF)
BF720	BF721	SOT-223	2	300	0.8	60	2.50
FZT458	FZT558	SOT-223	3	400	<5	50	0.60
KSC3503	KSA1381	TO-126	7	300	3.1	150	2.46
KSC2682	KSA1142	TO-126	8	180	4.5	180	2.08
MJE340	MJE350	TO-126	20	300	15	–	1.83
2SC4793	2SA1837	TO-220F	20	230	30	70	0.80
2SC4883A	2SA1859A	TO-220F	20	180	30	120	0.67
MJE15032	MJE15033	TO-229	50	250	120	>30	0.42
2SC5200	2SA1943	TO-220	80	250	360	30	0.29
FJP5200	FJP1943	TO-220	80	250	360	30	0.29
2STC5200	2STA1943	TO-264	150	230	225	30	0.71
MJL21196	MJL21195	TO-264	200	250	200	5	1.33

A. Circuit details

Referring to Figures 2x.89 and 2x.92, we started by replacing Q_7–Q_{10} with four sets of 180 V, 20 W 120 MHz *npn* and *pnp* power transistors in their (isolated) TO-220F packages. Five *npn* transistors Q_{11A}–Q_{11F} are driven by emitter follower Q_9, and likewise for five more *pnp* transistors (Q_{12A}–Q_{12F}); the driving pairs (Q_9Q_{10}) are biased at 60 mA. Only two C_{cb} capacitances (call it C_X) are presented to the voltage-amplifying stage (VAS) critical high-impedance node ("X"), thanks to the emitter followers Q_9Q_{10}. The Q_3Q_4 transistors are doubled pairs, running the VAS node at 80 mA, for 2.7× higher current and 9× higher power, compared with the PG508. We replaced the latter's D_1D_2 and R_4 with a pair of V_{BE} diodes (Q_7Q_8) to provide temperature-tracking class-AB bias for the 12 output transistors. One of the V_{BE} bias circuit resistors is adjusted for about 40 mA quiescent current in each output transistor.

The input transistors are TO-92 types, chosen for their task; Q_1Q_2 for high beta, and Q_3Q_4 to handle high currents. We clamped the collectors of the latter to 5 V with a cascode-connected power transistor (Q_5), which allowed us to run the gain stage at 160 mA peak during slewing. We added floating 5 V supplies to power each folded cascode stage (Q_6 for the upper half), allowing us to drive the output transistors almost to the supply rails. The full set of 10 video power transistors was soldered with bent leads under the PCB (Fig. 2x.93), and screw-mounted with plates to a single large heatsink with a large fan.

---

[55] The transistor spec C_{ob} (= $C_{cb}+C_{ce}$) is measured with the base grounded, and primarily consists of the C_{cb} term. To obtain the TO-126 parts, you may have to check with Rochester or Alibaba.

[56] H. Berg and L. Turner, "Torque Generated by the Flagellar Motor of E. coli," *Biophysical Journal* **65** 2201-16 (1993); N. Wadhwa et al., "Torque-dependent remodeling of the bacterial flagellar motor," *PNAS*, **116**, 24, 11764-11769 (2019).

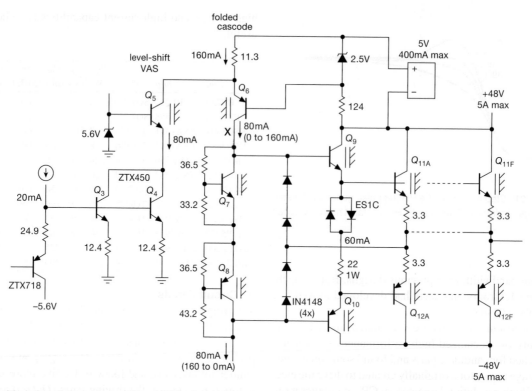

Figure 2x.92. Laboratory power amplifier, 100 V, 5 A, 10 MHz version. Simplified output-stage wiring, upper-side only. All *pnp* and *npn* power transistors are identical, and are mounted on one heatsink plate for temperature tracking. An emitter-follower stage isolates the paralleled output transistors and helps to keep the VAS-node capacitance low. Combined with high VAS currents, this allows for very fast slewing.

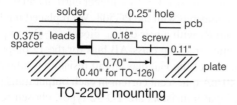

TO-220F mounting

Figure 2x.93. By leaving matching holes in the PCB, TO-220F power transistors can be dismounted from the underlying heatsink without unsoldering them from the PCB.

We chose 2SC4883A and 2SA1859A for our particular amplifiers, giving us 120 pF of total node capacitance, and 160 mA peak current, as shown. It's these numbers that are responsible for our amplifier's high-speed performance: the calculated slew rate is $S=I/C=160\text{mA}/120\text{pF}=1300\text{V}/\mu\text{s}$, and the VAS gain at 10 MHz is $G=g_m X_C=43$, where X_C is the reactance of the combined C_{cb}'s of Q_7–Q_{10}, and g_m is the combined transconductances of common-emitter amplifiers Q_3Q_4 and their mirror twins (not shown), which sums

to $g_m=4/R_E$ (check our arithmetic!). At high frequencies (above 1.3 MHz) the C_1R_2 network (in this case, empirically, $C_1=680$ pF and $R_2=170\,\Omega$) serves to increase the effective stage gain to $G_V=5\text{k}/170=29$, up to 10 MHz. To complete this "laboratory amplifier" we wrapped feedback[57] around the TIA power stage, setting an inverting voltage gain of $G_V=-5$.

B. Output protection

Users need to be aware of the frailty of amplifiers like this, given the high thermal resistance of the plastic-insulated TO220F package, and the poor SOA capability of BJT power transistors (see for example Figs 3.95 and 2x.94). When operating into a resistive load, the highest currents will occur at high output voltages (corresponding to the lowest V_{CE} across the output transistors) which helps to

---

[57] Specifically, $R_F=5$k, $R_{in}=1$k, and decorated with some additional small compensation R's and C's. For details, visit the project at www.dropbox.com/sh/dno89om1uezxs8a/AACoJsLyNazSQZvE9 _cTcH4Ja

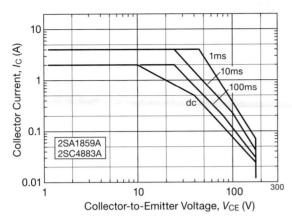

Figure 2x.94. Safe operating area (SOA) curves for the 2SA1859.

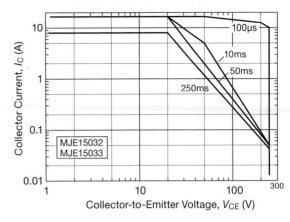

Figure 2x.95. Safe operating area (SOA) curves for MJE15032 and MJE15033.

keep their power dissipation and SOA in check. However, with a reactive load the highest currents can occur at high voltages across the transistors, and under those conditions you need to keep the output currents far below their ratings.[58]

C. Transistor choices

The TO-220F parts we chose have an attractive low capacitance among the power parts, but some may worry they are excessively wimpy, and may be tempted to use larger TO-220 or TO-264 parts. For example, the 50 W MJE15032 types have more attractive SOA limits, see Figure 2x.95. One possibility is to restrict their use to the ten output transistors. But their larger capacitance requires greater drive currents; for example, if they have $C_{cb}=50\,pF$, the emitter-follower driver will see a 500 pF load during slewing, so to achieve 1000 V/μs it would see a load current $i=C\,dv/dt=0.5\,A$. This might constrain the maximum allowable slewing repetition rate.

Some users may want to consider a set of power MOSFETs for the output transistors; here are some choices:

nMOS	pMOS	Pkg	P_{diss} (W)	V_{BR} (V)	C_{ob} (pF)	f_T (MHz)	FOM (W/pF)
2SK1058	2SJ162	TO-3P	100	160	400	35	0.25
IRF640	IRF9640	TO-220	125	200	375	50	0.33

Lateral MOSFETs (2SK1058, 2SJ162), with their negative temperature coefficient (see Fig. 3.118), can be safely paralleled, but ordinary VMOS power MOSFETs are at risk of thermal runaway. You'd want to keep as many fast BJTs as possible in the design, but the heatsink-mounted bias transistors likely would need to be matching MOSFETs. If you tamper with the emitter follower, your circuit modifications could include changing the floating 5 V supplies to 12 V to provide enough gate drive.

---

[58] With 48 V supplies, and taking into account the five output transistors on each rail, we can evaluate how far away from the rails (i.e., how much voltage across the transistors) the output voltage can go when driving a high-current load, while staying within the SOA limits. Without allowing for safety margins or quiescent temperatures, we can have single 10 ms 5 A current pulses for resistive loads at any output voltage, or single 100 ms 5 A in-phase (resistive) current pulses for load voltages above 13 V. For steady *dc* output the SOA limits are respected for currents up to 5 A with output voltages above 28 V, or for currents up to 1.5 A at any output voltage. The latter condition could create a power dissipation of up to 15 W in each of the five output transistors, raising the junction 94°C above the heat-sink temperature (which is pushing the limits, since $T_{J(max)}$ is 150°C). Out-of-phase currents (e.g., pulling up from near the negative rail) should have short duration, such as during slewing.

ADVANCED FET TOPICS

CHAPTER **3x**

In the third chapter of AoE3 we explored in some depth the field-effect transistor (FET), in both its flavors (JFET and MOSFET). These are 3-terminal devices, analogous to the BJT, but offering the important advantage of near-zero gate current. MOSFETs are the dominant species in digital circuits (logic and memory) and in power switching, but JFETs are valuable components in linear and low-noise design.

In this chapter we go both wider and deeper: first we explore JFET properties of transconductance and gain, and how to squeeze the most linearity out of a JFET amplifier or follower; then we turn our attention to MOSFETS, including their use in linear circuits, in a variety of current sources, in high-voltage and high-current pulse circuits, and in other applications. More than once we'll see the remarkable cascode configuration come to the rescue, as we continue our survey of FET properties and applications.

Review of Chapter 3 of AoE3

To bring the reader up to speed, we start this chapter with the end-of-chapter review from AoE3's Chapter 3:

¶A. FETs.
In Chapter 3 we explored the world of Field-Effect Transistors, or FETs. FETs have a conducting channel with terminals named *Drain* and *Source*. Conduction in the channel is controlled by an electric field created by a third *Gate* electrode (§3.1). As with bipolar transistors (BJTs), FETs are transconductance devices (see ¶G below), which means the drain *current* (assuming sufficient drain-to-source voltage) is controlled by the gate *voltage*.

¶B. *n*-channel and *p*-channel.
Like BJTs with their *npn* and *pnp* types, FETs come in both *n*- and *p*-channel polarities (§3.1.2). In either case the channel conductance increases if the gate voltage is taken toward the drain voltage. For example, for an *n*-channel FET with a positive drain voltage, the channel can be turned on with a sufficient positive-going voltage, and cutoff with a sufficient negative-going voltage. That's not to say the *n*-channel device requires positive and negative voltages to turn on and off. A threshold voltage V_{th} can be defined where the FET is just slightly turned on, and the channel responds to gate voltages above and below V_{th} for control.

¶C. Enhancement and Depletion Modes.
See Figure 3.8. Enhancement-mode devices have a high enough threshold voltage V_{th} that they are nonconducting (i.e., off) when their gate voltage is at $V_{GS}=0$ V. To bring such a FET into conduction, the gate is brought positive (if *n*-channel) or negative (if *p*-channel). Depletion-mode devices, by contrast, have their threshold voltage well into the "off" direction, thus they are conducting (i.e., on) with their gate-voltage at $V_{GS}=0$ V. Thus for example you must apply a considerable negative gate voltage V_{GS} to turn off an *n*-channel depletion-mode FET. See Figure 3.9 where drain current versus gate voltage is shown for a selection of *n*-channel devices. FETs can be fabricated with the transfer curve shifted left or right (more about this in ¶H below). Figures 3.10 and 3.11 show convenient maps of the FET types.

¶D. MOSFETs and JFETs.
In *metal-oxide* FETs (MOSFETs) the gate electrode is fully insulated from the channel, and can be taken positive or negative, typically up to ± 20 V. In junction FETs (JFETs) the semiconductor gate contacts the channel and acts as a diode junction, so it is insulated only in the reverse direction. Therefore JFETs are necessarily depletion mode devices; one cannot make an enhancement-mode JFET. Figures 3.6 and 3.7 show FET symbols.

¶E. FET Characteristics, Gate and Drain.
See Figure 3.13. A FET's channel conductance and current is controlled primarily by its gate voltage, but it's also affected by the drain voltage V_{DS}. At very low drain voltages the channel acts like a resistor, whose value is controlled by the gate (§3.1.2 and §3.2.7); this is called the *linear* region. At higher drain voltages the drain current levels off, being controlled by the gate voltage and only weakly dependent upon drain voltage; this is called the *saturated* region. In the saturated region the FET drain acts like a current source (or sink), and the device is characterized by its transconductance g_m (see ¶G below). MOSFETs are often used as switches. In this mode of operation a large gate voltage (e.g., 10 V) is applied to make the channel resistance low enough to approximate a closed switch. More on FET switches in sections ¶¶ O–Q below.

¶F. Square-law.
Over a large region of gate voltages greater than V_{th}, and for drain voltages above a volt or so (i.e., in the saturated region), a FET's drain current behaves like a square-law device; that is, its drain current is proportional to the square of the excess gate-drive voltage $(V_{GS}-V_{th})^2$, see Figure 3.14 and eq'n 3.2. This is sometimes called the *quadratic* region. The threshold voltage V_{th} is generally determined with an extrapolated $\sqrt{I_D}$ plot, as the figure shows. For V_{GS} below threshold the FET is in the subthreshold region; see ¶I below.

¶G. Transconductance and Amplifiers.
Transconductance g_m is the change in output drain current caused by a change in gate voltage: $g_m=i_D/v_{GS}$ (the lower-case i and v signify small signals). Common-source FET amplifiers (§3.2.3, Figures 3.28 and 3.29) have voltage gain $G=-g_m R_D$, where R_D is the drain load resistance. In contrast to BJTs (where $g_m \propto I_C$), the transconductance of FETs rises only as $\sqrt{I_D}$ in the important quadratic region; see Figures 3.53 and 3.54. As a consequence FET amplifiers with resistive drain loads have lower gain when designed to operate at higher current, because R_D is generally chosen

inversely proportional to drain current. The FET's internal output resistance also acts as a load resistance, thus limiting gain ("G_{max}") even with an ideal current-source drain load; see §3.3.2 eq'n 3.13, and Table 3.1.

When used as a *follower*, an FET has an output impedance $r_{out}=1/g_m$, see ¶K below.

¶H. Biasing JFET Amplifiers.
JFETs are well suited for making signal amplifiers (by contrast there are few viable small discrete *MOSFETs*), and they work especially well in low-noise amplifiers. But there's one very painful issue analog designers face: the uncertain value of the gate operating voltage for any given part. Scanning the min and max columns for $V_{GS(off)}$ in the JFET Table 3.1 on page 141, we see values for a particular JFET that range from -1 V to -7 V, or -0.4 V to -4 V. The latter is a 10:1 ratio! Figure 3.17 shows V_{GS} histograms for 300 parts, 100 each for three different JFET types in a family. Here we see gate voltage spreads of about 1 V, which you might rely upon if you buy a batch of parts from one manufacturer and measure them. But, *caution*: Figures 3.51 and 3.52 show how the same part type may vary when purchased from different manufacturers. To deal with the uncertainty, special biasing schemes are often required in FET amplifier circuits. Figures 3.25 and 3.41 show examples of the load-line concept for analyzing amplifier biasing.

¶I. Subthreshold Region.
The simple FET formula of eq'n 3.2 predicts zero drain current when the gate voltage reaches threshold ($V_{GS}=V_{th}$). In reality the drain current is not zero, and transitions smoothly to a subthreshold region (see Figure 3.16) where the FET looks more like a BJT, with its exponential Ebers-Moll characteristic (§2.3.1). In this region (where I_D rises exponentially with V_{GS}) we're glad to see a higher $g_m \propto I_D$; but sadly the FET proportionality constant is usually $2\times$ to $5\times$ smaller than for BJTs, see Figure 3.53.

¶J. Self-biased Amplifiers.
Depletion-mode MOSFETs (and all JFETs) operate with a reverse voltage on their gates, which allows them to be self-biased (§3.2.6A). The source terminal is "higher" than the gate terminal, so a source resistor connected between them sets the drain current to $I_D=V_{GS}/R$. This is also a convenient way to make a 2-terminal current source, but the tolerance will be poor due to the wide variability in V_{GS}, see ¶H. Alternately the V_{GS} voltage available at the source pin may be used to operate a current-setting IC like the LM334.

¶K. Source Followers.
Source followers (§3.2.6), Figure 3.40 have a nominal gain of 1, analogous to the BJT emitter follower. Because of their lower g_m, however, they have considerably higher output resistance, $r_{out}=1/g_m$, so the ideal unity gain is reduced by load resistance, see eq'n 3.7.

¶L. FETs as Variable Resistors.
At low drain voltages ($V_{DS} \ll V_{GS}$) FETs act as variable resistors programmed by the gate voltage. Because the slope varies with V_{DS}, however, the resistance is somewhat nonlinear. But there's a simple trick to linearize this resistance, by exploiting the quadratic behavior of FETs, see Figures 3.46 and 3.47.

¶M. FET Gate Current.
The gate of a JFET forms a diode junction with the channel; it's normally reverse-biased, with some non-zero dc leakage current (§3.2.8). This current roughly doubles for every 10°C temperature increase; furthermore it increases dramatically at high drain currents and drain voltages due to impact ionization, see Figure 3.49. MOSFET gates do not suffer from either of these leakage-current-increasing effects. In contrast to the generally negligible dc gate leakage, the input capacitance C_{iss} of FETs (which can be quite high, many hundreds of pF for large power MOSFETs) often presents a substantial ac load. Use a gate-driver chip (Table 3.8) to provide the high transient currents needed for rapid switching.

¶N. JFET Switches.
JFETs can be used as analog-signal switches, as in the n-channel switch of Figure 3.62. The switch is OFF when the gate is taken at least V_{th} below than the most negative input signal. To turn the switch ON the gate voltage must be allowed to equal the source. JFETs are symmetrical, so e.g., for an n-channel part, the "source" would be the most negative pin. Large-die JFETs work well as power switches up to 100 mA; Table 3.1 lists parts with R_{ON} as low as 3 Ω.

¶O. CMOS Switches.
CMOS signal switches are made with a parallel pair of complementary n- and p-channel MOSFETs. This reduces R_{ON} as shown in Figure 3.61, and beneficially causes cancellation of most of the injected charge transfer (§3.4.2E), see Figure 3.79. The injected charge scales roughly inversely proportional to R_{ON} (Figure 3.81), so there's a tradeoff between desirably low on-resistance and desirably low self-capacitance. As an example, Table 3.3 lists a switch with an impressive $R_{ON}=0.3$ Ω – but it's burdened

with a whopping 300 pF of self-capacitance. A T-switch configuration can be used to reduce the signal feedthrough at high-frequencies, see Figure 3.77.

¶P. CMOS Logic Gates.

See Figure 3.90. A series pair of complementary (*n*- and *p*-channel) small-geometry MOSFETs between the positive rail and ground forms the simplest logic inverter (Figure 3.90); more switches can be arranged to make CMOS logic gates (e.g., Figure 3.91, §3.4.4), with the attractive property of nearly zero static power, except when switching. CMOS logic is covered extensively in Chapters 10 and 12, and is the basis for all contemporary digital processors.

¶Q. MOSFET Power Switches.

Most power MOSFETs (§3.5) are enhancement type, available in both *n*- and *p*-channel polarities. They are very popular for use as high-current high-voltage power switches. A few relevant parameters are the breakdown voltage V_{DSS} (ranging from 20 V to 1.5 kV for *n*-channel, and to 500 V for *p*-channel); the channel on-resistance $R_{DS(on)}$ (as low as 2 mΩ); the power-handling ability (as high as 1000 W with the case held unrealistically at 25°C); and the gate capacitance C_{iss} (as high as 10,000 pF), which must be charged and discharged during MOSFET switching, see ¶S below. Table 3.4a lists representative small-package *n*-channel parts rated to +250 V and *p*-channel parts of all sizes to −100 V; Table 3.4b extends the *n*-channel selection to higher voltage and current; more complete tables are found in this chapter (Chapter 3x).

¶R. Maximum Current.

MOSFET datasheets list a maximum continuous rated current, specified however at an unrealistic 25°C case temperature. This is calculated from $I_{D(max)}^2 R_{DS(ON)} = P_{max}$, substituting a maximum power $P_{max} R_{\Theta JC} = \Delta T_{JC} = 150°C$ (see §9.4), where they have assumed $T_{J(max)} = 175°C$ (thus a 150°C ΔT_{JC}), and they use the value of $R_{DS(ON)}$ (max) at 175°C from an R_{DS} tempco plot (e.g., see Figure 3.116). That is, $I_{D(max)} = \sqrt{\Delta T_{JC}/R_{\Theta JC} R_{ON}}$. Some datasheets show the calculation for a more realistic 75°C or 100°C case temperature. Even so, you don't really want to run your MOSFET junction at 175°C, so we recommend using a lower maximum continuous I_D and corresponding P_{diss}.

¶S. Gate Charge.

The capacitances in power MOSFETs that slow down switching are most easily analyzed with gate-charge plots, like Figure 3.101. First consider turn-ON: as current flows

into the gate capacitance $C_{iss}+C_{rss}$ (dominated by C_{iss}) the gate voltage rises. There is a switching delay, because the FET's drain remains off until the gate voltage is high enough for the FET to sink the drain current. Then the drain voltage starts to fall, as seen in Figures 3.102 and 3.103. The falling drain creates a reverse gate current $I = C_{rss} \, dV_D/dt$ that prevents further increase in the gate voltage. Put another way, the falling slew-rate $dV_D/dt = I_G/C_{rss}$ is set by the gate current available to charge the feedback (Miller) capacitance C_{rss}. When V_{DS} reaches zero the gate resumes charging, now at a slower rate because the C_{rss} contribution to total gate capacitance is larger at $V_{DS}=0$, see Figure 3.100. The MOSFET does not reach its intended low value of $R_{DS(ON)}$ until the gate attains its full drive voltage. Turn-off proceeds similarly. MOSFET datasheets include values for C_{iss} and C_{rss}, but the latter is typically at $V_{DS}=25$ V, so you need to go to the datasheet plots of capacitances versus drain voltage.

¶T. MOSFET Gate Damage.

MOSFET gates typically have ± 20 V to ± 30 V maximum ratings, beyond which the very thin metal-oxide gate-channel insulator can be permanently damaged, see Figure 3.105. Be sure to discharge static charge before installation of discrete MOSFETs and MOS ICs.

¶U. FET versus BJT for Power Switching.

See §3.5.4H; see also ¶Z below.

¶V. MOSFET Switch Polarity.

Both *n*- and *p*-channel polarities of MOSFETs can be used to switch a voltage, see Figure 3.106 where most of the circuits show a conventional approach with a *p*-channel FET switching a positive voltage. But circuit E shows an *n*-channel FET doing the same task, with an additional voltage source powering the gate (the better-performing *n*-channel FET is preferred if it can be easily used, see §3.1.2). Figure 3.107 illustrates the use of photodiodes to power the high-side gates, to make "floating" switches.

¶W. Power MOSFET Amplifiers.

Unlike bipolar power transistors, power MOSFETs have a wide safe-operating area (SOA) and do not suffer from second breakdown (see Figure 3.95), which is due to a localized thermal-runaway heating problem. Figure 3.119 shows typical class-AB biasing techniques necessary for use in linear power amplifiers.

¶X. Depletion-mode Power MOSFETs.

Although most power MOSFETs are enhancement-mode types, *n*-channel depletion-mode types are available; §3.5.6D shows some applications. See also Table 3.6 on page 210.

¶Y. Paralleling Power MOSFETs.

When used as switches, yes, but when used in power amplifiers, no, at least not without high-value source-ballast resistors! Figure 3.117B shows an elegant active-feedback workaround for use with regulator pass elements.

¶Z. IGBTs.

IGBTs are an alternative to power MOSFETs, see §3.5.7 where we show a comparison between power MOSFETs, IGBTs and BJTs. They're primarily useful at voltages above 300 V and switching rates below 100 kHz, though there are some nice IGBTs for use at RF, for example the IRGB4045, good for 150 W or more at 20 MHz.

3x.1 A Guided Tour of JFETs

Back in Chapter 3 we included an abridged selection of *n*-channel JFETs for circuit design (Table 3.1, on page 141) and also an exhaustive listing (Table 3.7). The latter lists most available JFETs, and it includes both datasheet min/max specifications and measurements we've made on sample parts. In Chapter 3 and this chapter we've devoted considerable space to circuit design with JFETs, and for that reason we've included the full JFET table here (Table 3x.1 on the following page). But that still leaves the obvious question: how do you choose the best JFET for your application? You can choose from parts in the table with I_{DSS} ranging over a factor of more than 10,000, with transconductance ranging over nearly $\times 1000$, and with capacitances ranging over $\times 100$. How to decide?

Herewith a handy guide to Table 3x.1. As always, we caution you to study carefully the candidate part's datasheet, and be sure to check out the part's availability in the package style you want, before committing it to your design.

JFET Families; *n*- and *p*-channel types. Most of the JFETs are grouped by families of parts, e.g., the 2N5457, 58 and 59. These are manufactured in the same process, and they'll have similar specifications of parameters such as g_m and capacitance; but the parts in the series will offer overlapping ranges of I_{DSS} and $V_{GS(off)}$. Table 3.1 is sorted approximately by increasing JFET family transconductance g_m, but with exceptions for families such as the J105–J113, which are listed in numerical order.

Only two *p*-channel families are listed, the 2N5460-series amplifiers and the J174-series switches. Among *n*-channel parts, the 2N5457 family is a complement to the 2N5460 family, and the J111 family roughly complements the J174 family. The 2SK170 and LSK170 are very nice low-noise, low-gate-voltage *n*-channel JFETs. Their highly useful complement *was* the Toshiba 2SJ74, now sadly discontinued (RIP). If you find some on the surplus market, grab them!

Jellybeans. "Jellybeans" (third column) are perennially popular parts with good distributor inventory, low prices, and (generally) multiple manufacturers. Other listed parts that are not jellybeans are likely to have been more recently introduced; they often have superior properties, and may be preferred for your designs. They'll cost more, and they likely won't have a second source.

Packages. Classic JFET types came in metal TO-18 and plastic TO-92 "through-hole" packages. The latter are convenient for prototyping, but choose parts that are also available in SOT-23 (and smaller) surface-mount packages if you want some confidence that you will still be able to get them in a few years. There are two incompatible pinout configurations for both the through-hole and the SMT parts, shown in Figure 3x.1 and indicated in the table with corresponding letters. In many cases the SOT-23 package versions carry MMBF or PMBF prefixes in place of the 2N or PN for the TO-92 versions. In some cases the TO-92 versions will disappear from distributors' shelves, but you should still be able to find the SMT versions by searching with the numerical portion of the original part number, e.g., "5486," etc. Parts beginning with PMBF are made by NXP; the parts with MMBF prefix are usually made by ON Semiconductor (which acquired Fairchild Semiconductor in 2016).

V_{DS} max. It's surprising how many JFETs are limited to 20 or 25 volts, with only a few making it to 50 or 60 volts. You may find yourself using cascode circuits because of this issue alone. It's important to realize that the maximum voltage rating also applies to V_{DG} (the $V_D - V_{GS}$ difference), further exacerbating the low voltage ratings.

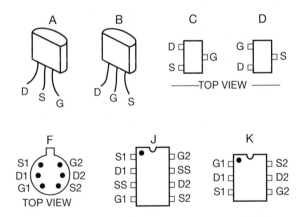

Figure 3x.1. Pin assignments ("pinouts") for the JFETs in Table 3.1. Some duals allow connection to the substrate ("**SS**"). We have found no exceptions to the general rule that JFETs are symmetrical – that is, **D**rain and **S**ource terminals are interchangeable.

Table 3x.1: Junction Field-Effect Transistor (JFETs)[a]

Part #[b]	N or P-channel	Jellybean?	TO-92: 2N,PN	SOT23: MMBF	SOT23: PMBF	MMBF_LT	V_{DSS} max (V)	I_{DSS} min (mA)	I_{DSS} max (mA)	I_{DSS} meas (mA)	R_{ON} max (Ω)	$V_{GS(off)}$[c] min (V)	$V_{GS(off)}$ max (V)	V_{GS} @ I_D measured (V)	V_{GS} @ I_D (mA)	g_m min (mS)	g_m max (mS)	g_m @I_D (mA)	g_m meas (mS)	g_m @I_D (mA)	G_{max}[e]	C_{iss} typ (pF)	C_{rss} typ (pF)
PN4117	N	•	A	C	-	-	40	0.03	0.09	0.07	-	-0.6	-1.8	-0.33	0.03	0.07	0.21	z	0.09	0.03	420[r]	1.2	0.3
'4118	N	-	A	C	-	-	40	0.08	0.24	0.20	-	-1	-3	-1.33	0.1	0.08	0.25	z	0.13	0.1	260[r]	1.2	0.3
'4119	N	-	A	C	-	-	40	0.20	0.60	0.30	-	-2	-6	0.0	0.3	0.10	0.33	z	0.18	0.3	140[r]	1.2	0.3
BFT46	N	-	-	C	-	-	25	0.20	1.5	0.63	-	-	-1.2	-0.16	0.3	1	-	z	1.7	0.3	190[s]	3.5	0.8
BF511	N	-	-	D	-	-	20	0.7	3	4.2	-	-	-1.5	-0.75	1	4	-	z	2.7	1	120	-	0.3
2N5457	N	•	A	C	-	-	25	1	5	3.5	-	-0.5	-6	-0.81	1	1	5	z	2.3	1	220	4.5	1.5
'5458	N	-	A	C	-	-	25	2	9	4.1	-	-1	-7	-0.97	1	1.5	5.5	z	2.2	1	190	4.5	1.5
'5459	N	-	A	C	-	-	25	4	16	9.9	-	-2	-8	-1.82	3	2	6	z	2.9	3	100	4.5	1.5
2N5460	P	•	A	C	-	-	25	-1	-5	3.4	-	0.75	6	+0.97	1	1	4	z	1.9	1	260	4.5	1.2
'5461	P	•	A	C	-	-	25	-2	-9	2.7	-	1	7.5	+0.67	1	1.5	5	z	2	1	210	4.5	1.2
'5462	P	•	A	C	-	-	25	-4	-16	5.9	-	1.8	9	+4.15	1	2	6	z	2.5	3	30	4.5	1.2
MMBF4416	N	•	A	C	C	C	30	5	15	5.9	-	-	-6	-0.19	5	4.5	7.5	z	3.9	5	70	4	0.8
2N5484	N	-	A	C	-	C	25	1	5	3.3	-	-0.3	-3	-0.73	1	3	6	z	2.3	1	230	10	2.2
'5485	N	-	A	C	-	-	25	4	10	6.6	-	-0.5	-4	-1.65	1	3.5	7	z	2.1	1	150	10	2.2
'5486	N	-	A	C	-	-	25	8	20	14	-	-2	-6	-2.61	1	4	8	z	2.1	1	75	10	2.2
2SK170BL	N	-	B	-	-	-	40	6	12	6.1	-	-0.2	-1.5	-0.04	1	22[t]		z	29	5	470	30	6
LSK170B	N	-	B	C	-	-	40	6	12	7.6	-	-0.2	-2	-0.17	3	10[t]		1	20	3	160	20	5
LSK170C	N	-	B	C	-	-	40	10	20	13	-	-0.2	-2	-0.26	5	10[t]		1	24	5	90	20	5
BF861B	N	-	-	C	-	-	25	6	15	8	-	-0.5	-1.5	-0.47	1	16	25	z	16	5	150	7.5	
BF545C	N	-	-	C	-	-	30	12	25	19	-	-3.2	-7.8	-1.80	5	3.0	6.5	z	3.7	5	30	1.7	0.8
BF862	N	-	-	C	-	-	20	10	25	12	-	-0.3	-1.2	-0.21	5	35	45[t]	z	26	5	270	10	1.9
PF5103	N	-	A	C	-	-	40	10	40	-	30	-1.2	-2.7	-1.00	5	7.5	-	2	10	5	160	16	6
PN4391	N	•	A	C	C	C	40	50	150	115	30	-4	-10	-7.15	5	12[t]		5	8.8	5	30	12	3.5
'4392	N	•	A	C	C	C	40	25	75	38	60	-2	-5	-1.67	5	16[t]		10	10	5	130	12	3.5
'4393	N	•	A	C	C	C	40	5	30	16	100	-0.5	-3	-1.25	1	13[t]		10	6.2	1	150	12	3.5
J105	N	-	A	C	-	-	25	500	-	-	3	-4.5	-10	-8.39	5	40[t]		5	37	10	60	160[m]	35[m]
J106	N	-	A	C	-	-	25	200	-	-	6	-2	-6	-2.42	5	53[t]		5	43	10	230	160[m]	35[m]
J107	N	-	A	C	-	-	25	100	-	-	8	-0.5	-4.5	-1.93	5	75[t]		5	48	10	340	160[m]	35[m]
J108	N	-	A	C	C	-	25	80	-	325	8	-3	-10	-5.83	5	37[t]		5	31	10	60	85	15
J109	N	•	A	C	C	-	25	40	-	201	12	-2	-6	-2.85	5	26[t]		5	32	10	160	85	15
J110	N	-	A	C	C	-	25	10	-	122	18	-0.5	-4	-1.80	5	20[t]		5	34	10	220	85	15
J111	N	•	A	C	C	-	35	20	-	115	30	-3	-10	-7.6	5	-	-		8.4	5	30	28	5
J112	N	-	A	C	C	-	35	5	-	47	50	-1	-5	-2.8	5	6.7[t]		1	9.5	5	100	28	5
J113	N	-	A	C	C	-	35	2	-	21	100	-0.5	-3	-1.0	5	8[t]		1	11	5	100	28	5
J174	P	-	B	C	C	-	30	-20	-135	26	85	5	10	+2.08	5	4.5	-	5	-	-	15	13	6
J175	P	•	B	C	C	C	30	-7	-60	13	125	3	6	+1.58	1	-	-	5	-	-	30	13	6
J176	P	-	B	C	C	C	30	-2	-25	6.1	250	1	4	+0.86	1	6.3	-	5	-	-	40	13	6
J177	P	•	B	C	C	C	30	-1.5	-20	4.2	300	0.8	2.5	+0.62	1	-	-	5	-	-	50	13	6
J308	N	-	A	C	C	-	25	12	60	35	-	-1	-6.5	-	-	8	-	10	12	5	120	4	2
J309	N	•	A	C	C	C	25	12	30	23	-	-1	-4	-1.2	5	10	20	10	11	5	300	4	2
J310	N	-	A	C	C	C	25	24	60	39	-	-2	-6.5	-2.4	5	8	18	10	8.9	5	100	4	2
dual JFETs											V_{OS} (mV)												
LS840-42	N		F				60	0.5	5	3.3	5[m]	-1	-4.5	-0.85	1	0.5	1	0.2	2.1	1	180	4	1.2
'843-5	N		F				60	1.5	15	-	1[m]	-1	-3.5	-	-	1	1.5[t]	0.5	-	-	-	8[m]	3[m]
LSK389A	N		F, J				40	2.6	6.5	-	20[m]	-0.15	-2	-	-	8	20[t]	3	-	-	-	25	5.5
'389B	N		F, J				40	6	12	12	20[m]	-0.15	-2	-0.24	8	8	20[t]	3	23	5	170	25	5.5
'389C	N		F, J				40	10	20	-	20[m]	-0.15	-2	-	-	8	20[t]	3	-	-	-	25	5.5
LS5912	N		F, J, K[p]				25	7	40	18	15[m]	-1	-5	-1.75	5	4	10	5	5.7	5	70	5	1.2
IFN146	N		F[v]				40	-	30	6	20[m]	-0.3	-1.2	-0.19	1	30	40[t]	z	25	5	660	75[m]	15[m]

(a) listed generally by increasing I_{DSS}, but also by part number within a family (e.g., J105–J113); see also Table 8.x for noise parameters.
(b) for families of related parts, **boldface** designates the family matriarch. (c) usually specified at I_D=1nA or 10nA, though sometimes at 10µA or even 200µA (e.g., for the J105–J113 "switches"); it doesn't much matter, given the wide range of specified $V_{GS(off)}$. (d) see the accompanying pinout figure; all JFETs appear to be symmetric (source and drain are interchangeable), but *italic* designates a datasheet pinout in which the S and D terminals are interchanged. (e) $G_{max}=g_m/g_{os}$, the maximum grounded-source voltage gain into a current source as drain load; listed values measured at I_D=1mA and V_{DS}=5V, unless noted otherwise. G_{max} is proportional to V_{DS}, and for most JFETs G_{max} is relatively constant over varying I_D. Use tabulated G_{max} to find $g_{os}=g_m/G_{max}$. (m) maximum. (p) several PDIP-8 pkgs available. (r) at I_D=30µA. (s) at I_D=300µA. (t) typical. (v) variant of "F" pinout: G and D terminals interchanged. (z) at I_{DSS}.

I$_{DSS}$ and V$_{GS(off)}$. For datasheet purposes the linear operating region for a JFET is bounded by a maximum current I_{DSS} corresponding to zero gate voltage V_{GS} (recall that JFETs are depletion-mode devices), and by a gate–source cutoff voltage $V_{GS(off)}$ at which the drain current is close to zero. (It's not truly zero, but rather a small current defined in the spec.) For the PN/MMBF/PMBF/SST4392, for example, you'll find from the datasheets of every manufacturer that $V_{GS(off)} = -2\,V$ (min), $-5\,V$ (max) at $I_D = 1\,nA$ and $V_{DS} = 20\,V$. But a caution: the "off"-current spec may be as high as $10\,\mu A$.[1] Both the I_{DSS} and $V_{GS(off)}$ specs have minimum and maximum values, defining the JFET's operating region, see Figure 3x.2.

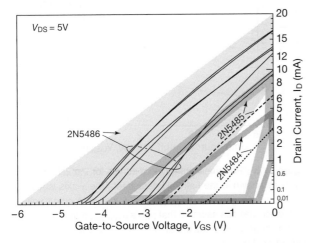

Figure 3x.2. Measured transfer characteristics of nine JFETs from the 2N5484–6 family, plotted on a $\sqrt{I_D}$ axis (thus straight lines in the quadratic region). The borders of the shaded contours connect the specified limits of I_{DSS} and $V_{GS(off)}$: in all cases the animals are kept in their cages.

The minimum and maximum specified I_{DSS} is useful when selecting a part for operation within some desired range of drain current with zero gate voltage. But it's often the case that dealing with the wide range of JFET datasheet min–max voltages and currents can be a real pain. For example, we might really like to know the min–max gate voltage at our chosen operating current, e.g., $I_D = 1\,mA$, rather than at the maximum (and poorly specified) current. Making a $\sqrt{I_D}$ plot like Figure 3x.2 for your candidate JFETs can help narrow the wide spec spread.

Or, if you're fond of equations (or hate graphs, or both) you can deal with this analytically (but don't confuse "analytical" with "accurate"!). The JFET's drain current in the quadratic region[2] can be written as $I_D = I_{DSS}(V_{GS}/V_{th} - 1)^2$, into which you can substitute the reasonable estimate $V_{th} \approx V_{GS(off)} + 0.2\,V$ for V_{th} at $I_D = 1\,\mu A$ (i.e., extrapolated from the cutoff voltage at a drain current of 1 nA; see Figure 3x.3). Repeat this process for the datasheet's min and max values. Review §3.3 and Figures 3.51, 3.52, and 3.55. Keep in mind that you can screen and bin your incoming parts for critical applications. You'll probably find reasonably tight matching of the measured I_{DSS} and $V_{GS(off)}$ parameters from a single batch of parts, as we did in the data plotted in Figure 3.17.

Lest we paint too rosy a picture, thereby leading you to believe that you can effortlessly design linear JFET circuits with confidence in their quiescent bias conditions, we hasten to add that the loose worst-case specifications of JFET I_{DSS} and $V_{GS(off)}$ can (and probably will) be your nemesis. And Nemesis is "the punisher of hubris." The situation is not *hopeless* (as the circuit examples in Chapter 3 demonstrate); it's just difficult, and requires extra care (and the use of circuit configurations that create predictable operating points, for example I_D-setting feedback).

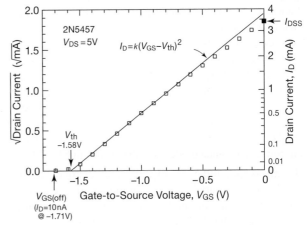

Figure 3x.3. Measured I_D versus V_{GS}, plotted on a square-root y-axis. The "quadratic region" is a reasonable approximation, not bad but not perfect. The "threshold voltage" V_{th} (or $V_{GS(th)}$) is the straight-line extrapolation to zero current, which is different from $V_{GS(off)}$, the latter conventionally specified at some small drain current (often 1 nA or 10 nA).

---

[1] It's not hard to extrapolate to a different off-current of your choice, knowing that a JFET's transconductance approaches that of a BJT in the subthreshold region: simply shift the $V_{GS(off)}$ according to the BJT's "60 mV/decade" rule. For example, a 180 mV shift of V_{GS} would translate a manufacturer's $V_{GS(off)}$ (at $I_D = 10\,nA$) to a new $V_{GS(off)}$ at $I_D = 10\,\mu A$ "spec," handy for use in a $\sqrt{I_D}$ versus V_{GS} plot.

[2] We emphasize *JFETs* specifically, because power MOSFETs used in linear circuits are usually operating in or near the (exponential) subthreshold region, rather than the quadratic region where JFETs typically operate.

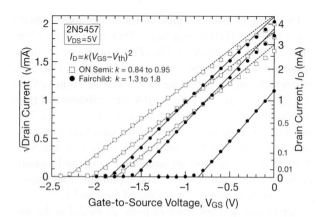

Figure 3x.4. Six parts from two manufacturers, plotted as in Fig. 3x.3, showing sample and manufacturer variability.

Switches versus amplifiers; R_{ON}. Any JFET can serve either as an amplifier or as a switch; but some are promoted as particularly well suited for switching. Such parts include R_{ON} specs in their datasheets[3]; these are shown in Table 3x.1, with values ranging from $3\,\Omega$ to $300\,\Omega$. Amplifier JFETs, when used as switches, will have roughly $R_{ON} \approx 1/g_m$, where g_m is the transconductance at I_{DSS} ($V_{GS} = 0\,\text{V}$). From the table you can see that low-R_{ON} switch JFETs tend to have rather high capacitances, with channel-to-gate capacitance C_{rss} values up to $35\,\text{pF}$ when back-biased to $V_{GS} = -10\,\text{V}$ (i.e., OFF), and even higher (according to the datasheet plots) when ON, as much as $60\,\text{pF}$ or so for the J105 family.

Some designers intentionally use a small, high-resistance amplifier JFET like a 2N4117 as a switch, because its low capacitance minimizes charge coupling during switching. Some applications work perfectly well with a $5\,\text{k}\Omega$–$20\,\text{k}\Omega$ ON-resistance. These parts also make excellent low-capacitance diodes, see the graphs in §1x.7.1.

Measured I_{DSS} and V_{GS} at I_D. We provide for each JFET type both a measured I_{DSS}, and a measured V_{GS} at some typical operating current I_D. These values are not meant to contravene the datasheet's max and min specs, rather they are representative of what we observed with delivered parts, and they could be used as starting points for trial designs. We have generally picked an I_D that's well below $I_{DSS(max)}$ and usually below $I_{DSS(min)}$, but high enough to yield a good transconductance gain (g_m) and low noise voltage[4] (e_n).

Low gate voltages are often helpful when designing JFET circuits. Looking at the table, you'll see that for a

given JFET family the lowest $V_{GS(off)}$ part is also the lowest I_{DSS} member of the family. High I_{DSS} parts can be used at low currents, but this may require an inconveniently high gate back-off voltage. For example, see Figure 3.51, where the low-I_{DSS} 2N5457 can be used with a V_{GS} of $-1\,\text{V}$ to $-2\,\text{V}$ to operate at 0.1–1 mA, but the 2N5458 may force you to a gate back-bias of $-3\,\text{V}$, and the 2N5459 (not shown) to as much as $-5\,\text{V}$. Some datasheets have "Parameter Interaction" graphs showing this effect, as for example those in Figure 3x.5.

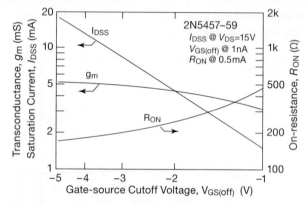

Figure 3x.5. "Parameter Interaction" curves for the 2N5457 JFET family, as shown in the Fairchild datasheet. The x-axis is not gate voltage; it is a given part specimen's actual gate cutoff voltage, from which you can estimate that part's I_{DSS}, g_m, and R_{ON}. The low cutoff-voltage '5457 parts are at the right, '5458 parts in the middle, and '5459 at the left.

Measured g_m at I_D. The table shows a measured value of g_m, generally at the same drain current as our V_{GS} measurement. A reasonable value for g_m at other currents can be had using equation 3.6's relationship $g_m/g_{m0} = (I_D/I_{D0})^{1/2}$. From the table's data you can see the interesting fact that there's not a large g_m variation among parts in a given family,[5] but there's a substantial variation in g_m between different families. For example, at a drain current of 5 mA we measured transconductances of 3.9 mS for a '4416 RF amplifier JFET, compared with 8.4 mS for a J111, and an impressive 29 mS (more transconductance is always better) for a 2SK170B low-noise JFET (a bipolar transistor would measure $g_m = 200\,\text{mS}$ at a collector current of 5 mA). Finally, note that the transconductance at their individual I_{DSS} values is nearly the same among the different members in a family, despite the large variation

---

[3] On some datasheets this is called $R_{DS(on)}$.

[4] See Chapter 8 for JFET noise data and measurements.

[5] This trend seems to hold even among different manufacturers of JFETs with a similar part number, as seen for example in the measured transconductance data of Figure 3.56.

in their I_{DSS}'s, as illustrated nicely in the measured data of Figure 3x.6.

There's plenty of interesting detail in the measured g_m-vs-I_D plots of dozens of JFETs in Figure 3x.12.

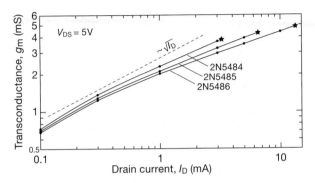

Figure 3x.6. Within a family of JFETs, parts with a lower rated I_{DSS} generally provide the greatest transconductance at a given drain current. Moral: in a circuit run at known I_D, select a JFET with the lowest I_{DSS} that is guaranteed to work, i.e., $I_{DSS} \geq I_{D(ckt)}$.

Maximum gain, G_{max}. We invented the term "G_{max}," in part as a way to parameterize a JFET's maximum possible gain, and in part as a convenient way to present our measured drain output-resistance values $r_o = 1/g_{os}$. Although g_{os} (and therefore r_o) changes with drain current, it scales with I_D in the same square-root fashion as g_m, so that the ratio $G_{max} \equiv g_m/g_{os}$ is roughly independent of the JFET's drain current (see §3x.4.1 and Figure 3x.20). That makes G_{max} an attractive parameter to tabulate.

For example, to find g_{os} at some current I_D, simply use $g_{os} = g_m/G_{max}$, with g_m scaled from its value at a known current I_{D0}; in the quadratic region, for example, you'd have $g_m = g_{m0}(I_D/I_{D0})^{1/2}$. In some cases we've formulated things directly in terms of G_{max} (rather than g_{os}), for example equations 3.11 and 3x.4.

Although G_{max} is relatively insensitive to drain current, it does show significant variation with drain *voltage*. The G_{max} values in the table were measured at $V_{DS} = 5$ V. In general, a FET's output resistance r_o increases with increasing drain voltage, so you get increasing G_{max}, as can be seen in the measured data of Figure 3x.23.

There's an interesting trend evident in the tabulated G_{max}: you'll see that for most JFET families the highest G_{max} is seen for the lowest I_{DSS} member of the family. Although that part has somewhat lower transconductance at its (lower) I_{DSS}, the G_{max} improvements can be impressive. For example, an LSK170C ($I_{DSS} = 13$ mA) has $G_{max} = 90$, whereas the less muscular LSK170A with half the I_{DSS} (6 mA) has five times the maximum gain ($G_{max} = 470$).

Looking up and down the table, you'll see this general trend repeated. While it's true that often you can design your way past a poor G_{max} (e.g., by using the cascode configuration), it's sometimes better simply to choose a more capable part.

Capacitance. Table 3x.1 lists typical JFET input and feedback capacitances, as provided on the manufacturers' datasheets. But a caution is in order: you'll see different measurement conditions used, including some that are quite unrealistic for common operation. For example, three popular parts have their capacitances specified at $V_{DS} = 0$ V and $V_{GS} = -10$ V; $V_{DS} = 15$ V and $V_{GS} = 0$ V; and $V_{DS} = 0$ V and $V_{GS} = 0$ V. The datasheets often include capacitance plots, but these sometimes fail to cover your circuit's operating conditions. Also, one might question the likelihood that a single set of tabulated or graphed capacitances accurately represents the behavior of an entire family's set of parts, which includes family members whose drain current capabilities range from low to high values of I_{DSS}; see also §3.3.5 and Figure 3.57.

Dual JFETs. Sadly, there's not much selection available in monolithic dual (matched) JFETs, but they're what you want when you design differential or low-offset amplifiers (as for example in Figures 3.36 and 3.37), or low-offset followers (Figure 3.43). The semiconductor industry has worked hard to create well-matched JFET pairs in the form of JFET-input *op-amps* (see Table 5.5, where some JFET-input op-amps have 100 μV maximum offset), but the available discrete monolithic JFET pairs in Table 3x.1 do less well, with maximum offset voltages typically in the range of 10 mV–20 mV (the LS843 does best, at $\Delta V_{GS} = 1$ mV, max).

If none of the available dual JFETs have the specifications you want, you'll have to use a pair of discrete JFETs. If you take literally what the datasheet says about the range of gate voltages, a "discrete matched pair" would seem hopeless; for example, the 2N5457 datasheet specifies $V_{GS(off)} = -0.5$ V min, -7 V max. But measurements we've made give a ray of hope. Figure 3.17 shows measured gate voltage histograms of 300 JFETs from the 2N5457–59 family. We see a narrowed 500 mV to 1500 mV spread for V_{GS} (for 1 mA drain current) for each of three family members. If you were to sort a shipment of 100 JFETs into 50 mV bins, you'd get a reasonable population in most of the bins. You could take any two parts from a bin and use them as a "50 mV matched pair."

3x.1.1 Gate current, I_{GSS} and I_G

Typical JFET datasheets include two kinds of gate current specifications: *Gate Reverse Current* (I_{GSS}), and *Gate Operating Current* (I_G). We did not include these in Table 3x.1, for reasons that will soon become evident. Let's look at the gate current scene.

Gate reverse current. Gate reverse current I_{GSS} is the dc gate leakage current with the gate held at a substantial back bias and with source and drain connected together. That's not the way you run a JFET, and it's useful mostly for evaluating a JFET as a low-leakage clamp diode. Even for this purpose the I_{GSS} values shown on datasheets are not terribly useful, being overly conservative (generally they are specified at large reverse bias, e.g., -15 V or -20 V), and often only a maximum value is given. Such a worst-case value tends to be outrageously conservative because of the practical difficulty of testing production parts (with "ATE" – automatic test equipment) at these low currents: it takes longer to measure low currents, and some ATE setups cannot even handle picoamp-scale currents. As an example, how useful is it to know that the 2N5460-series JFETs have $I_{GSS} = 3$ pA typ, 5000 pA max? (That's at 25°C; at 100°C the corresponding figures are 0.3 nA typ, 1000 nA max). Perhaps you should consider yourself lucky to have "typical" values (however unreliable they may be) as a guide; if you were using the *n*-channel complement (the 2N5457-series), you'd find only "maximum" values (1000 pA at 25°C, 200 nA at 100°C).

Gate operating current. When JFETs are in the active region the reverse gate bias is considerably less; but an effect known as "ionization-impact gate current" (§3.2.8) causes dramatic increases in gate current at moderate to large drain-to-gate voltages, particularly for *n*-channel types. Datasheets specify this (if at all) as a tabulated value of gate operating current (I_G) at some chosen operating condition (e.g., $I_G = 20$ pA typ at $V_{DG} = 10$ V and $I_D = 1$ mA for the 2N4416). The additional gate current is approximately proportional to drain current, the ratio rising rapidly with increasing drain-to-gate voltage V_{DG}. This can get pretty serious: for the same 2N4416 at a drain current of 1 mA, I_G rises from 20 pA (at 10 V) to 200 pA at 12 V, 2000 pA at 14 V, and 100,000 pA at 20 V! This suggests real caution when evaluating a JFET based upon tabulated gate operating current alone. There's hope, though: if you're lucky, you'll see gate operating current plotted as a family of curves of I_G versus V_{DG} at a set of drain currents, as in Figure 3x.7; there the proportionality is evident, and the gate current is seen to rise by more than a factor of a million.

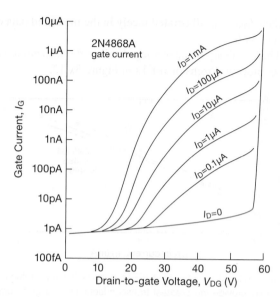

Figure 3x.7. The mother of all ionization-impact gate-current graphs: JFET gate leakage increases disastrously at higher drain-to-gate voltages, and is proportional to drain current.

We showed this latter effect earlier (Figure 3.49), illustrating how our favorite low-noise BF862 JFET suffers rapidly increasing gate currents starting at just 4 volts, if we run it at say 5 mA drain current for good low-noise performance.

To verify the claim that this gate current contribution is proportional to drain current, we made measurements of I_G at four decades of current, plotted in Figure 3x.9 as the *ratio* I_G/I_D versus V_{DG}. As the Mythbusters would say, this claim is "CONFIRMED." Also confirmed, by the way, is the claim that ionization-impact gate current is far smaller in *p*-channel devices. Note particularly that the two JFETs in Figure 3x.9 have a maximum V_{DG} rating of 25 volts! At voltages up to that maximum there is essentially *no* additional gate current in the *p*-channel 2N5461, in contrast to the *n*-channel 2N5457.[6]

In Figure 3x.8 we've plotted the measured ratio I_G/I_D versus V_{DG} for two dozen of the JFETs in Table 3x.1. This can be helpful in selecting JFETs for circuit situations where the effect is unavoidable. Be sure to check the JFET datasheet for detailed plots; if they aren't shown, assume the worst!

---

[6] We carried these measurements up to 60 V, way beyond the "absolute maximum" ratings in their respective datasheets. We did that because, curiously, datasheets often show gate current plots at voltages well beyond their own ratings (45 V and 50 V for these two parts). We remain mystified – can their message really be "don't do what we're doing"?

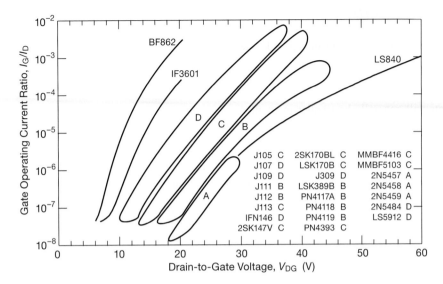

Figure 3x.8. Gate operating current for a selection of JFETs from Table 3x.1, plotted as a ratio I_G/I_D.

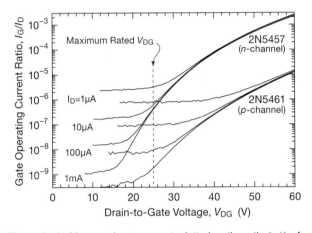

Figure 3x.9. Measured gate current, plotted as the ratio I_G/I_D, for a pair of popular JFETs. The flat tails are dominated by the measurement floor of $I_G \approx 1$ pA. With reckless abandon, these plots go well beyond the Absolute Maximum Rating of $V_{DG} = 25$ V.

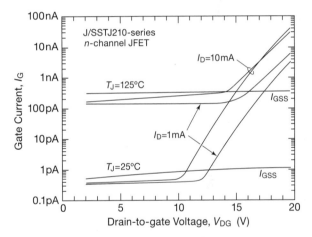

Figure 3x.10. When worrying about ionization-impact gate current, don't forget about the *other* big gorilla in the room – leakage doubles roughly every 10°C rise in temperature.

Gate current: a balanced view. When worrying about soaring gate current, don't forget about the exponential rise of gate current with *temperature*. Figure 3x.10, where both effects are shown, helps set the scale here.

Transistor die size also matters: larger JFETs will generally have larger gate leakage currents (as well as larger capacitance, hence larger dynamic gate current). For example, here's a comparison of Vishay's specified "typical" gate reverse current and capacitance specs for the small 2N4117-series, larger 2N4391-series, and very large J105-series:

| | I_{GSS} (typ) | | | |
| | @25°C | @150°C | C_{iss} | C_{rss} |
Type	(pA)	(pA)	(pF)	(pF)
2N4117	0.2	400	1.2	0.3
2N4391	5	13000	12	3
J105	20	10000*	120	20

* at 125°C

In summary, keep in mind that:

(a) gate reverse current is *leakage*, and increases rapidly with increasing temperature, typically doubling every ~10°C;

(b) when JFETs are operated at typical currents and at substantial drain-to-gate voltage, impact ionization creates actual gate operating currents I_G far larger than gate leakage current I_{GSS};

(c) dynamic gate currents ($C\,dV/dt$) can easily dominate over static gate current when signals are present;

(d) dc gate currents will generally be smaller in small-die JFETs.

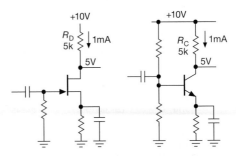

Figure 3x.11. Comparing JFET and BJT amplifiers with resistive load.

3x.2 A Closer Look at JFET Transconductance

3x.2.1 Dependence of g_m on I_D

Elaborating on the discussion in §3.3.3 (and see Figure 3.53), the transconductance of JFETs is most easily understood by considering the regions of drain current operation:

subthreshold ($I_D \ll I_{DSS}$) $I_D = I_o \exp(V_{GS}/nV_T)$,
 so $g_m = \partial I_D/\partial V_{GS} = I_D/nV_T$,
 where n is typically in the range $1 < n < 4$;
 alternatively, $g_m \propto \exp(V_{GS}/nV_T)$
quadratic $I_D = \kappa(V_{GS} - V_{th})^2$,
 so $g_m = \partial I_D/\partial V_{GS} = \kappa'\sqrt{I_D}$,
 where $\kappa' = 2\sqrt{\kappa}$ (with units of $\sqrt{I}/V$);
 alternatively, $g_m \propto (V_{GS} - V_{th})$

In the subthreshold region of drain current the transconductance of JFETs is linear in drain current (or exponential in gate voltage), and approaches that of BJTs (for which $n=1$). In the quadratic region the transconductance rises more slowly, being proportional to the square root of drain current (or linear in gate drive voltage $V_{GS} - V_{th}$). At these currents (more accurately, current *densities*) a JFET's transconductance falls well below that of a BJT. Most JFETs are still in the quadratic region at I_{DSS} (i.e., at $V_{GS}=0$), and continuing even to drain currents corresponding to forward gate bias voltages up to 0.5 V or so (beyond which thou shalt not go).[7]

To make our last remark quantitative, consider a JFET and a BJT, each operating at 1 mA. Imagine they are connected as common source (emitter) amplifiers, with a drain (collector) resistor of 5k to a +10 volt supply (Figure 3x.11). Let's ignore details of biasing and concentrate on the gain. The BJT has an r_e of 25 Ω, hence a g_m of 40 mS, for a voltage gain of −200 (which you could have calculated directly as $−R_C/r_e$). A typical JFET (e.g., a

---

[7] This is not true for MOSFETs, for which the drain current flattens out at some maximum value before you reach the maximum allowed (forward) gate voltage of 10 V–20 V; thus in this asymptotic region of drain current the transconductance ($\partial I_D/\partial V_{GS}$) in fact *decreases* with increasing drain current (or gate drive). See, for instance, the LND150 curve in Figure 3.9.

2N5457) has a g_m of 2 mS at a drain current of 1 mA, giving a voltage gain of −10. This seems discouraging by comparison. The low g_m also produces a relatively large Z_{out} in a follower configuration, where the JFET has $Z_{out} = 1/g_m$, which in this case equals 500 Ω (independent of signal source impedance); compare this with the BJT, which has $Z_{out} = R_{sig}/\beta + r_e$, or (at 1 mA) $R_{sig}/\beta + 25 \Omega$. For typical transistor betas, say ~ 100, and reasonable signal sources, say with $R_{sig} < 5 k\Omega$, the BJT follower is an order of magnitude stiffer ($Z_{out} = 25 \Omega$ to 75 Ω). Note, however, that for $R_{sig} > 50k$ the JFET follower will be better.

Of course, at much lower operating currents (down in the subthreshold region) the JFET's transconductance is much closer to (though still less than) the BJT's. But then you've got a circuit with much larger load resistors, thus higher output impedance and less bandwidth, and (as explained in Chapter 8) considerably higher noise.

As we discussed in Chapter 3, the problem of low voltage gain in FET amplifiers can be circumvented by resorting to a current-source (active) load (or a transimpedance stage), but once again the bipolar transistor will be better in the same circuit. For this reason you seldom see FETs used as simple amplifiers, unless it's important to take advantage of their unique input properties (extremely high input resistance and low input current).

Returning to Figure 3x.12, one might ask how well a JFET's transconductance versus drain current conforms to the ideal $\kappa'\sqrt{I_D}$ behavior? Although the curves of drain current versus V_{GS} by themselves look plausibly quadratic, a direct measurement of transconductance versus drain current reveals plenty of variation, as can be seen in the curves of Figure 3x.13 where we've plotted the "normalized" transconductance ($g_m/\sqrt{I_D}$) of three JFET types over three decades of drain current.

We measured the transconductance of many of the JFETs in Table 3x.1 over four decades of drain current, with the results plotted in Figure 3x.12. The topmost curve

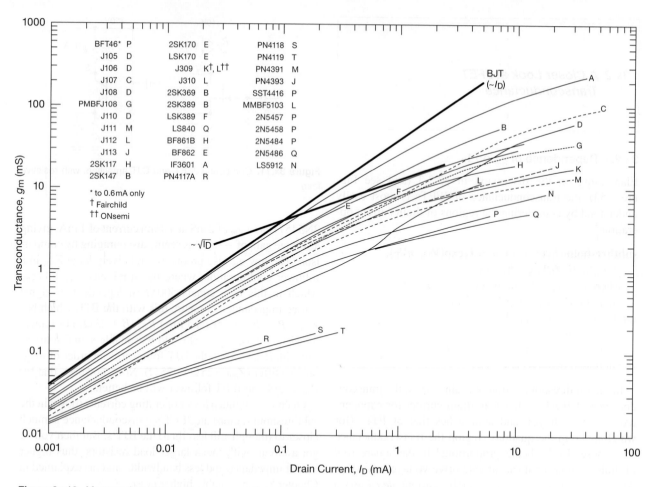

Figure 3x.12. Measured transconductance versus drain current, for most of the JFETs in Table 3x.1.

is the unique large-geometry IF3601 (also the quietest: $e_n \sim$ 0.3 nV/$\sqrt{\text{Hz}}$), a contender for the BJT transconductance look-alike award. At the bottom are the low-I_{DSS} 2N4117–19 family members, which are well into the quadratic region of drain current even at 1 μA. Noteworthy also, perhaps, are the several JFETs with anomalous shapes (most dramatically the J309); they get a "second wind" at higher drain currents, as if they were constructed of two paralleled JFETs such that a higher-current sibling kicks in at a less negative gate voltage.

3x.2.2 Dependence of g_m on V_{DS}

If you measure a JFET's transconductance at a given drain current (§3x.3 shows you how), you'll find that it's rather independent of drain voltage, once V_{DS} is above a few volts or so. But the transconductance falls off rapidly as you approach zero volts, as can be seen in the measured

data of Figure 3x.14. The voltage at which things get bad varies from one JFET type to another, but typically the transconductance turns seriously downward somewhere in the range of 200 mV to 1 V. For a given JFET the effect is more severe at higher drain currents, as can be seen in the normalized plots of Figure 3x.15.

The manufacturers' datasheets are not particularly revealing about this behavior, typically specifying the transconductance at some favorite drain voltage, usually 10 V or 15 V. That's not a lot of help when it comes to choosing V_{DS} for a JFET with a cascode in its drain. Best to aim for a volt or so, minimum, unless you've got good data for the part you're using.

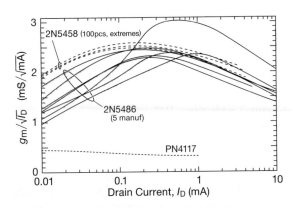

Figure 3x.13. Normalized JFET transconductance versus drain current. Although $g_m \propto \sqrt{I_D}$ is a good rule-of-thumb for JFETs above their subthreshold region, you do see deviations of order $\pm 25\%$ or so over a 1000:1 range of drain currents.

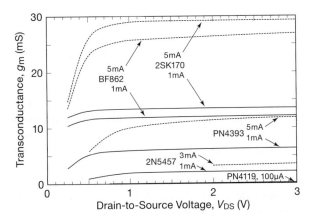

Figure 3x.14. Transconductance is flat for V_{DS} above a few volts, but decreases markedly at low voltages, as seen in these measured data.

3x.2.3 Performance of the transconductance enhancer

In §3.2.3 we introduced the "transconductance enhancer" (Figure 3.29F, repeated here as Fig. 3x.16), a configuration in which a JFET and a complementary-polarity BJT combine to produce a 3-terminal compound JFET with greatly enhanced transconductance. How well does it work, really?

To find out, we compared the measured transconductance of several JFET candidates with the transconductance of the same parts when combined in the enhancer configuration of Figure 3x.16; R_1 was chosen to produce a V_{BE} drop of ~ 0.65 V at the target drain current, and the gate bias was then adjusted to set $I_C = I_D$. For both measurements we used the test circuit of Figure 3x.18, in which the device under test is connected as a grounded-source amplifier

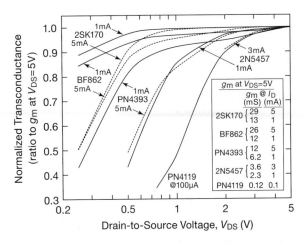

Figure 3x.15. Normalized transconductance (relative to g_m at $V_{DS}=5$ V) for the JFETs in Figure 3x.14.

with cascode (thus eliminating errors from finite drain output impedance, see §3x.4).

Here's what we found (all transconductances are in units of millisiemens, mS):

		Q_1 (JFET) g_{m1}	Q_2 (pnp) g_{m2}	$g_m' = g_{m1}R_1 g_{m2}$ calc	meas
	I_C, I_D				
2N5457	1 mA	2.26	39.5	56.6	52.9
LSK389B	1 mA	9.28	39.5	232	222
	5 mA	22.8	198	600	529
2SK170BL	1 mA	13.0	39.5	325	308
	5 mA	29.0	198	764	681

The measured values are in good agreement with expectations – the transconductance enhancer really enhances! Note also that most of the compound transconductance comes from the BJT, whose g_m dominates at these currents (where the JFET is in the quadratic region of drain current – recall Figures 3.53 and 3.54). Because most of the voltage gain comes from Q_2, it's important to select a low-noise BJT (see Chapter 8) if the amplifier is dealing with low-level signals. To achieve the full transconductance enhancement, the BJT should also be chosen with plenty of current gain (β), because the input impedance at the base diverts some signal current. This means that, ideally, $R_1 \ll \beta r_e = \beta/g_{m2} = \beta V_T/I_C$. This condition on β can be boiled down to a simple number by eliminating R_1, which equals V_{BE}/I_D; so, if you choose to run both transistors at equal current, as we did (i.e., $I_D = I_C$), then $\beta \gg V_{BE}/V_T \approx 650\text{mV}/25.3\text{mV} = 25$. The 2N4403 used for our measurements had a measured beta of 170, high enough so that the circuit's transconductance enhancement was little degraded from the ideal $g'_m = g_{m1}R_1g_{m2}$.[8]

The example in §3x.4.6 exploits the high transconductance of this configuration to implement a low-noise differential amplifier of predictable gain without the use of overall negative feedback.[9]

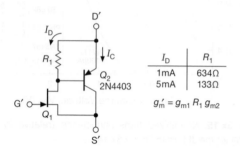

Figure 3x.16. Transconductance enhancer. The expression for the effective transconductance g'_m assumes sufficiently high beta for Q_2, such that the impedance looking into Q_2's base is much higher than R_1; i.e., $\beta \gg R_1g_{m2}$.

3x.2.4 Transconductance in the JFET source follower

The generally modest transconductance of FETs (compared with BJTs) limits the gain of single-stage common-source amplifiers, as we saw in §3.2.3. The addition of a "transconductance enhancer" helps considerably, as does the use of a current-source drain load (the latter requires feedback to establish its operating point). And ultimately the single-stage gain is limited by the FET's drain output conductance g_{os}: $G_{max} = g_m/g_{os}$, as discussed in §3.2.3 and 3x.4.

The FET follower is not immune to these same effects, even though it has no ambitions for greater-than-unity voltage gain. Look at Figure 3x.17, which shows three JFET follower circuits. Circuit #1 is a simple follower: its output impedance, seen looking into the source, is just $1/g_m$,

about $70\,\Omega$ as measured for this JFET at 6 mA operating current; that forms a voltage divider with the load impedance, with the calculated voltage gain shown (the "if $g_{os}=0$" column). Not bad, for high load resistance, but a significant departure from unity gain when driving a 1 kΩ load.

Circuit 2 adds some BJT muscle, raising the effective transconductance to $g'_m = g_{m1}R_2g_{m2}$, where g_{m2} is BJT Q_2's transconductance at its 3 mA collector current ($g_{m2} = I_C/V_T \approx 120$ mS). Think about it this way: assuming that R_2 is small compared with the impedance looking into Q_2's base (i.e., $R_2 \ll \beta r_e$), Q_1 generates a signal *voltage* of $g_{m1}R_2v_{in}$ across R_2; Q_2 converts that back to a signal *current* $g_{m1}R_2v_{in}g_{m2}$, thus a compound transconductance $g_{m1}R_2g_{m2}$.[10] This structure is analogous to the BJT "complementary Darlington" (Sziklai) connection. The enhanced transconductance reduces the output impedance (by a factor R_2g_{m2}, approximately 25), resulting in a voltage gain close to unity even into a 1k load: here $G=0.996$, once again if g_{os} is ignored.

We cannot continue to ignore g_{os}: Q_1's drain-to-source voltage is changing with input swing, reflected as a (slightly) changing drain current (at constant V_{GS}); or, equivalently, at fixed drain current V_{GS} changes slightly over the swing. That's a deviation from unity gain, as the equations and calculations in Figure 3x.17 indicate. Looking at the columns labeled "actual g_{os}," we see continued problems in paradise: though the transconductance enhancer greatly improved the low-R_L problem, it did little to bring the lightly loaded gain all the way up to unity. The solution here is to protect Q_1 (and also Q_2) from V_{DS} variation, by adding a cascode (circuit #3). The equations and calculated values tell the story, in symbols. The next section tells the story, in prose.

[8] Substituting a high-beta 2N5087 (measured $\beta = 440$) brought the g_m at 1 mA up to the theoretical value. But the 5 mA values declined, likely owing to internal bulk resistances in this low-current transistor that reduced its transconductance at the higher current density. Additionally, this transistor would not be a good choice here, because its relatively large "base spreading resistance" $r_{bb'}$ contributes excessive Johnson noise – see Chapter 8.

[9] Why on earth would anyone eschew negative feedback? Look to that section to find out why!

[10] If you chase the numbers through, assuming that $I_D=I_C$, you'll find that the JFET's transconductance is enhanced by a factor of 26, relative to its transconductance (without enhancer) at that same drain current. If we instead operate at the same *total* current (I_D+I_C), as we've done here, the transconductance enhancement factor is 18 (under the assumption that we're in the quadratic region of drain current, where $g_m \propto \sqrt{I_D}$).

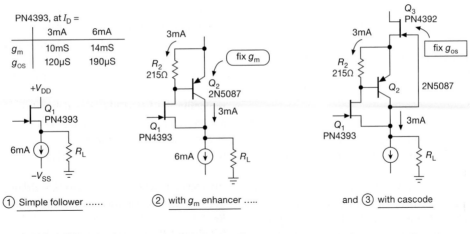

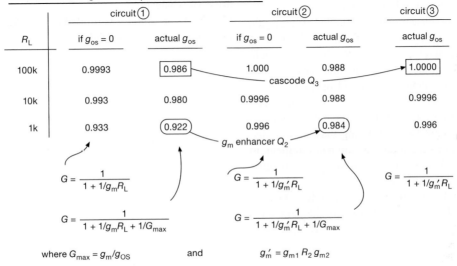

Figure 3x.17. JFET follower progression: raising the effective transconductance with a "g_m enhancer," then adding a cascode to reduce drain output conductance.

3x.3 Measuring JFET Transconductance

As we've seen, all JFETs are not created equal. Unlike BJTs (whose transconductance depends simply upon collector current, as $g_m = I_C/V_T = qI_C/kT$; thus, for example, 40 millisiemens at 1 mA), JFET transconductance at a given current may vary over two orders of magnitude, as seen for example in Figure 3.54. To develop the data for that graph, and for the entries in Chapter 3's JFET table, we used the circuit shown in Figure 3x.18.

The basic idea is to measure the ac-coupled voltage gain with the JFET biased to a known drain current. That current is set by a 5 V drop across drain load R_D. Simple enough. But it's tiresome to set the gate voltage manually for each choice of R_D; so we connected a biased integrator (U_{1a}), which slews to the correct bias (when S_2 is pressed), and holds that bias during the gain measurement. In practice we disconnect the input (via S_1) during biasing, and disconnect the integrator's input during gain measurement, to circumvent stability issues that arise from a pair of lagging phase shifts within the loop.

Continuing with the circuit description, we found by actual measurement that it's essential to wire the gain stage as a cascode (via Q_2); otherwise the measured gain is significantly lower (as much as 30%), owing to the JFET's finite output impedance seen at the drain (variation in I_D versus V_{DS} at constant V_{GS}, analogous to Early effect in a BJT). Here we biased the cascode to set the JFET's drain at a nominal +5 V, with a 2 V drop across Q_2. (It's interesting also to see how g_m varies with V_{DS}; that's easily done by replacing the lower bias resistor with a pot. Some results are plotted in Figures 3x.14 and 3x.15.)

Some additional wrinkles:

(a) You get a measurement of V_{GS} for free – at the output of U_{1b};

(b) You can measure I_{DSS} (zero-bias drain current), and the transconductance at I_{DSS}, by closing S_3 with an appropriate R_D;

(c) D_1 and R_1 prevent gate damage from outrageous overdrive, or whatever;

(d) CMOS op-amp U_1 is limited to 15 V total supply, thus the non-standard supply voltages.

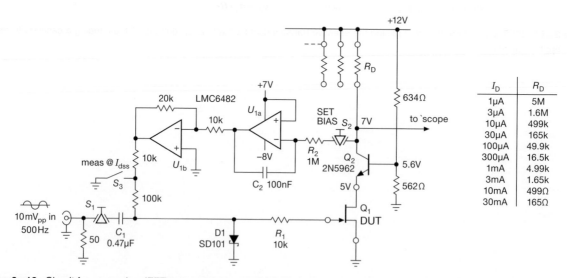

I_D	R_D
1µA	5M
3µA	1.6M
10µA	499k
30µA	165k
100µA	49.9k
300µA	16.5k
1mA	4.99k
3mA	1.65k
10mA	499Ω
30mA	165Ω

Figure 3x.18. Circuit for measuring JFET transconductance. Cascode stage Q_2 (which pins the JFET drain at +5 V) is essential; without it measured values of g_m can be in error by 30% or more.

3x.4 A Closer Look at JFET Output Impedance

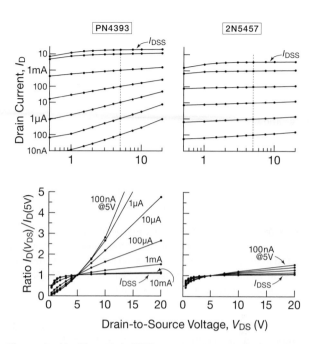

Figure 3x.19. Observing JFET output resistance. Top graphs: a family of measured I_D versus V_{DS} curves for two JFETs, with a set of fixed gate voltages chosen to produce decade steps in I_D at V_{DS}=5 V. Bottom graphs: the same data, normalized to the drain current at V_{DS}=5 V.

Ideally we'd like a JFET's drain current (in the saturation region) to depend only upon the gate drive voltage, and not upon the drain-to-source voltage V_{DS}; i.e., it should be an ideal transconductance device. In the quadratic region of drain current, for example, this is usually described by the classic behavior $I_D = \kappa (V_{GS} - V_{th})^2$, as we saw in §3.1.4 and §3x.1; in the subthreshold region the drain current is exponential (rather than quadratic), but, as in the quadratic region, it ideally does not depend on V_{DS}.

In real life, it does. This is analogous to Early effect in BJTs, and is usually described by the output conductance g_{os}, the small-signal ratio v_{ds}/I_D. Its inverse is r_o, the output resistance seen at the drain with fixed V_{GS}. Zero g_{os} would be the ideal; real JFETs (and MOSFETs) have non-zero g_{os}, typically in the range of $10\,\mu S$ to $1\,mS$ at drain currents of a few milliamps (equivalent to drain resistances of $100\,k\Omega$ to $1\,k\Omega$). Roughly speaking, g_{os} is commonly in the range of $g_m/100$ to $g_m/500$. The overall magnitude of the effect varies widely among different JFETs, as can be seen in the measured curves of Figure 3x.19. For a given JFET the effect is larger for smaller drain currents.

For a common-source amplifier, the drain resistance r_o is effectively in parallel with the external drain load. So, for example, with a current-source load ($R_D = \infty$) the voltage gain becomes $G = g_m r_o = g_m/g_{os}$. Let's call this quantity "G_{max}," which (unlike g_{os} or g_m separately) conveniently turns out to be relatively independent of drain current (see below). For an external drain resistor R_D the gain becomes $G = g_m(R_D \| r_o) = g_m R_D/(1 + g_{os}R_D)$.

The effect can be substantial: from the PN4393 data, its drain resistance at 0.1 mA is about $128\,k\Omega$; so a common-source amplifier with 10 V across a $100\,k\Omega$ drain resistor sees an effective $56\,k\Omega$ drain load, reducing its voltage gain from the ideal G=152 (i.e., $g_m R_D$, with a measured g_m of $1.57\,mS$ at 0.1 mA) to G=88. And, in the case of a current source in the drain (i.e., $R_D = \infty$), where you expect very high ideal gain, the actual gain becomes $G = g_m r_o$ (here ~200).

3x.4.1 A JFET's g_{os}-limited gain, G_{max}

Interestingly, g_{os} exhibits approximately the same variation with drain current as does g_m, so their ratio g_m/g_{os} shows little variation with drain current, even over four decades of I_D (Figure 3x.20).[11] For this reason we've listed the simple parameter G_{max} in Tables 3.1 and 3x.1 (based on measured g_m and g_{os}), from which you can find

$$g_{os} = g_m/G_{max}, \qquad (3x.1)$$

with the latter's scaling over current approximately the same as that of g_m.

Both g_m and g_{os} are important. Transconductance sets the output impedance of a follower; and it sets the voltage gain of a common-source amplifier when the drain load impedance R_D is much smaller than $1/g_{os}$. High transconductance is good. But g_{os} limits the high-R_D voltage gain, to a value $G_{max} = g_m/g_{os}$; so the latter might be considered one of a JFET's figures of merit. However, a simple cascode connection is remarkably effective in pre-

---

[11] Most JFETs exhibit relatively constant G_{max} over wide ranges of I_D, though you'll see some with 25%–50% falloff at high current densities. Behavior like that of the 2N5457–59 family (dashed lines in Figure 3x.20) is relatively rare.

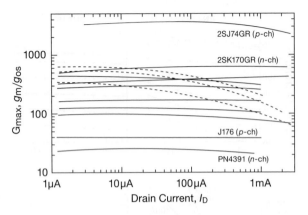

Figure 3x.20. For most JFETs the variation of output conductance g_{os} with drain current is similar to that of the transconductance g_m, leading to roughly constant G_{max}, as seen in these measured data (taken at V_{DS}=5 V). A small subset of JFETs (e.g., the 2N5457–59 family, dashed lines) depart from this rule-of-thumb.

venting gain limitations[12] caused by g_{os}; it also prevents Miller effect in a common-source stage, and it lets you set V_{DS} to a relatively low voltage,[13] preventing the precipitous rise of gate current described in §3.2.8. Once again, the remarkable cascode configuration comes to the rescue, clamping the drain voltage while passing through the full transconductance-limited current gain. The cascode is a wonderful thing; remember this always.

To see how this works, look at Table 3x.2, which summarizes a set of measurements we made on a selection of JFETs: both g_m and g_{os} at I_D=1 mA and V_{DS}=5 V, and then the voltage gain of each JFET in a common-source amplifier circuit operating at the same voltage and current, both with and without a cascode. With a cascode the measured gain is in good (almost *too* good!) agreement with the simple $G=g_m R_D$, and demonstrates why you want lots of transconductance. Without the cascode the gain is reduced, most seriously for the JFET with the poorest g_{os} (the PN4393, whose g_{os} corresponds to r_o=16 kΩ, seriously compromising the external 10 kΩ drain load resistor). At the other end, the very low g_{os} of the remarkable p-channel 2SJ74GR (which excels also in transconductance) hardly compromises the single-stage gain without cascode.[14]

Note that finite drain conductance affects also circuits other than common-source amplifiers, as seen in Figure 3x.17 where the non-zero variation of drain current with drain voltage (at constant V_{GS}) translates to a non-zero variation of gate–source voltage (at constant I_D), thus a departure from unit gain in the JFET follower. There (as here) the cascode gallops valiantly to the rescue; three cheers for the cascode!

3x.4.2 Source degeneration: another way to mitigate the g_{os} effect

Back in Chapter 3 we visited the common-source amplifier with source degeneration (Figure 3.29B,D), there featured as a way to reduce the voltage gain (and improve the linearity). Source degeneration (Figure 3x.21) also raises the JFET's output impedance r_o (reduced output conductance g_{os}), producing a closer approximation to an ideal current source output (and in some cases eliminating the need for a cascode stage). In particular, with a bit of algebraic fiddling you can convince yourself that the output conductance g_{os} is reduced to

$$g'_{os} = \frac{g_{os}}{1+g_m R_S+g_{os}R_S} \approx \frac{g_{os}}{1+g_m R_S} \qquad (3x.2)$$

i.e., by a factor of $1/(1+g_m R_S)$, where R_S is the source degeneration resistor, and where the (discarded) last term in the denominator is insignificant (by a factor of G_{max}). Source degeneration reduces the transconductance by the same factor, i.e.,

$$g'_m \approx \frac{g_m}{1+g_m R_S} \qquad (3x.3)$$

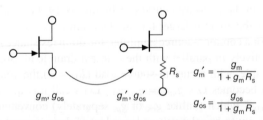

Figure 3x.21. Adding a "source degeneration" resistor R_S has the beneficial effect of reducing the output conductance g_{os}, but this comes at the expense of reduced transconductance g_m.

So, adding a source resistance R_S that is much greater than the impedance $1/g_m$ seen looking into the source increases the output impedance seen at the drain by that ratio. For example, a self-biased (degenerated) JFET current

---

[12] The effective G_{max} of a JFET with cascode is the product of the individual G_{max}'s: $G'_{max} = G_{max}(1)G_{max}(2)$.

[13] Don't overdo it – transconductance falls off sharply at very low V_{DS} (Figures 3x.14 and 3x.15), and g_{os} rises as well, a double whammy.

[14] That's the good news. The bad news is that this part is not currently being manufactured. But the good news is that you can still get them. We thought the same fate had befallen the fine n-channel 2SK170,

but, happily, they have risen from the grave (Mouser shows 4,000 in stock, 11,000 on order); there is also the almost-as-good second-source LSK170 from Linear Integrated Systems.

Table 3x.2: JFET Amplifier Gain Tests[a]

	Measured Parameters		Voltage Gain					Summary		
			with cascode		no cascode		Gmax			
	g_m (mS)	g_{os} (μS)	meas	$G=g_mR_D$	meas	$G=\dfrac{g_mR_D}{1+g_{os}R_D}$	$G=g_m/g_{os}$	g_m	g_{os}	Gmax
PN4393	6.31	62	61.6	63.1	36.4	39	102	good	poor	so-so
2N5457	2.28	10.95	22.6	22.8	18.7	20.5	210	poor	good	better
2SK170BL	13.1	15	130	131	105	114	870	excellent	OK	much better
2SJ74GR	14.9	4.2	152	149	147	143	3550	excellent	excellent	WOW!

(a) I_D = 1mA, V_{DS} = 5V, R_D = 10kΩ.

source will perform much better than the same JFET with its gate held at a fixed back-bias chosen to produce the same drain current; you can see the dramatic improvement in the pair of measured curves for a PN4393, configured as a 10 μA current sink, in Figure 3x.22.

For a common-source amplifier, the gain without degeneration

$$G = g_mR_D\,\frac{1}{1+g_{os}R_D}$$

becomes

$$G = g'_mR_D\,\frac{1}{1+g'_{os}R_D}$$

with source-degeneration-reduced values for g_m and g_{os}. This can be rearranged as

$$G = g_mR_D\,\frac{1}{1+g_mR_S+\dfrac{G_{prog}}{G_{max}}+g_{os}R_D} \qquad (3x.4)$$

where G_{prog} is the ideal gain (R_D/R_S), and R_S is the source degeneration resistance (R_S in Figure 3.29B, or $R_S \parallel R_{S'}$ in Figure 3.29D). As a numerical example, a PN4393 grounded-source amplifier running 1 mA into a 10k drain resistor would have its ideal gain g_mR_D=61.6 degraded (by g_{os}) to an actual gain G=36.4; by contrast, a source degeneration resistor R_S of 162 Ω (chosen equal to $1/g_m$, thus halving the ideal gain) results in an actual gain G = 19.3, with most of the gain reduction due to R_S, and g_{os} effects now in the minority. In this transaction we lost gain, but we gained linearity – the usual trade-off in source (or emitter) degeneration, which is, of course, a form of negative feedback. For sufficiently large R_S (i.e., $g_mR_S \gg 1$) the degeneration term dominates, and the gain approaches the simple $G=R_D/R_S$.

3x.4.3 Dependence of g_{os} on drain current density

As we remarked above, the variation of I_D with V_{DS} (at fixed V_{GS}) is greatest when operating at low current densi-ties, i.e., with drain currents far below the JFET's I_{DSS}. To see this, look at Figure 3x.22. We've plotted the same effect for four different JFET types, of widely varying zero-bias drain current (I_{DSS}), each with V_{GS} adjusted to produce a drain current of 10 μA at V_{DS}=5 V. The PN4391 is the part with the highest I_{DSS} (measured at 115 mA), trailed by the next three curves (with I_{DSS} of 16 mA, 6.1 mA, and 3.5 mA, respectively).

Note that these data are taken with a fixed gate-to-source bias voltage. By contrast, the drain current is largely independent of drain-to-source voltage if instead the JFET is self-biased with a source resistor R_S, chosen to produce the desired drain current. The stabilizing effect is seen clearly in Figure 3x.22, where the second-worst offender (PN4393) is magically transformed into the poster child. Don't fall into the trap, however, of thinking that this "fixes" the gain problem: the self-bias resistor degenerates the gain, in this case reducing the transconductance (at I_D=10 μA) from g_m=240 μS (for grounded source, with V_G held at -1.58 V) to a paltry g'_m=6.2 μS (158 kΩ source degeneration resistor, with V_G=0).

The situation is analogous to the effect of adding emitter degeneration to the common-emitter amplifier (or current source): you get improved linearity (or, for the current source, good constancy of current, i.e., high r_{out}), but at the expense of gain. Putting numbers on it, in spite of greatly increased r_0, the voltage gain into an R_D=500 kΩ drain resistor (5 V drop across R_D) would in fact *drop* – from G=55 (g_m times $R_D \parallel r_0$) down to G=3.2 (g'_mR_D). If you want lots of voltage gain into a resistive (or current source) drain load, what you really want is both high transconductance *and* high r_0 – and an appropriate figure of merit is their product g_mr_0, or, put in terms of output conductance, g_m/g_{os}. That product is, in fact, just G_{max}, the gain of a grounded-source amplifier with an ideal current source in the drain.[15]

---

[15] You can think of the effective drain load impedance as simply the paral-

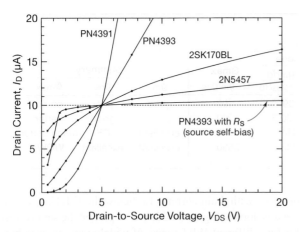

Figure 3x.22. Measured I_D versus V_{DS} for four JFETs, each with a fixed V_{GS} chosen to produce $10\,\mu A$ of drain current at $V_{DS}{=}5\,V$. Note by contrast the stabilizing effect of a $158\,k\Omega$ source self-bias resistor.

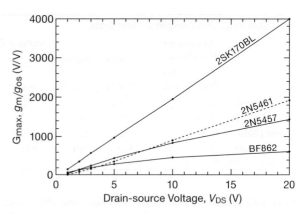

Figure 3x.23. Output conductance g_{os} usually increases with decreasing drain voltage, leading to greatly decreased G_{max} at low V_{DS}, as seen in these plots of four JFETs (derived from measured g_m and g_{os}). The 2N5461 (dashed line) is p-channel, the rest are n-channel.

3x.4.4 Dependence of g_{os} and G_{max} on V_{DS}

As Figure 3x.14 shows, a JFET's transconductance is roughly independent of drain voltage (except at very low V_{DS}, say less than 1 V). But the output conductance g_{os} does depend somewhat on V_{DS}, usually decreasing with increasing V_{DS}; this can be seen in the curvature of the I_D versus V_{DS} curves in Figures 3x.19 and 3x.22, particularly in the drain voltage region from $\sim$1 V to ten volts or more. As a result, G_{max} generally increases with increasing drain voltage, illustrated in the plots of G_{max} versus V_{DS} of Figure 3x.23. One consequence of this is that a JFET running at very low drain voltages will exhibit anomalously low G_{max} (thus the caution in the footnote 13 on page 176: don't bias a cascode so that the underlying common-source JFET is starved of adequate V_{DS}).

3x.4.5 A parting shot: g_{os} – sometimes it matters, sometimes it doesn't

A final comment: a JFET's finite output impedance ($1/g_{os}$) degrades an amplifier's gain *for high impedance drain loads* (large R_D, or a current source), but is insignificant if the impedance seen looking out of the drain is much less than $1/g_{os}$. That may be the case (if R_D is relatively low); or you can make it that way with the versatile cascode configuration, which clamps the drain voltage while passing the drain current through. With a cascode you get the full $g_m R_D$ voltage gain, where R_D is repositioned atop the cas-

---

lel combination of $1/g_{os}$ with the external impedance seen at the drain, the latter being infinite for an ideal current source.

code transistor. Likewise, a cascode greatly increases the output impedance of a current source.

3x.4.6 Example: A low-noise open-loop differential amplifier

Here's a good example circuit that pulls together these three advanced JFET topics (g_m enhancer, g_{os} mitigation with cascode, source degeneration for predictable open-loop gain). Many audio engineers believe that excessive negative feedback creates nonlinear artifacts that compromise fidelity. For example, in a low-level preamplifier stage you may have impulsive or RF interference that is outside the bandwidth of the feedback path, leading to transient distortion or intermodulation. A better solution, it is thought, would be an intrinsically linear open-loop input stage of predictable gain (switch-selectable over a wide range, in 3 dB steps, say). It should, of course, exhibit low noise, and ideally be fully differential with at least a few volts of common-mode range.

Figure 3x.24 shows how this can be done. The input stage is a JFET matched-pair differential amplifier with source degeneration, with current-mirror active load; it produces an output *current* (at Q_5's drain) that is converted to an output voltage via the transresistance amplifier U_1.

That's the view from 50,000 km. Zooming in, there's lots to see, and explain. First, to get predictable gain we boost the transconductance of the input pair $Q_{1a,b}$ (a high-g_m low-noise matched pair) via enhancers Q_2 and Q_3. We run each transistor at a substantial 5 mA (nearly the JFETs' specified minimum I_{DSS} of 6 mA), for maximum

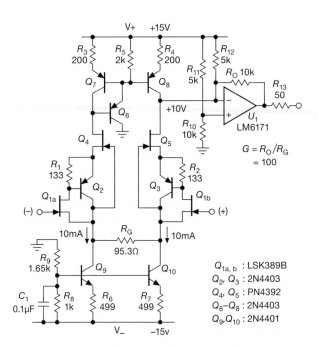

Figure 3x.24. A low-noise JFET differential amplifier of predictable gain. Source degeneration in the transconductance-enhanced input stage makes the gain relatively independent of JFET transconductance, without the use of negative feedback.

gain and minimum noise. The resulting high input-stage transconductance (measured at $g_m \approx 530$, see §3x.2.3) is then degenerated by a factor of ~ 25 by R_G, which makes the stage's gain predictable with good accuracy. Here's how this goes: the measured enhanced transconductance (530 mS at $I_D = I_C = 5$ mA) looks like an intrinsic resistance $r_s = 1/g'_m = 1.9\,\Omega$ (analogous to a BJT's r_e) at each JFET's source terminal; degeneration adds R_G to the sum of both r_s's, thus about $100\,\Omega$ total, setting the (degenerated) differential gain at 10 mA/V (i.e., 10 mS). This is rather insensitive to the JFETs' actual transconductance, because R_G dominates over r_s; for example, if $Q_{1a,b}$ exhibited just its minimum guaranteed g_m (10 mS at 5 mA, a factor of 2.3 less than actually measured), the stage's gain would decrease from 100 mS to 96.2 mS, an insubstantial 4% loss.

A few more details about the input stage:

biasing Current sink $Q_9 Q_{10}$ sets the operating currents at 10 mA for each side; the emitter resistors degenerate the current sinks, for better matching and high output impedance. We could have used a single 20 mA sink, with R_G split into a series pair; but dual current sinks permit gain selection via the single resistor R_G. R_1 and R_2 set the JFET drain currents to 5 mA, a value that drops a V_{BE} across the respective resistors R_1

and R_2. Current mirror Q_6–Q_8 is the drain load, with emitter degeneration (for improved balance and output impedance), assisted by follower Q_6, whose low output impedance (R_5 provides operating current) suppresses Miller effect at Q_8's output.

cascode JFETs Q_4 and Q_5 form the input stage cascodes, with all the benefits that accrue thereby: (a) eliminating Miller effect, (b) eliminating g_{os} effect, and (c) operating the input JFETs at low V_{DS} to prevent excessive gate leakage currents. Note that the cascode is configured *external* to the BJT enhancers, rather than being wrapped around $Q_{1a,b}$ only: this is the better configuration, because it suppresses both g_{os} effects (in the JFETs) and Early effects (the analogous effect in the BJTs). The cascode transistors have a relatively large V_{GS} (~ 2 V at 5 mA) at the operating current, chosen in order to set sufficient drain-source operating voltage for the input JFETs (V_{DS} is a diode drop less, at ~ 1.4 V).

The output stage is an op-amp (U_1) configured as a current-to-voltage ("transresistance") amplifier (see for example §§4.3.1 and 4x.3), with gain equal to R_O, i.e., 10 kV/A. The circuit's overall voltage gain is therefore $G_V = R_O/(R_G + 2r_s) \approx 100$, and is relatively insensitive to uncertainty in the input-stage JFETs as described above. The voltage divider $R_{10}R_{11}$ sets the operating voltage of Q_5's drain, via feedback through R_O,[16] with R_{12} canceling the output dc offset. The LM6171 is a wideband op-amp that runs from ± 15 V rails, with substantial output drive capability (± 100 mA), and with adequately low input noise. It's used also in the circuit examples of Figures 3.34 and 3.37.

---

[16] That's what op-amps do for a living, see Chapter 4 if this is new territory for you.

3x.5 MOSFETs as Linear Transistors

In Chapter 3 we treated extensively the subject of linear circuits built with *JFETs*, but in contrast MOSFETs were seen almost entirely in switching applications: analog switches (§3.4.1), logic switches (§3.4.4), and power switching (§3.5). But MOSFETs are useful also in numerous linear applications, for example in audio amplifier power stages, low-dropout voltage regulators, high-voltage piezo drivers, and so on.

Here we look at some of the properties of MOSFETs that matter in linear applications. As we'll discover, in linear applications we're usually running MOSFETs in their sub-threshold region of operation, a region that is often poorly specified (if at all) on manufacturers' datasheets. To illustrate how it goes, we've made lots of measurements of real MOSFETs, collected here in some sixteen graphical plots. Some detailed linear MOSFET designs are discussed elsewhere, for example in §3x.9 (high-voltage, high-Z probe), and in §4x.23 (precision high-voltage amplifier).

3x.5.1 Output characteristics and transfer function

A. Datasheet curves

To get started, we've picked a typical small MOSFET, but the principles and measurements we'll explore apply to the other power MOSFETs you might want to consider using in linear designs. Afterwards, we'll repeat some of

the measurements for a 1.2 kV MOSFET used in our high-voltage amplifier designs.

Look first at Figure 3x.25, which shows a datasheet graph of a set of drain currents versus drain voltage, for a selected set of gate voltages. This is usually called a plot of *output characteristics*, or ON-*region characteristics*; this one corresponds to the *n*-channel VN2222, a small power MOSFET rated at 60 V and 1 A.[17] It comes in the TO-92 small transistor package, and is quite popular,[18] with multiple manufacturers, and a price of $0.06 (qty 100).

This graph shows the expected increase of drain current with successively high gate voltages, and, for each curve, the resistive region at low drain voltages (extending to a drain voltage roughly half the gate drive, $V_{GS} - V_{th}$), leveling off to the flat region ("current saturation") at higher drain voltages. What the graph *doesn't* show, however, is the MOSFET's behavior at low currents, just the region where we would operate it in a linear circuit. Think about it: with a maximum power dissipation of ~ 0.5 W,[19] we're talking about drain currents down around 10 mA–100 mA (depending on our operating drain voltage), right at the bottom of the graph.[20]

Evidently a datasheet plot like this is aimed at power switching applications, where you want to know how much current the MOSFET can handle, and how much gate voltage you have to apply to drive it into voltage saturation (where the small V_{DS} permits relatively high currents while staying within permitted power dissipation) or to generate a short high-current pulse at higher drain voltages (where higher momentary peak power dissipation is permitted, see §§9.4.2B and 9x.25.8).

This emphasis on high current operation is evident also

---

[17] Pulsed; the maximum continuous drain current is considerably less, being limited by maximum rated junction temperature of 150°C.

[18] It may have acquired some of its popularity from its memorable part number, an echo of the ever-popular *npn* BJT 2N2222/2N4401 of comparable ratings. Similar MOSFETs of greater popularity are the 2N7000 and BS170 (in TO-92 package) and their even more popular SOT-23 twins (2N7002 and MMBF170), all comparably inexpensive, and with distributor stock up to a million pieces (check them out at the excellent resource www.octopart.com).

[19] As discussed in §9.4, the semiconductor's temperature rise above ambient is $\Delta T = P_{diss} R_{\Theta JA}$; here the VN2222's thermal resistance $R_{\Theta JA}$ is 170°C/W, thus a rise of 85°C when dissipating 0.5 W.

[20] In a *pulsed* application you can operate, briefly, at substantially higher instantaneous dissipation, relying on the transistor's thermal mass to absorb the dissipated energy. Datasheets provide plots of the "Transient Thermal Impedance" data needed to evaluate such pulsed operation, see §3x.13. For example, a VN2222 or 2N7000 starting at 25° can handle 10 W during a single 10 ms pulse.

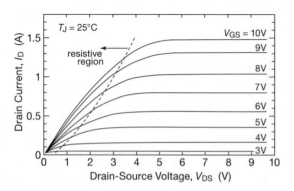

Figure 3x.25. VN2222 *n*-channel MOSFET output characteristics (a family of I_D versus V_{DS}, for chosen values of V_{GS}), adapted from the ON Semiconductor datasheet.

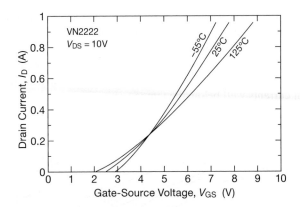

Figure 3x.26. VN2222 transfer characteristics (I_D versus V_{GS}), adapted from the ON Semiconductor datasheet, showing variation with junction temperature. The drain current tempco is positive at low currents, but negative at high currents.

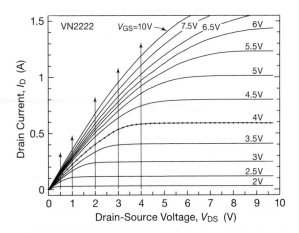

Figure 3x.27. Measured output characteristics of a VN2222 MOS-FET (TO-92 small transistor package). Datapoints are shown for one curve, to demonstrate authenticity. Vertical lines correspond to drain voltage choices of Figure 3x.28.

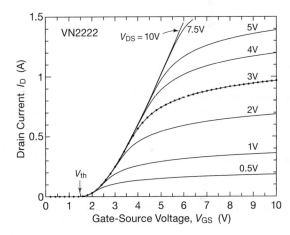

Figure 3x.28. Measured transfer characteristics (a family of I_D versus V_{DS}, for chosen values of V_{DS}) for the same VN2222 as in Figure 3x.27.

in the datasheet's plot of drain current versus gate voltage (*transfer characteristics*, Figure 3x.26). It does show nicely the stabilizing negative temperature coefficient of drain current at high currents, and (unhappily) the reverse at lower currents where we operate linear circuits (a subject discussed in §3.6.3). We're definitely in the latter region in any linear circuit: the crossover is around $I_D = 0.25$ A, where the dissipation at $V_{DS} = 10$ V is a whopping 2.5 W, far above this transistor's capability. And this plot, showing behavior at high currents, is not of much help in informing our design down in the ~ 10 mA regime.

B. Measured data

What we need are some good measurements and understanding of MOSFET behavior at much lower drain currents. Those are actually the easier measurements to make, because the power dissipation is so low that heating effects are generally negligible. We began by taking some measurements of a VN2222 sample at currents similar to those in the datasheet plots. Figure 3x.27 shows the resulting set of output characteristics, and Figure 3x.28 shows the corresponding transfer characteristics (the latter at a normal ambient temperature of 20°C only). These can be laborious measurements, unless you're lucky enough to have access to a dual-channel source-measure unit (SMU). That's what we used here, producing a nice set of datapoints (shown as dots in one of the curves in each figure). Because the peak power ranges up to 10 W or so, we used a pulsed measurement mode for these two figures: We took data at one point per second, but with the gate driven only for 125 μs per point.

Looking first at the measured output characteristics

(Fig. 3x.27), and ignoring for the moment the ebullient arrows, it's easy to spot the resistive region (also known as the *triode* region) where the drain current rises linearly with drain voltage. It's also evident that the resistive region extends to a higher drain voltage when driven with a higher gate voltage, as described back in Chapter 3 (Fig. 3.13), and that the drain current levels off at still higher drain voltages (the region of *current saturation*, also known as the *pentode* region, and not to be confused with saturation in a BJT, for which we recommend[21] the disambiguating term *voltage saturation*). Note that this graph (and the next) are

---

[21] Thanks to our colleague Nat Sokol for making a fuss about this!

plotted on linear axes, and, because they show no detail at low drain currents, they do not address our original complaint about low-current MOSFET analog circuits. We'll deal with that presently, after a look at the transfer characteristics, plotted on the same coarse linear current scale.

At first glance, Figure 3x.28 (transfer characteristics) bears some resemblance to Figure 3x.27: a family of lines, going up and to the right. Don't be confused! Here we're plotting drain current versus *gate* voltage, and the family members are fixed values of drain voltage, whereas in Figure 3x.27 the drain current is plotted versus *drain* voltage, and the family members are fixed values of gate voltage. Of course, it's the same transistor, doing its thing – and you can make the connection between these plots by imagining moving upward along one of the arrows (i.e., fixed drain voltage) in Figure 3x.27, reading off values of drain current as you cross the lines representing increasing values of gate voltage.

3x.5.2 Linear operation: hotspot SOA limitation

Back in §3.6.3 in the main volume we pointed out that MOSFETs, when run in their low-current regime (as in linear applications), exhibit a positive tempco of drain current, and thus cannot be directly paralleled; you need to include small source ("ballast") resistors. However, we blithely assumed that an individual MOSFET, with its isothermal silicon die, suffers no *internal* instabilities from positive I_D tempco.

As luck would have it, this positive I_D tempco, operating at the individual transistor cell scale, can also cause a reduction in a MOSFET's safe operating area, analogous to (but far less severe than) the "second breakdown" limitation that power BJTs suffer (see, e.g., Fig. 3.95). This "Spirito" effect[22] is more prominent in later-generation high trench-density MOSFETs. Figure 3x.29 shows this effect, which is being recognized in recently introduced MOSFETs where it becomes significant.[23] This does not affect operation at high current (where the I_D tempco is

negative), nor at low voltages (where the local dissipation is inadequate to cause thermal bunching), but it is significant in linear applications where the MOSFET is run at moderate current and higher voltages (in linear applications you are forced, by power dissipation limits, to run at currents well below the tempco crossover).

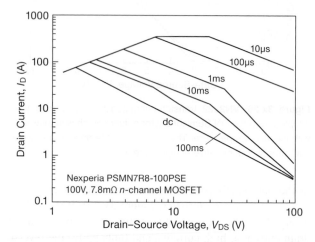

Figure 3x.29. The Spirito effect (local hot spots from positive I_D tempco) causes a reduction in SOA of high-density MOSFETs at higher voltages and moderate currents, seen here in the kink in 1 ms–100 ms curves.

3x.5.3 Exploring the subthreshold region

As we explained at the outset, we want to run these puppies in the linear region, and at much lower drain currents than these curves show, at a minimum to keep the power dissipation within allowable bounds (and probably for other circuit-related reasons as well). It turns out that "power MOSFETs," with their ≥ 1 A drain current ratings, work quite well down in the microamp region; it's just that the manufacturers don't make a big deal about it.

In fact, they don't show it at all on their datasheets. But you can measure it yourself, getting something like the plot of Figure 3x.30, which reveals the hidden life of our VN2222, plotted on a logarithmic axis spanning 11 decades of drain current. Just as with JFETs (§3.1.4A), there's a subthreshold region in which the drain current is exponential in gate voltage (and where, as with

[22] G. Breglio et al., "Electro-thermal instability in low voltage power MOS: experimental characterization," Proc. IEEE ISPSD'99, 233–236 (1999); P. Spirito et al., "Analytical model for thermal instability of low voltage power MOS and S.O.A. in pulse operation," Proc. IEEE 14th Intl. Sym. Power Semiconductor Devices and ICs, 269–272 (2000); P. Spirito et al., "Thermal instabilities in high current power MOS devices: Experimental evidence, electro-thermal simulations and analytical modeling," Proc. IEEE 23rd Intl. Conf. Microelectronics (MIEL2002), 23–30 (2002).

[23] See, for example, M.F. Thompson, "Linear mode operation of radiation hardened MOSFETs," International Rectifier App Note AN-1155;

"Power MOSFET frequently asked questions and answers," Nexperia Technical Note TN00008 (2018); H. Shah et al., "From planar to trench – evaluation of ruggedness across various generations of power MOSFETs and implications on in-circuit performance," 26th Annual IEEE Applied Power Electronics Conference (APEC, 2011).

BJTs, the transconductance is proportional to drain current); at higher drain currents the device enters the familiar quadratic and current-saturation regions seen in the previous linear plot (Fig.3x.28).

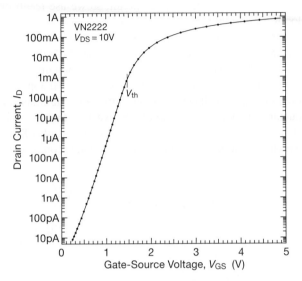

Figure 3x.30. Measured transfer characteristics for the same VN2222, extended to the deep subthreshold region. This 1A-class transistor works way down at the picoampere range (at room temperature, anyway)!

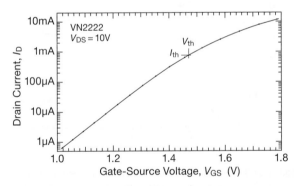

Figure 3x.31. Expanded view of Figure 3x.30's data, in the useful μA-to-mA range. The drain current at V_{th} is about 0.8 mA; currents less than that are in the subthreshold region.

For small-signal linear circuits the practical region of operating currents goes from the neighborhood of 1 μA to 10 mA, which is shown in more detail in Figure 3x.31. This includes the transition from subthreshold to quadratic regions, at the official gate threshold voltage V_{th}. Although datasheets usually define V_{th} somewhat arbitrarily (e.g., the gate voltage at which the drain current is 1 mA, with $V_{DS} = V_{GS}$), a better definition in terms of MOSFET oper-

ation is the extrapolation of the (quadratic) drain current onto the voltage axis, when the drain current is plotted on a square-root axis; see Figure 3x.32. For this MOSFET specimen the gate threshold voltage is 1.47 V, as indicated on the graphs of transfer characteristics; and from the expanded curves of Figure 3x.31 we can see that this corresponds to a drain current of 0.8 mA. We'll call the latter I_{th}.[24]

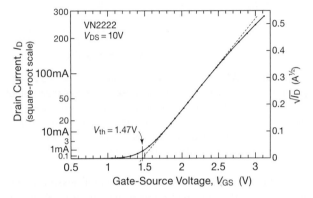

Figure 3x.32. Transfer characteristics of Figure 3x.30, plotted on a square-root vertical scale, to find the threshold voltage. Note the extensive "quadratic" drain-current region.

A. MOSFETs at low drain voltage

From a plot like Figure 3x.27 it is apparent that a MOSFET, when operated at low (subthreshold) currents, enters its region of current saturation at a low drain voltage; for example, in that plot the drain current levels off at about 25 mA for drain voltages above 0.5 V (the $V_{GS} = 2$ V curve). As with subthreshold operation generally, MOSFET datasheets are silent on this issue. But, voltmeter (actually, SMU) to the rescue: Figure 3x.33 shows a set of measured output characteristic curves, plotted over six decades of drain current, for two samples of VN2222. For each curve the gate voltage was chosen to put the drain current on a decade, spanning nominal drain currents from 1 μA to 100 mA (with pulsed measurements for the higher currents). The logarithmic V_{DS} axis expands nicely the low-voltage region (which is squashed to unusability in a plot like that of Figure 3x.25), and it shows that the MOSFET is quite usable in linear circuits operating with drain voltages down as low as 0.1 V when running at currents below 1 mA.

---

[24] It's the drain current I_{th} at V_{th} (rather than the value of gate threshold voltage V_{th} itself) that we can count on as a parameter characterizing a particular transistor type; the gate threshold voltage varies from part to part, and more widely from batch to batch.

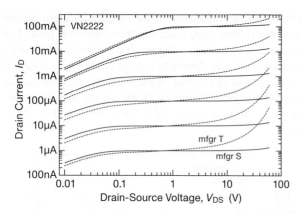

Figure 3x.33. All transistors are not created equal: VN2222 samples from two manufacturers exhibit markedly different drain output resistances (variation of drain current with drain voltage), as seen in these measured output characteristic curves (where the gate voltages were chosen to span decades of drain current). We used an "S" sample for all other VN2222 measurements.

Figure 3x.33 illustrates another essential lesson: Components with the same part number may differ markedly in characteristics that are not specified. Here we see that manufacturer T's VN2222 shows a large rise (as much as ×10) in drain current at higher voltages, compared with the textbook behavior of the VN2222 from manufacturer S; put another way, its dynamic output resistance ($r_o = dV_{DS}/dI_D$) is much smaller (analogous to Early effect in BJTs), an undesirable characteristic in a linear application.[25] Both transistors meet their datasheet specs – it's just that r_o (the drain resistance in current saturation) is not specified: These transistors are primarily intended for switching applications, where r_o is of no importance; instead the resistance that *is* specified is $R_{DS(ON)}$, i.e., the drain resistance in *voltage* saturation.

To see this behavior in greater detail, we normalized the curves and expanded the (still logarithmic) vertical scale (Figure 3x.34). You can see that the fractional effect is greater at lower drain currents in the region of drain voltages above a volt or so, where you're likely to run an amplifier circuit.

What can be done to circumvent this effect? There are two issues here: (a) the gain-spoiling effect of reduced r_o in a common-source stage, and (b) the variation of gate voltage with changing drain voltage, i.e., bias stability. Taking these in order, an excellent cure for low output resistance r_o

---

[25] Although transistor S exhibits "textbook behavior" (with its drain current not greatly rising at high drain voltages), we've found that most power MOSFETs behave more like transistor T. Maybe the textbooks need fixing.

is the ever-useful cascode circuit (see for example §2x.3 or Fig. 3.29), which clamps the gain-stage drain while passing the current to the drain load. As for the bias problem, we never know the exact V_{GS} needed by a MOSFET anyway, when designing a linear circuit, so we use feedback loops, or current sources, or source degeneration, or other schemes to set the operating current. In this respect both the "S" and "T" parts may be equally useful. And we can estimate the change of gate voltage easily enough: say we anticipate a 10× increase in drain current as we approach the rated drain voltage; from an estimate of the transconductance (say $n = 4$, see §3x.5.4B) we find that V_{GS} will change by about 250 mV (i.e., four times that of a BJT), which is what the circuit design will need to accommodate.

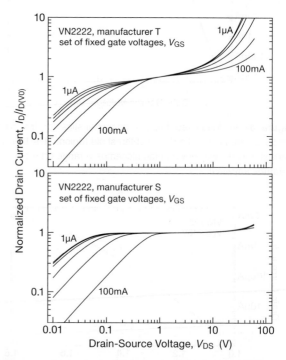

Figure 3x.34. Plot of normalized drain current versus drain voltage (at a set of fixed gate voltages), from the same data plotted in Figure 3x.33.

B. MOSFETs at high drain voltage

The upward curvature of drain current with increasing drain voltage seen in some MOSFETs (as in this example) is caused by channel-length modulation, and is analogous to Early effect in bipolar transistors (§2x.5). It is not leakage current, nor is it avalanche conduction (of the sort seen in zener diodes). To explore this a bit further we tortured a 2N7000 sample, measuring a set of output characteristic curves beyond its rated 60 V maximum drain voltage. It cried uncle at about 76 V (Fig. 3x.35), where its drain current increased abruptly, over just a few millivolts of increasing drain voltage, from leakage currents at the sub-nanoamp level to our preset 10 mA current limit.[26]

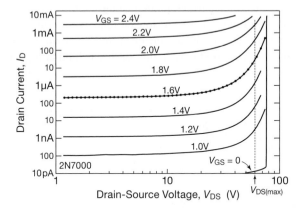

Figure 3x.35. The smooth upward curvature of drain current seen in Figures 3x.33 and 3x.34 differs from the abrupt "avalanche breakdown," seen here in measurements of a 2N7000 MOSFET (similar to the VN2222).

Avalanche. Although the very term "avalanche" evokes images of mayhem and destruction,[27] MOSFETs are not damaged by such insults as long as the peak and repetitive avalanche ratings of the device are respected. Most power-switching MOSFETs are explicitly avalanche rated, in many cases eliminating the need for external clamp diodes in switching power converters.

As we explain in §3x.13.1B, it turns out that those mysterious "single-pulse avalanche" and "repetitive-pulse avalanche" ratings correspond exactly to the power delivered during the avalanche (voltage times current) equaling the maximum power handling capability of the MOSFET, as shown in the datasheet's Transient Thermal Impedance plots. For example, the datasheet for the classic "fully avalanche rated" IRF540N specifies limits of avalanche current and repetitive avalanche energy of 16 A and 13 mJ, respectively, with the pulse width limited by the maximum allowable junction temperature of 175°C. See the further discussion in §3x.13.

3x.5.4 Exploring a high-voltage MOSFET

A small MOSFET, for example a 2N7002 or BSS84 in their diminutive SOT23-3 packages, are nice components for many purposes, particularly small switching applications; and there are dozens of similar parts, in both *n*MOS and *p*MOS flavors. But for voltages less than 40 V, and in the context of *linear* circuits, there's a wide selection of complete high-performance op-amps in the same small package.[28] So linear circuits with MOSFETs are more interesting when they exploit the capabilities of high-voltage parts (see Table 3x.3), to handle jobs not possible with standard ICs.

In Chapter 4x (§4x.23) we've worked out and tested a complete design for a precision high-voltage amplifier, of the sort you might use for driving deflection electrodes or piezo transducers; it can swing to ±500 V, or from zero to +1000 V, and works stably into load capacitances as large as 10 nF. That design exploits the nice properties of a 1200 V MOSFET, the IXTP1N120.

Here we explore the properties of that device, in particular its dc transfer function (I_D versus V_{GS}), the corresponding small-signal transconductance gain ($g_m = i_d/v_{gs}$), and the complexities of SPICE modeling in the subthreshold region.

A. IXTP1N120 transfer characteristics

Figure 3x.36 shows the measured transfer characteristics of an IXTP1N120 sample, here plotted on log axes (as we did for the small 60 V VN2222 in Fig. 3x.30) to display the relevant six decades of current (ignore for the moment the additional SPICE-model curves). The drain current exhibits the familiar exponential behavior in the subthreshold region below about 10 mA, with a smooth transition to the quadratic region at higher currents. As with the previous VN2222 example, the threshold voltage is most easily identified by re-plotting the drain current on a square-root axis (Fig. 3x.37). For this MOSFET the threshold voltage is $V_{th} \approx 4.18$ V; at that gate voltage the drain current is 7 mA.

To give some context, the high-voltage amplifier in §4x.23 runs from a total supply voltage of 1,100 V. To keep power dissipation within reasonable limits with surface-mount power packages (or TO-220 through-hole parts with clip-on heatsinks), we chose quiescent currents in the range of 0.25–0.5 mA, with dynamic current limits of 25 mA. For that application the MOSFETs are operating entirely in the

---

[26] The avalanche voltage is typically some 10%–30% above the part's rated breakdown voltage $V_{BR(DSS)}$.

[27] "Which bothers some men. The word itself makes some men uncomfortable."

[28] A 3-pin MOSFET in the tiny SOT-323/SC70 package (e.g., BSS123W or BSS84W) has a better chance competing against an op-amp in its (relatively!) huge SOT23-5 package.

subthreshold region of drain current, a situation that is not unusual in linear MOSFET applications.

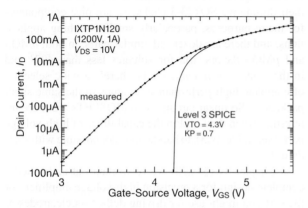

Figure 3x.36. Measured I_D versus V_{GS} for an IXTP1N120 high-voltage n-channel MOSFET (TO-220 power package). A best-fit SPICE level 3 model (with parameters shown) fails badly in the sub-threshold region of $I_D \lesssim I_{D(max)}/100$.

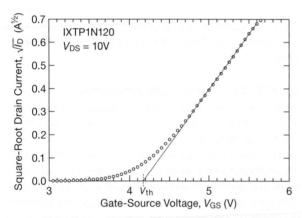

Figure 3x.37. Square-root plot of I_D versus V_{GS}, showing threshold voltage $V_{th} = 4.18\,V$ (at which $I_D \approx 8\,mA$).

B. IXTP1N120 transconductance

As we've remarked elsewhere, the exponential I_D versus V_{GS} in the subthreshold region is similar to the Ebers–Moll law for BJTs (eq'n 2.8), but with an exponential coefficient reduced by a scaling factor n (i.e., $I_D \propto \exp\left(V_{GS} - V_{th}\right)/nV_T$, where $V_T = kT/q$ is the familiar 25.3 mV at room temperature).

Transconductance gain g_m is the slope (derivative) of the transfer characteristic; i.e., $g_m = \Delta I_D/\Delta V_{GS}$ (i.e., i_d/v_{gs}). Because of the factor n in the denominator of the exponent, the transconductance gain of a MOSFET in the subthreshold region scales similarly to that of a BJT, but with

a gain degradation factor n; that is, a MOSFET's transconductance $g_m = I_D/nV_T$. We saw this same behavior earlier in connection with JFETs (in §3x.2, see particularly Fig. 3x.12), with values of n ranging from 2 to 6.

How do MOSFETs compare? From the measured transfer curve of Figure 3x.36 we calculated the transconductance versus drain current, plotted alongside the ideal gain of a BJT in Figure 3x.38. At low currents the MOSFET's transconductance is about a factor of four less than that of the BJT ($n \approx 4$), gradually falling further behind at higher drain currents.

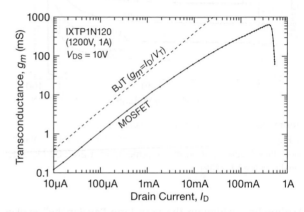

Figure 3x.38. Transconductance versus drain current of an IXTP1N120, calculated from the data of Figure 3x.36. At 1 mA the MOSFET's transconductance is about a quarter that of a BJT.

Because the factor n (ratio of $g_{m(BJT)}/g_{m(MOSFET)}$) is relatively constant over the subthreshold region, it's useful to plot n (rather than g_m) versus drain current. You get the plot in Figure 3x.39 (where we've also plotted its inverse, $1/n$, which you can think of as a figure of merit). You can see that the MOSFET's transconductance really falls behind the BJT for drain currents in the quadratic region and above. The gain factor $1/n$ begins falling at drain currents above about 0.4 mA, roughly 5% of the drain current I_{th} corresponding to the gate threshold voltage V_{th}. In Figure 3x.40 we've collected measured data for a selection of additional MOSFETs that you might choose for high-voltage amplifier applications. You can see that MOSFETs with higher current ratings maintain their transconductance to higher currents, owing to their larger die area; that is, they require a higher drain current to reach the same current density as a smaller part. Those parts are included in Table 3x.3, which lists the essential data (including measured transconductances) of a good selection of high-voltage MOSFETs.

Table 3x.3: High-Voltage Amplifier MOSFETs[a]

Part Number[x]	Mfg	TO-220	D2-Pak	D-Pak	TO-251	TO-247	insulated	V_{DSS} max (V)	P_{diss}^c max (W)	I_D^c max (A)	$R_{DS(ON)}$ max[v] (Ω)	g_m (mS)	@ I_D (mA)	V_{GS} (V)	g_m (mS)	C_{iss} typ (pF)	C_{oss} typ (pF)	C_{rss} typ (pF)	Cost[q] ($US)
IXTx02N450	IXYS	-	V	-	-	H	F	4500	113	0.2	480	150	50	-	-	246	19	6	18.00
IXTx02N250	IXYS	-	V	-	-	H	-	2500	57	0.2	450	145	100	3.9	3.1	120	9	3	18.60
2SK1412	Sanyo	•	-	-	-	-	-	1500	20	0.1	140	100	50	2.6	8.3	40	12	3	5.82
2SK1317	Renesas	-	-	-	-	-	y	1500	100	2.5	9	750	1000	-	-	990	125	60	5.93
STx3N150	ST	P	-	-	-	W	F	1500	140	2.5	6	2600	1300	4.3	7.4	940	102	13.2	4.64
STx4N150	ST	P	-	-	-	W	F	1500	160	4	7	3500	2000	3.8	7.8	1300	120	12	4.15
IXTx02N120	IXYS	P	-	Y	-	-	-	1200	33	0.2	75	200	100	3.5	6.5	104	8.6	1.9	2.92
IXTx06N120	IXYS	P	A	-	-	-	-	1200	42	0.6	32	450	300	3.3	7.7	270	19	3.2	1.72
IXTx1N120	IXYS	P	A	Y	-	-	-	1200	63	1	15[t]	920	500	3.9	7.9	550	25	5.4	2.46
IXTx3N120	IXYS	P	A	-	-	H	-	1200	150	3	4.5	2600	1500	4.9	7.1	1050	100	25	3.97
MTP3N120[d]	ON Semi	•	-	-	-	-	-	1200	125	3	4	3100	1500	3.2	9.3	1300	100	10	NLA
IXTx01N100	IXYS	-	-	Y	U	-	-	1000	25	0.1	80	160	50	3.9	6.7	54	6.9	2	0.93
IXTx05N100	IXYS	P	A	Y	U	-	-	1000	40	0.5	17	930	500	3.5	8.9	260	22	8	1.68
IXTx1N100	IXYS	P	A	-	-	-	-	1000	54	1.5	11	1500	1000	3.6	5.8	400	37	13	0.93
BUZ50B[d]	Philips	•	-	-	-	-	-	1000	75	2	8	1500	1500	2.8	6.2	1600	70	30	NLA
FQx2N100	Fairchild	-	-	D	U	-	-	1000	50	1.6	7[t]	1900	800	4.4	7.4	400	40	5	0.79
STx2NK100Z	ST	P	-	D	U	-	-	1000	70	1.9	8.5	2400	900	4.0	8.4	500	53	9	2.19
STx3NK100Z	ST	P	-	D	U	-	-	1000	90	2.5	6	2400	1250	4.2	8.0	600	53	12	2.17
IRFBG20	Vishay	•	-	•	-	-	-	1000	54	1.4	8[t]	1000[n]	840	3.8	7.3	500	52	17	1.02
FQx2N90	Fairchild	P	B	D	-	-	F	900	85	2.2	7.2	2000	1100	4.1	7.5	390	45	5.5	0.81
STx2NK90Z	ST	P	-	D	-	W	-	900	70	2.1	6.5	2300	1050	4.4	7.7	485	50	10	0.82
FQD1N80	Fairchild	-	-	D	U	-	-	800	45	1	20	750	500	4.3	5.8	150	20	2.7	0.52
SPx02N80	Infineon	P	-	-	-	-	A	800	42	2	2.7	1500	1200	3.3	9.6	290	130	6	1.13

(a) sorted first by voltage, then by die size and R_{ON}. (c) max pwr dissipation or drain current at T_{case}=25°C. (d) discontinued. (m) maximum. (n) minimum. (q) quantity 100; NLA = no longer available. (t) typical. (u) at V_{DS}=10V. (v) at V_{GS}=10V. (x) replace the symbol "x" with the corresponding upper-case package letter in the "Packages" columns. (y) TO-3P package.

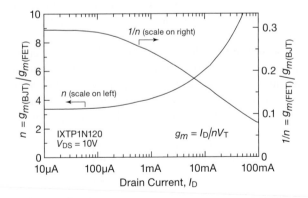

Figure 3x.39. Ratio of BJT to FET transconductance, n, versus drain current, calculated from the data of Figure 3x.38.

3x.5.5 SPICE models for power MOSFETs in the subthreshold region

Power MOSFET models are a complicated business. Many of the common parameters, such as C_{rss}, undergo highly nonlinear changes (i.e., variation with voltage, see Figs. 3x.41 and 3x.42) and are difficult to model. In addition, power-supply designers would like their SPICE models to work over large temperature excursions. These challenges often cannot be met with the basic "level-3" model behavior built into SPICE, so manufacturers create complex subcircuit models, with an array of components added around the intrinsic level 3 model. Many are messy and awkward beyond comprehension; but some, like those by Ronan and Wheatley,[29] are elegant and deserving of more attention. In this section we'll deal with a power MOSFET modeling problem of special interest to linear designers: operation in the subthreshold region.

Along with the IXTP1N120 measured data, Figure 3x.36 shows a simulated output characteristic curve for our SPICE

---

[29] Ronan and Wheatley, "Spicing-Up Spice II Software For Power MOSFET Modeling," Fairchild Semiconductor Application Note AN-7506, Feb. 1994.

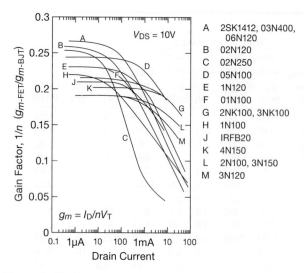

Figure 3x.40. Measured transconductance gain factors for a selection of high-voltage enhancement-mode MOSFETs. The leading digit(s) of the part numbers indicate the drain current ratings; thus 01N120 is a 1.2 kV 100 mA part.

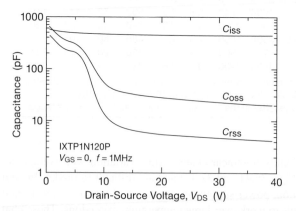

Figure 3x.41. Many power MOSFETs have highly nonlinear capacitances, as seen in these datasheet plots for an IXTP1N120.

model for the transistor. We used the standard level 3 MOSFET model, adjusting the V_{th} and beta parameters (VTO=3.3V and KP=0.7) to match our '1N120 measurements above 20 mA. This model implements the quadratic region formula $I_D = KP(VGS - VTO)^2$, which means that I_D plummets unrealistically to zero for all V_{GS} less than V_{th}, as seen in Figure 3x.36. Such models typically run into trouble below roughly $I_{max}/200$. This is devastating for any circuit modeling in the subthreshold region. This situation has not gone unnoticed by MOSFET experts, who suggest, as we do, that SPICE models should always be validated with bench measurements. For example, Tsividis

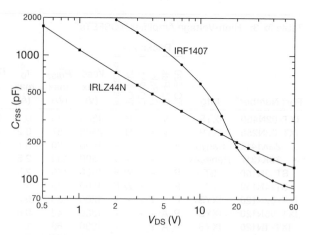

Figure 3x.42. Nonlinear capacitance versus drain voltage is not a universal characteristic, as seen in this pair of MOSFETs.

and Suyama wrote[30] "Rather than trying to convince the reader about the sad state of affairs when it comes to MOSFET modeling for analog work, we propose that the reader him/herself run a few tests that speak for themselves." We recommend this article.

Fortunately there are workarounds to SPICE's serious problems in modeling power MOSFETs in the subthreshold region. A special case occurs if you can model your circuit with a fixed drain current. In that case you can replace the intrinsic MOSFET with an "*E*" (voltage amplifier) or "*G*" (transconductance) element with a fixed g_m, as shown in Figure 3x.43. The MOSFET's capacitances are then added externally. We used this approach with the high-voltage amplifier in Figure 4x.137, in order to do a SPICE analysis of small-signal stability and settling to the quiescent state.

The precipitous drop in SPICE-modeled drain current seen in Figure 3x.36 corresponds to a spike in transconductance. Because we wanted a way to slow the change in I_D and thereby control g_m, we devised a second approach: Add BJTs in series with the source pin; for example, for transconductance factor $n = 5$ we would add five diode-connected BJTs by editing the MOSFET's subcircuit model. We set the intrinsic MOSFET's C_{iss} to zero, and added it back outside the BJTs. We reduced its V_{GS} by about five V_{BE}'s, and trimmed by simulating with V_{GS} sweeps. Amazingly, this kludge works.

But then we learned of a better approach by SPICE guru

---

[30] Tsividis and Suyama, "MOSFET modeling for analog circuit CAD: Problems and prospects," *IEEE J. Solid-State Circuits*, **29-3**, 210 (1994).

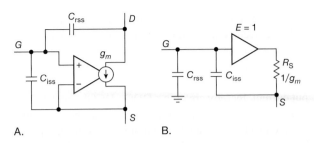

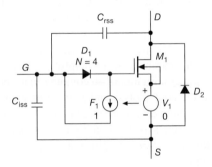

Figure 3x.43. Simplified fixed-current (i.e., fixed g_m) MOSFET models. A. Replace the MOSFET with a "G" element (VCCS: Voltage-controlled current source) and add the capacitances externally. B. For source followers you can use a $G=1$ gain element "E" (VCVS: Voltage-controlled voltage source) with an output resistor $R_S = 1/g_m$.

Steven Sandler.[31] He placed the Ebers–Moll correction element in the MOSFET's gate, where it doesn't squander V_{DS} at high drain currents; see Figure 3x.44.

Figure 3x.44. Sandler's VGS correction element added to the MOSFET's gate to create a valid subthreshold model. Capacitances and diodes are added externally.

Sandler used a diode element, D, with settings for N and IS, fed from an F element (a CCCS that mirrors I_D), as shown in Figure 3x.44. We tried this for the '1N120 in the figures above, using $N = 4$ (like four ordinary diodes in series) and IS=1u (the gain of the F element is set to unity, so the diode has lots of current going through it). We also changed the intrinsic FET's VTO from 4.3 V to 3.15 V. These values were determined empirically, and now the adjusted model matches quite reasonably our measurements down to about 5 mA.

You can apply these model corrections even if you lack subthreshold measurements for your MOSFET: Pick a reasonable value for n, say from 3 to 8, and adjust the parameters until the new plot both matches the manufacturer's

model at currents about three times V_{th}, and also looks more or less like the measured data in Figure 3x.36, with a smooth transition to the exponential low-current region.

3x.5.6 Typical SPICE model for a power MOSFET

When you look at a SPICE model for a MOSFET, for example, what you see are just lists of parameters and node connections. It's instructive, though, to see what the circuit diagram of a typical SPICE model looks like. Here's the (text) model for the classic IRF740, a 400 V 10 A n-channel power MOSFET:[32]

```
     10 20 30
*      TERMINALS:    D   G   S
M1    1    2    3    3    DMOS L=1U W=1U
RD    10   1    0.26
RS    40   3    14.8M
RG    20   2    15
CGS   2    3    1.17N
EGD   12   0    2    1    1
VFB   14   0    0
FFB   2    1    VFB   1
CGD   13   14   1.03N
R1    13   0    1
D1    12   13   DLIM
DDG   15   14   DCGD
R2    12   15   1
D2    15   0    DLIM
DSD   3    10   DSUB
LS    30   40   7.5N
.MODEL DMOS NMOS (LEVEL=3 VMAX=833K
+ THETA=58.1M ETA=2M VTO=3.1 KP=13.4)
.MODEL DCGD D (CJO=1.03N VJ=0.6 M=0.68)
.MODEL DSUB D (IS=41.5N N=1.5 RS=0.125
+ BV=400 CJO=946P VJ=0.8 M=0.42 TT=390N)
.MODEL DLIM D (IS=100U)
.ENDS
```

A. Equivalent circuit

We followed the node numbers in this IRF740 model listing to generate the schematic drawing in Figure 3x.45. Here M1 is the basic dc transconductance NMOS model, with only five parameters specified here. RG = 15 Ω is the gate spreading resistance, which we find is often set too

[31] Sandler and Hymowitz, "SPICE Model Supports LDO Regulator Designs," *Power Electronics Technology*, pp. 26–31, May 2005.

[32] The IRF740 model and some of this discussion is based on information (or quoted) from IntuSoft's SPICE, and an application note (on IGBTs!) by A.F. Petrie and C. Hymowitz.

high.[33] RS and LS are passive elements in series with the source pin, and RD is the minimum $R_{DS(on)}$ resistance.

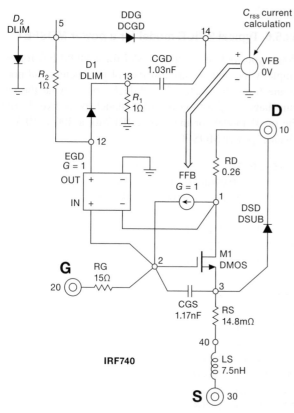

Figure 3x.45. Schematic diagram of an IRF740 MOSFET SPICE model, showing custom C_{rss} modeling. C_{oss} is modeled with DSUB.

There are nine parts that replace the gate–drain capacitor (C_{rss}), in an attempt to more accurately model the change in capacitance with gate and drain voltage. EGD is a voltage generator equal to M1's gate-to-drain voltage, which is used to supply voltage to the feedback

capacitance-emulating subcircuit.[34] FFB is a current controlled current source used to inject the C_{rss} feedback current back into M1's gate. VFB is a voltage generator shorting node 14, used to monitor the current in the feedback capacitance-emulation subcircuit, and control FFB's output current, thereby creating a DDG–DCGD modeled C_{rss}. EGD, VFB, and FFB provide the necessary power to drive the feedback components in parallel without loading M1. They also permit ground connections in the subcircuit, improving convergence and accuracy.

DDG is a reverse-biased diode, whose DCGD model emulates the gate-to-drain capacitor at high voltages. R2 and D2 prevent forward biasing of DDG, by limiting its operation to the region where the drain voltage exceeds the gate voltage. When the drain voltage is less than the gate voltage, R1 and D1 turn on CGD.

B. Model capacitances

In this IRF740 model, CGS ($= C_{iss} - C_{rss}$) is permanently set to 1170 pF. $C_{oss} = C_{ds} + C_{rss}$, where C_{ds} is given by the DSUB model. C_{rss} is pegged at 1030 pF until V_{DS} exceeds V_{GS} (around 5 V, depending on I_D), then it follows the DCGD model's parameters. It's helpful to consider capacitances three ways. Figure 3x.46 combines datasheet plots, bench-measurement plots[35] and SPICE-model plots[36] for C_{oss} and C_{rss}. Note that in all three cases C_{rss} is measured with $V_{GS} = 0$ V, the standard condition. This means that for SPICE, with $V_{DS} > V_{GS}$, the plot only shows the model's DCGD region.

We consider bench measurements to be the "gold standard" against which other data is to be compared. The SPICE plots more or less follow the datasheet plots, which was probably the goal of the model's creator. But the SPICE plots don't show the typical precipitous capacitance drop in the 15 V to 35 V region, as seen in the datasheet plots and in our data. The squares at 25 V are the datasheet's spec values. These agree well with our measurements, but not so well with the datasheet plots, which show the transition at about 17 V (a more common value). The model's diode capacitance element entirely fails to see the drop, and also

---

[33] For turning ON purposes, a significant portion of the MOSFET's gate-channel area is rapidly accessed by the gate, and deserves a low RG value. On the other hand, when turning OFF, some of the MOSFET's area is remote and accessed slowly by the gate's spreading resistance, and needs a high RG value to model its slow turnoff. A more accurate model for this situation might be to break M1 into several parts, so the different I_D rise- and fall-time situations can be better modeled. One further note on this subject: we have observed that some types of MOSFETs display much less of this effect than others. What's more, datasheets with only a single RG value are of little help to learn which ones are which. As is often the case, only bench measurements will reveal the truth.

[34] Reader Tim Williams suggested the model might better use EDS, rather than EDG for CDG.

[35] Measured with an HP 4280A C–V meter (this instrument is rack-mounted, no bench).

[36] To measure the SPICE model's capacitances, wire up the MOSFET in the prescribed fashion. For example, for C_{oss} you'd ground the gate and source, apply a fixed current to the drain, and measure V_{DS} versus time. Using $dV/dt = I/C$, take dV/dt and calculate $C = I/(dV/dt)$. Likewise, for C_{rss} measure the gate current to ground, to find $C = I_{GS}/(dV/dt)$. We find it convenient to do the calculations and plotting in a spreadsheet.

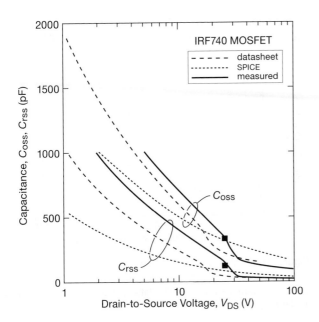

Figure 3x.46. Comparing MOSFET capacitances versus drain voltage: Datasheet curves, SPICE simulation, and measured values. The black squares are datasheet "typical" tabulated values at $V_{DS} = 25$ V.

fails to show the low C_{oss} and C_{rss} values from there up to the rated voltage. This means SPICE will overstate switching losses ($P = fCV^2$) above the transition, and understate them below.

C. Other models

This SPICE model is only one example of the complicated schemes used to model power MOSFETs. The model uses SPICE level 3, and doesn't have switches often seen in level 1 models. Accurately modeling the three MOSFET capacitances is particularly painful. Perhaps the Ronan and Wheatley approach we mentioned in §3x.5.5, with its cascode JFET, is an exception. There are plenty of far more complicated models. Some, such as BSIM3 versions commonly found, will be more accurate. Adding transient thermal effects, with thermal resistance and mass, is a necessary complication for many users.

Specialized models will often succeed in some aspects, but may fail badly in others. It's best to examine what aspects matter in a particular circuit design, and carefully test the SPICE model against bench measurements. For example, in our use of power MOSFETs as linear elements in high-voltage amplifiers, we need a good model of g_m and V_{GS} versus I_D (see §3x.5.5), but we can get by with fixed values for the three capacitances, because we can ignore operation with V_{DS} below 25 to 50 V (knowing that circuit speed will be degraded near the supply rails).

For high-current MOSFET switching applications with rated V_{DSS} voltages of 500 to 800 V, it's often preferable to use a "super-junction" part (see §3x.11.2). These have much higher C_{oss} and C_{rss} transition voltages, so the datasheet typically specs them at $V_{DS} = 100$ V,[37] where the values are often astonishingly low. Would you believe $C_{rss} = 0.8$ pF?! And, some manufacturers are omitting the C_{rss} spec entirely.

The super-junction capacitance-versus-voltage scene is so messy that manufacturers have introduced effective-capacitance specs for calculations. For example, "effective output capacitance, energy related" ($C_{O(er)}$) for calculating switching losses. This is commonly defined as a constant equivalent capacitance giving the same stored energy as C_{oss} when V_{DS} increases from 0 to 80% of V_{DSS}. There's also a time-based value that matches charging times.

3x.5.7 An unusual low-voltage MOSFET

Back in Chapter 3, when introducing enhancement and depletion modes (§3.1.2C), we made the bold statement that the distinction was somewhat arbitrary because it is possible to manufacture "in-between" MOSFETs. Possible, indeed, but does anyone do it? And if so, can you actually *buy* these things?

The answer is yes and yes, the "anyone" is Advanced Linear Devices, and the devices are called Zero Threshold™ MOSFETs. They come packaged in duals and quads, for example the ALD210800A quad MOSFET that houses two n-channel matched pairs. These devices belong to ALD's electrically programmable analog device family (EPAD®), which allow field- or factory-programming and matching of threshold voltages. The zero-threshold devices are factory trimmed to a threshold voltage of $V_{th} = 0 \pm 10$ mV, where the threshold voltage is defined for a drain current of $10\,\mu$A at a drain voltage of 0.1 V. This series of MOSFETs includes members with selected threshold voltages near zero and of both polarities, as seen in the set of transfer curves of Figure 3x.47.

These are most unusual parts, certainly among the choices of available *discrete* MOSFETs: they're neither depletion nor enhancement, and they are specified for operation at very low drain voltages, evidenced among other things by the exceedingly low drain voltage at which V_{th} is specified. We jumped at the chance to measure some samples, with the results shown in Figures 3x.48–3x.50

The transfer curves in Figure 3x.48 exhibit the now-familiar subthreshold exponential, but with an admirably

---

[37] Some manufacturers are specifying them at 80% of V_{DSS}.

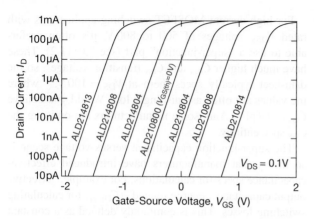

Figure 3x.47. Low-current (subthreshold) transfer characteristics for the ALD family of EPAD MOSFETs, as shown on their datasheet.

high transconductance (about 60% that of a BJT, $n = 1.6$), and this at a drain voltage of just 100 mV.

This suggests the obvious next measurement, namely the behavior over a range of drain voltages (the output

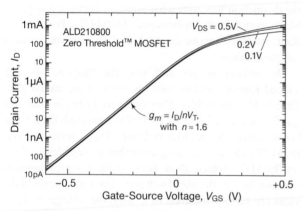

Figure 3x.48. Measured subthreshold transfer characteristics of the ALD210800 Zero Threshold matched quad MOSFET, at $V_{DS} = 0.1$ V. The drain current rises markedly at higher drain voltages, reaching approximately ×10 at $V_{DS} = 5$ V.

characteristics). Figure 3x.49 shows the low-voltage end, where we've plotted the data for an LND150 depletion-mode MOSFET for comparison. The ALD part is fully in current saturation by $V_{DS} = 0.1$ V, whereas the LND150 is just getting started. Figure 3x.49 extends these curves up to the voltage limit of both parts (10 V and 500 V, respectively), from which it can be seen that the ALD part shines at low drain voltages, but its drain current rises 50-fold at its 10 V drain limit; by comparison, the depletion-mode part reaches current saturation only after 1.5 V, but above that it is exemplary in constancy of drain current (i.e., high

output impedance). We explore the use of depletion-mode MOSFETs as current sources in the next section (§3x.6).

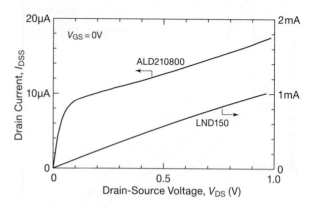

Figure 3x.49. Measured I_{DSS} (I_D with $V_{GS} = 0$) versus V_{DS}, at very low drain voltages, for the "Zero Threshold" ALD210800 and the depletion-mode LND150. The latter's drain current flattens at 1.54 mA above 1.8 V.

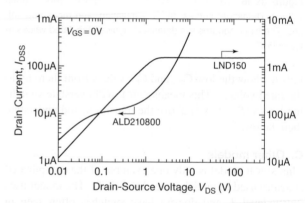

Figure 3x.50. Data of Fig. 3x.49, plotted on log scales and spanning the full specified drain voltage range of each transistor.

3x.6 Floating High-Voltage Current Sources

A simple pull-up resistor is commonly used within circuits, for example to bias a circuit, to act as the load for an amplifier stage, to kickstart a regulator, etc. But the resistor's current is linear in the voltage drop, which can be inconvenient when it has to operate over a range of voltages: The available current falls as the signal rises toward the supply voltage (when you may need it most to drive the output stage, say). Its finite impedance also limits amplifier gain, according to $G = g_m R_{pullup}$. A good current source provides an excellent solution to these shortcomings; it is often a superior replacement for the humble pull-up resistor.

Most of the current sources, current sinks, and current mirrors we've discussed required using the ground bus or a power-supply rail; that is, they were three-terminal circuits. But for many applications what you need is a "floating" (two-terminal) current source. It's a bit tricky, though, because a two-terminal current source requires an active component that's always on, powered from the same current that it's regulating.

You can make a simple 2-terminal current source with a self-biased depletion-mode FET (JFET or MOSFET), a subject we introduced in §3.2.2 (there built with JFETs), and which makes a later appearance in connection with voltage regulators in §9.3.14 (there built with depletion-mode MOSFETs). In §3.2.2B we discussed the problem of predictability of I_D caused by poorly specified transfer characteristics (I_D versus V_{GS}), and in §3.2.2D we introduced the JFET cascode (Fig. 3.27) to mitigate the variation of I_D with drain voltage.

The unpredictability of the simple two-terminal self-biased current source (§9.3.14, Figure 9.36D,E; repeated here later as Figure 3x.52A) can be addressed by combining a regulating IC with a depletion-mode MOSFET. In the main volume we hinted at this technique (Figures 3.114 and 9.104), but there's a gotcha lurking in those circuits, because they depend on the relatively large gate–source back-bias of about $-2\,V$, for the suggested IXTP08N50 depletion-mode MOSFET at the operating current, to power the IC. This problem can be circumvented by boosting Q_1's gate voltage with a zener, raising the MOSFET's source terminal well above the dropout voltage

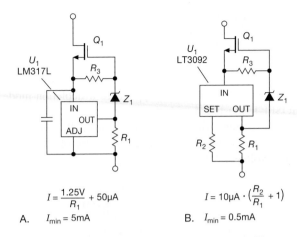

$$I = \frac{1.25V}{R_1} + 50\mu A$$
A. $I_{min} = 5mA$

$$I = 10\mu A \cdot \left(\frac{R_2}{R_1} + 1\right)$$
B. $I_{min} = 0.5mA$

Figure 3x.51. Floating (two-terminal) current sources. The negative gate–source voltage V_{GS} of depletion MOSFET Q_1 creates a current through R_3 to power active zener Z_1.

of regulator U_1; see Figure 3x.51. These circuits require a minimum operating current (to bias the zener and power the regulator), but they are quite effective at high currents, particularly if the voltages aren't too high. A similar approach is shown in Figure 3x.56, and later in Figure 3x.68 in the context of high-voltage precision two-terminal current sources.

In this section we're going to concentrate on low-current high-voltage two-terminal current sources, with an emphasis on circuits with low capacitance, and aimed particularly at achieving better constancy of current with variations in voltage (i.e., raised output impedance), and the extension of operating voltage to the kilovolt scale. In a later chapter (in §4x.23) we describe an amplifier whose performance relies on a high-voltage current source with low capacitance; that application does not require precision, or unusually high dynamic impedance. In this section, however, we'll see how well one can do with circuits that are relatively simple.[38]

3x.6.1 Raising output impedance with a cascode

Figure 3x.52A shows the simple self-biased depletion-mode current source, where R_1 sets the operating current I_D to a value at which the voltage drop $I_D R_1$ equals the gate–source voltage V_{GS} appropriate to that current (refer back to Figure 3.23). For a given value of R_1 the current will show some variation with changes in drain voltage, because the

---

[38] In some respects we'll succeed *too* well – for example producing dynamic impedances greater than $10^{10}\,\Omega$, not needed for any application we can think of!

FET with fixed V_{GS} is not a perfect current source, that is, it does not have infinite output resistance (Figure 9.41 illustrates this problem).

The cascode does its magic by protecting the current-setting transistor from drain voltage variations, as seen in Figure 3x.52B (where it is applied to a depletion-mode MOSFET, illustrated with the handy LND150). Here Q_1 sets the current with self-biasing resistor R_1, while Q_2 clamps Q_1's drain-source voltage to the value of Q_2's gate–source voltage;[39] that is, $V_{DS1} = V_{GS2}$.

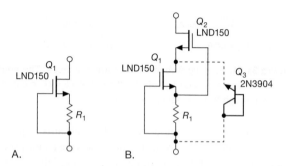

Figure 3x.52. Depletion-mode MOSFET 2-terminal current source. A. Simplest self-biased circuit. B. Cascode Q_2 isolates Q_1 from voltage swings, raising output impedance. Further refinements follow in Fig. 3x.54, where they are labeled C and D.

We set up both circuits, selecting values of R_1 to produce $100\,\mu A$ or $500\,\mu A$ drain currents, and measured the current variation over the full specified 500 V drain-source voltage range. Figure 3x.53 shows the results, along with approximate values of the equivalent output resistance.[40] The cascode improves constancy of current by a factor of five to ten, with quite respectable output impedances ($1.7\,G\Omega$ at $100\,\mu A$, $90\,M\Omega$ at $500\,\mu A$).

You might well wonder why there's any variation at all. A good question. And the answer is that the V_{GS} of Q_2 varies a bit with drain voltage, and, because it sets Q_1's drain voltage, it causes Q_1's drain current to vary. This second-order effect would ordinarily be insignificant, but it matters here because Q_1 is operating at very low V_{DS} (typically less than a volt), where its drain current has not yet

plateaued to its saturation value, and is thus somewhat sensitive to drain voltage (see Fig. 3.13). Ideally you'd like Q_2 to be operating with somewhat more than 1 V; you could use a MOSFET with larger $V_{GS(off)}$, perhaps[41] a BSS126 (with $V_{GS} \approx -1.45$ V at $I_D = 0.5$ mA), but it's best to avoid a large-geometry MOSFET because the latter generally have undesirably large capacitances.

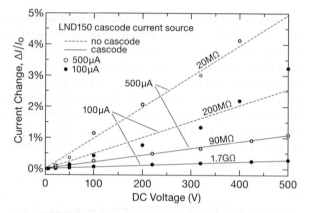

Figure 3x.53. Measured current variation for the circuits of Fig. 3x.52. Points represent equilibrium values of dc current in still air. The output resistance is dominated by thermal effects; the *dynamic* impedances are 10×–100× higher.

There's another effect at work here, namely thermal drift from $P_{diss} = I_D V_{DS}$ self-heating in the transistor (recall the tempco of I_D, seen in Fig. 3.14, reflected in the measured thermal drift of Fig. 3.22A). This affects particularly the simple current source (Fig. 3x.52A), where the current-setting transistor Q_1 bears the full brunt of the power dissipation. When measuring the currents for Figure 3x.53, we typically saw very little initial current change when the voltage was stepped, followed by a slow rise (on a time scale of a minute or so) to a new equilibrium value. Put another way, the *dynamic* resistance of the current source is far larger than its dc resistance, by an estimated factor of 10×–100×. That puts the signal-frequency impedance up in the 10's of gigaohms, far better than needed for any real-world application.[42] We'll pick up the small-signal thread presently (§3x.6.3 on the facing page), after a short digression on power dissipation.

---

[39] BJT Q_3, wired as a low-capacitance ~ 6 V zener, protects Q_2's gate in the event of drain-source breakdown.

[40] A note about this graph, and those following: For the common case of a current source pulling up to a positive supply rail, increasing voltages across the (two-terminal) current source correspond to more *negative* node voltages within the components of the current source. And the highest voltage drops correspond to the (low-side) output near the negative supply rail.

[41] We reconfigured the circuit of Figure 3x.52B, replacing Q_2 with a BSS126. That raised Q_1's V_{DS} from 1.13 V to 1.56 V (for $100\,\mu A$), and from 0.73 V to 1.32 V (for $500\,\mu A$). This change doubled the measured static (dc) output resistances, increasing them to 3 GΩ and 230 MΩ (for $100\,\mu A$ and $500\,\mu A$, respectively).

[42] Are the authors merely lacking in imagination? Send us your suggested applications.

3x.6.2 Reducing power dissipation

Happily, these thermal effects can be mitigated significantly by bridging the drain-source terminals with a resistor (R_2 in Fig. 3x.54C), chosen to bypass a bit less than the full current at maximum voltage. Your first reaction might well be "That can't work, because a resistor is not a constant current source." Your second thought, if you stick with it, is a resounding "Yes! It works just fine, as long as the transistor is still in conduction, because R_1 sees the total current, and adjusts Q_1's drain current as needed." Good thinking.

Another benefit is the ability to operate at a higher current without exceeding the transistor's maximum temperature. For the single transistor circuit of Figure 3x.54C, for example, the transistor would dissipate 250 mW at the maximum voltage of 500 V, raising the junction temperature[43] (T_J) an impressive $\Delta T_{JA} = R_{\theta JA}P_{diss} = 42°C$ above ambient[44] (see §9.4 for for a discussion of thermal calculations). That's not a good situation, because the inside of a box full of powered electronics may be around $\sim 50°C$, or even $60°C$, thus a transistor temperature of as much as $125°C$, which is our nominal design limit for reliable power electronics; and if we were to try to go up to 1 mA, we'd see a scary $125°C$ rise to $180°C$.

It's easy to figure out the reduction in transistor dissipation, which we've plotted for several current-bypass ratios in Figure 3x.55. With the values shown we've bypassed a fraction $\eta = 0.66$ of the current at the maximum operating voltage of $V_{max} = 500$ V, reducing the worst-case dissipation of a SOT-23 to 33% of 250 mW, or 83 mW. So the SMT transistor's maximum heating above ambient amounts to $\Delta T_{JA} = 21°C$ (thus a very reasonable $T_J \approx 81°C$ even at a tropical $60°C$ ambient temperature).

You can exploit this same benefit with the cascode current source, as in Figure 3x.54D. An analogous thermal calculation for the cascode transistor (that's the one that gets hot, here) shows that the peak dissipation is reduced from 250 mW to 80 mW, lowering the transistor's worst-case heating from $\Delta T_{JA} = 63°C$ down to $20°C$. (See Figure 3x.66 for helpful plots of I_D versus V_{GS} for the BSS126 and LND150 small depletion-mode MOSFETs.)

Readers in a cranky frame of mind may object that we're not accomplishing much here, because the same total amount of heat has to be dissipated either way. True

enough. But keep in mind that (a) resistors are allowed to get hotter than semiconductors; and, perhaps most important, (b) we want to mitigate the degradation of circuit performance that is caused by heating of the *active* devices (i.e., semiconductor temperature coefficients of I_D, etc.), so it's better to offload the heat to the simple-minded (and non-critical) resistors.[45]

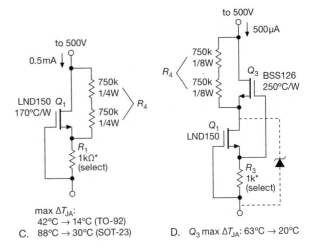

max ΔT_{JA}:
42°C → 14°C (TO-92)
C. 88°C → 30°C (SOT-23)   D. Q_3 max ΔT_{JA}: 63°C → 20°C

Figure 3x.54. Reducing power dissipation with a shunt resistor, implemented as a series pair to respect maximum voltage ratings of the resistors (see Fig. 9.112). Thermal resistances shown are for Q_1 in a TO-92 package, and for Q_2 in SOT-23 with minimum pad sizes. These "C" and "D" circuit refinements follow on the simple current sources of Fig. 3x.52A and B. See Fig. 3x.57 for extending this trick to higher voltages. ***Warning:*** the LND150 has a non-standard pinout; it is *not* the same as the BSS126!

3x.6.3 Small-signal output impedance

Plots of current versus voltage, like those in Figure 3x.53, can be characterized by an output resistance r_o, as we've done there. But there's another way of thinking about the origin of the current change that is helpful in revealing the important factors that limit the ideal of infinite r_o. To keep things simple, imagine the case of a single FET with grounded source terminal.

---

[43] Well, it's not really a *junction*, but that's the conventional language, given the historical fact that junction transistors (BJTs) predated the oxide-insulated field-effect transistors (MOSFETs) used here.

[44] For the through-hole TO-92 package; for the BSS126's surface-mount package it would be 63°C.

[45] An important note: Most of the measurements above were made with though-hole parts mounted on a solderless breadboard. This had the effect of isolating the current-programming semiconductor from the heat-producing resistors. To achieve similar results with SMT components on a PCB, you may need to isolate the critical part (for example in its SOT-23 package) from the hot parts, with a slot cutout in the PCB. A 1 mm-wide slot curved around the part should work well. Read on, for analogous performance results with components mounted on a PCB.

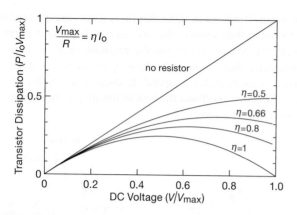

Figure 3x.55. Fractional reduction of transistor power dissipation for the circuit technique of Figure 3x.54. The parameter η is the fraction of total current that is bypassed through the shunt resistor at maximum voltage; $\eta = 0.66$ is a good choice, with maximum transistor dissipation just 34% of the maximum total.

Written in terms of small-signal variations, the meaning of a drain output resistance r_d is that a drain voltage variation v_{ds} produces a drain current variation i_d according to $i_d = v_{ds}/r_d$. Recalling that $i_d = g_m v_{gs}$, we can substitute to find that a drain voltage variation produces an effective gate–source change (analogous to Early effect) of $v_{gs} = v_{ds}/g_m r_d$, or, effectively, a "feedback factor"

$$\mu \equiv v_{gs}/v_{ds} = 1/g_m r_d,$$

where μ represents the effective change of gate–source voltage as a fraction of change in drain voltage.

Now, here's the interesting point: In the subthreshold region of drain current the quantity μ is relatively constant, because g_m is approximately proportional to drain current, and r_o is approximately proportional to $1/I_D$. That is, the feedback factor μ is mostly a property of the particular FET (as discussed in detail in §3x.4); typically μ is 1/250 or less. And for our simple depletion-mode current source it is the changes in v_{gs} (proportional to μ) that, when combined with the source resistor R_1, degrades the constancy of current. To improve the constancy of current we need to reduce μ; that is accomplished nicely by the cascode, effectively replacing μ with μ^2.[46]

3x.6.4 Low-cost predictable current source

Current sources that depend on the I_D-vs-V_{GS} transfer characteristic of a depletion-mode FET (whether JFET or

---

[46] As will become evident, however, at higher currents and voltages changes in V_{GS} caused by local heating can far exceed changes due to V_{DS}, forcing us to seek additional circuit improvements.

MOSFET) to program the output current are inherently inaccurate, owing to their manufacturing spread; see for example the measured histogram in Figure 3.17, or the measured transfer characteristics in Figure 3.25. Although you can generally expect a smaller spread within a given manufacturing batch of MOSFETs, you cannot depend upon finding the same characteristics in some other batch, or in the parts made by a different manufacturer. In other words, if you need to create a current with an accuracy better than, say, ±50%, you'll have to select the value of the source bias resistor to match the particular FET.

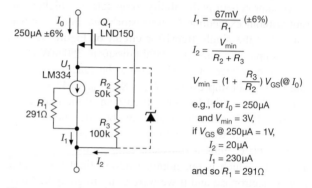

Figure 3x.56. Cascode Q_1 raises the output impedance (and extends the voltage range) of the LM334 programmable (and predictable) current source.

A different approach is shown in Figure 3x.56, where an inexpensive programmable current-source IC[47] is used as the current-setting portion of a cascode (analogous to Fig. 3x.52B, and see Fig. 2x.13). The LM334's current is programmed (over a range of $1\,\mu$A to $10\,$mA) with a single resistor R_1, according to $I = 0.067/R_1$, where the coefficient of $67\,$mV is the nominal value at 20°C (it increases proportional to absolute temperature, "PTAT").[48] The programming accuracy is ±6% for the least expensive grade ($0.60 in unit quantity), and ±3% for the intermediate grade ($1 in unit quantity). The LM334 by itself operates over a voltage range of 1 V to 40 V; here the cascode allows circuit operation to the breakdown limit of Q_1, or 500 V. Because the cascode transistor's gate–source voltage may be less than the minimum operating voltage of the LM334, the divider $R_2 R_3$ is arranged to multiply V_{GS} by

---

[47] See Figure 2x.13 for its elegant design.

[48] The programmed current I_1 includes the bias current through the programming resistor R_1, the latter contributing about 6% to the total. Note: when used at higher currents and voltages, adjacent SMT parts on the PCB will raise U_1's temperature, increasing its current by 25% per 100°C. Consider 1 mm slots for thermal isolation.

a modest factor; here we've chosen to run the LM334 at three times V_{GS}.

You can extend the current or voltage range of this circuit, and of the other current sources in this section, either by using a transistor of higher voltage (Table 3.6), or by the series-stacking technique described next.

3x.6.5 Current sources for higher voltages

You can get depletion-mode MOSFETs with voltage ratings to several kilovolts (e.g., the IXTH2N170, 1700 V; see also Table 3.6), allowing variations of these circuits to operate into the kilovolt range. But such MOSFETs tend to be expensive, hard to get, and usually of high current ratings with corresponding (and undesirable) large capacitances. For example, the IXTH2N170 is rated at 2 A, with an output capacitance C_{oss} of 200 pF at 25 V (compared with 2 pF for the LND150, and 2.4 pF for the BSS126); it costs \$16 in unit quantity.

Instead, you can extend the voltage range of MOSFETs with a series string, rigged up to ensure that the total voltage distributes itself without damaging the individual transistors.

A. A simple scheme

The simplest (and least elegant) scheme is shown in Figure 3x.57A: As the total voltage is increased, the transistor with the lowest current initially rules, until it reaches its avalanche voltage, or until its (imperfect) gradually rising current brings the transistor with the next higher current into play. The circuit works – but we'll see presently that adding a few components (circuit B) makes a circuit with better properties.

Figure 3x.58 shows the measured current of Figure 3x.57A, with three 900 V MOSFETs. With equal 2 kΩ resistors for R_1–R_3 the current rises smoothly, with an equivalent output resistance of ~ 25 MΩ. Evidently these three transistors have well-matched transfer characteristics. When the source resistors are mismatched, to simulate a ± 200 mV spread in V_{GS}, the curve of current shows soft steps at voltages corresponding to multiples of the 1.0kV–1.2kV avalanche voltage.

Figure 3x.59 shows the performance of an analogous circuit, this time with a set of five LND150's from a single manufacturing batch. This time we're not so lucky: the current exhibits a set of soft steps, the steepest of which corresponds to an incremental resistance of 60 MΩ (compared with an overall slope of 250 MΩ).

Comparing these two simple approaches, the set of five LND150s was operated at about 0.11 mA, with each dissi-

pating up to 70 mW; the net current-source capacitance was about 2.5 pF. By comparison, the set of three IXCP10M90s in TO-220 packages was operated at 2 mA, with each transistor dissipating up to 2 W near its avalanche voltage; a clip-on heatsink was required. The C_{oss} for these parts is about 40 pF (presently we'll see how to reduce the overall capacitance to less than 15 pF while retaining the 6 W capability[49]).

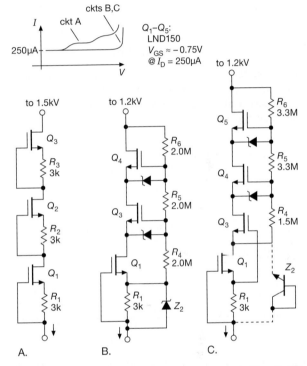

Figure 3x.57. Extending the voltage range of Fig. 3x.52's depletion-mode current sources. A. Simple series string. B,C. Cascode with series stack.

B. Distributed series string

A better way to extend the operating voltage is shown in the circuits of Figure 3x.57B and C, where a string of resistors R_4–R_6 distributes the voltage span across a series-connected group of transistors Q_3–Q_5. The series stack serves as a high-voltage cascode transistor for current-setting Q_1. As with Figure 3x.54, the shunt resistors do not compromise the high output impedance, because their (resistive) current is sampled by current-sensing resistor R_1.[50]

---

[49] Or, with IXTP01N100D depletion-mode FETs (see Table 3.6) reduce the capacitance further, to about 10 pF.

[50] In this circuit R_4 is deliberately chosen smaller than the other resistors, to limit the maximum voltage across Q_3 in order to reduce its V_{GS} vari-

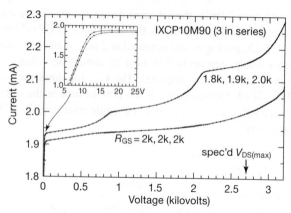

Figure 3x.58. Measured current variation for three 900 V IXCP10M90's configured as in Fig. 3x.57A. The unmatched resistors (upper curve) illustrate the effect of a ±0.2 V scatter among the transistors' V_{GS}.

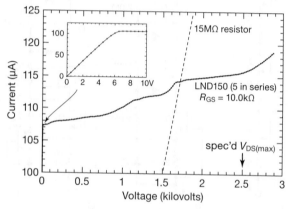

Figure 3x.59. Measured current variation for five 500 V LND150's configured (with equal 10k resistors) as in Fig. 3x.57A, illustrating current variation from V_{GS} scatter.

Circuit B is inferior to C, first because Q_1 runs at a high voltage, thus hot; and second because R_4 robs current from Q_1. Both of these effects raise Q_1's V_{GS}, and thus the programmed current.[51]

The series-stack configurations (Fig. 3x.57B and C) eliminate entirely the current steps of Figure 3x.57A, as seen in the measured current error of Figure 3x.60. Because the current hardly varies with applied voltage, we've here plotted the current *error*, on a greatly expanded (×250,

compared with Fig. 3x.59) scale. The equivalent output resistance is an impressive 15 GΩ.[52]

There's more to like about the series-stack circuits B and C: (a) you can use a selected low-voltage transistor for Q_1; (b) the effective capacitance is C_{oss}/n, for n transistors in the string; (c) the power dissipation is spread equally across Q_3–Q_5 (unlike the situation in circuit A), and at high voltages the gate equalizing resistors take more of the dissipation; and (d) you can replace Q_1 with an integrated current sink, for example an LM334 as in Figure 3x.56.

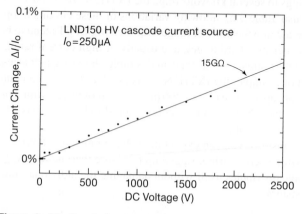

Figure 3x.60. Greatly improved constancy of current is seen in the measured current of seven LND150's, configured as a cascode with a 5-transistor series stack (Fig. 3x.57C). Compare with Figure 3x.59: Here the vertical scale is expanded ×250. See caution in the footnote 52 on this page.

C. Some applications: HV amplifier; HV probe

We've used these techniques to make current-source loads for high-Z high-voltage probes (§3x.9), and to make current-source pull-ups for MOSFET driver stages in high-voltage linear amplifiers, for example of the kind described in §4x.23. For these kinds of applications we favor low-current depletion MOSFETs, owing to their very low capacitance; for example, the BSS126 (600 V, I_{DSS}=7 mA, min) has typical capacitances of 21 pF, 2.4 pF, and 1.0 pF (C_{iss}, C_{oss}, and C_{rss}, respectively) at V_{DS} = 25 V. Nothing comes for free, and the price you pay for low capacitance is very limited power dissipation: the BSS126 comes only

---

ation, and thus the current variation of Q_1. Be sure to use resistors rated for 500 V (or several lower-voltage resistors in series) for R_5 and R_6.

[51] In circuit C, R_4 robs current from Q_3, changing both its V_{GS} and Q_1's V_{DS}. Happily, though, a MOSFET's V_{GS} is not sensitive to changes in V_{DS}, see §3x.5.1.

[52] As noted earlier, our measurements were made in a solderless breadboard with a test circuit using TO-92 transistors and axial-lead resistors. In addition to style C's superior configuration, this served to isolate Q_1 and its positive gate-voltage tempco from as much as 0.6 W generated by the resistors and other transistors. Surface-mount parts on a PCB are not so fortunate, suffering the effects of adjacent hot parts, especially at higher currents.

in a SOT-23 package, with a pathetic thermal resistance of $R_{\Theta JA}=250°$C/W (on a PCB with "minimal footprint").

As we'll see presently, the simple circuits of Figure 3x.57 are adequate for non-critical applications, but they suffer from non-constancy of current, which varies somewhat as the voltage varies. That is because the (changing) resistor-string current changes the drop across R_1, thus the bottom transistor's V_{GS}, and thus its drain current.

A second (and even larger) effect is caused by heating of Q_1 by adjacent hot parts, particularly when there's a large voltage drop across the current source. These may not be important, if the goal is simply to provide a driver load that is more like a current source than a resistor. But as we seek to run our current sources at higher currents and voltages, these increases make the power-dissipation problems that much harder to solve. In the following sections we take a deep dive into perfecting the high-voltage current source.

D. High-voltage current sources: 250 μA

Let's review the choices so far, all of which use small SOT-23 MOSFETs. Starting with Figure 3x.52, the simplest current-source (CS) in our lineup is circuit A, a single depletion-mode MOSFET with a current-setting source resistor. With available parts this circuit is limited to 500 to 600 V, and to currents no more than 0.5 mA for voltages over 400 V due to excess heating. At power dissipations above about 200 mW the V_{GS} tempco of Q_1 causes the current to increase beyond its already uncertain value. Next in circuit B we can go to somewhat higher currents, having moved the power dissipation to Q_2, but we're still limited by using a single transistor. In Figure 3x.54C we added resistors to handle half or more of the power dissipation. But circuit C still suffers from the same V_{GS} tempco degradation of the current-setting transistor Q_1. This effect is greatly reduced in circuit D by voltage-isolating cascode Q_3, assuming that Q_1 can be thermally isolated from the other hot parts.

In Figure 3x.57 we stacked up series stages, allowing us to go to higher voltages. But in circuit A only one MOSFET is working at a time, until it "fills up," and we suffer also from the same tempco problems we had with the single FET in Figure 3x.52A. It's far better to partition the power dissipation among the series-stacked parts, so they'll all run cooler, as in Figure 3x.57B and C.

Figure 3x.60 shows how beautifully circuit C worked in a breadboard, maintaining a steady 250 μA with less than 0.1% variation all the way to 2.5 kV. This excellent performance comes from having Q_3–Q_5 doing the heavy lifting (with cascode Q_3 helped by reduced shunt resistor R_4), and

also having isolated current-setting transistor Q_1 (a TO-92 through-hole part, standing tall in the breadboard).

In a PCB implementation, though, it's difficult to prevent thermal coupling to Q_1, so it's tempting to try a different approach – namely using a current-reference IC (or equivalent) to set the pull-down current at the bottom of the string. We saw this in Figure 3x.56, where U_1 set most of the CS current; but in that circuit a portion of the total current flows through R_2R_3, and that is affected by Q_1's tempco, so there is residual current variation with heating (i.e., with total voltage across the CS). Higher resistor values would reduce the effect, but then we get to worry about Q_1 responding quickly to rapid changes in V_{DS}. (We may be tempted, but we're not allowed to add a capacitor across the resistors!) And there's another degrading effect in that simple circuit, because the current-reference IC (U_1) happens to have a deliberate PTAT tempco (proportional to absolute temperature), in which serious heating could cause up to 25% increased currents.

Why all the fuss? At this point the reader can be excused for wondering why they are being dragged through so many circuits? Why not just show the best HV current-source circuit, and leave it at that? Well, as in engineering generally, there is no overall "best" circuit. Many applications will be fine with simple circuits, but others require more effort. The story of improving HV current-source performance is the story of adding more and more parts. But there's no point in using a complex part-laden design for a non-critical application. We do hope you'll enjoy exploring the options (as we did, designing them!) as we continue with higher-voltage, higher-current versions.

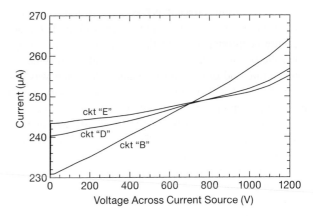

Figure 3x.61. Measured current versus voltage for the 250 μA current-source circuits of Fig. 3x.57B and Fig. 3x.62D and E, implemented as surface-mount parts on a PCB. Even though the maximum power dissipation was only 300 mW, all three reveal increasing thermal effects as they approach 1.2 kV.

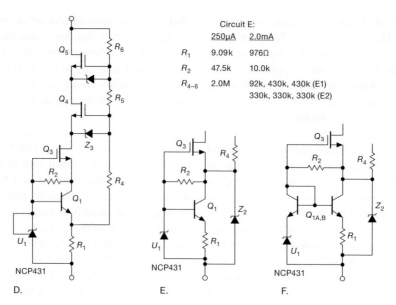

Figure 3x.62. Series-stacked HV current sources, with a 2.5 V voltage reference (ON Semi's NCP431A) setting the current. All have identical Q_4Q_5 and R_5R_6 series-stack wiring as shown in circuit D. The zeners are standard gate-protection diodes.

To explore issues with these circuits, we created production PCBs, with small surface-mount parts. Figure 3x.61 shows measurements for three 250 µA versions (two of which we describe below). Plot B is for Figure 3x.57's circuit B, whose current is set by Q_1's V_{GS}. No surprise here to see a 12% increase at 1.2 kV relative to that at 200 V. If this 2-terminal current source were used as a pull-up in a high-voltage amplifier, there would be an unsatisfactory decrease of pull-up current as the signal swings from full negative to full positive output.

Next step: stable current reference. In the preceding collection of circuits, depletion-mode MOSFET Q_1 set the current through R_1. But this depends on its (poorly-controlled) gate–source voltage, along with sensitivities to temperature and drain-source voltage. The result was undesirable current variation with total voltage drop, which is less than optimal in a current source, and additionally not a good thing when we were already struggling with excessive power dissipation. The next set of circuits addresses this shortcoming by using IC voltage references (2-terminal shunt type) to set the R_1 bias voltage, thus the set-point current.

Look at Figure 3x.62, where we've used the NCP31A 2.50 V active shunt-mode (zener-like) reference, biased at 40 µA by R_2 (47.5k for the 250 µA version). Emitter follower Q_1 loaded with R_1 then sets the major portion current of the total current, about 210 µA ($I_{U1} = [2.5\text{V}-V_{BE}]/R_1 + V_{GS}/R_2$). Figure 3x.61 shows circuit Figure 3x.62D im-

proving the constancy of current, in this case a rise of 6% (compared with 12% for Fig. 3x.57B); of this, 4% is due to Q_1 heating (40°C rise with -2.2 mV/°C tempco), and 2% blamed on Ebers–Moll (Q_1's V_{BE} drops by 37 mV due to current dropping from 230 µA to 57 µA, see Fig. 2x.11).[53]

The next refinement (Fig. 3x.62E) was to eliminate the 2% Ebers–Moll effect by moving R_4 to Q_3's source terminal, so we now see the overall current rise reduced to 4%.

Where can we go from here? Well, how about a BJT matched pair,[54] to eliminate the V_{BE} tempco effect? That's the circuit of Figure 3x.62F, whose current is given by $I = (2.50\text{V}-60\text{mV})/R_1 + V_{GS}/R_2$.

As we refine these circuits, we are using more parts! Checking the parts counts for our high-voltage current sources, circuits A–C (Fig. 3x.57) used 6, 10, and 11 parts, respectively. After we added the U_1 voltage reference, circuits D–F (Fig. 3x.62) had 12, 13, and 13 parts. But hey, that's not so bad – soon we're going to top out at 16 parts with our "perfect" version! But that's OK, the various circuits all have eight required parts, associated with Q_3 to Q_5, to handle the high voltage and high power, so the best cir-

---

[53] To address these drifts, reader "whit3rd" suggests moving U_1's adjust pin to Q_1's emitter in circuit D of Figure 3x.62. This is a good idea, but note that U_1 responds slowly to changes in Q_1's V_{BE}. Making this change in circuit E, where R_4 has been moved to Q_3, would isolate Q_1 from fast current changes, and hence V_{BE} changes; and adding Q_2 as in Figure 3x.68H would also isolate U_1 from voltage changes.

[54] Such as a BCM847DS, a dual version of the BC847, see Table 8.1b.

cuit uses only six more. And keep in mind, you'll need another four parts per kilovolt if you want to go to yet higher voltages. This is what comes with the territory.

E. High-voltage current sources: 2 mA

These basic circuits can run at higher currents, with appropriate component scaling and with attention to maximum power dissipation. Figure 3x.63 shows measured performance for the circuits of Figure 3x.62, configured for 2 mA with the resistor values shown. First we set the series resistance of R_4–R_6 to about 1 M, to take up about 1.25 mA at 1.25 kV, thus limiting the MOSFETs to less than 40% of the total current at 1.25 kV. In circuit D R_4 is returned to Q_1's emitter, giving a 9% rise in total current as the total voltage goes from 200 V to 1.2 kV. As with the 250 μA circuits earlier, in circuit E we return R_4 to Q_3's source, producing a smaller 5% rise (when the resistor string is partitioned as indicated in the circuit: version "E1"). Circuit F offers further improvement with the same parts count.

The current sharing and power dissipation of the transistors and of the resistors is plotted in Figures 3x.64 and 3x.65, assuming equal sharing among each component type. The worst-case MOSFET dissipation is 0.35 W with equal current-sharing resistors R_4–R_6 (but in our "E1" circuit the upper two MOSFETs peaked at about 0.5 W).[55]

F. Current sources in high-voltage amplifiers

These HV current sources find application as pull-ups in high-voltage amplifiers. For example, the high-voltage amplifier described in Figure 4x.137 uses a 250 μA current-source pull-up of either form A or B (Fig. 3x.57). We're often quite parsimonious with current drawn from high-voltage supplies,[56] and even 0.25 mA can seem quite a

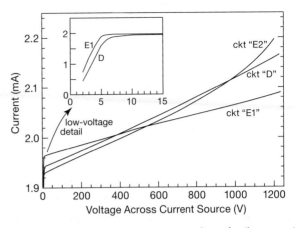

Figure 3x.63. Measured current versus voltage for the current-source circuits of Fig. 3x.62.

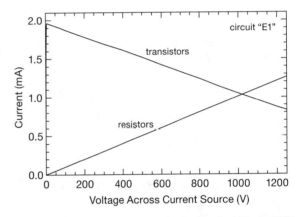

Figure 3x.64. Current in Q_1 and R_4 for the circuit of Fig. 3x.62E1.

lot. For that circuit, the 0.25 mA current-source pull-up allowed the amplifier to have a reasonable +7.5 V/μs slew rate. But we often need higher slew rates;[57] for example, in §4x.23.9 we required 10 mA pull-up currents to achieve $S = I/C = 10\text{mA}/20\text{pF} = +500\,\text{V}/\mu\text{s}$ slew rates. Clearly there's benefit to be had from further stepping up our current-source game to work to higher voltages and higher currents.

The critical high-impedance low-capacitance node in the HV amplifier of Figure 4x.137 is marked "X," and the capacitance of that node is $C_X = C_{oss} + C_{rss} + C_{pullup} + C_{stray}$. The MOSFETs used in these amplifiers can be chosen from Table 4x.5 in §4x.23.9. For the fastest possible amplifiers, we'd be choosing parts near the top, where $C_{oss} + C_{rss}$ is 15 pF or less. We'd like C_{pullup} of our current source to

---

[55] In the E1 test we reduced R_4 to a tenth of the total resistor string, to put 90% of the resistor heating power to the parts farthest from current-setting Q_1. With Q_3 and R_4 running cooler, Q_1's tempco effects were reduced, leading to the improved result. But we shouldn't celebrate, because the four remaining hot parts pushed the surface of the PCB to a toasty 150°C! The data looked fine, but the FLIR thermographic image (see §9x.24.2) said otherwise. We'll have more to say about those issues later. The E2 test was like E1, except we used three 232k resistors, for 1.8 mA max at 1.2 kV, thinking it might be a good idea to reduce the MOSFET currents to 20% of the total. But the resistor bank heated Q_1, causing the current to climb an extra 10% at 1.2 kV, and the 0.2-to-1.2 kV current rise degraded to 12%.

[56] One of our instruments had twenty-four 1.2 kV amplifiers, supplying precision voltages for deflector and shim electrodes in a molecular beam with a 10 kV trap. With such a collection of amplifiers, the accumulated current drains can really add up, so you need to practice restraint.

[57] In circuits like this, *negative*-going slew rates can be very fast, as much as −500V/μs, due to Q_5's very high current-sinking capability.

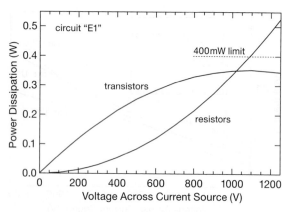

Figure 3x.65. Average power per MOSFET and per resistor for the circuit of Fig. 3x.62E2.

be a small fraction of that, say no more than 3 pF. The 3-transistor stacks we've been evaluating here should contribute no more than 1 pF, if made with the small SOT-23 LND150 or BSS126 transistors. By contrast, the larger-die (higher-power) depletion-mode MOSFETs from Table 3.6 have C_{oss} of 6 pF to 45 pF, so a series stack of three MOSFETs would add 2 pF to 15 pF to the C_X node. Of these larger MOSFETs, the 400 V 150 mA DN2540N5 (with $C_{oss} = 12$ pF) is the most attractive TO-220 part. A stack of three would add 4 pF to C_X, perfectly reasonable given the 15 pF or more of existing node capacitance. This MOSFET is well suited for use with 3 W to 5 W heatsinks, each taking 2 to 4 mm^2 of space.

Issues with small depletion-mode MOSFETs at high currents. Looking at the role of Q_3 in the circuits above, it's clear that depletion-mode MOSFETs are handy for creating bias voltages to power circuitry in 2-terminal current sources. The BSS126 is a favorite part for use in the 0.5 mA to 10 mA region. Like the lower-current LND150, it has low C_{oss} capacitance (2.4 pF versus 2.0 pF), but can work at higher currents, as seen in the datasheet curves of Figure 3x.66. It has high transconductance at low currents, for example at 250 μA, where V_{GS} appears nearly independent of current changes. But then notice the high V_{GS} changes at 5 mA (a 20× higher current). Also notice the dramatically higher I_D versus V_{GS} tempco at 5 mA, nearly +200 mV, or about +25% per 100°C. This means its usefulness as a voltage reference within a current source will be poor at higher currents and high temperatures, and we'll have to work harder to create effective current sources. Alternately we could choose depletion-mode MOSFETs with higher minimum I_{DSS} currents, but these will have higher V_{GS} voltages at 5 mA and 10 mA. A good choice would be the 800 V

20mA CPC3982 in a SOT-23 package, with a C_{oss} of 2.2 pF and a typical V_{GS} of −1.25 V at 10 mA.[58]

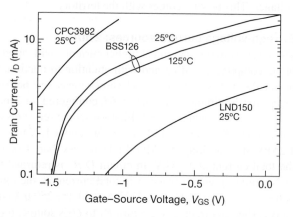

Figure 3x.66. Transfer curves (typical) of the only available depletion-mode MOSFETs in SOT-23 packages, as shown in their respective datasheets.

Also, before designing in a BSS126 at 10 mA or above, note that its datasheet's minimum I_{DSS} is 7 mA (with no "typical"[59] or maximum specified!). That means we might need a slightly *positive* gate bias to reach 10 mA. A nice way to finesse this is to use a high-current JFET (e.g., an MMBF4392, see Table 3x.1) for Q_2, for the critical depletion-mode function. The JFET's modest voltage rating means we'll have to cascode it with a high-voltage MOSFET (Q_3 in Fig. 3x.68G). Then we add U_2, a 2.5 V active zener (an LM385-2.5, that works to 15 μA) to ensure Q_2 has enough headroom. Depletion-mode MOSFETs Q_3 and up are allowed to work with slight positive voltages, up to 0.5 V or so with typical TVS protection zeners like the D1213A, so the BSS126 should be OK at 10 mA or even 15 mA.

G. High-voltage current sources: 5 mA and more
Going to higher currents means rising thermal dissipation... and *problems!* Figure 3x.67 shows measurements on two PCB-built 5 mA 550 V (3 W) current sources, both with 68k resistors for R_4 to R_6;[60] for these tests we initially used the circuits of Figure 3x.57B and Figure 3x.62D. Circuit B relies on Q1's V_{GS} voltage, with 274 Ω for R_1. A

---

[58] Other examples of MOSFETs with low capacitance and higher minimum I_{DSS} are the 800 V 100 mA CPC3980 ($C_{oss} = 5$ pF), the 350 V 180 mA 350-volt DN3135 ($C_{oss} = 6$ pF), or the 450 V, 120 mA DN3145 ($C_{oss} = 15$ pF).

[59] There's a hint in the graphical transfer plot, which shows a zero-V_{GS} drain current of 22 mA.

[60] And a third "perfect" current-source circuit, see §3x.6.5H below.

better choice would have been circuit C, because in circuit B R_4 of the series stack of resistors is returned to the source of Q_1 (instead of to Q_3). Although R_1 still monitors and controls the current, at increasing voltages R_4 robs Q_1's current, decreasing its V_{GS} and thus increasing the current through R_1. The effect is exacerbated with Q_1's heating from adjacent components, and this resulted in an unacceptable 34% current increase from 100 to 500 V, with consequent severe rising temperatures and currents at 550 V.

By contrast, circuit D is active-zener based. We set $R_1 = 402\,\Omega$ (for 5 mA current) and $R_2 = 10\mathrm{k}$ (for $80\,\mu\mathrm{A}$ zener bias). As before, the R_4 stack still robs current from Q_1, but, owing to the 2.5 V bias at its base (large compared even with a 60 mV change in V_{BE} that would result from a $10\times$ change in current), the effect of R_4's changing current is minor. In a quick measurement (to keep it from overheating) we saw circuit D's current increase by 6.6% as we ramped the voltage to 500 V (plus some more as the PCB warmed up); circuit E, with R_4 returned to Q_3's source, would have done better. It's evident that more complex circuits are appropriate at higher currents. But, at these voltages and currents, both versions have three very hot resistors and three hot transistors: in the case of a 5.5 mA current source with 550 V across it, these six surface-mount components are dissipating 3 W. Although the voltage ratings of the transistors are sufficient for a 1.5 kV current source, the voltage is limited by dissipation, not breakdown; that's why we needed three transistors to distribute the heat. Even so, we measured surface temperatures of $+130\,°\mathrm{C}$ before turning off the supply. Evidently additional thermal mitigation will be required – for example by mounting the hot components on a small daughterboard with modest airflow (a small fan).

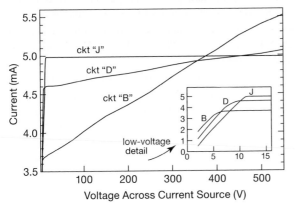

Figure 3x.67. Measured current versus voltage for the circuits of Figs. 3x.57B, 3x.62D, and 3x.68J.

There's more than just overheating to worry about in these circuits: Circuits D and E suffer from a substantial positive tempco, (Q_1's V_{BE} drops by 220 mV for a 100°C rise, a +13% effect). This overshadows the stable U_1 voltage reference. Moving R_4 to Q_3's source (circuit E) helps, but Q_1 still suffers from its temperature coefficient; that's where circuit F (with balanced V_{BE}s) would show its stuff.

H. Perfect high-voltage current source

We'd like to have a "perfect" 2-terminal current source, with its current set by precision resistors, etc., and independent of total voltage. But instead we've been struggling with high-voltage current sources whose currents change with voltage and temperature. Thinking through the accuracy and stability problems of our designs, one thing that stands out is their need for a bias current to power the current-setting reference. What we need is a self-contained reference that is self biased. Ideally it shouldn't require an output capacitor, so that we can make a current regulator by simply adding a load resistor. This ideal leads to circuits H and J in Figure 3x.68.

We first looked at 3-terminal voltage references (see Table 9.8) and found several that didn't require output capacitors; but they were all low-dropout (LDO) types, which means they have high-impedance collector or drain outputs, rather than a low-impedance emitter or source-follower. They've managed to internally stabilize the circuit without an output capacitor, but the price you pay is having to live with a slow control loop. Since our fastest high-voltage amplifiers can complete their slewing in a microsecond, the current source needs to respond as fast. In Figure 3x.68H, R_2 and cascode Q_2 isolate U_1 with its stabilizing C_1. Then we trust Q_3 to provide the required fast response. We're rewarded with a perfect current source: $I = I_Q + 2.5\mathrm{V}/R_1$, where $I_Q = 60\,\mu\mathrm{A}$.

An alternative is to find a 3-terminal low-voltage *non-*LDO with very low quiescent current; the only one we could find (in SOT-23 or smaller package) was the ZMR250 from Zetex (Diodes Inc.), which turns out to be ideal (Fig. 3x.68J). It's well characterized, and in normal use does not require input or output capacitors. It draws a low $30\,\mu\mathrm{A}$ quiescent current, and it should be tolerant of six or more adjacent hot components, since its 2.5 V output is specified to change by 1% (typ) per 100°C. It needs 1.5–2 V overhead to operate, so we added Z_1 (an NCP431A active reference) to bias Q_2's gate. Now Q_3 becomes the high-voltage MOSFET, with R_4 returned to its source terminal.

You might worry that there's a small error from Z_1's uncertain current (set by Q_2's uncertain gate voltage across

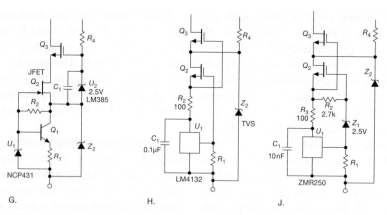

Figure 3x.68. High-voltage current-sources suitable for 5 mA and up. In circuit G a JFET biases reference U_1 at high currents. Circuits H and J sport a perfect formula, $I = I_Q + 2.5V/R_1$, excellent at any current well above U_1's quiescent current I_Q. Circuit H uses a precision 3-terminal shunt voltage reference; it's LDO, and needs little overhead to operate. Circuit J uses a linear-regulator IC that needs several volts to operate; we found that a small capacitor C_1 is required for stability.

R_2); but in fact there is no error because that current is metered by R_1. We measured the current of this circuit (curve J in Fig. 3x.67), and found it constant at 4.986 mA from 12 V clear up to 550 V. (When configured as a 2 mA current source it measured a constant 1.980 mA from 10 V to 1.25 kV, having reached within 1% of that value by 6.5 V.) The minimum compliance voltage for circuit J may be as high as 10 V; so, though it's "perfect," it's not best for every situation.

Higher currents. The examples we've shown are aimed at high-speed applications, where low capacitance is paramount. But if you don't need speed, you can take advantage of larger MOSFETs, in power packages, to create higher current versions of these circuits. For example, you could use a string of DN2540N5 (400 V, 150 mA, 12 pF) or IXTP02N50D (500 V, 200 mA, 25 pF) MOSFETs, both depletion mode types in TO-220 power packages.

Enhancement-mode MOSFETs. These circuits do not *require* depletion-mode MOSFETs, except for the biasing and current-setting transistors (e.g., Q_3 in Figure 3x.62, or Q_2 and Q_3 in Figure 3x.68); all the others can just as happily be enhancement mode. For example, the enhancement-mode IXTY01N100 comes in a 25 W DPAK, with a C_{oss} of just 7 pF, and the IXTA08N100P comes in a 40 W TO-220 with just 18 pF of C_{oss}. These are 1 kV parts, chosen not for their high voltage rating, but for their low capacitance. Also, keep in mind they'll work fine at low voltages.

Depletion-mode power MOSFETs. For readers anticipating a discussion of *depletion*-mode power MOSFETs, we're sorry not to have a section about these amazing and

super-useful devices in *The x-Chapters*. But we have not ignored them, see §3.6.2 in the main volume, the high-voltage discharge circuit in §9x.12, and the detailed Table 3.6. There has not been much action in that scene, so that table is up-to-date. It was missing the DN2470 DPak, and also the DN2530, which is a useful 13 pF part in SOT-89 and TO-92 packages. The table is also missing the 800 V 20 mA CPC3982 in SOT-23 mentioned earlier (but it does have the *discontinued* CPC3720, 3730, and 3714!).

A parting shot. It's important to realize that, behind an innocent-looking and perfectly constant output current curve, the devil of excessively high temperatures may be hiding. It's essential to understand and address thermal problems if you want to avoid premature component failure. If significant power dissipation is anticipated, be sure to couple surface-mount transistors and resistors to a heatsink, using compliant thermally-conductive sheet material such as Bergquist Gap Pad VO, or 3G Shielding ThermaWorx T600.

For circuits with high currents at high voltages, you'll need to use larger MOSFETs with higher power capabilities, even though they have higher capacitance. For example, a stack of three of the 12 pF or 25 pF parts mentioned above will result in a 4 pF or 8 pF current source. Let's say this is used in a high-voltage amplifier, e.g., driving node X in Figure 4x.137, with MOSFETs like the IXTP06N120 (1200 V, 0.6 A). The node's $C_{oss} + C_{rss}$ capacitance will be 18 pF already, so adding another 8 pF or 45% will be OK, as long as you can compensate by increasing the current by 45%. But you might be increasing the current by a factor of five, to achieve slew rates of 25 mA/26 pF=900 V/μs! At this point it's clear that heatsinks alone will not do the job – a fan will have to be added. Once you have the fan, its airflow will cool the hot current-source heatsinks as well.

3x.7 Bandwidth of the Cascode; BJT versus FET

In previous examples here and in Chapter 3 we've happily exploited the cascode configuration, celebrating its triple virtues (Miller killer; g_{os} killer; low gate current via low V_{DS}). But what about its speed?

The cascode transistor is, in essence, a *common-gate* amplifier, with its input being a signal current applied to the source terminal, and the output being the current that appears at the drain. It's analogous to the BJT common-base configuration, and it shares the characteristic of having voltage gain, but no current gain. Let's look first at this amplifier configuration.

3x.7.1 The common-gate/common-base amplifier

Common-base BJT amplifiers are widely used at radiofrequencies, owing to their freedom from Miller effect, and their good match to 50 Ω signal sources. But you rarely see common-gate JFET amplifiers (with the exception of the JFET cascode, discussed in the next section), and for good reason. Look at Figure 3x.69, which shows the essence of a grounded-base and a grounded-gate amplifier stage, driven by a signal voltage in series with an external resistor (R_E and R_S, respectively, which include the signal's Thévenin impedance). In each case the (non-inverting) voltage gain is the ratio of the load impedance to the input signal current; and the latter is[61] the input signal voltage divided by $R_{ext}+1/g_m$.

Compared with the BJT, the JFET's low transconductance (in the range of $\sim 10\,\mathrm{mS}$ at typical drain currents of a few milliamps) is a serious problem, greatly limiting the single-stage voltage gain. This is especially true with radiofrequency signal sources, which usually have a Thévenin signal impedance of 50 Ω, and which expect to be loaded with that same impedance as a termination. One possible solution is the transconductance enhancer... but what's the point? In a common-gate amplifier the input signal supplies the full output current, and so there's no benefit to be gained from the JFET's extraordinarily low (es-

---

[61] Except at frequencies approaching the transistor's unity-gain frequency f_T, where the common-base or common-gate current gain is 3 dB down from unity, see §2x.11 for BJTs and §3x.7.3 for FETs.

sentially zero) gate current. For this kind of application – a relatively low impedance signal source of wide bandwidth – use the BJT common-base configuration.

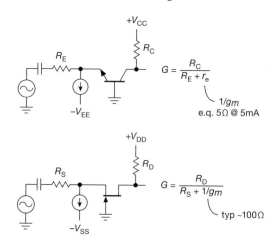

Figure 3x.69. Grounded-base and grounded-gate amplifiers. Zero gate current is a wonderful thing – but of no benefit here. The BJT's higher transconductance wins the day.

3x.7.2 Cascode as common-gate/common-base amplifier

Though the cascode transistor is indeed a common-gate (or -base) amplifier, the situation is somewhat different, because the input signal is a current (approximately) from the underlying drain. That is, the cascode transistor is driven by a high-impedance signal. Here the limited transconductance of the JFET cascode (Q_2 in Figure 3x.70) is less important: what matters is that (a) the impedance $1/g_{m2}$ seen looking into the cascode source terminal be much smaller than the output impedance $1/g_{os}$ of Q_1 (to mitigate the g_{os} effects described in §3x.4), and (b) the transconductance of the cascode transistor g_{m2} be at least comparable to the transconductance of the gain stage g_{m1} (to suppress Miller effect, by minimizing drain voltage swing).

The JFET cascode has the benefit of simple biasing: Q_2's gate–source voltage sets Q_1's drain-source operating voltage (and this works for a follower, too, by connecting the Q_2's gate to Q_1's source). This simplicity is tempered somewhat, though, by the loose V_{GS} specifications, which creates considerable uncertainty about the actual V_{DS} of Q_1. And, as we'll see presently, if you choose a high I_{DSS} part for Q_2 (to ensure adequate V_{GS} margin, when run at low current), you'll wind up with degraded bandwidth.

The BJT cascode takes a few more parts (Figure 3x.70), but has pleasantly high transconductance ($g_m = I_C/V_T$), and

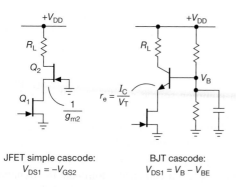

Figure 3x.70. Cascode choices for a common-source JFET amplifier.

it lets you have confidence in setting Q_1's drain voltage (a diode drop below the base voltage from the bias divider). More bandwidth, too, as we'll see next. In fairness to the JFET, we should point out that it is somewhat more difficult to bootstrap the BJT cascode, in a situation where Q_1's source terminal is not at a fixed voltage (e.g., in a follower or differential amplifier).

3x.7.3 Estimating cascode bandwidth

We haven't forgotten our question... we've just been warming up for the kill: what is the bandwidth of the cascode? Figure 3x.71 shows how to think about it, for both JFETs and BJTs. An applied signal current generates some swing at the source [emitter] terminal, of amplitude $v_{in} = i_{in}/g_m$. But this causes some signal current to be diverted to ground by the gate–source [base–emitter] capacitance C_{iss} [C_{be}]. The current gain (nominally unity) is down 3 dB when the current so diverted is equal in magnitude to the current entering the source [emitter], i.e., when the impedance of C_{iss} [C_{be}] equals $1/g_m$; this gives the gain–bandwidth product (f_T, or "GBW") expressions:

$$f_T = \frac{1}{2\pi r_e C_{be}} \qquad \text{(BJT)} \qquad (3x.5)$$

$$f_T = \frac{g_m}{2\pi C_{iss}} \qquad \text{(FET)} \qquad (3x.6)$$

Low transconductance and high capacitance are the enemies of bandwidth. That's why a large JFET, run at a tiny fraction of its considerable I_{DSS}, would be a poor choice. To see this clearly, look at the following comparison:

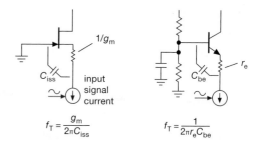

Figure 3x.71. Estimating bandwidth in the cascode.

Type	$I_{DSS(min)}$	I_C, I_D	g_m (mS)	C_{iss}, C_{be} (pF)	f_T (MHz)
BF862	10 mA	1 mA	12.1	9	214
		5 mA	26	9	460
J107	100 mA	1 mA	8.24	70	19
		5 mA	30.2	70	69
2N3904	–	1 mA	40	29*	220
		5 mA	200	80*	400

* inferred from datasheet f_T at I_C values

The J107 is a large JFET (100 mA minimum I_{DSS}, no maximum specified!), and its large capacitance reduces the bandwidth considerably, compared with the smaller BF862 (whose transconductance is comparable). Bipolar transistors operate with their base–emitter junction forward biased, and datasheets do not specify a capacitance value (which would be a strong function of I_C). Instead they specify f_T versus I_C, a curve that has a broad maximum at intermediate currents,[62] dropping at both high and low currents.[63] From these data you can infer C_{be} (as we've done in the table) – but what you ultimately want to know is how that capacitance affects transistor bandwidth, which is why f_T is specified (rather than C_{be}).

There are some situations where you cannot follow the advice of using a low-current (therefore low-capacitance) FET in a cascode, notably in high-voltage (kilovolt-level) circuits. There you've got to use discrete "power" MOSFETs, the smallest of which have I_{DSS} ratings of at least a few hundred milliamps – and the capacitance to prove it! See the circuit example in §3x.9

---

[62] See for example Figure 2x.80 on page 143.

[63] This suggests that the current-dependent ("nonlinear") portion of C_{be} is roughly proportional to I_C; the f_T dropoff at low currents is due to the effects of the fixed portion of C_{be}, and the f_T dropoff at high currents is due to residual bulk resistances in the emitter and base.

3x.7.4 What about MOSFETs?

High-voltage MOSFETs and power MOSFETs, apart from their much larger capacitances, exhibit the same capacitive current-robbing behavior as JFETs, with a common-base (or cascode) bandwidth given by

$$f_T = \frac{g_m}{2\pi C_{iss}}. \qquad (3x.7)$$

At a frequency of f_T the signal current flowing through the gate–source capacitance robs half the current that should be flowing through the source terminal; in other words, f_T is the frequency at which the FET has unity current gain.[64] As an example, the IXTA1N120 n-channel power MOSFET (1200 V, 1 A, used in the precision high-voltage amplifier in §4x.23) has a gate–source capacitance of 450 pF and a transconductance of 8 mS at a drain current of 1 mA. At that current, which is not unreasonable for a high-voltage amplifier intended for low-current applications, the bandwidth (i.e., the frequency at which the current gain is unity) is $f_T = 2.8$ MHz.

You are probably asking yourself why anyone would choose a power MOSFET for a small-signal amplifier application, given its poor bandwidth (a few megahertz) compared with the much greater bandwidths we just calculated for a BJT or JFET (tens to hundreds of megahertz). One answer is that MOSFETs are the only game in town for voltages upward of 500 V; and discrete MOSFETs are only available in large geometry types with ampere-scale current ratings (and correspondingly large C_{iss} capacitances). So you have to contend with quite limited bandwidths; that's not a killer, though, if what you need is a high-voltage amplifier of only modest bandwidth, of the kind shown in §4x.23.

You *can* do better, though, by seeking small-die MOSFETs with their lower capacitance; see Table 4x.5 in §4x.23.9. A good choice, for example, might be the IXTY01N100, with $C_{iss} = 54$ pF, or 8× less than the IXTA1N120; its f_T is over 50 MHz at 2 mA.

3x.7.5 Bandwidth of the source follower

You can similarly define the frequency at which the current gain of an FET operating as a source follower (with either a resistive or current-sinking load) falls to unity; it is approximately the same as the f_T of the common-gate (or cascode) configuration: in either case it is set by the diversion of source-terminal current comparable to I_S through the gate–source capacitance. This sets an upper bound on the follower's actual 3 dB signal bandwidth, which (just as with a common-source amplifier) will often be less, especially with a capacitive load (discussed next). For the source follower there is an additional current diversion through the gate-drain capacitance C_{rss}; although the latter is always significantly smaller than the gate–source capacitance C_{iss},[65] it can easily dominate the dynamic input current, because the voltage across C_{iss} is bootstrapped by the near-unity follower gain. A capacitive load reduces the bootstrap action at high frequencies, however, causing the large bandwidth-killing C_{iss} to reassert itself, as we'll see next.

Capacitive loads. We've considered only resistive loads (or current-source loads) in the preceding discussion. Interesting things happen when a source follower drives instead a *capacitive* load, a situation you encounter for example when sending signals through coax cable. We visit this in the next section of the chapter.

---

[64] A rude shock to those accustomed to thinking of an FET as a device with near-infinite current gain! Indeed, MOSFETs in particular, with their picoamp-scale gate leakage currents, have current gains *at dc* in the range of 10^{12} or more. But at signal frequencies the capacitance seen at the gate causes dynamic gate currents, which at frequencies above f_T render it unusable as an amplifier.

[65] Often by an order of magnitude or more. For the IXTA1N120, for example, $C_{rss} \approx 5$ pF at typical drain voltages – nearly a hundred times smaller than the 450 pF value of C_{iss}.

3x.8 Bandwidth of the Source Follower with a Capacitive Load

As we remarked in the previous section, a capacitively loaded source follower has some quirky behavior, and is worthy of its own discussion. When driven with a finite source resistance, for example, it exhibits gain peaking (voltage gain $G > 1$) and an ultimate 12 dB/octave roll-off. We'll consider first the situation with a resistive signal source, then with an input signal that is a current (i.e., a capacitively loaded transresistance stage). Let's see how it goes.

3x.8.1 Follower with resistive signal source

Figure 3x.72 shows the circuit model, in which the input signal source has Thévenin resistance R_1, and the FET is modeled by a unity gain voltage amplifier with output impedance $1/g_m$ and input-to-output capacitance C_{iss}; we'll call these R_2 and C_1, to keep the equations uncluttered. Because we're dealing with capacitive impedances, we'll adopt the widely used abbreviation $s \equiv j\omega$; thus, for example, the impedance of a capacitor is $\mathbf{Z} = 1/sC$, and a signal voltage v across a capacitor C produces a current $i = vsC$.

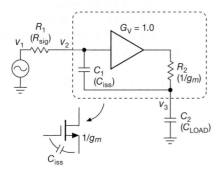

Figure 3x.72. Model of a source follower with capacitive load, where C_1 represents the gate–source capacitance, and R_2 represents the output resistance $1/g_m$ at the source terminal. The effects of the (omitted) gate-drain and source-drain capacitances are negligible.

We're interested in the behavior of the voltage gain $G_V \equiv v_3/v_1$ with frequency. With the small-signal voltages as

shown, we have for the current at the output

$$v_3 sC_2 = (v_2 - v_3)\left(sC_1 + \frac{1}{R_2}\right)$$

which can be grouped and written

$$v_3(1 + s\tau_2 + s\tau_3) = v_2(1 + s\tau_3), \tag{3x.8}$$

where any RC products that appear are written as time constants: $\tau_1 = R_1 C_1$, $\tau_2 = R_2 C_2$, and $\tau_3 = R_2 C_1$. We've got two unknowns (v_2 and v_3) and only one equation; so to eliminate the pesky v_2 we need an independent equation, which we can get by writing v_2 as a divider between v_1 and v_3:

$$v_2 = v_3 + (v_1 - v_3)\frac{1/sC_1}{R_1 + 1/sC_1}$$

which can be grouped and written

$$v_2 = v_1\left(\frac{1}{1 + s\tau_1}\right) + v_3\left(1 - \frac{1}{1 + s\tau_1}\right). \tag{3x.9}$$

Substituting v_2 from eq'n 3x.9 into eq'n 3x.8 and rearranging, we get finally

$$G = \frac{v_3}{v_1} = \frac{1 + s\tau_3}{1 + (1 + s\tau_1)(s\tau_2 + s\tau_3)} \approx \frac{1}{1 + s\tau_2 + s^2\tau_1\tau_2}, \tag{3x.10}$$

where we've discarded τ_3 (which will often be insignificant) to arrive at the right-hand expression.[66]

Aficionados of the s-plane will instantly recognize this as a flat low-frequency response, rising to a peak near a frequency at which the (real) s^2 term equals -1, then falling off at increasing frequency, ultimately at 12 dB/octave (from the s^2 term in the denominator, a "pole pair" that represents a pair of cascaded lagging RC phase shifts). The peaking frequency is given approximately by $\omega \approx 1/\sqrt{\tau_1\tau_2}$, i.e., the geometric mean of the characteristic frequencies of the input and output RC pairs (do the arithmetic!). Framed in terms of the MOSFET's f_T, we can write the approximate peaking frequency (or, to comparable approximation, the bandwidth) as

$$f_{\text{peaking}} \approx \sqrt{\frac{f_T}{2\pi R_1 C_{\text{LOAD}}}}. \tag{3x.11}$$

Note that the bandwidth is inversely proportional to the square root of load capacitance, and proportional to the square root of transconductance (because $f_T = g_m/2\pi C_{\text{iss}}$); in the subthreshold region (where we operate power MOSFETs) $g_m \propto I_D$, so the bandwidth is proportional to the square root of drain current. The equation also appears to

---

[66] The case of a *resistively-loaded* source follower is treated in §4x.26; see particularly eq'n 4x.13.

say that the "follower" has an *inverted* output at frequencies well above the peak (because $s^2 = -\omega^2$) – and in fact this is exactly what you would expect from two 90° phase shifts (more on this topic in the next section).

To get some "ground truth" we measured the response of an IXTA1N120 MOSFET ($C_{iss} = 450$ pF, $R_2 = 1/g_m = 250\Omega$ at $I_D = 0.5$ mA), connected as a source follower with a load capacitance $C_2 = 10$ nF and with a signal source resistance $R_1 = 200$ kΩ.[67] For this set of parameters the time constants $\tau_1 - \tau_3$ have values 90 μs, 2.5 μs, and 0.11 μs, respectively (thus justifying the omission of the latter terms in eq'n 3x.10). Figure 3x.73 shows the results, in good agreement with calculations. Note particularly the 12 dB/octave falloff above the peak, and the corresponding phase shift approach to $-180°$.

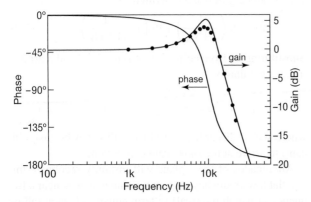

Figure 3x.73. Calculated (solid curves) and measured response (dots) of an IXTA1N120 MOSFET source follower operating at 0.5 mA drain current, with a load capacitance of 10 nF and a signal source resistance of 200 kΩ.

3x.8.2 Follower driven with a current signal

Sometimes a FET follower is used to buffer the high-impedance signal at the drain of a driver stage (we'll see a nice example in §4x.23), in which case the input signal to the follower's gate is a current. The output signal at the FET's source terminal is a voltage, so it's properly a transimpedance stage (current-to-voltage), rather than a voltage follower. We can model the circuit as in Figure 3x.74A and write down a pair of equations for the currents at nodes v_1 and v_2, analogous to what we did in §3x.8.1. For discrete MOSFETs the input shunt capacitance C_0 is usually small compared with the gate–source capacitance C_{iss}, as seen in the values listed in the figure.

---

[67] Similar to its operation in the precision high-voltage bipolar amplifier in §4x.23.

Sparing the reader the gory details, we find that the transimpedance gain (v_2/i_1) is given, approximately,[68] by

$$g_m = \frac{v_2}{i_1} = \frac{1}{2\pi f} \left\{ \frac{1}{C_0\left(1 + \frac{C_2}{C_1}\frac{1}{1+f_T/f}\right) + C_2\frac{1}{1+f_T/f}} \right\}, \tag{3x.12}$$

and the input capacitance C_{eff} seen by the source signal (whatever its Thévenin source impedance) is given by

$$C_{eff} = C_0 + C_1 \left\{ \frac{1}{1 + \frac{C_1}{C_2}\left(1 + \frac{f_T}{f}\right)} \right\}, \tag{3x.13}$$

which is useful because it's C_2 (the load) that you know about. Equivalently, you can write C_{eff} as

$$C_{eff} = C_0 + C_2 \left\{ \frac{1}{1 + \frac{C_2}{C_1} + \frac{f_T}{f}} \right\}, \tag{3x.14}$$

which looks awful, but happily reduces to reasonable expressions in the regions of small and large load capacitance.

The C_{eff} formulas are particularly important when designing voltage amplifiers with MOSFET-follower output stages, for example like that in §4x.23 (Fig. 4x.137), where the open-loop voltage gain of Q_5 depends directly on the load capacitance seen at its drain terminal.

For no load ($C_2 = 0$), or in fact for capacitive loads for which $C_2 \ll C_1 f_T/f$ and $C_2 \ll C_0 f_T/f$, $C_{eff} \approx C_0$ and therefore the transimpedance gain is just $G = 1/sC_0$. This is simply understood as a voltage follower replicating the voltage across C_0 that is produced by the input current; no input signal current flows through C_1, which is bootstrapped to nothingness by the unity-gain follower. At the other extreme of load capacitance, $C_2 \gg C_1$, the effect of a large C_2 is to cause the unloaded gain (which was dropping at 6 dB/octave) to begin dropping at 12 dB/octave (2 poles) at a frequency given by $f_1 = f_T(C_0/C_2)$, where $f_T = g_m/2\pi C_1$ is the bandwidth of the FET follower as previously derived (eq'n 3x.6). That is, the gain begins falling at 12 dB/octave at a frequency[69]

$$f_1 = f_T\frac{C_0}{C_2} = \frac{g_m}{2\pi C_{iss}}\frac{C_0}{C_{LOAD}} \tag{3x.15}$$

where the last term follows because $g_m = sC_1$ at the FET's f_T. Figure 3x.75 shows the situation graphically, for a resistive-source follower, with characteristic frequencies

---

[68] We have swept phase under the rug; i.e., we've let $s \to 2\pi f$.

[69] We can ignore the source-follower's load on the v_1 node for frequencies below $f_T C_1/C_2$ and $f_T C_0/C_2$, where $C_2 = C_{load}$.

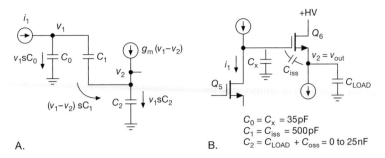

Figure 3x.74. A. Model of a capacitively loaded source follower with input signal current i_1 and output signal voltage v_2. B. Application of the model to the high-voltage amplifier of Figure 4x.137.

corresponding to the C_0 and C_1 capacitances shown in Figure 3x.74B.

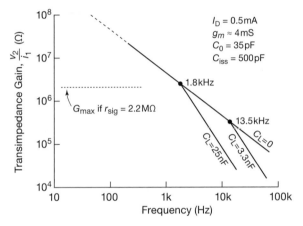

Figure 3x.75. Transimpedance gain of the capacitively loaded source follower, with a MOSFET of bandwidth $f_T = g_m/2\pi C_{iss} = 1.3$ MHz, and with an input gate-to-ground capacitance $C_0 = 35$ pF.

Figure 3x.75 might seem to imply that MOSFET amplifiers are limited to low frequencies, but it's not necessarily so. You can select a small-die MOSFET: for example, the IXTY01N100, with its $C_{iss} = 54$ pF, when operated at 1.5 mA has an f_T of 30 MHz, more than twenty times better. Or you may be able to add a pole-canceling zero inside your amplifier. Also, it's always helpful to add a series resistor at the amplifier's output, to isolate the capacitive load.

Low-frequency gain limited by signal's source impedance. The gain would be divergent at low frequencies (recall $G = -1/2\pi f C_0$) if the input signal were a perfect current source, i.e., if $Z_{sig} = \infty$. In the real world, as represented for example by the circuit of Figure 3x.74B, MOSFET Q_6's gate sees a finite driving impedance, set by the output resistance of common-source driver Q_5 in

parallel with the finite (i.e., imperfect) resistance of drain pull-up current source. This same circuit fragment is a part of the precision high-voltage amplifier of §4x.23, where these effects contribute to an estimated driving impedance of ~ 100 MΩ, dominated by the current source (for a preview, look at the measured current-source data in Figure 3x.59). That's the reason for the suggestive dashed line at the low-frequency end of the gain curve.

You could, of course, replace Q_5's drain-load current source with a simple pull-up resistor. For the amplifier of §4x.23, where the +HV rail is +550 V and Q_5's drain current is 0.25 mA, a suitable pull-up resistor would be 2.2 MΩ; for that situation the low-frequency transimpedance gain is just R_{pullup}, as indicated in Figure 3x.75. You get a lot more gain at low frequencies with a current-source load; and, as a bonus, you get uniform slewing over the full swing.

To see quantitatively the advantage of a current-source pull-up, consider the situation with a 2.2 MΩ pull-up to +550 V, with the output swinging between 5 V and 525 V: the resistive pull-up current drops from about 250 μA down to only 11 μA, a factor of 22. In addition to greatly slowing the I/C_x slew rate, the transconductance of Q_5 will fall by the same factor, reducing its gain and f_T, and thus its frequency response in the feedback loop. Bottom line: a current-source pull-up is a huge benefit; consider it essential.

Stability inside a feedback loop. The gain curve (Fig. 3x.75) of the capacitively loaded transimpedance follower falls off at 12 dB/octave above the f_1 breakpoint (eq'n 3x.15); the corresponding lagging phase shift goes from $-90°$ (for $f \ll f_1$), through $-135°$ (at f_1), and approaches $-180°$ (for $f \gg f_1$).

That's bad news inside a feedback loop! One solution is to ensure that the loop gain falls to unity by f_1, i.e., with conservative "dominant-pole" compensation. But a nice way to ensure stability without such a large reduction in bandwidth is to isolate the load capacitance at frequencies of f_1 and above, with a series resistor at the output. At first you might think that the resistor value must be tailored to a particular value of load capacitance. But, happily, you would be wrong: the condition is simply that the impedance of the series resistor (call it R_{20}) should equal the reactance of C_{LOAD} at f_1. That is, $R_{20} = 1/2\pi f_1 C_{LOAD}$; but substituting f_1 from eq'n 3x.15 eliminates C_{LOAD}, and we find

$$R_{20} = \frac{1}{2\pi f_T C_0}. \tag{3x.16}$$

In other words, this single choice of isolation resistor cancels the capacitively loaded breakpoint at f_1 for all values of load capacitance.

Sharp-eyed readers may have two complaints at this point, namely (a) the output RC (i.e., $R_{20}C_{LOAD}$) slows the response, with a 3 dB loss of gain at f_1 and -6 dB/octave thereafter, and (b) the series resistor R_{20} causes a dc error, given that the feedback is taken upstream of R_{20} (i.e., from Q_6's source terminal). A good eye! But not to worry – to the first point we respond that you get more transimpedance gain at any frequency than you would with dominant-pole compensation; and to the second point we direct the reader to the implementation in Figure 4x.137, where a split feedback path ensures dc accuracy in the presence of load currents. The latter is a common technique for dealing with capacitive loads, see for example Figure 4.78B.

For formulas dealing with the bandwidth of a source follower with a resistive load, see §4x.26, eq'n 4x.13.

3x.9 High-Voltage Probe with High Input Impedance

Problem: Measure voltages or waveforms in a high-voltage circuit where the loading from even a $10\,M\Omega$ or $100\,M\Omega$ probe would be unacceptable. Solution: Come up with a circuit that can handle high voltages (and produce a $V_{in}/100$ "monitor" output), and which has input currents well below a microamp (and ideally down in the picoamp range). Happily, high-voltage MOSFETs provide a good solution. Here we'll evolve two approaches to the problem.

3x.9.1 Compensated-offset MOSFET follower

Figure 3x.76A shows a first stab, just a high-voltage MOSFET follower with 100:1 divider in the source terminal. It will work – sort of. But it has some serious problems: most obviously, the output drops to zero for inputs below some effective gate threshold voltage V_{th}; that is, there is no compensation for the non-zero error caused by V_{GS} offset. Furthermore, it suffers from a variable offset voltage (i.e., the MOSFET's changing V_{GS}) as the drain current goes from its maximum (here $100\,\mu A$ at $V_{in} = 1\,kV$) down to a microamp or less when the input is down around $10\,V$ or less.

The basic source-follower idea can be improved considerably with the circuit modifications of Figure 3x.76B. A brief run-through of its operation goes like this: (a) We replace the source load of follower Q_1 with a current sink that returns to a low-voltage negative rail (to permit inputs down to zero), with the current set by low-voltage components A_3 and Q_3, assisted by high-voltage cascode MOSFET Q_2; next (b) we add to Q_3's emitter a correction current (via Q_4) that tracks the current diverted by the 100:1 output divider, so that follower Q_1 operates at a constant drain current (nominally $50\,\mu A$); next (c) we offset the bottom of the 100:1 divider with follower A_2 and Q_4 to cancel the (approximately constant) V_{GS} of Q_1; and finally (d) we calibrate the gain by trimming the divider ratio with R_6.

A few fine points: R_1 limits fault currents through protection diodes D_1–D_3; gate resistors R_2 and R_{13} suppress oscillation tendencies in the high-voltage MOSFETs; D_3 lowers the effective capacitance of zener D_2; and A_1's $50\,\Omega$ output resistor suppresses oscillation from external cable capacitive loading. See also §3x.6.5 for the use of series

MOSFET stacks to go to higher voltages, and the use of higher currents for faster negative-slewing speeds.

3x.9.2 Bootstrapped op-amp follower

Here's an unusual circuit design (Fig. 3x.77) that puts an op-amp follower at center stage, assisted by high-voltage MOSFETs to bootstrap its supply rails to follow the input voltage. You can think of this as an op-amp-centric design, as contrasted with the MOSFET-centric design of Figure 3x.76. It's pretty daring – a low-bias 5 V CMOS op-amp handling input signals (and even abrupt steps) over nearly a kilovolt range. Scary enough in fact that we built and abused it mightily, and it just kept working.

The idea here is to leverage the accuracy and very low input current of a precision CMOS op-amp to achieve performance superior to that of the MOSFET follower. That's not hard to imagine – the LMP7721, for example, has a maximum input offset voltage of 0.15 mV, and a maximum bias current (at room temperature) of 20 *femto*amps (it claims *typical* values of $26\,\mu V$ and 3 fA, damn impressive!).

So the basic circuit is just unity-gain follower U_1, with both full-swing and 100:1 outputs. Very clean, no offset trims, etc. The hard part is ensuring that the op-amp stays in its operating range and is not damaged by the worst possible insults at the input terminal. The op-amp's positive supply rail is created by depletion-mode follower Q_1, bootstrapped from the op-amp's output; that puts the rail typically 1–2 V above the op-amp output (it's a rail-to-rail op-amp, so there's no worry about headroom). Zener Z_1 sets the op-amp's total supply voltage, biased by depletion-mode current sink Q_2. The latter's current, nominally 1.5 mA, sets Q_1's operating current, and must be somewhat greater than the op-amp's quiescent current.[70]

You can kill a high-voltage circuit in an eyeblink, so we took pains to anticipate the worst: Diode clamps D_1 and D_2, with input current limited to less than a milliamp by the $1\,M\Omega$ input resistor, bound U_1's input voltage to no more than a half volt beyond the supply rails (and with any input current through U_1's input protection diodes kept below 0.1 mA by series resistor R_2). We've used the gate-channel diode of a small-geometry JFET for these clamp diodes, to minimize their contribution to the input current; at room temperature we measured leakage currents of just $\sim 20\,fA$ with an applied voltage less than 3 V (see Figure 1x.120).

---

[70] We'd like to run at lower currents to minimize power dissipation in the MOSFETs, but we found that the low bandwidth of a micropower op-amp like the AD8603 ($I_Q \approx 40\,\mu A$) caused oscillation in this circuit arrangement.

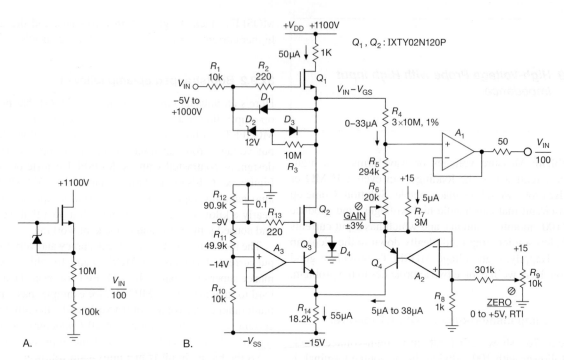

Figure 3x.76. High-impedance probes for input voltages to +1 kV. A. Simplest MOSFET follower. B. Cascode current sink biases MOSFET follower Q_1, with trim to cancel the V_{GS} offset of the latter.

Zeners Z_2 and Z_3 protect the MOSFET gates; note the polarity, appropriate for the depletion-mode's negative V_{GS}.

Finally, we added a 1 GΩ input bias path through R_7, so that an open input produces an output near ground. It can be returned to ground, if 1 GΩ is an acceptably high input resistance. Alternatively, R_7 can be bootstrapped as shown to produce an input resistance 1000 times higher (i.e., 1 TΩ). In the latter case the float voltage (referred to the input) is

$$V_{\text{float}} = 10^3 V_{os} + 10^9 I_{\text{bias}} \qquad (3x.17)$$

(replace the factor 10^3 by unity if R_7 is returned to ground). For the component values here, and with an LMP7721 opamp for U_1, that amounts to a worst-case float voltage (referred to the input) of ±150 mV (with bootstrapped R_7), or ±200 μV (with R_7 returned to ground).

In either case the probe's input current is

$$I_{\text{in}} = I_{\text{bias}} + \frac{V_{os}}{10^9} \qquad (3x.18)$$

which amounts to a worst-case input current of ±200 fA.

Torture tests. This circuit worked nicely, doing what it was supposed to do. With a step input, measured rise and fall times were around 15 μs with the input resistor R_1 set at a conservative 1 MΩ; reducing R_1 to 100 kΩ speeded up the step response, to ~2 μs for small steps (<

±100 V), but slew-rate limited for large steps (+100 V/μs and −50 V/μs).

We tried our best to destroy the prototype, by switching the input abruptly over a full swing (from zero to ±400 V); to ensure the fastest input slew rate we used a mercury relay, and banged the input repeatedly to a set of stiff positive and negative voltages. Figure 3x.78 shows some measured single-shot waveforms. These tests were a failure – we couldn't break the thing!

Extension to higher voltage. The voltage range of the circuit of Figure 3x.77 is limited by the voltage ratings of the depletion-mode MOSFETs, which for currently available parts top out at 1 kV.[71] You can extend the circuit to higher voltages, however, by spreading the voltage across several MOSFETs, a technique we've used earlier (e.g., see §3x.6.5 and Figures 3x.57 and 3x.61). And you can take advantage of the higher voltage ratings available in enhancement-mode MOSFETs – standard *n*-channel parts from IXYS go as high as 4500 V (see Table 3x.3).

Figure 3x.79 shows a simple extension of our probe circuit, to accommodate inputs to ±1 kV. Here Q_1 operates as before, bootstrapping the op-amp's positive rail; but di-

---

[71] We ran this circuit at somewhat lower voltage, because the measured drain curves rise rapidly as V_{DS} approaches 1 kV.

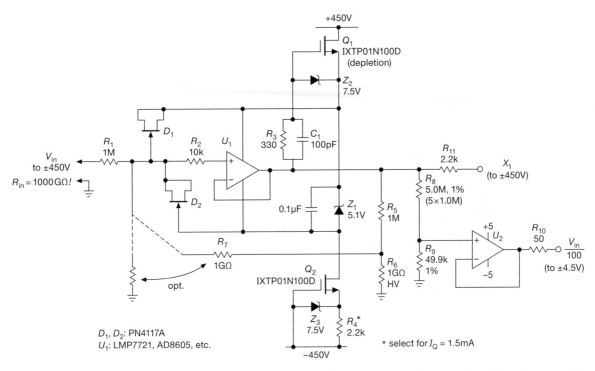

Figure 3x.77. Another high-voltage probe circuit, exploiting the accuracy of a low-I_b op-amp. The unusual bootstrap arrangement keeps the input signal within the op-amp follower's 5 V total supply.

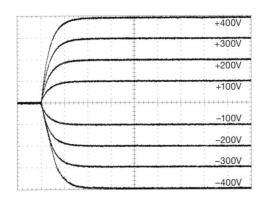

Figure 3x.78. Measured X1 output waveforms of the circuit of Figure 3x.77 when driven with abrupt steps to the voltages indicated. Vertical: 100 V/div; Horizontal: 20 μs/div.

rent, with Q_2 acting as a cascode (a technique used earlier, in §3x.6.4). This scheme can be extended to higher voltages by stacking additional MOSFETs atop Q_3 and Q_4, or by substituting parts of higher voltage ratings, or both.

vider $R_{12}R_{13}$ splits the span up to the positive HV rail, with enhancement-mode MOSFET Q_3 sharing the voltage burden. Down at the negative HV rail Q_2 is a current sink, with the voltage span assisted by Q_4. This time we've used a different scheme to set the sink current: instead of the not-very-predictable source self-bias arrangement (Q_2R_4 in Fig. 3x.77), we've used the LM334 programmable 2-terminal low-voltage current source (U_3) to set the sink cur-

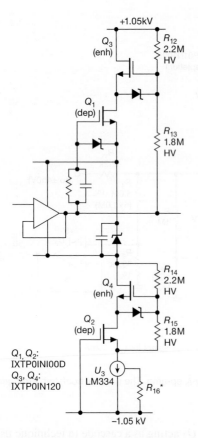

Figure 3x.79. Stacking a pair of cascode MOSFETs to Figure 3x.77 extends the voltage range to ±1 kV.

3x.10 CMOS Linear Amplifiers

CMOS inverters – and indeed all CMOS digital logic circuits – are intended to be used with *digital* signal levels (§10.1.2). Except during transitions between defined HIGH and LOW voltage states, therefore, the inputs and outputs are close to ground or V_{DD} (commonly +2.5 V, +3.3 V or +5 V). And except during those transitions (with typical durations of a few nanoseconds), there is no quiescent current drain.

The CMOS inverter turns out to have some interesting properties when used with *analog* signals. Look back at Figure 3.93. You can think of Q_1 as an active (current-source) load for inverting amplifier Q_2, and vice versa. It's a push–pull inverting amplifier stage. When the input is near V_{DD} or ground, the currents are grossly mismatched, and the amplifier is in saturation (or "clipping") at ground or V_{DD}, respectively. This is, of course, the normal situation with digital signals. However, when the input is near half the supply voltage, there is a small region where the drain currents of Q_1 and Q_2 are nearly equal; in this region the circuit is an inverting linear amplifier with high gain. Its transfer characteristic is shown in Figure 3x.80. The variation of R_{load} and g_m with drain current is such that the highest voltage gain occurs for relatively low drain currents, i.e., at low supply voltages (say 2.5 V).

This circuit is not a good amplifier; it has the disadvantage of very high output impedance (particularly when operated at low voltage), poor linearity, and unpredictable gain. However, it is simple and inexpensive (this specimen comes six to a package, for about $0.25), and it is sometimes used to amplify small input signals whose waveforms aren't important. Some examples are proximity switches (which amplify 60 Hz capacitive pickup), crystal oscillators, and frequency-sensing input devices whose output drives digital counting logic.

To make one of these amplifiers you can use an "unbuffered" inverter from any of the CMOS logic families (see Chapter 10). We rigged up several of these, and measured the gain versus frequency. Figure 3x.81 shows measured data for the high-voltage (to +15 V) 4000B-series unbuffered inverter. You can see the trade-off of bandwidth versus gain. For operation at lower voltages (down to 1 V) you could use something like a 74LVC1GU04 (+1.8 V to +5.5 V) or a 74AUP1GU04 (+0.8 V to +3.6 V). Figure 3x.82 plots their measured −3 dB bandwidths over the full supply voltage range.

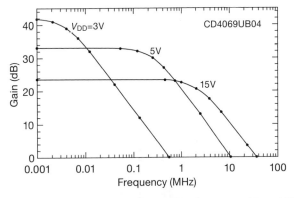

Figure 3x.81. Measured gain versus bandwidth for the CD4069UB unbuffered inverter, wired as a linear amplifier with 10 MΩ feedback.

To use a CMOS inverter as a linear amplifier, it's necessary to bias the input so that the amplifier is in its active region. The usual method is with a large-value resistor from output to input, as shown in Figure 3x.83. That puts us at the point $V_{out}=V_{in}$ in Figure 3x.80. Such a connection (circuit A) also acts to lower the input impedance, through "shunt feedback," making circuit B desirable if a high input impedance at signal frequencies is important. The third circuit is the classic CMOS crystal oscillator, discussed in §7.1.6D. Figure 3x.84 shows a variant of circuit A that we've used to generate a clean 10 MHz full-swing square wave (to drive digital logic) from an input sinewave. The

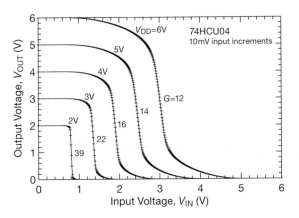

Figure 3x.80. Measured transfer characteristics (V_{out} vs V_{in}) for a 74HCU04 "unbuffered logic inverter" over its full range of allowed supply voltages. Linear gain increases with decreasing supply voltage.

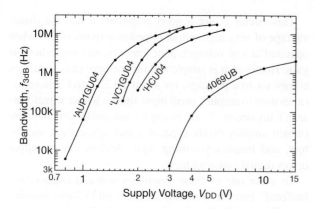

Figure 3x.82. Measured $-3\,dB$ bandwidths for four popular logic families, covering the full range of specified operating supply voltages.

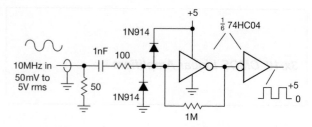

Figure 3x.84. Creating a clean digital replica of a 10 MHz sinewave "reference clock" input.

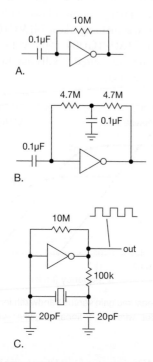

Figure 3x.83. CMOS linear amplifier circuit examples.

circuit works well for input amplitudes from 50 mVrms to 5 Vrms. This is a good example of an "I don't know the gain, and I don't care" application. Note the input-protection network, consisting of a current-limiting series resistor and clamping diodes.

family (by size):	'x10	'x20	'x30	'x40
die size (mm^2):	8	16	33	
IRF5x0 (100V)	'510	'520	'530	'540
IRF6x0 (200V)	'610	'620	'630	'640
IRF7x0 (400V)	'710	'720	'730	'740
IRF8x0 (500V)	–	'820	'830	'840

3x.11 MOSFETs Through the Ages

MOSFETs are the dominant transistor species on the planet, having displaced bipolar technology in logic, microprocessors, and microcontrollers. Early MOS devices were slow, and generally unsuitable for power applications, but all that changed long ago.

In other sections of the book we've treated the logic and processor scene (see, for example, §12.1.1 for a brief history of BJT and CMOS logic, and §11.1 for the programmable logic story); here we tell the story of *power* MOSFETs – their steady development through a half century to the excellent contemporary performers; of the latter we show graphs and scatterplots to illustrate the continual improvement in various figures of merit.

3x.11.1 A MOSFET Saga: the First 30 Years

International Rectifier Corporation, usually called "IR," was a force in the early history of power MOSFETs.[72] Alexander Lidow, one of the owner's sons, experimented with MOSFET technology in 1976 a year before receiving his physics PhD from Stanford. His father gave Alexander a modest budget of $100k of equipment and the aid of one engineer. With this he developed the HEXFET, so called because of the hexagonal shape of each of its small cells. The HEXFET was expensive to make, so to reduce manufacturing costs they created a huge fab, large enough to supply the "entire world need" for MOSFETs. Alexander put it this way: "We bet the company, bet again, and borrowed to bet some more on HEXFET." In 1979 they introduced the IRF500 to IRF800 HEXFET series, in inexpensive plastic TO-220 packages.

A coherent family of power MOSFETs. The part-numbering scheme handled 16 possible types, with voltages from 100 to 500 V and power levels from 36 to 125 watts.

The parts had simple names, IRF for IR FET, plus three digits. The first digit, from 5 to 8, indicates the maximum drain voltage, and the last two digits, 10 to 40,[73] indicates the part's die area, in factors-of-two steps.[74] We circuit designers thought of it as the current- and power-handling capability, as the actual die size was rarely mentioned.

With its well-organized parameter set, this family of MOSFETs provided engineers with the parts they needed to explore the exploding power-MOSFET design scene in the early 1980s. It was a brilliant approach, and the parts were immediate smash hits. They were rapidly second-sourced by many other companies, and the dropping prices became very attractive.[75] As we'll see, these widely sourced parts are also enjoying a long product life cycle, making them good choices for designs that need a long manufacturing life. Recently IR's original IRF500 to IRF800 die design and rights were purchased by Vishay. They updated the datasheets in 2010, and seem determined to provide an extended available life for these legacy parts. However, most of the family's parts have morphed into cheaper, improved modern versions, such as IRF830N replacing the IRF830, etc. We'll return to this story shortly.

The IRF500 family of HEXFETs represents an opportunity to look at a single series of parts, all made using the same fab process and design techniques, to see how die size and voltage rating can affect a part's capabilities.

Small and large MOSFETs. Many important parameters are primarily influenced by die size. Gate capacitance, C_{iss}, and total gate-charge, Q_G, are directly proportional to the number of identically sized HEXFET hexagonal cells and hence to the die area, as the table below shows.[76] Thermal conductance from the die to the case decreases with active die area, although not quite proportionally, as the $R_{\Theta JC}$ parameter shows. This translates directly into the maximum power dissipation (P_D @$T_{case}=25°C$), namely $P_{D(max)}=(T_J-T_C)/R_{\Theta JC}$.

	'x10	'x20	'x30	'x40	units
C_{iss}	163	348	696	1425	pF
Q_G	5.7	13.0	23.3	46.8	nC
$R_{\Theta JC}$	3.5	2.5	1.7	1.0	°C/W
P_D	36	50	75	125	W (T_C=25°C)

---

[72] Though IR did not *invent* the power MOSFET: the first edition of *The Art of Electronics*, finished in 1979, has Table 6.4 (Power MOSFETs) showcasing 47 different parts; they had voltage and power capabilities ranging from 35 to 400 V, and 0.6 to 90 W. The MOSFET pioneer Siliconix had the most entries, but IR, Intersil, Hitachi and Supertex were present. Most MOSFETs were in metal packages, but a few TO-92 and TO-220 plastic packages were present. IR's HEXFET was there, mostly in TO-3 packages, but the brand new IRF530 and IRF730 in TO-220 plastic made it into the table. Many power engineers were initially hesitant to use the plastic packages, but the low prices and apparent reliability won them over.

[73] The highest-current '40 series was introduced a few years later.

[74] International Rectifier, HEXFET Databook, HDB-1, 1981, page 96. The "mm^2" area notation means, e.g., 8×8 mm for "8 mm^2," etc.

[75] Curiously, the place reserved for a small 500 V part (IRF810) was never filled.

[76] The values are averages – specific different-voltage parts may vary.

$R_{DS(ON)}$ (mΩ, typical, at T_C=25°C)				
	family:	'x10	'x20	'x30	'x40
100V	IRF5x0	410	200	100	50
200V	IRF6x0	1250	550	220	130
400V	IRF7x0	3100	1300	740	435
500V	IRF8x0	–	2500	1200	800

The second table shows the typical ON-resistance, $R_{DS(ON)}$, for the 15 parts in the family. The 100-volt IRF510 to '540 parts show factor-of-two ON-resistance steps, nicely decreasing with the die area as we scan across the 100-volt row. Looking down any of the columns, we see the ON-resistance increasing dramatically with drain-voltage capability, by more than a factor of 10 for a 5× increase in voltage. For example, the 100 V IRF520 with 0.2 Ω versus the 500 V IRF820 with 2.5 Ω. $R_{DS(ON)}$ scales approximately as $(V_{DSS})^{1.5}$. We'll revisit this table shortly to look at the relationship of $R_{DS(ON)}$ to Q_G for different parts.

MOSFET datasheets highlight a current spec, $I_{D(max)}$ @ T_C = 25°C, which is the maximum-continuous current allowed with the junction temperature at 175°C and the case temperature somehow unrealistically clamped to 25°C. This is a calculated value, $I_D = (P_D/kR_{DS(ON)})^{1/2}$, where k is the increase in ON-resistance with temperature, a factor of ~ 2.4 at $T_J = 175$°C for these parts. It's a best-case continuous current capability, good for boasting in ads and on the front page of the datasheet. The value ranges from 2A to 28A for the 15 parts in the table below. But it is not realistic, because we can't keep the metal tab at 25°C, and we don't really want to run the junction at 175°C.

This current spec increases with die size, but not quite proportional to die area. Although $R_{DS(ON)}$ does decrease inversely, $R_{\Theta JC}$ fails to improve inversely with die area. Following the observed increase of $R_{DS(ON)}$ with voltage rating, the I_D spec decreases with the drain voltage rating, $V_{DS(max)}$.[77]

$I_{D(max)}$ (A, at T_C=175°C)				
	family:	'x10	'x20	'x30	'x40
100V	IRF5x0	5.6	9.2	14	28
200V	IRF6x0	3.3	5.2	9	18
400V	IRF7x0	2	3.3	5.5	10
500V	IRF8x0	–	2.5	4.5	8

IR also provided an additional conservative spec in its datasheets, maximum drain current at a case temperature T_C = 100°C (also a calculated value). Substituting P_D in the I_D formula above, $I_D = (P_D/kR_{DS(ON)})^{1/2}$, we see a $(150/75)^{1/2} = 1.41\times$ reduction in I_D when we increase T_C from 25°C to 100°C.

$I_{D(max)}$ (A, at T_C=100°C)				
	family:	'x10	'x20	'x30	'x40
100V	IRF5x0	4	6.5	10	20
200V	IRF6x0	2.1	3.3	5.7	11
400V	IRF7x0	1.2	2.1	3.5	6.3
500V	IRF8x0	–	1.6	2.9	5.1

[77] In these tables IR uses the maximum $R_{DS(ON)}$ spec at 175°C.

We don't favor bumping up against the $T_J = 175$°C limit. Taking instead a value of $T_J = 150$°C, we have a $(2/3)^{-1/2} = 1.22\times$ reduction from the datasheet's overly-optimistic $I_{D(max)}$ spec. We can simplify our lives and sleep better at night if we just divide the 25°C datasheet values by a factor of two, giving us conservative numbers we can actually use. It's then our job to figure out how to keep the case temperature under 70°C to 100°C.

Conservative current ratings. Now the conservative current capability of our 15 parts looks like this.[78]

$I_{D(max)}$ (A, 50% derated from T_C=25°C)				
	family:	'x10	'x20	'x30	'x40
100V	IRF5x0	2.8	4.6	7	14
200V	IRF6x0	1.7	2.6	4.5	9
400V	IRF7x0	1	1.7	2.8	5
500V	IRF8x0	–	1.3	2.3	4

The IRF530 can safely switch 7 amps, and remain on 100% of the time. Ditto for the IRF540 switching up to 14 amps.

Returning to our story, IR's IRF500 series was a smash success, so much so that four or five manufacturers were offering MOSFETs with the same part numbers and specifications within a year or two. IR also introduced logic-level versions of the 100 V and 200 V MOSFETs, the IRL5xx and IRL6xx parts. The other manufacturers immediately offered these other parts as well.

Low voltage parts. IR brought out parts using the same dies but with degraded current or voltage specs, e.g., the IRF531 was a 60-volt version of the IRF530. It's doubtful that a significant portion of the fab's output failed to meet the full specs, but this was a good way to offer cheaper parts without seriously damaging the market for more expensive parts.

The popularity of the 60-volt HEXFET parts wasn't simply because they were cheaper; IR had in fact found a sweet spot for MOSFET switches: 55 V to 60 V parts are especially useful with low voltage supplies and in the automotive industry.[79] To capitalize on this market, IR introduced higher current parts specifically designed for lower voltages. Enter the IRFZ14 to IRFZ44 series.

	'z14	'z24	'z34	'z44	units
C_{iss}	300	640	1200	1900	pF
Q_g	9.7	19	30	42	nC
$R_{\Theta JC}$	3.5	2.5	1.7	1.0	°C/W
P_D	43	60	88	150	W (T_C=100°C)
$R_{DS(ON)}$	135	68	42	24	mΩ
$I_{D(max)}$	7.2	17	21	36	A (T_C=100°C)

This series had a single voltage spec, $V_{DSS(max)} = 60$ V. Comparing the unchanged $R_{\Theta JC}$ ratings, it appears that IR used the

[78] Note we're discussing only the TO-220 package versions. The surface-mount ("tabless") packages – I²PAK (TO-262) and D²PAK (TO-263) – use the same die, so most of the specs are unchanged, providing you can get the heat off the metal back.

[79] Even in a system powered by a 13.8-volt battery, a higher voltage rating is needed, to safely handle surges and spikes.

same set of die sizes for the IRFZ14 to 'Z44 series as for the older IRF5x0-series. However, the C_{iss} values have nearly doubled, so they may have increased the cell density. At any rate IR improved upon the $R_{DS(ON)}$ values: 410 mΩ down to 50 mΩ (for the older 100 V parts) was lowered to 135 mΩ down to 24 mΩ (for the 60 V series). And the $I_D @ T_C = 100°C$ currents were likewise improved, by about 2×: from 4 A–20 A (100 V parts), to 7 A–36 A (60 V parts). If we apply our conservative factor of two, the IRFZ44 part can safely switch a 30 A load, a welcome improvement over 14 A.

Improved MOSFET die structures. In a switching MOSFET's vertical structure, the source-to-drain path is formed on the sides of a groove in the silicon, in an array of identically sized cells (Figure 3x.85). The current flow is vertical, from source metalization on the top, through a channel next to the gate formed on the groove, down to the drain substrate, to the thick metal drain tab. If the cell geometry is reduced, with finer lines and smaller cells, more cells fit onto a given die area, or even into a reduced die area. The smaller die may degrade the part's thermal resistance $R_{ΘJC}$, but the larger number of cells in parallel can greatly reduce its $R_{DS(ON)}$ value. If $R_Θ \cdot R_{DS(ON)}$ is maintained, or improved, the resulting new smaller part will have the same $I_D @ T_C$ spec, and can be used in place of the older larger part. Often the new smaller-geometry design will also have lower C_{iss}, and reduced total gate-charge, Q_G. The manufacturer's motivation is to make more parts from each silicon wafer (at the same per-unit price); the engineering customer's motivation is to move to a part with lower gate-charge; and the purchasing department's motivation is to move to a part with a lower price. We've found a guiding rule for the MOSFET industry.

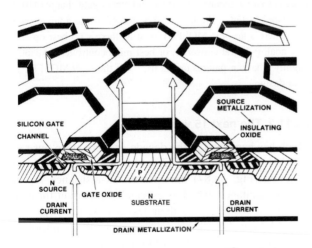

Figure 3x.85. International Rectifier's HEXFET® vertical power MOSFET structure. Reproduced with permission of Infineon Technologies AG.

Why we win twice. We users have another motivation for the later-generation parts. Our usual formula for $I_{D(max)}$ has to include the total thermal resistance from junction to ambient (i.e.,

$R_Θ \equiv R_{ΘJA} = R_{ΘJC} + R_{ΘCS} + R_{ΘSA}$), not simply the manufacturer's portion of the thermal job (i.e., $R_{ΘJC}$). So if the product $R_Θ \cdot R_{DS(ON)}$ can be maintained (or improved) by reducing $R_{DS(ON)}$, then the in-system relevant product ($R_{ΘJC} + R_{ΘCS} + R_{ΘSA}) \cdot R_{DS(ON)}$ will be greatly improved. We can't make a perfect heat sink, and thus for us the case-to-ambient thermal resistance $R_{ΘCS} + R_{ΘSA}$ may greatly dominate over the part's junction-to-case $R_{ΘJC}$. We experience lower I^2R dissipation, proportional to the extent that the $R_{DS(ON)}$ spec is improved. So we benefit twice from the improved part.

Reduced gate drive, and $R_{DS(ON)}Q_G$ figure-of-merit. As discussed elsewhere, it's usually desirable to use parts with lower values of Q_G to reduce gate-drive power losses. Advanced designs using improved fab processes often have lower Q_G values for a given MOSFET capability. From the $R_{DS(ON)}$ and Q_G values in the earlier IRF500-series table, we see that $R_{DS(ON)}$ decreases and Q_G increases with increasing die size. So $R_{DS(ON)} \cdot Q_G$ is a convenient figure-of-merit (FOM) for evaluating a design and its fab process. In this FOM table we use the actual Q_G values for each part, and we've included the 60 V IRFZx4 parts. The units of this FOM are nC-Ω. As in golf, low scores are better.

Figure-of-Merit, FOM (nC-Ω)					
	family:	'1	'2	'3	'4
60V	IRFZx4	1.31	1.29	1.26	1.01
100V	IRF5x0	2.13	2.06	1.62	2.35
200V	IRF6x0	7.9	7.9	5.9	5.9
400V	IRF7x0	17.7	19.5	17.8	18.7
500V	IRF8x0	–	40.0	31.2	41.6

Note that closely similar FOMs are obtained for parts with the same voltage rating. To compare parts with different voltage ratings, divide the FOM by $V_{DS(max)}^p$, where the power p is in the range of 1.5–1.6. This helps to wade through the huge range of available parts and determine which have newer designs and are made in advanced fabs, and which are not. Let's return to our story about the evolution of IR's HEXFETs.

A progression of steady improvements. The IRFZ34 and IRFZ44 were also smash hits, with equivalents quickly offered by the competition. Newer parts with lower prices and better specs could win out over older parts. By the early 1990s MOSFET manufacturers had learned the design improvement tricks discussed above, and this scene included MOSFETs like the IRFZ44. To explore this, let's follow IR's 'Z44 product line, in the convenient TO-220 package. IR has since offered at least five variants beyond the IRFZ44, identified by a suffix. As before, we use typical Q_G and $R_{DS(ON)}$ values, and $I_{D(max)}$ is the maximum continuous current with the case at 100°C. For comparison we've included the older IRF540 in the table below.

At first glance the table is just a forbidding mass of numbers – but there's a lot to learn with a closer look. Scan across to compare a specific parameter. For example, the $I_{D(max)}$ values stick pretty much to the original 36 A for all the parts; that's the parameter that lets these parts share the '44 label.

	'540	'Z44	'Z44E	'Z44V	'Z44N	'Z44VZ	'Z44Z	units
V_{DS}	100	60	60	60	55	60	55	V
$R_{\theta JC}$	1.0	1.0	1.4	1.3	1.5	1.64	1.87	°C/W
P_D	150	150	110	115	94	92	80	W*
$R_{DS(ON)}$	52	24	20	14	14	9.6	11	mΩ
$I_{D(max)}$	20	36	34	39	35	40	36	A*
Q_g	47	42	38	55	41	43	29	nC
Q_{gd}	17	17	16	22	16	18	12	nC
C_{iss}	1700	1900	1360	1812	1470	1690	1420	pF
C_{oss}	560	920	420	393	360	270	240	pF
C_{rss}	120	170	160	103	88	130	130	pF
E(10μs)	136	150	65	65	54	68	79	mJ
FOM	2.44	1.05	0.76	0.77	0.57	0.41	0.32	nC-Ω

* at T_C=100°C

Look next at the $R_{\theta JC}$ data, which starts at 1.0°C/W but rises to 1.87°C/W; this shows us that IR has dramatically reduced the die size of the newer offerings. This means they can make more MOSFETs per wafer and reduce the selling price without damaging profits. In some cases C_{iss} and Q_G are lower, reducing switching losses. This also shows in the calculated FOM, which improves by more than 3× for the IRFZ44Z.[80,81] Looking at the C_{iss} and C_{oss} data, we see another hidden improvement. The newer parts show a dramatic reduction in C_{oss}, even if less reduction was achieved with C_{iss}. This means the FET's design has been altered to favor lower output-node capacitive switching losses, and the newer part is better suited to work at higher switching frequencies. This improvement is hidden deep within the datasheets, but we're confident that IR salespeople did not fail to mention it to prospective customers.

The table shows the FOM improving from 1.05 nC-Ω for the IRFZ44 down to 0.32 nC-Ω for the IRFZ44Z. That means IR has made dramatic improvements in their designs and the fab process. But the competition has also been making dramatic progress. And as we will soon see, reducing the required gate-drive voltage from 10V to 5V helps a lot, and further reducing the drain-voltage capability also helps greatly. We'll return to explore this scene presently, and no doubt we'll find some parts that improve upon the performance of the IRFZ44 series.

The original IRFZ44 was a big improvement over the older IRF540, with $R_{DS(ON)}$ reduced from 52 mΩ to 24 mΩ, but these IRFZ44 variants show further dramatic improvement, down to 11 mΩ (and even 9.6 mΩ for the IRFX44VZ). This is helpful for dc switching applications, as seen earlier. If we apply our conservative factor of two to the T_C = 25°C value, we get 33 A for the IRFZ44VZ, not much of an improvement. But even if we're stuck with a poor heat-removal design, I^2R tells us that an 11 mΩ part can carry 1.5× more current than a 24 mΩ part.

Ironically, high-power linear applications suffer with newer

parts. The lower $R_{DS(ON)}$ values in "improved" versions are of little benefit for linear applications, which don't saturate anyway (saturation is the antithesis of linearity). And the degraded thermal conductance and thermal mass seen in later-generation "shrink die" MOSFETs are a real disadvantage in a linear circuit that is producing substantial heat that must be removed. Aware of this issue, IR offers the IRFZ44R (not shown above), made with the old IRFZ44 specs, and called a "drop-in replacement for linear/audio applications." The original IRFZ44 is also still available from Vishay. Be sure to look carefully at datasheet $R_{\theta JC}$ and P_D values when selecting power MOSFETs for linear applications.

One other parameter is dramatically degraded, and that's the thermal mass, because the dies are smaller. This shows up on the datasheet's "Single-pulse" curve in the Transient Thermal Impedance graphs. We use the 10 μs value on the curve to calculate the maximum energy E(10 μs) that can be absorbed in 10 μs, raising the junction temperature by 150°C, from 25°C to 175°C. This parameter is closely related to EAS, the maximum Single-Pulse Avalanche Energy specification,[82] with related discussion in §§3x.13 and 9x.25.8. Parts with reduced EAS are arguably less robust. We see this parameter degraded from 150 mJ down to as low as 54 mJ.

Bigger is usually *not* better. Often when choosing a power MOSFET, it's tempting to simply pick the biggest baddest one, the one with the lowest ON-resistance, or the highest P_D power-handling spec, or the highest I_D @ T_C = 25°C spec. But there are penalties for using unnecessarily large-die high-performance parts, as can be seen by examining Table 3.4 on page 188 of AoE3. We see higher prices (up to $5 compared to $0.25), higher gate capacitances (13 nF compared to 50 pF), and higher gate charge (410 nC compared to 1 nC). We may also be forced to use a larger package. There's no *best* MOSFET (just as there's no best op-amp, best analog-digital converter, or best microcontroller). Adapting slightly what our parents taught us,[83] use the *right* MOSFET for your design.

3x.11.2 The next 15 years

Much has happened in the decade or more since our saga concluded. Silicon MOSFETs with "super-junction" construction (Fig. 3x.87) have lower ON-resistance for the same capacitance, which is especially important for high-voltage MOSFETs with their less-than-stellar R_{ON}. And MOSFETs made with silicon carbide or gallium nitride are the hottest new item in the MOSFET zoo.

The power MOSFET industry's recent progress is seen in two main areas: low-voltage parts, mostly with maximum voltage ratings below 200 V, and high-voltage parts,

---

[80] IR calls the Z and VZ parts "automotive MOSFETs."

[81] IR made a few changes in the datasheet, such as showing the typical data values favored by competitors, and extending the Transient Thermal Impedance curve from 10 μs down to 1 μs.

[82] The EAS value is measured for a longer pulse time, during which heat can spread further into the leadframe, so it's a higher value.

[83] "Use the right tool for the job."

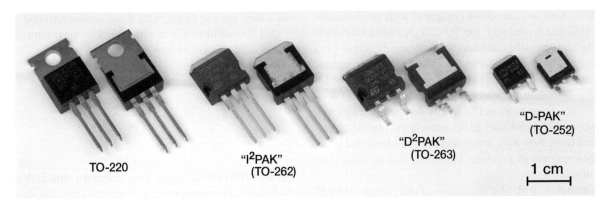

Figure 3x.86. Evolution of the D²Pak and DPak surface-mount packages.

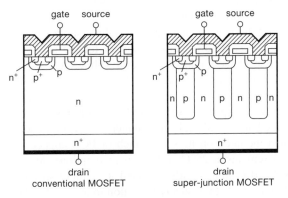

Figure 3x.87. A super-junction MOSFET's vertical p–n junctions result in low ON-resistance (R_{ON}) and reduced gate charge (Q_g) compared with the conventional planar MOSFET, especially at high voltages.

mostly 500 V to 1200 V. Both categories have seen a continuing trend of smaller feature sizes and improved geometries (Fig. 3x.87), leading to lower $R_{DS(ON)}$ for a given die size, and, usually, lower capacitances. In the case of low-voltage parts, there's been an explosion of package variations, especially for surface-mount parts (SMT), with some of the highest-performance parts appearing in some of the smallest and most interesting packages. But a major struggle has always been how to get the heat out of the transistor and into something else – the circuit board, the heatsink, or whatever!

A. Logic-level gates

A gate switching voltage spec of 10 V was adopted in the 1970s, and a decade later we got a convenient "logic-level" spec of 4.5 V (only available for parts with V_{DSS} of 400 V or less). After another decade the low-voltage MOSFET scene began providing lower-threshold parts with gate-

drive specs of 2.5 1.8 V, and then 1.5 V, 1.2 V and 0.9 V.[84] Low gate-drive voltage requirements not only simplify the driver electronics, but, given constant C_{iss}, they also lower the switching gate charge Q_g. This is doubly important, given that gate-drive power $P = fV_gQ_g$.

B. Packages

One of the oldest classic surface-mount packages, the TO-263 or "D2Pak" (short for DDPak, larger than the DPak), as it was called – was not much of an innovation. It was merely the long-lived TO-220 bolt-down package with the tab end truncated to 1.65 mm (becoming the sole drain contact), with the middle lead now useless and chopped off at 1.4 mm, and the outer two leads (gate and source) shortened to 3.0 mm and bent to line up with PCB traces, see Figure 3x.86. The TO-220's large metal back remains, with the 1.65 mm extension, and takes on the responsibility for heat transfer to the PCB.

Given the package similarity, many parts are available in both TO-220 and TO-263/DDPak packages. If you look at the power spec (in watts), which derives from the $R_{\Theta JC}$ spec (note the J-to-C, the thermal resistance from the junction to the case), you'll see that these are identical for the two packages. But that's really an unuseful spec, based solely on the nearly-identical pad of metal on the two parts: in the case of the TO-220 we can efficiently deliver the heat to an attached heat sink, suffering only a small additional $R_{\Theta CS}$ (case-to-sink) through thermal grease or thermal pad. The situation is not so rosy for the DDPak. Here, though the case-to-sink thermal resistance is very low (a large-area solder connection), the sink-to-ambient part of the equation is the circuit board itself, which doesn't conduct and

---

[84] For example the Rohm RYC002N05, a 50 V MOSFET in a SOT-23 package with $R_{DS} - 3\,\Omega$ at 0.9 V gate drive.

dissipate heat well, certainly compared with an aluminum heat sink that is used with the TO-220 packaged MOSFET.

The unimpressive datasheet specs typically admit (in a fine-print footnote) that they're assuming a one square-inch copper surface (who has room on their PCB layout for that?!). The resulting power capability drops from 50 W or more (for a large heatsink) to 3.75 W for the square inch of copper in an (unrealistic) 25°C ambient environment (calculated from their $R_{\Theta JA}$ spec), and even less in a final product that operates in a realistic environment at elevated temperature.

Continuing with the evolution of the hoary TO-220 (which evolved to the SMT DDpak), the TO-252 ("DPak") package is about 60% smaller, but the same thermal principles apply. Some parts are also available in IPak and I^2Pak versions. These packages keep the truncated tab, but leave three legs in place for through-hole mounting. That allows them to be installed standing up on the board, so small clip-on heatsinks can be attached to get you into the $\sim$3 W territory.

Figure 3x.88. The SO-8 power package has a 3.8×4 mm bottom-side thermal pad. The PowerFLAT DSC package from ST adds a 2.4×3.8 mm topside pad. Both are *small* – even the diminutive DPak dwarfs these puppies.

A welcome development, for those seeking to design smaller electronic devices, was the power MOSFET adoption of the SOIC-8 IC package footprint, see Figure 3x.88. As with the DPak, a large heat-transfer pad on the bottom

was added for the drain, with three source-terminal pins allotted for additional heat removal. The maximum drain current at 25°C case temperature is

$$I_{D(max, 25°C)} = \left(\frac{T_{Jmax} - 25°C}{R_{DS(on)} R_{\Theta JC}} \right)^{\frac{1}{2}} \qquad (3x.19)$$

If the MOSFET designers have reached a limit in driving down $R_{\Theta JC}$ (or worse, have *lost* ground from the smaller "shrink" dies being produced by the semiconductor fabrication teams), they can turn their attention to $R_{DS(on)}$. The industry's answer to the "heatsink problem" has been to aggressively drive down the ON resistance, so the MOSFET doesn't dissipate as much power in the first place. This approach became especially important for the new smaller power packages. Table 3.4b in AoE3 shows $R_{DS(on)}$ values as low 1.65 mΩ for 60 V parts. But we could have shown 25 V parts with $R_{DS(on)}$ as low as 0.48 mΩ!. One such part, Vishay's inexpensive SiRA20DP, in their SO-8 PowerPAK, is rated at 6.25 W, and can conduct 81.7 A for 10 seconds when mounted on its square inch of PCB copper (if the PCB remains at 25°C, a dubious assumption). At a more likely 70°C ambient, it's still rated at an impressive 65.3 A. It's evident that the steady reduction in $R_{DS(on)}$ has been a boon to the designer of small circuits that must handle high current.

What more to say? Well, once you've got a power MOSFET in its surface-mount package safely soldered to the PCB, you might find yourself wishing you could slap a heatsink on the *top*side, to get more heat-removal area in play. Happily, the semiconductor industry has responded, with SMT MOSFETs featuring dual-side cooling. Figure 3x.88 shows ST's "PowerFLAT DSC" package, with its topside 2.4×3.8 mm pad along with the powerPAK SO-8 bottom thermal pad and pinout. For example, the STLD200N4F6AG, a 40 V, 120 A part, with an $R_{DS(on)}$ of 1.27 mΩ; its $R_{\Theta JC}$ is 0.95°C/W on the bottom and 2.9°C/W on the top. Hey, we'll take the 2.9°C/W value and go with it! With a reasonable 3°C/W top-side thermal pad and heatsink, this part could dissipate up to 25 W, breaking us away from the painfully-limited PCB thermal tyranny.

Circuit designers are increasingly interested in small packages, as they shrink their industrial and consumer widgets, and the industry has responded with a multitude of even smaller power packages; there's a nice sampling in the photographs in Figures 9.46 and 9.47 in the main volume.

C. P-channel MOSFETs

There's a *p*-channel power-MOSFET Table in AoE3 (Table 3.4a), but it goes only to 100 V. However, in a big spreadsheet we keep track of what's available, and we're sorry to report that there are only six entries at 600 V, all from IXYS, and not that many more in the 350–500 V region. What's more, most of these are older parts; evidently the manufacturers don't seem interested. This is not good news for those of us who like to make high-voltage power amplifiers using standard silicon power MOSFETs. There's more new interest in the under 100 V region, but Table 3.4a is still reasonably up to date, and a good place to start exploring (most of the new parts are in unusual and attractive small packages, and are for use below 50 V). There's a nice assortment of parts with one *n*-channel and one *p*-channel MOSFET per package.

D. High-voltage parts

Our Table 3.4b (pages 189 to 191 in the main book) lists *n*-channel silicon MOSFETs all the way up to 4.5 kV, and we're happy to report seeing more new entries in the 2.5 kV to 4.7 kV region – once again all from IXYS. Several of their parts have been discontinued, but the scene is settling down. There are new manufacturers paying attention to the 1.7 kV and lower range, and interesting new parts to explore. A dramatic development in the last decade has been the explosion of super-junction MOSFETs in the 500 V to 900 V range. These feature dramatically-lower capacitances, and can be used at higher switching frequencies, which allows smaller magnetics. There are also new MOSFETs made from other materials, like silicon-carbide. There's a 1.2 kV SiC part that some of us have found quite useful – see for example §3x.15.2. We'll discuss those later.

The story about 600 V parts . . .

In the early years when the power MOSFET market was exploding, and engineers were discovering amazing new things to do with them, the highest voltage parts you could get were rated at 400 V. Then IR introduced the 500 V IRF820, 830 and 840. The latter was rated at 125 W, and was good for 5 A switching; then Motorola and other manufactures jumped on the bandwagon. And power-supply designers jumped at the chance to make universal-input-voltage supplies, working from 95 to 120 Vac, or 230 to 250 Vac (see §9.7.1). The drill goes like this: rectify the ac, then cycle-by-cycle boost convert the half-sine at 400 Vdc, charge a "bulk" storage capacitor (perhaps using PFC, see Figs. 9.77 and 9.78), then add dc–dc converters with transformers for whatever lower voltages you need. But designers quickly realized that 500 V MOSFETs didn't provide

enough safety margin for their ambitious ideas (with 400 V and 5 A, we're talking 2 kilowatts!), so the pressure was on to provide 600 V parts. When these came out in the early 1990s, everyone was in a good place. And in the mid-2000s 600 V super-junction MOSFETs became available, and we were in a very, very good place.

The "space" of MOSFET parameters is huge (voltages from 15 V to 4700 V, currents from milliamps to kiloamps, etc.), and a full exploration of their evolution over the decades is, as they say, well beyond the scope of this book. So, to keep this discussion within the range of available pages we've picked one region to explore: MOSFETs rated at 600 V.

E. Capacitances

One of the most important parameters of a power MOSFET is its capacitance, which has a direct effect on switching speed, bandwidth, and required gate drive current. There are three primary MOSFET capacitances to consider.[85] There's the gate-input capacitance C_{iss}, the drain-to-gate capacitance C_{rss}, and the output capacitance C_{oss}. The gate is grounded for the latter, so it includes C_{rss}. We often pay special attention to C_{oss}, to help calculate switching losses; see also §3x.5.6.

We will look at these capacitances individually.

Input capacitance, C_{iss}

Intuitively the drain-source ON-resistance is inversely proportional to the FET's effective channel surface area (imagine a bunch of FETs in parallel). And we know that the FET's capacitances must be proportional to area. So we might expect the gate capacitance C_{iss} to vary inversely with $R_{DS(on)}$.

And, indeed, the scatterplot in Figure 3x.89 shows C_{iss} varying inversely with $R_{DS(on)}$. Note the close adherence of datapoints to the straight line of constant product $C_{iss}R_{DS(on)}$, over three orders of magnitude; the best parts hug the trend line, with other less-successful parts with the same $R_{DS(on)}$ having two or three times higher input capacitance. Since C_{iss} and $R_{DS(on)}$ are roughly inversely related, we can define a Figure-of-Merit, $FoM = C_{iss}R_{DS(on)}$. Calculating this FoM is an excellent way to evaluate parts you're

---

[85] The capacitances are generally measured with a low-level signal (100 mV) at 1 MHz. Most manufacturers use HP's classic 4280A, which can measure capacitance to ground, suitable for C_{iss} and C_{oss}, and a floating capacitance between two pins, e.g., gate and drain, as required for C_{rss}. The standard voltages are $V_{GS} = 0$ V and $V_{DS} = 25$ V (although the 4280A lets you adjust the drain voltage from -100 V to $+100$ V, with $+100$ V commonly used for super-junction parts).

considering, either from their datasheet specs, or from values in Table 3.4b.[86]

Closely related to C_{iss}, and even more interesting, is the total gate charge, Q_{gd}. This is the charge you have to pump into the gate's capacitance to raise its voltage up to the ON value, typically 10 V. In a naive view, gate charge would be $Q = C_{iss}V_{on}$. But as you drive the gate of an active MOSFET, the drain voltage descends to zero, sinking current from the gate through the feedback capacitance C_{rss} (Miller effect). Although the feedback capacitance is much smaller than the gate capacitance,[87] this effect can nevertheless be serious: imagine a drain voltage changing by, say, 400 V in an offline power converter. The charge from the small feedback capacitance, dropping 400 V, can easily match the charge from a 40× larger gate input capacitance that's changing 10 V. This is, of course, the Miller plateau at work, which we saw in the main volume in §3.5.4A; see also §3x.12 for greater detail, with measurements.

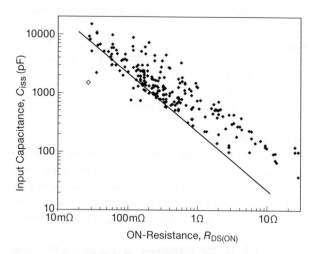

Figure 3x.89. Input capacitance versus ON-resistance for a selection of 600 V and 650 V n-channel MOSFETs.

Capacitance versus drain voltage

The MOSFET C_{rss} and C_{oss} capacitances change dramatically with drain voltage. But we defer discussion and graphs to §3x.11.3 below, where we compare four MOSFET technologies. See also §3x.5.6, where we discuss

MOSFET SPICE models, with further discussion of MOSFET capacitances.

Feedback capacitance, C_{rss}

This parameter, C_{dg} or C_{rss}, can be important in evaluating gate drive requirements, and also other switching aspects, such as problems caused by high drain-voltage slew rate (dV/dt). The latter can cause damaging voltage spikes: the drain slewing causes a transient current $I = C\,dV/dt$, which in turn interacts with parasitic inductances to create a voltage transient $V = L\,dI/dt$.[88]

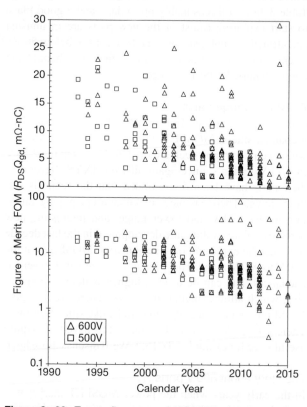

Figure 3x.90. Twenty five years of MOSFET evolution, showing improvements in gate charge figure-of-merit (closely related to feedback capacitance) for a selection of 500 V and 600 V MOSFETs.

MOSFET data in Table 3.4b (and in our contemporary spreadsheet) shows that Q_{gd} (the portion of gate charge due

---

[86] We haven't prepared a scatterplot of this FOM, but you can look ahead to Fig. 3x.90 for a related scatterplot showing improvements over 25 years in another FoM – the product of $R_{DS(on)}$ and total gate charge Q_{gd}.

[87] And highly nonlinear in drain voltage, as we saw in Fig. 3.100; see also Fig. 3x.92.

[88] High drain-voltage dV/dt, such as from body-diode reverse-recovery snap-off, combined with the feedback capacitance, can create unintended and damaging gate voltages. Half-bridge and full-bridge configurations can have both polarities of snap-off spikes. This can be the result of excessive switching deadtimes, hard (rather than soft) freewheel diodes, activation of the MOSFET's parasitic *npn* transistor, etc. You can have fun Googling this huge can of worms.

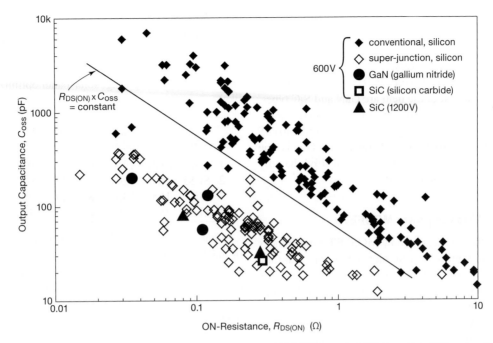

Figure 3x.91. Scatterplot of C_{oss} versus $R_{DS(ON)}$ for some three hundred MOSFETs rated at 600 V, and two SiC parts rated at 1200 V.

to C_{rss}) can amount to nearly half of the total gate charge Q_g for many MOSFETs. Figure 3x.90 shows the improvements in C_{rss} over 25 years of 500 V and 600 V MOSFET development. It's nice to see how designers have been successful at reducing feedback capacitance, and thereby the gate-charge fraction Q_{gd}. The data includes a host of super-junction MOSFETs, plus a few 600 V silicon carbide parts, revealing improvement by factors of five to ten. It's not an exaggeration to assert this is one of the reasons power-design engineers have been having a good time lately.

Output capacitance, C_{oss}

Look at Figure 3x.91, where we've made a scatterplot of C_{oss} versus $R_{DS(on)}$ for a large collection of available high-voltage (600 V and 650 V, but 1200 V(!) for the silicon carbide parts) MOSFETs. New GaN and SiC MOSFETs have dramatically lower capacitance specs than conventional MOSFETs, but many of the super-junction (silicon) parts are competitive. The C_{oss} capacitance is especially important, in part because of its switching-loss contribution[89] $P = fCV^2$. As we've seen, this capacitance is inversely proportional to ON-resistance, the latter

roughly proportional to the effective die size. Note the line $R_{DS(ON)}C_{oss} = $ constant, which can be considered a figure of merit (lower is better). This is one reason you shouldn't select a "bigger" part than needed – bigger isn't better, if you want to minimize switching losses caused by excessive capacitance.

Capacitances for conventional MOSFETs with similar $R_{DS(ON)}$ vary within a 4:1 band (upper black symbols); their C_{oss} is usually specified at 25 V, because it doesn't drop much more at higher voltages (see curve 1 in Figures 3x.92 and 3x.93). By contrast, the new super-junction MOSFETs[90] are specified at 100 V or above (and sometimes even at 480 V or 1000 V), because their capacitance is much higher below those voltages (see curve 2 in Figures 3x.92 and 3x.93). They also vary within a 4:1 band (open symbols), but the C_{oss} at high V_{DS} can be as much as $10\times$ lower than conventional MOSFETs.

Infineon's IPZ60R017C7 is a very low capacitance silicon super-junction part. One thing that's impressive is that many 1.2 kV SiC parts (black triangle) compare favorably with 600 V parts, so they're worth considering for lower-voltage applications. Transphorm's TPH3207WS and TPH3208PS are GaN parts incorporating cascode

---

[89] An additional important switching-loss term (sometimes called "class-A" switching loss) relates to gate drive and switching speed: during a slow transition from ON to OFF (and vice-versa), the MOSFET is burdened with an instantaneous power dissipation $P = I_D V_{DS}$.

[90] Super-junction MOSFETs are not available for voltages below 500 V. Also, logic-level gate operating voltages are not available for voltages above 400 V.

JFETs, giving them C_{oss} capacitances near the bottom of the pack.

Wide bandgap (WBG) devices

The half-dozen parts in Figure 3x.91 with dramatically lower C_{oss} than the rest of the field are GaN and SiC types, except for the unique $27\,m\Omega$ best-in-class part (open-square), which is an SiC type – the UJ3C065030T3S by United Silicon Carbide, Inc. (UnitedSiC). This is one strange beast, including an explicit cascoded JFET, happily packaged in a convenient TO-220.

3x.11.3 Four kinds of power MOSFETs

We've put off the business of how MOSFET capacitances depend on drain voltage, having mentioned only that they decrease significantly with increasing voltage. As it turns out, the four contemporary technologies of MOSFET construction – [1] conventional silicon, [2] super-junction silicon, and the two wide-bandgap materials [3] silicon carbide (SiC) and [4] gallium nitride (GaN) share characteristics that set them apart from the others.[91] Here's a minitable of properties of the three semiconductor types:

#	type	E_G (eV)	μ (cm^2/V·s)	E_B (MV/cm)	k (W/cm°C)
1,2	Si	1.14	1200	0.3	1.5
3	SiC	3.2	900	3.0	4.9
4	GaN	3.4	1500	3.0	0.5

The bandgap voltage E_G for SiC and GaN is much higher than silicon, which is why they're called wide-bandgap or WBG devices. Silicon's electron mobility μ is midway between the WBG materials; the latter have an order of magnitude higher breakdown field E_B than silicon. As a consequence WBG MOSFETs can have a very thin drift layer and/or higher doping concentration. The thermal conductivity k of SiC is much higher than silicon, while GaN is much lower. It's clear that we can expect designs made with these semiconductors to behave differently.

In the minitable below we compare the electrical properties of four 600 V MOSFETs of these differing technologies,[92] all with an R_{on} of 120 mΩ. That ON-resistance puts

these parts squarely in the muscular region for 600 V MOSFETs, easily able to manage kilowatts of power (compared with, say, 250 mΩ to 600 mΩ parts that would be commonly selected for offline ac–dc power supplies, inverters, small motors, lighting, solar-panel optimizers, and the like).

#	type	P_D (W)	R_{on} (mΩ)	@ V_{GS} (V)	I_{D25C} (A)	Q_g (nC)	C_{iss} (pF)
1	Si	1040	125	10	50	94	6300
2	Si-SJ	165	120	10	36	61	1200
3	SiC	165	130	18	29	48	1680
4	GaN	96	110	8	20	10	760

What's striking about this table is the dramatic *reduction* of maximum power dissipation in the newer technologies, compared with classic silicon MOSFETs. This is primarily due to the smaller semiconductor die sizes, with their correspondingly increased thermal resistance (recall eq'n 3x.19, where $R_{\Theta JC}$ appears in the denominator). So, even though these representative parts have comparable ON-resistance, the newer technologies cannot handle anywhere near as much power as conventional silicon: a kilowatt for the silicon part, versus 100–170 W for the others. To be able to handle comparable current (and power), the die sizes for the newer technologies would have to be increased (raising their gate charge and capacitance from their lower initial values).

Silicon carbide MOSFETs require much higher gate-drive voltages than the 10 V spec of conventional MOSFETs (early SiC parts required +20 V and −5 V), but, happily, most GaN parts are happy with less than 10 V (which is unusual for high-voltage conventional MOSFETs). When all factors are considered, the three newer types generally require less semiconductor area to get the job done, potentially benefitting the manufacturer. It's amusing to see tiny SiC dies in huge TO-247 packages, rather than a TO-220 or smaller that would work fine, evidently done to impress the buyers. See further discussion of SiC and GaN MOSFETs in §3x.17.3, and some application circuits in §3x.15.

A. Comparison of capacitances

Revisiting the MOSFET parameter of capacitance, which figures importantly in switching speed and switching losses, let's see how these technologies compare.

Input capacitance, C_{iss}

Input capacitance C_{iss} does not vary much with drain voltage (see for example Fig. 3x.41), so a look at the minitable

---

[91] Of course, for any given type of MOSFET you can expect individual designs to show considerable variation among different manufacturers' parts, or even different products from the same manufacturer.

[92] The part numbers are IXFH50N60P3 (silicon, IXYS/Littelfuse); TK20E60W (Si-SJ, Toshiba); SCT2120AF (SiC, Rohm); and TPH3208PS (GaN, Transphorm).

above tells the story: for comparable ON-resistance and MOSFET voltage rating, the input capacitance of super-junction and SiC MOSFETs is reduced (improved) by a factor of four to five, and for gallium nitride a factor of eight, relative to conventional silicon MOSFETs.

Feedback capacitance, C_{rss}

Feedback capacitance C_{rss} is another story altogether (one that needs to be told with graphs!), because there's a strong dependence on drain voltage, as was seen in Figure 3x.41 (where the feedback capacitance dropped by a factor of 100 as the drain voltage went from zero to 40 V). And here the different technologies differ widely, as seen in Figure 3x.92's plot of C_{rss} versus drain voltage.

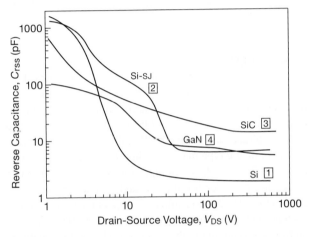

Figure 3x.92. Miller capacitance (C_{dg} or C_{rss}) versus drain voltage, for MOSFETs with similar $R_{DS(on)} = 120\,\text{m}\Omega$, made with four different technologies.

The ordinary silicon type (an IXFH50N60P3), trace 1, has the highest capacitance below 5 V, but quickly drops (by a factor of 1000!) to only a few pF, and is the clear winner.[93] It will therefore have the lowest total energy required to charge the C_{rss} capacitance from 0 to 400 V (80% of V_{DSS}), and C_{rss} will make only a small contribution to gate charge Q_g. However, this MOSFET's high input capacitance C_{iss} makes it last in that category. Usually the super-junction parts (Si–SJ, trace 2) will win in the C_{rss} category, above 100 V anyway, and their lower Q_g wins over the silicon part. The SiC and GaN parts, traces 3 and

---

[93] Perhaps it's unfair to select an IXYS part – for a few years they have been touting "Ultra Junction X2-Class" parts with superior properties. We didn't choose one of their X2-class parts, but something unusual is still going on. In our defense we like to show what manufacturers can do, even if they don't always do it.

4, do poorly, even above 100 V, but their low C_{iss} comes to the rescue, and they win in the Q_g category. However, don't forget that you'll need to select a lower R_{on} (larger die, thus larger capacitance and gate charge) part to equal the current capabilities of silicon parts.

Drain output capacitance, C_{oss}

Finally, in the C_{oss} plots in Figure 3x.93 the potential advantages of the new technologies become evident. The silicon part (trace 1) is the overwhelming loser above 30 V. The silicon super-junction part (trace 2) is the winner above 40 V, but not by much, and it has to overcome a huge deficit built up below 25 V. The SiC and GaN parts (traces 3 and 4) look pretty good, but as we'll see below in Figure 3x.94, an integrated energy calculation will hand the prize to the super-junction MOSFET.

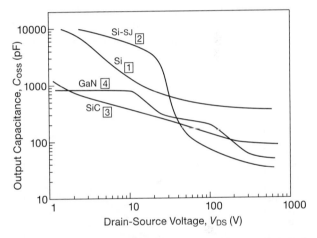

Figure 3x.93. Drain output capacitance (C_{oss}) versus drain voltage, for parts of comparable ON-resistance made with four different MOSFET technologies.

B. Energy: what does all this capacitance stuff mean?

As you might suspect, examining the tortured plots of capacitance versus voltage for these power MOSFETs makes it painful to estimate the energy and power dissipation of circuits using them. In the good ol' days of ordinary silicon MOSFETs, most of the capacitance drop occurred in the 5–20 V region, and we could simply use the datasheet's capacitance spec at 25 V for rough calculations. But manufacturers of the new high-voltage super-junction MOSFETs were proud of the remarkably low capacitance their parts could achieve at high voltages, and they weren't happy showing the unflattering high capacitance at the conventional 25 V. So they changed their specification to list instead the capacitance at 100 V. Now, while presenting nice small values, they were hiding the high energy levels involved in charging the much higher capacitance for the first 75 V. This was an especially significant issue for C_{oss}, which is important in calculating switching losses.

Equivalent energy. The solution to this issue was to introduce new energy-related variables. The simplest (and least useful) energy variable is "time-related equivalent energy." You apply a charging current to the drain, and see how much time it takes to charge to each x-axis drain voltage point; then you calculate the effective capacitance, assuming that it maintained a fixed value: $C_{eff} = It/V$. Then use those effective capacitances to calculate the effective energy for each capacitance and voltage: $E_{eff} = \frac{1}{2}C_{eff}V^2$. The resulting energy values may not be that useful, but the curve shapes do reveal something about what's going on at low voltages. On datasheets it's common to see the energy values at 80% of the maximum drain voltage at 25°C ($0.8V_{DSS}$), to be used when comparing competing parts.

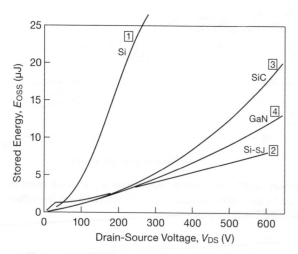

Figure 3x.94. Stored energy (E_{oss}) versus drain voltage, for four representative 120 mΩ MOSFETs. The plotted values show the integrated energy from zero volts to the x-axis voltage.

Stored energy. The second (and more useful) method is to calculate the energy required to *change* the voltage (in small steps) up to a final drain voltage V_{DS}; in other words, integrating IdV to form a plot of stored energy, E_{OSS}, versus final drain voltage V_{DS}. Figure 3x.94 shows that plot for our four candidate 600 V MOSFETs of comparable ON-resistance, revealing the dramatic differences between the various technologies. Classic silicon power MOSFETs (trace 1) appear dramatically worse than the rest (lower

is better), shooting clear off the chart.[94] The SiC and GaN parts (traces 3 and 4) look much better – but remember that you may need a part with lower ON-resistance (thus a larger die, with more capacitance) to get the needed low thermal resistance if your circuit operates the MOSFET at high power dissipation.

In the figure, the silicon super-junction part (trace 2) looks the best of all. It wasn't supposed to be that way (secretly we were rooting for silicon carbide), and our preliminary experience with the new WBG technologies leaves us very impressed with certain SiC parts, compared with some super-junction parts. But this may simply illustrate an important point, that the specific parts being compared must be examined in detail, with all their parameter interactions included, and that you may discover gems in the special offerings of particular manufacturers.

C. Conclusion

Our comparison of MOSFETs made with the four different technologies is revealing. But we have to confess that the choice of parts with comparable R_{on} was a bit unfair (though logical): Our giant MOSFET spreadsheet is sorted first by voltage, and second by R_{on}; this makes sense, given that a single die type may come in different packages, with different P_{diss} and I_{D25C} ratings, and it's also useful when evaluating parts for use in, say, high-voltage pulse generators with 50 Ω back-terminated outputs (see §3x.15). For such applications a low R_{on} compared with 50 Ω is important, whereas having high power or high current capability usually is not.

But for power conversion the maximum current I_{D25C} is a more important parameter. So by way of recompense, we've selected different silicon parts[95] for an updated comparison mini-table in which the contenders have comparable I_{D25C} ratings of ~20 A (rather than comparable R_{on} of 120 mΩ). The two WBG devices remain the same, but now their R_{on} values are much less than that of the silicon types, and silicon doesn't look so bad.[96]

---

[94] But, hey, the silicon MOSFET is a BIG part! Recall its 1040 W rating, compared with only 165 W or less for the others.

[95] The part numbers are STP20NM60 (silicon, ST Microelectronics); FCP190N65F (Si-SJ, Fairchild/ON Semi); SCT2120AF (SiC, Rohm); and TPH3208PS (GaN, Transphorm).

[96] The silicon super-junction part's parameters aren't as good as we're used to seeing, compared with the others, but that was the best we could find in the 20 A region. For a better-looking part, check out the 44 A FCP067N65S3.

#	type	P_D (W)	R_{on} (mΩ)	@ V_{GS} (V)	I_{D25C} (A)	Q_g (nC)	C_{iss} (pF)
5	Si	192	250	10	20	39	1500
6	Si-SJ	208	168	10	21	60	2425
3	SiC	165	130	18	29	48	1680
4	GaN	96	110	8	20	10	760

For this new comparison set of MOSFETs, Figure 3x.95 replaces the earlier plot (Fig. 3x.93) of C_{oss} versus drain voltage. And the corresponding stored energy E_{oss} values at $V_{DS} = 480$ V become 25 μJ for the silicon part[97] and 7.5 μJ for the silicon super-junction part; the latter is still the winner in the stored-energy contest.

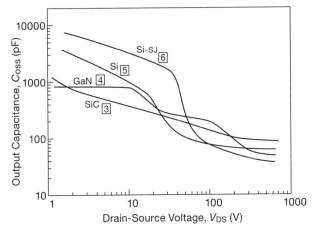

Figure 3x.95. Drain output capacitance (C_{oss}) versus drain voltage, for parts of comparable maximum drain current, made with four different MOSFET technologies.

The wide-bandgap team responds. You guys were unfair to our team's excellent WBG MOSFETs! For one thing, you could have chosen higher voltage parts, where our silicon carbide technology shines, for example the Cree C2M0280120D[98] (1200 V, 10 A, $5), or the STMicroelectronics SCT20N120 (1200 V, 20 A, $10). Or you could have included higher current parts, like the Rohm SCT2120AF (650 V, 29 A, $10), which you evidently excluded because it's *too good* (i.e., 29 A, versus the 20 A competitors).

And then, in the gallium nitride arena, our lower voltage technology is most impressive, almost magical in its fast-switching instant-off performance. For example,

[97] Calculated as $E_{oss} = 0.5C_{oss(eff)}V^2$ from the datasheet's $C_{oss(eff)}$ value of 215 pF.

[98] Authors' note: a favorite of ours (even for 600 V applications), see Fig. 3x.111.

a 48 V–to–1.0 V dc–dc converter ordinarily requires two stages (for example, converting to 12 V, then down to 1.0 V), but Texas Instruments is touting single-step designs with their LMG5200 80 V, 10 A, $10 half-bridge GaN IC. And they're using their 600 V, 40 A, $25 LMG3410 GaN integrated switch for smaller, more efficient 380 V dc converters. Both of these parts integrate high voltage level-shifting, gate drive, protection, and power GaN MOSFETs. Besides taking less space, they benefit from low-inductance internal connections (gate-driver to MOSFET source), reducing deadtime and minimizing reverse-recovery problems. And the fast GaN technology lets them run their 380 V PFC stage at 1 MHz, dramatically reducing the size and weight of the magnetics. But, showing your silicon favoritism, you guys blocked that 600 V part from the contest, because it wasn't a discrete transistor! Once again, we protest this unfair treatment, and we're filing a complaint with the referee.

Wrapup. As we conclude this lengthy section, rich with intricacies of the history of power MOSFETs, we have to admit the omission of a few important MOSFET topics:
(1) RF power MOSFETs. These are specialized beasts that work to high power levels (750 W) and to high frequencies (tens of GHz). This is an active and fast-growing area.
(2) Lateral power MOSFETs. These avoid using V-grooves and other high channel-density schemes. They feature a negative tempco bias property, see §§3.6.4 and 3x.18, required for high-power linear audio power amplifiers. ProFusion sells parts made by Exicon. Hitachi abandoned the market, and now users pray these parts won't disappear.
(3) Depletion-mode MOSFETs. Unlike enhancement-mode MOSFETs, which are normally OFF and require a positive bias to turn ON, these are normally ON, and require a reverse-bias to turn OFF. See the discussion in §3.6.2 and Table 3.6 of the main book, and also §9.3.14 for applications. In this volume see §§3x.6, 4x.23, 9x.3, and 9x.12. Very nice depletion-mode parts are available, both small and large, and seem to have a stable marketplace, however we aren't seeing new parts being introduced. And only *n*-channel types available.
(4) Pioneered by companies like Agilent (spun off to Avago, and purchased by Broadcom), super-fast high-frequency small-signal MOSFETs flourished. With technologies like pHEMT (pseudomorphic high-electron-mobility transistor) and E-pHEMT (enhancement-mode pHEMT), we got inexpensive discrete parts like Avago's ATF-38143, a 10 GHz 4 V 200 mA FET in a convenient 4-lead SC-70 package.[99] But these handy parts have disap-

[99] John Larkin wrote "The really sad demise is the ATF-50189, a power

peared, as semiconductor manufacturers created IC-based solutions more attractive to the telecom customer base.[100] Now we are left with parts like the SAV-551+, made by Mini-Circuits (thank you!) and not even appearing on Octopart. It substitutes for Avago's ATF-55143, and features an ON-resistance of $2\,\Omega$, and an output capacitance of $0.4\,pF$, numbers we were getting used to in the good-old-days.

---

pHEMT in a SOT-89 package. It was rated for 7 V but, in the tradition of RF parts, was good for a lot more. I have/had a beautiful pulse generator output stage, and some great laser drivers, that use them. There were nice SOT-89 MESFETs, long gone. I'm having to transition some designs to GaN, which needs a lot more gate drive, but can switch a lot more volts."

[100] Of the parts on Larkin's 2017 small pHEMT list (with pinouts and SMT labels), only CEL's CE3514 is still available, and Mouser has a pile of SKY65050 parts left in stock.

3x.12 Measuring MOSFET Gate Charge

Reviewing the discussion in Chapter 3. When MOS-FETs are used in power switching applications, the parameter of *gate charge* becomes especially important: you've got to supply (and later remove) substantial current to charge (and discharge) the capacitances seen at the gate. This has to be done quickly enough to keep the losses sufficiently small during partial conduction ("class-A conduction") as the drain current is flowing at the same time as the drain is swinging between saturated and OFF states.[101] For a MOSFET switching circuit the capacitance seen at the gate changes markedly during the switching cycle: initially (transistor OFF) it is $C_{GS} + C_{DG}$ (officially called C_{iss} and C_{rss}), with the latter "feedback capacitance" depending nonlinearly on V_{DS}, but generally about 10% of C_{GS} when there's plenty of voltage on the drain; see Figure 3.100. By the time the MOSFET has brought the drain down to the source voltage the feedback capacitance C_{DG} has increased, to about 30% of C_{GS}. On the way down, though, the gate sees an enhanced drain-to-gate capacitance: the infamous Miller Effect. So, for instance, if we supply a constant current to the gate of an initially OFF MOSFET (e.g., as shown in Figure 3.102), we'll see the gate voltage climb linearly to the onset of conduction, then hover while the drain comes bareling down to saturation, then continue with a linear rise at somewhat reduced slope (because of the increased value of nonlinear feedback capacitance C_{DG}). Because $Q = It$, the horizontal axis (time) is also a proportional measure of charge: if you supply 1 mA of gate current, for example, it's 1 nC per μs.

We introduced this in §3.5.4B. Here we continue with a look at some actual measured gate charge curves, expanding the discussion to include both constant-current loads and resistive loads. We'll see some pitfalls (and ways to circumvent them) in making these measurements. Along

---

[101] This is one contribution among several mechanisms of "switching losses." Another is the power dissipated by the gate driver; and a third is the repetitive dissipation of energy stored in the MOSFET's drain capacitance (plus associated load capacitances) when its voltage is brought to zero. The latter amounts to $P = fC_D V_{DD}^2/2$, for "hard switching," in which the combined drain and load capacitance C_D is brought abruptly from V_{DD} to zero.

the way we'll reveal some fictions that have been perpetuated in the gate charge curves of MOSFET datasheets.

3x.12.1 The gate charge curve depends on load current

Look first at Figure 3x.96, an overlay of measured gate and drain voltages for several values of drain load resistance R_D; the initially OFF MOSFET was driven with a current of 0.5 mA applied to the gate. The drain supply voltage was +30 V, so those loads correspond to ON-state currents of 0.1 A, 1 A, 3 A, 10 A, and 25 A. You can see that the required gate charge does not much depend on load current, being ~3 nC to the onset of conduction, another ~15 nC to complete what we might call the "Miller plateau," and an additional ~10 nC to bring the gate up to a full +5 V forward bias. These are officially called the gate–source charge Q_{gs}, the gate-drain charge Q_{gd}, and the total gate charge Q_g, respectively.

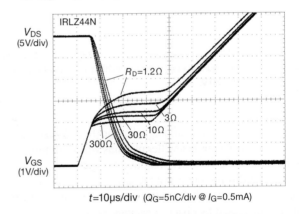

Figure 3x.96. Measured gate and drain voltage for an IRLZ44N n-channel power MOSFET, when switching several values of load resistance returned to +30 V. The gate was driven with 0.5 mA step.

From the figure you can see that smaller load resistances (higher drain currents) require larger gate voltages; you can see also that the gate voltage trace rises as the drain voltage is brought down to ground, because the drain current through the fixed drain resistor goes from zero to maximum. This is particularly evident at the highest currents, where the MOSFET's transconductance is lowest.

3x.12.2 Gate charge curves at constant load current

For applications where the load is resistive, the curves of Figure 3x.96 tell the story. It's more common, though, to see gate charge graphs generated at a constant value of drain *current* (rather than resistance); often that drain current is chosen to be close to the part's rated maximum continuous current. Figure 3x.97 shows a set of measured gate and drain voltages for the same MOSFET sample that was used for Figure 3x.96, but this time with a current source as drain load (with the set of drain currents chosen to be the same as the ON-state currents in the previous figure).

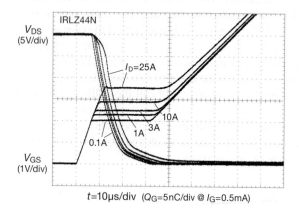

Figure 3x.97. Measured gate and drain voltage for the same MOSFET as in Figure 3x.96, when switching a constant-current load returned to +30 V.

Here you can see textbook behavior (hey, this *is* a textbook, sort of!), with flat Miller plateaus whose altitudes are a measure of the gate drive required for each value of drain current. In this figure (and in the last) the gate voltage slew rate is lower after drain voltage saturation, owing to the greatly increased feedback capacitance C_{DG}.

3x.12.3 The gate charge curve depends also on drain voltage

The Miller plateau gets longer if the drain starts at a higher voltage, as we saw in the datasheet extracts in Figure 3.101. Datasheet plots are good to have, but it's always better if you can measure things for yourself. Figure 3x.98 shows a family of gate charge curves for the same MOSFET sample that was used to generate Figure 3x.97, this time run at a constant drain current of 1 A, but with starting drain voltages going from 5 V to 60 V in 5 V steps.

Higher drain voltages require additional gate charge, as the drain has to swing further; but the reduced nonlinear capacitance C_{rss} at higher voltages reduces the incremental effect as the drain voltage is increased in equal steps. At the lowest drain voltage (5 V) C_{rss} becomes large enough to produce a visible reduction of the initial slope.

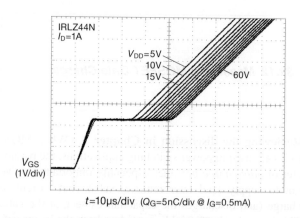

Figure 3x.98. Same MOSFET: Measured gate charge curves for I_D=1 A, with a set of initial drain voltages from 5 V–60 V, in 5 V steps.

3x.12.4 Gate charge test circuit

In many MOSFET datasheets you'll find the right-hand circuit of Figure 3x.99 stuffed somewhere in the back pages, without explanation. It has the germ of elegance – the matched "flying nMOS current source" as drain load for the device under test (DUT). But it's seriously incomplete: no detail of the gate current source, no bypassing of the supply voltage (in a circuit that runs with pulsed currents of 50 A or more!), and, as we'll see, a propensity for serious oscillation.

Starting with the kernel idea, we evolved the circuit shown on the left. The device under test is Q_1, with current source Q_2 as drain load (or you can select instead a load resistor R_L). The supply voltage V_{DD} is heavily bypassed directly to the DUT's source terminal, taking account of the high pulsed-current path (indicated in heavy lines).

For the gate current source we use a small JFET, with source self-bias to set the operating current (in the range of ∼0.1 mA–10 mA). The current is clamped to ground by Q_6, but released into the DUT's gate during the input pulse that brings Q_5 into conduction. You set I_G by shorting Q_5's collector to ground, reading the current at Q_6's collector as shown (keep V_{DD} unenergized, or you'll fry something – at these currents you've got to make short pulsed measurements).

Up at the drain end we've got the classic flying-MOSFET, with its battery-driven gate voltage. We don't know how the classic circuit is expected to work properly, because MOSFETs (especially high-voltage types) tend to robust oscillation unless you add some series resistance (and perhaps a ferrite bead) at the gate terminal. Here we used both, selecting the series resistance R_g for a clean current step as observed across sense resistor R_{sense} down at

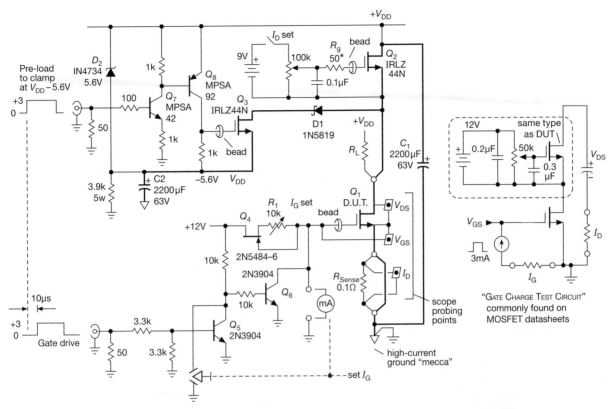

Figure 3x.99. Test circuit for measuring gate charge. The ferrite beads (J.W.Miller FB73-110-RC) fit over the MOSFET's gate lead, and are effective at killing oscillations.

Q_1's source. Choose R_g large enough to prevent instability, but not so large that the current overshoots when Q_1 is pulsed into conduction. After the pulsed drain current has been set, short out R_{sense} – it's used only for the fussy current-source adjustment.

Finally, we added a drain voltage clamp circuit (Q_3 and the stuff to its left), because the gate charge curves exhibited a bit of transient overshoot at the onset of DUT conduction, caused by the abrupt startup of the Q_2 current source circuit. The idea is to get Q_2 supplying current in advance of driving Q_1. That is done by generating a bypassed rail a few volts below V_{DD}, to which the current source is invited (via D_1 and Q_3) to source its current a few microseconds early. This circuit works well – the gate charge traces in the figures are unretouched authentic originals straight from our favorite TDS3044B "lunchbox" scope. To get clean traces it's important to probe directly across the gate–source (or gate-drain) terminals, as indicated: with 50 A currents switching in microseconds you get plenty of probing artifacts if you use the probe's ordinary 10 cm ground clips – see for example Figure 12.32 and its associated discussion.

3x.12.5 The Miller plateau

For a given gate voltage, the MOSFET used for Figure 3x.97 exhibits excellent constancy of drain current under wildly changing drain voltage. It has a Miller plateau flat enough to use as a landing strip. Don't expect such textbook behavior from every power MOSFET, however. Compare its gate charge curves with those in Figure 3x.101, a different MOSFET measured under identical conditions. Here the gate voltage rises significantly to maintain constant drain current as the drain voltage undergoes the same switching cycle from +30 V to ground.

In practice there's nothing really wrong with this, in a switching application, as long as you drive the gate to a high enough voltage to ensure that the MOSFET is brought into full conduction. But it does suggest, for example, that some MOSFETs are much better than others when rigged up as current sources like Q_2 in our gate charge test circuit.

It also suggests that the gate charge graphs shown in datasheets are not to be taken literally. Figure 3x.102 compares the datasheet's idealized gate charge graph with our measured waveform (the uppermost V_{GS} line in Fig-

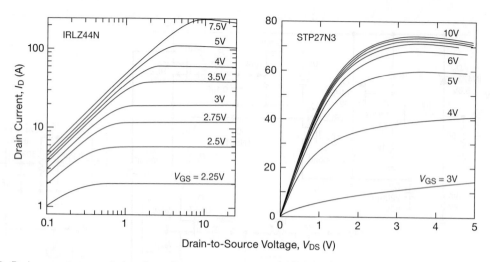

Figure 3x.100. Drain current versus drain voltage ("output characteristics") from the datasheets for the two MOSFETs of the previous figures.

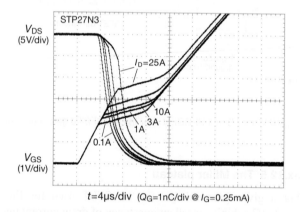

Figure 3x.101. Gate charge waveforms of a different MOSFET, measured under the same conditions as Figure 3x.97.

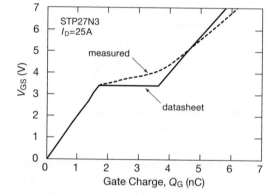

Figure 3x.102. The idealized gate charge graph on the STP27N3's datasheet gives quantitatively correct Q_g values, but seriously misrepresents the shape of the Miller plateau.

ure 3x.101). The datasheet graph does, however, provide correct values of Q_{gs}, Q_{gd}, and Q_g.

Interestingly, the "Output Characteristics" graph (I_D versus V_{DS} for a family of V_{GS} values) from the same datasheet predicts the observed sloped Miller plateau (and disagrees with the datasheet's idealized gate charge plot). Look at Figure 3x.100, which compares these curves from the respective datasheets. Even though they are plotted differently (log-log, linear), it is nevertheless clear that the flat curves of the IRLZ44N are consistent with the observed flat Miller plateaus, and conversely for the sloped curves of the STP27N3.

Looking deeper into what's going on with these two example MOSFETs, one only has to take a look at the *Capacitance versus Drain-to-Source Voltage* plots in their

datasheets. The IRLZ44N has a very high C_{rss} capacitance (700 pF at low voltage, dropping to 220 pF at 10 V and to its 150 pF spec at 25 V). This high C_{rss} gives it a strong Miller effect and a nice "textbook" flat Miller plateau. By contrast, C_{rss} for the STP27N3 starts at a much lower 220 pF and drops rapidly to 70 pF by 5 V, 30 pF by 10 V and 19 pF at 25 V. Its low C_{rss} means it has a very weak Miller effect, and a barely-visible Miller plateau.

Although these two MOSFETS in TO-220 packages have similar voltage, current and $R_{DS(on)}$ ratings (20 and 21 mΩ, with 5 V gate drive), they have dramatically different gate-charge specs: the total gate charge (Q_g) is 48 nC for the IRLZ44N versus 4.8 nC for the STP27N3, and the Miller charge (Q_{gd}) is 25 nC versus just 1.9 nC. Amazing!

The former may look like a better classic MOSFET, but the latter is the one that you'll want to choose for fast switching, because it's $10\times$ easier to drive.[102]

[102] The oldest datasheets stored in our computer have dates of 1997 for the IRLZ44N and 2009 for the STP27N3. MOSFET designers have been making steady progress – see the scatterplot in Fig. 3x.90.

3x.13 Pulse Energy in Power MOSFETs

MOSFETs – and many other power devices – can withstand very high instantaneous power dissipation, if the duration is short enough. As we'll see, the numbers are impressive – ten kilowatts or so for an ordinary power MOSFET. And this applies, perhaps surprisingly, to a transistor that is undergoing avalanche breakdown (that is, uncontrolled conduction when the drain-source voltage drop exceeds the breakdown rating $V_{(BR)DSS}$), as well as to ordinary "controlled" power dissipation (that is, drain current that is responsive to applied gate–source voltage, when the drain-source voltage is in the legal range). The ability to withstand high avalanche and pulse energy is particularly important in inductive power switching, for example in switching power converters (Chapter 9). You shouldn't feel as if you're doing something wrong by clamping inductive spikes with MOSFET breakdown: it's perfectly good design practice to take advantage of this pleasant property of power MOSFETs, which the datasheets advertise with language like "guaranteed to withstand a specified level of energy in the breakdown avalanche mode of operation." However, those same datasheets can be very misleading on this subject. Let's see what's going on.

3x.13.1 Limited only by maximum junction temperature

The basic fact is that allowable power dissipation in MOSFETs is limited essentially by $T_{J(max)}$. For dc or steady-state operation this means that the power dissipation is limited by the thermal resistance and the ambient temperature, namely $T_J = T_A + P_{diss}R_{\Theta JA} \leq T_{J(max)}$. So, for example, a power MOSFET in a plastic TO-220 power package ($T_{J(max)} = 175°C$), attached to a reasonably hefty heatsink, might have $R_{\Theta JA} = 2°C/W$, limiting its steady-state power dissipation to 62.5 W at an ambient temperature of 50°C (see Chapter 9 for more detail).

Things are different for a short duration pulse (less than a millisecond, say), because the heat capacity (the "specific heat") of the transistor itself absorbs the energy and limits the temperature rise. Datasheets usually characterize this as an "effective transient thermal impedance" $Z_{\Theta JC}$, which

is a function of pulse duration,[103] and from which you can calculate the maximum pulse power for any pulse width. You can do this for ordinary (controlled) pulsed operation, or for inductive avalanche breakdown. As we'll see, for the latter you need to know some circuit parameters, namely the inductance, the peak current, and the power supply voltage.

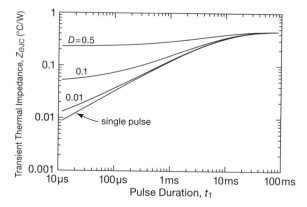

Figure 3x.103. Effective thermal impedance $Z_{\Theta JC}$ versus pulse width and duty cycle for an IRF1405 power MOSFET.

A. Controlled Conduction

To get started, look at Figure 3x.103, extracted from a datasheet for a representative power MOSFET (the IRF1405). The effective thermal impedance from junction to case $Z_{\Theta JC}$ is plotted versus pulse duration, for solitary pulses and for repetitive pulses at three values of duty cycle (sometimes called "duty ratio," or "duty factor"; it's simply the fraction of time the switch is ON). This transistor has a specified thermal resistance (steady-state) of 0.45°C/W, and the curves reach that value for pulse durations greater than about 10 ms. So, for example, if the heatsinking is adequate to hold the case at 50°C, then the transistor can continuously dissipate $(T_{J(max)} - T_C)/R_{\Theta JC} = 275$ W.

For much shorter pulses, however, the instantaneous power dissipation can be larger, as reflected in the lower effective thermal impedances seen in the figure. For example, during a 100μs pulse (for which $Z_{\Theta JC} = 0.032°C/W$) you could dissipate up to 3.9 kW at the same case temperature (for solitary or sparse pulses). The roughly square-root behavior of $Z_{\Theta JC}$ versus pulse duration is a reflection of the diffusive nature of heat conduction. For high duty-cycle operation the peak dissipation ultimately becomes limited

---

[103] And, for repetitive pulses, the "duty cycle" D, which is the ratio of pulse width to pulse period.

by the high average power. So, for example, the thermal impedance for short-period square-wave conduction (50% duty cycle) rises to half the dc value, or 0.23°C/W.

Exercise 3x.1. Verify the 3.9 kW calculation above. Then calculate the maximum pulse power for a kilohertz train of $10\mu s$ pulses, at 75°C case temperature.

B. Avalanche Mode

Contemporary power MOSFETs are not damaged by drain-source breakdown, as long as the energy delivered to the junction during such uncontrolled conduction stays within certain bounds. This is sometime specified on the datasheet: the IRF1405, for example, specifies a maximum E_{AS} ("single pulse avalanche energy") of 560 mJ. However, this figure depends on the time duration of the delivered energy (just as with maximum controlled pulse power or pulse energy), which depends in turn on the external circuit. The usual applications in which you produce avalanche breakdown are in inductive switching, in which energy stored in the inductor during conduction is delivered to the transistor when current is switched OFF. So, for example, the IRF1405 datasheet has a footnote saying that the specified E_{AS} limit is for a starting junction temperature of 25°C, an inductance of $110 \mu H$, and a peak inductor current of 101 A.

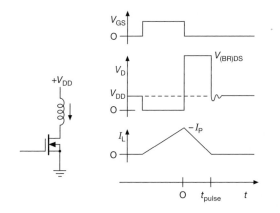

Figure 3x.104. A MOSFET switch ramps inductor current to I_p, then undergoes avalanche breakdown, absorbing power for the duration of the current ramp-down, t_{pulse}.

This is certainly interesting. But what if you want to use the MOSFET in a different circuit? The simple unifying fact here is that *the maximum allowable pulse avalanche energy is set by $T_{J(max)}$ alone* (just as with controlled pulse conduction).[104] So, for any given circuit you can use the

---

[104] This holds true for pulse widths as short as $10\mu s$; for much shorter pulses there may be other failure modes.

datasheet's graph of $Z_{\Theta JC}$ to figure out the limits of safe avalanche conduction.

We can verify this, first, by checking to see if T_J for the specified avalanche conditions is in fact $T_{J(max)}$ (175°C for this transistor). To do this we'll need to know two more quantities, namely the power supply voltage V_{DD}, and the assumed MOSFET breakdown voltage. It's reasonable to assume that the latter is simply $V_{(BR)DSS}$, here 55 V; there's no V_{DD} supply voltage mentioned, but we can take a wild guess (which will turn out to be right!), figuring that an "automotive MOSFET" (the datasheet's self-proclaimed name) is supposed to run from the normal car battery voltage of 14 V.

The setup is shown in Figure 3x.104. We imagine the MOSFET is held ON for a time that ramps the inductor current to the 101 A value (according to $V = LdI/dt$), followed by an OFF state during which the current ramps down to zero. During ramp-up the inductor has V_{DD} across it; during ramp-down it has $V_{(BR)DSS} - V_{DD}$. Ramp-down ends when the inductor current reaches zero; the average current during the pulse is half the peak inductor current. Let's do it as an exercise:

Exercise 3x.2. Find the junction temperature corresponding to the datasheet's specified circuit conditions for maximum avalanche energy, by doing the following calculations: (1) Assume that the peak inductor current (call it I_p) is 101 A when the switch is turned OFF (time $t = 0$); find the time t_{pulse} for the current to ramp down to zero. (2) From Fig. 3x.103 find the corresponding single-pulse transient thermal impedance. (3) Calculate the energy stored in the inductor, $E_L = \frac{1}{2}LI_p^2$, at its peak current; compare with the specified "single pulse avalanche energy" spec of 560 mJ. (4) Assuming E_L is the energy that is delivered to the MOSFET, initially at 25°C, find the average power delivered to the junction during the pulse, and the resulting junction temperature; compare with $T_{J(max)} = 175$°C.

You should have found that the inductor's stored energy indeed matches the 560 mJ specification, but the resulting junction temperature (≈ 140°C) is well below the $T_{J(max)}$ of 175°C. However, part (4) of the exercise is flawed, because the high end of the inductor is tied to V_{DD}, so additional power is delivered to the transistor as the inductor current ramps down. It's easy enough to do the correct calculation:

Exercise 3x.3. Calculate correctly the total power delivered to the junction in the above exercise, by including the additional average power supplied from V_{DD}, namely $\frac{1}{2}V_{DD}I_p$. From that calculate T_J, assuming as before a starting temperature of 25°C.

You should have found $T_J \approx 175$°C, in agreement with the specified $T_{J(max)}$. The additional energy from V_{DD} is not negligible, and increases rapidly as the supply voltage approaches the breakdown voltage; it's easy to show that the energy delivered to the transistor during the ramp-down is given by

$$E_J = \frac{1}{2}LI_p^2\left(1 + \frac{V_{DD}}{V_{(BR)DSS} - V_{DD}}\right),$$

with the additional term within parenthesis representing the energy delivered from the supply. For a supply voltage of half the breakdown voltage, the pulse energy is doubled compared with its low voltage value; furthermore, the ramp-down time doubles, which raises the transient thermal impedance. Both effects work to reduce the peak inductor energy you can switch.

There are several important lessons here:

- The "avalanche energy" specification (E_{AS}) given in datasheets assumes a particular test circuit and component values (inductance; peak current; power supply voltage); so you can't just compare these numbers when choosing a MOSFET.
- Even if your circuit is similar to the datasheet's, you have to be careful because the E_{AS} value may (as in this case) correspond to the inductor's stored energy, not the energy actually delivered to the transistor.
- For these reasons you should probably ignore the E_{AS} spec altogether, and just calculate T_J for your actual circuit configuration, using the datasheet's transient thermal impedance graph. You can rely on this as long as the calculated avalanche pulse duration is greater than approximately ten microseconds.

3x.13.2 Alternative graphs

There are other useful ways to plot pulse and avalanche limits. In Figure 3x.105 we've used the transient thermal impedance data from Fig. 3x.103 to generate plots of maximum pulse *power* and maximum pulse *energy* versus pulse duration. You've got to assume some case temperature T_C, in the conversion from thermal impedance to peak power/energy; here we've made plots for two choices ($T_C = 25°C$ and $75°C$), calculating power and energy values that produce a maximum junction temperature of $175°C$.

Plotted also are maximum values of avalanche current (averaged over the inductor's linear ramp from I_p to zero; that is, $I_p/2$). You've got to make some assumption about the actual breakdown voltage, since avalanche power depends on it; datasheets sometimes use a value of 1.3 times the rated $V_{(BR)DS}$, which we've done here. To use this plot,

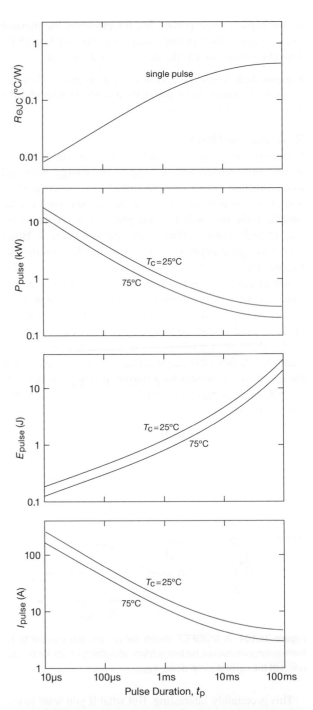

Figure 3x.105. Pulse and avalanche limits for the IRF1405 MOSFET, derived from the datasheet's single-pulse transient thermal impedance $Z_{\Theta JC}$, with $T_{J(max)} = 175°C$, for two choices of case temperature. The values plotted for current are averaged over the pulse, and assume avalanche clamping at 130% of rated drain-source breakdown voltage.

of course, you need to know the ramp-down time, which depends on L, I_p, V_{DD}, and $V_{(BR)DS}$.

Exercise 3x.4. What is the ramp-down time, t_{pulse}, in terms of these variables?

3x.14 MOSFET Gate Drivers

We introduced MOSFET gate drivers in Chapter 3 (see for example Fig. 3.96), and we listed a selection in Table 3.8 on page 218 of AoE3 (repeated here as Table 3x.4 on the next page). These drivers are powered from a single positive rail (typically 4.5 V–18 V), accept logic-level input, and produce full-swing outputs with substantial peak current capability (up to 10 A for some parts) from their internal push–pull MOSFET output stage. They made cameo appearances later, e.g., in Figures 12.42B, 12.44H, and 15.10. Those figures illustrate some circuit basics, but there is more to know. Here we gather together some useful circuit tricks.

which drops 0.1 V while charging the gate. Diode D_1 protects the driver from latchup, which can occur with very rapid drain slew-rates in combination with some source-terminal inductance; something like a 1N5819 (1 A Schottky) is good. The series gate resistors limit the transient gate current, slowing the switching time; this is useful to reduce system transients when you don't need the fastest switching speeds. Typically R_2 is smaller than R_1 (and is often zero), for faster turn-off; typical resistor values are in the range of 5 Ω–100 Ω.

The circuits in Figures 3x.106B and C show some variations: Transistor Q_2 in circuit B boosts gate sinking current for dV/dt protection, for example in bridge configurations where the upper MOSFET's slew rate can source enough current ($C_{rss}\,dV/dt$) to turn back ON the lower MOSFET, thus creating rail-to-rail shoot-through; and circuit C shows how drivers with separate sourcing and sinking outputs[106] let you tailor source and sink gate resistances without an extra diode.

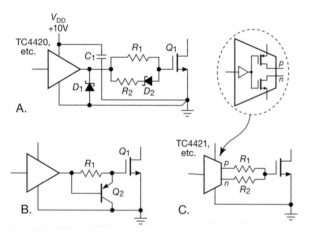

Figure 3x.106. MOSFET gate driver configurations. A. Full configuration: Capacitor C_1 supplies the transient sourcing current during turn-on, while D_1 protects the driver from dV/dt latchup; the gate resistors reduce turn-on and turn-off slew rates (R_2 is often omitted). B. Q_2 boosts turn-off gate drive. C. Driver with separate outputs eliminates need for D_2.

Figure 3x.106A shows the general case. Bypass capacitor C_1 is essential, given the substantial peak drive currents during MOSFET turn-on (e.g., driving a gate capacitance[105] C_{iss} of 1000 pF through 10 V in 10 ns requires one ampere); note the return path of both the driver and of C_1 to the MOSFET's source terminal. C_1 is usually 0.1 μF,

---

[105] Better stated in terms of total gate charge, i.e., Q_g=10 nC. See §3.5.4B and Figs. 3.101 and 3.102.

[106] E.g., TC4421/22, TC4431/32, TC4451/52, LM5114, MAX5048, MAX15070, ZXGD3002/04, and UCC27531/38.

Table 3x.4: Low-side MOSFET Gate Drivers[a]

Part #	Mfg[d]	# channels	V_{min} (V)	V_{max} (V)	I_{pk} (A)	$t_d+0.5t_r$ (ns, typ)	C_{load} (nF)	logic thresh[p]	source below gnd?	current limit	UVLO?	enable?	inv?	non-inv?	output rail-to-rail?	TO220, Dpak	DIP	SOIC, MSOP	SOT23	smaller	Comments
TC4426-28	MC+	2	4.5	18	1.5	55	1	T	-	-	-	-	n	n	•	-	•	•	•	•	G,H
TC4423-25	MC+	2	4.5	18	3	70	1.8	T	-	-	-	-	n	n	•	-	•	•	•	•	G,H
TC4420,29	MC+	2	4.5	18	6	80	2.5	T	-	-	-	-	•	•	•	•	•	•	•	-	G
TC4421-22	MC+	1	4.5	18	9	85	10	T	-	-	-	-	•	•	•	•	•	•	-	•	G,J
FAN3111	F	1	4.5	18	1	20	0.5	C	-	-	-	c	c	c	•	-	-	-	•	-	
FAN3100C,T	F	1	4.5	18	2	20	1	C,T	-	-	•	c	c	c	•	-	-	-	•	•	A
FAN3180	F	1	5	18	2	30	1	T	-	-	•	-	-	•	•	-	-	-	•	-	B
FAN3216-17	F	2	4.5	18	2	25	2.2	T	-	-	•	-	•	•	•	-	-	•	-	-	D
FAN3226-29C,T	F	2	4.5	18	2	25	1	C,T	-	-	•	c	c	c	•	-	-	-	•	•	C,E
FAN3213-14	F	2	4.5	18	4	20	2.2	T	-	-	•	-	•	•	•	-	-	•	-	-	C
FAN3223-25C,T	F	2	4.5	18	4	25	2.2	C,T	-	-	•	c	c	c	•	-	-	-	•	•	E
FAN3121-22	F	1	4.5	18	9	21	10	T	-	-	•	•	•	•	•	-	-	-	•	-	
IRS44273L	IR	1	12	20	1.5	50	1	T	-	-	•	-	-	•	•	-	-	-	•	-	
IR25600	IR	2	6	20	1.5	75	1	T	-	-	•	-	-	•	•	-	-	•	-	-	
MAX17600-05	MA	2	4	14	4	15	1	C5,T	-	-	-	•	n	n	•	-	-	-	•	-	H
MAX5054-57	MA	2	4	15	4	38	5	C,T	-	-	-	c	c	c	•	-	-	•	-	-	
MAX5048A,B	MA	1	4	12.6	7.6[h]	18	1	C,T	-	-	•	c	c	c	•	-	-	•	-	-	
UCC37323-25[k]	TI	2	4.5	15	4	47	1.8	T	-	-	-	•	•	•	•	-	-	•	•	-	
UCC27517	TI	1	4.7	20	4	17	1.8	T	-	-	•	c	c	c	-	-	-	-	•	-	
UCC27516-19	TI	1	4.7	20	4	17,21	1.8	T,C	-	-	•	•	•	•	-	-	-	-	•	-	
UCC27523-26	TI	2	4.7	20	5	17	1.8	T	-	-	•	•	•	•	•	-	-	-	•	•	E,H
UCC37321-22[k]	TI	1	4	15	9	50	10	T	-	-	•	•	•	•	•	-	-	•	•	-	
MIC44F18-20	MI	1	4.5	13.2	6	24	1	T	-	-	•	•	•	•	•	-	-	•	•	-	
ADP3623-25	A	2	4.5	18	4	28	2.2	T	-	-	-	•	•	•	•	-	-	•	-	-	H,P
LM5110	TI	2	3.5	14	5[f]	38	2	T	•	-	•	-	n	n	•	-	-	•	-	-	H,L
LM5112	TI	2	3.5	14	7[g]	38	2	T	•	-	•	-	n	n	•	-	-	-	•	-	H,L,M
LM5114	TI	1	4	12.6	7.6[h]	16	1	C	-	-	•	c	c	c	•	-	-	-	•	-	
ISL89367	IN	2	4.5	16	6	45	10	F	•	-	•	n	o	o	•	-	-	•	-	-	N
ISL89160-62	IN	2	4.5	16	6	45	10	C5,T	-	-	•	-	•	•	•	-	-	•	-	-	O
MC34151	O	2	6.5	18	1.5	50	1	T	-	-	•	-	-	•	•	-	-	•	-	-	
IR2121	IR	1	12	18	2[e]	200	3.3	T	•	•	-	-	-	•	•	-	-	•	-	-	F
UC3708	TI	2	5	35	3	37	1	T	-	-	•	-	•	•	•	-	-	•	-	-	
IXDD602	IX	1	4.5	35	2	50	1	C5	-	-	•	•	•	•	•	-	-	-	•	-	H,R
IXDD604	IX	1	4.5	35	4	40	1	C5	-	-	•	•	•	•	•	-	-	-	•	-	H,R
IXDD609	IX	1	4.5	35	9	60	10	C5	-	-	•	•	•	•	•	•	-	•	-	-	R
IXDD614	IX	1	4.5	35	14	70	15	C5	-	-	•	•	•	•	•	•	-	•	-	-	R
IXDD630	IX	1	10	35	30	65	5.6	C5	-	-	•	•	•	•	-	•	-	•	-	-	K,R
ZXGD3002-04	D	1	-	20,40	9,5	11	1	-	-	-	-	-	-	•	•	-	-	-	-	•	M,S

Notes: (a) sorted by family, within family sorted by Iout; except for ZXGD3000-series, all devices swing rail-to-rail, or nearly so. (b) into Cload at V_S=12V. (c) input gate with inv and non-inv inputs. (d) A=Analog Devices; D=Diodes,Inc; F=Fairchild; IN=Intersil; IR=International Rectifier; IX=Ixys/Clare; L=LTC; MA=Maxim; MC=Microchip; MI=Micrel; O=OnSemiconductor; S=STMicroelectronics; TI=Texas Instruments. (e) 1A source, 2A sink. (f) 3A source, 5A sink. (g) 3A source, 7A sink. (h) 1.3A source, 7.6A sink. (k) 37xxx for 0 to 70°C, 27xxx for -40°C to 105°C. (n) see part-specific comments. (o) XOR input sets optional invert. (p) C=CMOS; C5=5V CMOS; F=flexible, set by V_{ref-} and V_{ref+} input pins; T=TTL.

Comments: (A) suffix specifies logic threshold. (B) includes 3.3V LDO output. (C) 2ns td channel match. (D) 1ns t_d channel match. (E) dual inv+en, dual non-inv+en, dual inputs. (F) source-resistor current-sense input terminal, suitable for driving an IGBT. (G) industry std, many mfgrs. (H) dual inv, dual non-inv, or one each. (J) for 8-pin pkgs, n- and p-ch drains on separate pins. (K) t_r, t_f = 50ns into 68nF. (L) output swing to neg rail, can be 5V below logic gnd. (M) n- and p-ch drains on separate pins. (N) resistor-programmed edge-delay timers; 2-input AND signal inputs. (O) ISL89163-65 same, but include enable inputs; ISL89166-68 same, but include resistor-programmed edge-delay timer inputs. (P) overtemp protection and output. (R) full p/n is IXDx6..., where x = N, I, D and F for non-inv, inv, dual non-inv+en, or one of each. (S) series is one each high-current high-gain *npn* and *pnp* transistor emitter-followers for pull-up and down.

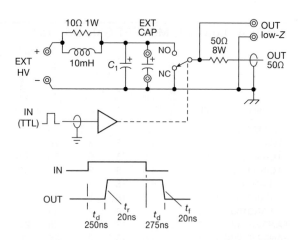

Figure 3x.107. Block diagram of the pulser of Figure 3x.109.

3x.15 High-Voltage Pulsers

High-voltage power MOSFETs are widely available; and, happily, they are quite inexpensive. The FQPF8N80, for example, rated at 800 V and 8 A, costs just $1 in single quantities. We owe our good fortune here to a host of important commercial applications in power control that provide incentive to the semiconductor manufacturers: line-powered ac–dc conversion, inverters for three-phase variable-speed motor drive, fluorescent lighting, and so on. For many of these applications there's a need also for an isolated high-side driver, to provide gate drive to the "flying" upper n-channel MOSFET switch of a push–pull pair, as for example in the classic half-bridge circuit (Fig. 9.73C). Here too the semiconductor manufacturers have responded, with inexpensive offerings that do the job (many of which cost less than a dollar, even in single quantities); some examples are listed in Table 3x.5 on page 250.

Quite apart from such commercial power applications, these high-voltage MOSFETs and drivers can be used to make impressive pulse generators for laboratory applications. We've designed pulsers for applications including particle trapping, electro-optic light modulation, generation of terahertz radiation, electroporation in cell biology, rapid bubble production in nanopores, and, at a more mundane level, testing the pulse-energy endurance of electronic components such as resistors and transient voltage suppressors.

In this section we present several useful designs for laboratory-scale high-voltage pulsers.

3x.15.1 Two-switch +600 V pulser

Figure 3x.107 shows the basic form of a unipolarity (positive only) pulse generator. A pair of n-channel[107] MOSFET switches (shown notionally as an ordinary SPDT switch) are used to connect the output either to ground or to a positive dc voltage. There are two outputs: For applications where you want to drive a hefty current into the load, the output is taken directly from the switch; but for ap-

---

[107] We use n-channel switches because the highest voltage p-channel MOSFETs top out at 600 V, and we want to go higher; p-channel parts of comparable voltage and current ratings also have poorer performance characteristics (higher R_{ON} and C_{iss}), and they are quite a bit more expensive.

plications where you want to deliver a clean pulse at the open (unterminated) end of a length of coax, there's a 50 Ω "back-terminated" output (see Appendix H). There's provision to attach a large external energy-storage capacitor, if you want to deliver high-current pulses without droop. The MOSFET switches respond to a logic-level input, with control circuitry to ensure non-overlapping conduction. The switch transitions are fast ($\sim$20 ns), but the drive circuitry imposes significant delay time ($\sim$250 ns), as indicated.

Before moving on to the full circuit, let's briefly visit the business of high-side drivers, a topic seen earlier in §3.5.6 and revisited in §9x.10 in the context of lower voltages. We need to generate a $\sim$10 V gate–source drive to the upper n-channel MOSFET switch, whose drain is tied to the positive rail and whose source terminal is at the output potential. This is a "flying MOSFET" with flying gate, for which the driver has to generate its V_{GS} (of 0 V or +10 V) relative to the source terminal, the latter jumping between ground and +HV at a prodigious rate. To get a sense of the problem, note that a 500 V step in 20 ns is a slew rate of 25 *kilo*volts per microsecond! And, as with MOSFET drivers generally, the high-side driver may need to supply dynamic drive currents in the range of an amp, in order to charge and discharge the gate capacitance on timescales of 10 ns; after all, a power MOSFET gate charge value of, say, $Q_g \approx 25$ nC requires a gate current of 2.5 A to switch in 10 ns. For our circuit it's important also to have good control over the relative timing of the high-side and low-side switches, to prevent conduction overlap; that's why a half-bridge driver chip, with its paired outputs, is just what we need here.

For this circuit we chose the IR2113 driver, for which a simplified diagram is shown in Figure 3x.108. The low side has a straightforward push–pull output stage, able to source or sink 2 A, and powered from a +10 V to +20 V supply (V_{CC}) referenced to an output COM terminal that must be within a few volts of input logic ground. The high-side push–pull driver floats atop the high-side MOSFET's source terminal (V_S), powered also from +10 V to +20 V relative to its common terminal V_S; i.e., $10V \leq V_B - V_S \leq 20V$. Both the low-side and high-side drivers are shut down if their supply voltages fall below a preset undervoltage lockout (UVLO) threshold of approximately 8 V; this ensures that the MOSFETs are fully driven (or not at all), thus preventing partial conduction that would cause excessive dissipation and overheating. An interesting design feature of this class of driver IC is the use of short pulses (rather than levels) to signal the intended high-side state; this minimizes power dissipation in the high-voltage level-shifters. This is quite effective, except at high switching frequencies where the dissipation rises: The datasheet graphs show minimal heating effects below ~ 10 kHz, but rising to maximum permissible junction temperatures at frequencies in the 100 kHz–1 MHz range. The heating effect is linear in switching frequency and in high-side supply voltage V_B, but it depends also upon the value of series gate resistor (R_1 and R_2 in Fig. 3x.109).

The IR2113/2213 drivers do their level shifting with internal high-voltage transistors, with ratings to 1.2 kV. It's also possible to do the job with transformer coupling, as in the ADuM6132 (see Table 3x.5 and Fig. 12.44G,H), or with optical coupling (see Fig. 3x.113).

Now for the full circuit (Figure 3x.109). Working from left to right, we generate true and inverted input signals for the high-side and low-side driver-chip inputs, matching delays with a pair of XOR gates. The IR2113/2213 itself creates a non-overlap interval of approximately 25 ns, which is the difference between its ON and OFF propagation delay times (120 ns and 94 ns, typ, respectively), but we have added ~ 15 ns of additional safety factor with the slow-on/fast-off gate circuits shown. We've done this because we've seen power MOSFETs take as long as 50 ns to completely cease drain conduction after V_{GS} is brought to zero.[108]

The output switches Q_1 and Q_2 are driven with small series gate resistors R_1 and R_2, paralleled with Schottky diodes to reduce turn-off time (for further overlap preven-

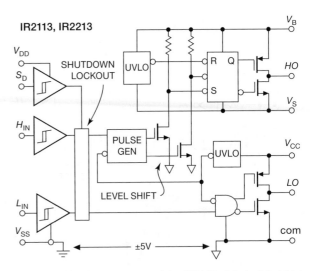

Figure 3x.108. Block diagram of the IR2113-style isolated high-side driver with matched non-isolated low-side driver. The high-side driver is implemented as a flip-flop, with level-shifted pulsed SET and RESET inputs. Both drivers implement undervoltage lockout.

tion). The 1 kΩ gate–source resistors prevent switch conduction if, for example, the high voltage is applied before the low voltage supplies. The bulk capacitor C_1 provides the low impedance path of pulsed output current; it can be supplemented with additional external capacitance. The power inductor L_1 (with 10 Ω damping resistor) isolates the HV supply from abrupt high-current transients.

Diode D_3 deserves some comment: The high-side driver requires its own 12 V dc supply, whose common terminal necessarily flies with the output. That could be an ac-powered dc supply – but beware, such a supply will have to slew at kilovolts per microsecond, no easy task for a conventional line-powered supply because of the interwinding capacitance of the transformer.[109] The trick in Figure 3x.109 is to use a relatively large storage capacitor (C_2) on the high-side supply, and let the low-side +12 V supply replenish its charge through D_3 during times when the output is low. That works fine if the output is low most of the time, or if the pulse rate is high enough. It would not work, though, if the output sits high for long intervals; in that case you'd have to arrange a flying supply. Alternatively you could take advantage of a half-bridge driver like the ADuM6132 (used in our three-switch design, §3x.15.5), which integrates its own flying high-side supply and requires only an external storage capacitor.

---

[108] The MOSFET channel itself switches very fast in response to the overlying gate electrode; but the resistance of the gate runners combines with the gate capacitance to cause an RC signal delay to the distant MOSFET cells. This varies widely among different MOSFETs and manufacturers, and is generally not specified.

[109] Which must be insulated to withstand the full output voltage. We've successfully used the PCP-series of miniature PCB-mounting transformers, rated at 2500 V isolation.

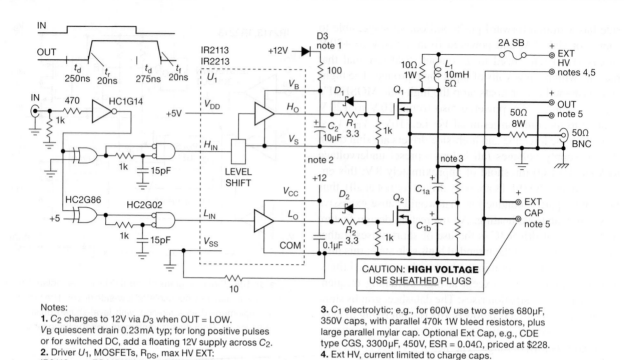

Figure 3x.109. High-voltage pulse generator. The *RC* delays at the input gates ensure non-overlapping conduction ("shoot-through"). The back-terminated BNC output drives unterminated 50 Ω coax, delivering full-swing pulse or dc output to a high-impedance load; the direct output delivers high-current short-pulse output. Bold lines indicate the output current path.

Notes:
1. C_2 charges to 12V via D_3 when OUT = LOW.
V_B quiescent drain 0.23mA typ; for long positive pulses or for switched DC, add a floating 12V supply across C_2.
2. Driver U_1, MOSFETs, R_{DS}, max HV EXT:
IR2113   FQA19N60   0.3Ω   600V
IR2213   IXFK26N120   0.46Ω   1200V

3. C_1 electrolytic; e.g., for 600V use two series 680µF, 350V caps, with parallel 470k 1W bleed resistors, plus large parallel mylar cap. Optional Ext Cap, e.g., CDE type CGS, 3300µF, 450V, ESR = 0.04Ω, priced at $228.
4. Ext HV, current limited to charge caps.
5. Banana jacks spaced 0.75" -- DANGER: High Voltage! Use spring-loaded sheathed plugs.

3x.15.2 Two-switch +500 V 20 A fast pulser

For several exotic applications we needed a faster HV pulser (<20 ns) with lots of peak current capability (to 20 A) and the ability to operate to several megahertz. For these requirements the IR2113 or IR2213 drivers are hopeless, with their sluggish ~100 ns or ~250 ns respective propagation times. Happily, there are some faster non-isolated drivers, for example the UCC2753x series; the UCC27538 we chose has typical propagation times of 17 ns (and that into a whopping 1.8 nF load, nearly twice the 1 nF that the IR2113/2213's specify). And to keep the load capacitance low we chose silicon carbide (SiC) MOSFETs, which are superior to silicon in several important parameters: Compared with silicon, SiC has an order of magnitude higher electric-field breakdown strength, three times higher thermal conductivity, and triple the bandgap. As a consequence, SiC MOSFETs of equivalent voltage rating and ON-resistance can be made much smaller (see Fig. 3x.110), with correspondingly lower gate capacitance. That's helpful for our application, permitting significantly greater switching speed with the drivers we've chosen.

The circuit is shown in Figure 3x.111, where we've omitted some details that only a nerd could love. As with the previous pulser, we've tailored the drive signals for delayed-ON/fast-OFF (both with the frontend gates and with the MOSFET gate drive resistors). For the high-side driver we chose the fast Si8610 digital isolator (8 ns propagation time, dc–150 Mbps, and isolation rated to 5000 Vrms and 60 kV/μs). High-side dc power comes from U_4, an isolated dc–dc module intended for SiC gate drive, where it's common to use gate swings to +20 V and −5 V. These are available from several manufacturers (two are listed on the drawing), with isolation ratings of 3.5 kVac (and 6 kVdc), and an astonishingly low isolation capacitance of just 3.5 pF. For convenience we used the same converter type for the low-side driver (where no isolation is needed).

Some details: (a) This circuit runs *hot*, particularly when switching fast, and into a low-impedance load. So we mounted the MOSFETs and the output terminating resistor R_3 (a non-inductive TO-247 type) onto a microprocessor-style heatsink with blower. (b) Note the high-current path at the output, best implemented with generous copper pours, and of course storage capacitors of low inductance.

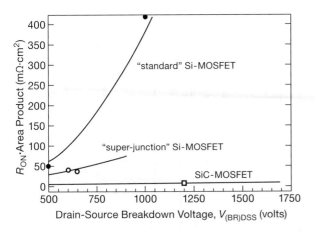

Figure 3x.110. Owing to silicon carbide's higher breakdown strength, the die size of SiC MOSFETs are far smaller than conventional silicon MOSFETs of comparable ON-resistance, as seen in this plot of $R_{ON}A_{die}$ figure-of-merit versus rated breakdown voltage. (Adapted from Rohm App Note 14103EBY01, to which we've added measured datapoints.)

(c) These particular gate drivers provide separate pull-up and pull-down outputs, convenient for setting ON and OFF delays; most drivers don't let you do that, so you use a resistor/diode combination, as in Figure 3x.109. (d) Because the low-side MOSFET driver U_3 rides on a $-4\,\mathrm{V}$ "ground," its input signal is level-shifted by Q_1 (see for example Fig. 12.44A,C).

The good news is that this circuit works well, at least for pulse rates to $\sim 2.5\,\mathrm{MHz}$: Figure 3x.112 shows a 500 V 5 A pulse with FWHM $\approx 20\,\mathrm{ns}$. But things go bad when you try to push it higher: we cranked it up to 10 MHz, and the gate drivers *exploded*, taking out a bunch of other components with them! The problem is that those fast drivers come only in tiny SOT-23 packages, an odd situation given that they are rated to peak currents of 2.5 A (sourcing) and 5 A (sinking). Drivers you can get in power packages (e.g., TC4422, in a TO-220) are slower ($t_p=30\,\mathrm{ns}$, typ), and their greater self-capacitance requires more supply current (e.g., 170 mA at 2 MHz with no load) than the isolated converters U_4 and U_5 can supply (maximum dc output $\pm 100\,\mathrm{mA}$). Desperate for operation to 10 MHz, we reconfigured the circuit with TC4422 heatsink-mounted drivers, powered by three paralleled dc–dc converters. Pretty crude, but it works.

3x.15.3 Two-switch reversible kilovolt pulser

Positive pulses are fine... but sometimes you want *negative* high-voltage pulses. With the previous circuit you're stuck with positive polarity only, because the low-side driver's COM terminal must be close to logic ground. But you can circumvent this limitation by using a dual driver in which both outputs are fully HV isolated, as in Figure 3x.113. Here we've used a dual optocoupler from Avago, with logic-level outputs (5.5 V maximum), pretty good speed ($t_p \approx 40\,\mathrm{ns}$), isolation to 1 kV, and slew rates[110] to 20 kV/μs. Because the optocoupler's output is intended for driving logic (i.e., 5 V swing, and 10 mA maximum), we add TC4420 gate drivers to bring the gate-drive swing to 10 V, and with sink or source peak currents to 6 A. That should do the job!

Because both drivers and their floating supplies are fully isolated, polarity reversal is as simple as switching which terminal of the HV supply is connected to ground. Here we've used a mechanical relay for the job, energized by the polarity input choice. Relays are slow ($\sim$ milliseconds), and so this circuit is not intended for rapid polarity reversals (for example, in a train of pulses of alternating polarity). We'll address that challenge in the last example of this section (§3x.15.5 on page 249).

The rest of the circuit is straightforward, and to save space in the drawing we've omitted some details: the gate drive resistors and diodes, the HV fuse and peak current limiting inductor, the direct (low-Z) output, and provision for an external HV capacitor.

3x.15.4 Output monitor

For most applications of these pulsers you'd like to know the actual output waveform, delivered at manageable 'scope voltages (say within a range of $\pm 10\,\mathrm{V}$), and with enough fidelity to reveal details at the rapid timescales of the pulse waveform ($\sim 10\,\mathrm{ns}$). A simple 100:1 resistive divider will not do the job, because (just as with a 'scope probe) stray capacitance distorts the waveform. But, as with a 'scope probe, we can compensate the stray shunt capacitance of the upper resistor with a trimmable capacitor across the lower resistor, as in Figure 3x.114.

Here we've chosen a precision high-voltage resistor for R_1, forming a 100:1 divider with R_2. Compensation capacitor C_2 combines with the estimated ~ 0.05–0.1 pF self-capacitance of R_1 to maintain the 100:1 divider at high frequencies; you trim it for best waveform fidelity, in the manner of 'scope probes. The 2 MΩ value of R_1 may seem surprisingly low – for example, it dissipates 0.5 W at $V_{out}=1\,\mathrm{kV}$. But, as in engineering generally, its value

---

[110] This specification is the minimum common-mode output slew rate for which the output state is guaranteed to be correct.

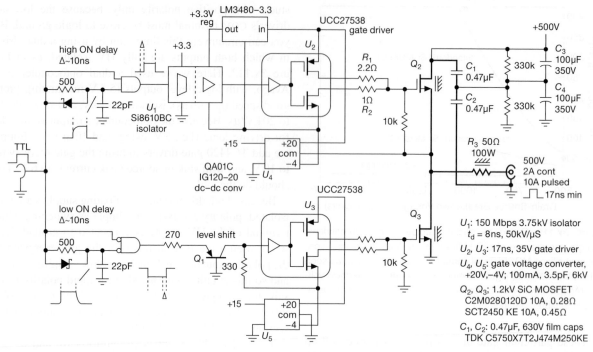

Figure 3x.111. Fast 500 V 10 A unipolarity pulser. The low capacitance of silicon carbide MOSFETs Q_2 and Q_3 allows the use of fast low-power drivers U_2 and U_3 and associated isolated gate supplies U_4 and U_5.

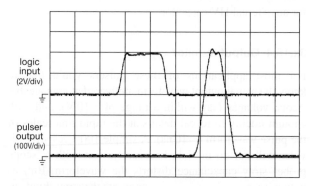

Figure 3x.112. Output waveform (upstream of $50\,\Omega$ series resistor) when driving $50\,\Omega$ power load. Horizontal: 20 ns/div.

is a compromise, in this case balancing resistor dissipation against the voltage error produced by U_1's input current.

The latter buffers the divided signal (whose source impedance is $20\,\text{k}\Omega$, thus unsuitable for any cable or capacitive loading, even that of a high-Z 'scope probe). The LT1363 was chosen for its high speed ($>500\,\text{V}/\mu\text{s}$) combined with relatively low offset current (120 nA, typ), and ability to operate from $\pm15\,\text{V}$ supplies. A 10 V swing into the $100\,\Omega$ of the terminated output requires 100 mA, here provided by the wideband unity-gain BUF634 (160 MHz,

$2000\,\text{V}/\mu\text{s}$). The latter is muscular (output currents to $\pm250\,\text{mA}$), but inaccurate ($V_{\text{os}} = 100\,\text{mV}$, would you believe? It's a muscle-car, with a terrible driver!), so we close the loop around U_1 as shown.

To minimize dc error we match source resistances into U_1 with a 20k feedback resistor (R_3), shunted at signal frequencies by C_4 (crossover at 800 Hz); series resistor R_4 allows C_3 to take over at frequencies above 30 MHz, forming a direct signal path around U_1 for high-frequency stability. The final result is a 200:1 monitor output into a $50\,\Omega$ termination that can follow a full-swing 1 kV HV output step in 20 ns ($500\,\text{V}/\mu\text{s}$), and with a maximum untrimmed dc offset of 0.7 V (referred to the HV signal); the latter is caused mostly by U_1's worst-case offset current of 350 nA, and could be trimmed to zero with its NULL pins.

This is pretty good, and simple enough. But you can do better, if you want, by separating the low- and high-frequency paths, processing them with optimized amplifiers (for LF: slow, low-bias, accurate; for HF: just fast), and combining the signals into the output buffer. Such an arrangement is some $50\times$ less vulnerable to errors from stray capacitive coupling, and in addition has far better untrimmed dc accuracy.

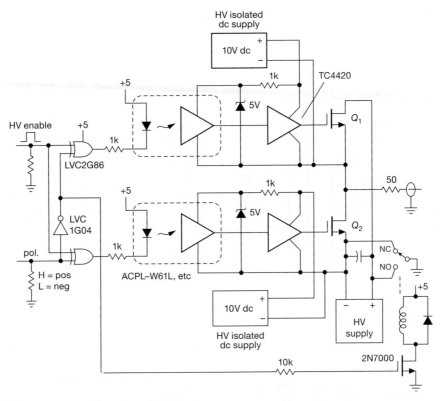

Figure 3x.113. By using a pair of high-voltage optocouplers, each with its own floating gate driver supply, the simple topology of Figure 3x.109 can be adapted to generate pulses of either polarity.

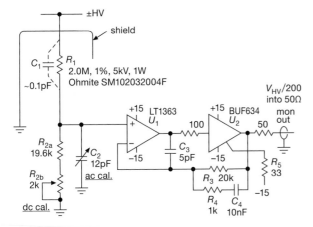

Figure 3x.114. High-voltage output monitor circuit with 30 MHz bandwidth and 10 ns pulse response, suitable for pulses of either polarity to ±1 kV.

3x.15.5 Three-switch bipolarity kilovolt pulser

Finally, with the addition of a third switch element we can make a high-voltage pulser with output pulses of either po-larity, switchable at full pulser speed. Figure 3x.115 shows the scheme. Here we've used the elegant ADuM6132 half-bridge driver,[111] which incorporates internal transformer-coupled high-side isolated dc supply and driver, with 15 V output swing and ±200 mA drive capability. For faster switching we've appended TC4420 gate drivers, powered by the same isolated dc.

The middle switch may confuse. The *n*-channel MOS-FET series pair Q_3Q_4 act as a bidirectional switch: when the gate is driven high, both switches are in conduction, bringing the output back to ground from its previous residence at either polarity. The body diodes (present in all power MOSFETs) are drawn explicitly here, so you can see that nothing bad happens when one of the series pair experiences reverse polarity: The body diode carries the reverse current if $I_{out}R_{ON}$ would be greater than a diode drop, otherwise the MOSFET is quite happy to conduct in the reverse direction. As with Figure 3x.113, we've omitted details such as the non-overlap delay circuits at the input, the

---

[111] It has also a matched non-isolated low-side driver, sitting idle in this application.

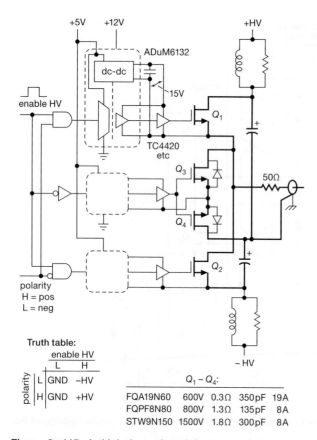

Figure 3x.115. A third electronic switch to ground (series pair $Q_3 Q_4$) permits rapid generation of pulses of either polarity (or from one rail to the other), to be contrasted with the slower polarity switching (via mechanical relay) of Figure 3x.113.

Truth table:

	enable HV	
	L	H
polarity L	GND	−HV
polarity H	GND	+HV

polarity
H = pos
L = neg

$Q_1 - Q_4$:

FQA19N60	600V	0.3Ω	350pF	19A
FQPF8N80	800V	1.3Ω	135pF	8A
STW9N150	1500V	1.8Ω	300pF	8A

Table 3x.5: High-Voltage Half-Bridge Drivers[a]

Part #	V_s^m HV (V)	I_{out}^t pos (A)	neg (A)	Delay t_{on} (ns)	t_{off} (ns)	UVLO[m] (V)	HI & LO inputs	Shutdown	Separate GNDs[v]
IR2113[p]	600	2.5	2.5	120	94	9.7	•	•	±5V
FAN7392	600	3.0	3.0	130	150	9.9	•	•	±5V
FAN7390M1	600	4.5	4.5	140	140	9.8	•	-	±7V
IR2101[i]	600	0.13	0.27	160	150	9.8	•	-	no
NCP5111	600	0.25	0.5	750	100[d]	9.9	-	-	no
FAN7382	600	0.35	0.65	170	200	10.0	•	-	no
IRS2108[e]	600	0.12	0.25	220	200[e]	9.8	•	-	±5V
IRS2109[e]	600	0.12	0.25	750	200[e]	9.8	-	•	±5V
IR2213	1200	2.0	2.5	280	225	11.7	•	•	±5V
with de-saturation detectors									
IR22141	1200	2.0[s]	3.0	440	440	11.4	•	•	±5V
self-oscillating									
IR21531	600	0.21	0.42	osc1[o]		9.9	-	•	-
NCP1392	600	0.5	1.0	osc2[o]		12.0	-	•	-
transformer-coupled									
ADuM6132	2500	0.2[u]	0.2	60	60	12.3	•	-	no

Notes: (a) all have isolated high-side driver; all accept "TTL" input logic levels, and most have Schmitt-trigger inputs. (d) 650ns deadtime. (e) IRS2108 and IRS2109 have 540ns deadtime; the '21084 and '21094 variants are deadtime programmable to 5µs. (i) IR2102 for inverted logic. (m) maximum. (o) osc1: 555-type oscillator, to 100kHz; osc2: to 480kHz. (p) IR encourages use of their IRS-series, rather than IRxxx parts. (s) 8V desaturation detector, causes shutdown after 10µs. (t) typical. (u) includes 15V 22mA floating supply, add capacitor and gate-driver IC. (v) for logic input and LOW driver output; i.e., separate V_{ss} and COM pins.

gate resistors and diodes, the HV fusing, the direct (low-Z) output, and provision for an external HV capacitor. For more details, inquire about our RIS-688.

A high-voltage switch with three states is unique. One of our colleagues (Gabriel Hosu) made a 1.5 kV version of this circuit for his experiment. He was using an EO (Electro-Optic) deflector, with a pinhole, to make a fast laser-light shutter. For the required deflection he needed to apply 1500 V, but the EO restricted him to 750 V, so he switched it with ±750 V. The manufacturer also warned against a shortened EO lifetime if it was kept at high voltage for long durations, so he had his program set the 3-state switch to zero except when his experiment momentarily needed fast shutter operation (and a mechanical shutter was used the rest of the time).

3x.16 MOSFET ON-*Resistance versus Temperature*

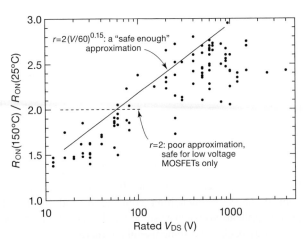

Figure 3x.117. Scatter plot of MOSFET ON-resistance ratio $r \equiv R_{ON}(150°C)/R_{ON}(25°C)$ versus rated voltage.

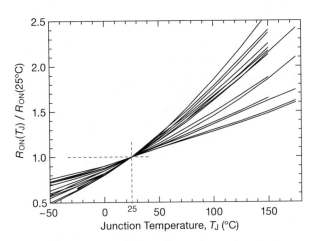

Figure 3x.116. Normalized MOSFET ON-resistance versus junction temperature for a selection of power MOSFETs with voltage ratings from 20 V to 600 V.

When designing circuits with power MOSFET switches, don't make the mistake of ignoring the rise of MOSFET ON-resistance with temperature. Although it's true that the R_{ON} value that's proclaimed on a datasheet's front page banner is indeed a "maximum" (worst-case) value, it's also true that it invariably is the $T_J = 25°C$ value. That's good enough for a rough estimate, and for comparison purposes among candidate MOSFETs, but it's not good enough for safe thermal design – power switches get *hot*, and you may find yourself running at junction temperatures of 70°C–100°C or more.

As always, the datasheet must be your ultimate guide. But there are some approximate trends and rules-of-thumb that can guide you.

Look first at Figure 3x.116, where we've plotted datasheet curves of R_{ON} versus T_J (normalized to R_{ON} at 25°C) for a dozen power MOSFETs (selected to span a wide range of rated voltages). You can see that the curves have a family resemblance; but you can see also that the increase of R_{ON} at some elevated temperature varies widely among parts (more on this, below), which is why you must consult the datasheet's plot of R_{ON} versus T_J. If you're unhappy with a plot, and want some analytical expression for R_{ON} at some junction temperature T_J, a straight-line ap-

proximation may be good enough; that is,

$$R_{ON(T_J)} \approx R_{ON(25)} + \frac{R_{ON(150)} - R_{ON(25)}}{R_{ON(150)}}(T_J - 25).$$

This could be useful in a thermal calculation where the junction temperature T_J depends in the usual way on heatsink thermal resistance and on power dissipation $I^2 R_{ON}$, but R_{ON} in the latter expression depends upon T_J itself. Then again, there's a certain uneasiness in depending on a closed-form solution in a situation like this. Better to take a conservative worst-case value from the datasheet and go with it.

High-voltage MOSFETs exhibit greater variation. It turns out there's some regularity in the range of R_{ON} variation in Figure 3x.116. Look at Figure 3x.117, a scatter plot in which the ON-resistance ratio $r \equiv R_{ON}(150°C)/R_{ON}(25°C)$ for 100 MOSFETs is plotted against their rated maximum V_{DS}.

Low-voltage parts exhibit small increase ($r \sim 1.5$), whereas high-voltage parts range up to $r \sim 2.5$ or more. Roughly speaking, you can safely assume an ON-resistance ratio $r = 2$ for MOSFETs rated at $V_{DS} \leq 60$ V, and $r = 3$ for higher voltage parts; but these figures are overly conservative for many MOSFETs, and it's always best to go back to the datasheet.

Note that we're talking about R_{ON} *ratios* here. When you consider that high-voltage MOSFETs have much higher R_{ON} to begin with, you see that they are doubly limited in their switched current capability.[112] For high current switching at voltages above 300 V or so, consider using IGBTs instead.

---

[112] Less so for pulsed operation at low duty cycle, where T_J is not elevated.

3x.17 Thyristors, IGBTs, and Wide-bandgap MOSFETs

		MOSFET	**IGBT**
Part #		IRFPG50	IRG4PH50S
V_{max}		1000 V	1200 V
I_{max}	dc	6.1 A	57 A
	pulse	24 A	114 A
R_{ON}(typ)	25°C	1.5 Ω	
	150°C	4 Ω	
V_{ON} (15A)	25°C	23 V	1.2 V
	150°C	60 V	1.2 V

The contemporary power MOSFET is a versatile transistor, for both power switching applications (e.g., dc power control, or dc–dc switching converters), and for linear power applications (such as audio amplifiers). But there are some drawbacks, and some useful alternatives.

3x.17.1 Insulated-gate bipolar transistor (IGBT)

The IGBT is an interesting MOSFET-bipolar hybrid, most simply described as an integrated complementary-Darlington-like (Sziklai) connection of an input MOSFET with a power bipolar transistor (Fig. 3.110). So it has the input characteristics of a MOSFET (zero dc gate current), combined with the output characteristics of a power bipolar transistor; note, however, that it cannot saturate to less than V_{BE}.

Nearly all available IGBTs come in the NMOS-*pnp* polarity only, thus behave as an *n*-type device.[113] They are generally high-voltage and high-power devices, available in discrete transistor power packages like the TO-220, TO-247, and in surface-mount packages like the D²PAK and SMD-220, with ratings to 1200 V and 100 A. For higher currents you can get them in larger rectangular power "modules," with the same high voltages and with current ratings to 1000 A or more.

IGBTs excel in the arena of high-voltage switching, because high-voltage MOSFETs suffer from greatly increased R_{ON}: an approximate rule-of-thumb for MOSFETs is that the R_{ON} increases as the square of the voltage rating.[114] For example, here's a comparison of two power products from International Rectifier:

These are comparably priced (about \$5) and packaged (TO-247), have similar input characteristics (2.8 nF and 3.6 nF input capacitance), and the resulting saturation voltages V_{ON} when switching 15 A are shown for the same full input drive of $V_{in} = +15$ V. The IGBT is the clear winner in this high voltage and current regime.[115] And, when compared with a power BJT, it shares the MOSFET advantage of high static input impedance (though still exhibiting the drastically reduced dynamic input impedance during switching, as we saw in §§3.5.1A and 3.5.4A). The BJT does have the advantage of lower saturation voltage (the IGBT's V_{ON} is at least V_{BE}), at the expense of high static driving current; the latter drawback is exacerbated at high currents, where BJT beta drops rapidly. A saturated BJT also suffers from slow recovery, owing to stored charge in the base region.

With the very high voltages and currents encountered with IGBTs, it is mandatory to include fault protection in the circuit design: an IGBT that is switching a 50 A load from a 1000 V supply will be destroyed in milliseconds if the load becomes short-circuited, owing to the 50 kW(!) power dissipation. The usual method is to shut off the drive if V_{CE} has not dropped to just a few volts after 5 μs or so of input drive.

3x.17.2 Thyristors

For the utmost in *really* high power switching (we're talking kiloamperes and kilovolts) the preferred devices are the *thyristor* family, which include the unidirectional "silicon controlled rectifiers" (SCRs) and the bidirectional "triacs." These 3-terminal devices behave somewhat differently from the transistors we've seen (BJTs, FETs, and IGBTs): once triggered into conduction by a small control current (a few mA) into their control electrode (the *gate*), they remain ON until external events bring the controlled current (from *anode* to *cathode*) to zero. They are used

---

[113] Currently the only *p*-type IGBTs we know of are the Toshiba GT20D200 series.

[114] You find exponents from 1.6 to 2.5 in the literature; the lower end of this range is likely to be more accurate.

[115] Where it excels also in maintaining high transconductance, compared with the MOSFET.

universally in house-current lamp dimmers, where they are switched ON for a fraction of each half-cycle of ac line voltage, thus varying the *conduction angle*.

Thyristors are available in ratings from 1 A to many thousands of amperes, and voltage ratings from 50 V to many kV. They come in small transistor packages, the usual transistor power packages, larger modules, and really scary "hockey puck" packages that are capable of switching megawatts. These are hefty devices; you can hurt yourself just by dropping one on your foot.

3x.17.3 Silicon carbide and gallium nitride MOSFETs

These recent devices are impressive alternatives to silicon MOSFETs, particularly suited to high-voltage power applications. We saw the silicon carbide (SiC) MOSFET back in §3x.15.2, where its significantly lower capacitance (compared with a conventional MOSFET of equivalent voltage and current ratings) let us make a faster-switching pulse generator.

Both silicon carbide and gallium nitride (GaN, a "III–V" semiconductor) are examples of wide-bandgap semiconductors, potentially permitting operation at higher temperatures than silicon devices (discussed at length in §3x.11.3). They also excel in breakdown voltage, thermal conductivity, carrier mobility, and low capacitance, leading to better figures of merit such as $R_{ON} \cdot$Area product (Fig. 3x.110). As we saw in the pulser circuit (Fig. 3x.111), however, their gate-drive requirements can be fussy: SiC MOSFETs often want -5 V (OFF), $+20$ V (ON), whereas currently available GaN devices have low thresholds (~ 1.5 V) and relatively low ON-state drive (~ 6 V).

Happily, semiconductor manufacturers are responding to the challenge, with integrated driver–MOSFET ICs. In the GaN arena a contemporary[116] example is the Navitas NV6113 (nMOS GaN, 10–24 V rail, logic-level input, 650 V 5 A, zero reverse recovery charge, 12 pF C_{oss}, dV/dt immunity of 200 V/ns, and 10 ns propagation time). You can also get standalone GaN drivers, for example the LMG1210 half-bridge driver, whose dual outputs can drive a pair of GaN MOSFETs (with the high-side FET operating up to 200 V), limiting each transistor's ON-state gate drive to 5 V. And, combining the features of the latter two ICs, you can get integrated GaN half-bridges with driver, for example TI's LMG5200; it includes a pair of 80 V 10 A

GaN FETs ($R_{ON}{=}15\,m\Omega$) with fast gate drivers that can operate up to 10 MHz switching frequency.

For SiC MOSFETs (which generally like a few volts of negative bias in the OFF state) you can use single and half-bridge drivers intended for silicon MOSFETs, with a level translator or opto-coupler to allow the output to go negative; see Figure 3x.111, where we played this trick on the UCC27538 drivers.[117] Other fast drivers intended for ordinary silicon MOSFETs include the IXDx609 and the opto-isolated ACPL-P346. But you can dispense with the level-shifting schemes by using a gate driver intended for SiC, for example ONsemi's NCP51705, which includes an on-chip charge pump to generate a -3 V to -8 V OFF-state drive level.[118] And you can get SiC half-bridges with integrated drivers, for example the Apex SA110 in its power package (400 V, 20 A).

---

[116] You can expect these pioneering devices to be rendered obsolete in a shockingly short time.

[117] TI offers other isolated drivers, for example the UCC5320 or UCC21732, intended for IGBTs or for Si or SiC MOSFETs.

[118] Although SiC MOSFETs permit 0 V OFF-state drive, a negative level speeds turn-off and enhances dV/dt immunity.

3x.18 Power Transistors for Linear Amplifiers

In Chapter 3 we showed typical audio power-amplifier output stage circuits (Figure 3.119), consisting of complementary positive and negative power transistors and a class-AB voltage-bias network. It's interesting, and instructive, to compare the various types of power transistors you might use.

Table 3x.6 shows candidate power transistors, chosen to be best in class for handling at least ±80 V output swings with up to 10 A of output current and 150 W of power dissipation;[119] for comparison we've added a few parts with different voltage or power ratings. In general the selection of p-channel parts is more limited than n-channel, so in most cases an available p-channel part dictates the n-channel choice, and thus the specifications for the complementary pair.

Topping the list we have the classic type for the job, a bipolar junction transistor (BJT). We've selected parts

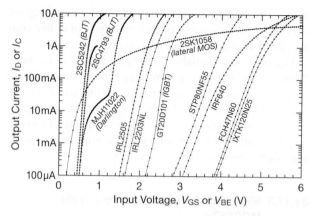

Figure 3x.118. Measured output current (I_D or I_C) versus drive voltage (V_{GS} or V_{BE}) for the n-type power semiconductors in Table 3x.6: BJT (single and Darlington), MOSFETs (vertical and lateral), and IGBTs. These were measured with V_{DS} or V_{CE} set to 5 V, with 1 ms pulsed measurements taken at 10 measurements/second to prevent thermal drifts. See §3x.18 for further discussion.

Table 3x.6: Representative Power-Amplifier Transistors[a]

Technology	Polarity N-type	Polarity P-type	V_{max} (V)	P_{diss}[b] (W)	C_{out} (pF)
Bipolar (BJT)	2SC5242	2SA1962[c]	230	150	360
– driver	2SC4793	2SA1837	230	20	30
– Darlington	MJH11022	MJH11021	250	150	600
MOSFET, lateral	2SK1058	2SJ162	160	100	400
VMOS – classic	IRF640	IRF9640	200	125	400
– high power	IXTK120N25	IXTK90P20	200	700	2210
– low-voltage	STP80NF55	STP80PF55	55	300	1130
– low-V_{th}	IRL2505	SUP90P06-09	55	200	975
– low-V_{th}, V_{DS}	IRL2203N	SUP75P03-07	30	180	1565
– high voltage	FCH47N60	IXTH10P60	600	300	430
– SOT-227 pkg	IXTN60N50	IXTN40P50	500	735	1325
IGBT, audio	GT20D101	GT20D201	250	180	450

(a) where the specifications differ for N- and P-type parts, the lower V_{max}, lower P_{max}, and higher C_{out} is shown. (b) at T_{case}=25°C. (c) for Fairchild 2SC5242 and 2SA1962 types, use their FJA4313 and FJA4213 parts, in a TO-3P package: 250V, 130W.

---

[119] It's important to note that the "maximum power dissipation" values you see on power transistor datasheets are based upon the very unrealistic assumption that the transistor case is held at 25°C; in real life you might be able to dissipate 1/3 of that – see §9.4.1 and the discussion in §3x.11.1.

first introduced by Toshiba,[120] which feature higher f_T (> 30 MHz) than typical audio power transistors (3 MHz).

First question: how do these parts compare in terms of output current versus input voltage? Figure 3x.118 shows the measured "transfer characteristic" curves for these parts, extending from 0.1 mA up to 10 A, with the drain (or collector) held at +5 V. (At the high current end you're talking 50 W dissipation, so you have to make brief pulsed measurements to prevent self-heating errors; for these particular measurements we used an Agilent B2902A "source-measure unit," set for 1 ms pulses every 100 ms.) As might be expected, the BJTs have the highest transconductance (i.e., slope), and the usual $V_{BE} \approx 0.6$ V operating characteristic (the Darlington's output transistor kicks in at ~ 1.2 V around 10 mA, when the drop across its internal 60 Ω base–emitter resistor reaches 0.6 V). The IRL-series "logic-level" (low threshold) vertical MOSFETs are next, with the other four (normal threshold) vertical MOSFETs off to the right. Compared with the vertical MOSFETs, the GT20D101 IGBT has a higher transconductance, and a decently low threshold. Finally, the 2SK1058 lateral MOSFET has the lowest threshold of all, but it loses steam at higher currents, where its transconductance falls markedly. Its redeeming virtue is its stabilizing negative tempco of drain current.

From the data of Figure 3x.118 it's easy to extract curves of transconductance – which we've plotted in Fig-

---

[120] They're also available from ON Semiconductor (Fairchild), who offers them in TO-3P cases (part numbers FJA4313 and FJA4213).

ure 3x.119. Bipolar junction transistors have much higher transconductance than FETs and IGBTs, especially at high currents. This leads to higher loop gain, $G = g_m R_L$, and the possibility of lower distortion. These transistors have gain specs as low as $\beta = 35$ at high currents, i.e., $I_B \approx 140\,\text{mA}$ at 5 A output, so they require driver transistors. The driver transistors shown on the second row of Table 3x.6 have $f_T > 70\,\text{MHz}$. One attractive possibility with separate driver and output transistors is that they can be used in the Sziklai configuration (Figure 3x.120). This allows improved quiescent bias-current temperature stability over the usual Darlington configuration, and allows using smaller current-sense resistors, in part because the driver transistor dissipates much less power, and in part because the thermal-compensation network can be made from the same parts, thermally coupled.

Darlington power transistors (3rd row of the table) dedicate about 10% of the die area to incorporate the driver on the same die. They also have built-in bias resistors to help in rapidly turning off the driver transistor. The MJH11000-series (ON Semiconductor) listed parts have $\beta > 1000$ (typical) at 5 A, though the beta drops to as little as 50 at lower currents like 30 mA (due to the built-in resistors); see the measured data of Figure 3x.121. The minimum required current drive reaches $I_B = 10\,\text{mA}$ at 10 A.

All of the bipolar types suffer from second breakdown (see Figure 3.95 in §3.5), which severely limits their maximum safe output current into low load resistances (high V_{CE}) or capacitive loads, and forces the use of multiple paralleled output transistors with current-equalizing emitter resistors. This is especially a problem for amplifiers with high supply voltages, and threatens the amplifier's reliability.

MOSFET and IGBT types do not suffer from second breakdown, making them attractive for high-voltage high-power linear amplifiers. However ordinary "vertical" power MOSFETs suffer from a positive drain-current tempco (at fixed gate voltage), see Figure 3.115 in §3.6.3, which makes them hard to set up for low class-AB bias currents and requires awkward high-value current-equalizing resistors when they're used in parallel.[121]

This can be addressed with a "V_{BE} multiplier" bias arrangement, in which several BJTs are used (see, e.g., Fig. 3x.120A) to generate a gate bias with negative tempco adequate to compensate the positive tempco of the MOS-FETs. A typical V_{GS} tempco is $-4\,\text{mV/°C}$ to $-7\,\text{mV/°C}$ (e.g., for the 125 W IRF640, estimated from the datasheet's

---

[121] However, high-density vertical MOSFETs *are* susceptible to a form of hotspot formation, the "Spirito effect," which limits their safe operating area at high voltage and moderate current; see the discussion in §3x.5.2.

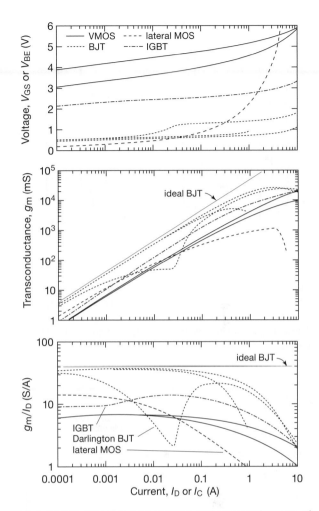

Figure 3x.119. Transfer characteristics and transconductance of the "200 W" power transistors from Table 3x.6, at V_D or V_C of 5 V, from measured I_D (or I_C) versus V_{GS} (or V_{BE}) data plotted in Figure 3x.118.

"Gate Threshold Voltage and Transfer Characteristics" plots), at a modest quiescent setpoint current of 50 mA to 100 mA. Combining this with the *p*-channel MOSFET's smaller $-2.5\,\text{mV/°C}$ to $-4\,\text{mV/°C}$ tempco, the overall positive quiescent current tempco of the push–pull pair can be crudely compensated with the V_{BE} multiplier. Although imperfect, this is in fact a widely used technique.

Hitachi nicely solved this problem by offering lateral MOSFETs, intended especially for audio amplifiers. Their parts also feature attractively low gate-threshold voltages, the lowest of all contenders shown in Figure 3x.119. The low V_{th}, along with their negative drain-current tempco (Figure 3.118), simplifies setting low class-AB bias currents, and allow us to reduce wasteful quiescent power heating. Lateral MOSFETs have poorer maximum current

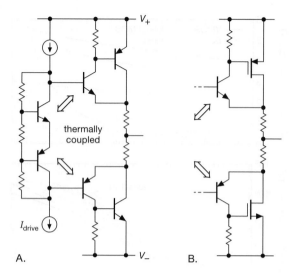

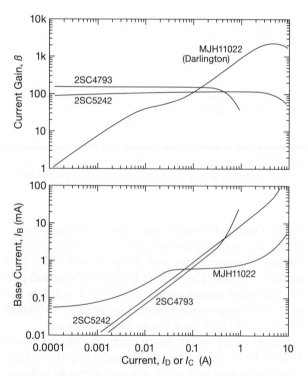

Figure 3x.120. Push–pull output stage with (A) complementary Sziklai-connected BJTs, and (B) complementary BJT + MOSFETs. Thermal stability is promoted by their single-V_{BE} drop (compared with a Darlington's $2V_{BE}$ drop), thermally coupled to the similar-size drivers.

Figure 3x.121. Measured base current, and current gain (β), versus collector current, for the bipolar transistors in Table 3x.6. $V_{CE} = 5\,\text{V}$.

and power-handling capability than ordinary power MOS-FETs, requiring us to use more parts in parallel. However, they can be paralleled without using any current-equalizing source resistors. Some high-power lateral MOS-FETs are available, made by Exicon (and sold by their Profusion Audio Semiconductor outlet), for example the ECW20N20 and ECW20P20 (200 V, 250 W) in TO-264 packages (change ECW prefix to ECF for TO-3). Ordinary vertical MOSFETs do have some advantages for use in linear power amplifiers: they have much higher transconductance than lateral MOSFETs, especially at currents above 2 A, they can handle higher currents (± 20 A), higher voltages (± 300 V) and higher power levels (an honest 300 W). They have multiple manufacturing sources, and they're much less expensive.

For our first ordinary vertical-structure power MOS-FETs, we've picked an old classic, the 200-volt IRF640 family. MOSFETs designed in the 1980s had a low-density groove geometry, so they generally require a larger die area for a given $R_{DS(ON)}$ spec than their modern replacements.[122] These are widely available, inexpensive, and come in convenient TO-220 packages. However they have limited power handling capability. For example, the IXYS parts listed next in the table are five times better in that regard. They're also five times more expensive.

Although MOSFETs have zero dc input current, they have substantial input capacitance, e.g., 400 pF to 2200 pF for the two types mentioned above. This means that for use at high audio frequencies they require high drive-current circuits, similar to fast-slewing BJT circuits. Let's do a quick calculation using the IXTK120N25. On each ac cycle it's important to get the trailing MOSFET turned OFF before the leading one is fully turned ON, to keep current shoot-through to an acceptable level. For the '120N25 running at 4 A we measured an attractively high $g_m = 10\,\text{S}$, or about 4A per 400 mV. The combined C_{iss} for the complementary parts is 20 nF. Let's do the calculation for $f = 200\,\text{kHz}$. If we assume a 1.6 V total gate-voltage change for both devices, transitioning in 1 μs, then $i = CdV/dt = 32\,\text{mA}$ peak. That's as high or higher than you might find in most BJT amplifier designs.

Toshiba's audio IGBT transistors are last on the table. One interesting thought: IGBTs come in a Sziklai configuration, with an n-channel MOSFET combined with a

---

[122] Generally speaking, an "improved" smaller die is usually a good feature, because of lower capacitance, and lower threshold voltages. Ironically, however, larger die area is actually a *very good* feature for circuits with lots of power dissipation, owing to lower thermal resistance $R_{\theta JC}$ thus permitting greater power dissipation. But modern '640 parts seem not to have fallen prey to this effect.

pnp BJT. Note their high transconductance in the 0.1 A to 1 A region (Figure 3x.119). Were they the best or the worst of both worlds? They enjoyed only limited acceptance and were finally discontinued by Toshiba. You can find *n*-channel IGBT candidates from other manufacturers, but without *p*-channel parts you're out of luck.

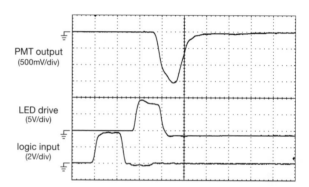

Figure 3x.122. Fast LED driver, for photodetector characterization. The ISL55110 dual gate driver can sink and source 7 A, with 1.5 ns risetime into 100 pF.

3x.19 Generating Fast High-Current LED Pulses

We've encountered some nice LED-driving challenges over the years. Here we collect some interesting solutions.

3x.19.1 10 ns pulser

For our optical SETI project we needed to generate short test flashes, ten nanoseconds or less in duration. We found that you can do this, with off-the-shelf LEDs, but it's important to drive it hard in the forward direction, then abruptly reverse the drive polarity to quench the LED's emission.

Figure 3x.122 shows a charmingly simple solution. We used a wicked-fast dual gate driver IC (3.5 A peak current per section, rise/fall time of 1.5 ns into 100 pF), and we coupled it to the LED with the $R_S C_S$ series network whose time constant is comparable to the pulse width. So the capacitor partially charges up during the ON portion, which creates a reverse bias when switched OFF. Figure 3x.123 shows multipulse average waveforms when we ran this thing in a dark box (with the LED set back 30 cm from the PMT, and attenuated with an ×1024 ND filter). The pulser is socking the LED, hard, with 10 V. Works great!

3x.19.2 High-power pulser

For this project we needed a powerful short intense flash of blue light for calcium fluorescence whole-brain imaging of freely-swimming zebra fish.[123] We chose an Osram 20 A high-power theater-projection LED (similar to the Luminus device in the lower lefthand corner of Fig. 9x.97). Its "surge current" rating is 40 A,[124] but to get the intensity

Figure 3x.123. Measured waveforms for the circuit of Fig. 3x.122, from a Rohm SLA580BCT blue LED, detected by a Hamamatsu R7400 PMT (0.78 ns risetime) terminated in 50 Ω. Horizontal: 10 ns/div.

we wanted, we had to drive it, during the 45 μs pulse, at 200 A![125]

We started with a simple MOSFET switch plus resistor (Fig. 3x.124A), but the rising current waveform was disappointingly sluggish (probably owing to distributed induc-

---

[123] D.H. Kim et al., "Pan-neuronal calcium imaging with cellular resolution in freely swimming zebrafish," Nature Methods, 2017 **14** 11, 1107–14 (2017).

[124] Some manufacturers recommend limiting the current's slew-rate or risetime; for example, the Luminus CBT-120's datasheet warns "In pulsed operation, rise time from 10–90% of forward current should be larger than 0.5 μs."

[125] That's serious mistreatment, and in practice we had a failure every few months; but you have to buy these puppies in significant minimum quantities, so we always had replacements to spare. Of course, we can't claim we weren't warned: on the Luminus datasheet, for example, we are told "Product lifetime data is specified at recommended forward drive currents. *Sustained operation at or beyond absolute maximum currents will result in a reduction of device lifetime compared to recommended forward drive currents.*" (emphasis added)

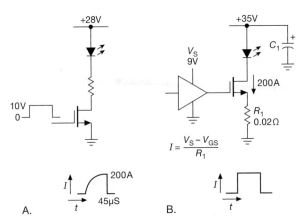

Figure 3x.124. Driving a high-current LED with a current-limited switch (A) produced a slow rise of current; using a current sink from a higher-voltage rail (B) solved the problem.

tance that reduced the current slew-rate). So we decided to use a current-switching version with a higher voltage rail (circuit B), which is analogous to the standard BJT current source (but afflicted with the less predictable V_{GS}, compared with the BJT's V_{BE}). We mounted the 460 nm power-LED on a water-cooled heatsink, and coupled it to a complex optical system, including a TI digital micromirror array (to track and illuminate the swimming fish's brain). The LED has an intrinsic forward voltage of about 3.5 V, plus an internal resistance of about 25 mΩ; so at currents of 200 A it may need as much as 10 V drive. For C_1 we chose a large 22,000 μF electrolytic, with an ESR of $\leq$18 mΩ; during the planned 45μs 200 A pulse the voltage droop will be $dV = I dt / C = -0.4$ V. We mounted the 515 W MOSFET (in a hefty TO-247 package) on a small fan-cooled heatsink of the sort used for desktop microprocessors.

Figure 3x.125 shows the complete LED driver circuit. Its simplicity conceals some important design choices; let's take a closer look.

A. Wiring

We closely wired one lead of a low-inductance 20 mΩ, 5 W axial resistor (Ohmite 15FR020) to the MOSFET's source pin. At the resistor's other end we made a close Kelvin (4-wire) connection to the gate driver, with a second heavy wire to C_1's negative screw lug. We chose a MOSFET driver with CMOS rail-to-rail outputs and made tight connections to its supply capacitor C_2. When you have hundreds of amps banging around in a circuit like this, with the likelihood of large $V = L dI/dt$ spikes, it's a good idea to isolate the logic-level input circuit, thus the fast 20 V optical isolator to couple the 45 μs input signal.

B. Gate voltage

As we've often complained (see, for example, §3.1.5), FETs (both JFET and MOSFET) have loosely specified gate turn-on voltages. And the power MOSFET in this circuit is no exception, with its $V_{GS(TH)}$ specified as 2 V min, 4 V max. For a laboratory circuit like this, it's OK to trim the gate drive (LM317 trimmer) to set the current you want. But what about the change of required gate voltage as the temperature changes (and it *will!* – see below)?

As luck would have it, we're in a good place, here: MOSFETs at high currents exhibit a favorable convergence of transfer curves, seen here in Figure 3x.126 (and, for example, in Fig. 3.115 in the main volume). MOSFET "Transfer Characteristics" (I_D versus V_{GS} curves) are usually given for three temperatures, $-55°$, $+25°$, and $+175°$C. At extremely high currents, far above common linear operation, the three temperature curves merge, and I_D versus V_{GS} has a very low temperature coefficient. For our chosen Q_1 (an HUF75652), that happens at about 200 A(!), where a temperature change of 230K requires a gate voltage change of just 0.2 V. For that current the nominal gate voltage is 5.2 V (with expected variation of $\sim$1 V over manufacturing batches). We chose $R_1 = 20$ mΩ, for a 4 V drop at 200 A; so a 9.2 V pulse at the MOSFET's gate will generate a sink current of 200 A.[126] That current, combined with a high positive rail voltage (30 V here) causes the LED to emit promptly. And, owing to the low tempco of gate voltage, the LED current will remain substantially constant as the MOSFET heats up.

C. Power dissipation

According to the datasheet, the HUF75652 power MOSFET can handle up to 1000 A for short pulses ($\tau < 50 \mu$s) at low duty cycle – but that's for (saturated) switching applications, where there's very little voltage across the MOSFET. Here we're operating it in linear mode, and we've chosen the 30 V rail voltage to accommodate all the voltage drops: there are the resistive drops (about 5 V), but then there are *inductive* drops. To get a sense of the latter, consider a wiring inductance of 100 nH and a current slew rate of 100A/μs – that's an $L dI/dt$ of 10 volts!

That's why we chose a relatively high power supply voltage (30 V) – which is exactly the reason that the LED is brought to full current rapidly – but as a consequence the MOSFET might have $\sim$15 V across it during the 200 A conduction. That 3 kW, yowie!

Sounds like a serious problem... but, happily, we're

---

[126] The adjustable gate voltage lets us set the amplitude of the current-pulse.

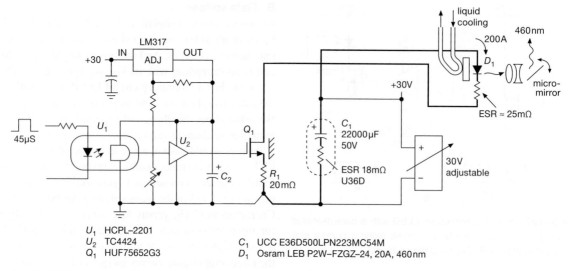

Figure 3x.125. Driving a power LED with 200 A pulses. The power MOSFET is operated in linear mode to create a fast and stable 200 A current sink. Heavy lines indicate the pulsed high-current path; for reduced inductance use a twisted pair to the remote LED.

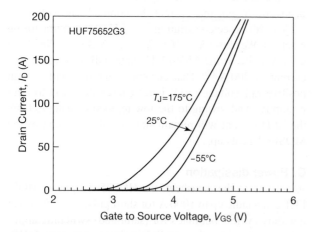

Figure 3x.126. MOSFET transfer characteristic plots show a tempco minimum at high currents.

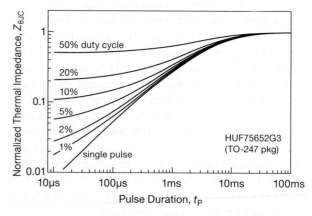

Figure 3x.127. Normalized Transient Thermal Impedance plots for a HUF75652G3 MOSFET.

rescued by the "thermal mass" of the MOSFET and its heat-spreading substrate: Figure 3x.127 shows the effective multiplier, for short pulses, of the part's steady-state (i.e., dc) thermal impedance (see also §§3x.13 and 9x.25.8). For 45 μs single pulses the normalizing factor is 0.03, and for a train of pulses with a 2% duty cycle (i.e., ~ 500 Hz) the factor is 0.05. Applying the latter to the datasheet's $R_{\Theta JC} = 0.29°$C/W, we have an effective thermal resistance ("junction" to case) of 0.0145°C/W, and thus a temperature rise of 44°C, which is well within safe limits.

Single pulses are absorbed entirely by the junction and

frame's thermal mass, but multiple pulses are another matter.

Figure 3x.127's graph shows the thermal mass of our MOSFET and supporting mass for pulses up to a few milliseconds in length, but for a 2% duty-cycle (i.e., a 500 Hz frame rate), the heat moves from the MOSFET to the heatsink, which should be able to dissipate the 60 W average power. For more information and free prototyping PCBs, inquire about RIS–796.

For another example of fast MOSFET gate-voltage current-source driving, see our piezo-actuator charge-control pulsed-step circuit, §3x.22.

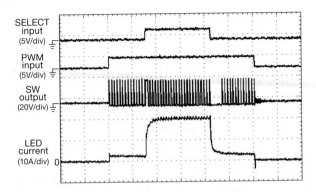

Figure 3x.128. The LT3743 steps rapidly between a pair of resistor-programmed currents: 0 to 2 A to 20 A. Horizontal: 20 μs/div. Taken from the datasheet for LT3743, and used with permission. All images, icons, and marks in said datasheet are owned by Analog Devices, Inc. ("ADI"), © 2019. All Rights Reserved. These images, icons, and marks are reproduced with permission by ADI. No unauthorized reproduction, distribution, or usage is permitted without ADI's written consent.

3x.19.3 Integrated LED Drivers

Semiconductor manufacturers recognize the market for high-intensity LED drivers, and have responded enthusiastically. For example, high-power projectors, using three colors of LEDs, require tightly-controlled switchmode *current-source* drivers that allow fast ON/OFF rates for PWM dimming. We've used Linear Technology's LT3743 buck converter to drive external half-bridge MOSFETs and an inductor, delivering up to 30 A at up to 36 V, with a 3000:1 PWM dimming ratio. With large externally-switched load capacitors, it's capable of changing regulated LED current levels within several microseconds, providing accurate, high-speed PWM dimming between two current levels (Fig. 3x.128).

The LT3743 is a serious 28-pin part in a power package. A simpler approach is exemplified by Maxim's MAX16819, an inexpensive ($1.34) 6-pin part intended for illumination applications. Driving a single MOSFET, it handles 30 A LEDs up to 28 V. It delivers a 5000:1 dimming range when operated at 20 kHz PWM. And Maxim's MAX16818 is a 28-pin part that's the basis for Luminus DK-136M, a 36 A driver with heatsink, for PhlatLight's 20 W LEDs.

All of these ICs are buck converters, with high-side current sensing to create a constant inductor current into a grounded load. Dimming in multi-color LED projectors requires a constant LED current when ON, independent of duty cycle. Because the LED drive is constant current, you can simply connect a MOSFET switch across the LED string to create PWM dimming, as suggested on the MAX16818 datasheet's front page. This method should work well with any buck converter that can maintain an unchanging inductor current despite output-voltage changes. For such a configuration it is best to use high-frequency switching, with a fast control loop.

3x.20 Precision 1.5 kV 1 μs Ramp

Here's a circuit that can generate a negative-going ramp, with up to 2 kV/μs slew rate, precise to within 1%. Figure 3x.129 shows the circuit, which was used to read-out Rydberg-atom energy levels. We'll relate the design in short-story format.

This originated in our instrument design lab, where requests for instruments that perform somewhat beyond the state of the art are not uncommon. In this case the request was for a precision 10 kV ramp in 0.25 μs (a precise 40kV/μs ramp – is that even possible?); while trying to do the impossible, however, we learned that the region of interest was 1.5 kV and below, and that a negative-going ramp would suffice.[127] More important, they wanted the thing right away!

Thus a quick-and-dirty temporary[128] version was the goal. This sounded like a MOSFET job, so we chose an STP4N150 (1500 V, 4 A, TO-220 or TO-247 power packages, about $6). The capacitance-load budget for coax and electrode was 180 pF, and the FET's C_{oss} of 120 pF was modestly in the same region. A quick calculation showed that $I = C\,dV/dt = 300\,pF \cdot 1.5\,kV/0.5\,\mu s = 900$ mA, another modest number (the 10 kV version would have required 6 A[129]). At this current the STP4N150's transconductance

is 2.3 S, and, with its C_{iss} of 1300 pF, its f_T would be 280 MHz (eq'n 3x.6), more than fast enough for our task.

So, the plan was to use Q_1 as a current sink, programmed with a gate voltage, with 7.5 Ω from source to ground, thus 135 mA/V. The usual scheme is to close the loop to accurately control the source voltage (thus the current). But that won't work here, because it's not constant *current* we want, it's constant *slew rate dV/dt*: the fly in the ointment is the nonlinear drain capacitance C_{oss}, which changes with drain voltage (see, for example, Fig. 3x.41).

What to do? The solution was to use a small high-voltage capacitor to generate a feedback *current*, which gets combined in a current (rather than voltage) summing junction. We had some 3 pF 5 kV capacitors in our inventory (thus 9 mA feedback current for the fastest ramps), and we used an operational transconductance amplifier (OTA) to program the ramp.

The Burr–Brown/TI OPA860 looked good with its 80 MHz bandwidth, 10 mA maximum OTA-node operating range, and 60 mA buffered node output. We added a BUF634 buffer, with its 250 mA 30 MHz comfort zone, to drive the MOSFET (are you getting the picture, the readily available resources were all much faster than needed; but they were *readily available!*). This worked very well when programmed at the slower ramp speeds, but there was too much loop gain at the fastest speeds, and the output oscillated at radiofrequencies as it ramped (making for an amazing scope trace!). We fixed this misbehavior by attenuating the feedback current with a capacitive divider (C_2) to ground. We used the usual 50 Ω back termination to the coax to kill any reflections.

In the instrument version put into "temporary" use, Figure 3x.129, attenuating capacitor C_2 has to be changed for different operating speed ranges (hey, we said this was a quick-and-dirty design).[130] In our research institute[131] the motto is "keep science moving forward as fast as possible." For more information about this project, ask about RIS-633.

---

[127] Details: what they actually wanted was a negative-going ramp, starting at about 5 kV, and going down to about 3.5 kV. Given the brief nature of the ramp, we could capacitively couple to a ground-based circuit. It looked well suited to a MOSFET-based current sink, with the MOSFET's control circuitry conveniently at ground.

[128] We've learned from experience that "temporary" versions have a way of becoming permanent.

[129] Check out the 5D21 vacuum tube, rated 20 kV, 15 A. This was used with pulse-forming transformers (see Appendix H.3 and Fig. H.16) to drive magnetron radar transmitters. Four tubes were used to make 500 kW pulses in the WWII submarine "SV" S-band radar; see maritime.org/tech/radiocat/sv.htm. The 5D21 was also used in the earlier "SJ" radar, which was decisive in allied submarine patrols – see, for example, the fascinating story of the USS Tang at en.wikipedia.org/wiki/USS_Tang_(SS-306), where the 5D21 is mentioned in the 2nd patrol report. It credited its radar to its success in sinking 33 ships totaling 116,454 tons; as fate would have it, the Tang succumbed to an ignominious end – sunk by its own errant torpedo that circled back. The 5D21 has found a second life, evidently: check out www.youtube.com/watch?v=re_ITPnM2TU

[130] The next version has an AD734 multiplier between U_1 and U_2, wired as a divider, to reduce the loop gain in proportion to the ramp slew-rate programming voltage.

[131] The Rowland Institute at Harvard, founded by Edwin Land.

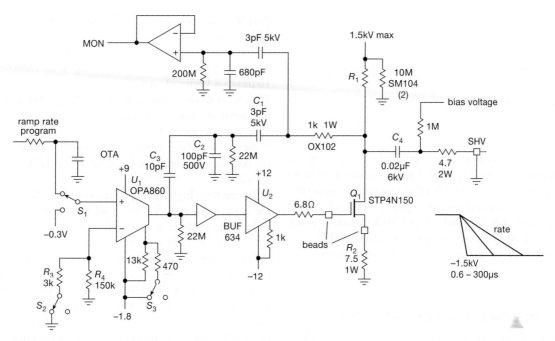

Figure 3x.129. Programmable ramp generator: 1.5 kV, 7.5–2000 V/μs. Switches S_1–S_3 are a 74VHC4053; S_1 is the ramp start pulse, and S_2 and S_3 are range switches.

263

3x.21 Fast Shutoff of High-Energy Magnetic Field

For atom- or ion-trapping experiments it's necessary to generate and then quickly quench magnetic fields. Here we describe a circuit (Fig. 3x.130) that can quickly shut off a 50 A to 100 A current in a set of magnetic-field coils. It uses a high-current contactor and a 1.2 kV IGBT module, which transfers the coil's stored magnetic-field energy ($LI^2/2$) to a 10 μF 1.5 kV oil capacitor with a resonant quarter-sine cycle.

3x.21.1 Helmholtz coils, rapid field shutoff

Helmholtz coils (see Fig. 1x.21) are a simple configuration of two coils that creates a magnetic field that is quite uniform over some interior volume; they are used for many purposes, including the trapping of ions. They can be quite large, and require high currents to create even modest fields. The large size also means these coils have a fairly high inductance. For our experiment the working currents were 50 A with a 5 mH coil, and 100 A with a 1.25mH coil (simply the same coils, in series or in parallel). The magnetic stored energy was $E = LI^2/2$, or 12.5 Joules in either configuration (as it had to be, since the magnetic field was the same in either configuration).

When the magnetic field is used to trap atoms or ions, it's often necessary to open the magnetic field rapidly, by turning off the current. But inductors hate to have their current change abruptly ($V = LdI/dt$), so the act of quenching the field produces a large "flyback" voltage across the coil's terminals, attempting to maintain the current (see for example Fig. 1.83); this would destroy the contacts of a relay or mechanical switch. But it's necessary to let the flyback voltage go as high as possible to get the current down to zero rapidly.

For example, if we let the voltage across the just-opened switch go quickly to 1 kV, the current would ramp down at $dI/dt = V/L$, so it would ramp to zero current in a time $t = IL/V = 125\,\mu$s. To make this work, though, we have to clamp the flyback voltage at 1 kV, otherwise it will rise as far as necessary to maintain the initial current; this usually results in destruction of some component, from arcing or avalanche. And, of course, the 1 kV clamp has to handle

the initial current (100 A) and its downward ramp to zero, taking place over the 125 μs.

In another fast magnet-quench experiment (Fig. 9x.117) we dumped an 875 A current, flowing in an inductance of 10 μH (3.8 Joules of energy), into a stack of eight 5KP26A TVS zener diodes, with each one handling 110A and breaking down at about 40 V. But for our Helmholtz problem here the energy is 3.5× higher and the voltage is 28× higher, so we chose a different approach, namely to transfer coil's energy into a substantial 10 μF 1.5 kV oil-filled capacitor.

To see how high the capacitor voltage will rise, equate the capacitor's energy $CV^2/2$ with the inductor's energy $LI^2/2$ (the transfer is essentially lossless). The result is $V = I\sqrt{L/C} = 1120$ volts (with either coil configuration). When the connection is opened, the coil's current will start falling with a quarter cosine waveform, and the capacitor's voltage will rise with a quarter sine waveform,. The resonant frequency of the LC pair is $f = 1/2\pi\sqrt{LC}$, or 712 Hz for a 5 mH coil (series connection of the Helmholtz coils) or 1424 Hz for a 1.25 mH coil (parallel connection); so the calculated quarter-cycle ring-down times are 175 μs and 350 μs. The switch (or equivalent) must be opened at the time the current falls to zero, when the coil's energy has collapsed to zero (and its stored magnetic energy resides entirely in the capacitor).

3x.21.2 High voltage, high current switches

The final circuit is shown in Figure 3x.130, and features an IGBT (insulated-gate bipolar transistor, see §§3.5.7 and 3x.17). The 1.12 kV flyback value was compatible with some 1.2 kV 300 A dual-IGBT half-bridge switch modules (Mitsubishi CM300DY-24H, \$450 at retail, a real monster – see Fig. 3x.131), which fortunately we had in stock from a lucky eBay purchase. The LC resonant activity must stop when the current reaches zero, otherwise it will simply oscillate, with the field reversing sinusoidally, so we placed a diode in series with the capacitor. This has to be one substantial diode – it must have a 100 A 1.2 kV rating; but we were only using one of the IGBTs in the module, and each one included a discrete reverse freewheel diode, so we used the second spare diode for the job. Diodes have "stored charge," though, so after the diode's anode went flying to zero volts, it experienced a severe snap-off spike (see §9x.6) despite being a "Super-Fast Recovery" type. To damp this we added a 0.1 μF 1.5 kV capacitor and 220 Ω snubber, creating a modest tamed oscillation. Things were coming together nicely.

But there was one further fly in the ointment: although

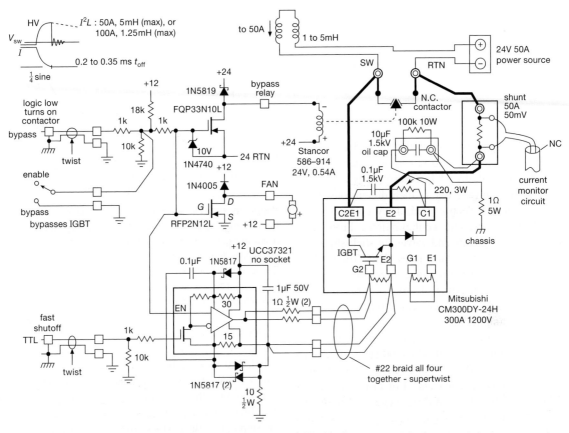

Figure 3x.130. This circuit turns off a 100 A magnet in a fraction of a millisecond.

1.2 kV IGBTs have respectably low ON voltages at high currents (lower than 1.2 kV MOSFETs, see §3x.17.1), the IGBT's ON voltage drop at 100 A was expected to be about 2 V, so it would dissipate 200 W, and some serious cooling would be required. We knew, however, that most of the time the magnetic field would be continuously ON, with the programmed rapid shutoff being an occasional event. So we added a *contactor* to carry the high current most of the time. Contactors are like huge relays, and they come in normally-open and normally-closed versions. We chose a normally-closed contactor (Stancor/White-Rodgers 586-914), thus saving the 10 W dissipation required to keep a normally-open contactor closed. In use, the computer controlling the experiment first turns ON the IGBT, then opens the contactor, and when the critical moment comes it turns OFF the IGBT and prepares to take measurements after the field has collapsed to zero.

For more information on this project, ask about RIS-614.

Figure 3x.131. For real power jobs you need a substantial module, able to dissipate a kilowatt or more. This dual IGBT module has a forward voltage of 2.5 V at its rated 300 A; that's 750 W dissipation per section!

265

3x.22 Precision Charge-dispensing Piezo Positioner

Piezoelectric actuators exploit the piezoelectic effect (production of an electric field in certain materials in response to mechanical stress) in reverse: they produce a strain (fractional change in one of their dimensions) in response to an applied electric field. They come in many different shapes, sizes, and configurations. From lens-focusing motors, to pumps and valves, to stick-slip walkers, and so on, they fill many roles. NEC/Tokin offers this description: "Multilayer piezoelectric actuators are ceramic elements for converting electrical energy into mechanical energy, such as displacement or force, by utilizing the piezoelectric longitudinal effect."

Tokin's high-sensitivity longitudinal actuator sticks have a maximum operating range of 500 to 900 ppm (fractional displacement) of their length. For example, their 20 mm-long AE0505D16 has a displacement of 11.6 μm at 100 V and 17.4 μm at 150 V. It has a capacitance of 1.4 μF, and an unloaded mechanical resonant frequency of 69 kHz (due to its stiffness). Piezo actuators are powerful, and can exert a substantial force against a resisting object. For our experiments we used a similar Thorlabs AE0505D18 (made by Tokin), with 1.6 μF capacitance and 15 μm motion at 100 V.

Because of nonlinearities, a piezoelectric actuator's step size depends somewhat nonlinearly on applied voltage, but rather accurately on delivered *charge*.[132] That is, charge-control of piezo elements provides linear and hysteresis-free displacement, as opposed to the conventional voltage-control approach. Figure 3x.132 shows the comparison graphically, where we've measured the deviation from linearity of the voltage steps across a 10 μm piezo actuator (of 1.5 μF capacitance) when driven with pulses of constant charge.

We needed to generate equally spaced actuator steps, to rapidly move the focal plane of a high-speed camera (200 μs per image), obtaining images of a white blood cell ("WBC") at ten depth levels as it streamed past. We first tried a high-voltage amplifier made with an Apex PA-15

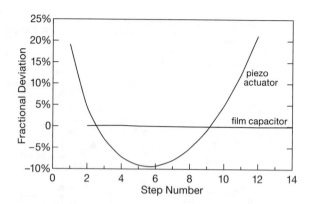

Figure 3x.132. Deviation of measured piezo-voltage step size from average, when driven with constant-charge steps. Data from a similar experiment, but substituting a 1.5 μF film capacitor, shows identical voltage steps and validates the measurement technique. Because piezo step size is known to be accurately set by the delivered *charge*, the curve represents the step-size error that would be incurred if driven with steps of constant voltage.

power IC, wired to handle a 150 V range with 150 mA maximum output currents.[133] But it was too slow: when commanded to make a 10 V step driving the 1.5 μF piezo, it took $t = C\Delta V / I = 100\,\mu$s to make the step. With this sluggish response there would be image smearing for 50% of the 200 μs imaging time; we wanted to do better, making the step in less than 10% of the imaging time.

The solution was a circuit to dispense precision packets of charge to the piezo-positioning stick. See the block diagram in Figure 3x.133, which shows a pulsed 2.5 A MOSFET current source, driven by a timer to precisely control the delivered *charge* per pulse. A second MOSFET resets the piezo element after a series of steps. This charge stepper circuit took 10 μs to move the focal plane (just 5% of the camera's 200 μs exposure), the rest of the exposure being motion free.

3x.22.1 Fast MOSFET pulsed charge dispenser

Figure 3x.134 shows the full circuit of the metered charge dispenser circuit. It consists of a fast-response MOSFET current source Q_1, switched with U_1, an IR2113 600 V MOSFET half-bridge driver (i.e., both high-side and low-side gate drivers, see Fig. 3x.108), and a 74HC221 adjustable monostable timer to make pulses, typically 2 μs to 10 μs long. The timer's initial trigger occurs at the beginning of a camera frame, coincident with the arrival of a new WBC. The IR2113 responds with roughly 100 ns delays,

---

[132] See, for example, A.J. Fleming and K.K. Leang, "Charge drives for scanning probe microscope positioning stages," *Ultramicroscopy*, **108**, 1551–1557 (2008).

[133] For more about this project ask about RIS-683-3.

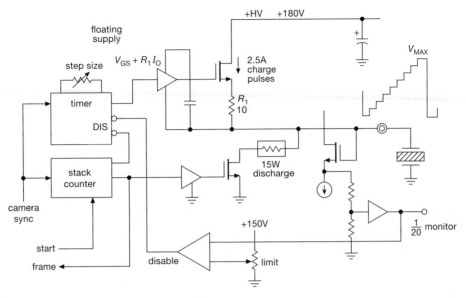

Figure 3x.133. Block diagram of charge-dispensing piezo driver.

and can provide 2.5 A gate drive with 25 ns risetime, capable of switching Q_1 in 15 ns. The MOSFET contributes a small delay, but all of these delays remain relatively stable throughout the actuator's step-by-step motion.

Circuit details. The output current is $I_o = (V_{DD} - V_{GS})/R_1$, with adjustable MOSFET gate drive V_{DD} via trimpot R_3, to trim the current-pulse and charge-packet size. MOSFET Q_3, thermally coupled to Q_1, tracks the latter's V_{GS}, and is included in the V_{DD} regulator voltage. Although the two V_{GS} values differ, we can write, approximately, $I_o = V_{FB}(1 + R_3/R_2)/R_1$, where V_{FB} is the regulator U_2's feedback reference voltage of 1.242 V. Cascode transistor Q_2 clamps Q_1's drain voltage to a few volts, to limit power dissipation and heating, and to remove any (already very low) influence from changing V_{DS}. In this fashion the gate drive voltage V_{DD}, and hence the R_1 current, remain reliably constant at its setpoint. In our application we trimmed R_3 for an output current of 1.25 A or 2.5 A.

Both Q_2's gate drive and the programming voltage V_{DD} are floating, and referenced to the piezo-actuator's standing voltage. The entire assembly is isolated from ground, with low leakage currents,[134] to minimize discharge of the piezo element. The high-voltage power supply may range from

160 V to 200 V, and it must have a large storage capacitor to provide the charge used for the piezo stepping. The 12 V isolated supply is a small dc–dc converter, e.g., muRata's NKE series.

The piezo-stick's length responds rapidly to its new charge level (without ringing[135]), and the bulk of the $200\,\mu s$ frame's imaging time occurs without further movement.[136] After the initial trigger the timer IC is triggered for each new $200\,\mu s$ camera frame, dispensing identical charge packets to the piezo element, and moving the camera focus in fixed steps for a set of z-axis stacked images. Comparator U_3 monitors the piezo voltage, and inhibits further charge packets when the piezo reaches a settable limit, usually between 100 V and 150. At this point the

---

voltage MOSFET driver U_1 uses short current pulses to change the state of the high-side MOSFET driver (see Fig. 3x.108) it may have modest leakage current, which could be detrimental in slow packet-based applications. In that case you'd want to use an optically-coupled high-side driver like those used in Figs. 3x.111 and 3x.113. The muRata NKE isolated dc–dc converters have a claimed $10\,\text{G}\Omega$ isolation resistance.

[135] See the many references in the literature, e.g., L. Huang et al., "Switched capacitor charge pump reduces hysteresis of piezoelectric actuators over a large frequency range," *Rev. Sci. Instrum.*, **81**, 094701 (2010)

[136] Each charge packet can increase the piezo voltage from 6 to 12 V, depending on its place in the non-linear response (see Fig. 3x.132), but the fixed charge packet results in a corresponding fixed displacement. There are some fifty papers dealing with piezo charge-drive, all of which accept the perfection of charge (versus voltage) drive; however, most deal with exotic aspects.

---

[134] We can expect the leakage currents of power MOSFETs Q_2 and Q_4 (both of which bridge the isolated piezo and its driver to the outside world) to be exceedingly low. Although maximum leakage spec for our chosen part, an FDP7N60NZ, is 1 μA, in practice we can expect the actual values to be 50 to 1000 times smaller than that at room temperature (as we saw, e.g., in §2x.1). Although the level-shifting circuit in high-

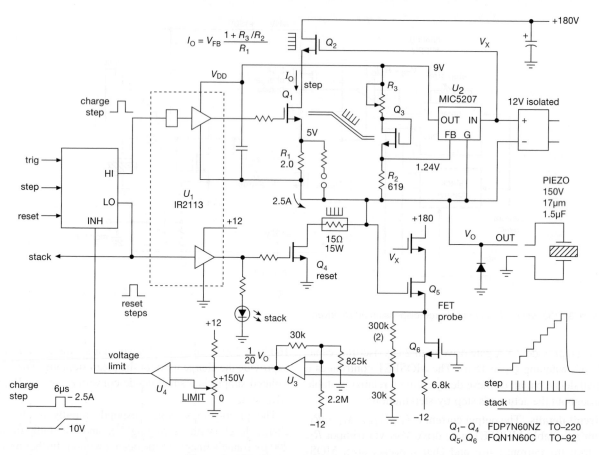

Figure 3x.134. Charge-dispensing piezo driver.

control circuit creates a STACK pulse, which turns on U_1's low-side, rapidly discharging the piezo element (through Q_4) in less than $200\,\mu s$ (time constant of $22\,\mu s$) so it's ready for a new set of image frames. The camera's image-storage program uses the STACK pulse to mark the end of a set of frames.

Here we deliberately avoided closed-loop control of the MOSFET current sources, preferring instead to drive the gate with a trimmable (and compensated) voltage, and with switchable current-setting source-terminal resistors. An alternative approach is to use a fast op-amp with closed-loop control. See §4x.26 for an analysis of op-amp-controlled MOSFET current sources. For more information about this project, ask about RIS-741.

3x.22.2 Analog charge dispenser

For comparison with the charge-dispensing system above, it's instructive to look at the more conventional tech-

nique[137] using op-amps (Fig. 3x.135), in which the piezo element C_P is in the feedback path. Capacitor C_1 sets the current-programming gain; an input-voltage change forces charge into C_1, which the op-amp then imposes onto C_P, so that the piezo charge is $Q_P = V_{in}C_1$. Resistor R_2 provides dc drift removal for the piezo voltage, and a time-constant-matching resistor R_1 is suggested, $R_1 = R_2C_P/C_1$. The authors of the referenced paper used an R_2C_P time constant of $1.4\,s$ ($0.1\,Hz$), with a reported linear frequency-response range of $1\,Hz$ to $20\,Hz$ (others have reported these circuits working to $1\,kHz$ or more). The authors struggled with A_3 driving a $C_P = 7.2\,\mu F$ load, and added pre-compensation. A_1 is also driving a capacitive load (which could be dealt with by adding a series resistor to create a zero above the maximum frequency).

We find the dc drift-removal concept unsatisfying, as it

---

[137] See, for example, K.A. Yi and R.J. Veillitte, "A charge controller for linear operation of a piezoelectric stack actuator," *IEEE Trans. Control Sys.*, **13**, 4, 517–526 (2005).

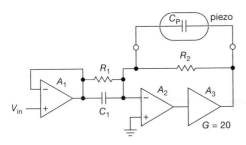

Figure 3x.135. Analog charge-drive circuit, for a piezo actuator in the feedback path of HV op-amp A_3.

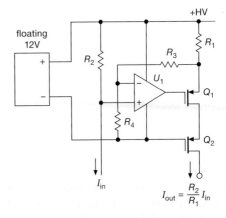

$$I_{out} = \frac{R_2}{R_1} I_{in}$$

Figure 3x.136. Charge drive for a piezo actuator, suited for small steps. R_4 offsets U_1's inputs enough to overcome the op-amp's V_{os}, ensuring that Q_1 is off when I_{in} is zero.

degrades the charge-linear ideal at low frequencies; we prefer instead to keep the drift low and apply an occasional reset (recall the simple op-amp integrator, §4.2.6, which can be stabilized with a resistor across the feedback capacitor, or you can omit the resistor to preserve accurate integration at low frequencies, resetting instead with a switch across the capacitor). Another issue with this circuit is that capacitor C_1 must be of a type that is free from dielectric absorption (capacitor "memory," see §§1x.3.10 and 5.6.2); since C_1 is typically ten times the piezo's capacitance, this can be a serious problem.[138] And, after our experience with pulsed charge packet drive, we prefer the latter technique, in spite of its circuit complexity.

3x.22.3 Small-step pulsed charge dispenser

The fast MOSFET pulsed charge dispenser of Figure 3x.134 was designed to work at currents of 1.25 A to 2.5 A. But if you want small charge packets to make small scanning steps, say 1/2000 of the piezo range, instead of 5 or 10%, you cannot practically shorten the pulse length to, say, 25 ns (it *might* work, but you couldn't rely on a stable charge-packet size). Instead one approach would be to modify its simple fixed-gate-voltage drive to make low current pulses, say 25 mA, by raising R_1 from 2 Ω to 2k.

But for a high-performance charge-based microstepping scanner, it would be nice to have more flexibility, for example programmable current to make the charge packets, analogous to the analog high-voltage current source of Figure 4x.152. But we like the idea of keep-

ing our speedy power MOSFETs, rather than the wimpy 50 mA BJTs in that figure. Figure 3x.136 does the job. It requires a serious op-amp to drive the MOSFET's gate capacitance (i.e., one with high output capability and more than 0.3 mA quiescent current), which is the reason for the floating 12 V dc–dc converter.

So far this is a pretty conventional current mirror; but it's important to ensure that there's no output current when I_{in} is zero. The problem is that the op-amp's offset voltage would produce an output current of V_{os}/R_1 (for one sign of offset voltage); adding R_4 creates a small threshold voltage (chosen slightly larger than $V_{os(max)}$) that $I_{in}R_2$ must overcome.

Scanners need to operate smoothly in both directions, so a mirror-image version of the current mirror (a "mirror of a mirror"?!) is needed to complete the system. Also, it's a good idea to monitor the piezo voltage without introducing a current error; for this a precision high-impedance (zero-current) voltage probe is needed, see the circuits in §3x.9. It's also a good idea to have a mid-scale initializer reset, using something like the floating-MOSFET switch configuration found in Figure 3.107 of the main volume, or the middle switch configuration of Figure 3x.115. To make this work, you'd connect the reset switch to an intermediate voltage, say 50 V, to center the scanning around the middle of the piezo actuator's range.[139] However, as Figure 3x.132 shows, you can comfortably work to the edges

---

[138] As summarized in the article by L. Huang et al., *ibid.*, "for a traditional charge driver, the two factors that may influence its performance are the resistor network and the dielectric absorption of the input capacitor. It is not an absolute charge driving system and has inevitable influence on the performance of the actuators. The existence of the resistor network worsens displacement linearity at low frequencies, while the dielectric absorption causes nonlinear displacement response over all the frequency ranges." Their solution: "switched-capacitor charge pump."

[139] Or you might prefer to have a reset voltage that puts the piezo at the start of the desired scanning region; this could be arranged by using an amplifier that can drive the storage capacitor to that reset voltage, for example something like the amplifier in Fig. 4x.137.

of the range without suffering dramatic nonlinearities or changes in sensitivity.

An important benefit of charge-based piezo drive is the high-speed scanning that's made possible with a linear open-loop charge-based system, when compared to conventional voltage drive that's buried inside a position-sensor feedback loop.[140]

Finally, keep in mind that the pulsed-charge piezo actuator drive does not create vibration, unlike the situation with pulsed *voltage* steps. It has been reported that, in some systems, charge pulses can cause noise in associated capacitive position sensors. Whereas in voltage-drive piezo systems the dramatic non-linearity makes such sensors necessary, they are not needed in charge-based systems, as many authorities suggest.

Figure 3x.134 is a switched-MOSFET current source. For an exploration of the design of a pulsed *op-amp controlled* current source, see §4x.26. There we show a 2.5 A version of Figure 3x.134, with speeds of 25 ns. We also discuss the issues involved in making fast versions that operate at low currents.

---

[140] If desired, you could retain the position sensor in a charge-based system, taking periodic measurements to check for creep or reduced hysteresis, and making slow corrections to the fast charge-based scanning.

ADVANCED TOPICS IN OPERATIONAL AMPLIFIERS

CHAPTER **4x**

In this chapter supplement we treat a number of more advanced topics dealing with operational amplifiers and their applications. In keeping with the objective of these "x-chapters," we have intentionally kept the discussions terse. The topics are presented in no particular order, and with no attempt to connect them logically. Think of them as a collection of short stories.

Among the rich collection of diverse topics, you'll find (a) feedback stability, phase margin, Bode plots, and the like; (b) a thorough discussion of transimpedance amplifiers (TIA) – stability, bandwidth, noise, and some example designs; (c) unity-gain buffers; (d) high-speed op-amps, both voltage feedback (VFB) and current feedback (CFB); (e) sections discussing slew rate, settling time, bias-current cancellation, and rail-to-rail op-amps; (f) analog "function" circuits; (g) normalizing TIAs and wide-range linear TIAs; (h) capacitive loads and stability; (i) precision and fast high-voltage amplifiers; and more...

Review of Chapter 4 of AoE3

To bring the reader up to speed, we start this chapter with the end-of-chapter review from AoE's Chapter 4:

¶A. The Ideal Op-amp.

In Chapter 4 we explored the world of Operational Amplifiers ("Op-amps"), universal building blocks of analog circuits. A good op-amp approaches the ideal of an infinite-gain wideband noiseless dc-coupled difference amplifier with zero input current and zero offset voltage. Op-amps are intended for use in circuits with negative feedback, where the feedback network determines the circuit's behavior. Op-amps figured importantly in the topics of Chapter 5 (Precision Circuits), Chapter 6 (Filters), Chapter 7 (Oscillators and Timers), Chapter 8 (Low-Noise Techniques), Chapter 9 (Voltage Regulation and Power Conversion), and Chapter 13 (Digital Meets Analog).

¶B. The "Golden Rules."

At a basic level (and ignoring imperfections, see ¶¶K–M below), an op-amp circuit with feedback can be simply understood by recognizing that feedback from the output operates to (I) make the voltage difference between the inputs zero; and, at the same level of ignoration, (II) the inputs draw no current. These rules are quite helpful, and for dc (or low-frequency circuits) they are in error only by typical offset voltages of a millivolt or less (rule I), and by typical input currents of order a picoamp for FET types or tens of nanoamps for BJT types (rule II).

¶C. Basic Op-amp Configurations.

In §4.2 and §4.3 we met the basic linear circuits (detailed in ¶¶D–F below): inverting amplifier, noninverting amplifier (and follower), difference amplifier, current source (transconductance, i.e., voltage-to-current), transresistance amplifier (i.e., current-to-voltage), and integrator. We saw also two important *non*-linear circuits: the Schmitt trigger, and the active rectifier. And in §4.5 we saw additional circuit building blocks: peak detector, sample-and-hold, active clamp, active full-wave rectifier (absolute-value circuit), and differentiator.

¶D. Voltage Amplifiers.

The *inverting amplifier* (Figure 4.5) combines input current V_{in}/R_1 and feedback current V_{out}/R_2 into a summing junction; it has voltage gain $G_V = -R_2/R_1$ and input impedance R_1. In the *noninverting amplifier* (Figure 4.6) a fraction of the output is fed back to the inverting input; it has voltage gain $G_V = 1 + R_2/R_1$ and near-infinite input impedance. For the *follower* (Figure 4.8) the feedback gain is unity, i.e., the resistive divider is replaced by a connection from output to inverting input. The *difference amplifier* (Figure 4.9) uses a pair of matched resistive dividers to generate an output $V_{out} = (R_2/R_1)\Delta V_{in}$; its input impedance is $R_1 + R_2$, and its common-mode rejection depends directly on the accuracy of the resistor matching (e.g., ~ 60 dB with $\pm 0.1\%$ resistor tolerance). Difference amplifiers are treated in greater detail in §5.14. A pair of input followers can be used to achieve high input impedance, but a better 3-op-amp configuration is the *instrumentation amplifier*, see §5.15.

¶E. Integrator and Differentiator.

The *integrator* (Figure 4.16) looks like an inverting amplifier in which the feedback resistor is replaced by a capacitor; thus the input current V_{in}/R_1 and feedback current $C\,dV_{out}/dt$ are combined at the summing junction. Ignoring the imperfections in ¶K below, the integrator is "perfect," thus any non-zero average dc input voltage will cause the output to grow and eventually saturate. The integrator can be reset with a transistor switch across the feedback capacitor (Figure 4.18); alternatively you can use a large shunt resistor to limit the dc gain, but this defeats the integrator operation at low frequencies ($f \lesssim 1/R_f C$). The integrator's input impedance is R_1.

The op-amp *differentiator* (Figure 4.68) is a similar configuration, but with R and C interchanged. Without additional components (Figure 4.69) this configuration is unstable (see ¶O, below).

¶F. Transresistance and Transconductance Amplifiers.

By omitting the input resistor, an inverting voltage amplifier becomes a *transresistance amplifier*[1], i.e., a current-to-voltage converter (Figure 4.22). Its gain is $V_{out}/I_{in} = -R_f$, and (ignoring imperfections) the impedance at its input (which drives the summing junction) is zero. Capacitance at the input creates issues of stability, bandwidth, and noise; see §8.11 and the discussion in this chapter (Chapter 4x). Transresistance amplifiers are widely used in photodiode applications.

---

[1] Or *transimpedance* amplifier, TIA.

A *transconductance amplifier* (Figures 4.10–4.15) converts a voltage input to a current output; it is a voltage-controlled current source. The simplest form uses an op-amp and one resistor (Figure 4.10), but works only with a floating load. The Howland circuit and its variations (Figures 4.14 and 4.15) drive a load returned to ground, but their accuracy depends on resistor matching. Circuits with an external transistor (Figures 4.12 and 4.13) drive loads returned to ground, do not require resistor matching, and, in contrast to the other circuits, benefit from the intrinsically high output impedance of the transistor. In this chapter (Chapter 4x) we describe a nice variation on the transistor-assisted current source that achieves both high speed and bipolarity output (i.e., sinking and sourcing)

¶G. Nonlinear Circuits: Peak Detector, S/H, Clamp, Rectifier.

Because of their high gain, op-amps provide accuracy to nonlinear functions that can be performed with passive components alone; in these circuits one or more diodes select the regions in which feedback acts. The *peak detector* (Figure 4.58) captures and holds the highest (or lowest) voltage since the last reset; the *sample-and-hold* (S/H) circuit (Figure 4.60) responds to an input pulse by capturing and holding the value of an input signal voltage; the *active clamp* (Figure 4.61) bounds a signal to a maximum (or minimum) voltage; the *active rectifier* creates accurate half-wave (Figures 4.36 and 4.38) or full-wave (Figures 4.63 and 4.64) outputs. In practice the performance of these circuits is limited by the finite slew rate and output current of real op-amps (see ¶ M, below).

¶H. Positive Feedback: Comparator, Schmitt Trigger, and Oscillator.

If the feedback path is removed, an op-amp acts as a *comparator*, with the output responding (by saturating near the corresponding supply rail) to a reversal of differential input voltage of a millivolt or less (Figure 4.32A). Adding some positive feedback (Figure 4.32B) creates a *Schmitt trigger*, which both speeds up the response and also suppresses noise-induced multiple transitions. Op-amps are optimized for use with negative feedback in linear applications (notably by a deliberate internal -6 dB/octave rolloff "compensation," see ¶ O below), so special comparator ICs (lacking compensation) are preferred, see §12.3 and Tables 12.1 and 12.6. A combination of positive feedback (Schmitt trigger) and negative feedback (with an integrator) creates an *oscillator* (Figure 4.39), a subject treated in detail in Chapter 7.

¶I. Single-Supply and Rail-to-rail Op-amps.

For some op-amps both the input common-mode range and the output swing extend all the way down to the negative rail, making them particularly suited for operation with a single positive supply. Rail-to-rail op-amps allow input swings to both supply rails, or output swings to both rails, or both; see Table 4.2a. The latter are especially useful in circuits with low supply voltages.

¶J. Some Cautions.

In linear op-amp circuits, the Golden Rules (see ¶ B, above) will be obeyed only if (a) feedback is negative and (b) the op-amp stays in the active region (i.e., not saturated). There must be feedback at dc, or the op-amp will saturate. Power supplies should be bypassed. Stability is degraded with capacitive loads, and by lagging phase shifts in the feedback path (e.g., by capacitance at the inverting terminal). And, most important, real op-amps have a host of limitations (¶¶K–N, below) that bound attainable circuit performance.

¶K. Departures from Ideal Behavior.

In the real world op-amps are not perfect. There is no "best" op-amp, thus one must trade off a range of parameters: input imperfections (offset voltage, drift, and noise; input current and noise; differential and common-mode range), output limitations (slew rate, output current, output impedance, output swing), amplifier characteristics (gain, phase shift, bandwidth, CMRR and PSRR), operating characteristics (supply voltage and current), and other considerations (package, cost, availability). See §4.4, Tables 4.1, 4.2a, and 4.2b, the more extensive tables in Chapters 5, and 8, and ¶¶L–N below.

¶L. Input Limitations.

The *input offset voltage* (V_{os}), ranging from about $25\,\mu$V ("precision" op-amp) to 5 mV, is the voltage unbalance at the input terminals. It's an important parameter for precision circuits, and circuits with high closed-loop dc gain; the error seen at the output is $G_{CL}V_{os}$. Some op-amps provide pins for external trimming of offset voltage (e.g., see Figure 4.43).

The *offset voltage drift*, or *tempco* (TCV$_{os}$, or $\Delta V_{os}/\Delta T$), is the temperature coefficient of offset voltage; it ranges from about $0.1\,\mu$V/$^\circ$C ("precision" op-amp) to $10\,\mu$V/$^\circ$C. Even if you're lucky and have an op-amp with low V_{os} (or you've trimmed it to zero), TCV$_{os}$ represents the growth of offset with changing temperature.

The *input noise voltage density* (e_n) represents a noisy voltage source in series with the input terminals. It ranges from about $1\,$nV/$\sqrt{\text{Hz}}$ (low-noise bipolar op-amp) to

$100\,\mathrm{nV}/\sqrt{\mathrm{Hz}}$ or more (micropower op-amps). Noise voltage is important in audio and precision applications.

The *input bias current* (I_B) is the (non-zero) dc current at the input terminals. It ranges from a low of about 5 fA (CMOS low-bias op-amps, and "electrometer" op-amps) to 50 nA (typical[2] BJT op-amps) to a high of $10\,\mu\mathrm{A}$ (wideband BJT-input op-amps). Bias current flowing through the circuit's dc source resistance causes a dc voltage offset; it also creates a current error in integrators and transresistance amplifiers.

The *input noise current* (i_n) is the equivalent noise current added at the input. For most op-amps[3] it is simply the shot noise of the bias current ($i_n=\sqrt{2qI_B}$); it ranges from about $0.1\,\mathrm{fA}/\sqrt{\mathrm{Hz}}$ (CMOS low-bias op-amps, "electrometer" op-amps) to $1\,\mathrm{pA}/\sqrt{\mathrm{Hz}}$ (wideband BJT op-amps). Input noise current flowing through the circuit's ac source impedance creates a noise voltage, which can dominate over e_n. The ratio $r_n=e_n/i_n$ is the op-amp's *noise resistance*; for signal source impedances greater than r_n the current noise dominates.

Op-amps function properly when both inputs are within the *input common-mode voltage range* (V_{CM}), which may extend to the negative rail ("single-supply" op-amps), or to both rails ("rail-to-rail" op-amps). Beware: many op-amps have a more restricted *input differential voltage range*, sometimes as little as just a few volts.

¶M. Output Limitations.

The *slew rate* (SR) is the op-amp's dV_{out}/dt with an applied differential voltage at the input. It is set by internal drive currents charging the compensation capacitor, and ranges from about $0.1\,\mathrm{V}/\mu\mathrm{s}$ (micropower op-amps) to $10\,\mathrm{V}/\mu\mathrm{s}$ (general purpose op-amps) to $5000\,\mathrm{V}/\mu\mathrm{s}$ (high-speed op-amps). Slew rate is important in high-speed applications generally, and in large-swing applications such as A/D and D/A converters, S/H and peak detectors, and active rectifiers. It limits the large-signal output frequency: a sinewave of amplitude A and frequency f requires a slew rate of $\mathrm{SR}=2\pi A f$; see Figure 4.54.

Op-amps are small devices, with *output current* deliberately limited to prevent overheating; see for example Figure 4.43, where $R_5 Q_9$ and $R_6 Q_{10}$ limit the output sourcing and sinking currents to $I_{lim}=V_{BE}/R \approx 25\,\mathrm{mA}$, illustrated in Figure 4.45. If you need more output current, there are a few high-current op-amps available; you can also add an external unity-gain power buffer like the LT1010 (I_{out} to $\pm150\,\mathrm{mA}$), or a discrete push-pull follower.

The open-loop *output impedance* of an op-amp is generally in the neighborhood of $100\,\Omega$, which is reduced by the loop gain to fractions of an ohm at low frequencies. Because an op-amp's open-loop gain G_{OL} falls as $1/f$ over most of its bandwidth (see ¶O below), however, the circuit's *closed-loop* output impedance rises approximately proportional to frequency; it looks inductive (Figure 4.53).

In general the *output swing* for an op-amp like Figure 4.43 extends only to within a volt or so from either rail. Many CMOS and other low-voltage op-amps, however, specify unloaded rail-to-rail output swings, see Figure 4.46.

Op-amps can be grouped into several *supply voltage* ranges: "low-voltage" op-amps have a maximum total supply voltage (i.e., $V_+ - V_-$) around 6 V, and generally operate down to 2 V; "high-voltage" op-amps allow total supply voltages to 36 V, and generally operate down to 5–10 V. In between there is a sparse class of what might be called "mid-voltage" op-amps, with total supply voltages in the neighborhood of 10–15 V. See Table 5.5. There are also op-amps that are truly high-voltage (to hundreds of volts), see Table 4.2b.

¶N. Gain, Phase Shift, and Bandwidth.

Op-amps have large dc *open-loop gain* $G_{OL(dc)}$, typically in the range of 10^5–10^7 (the latter being typical of "precision" op-amps, see Chapter 5). To ensure stability (see ¶O, below) the op-amp's open-loop gain falls as $1/f$, reaching unity at a frequency f_T (see Figure 4.47). This limits the closed-loop *bandwidth* to $\mathrm{BW}_{CL}\approx f_T/G_{CL}$. Over most of the operating frequency range the op-amp's open-loop *phase shift* is $-90°$, eliminated in the closed-loop response by feedback.

¶O. Feedback Stability, "Frequency Compensation," and Bode Plots

Finally, negative feedback can become *positive* feedback, promoting instability and oscillations, if the accumulated phase shift reaches $180°$ at a frequency at which the loop gain is ≥ 1. This topic is foreshadowed in §4.6.2 in connection with capacitive loads, and it is discussed in detail in §4.9. The basic technique is *dominant-pole compensation*, in which a deliberate $-6\,\mathrm{dB}$/octave (i.e., $\propto 1/f$) rolloff is introduced within the op-amp in order to bring the gain down to unity at a frequency lower than that at which additional unintended phase shifts rear their ugly heads (Figure 4.99). Most op-amps include such compensation inter-

---

[2] The input current of "bias-compensated" BJT op-amps is typically around 50 pA.

[3] But not "bias-compensated" BJT op-amps, see §8.9.

nally, such that they are stable at all closed-loop gains (the unity-gain follower configuration is most prone to instability, because there is no attenuation in the feedback path). "Decompensated" op-amps are less aggressively compensated, and are stable for closed-loop gains greater than some minimum (often specified as $G>2$, 5, or 10; Figure 4.95). Compensated op-amps exhibit an open-loop lagging phase shift of 90° over most of their frequency range (beginning as low as 10 Hz or less). Thus an external feedback network that adds another 90° of lagging phase shift at a frequency where the loop gain is unity will cause oscillation.

A favorite tool is the *Bode Plot*, a graph of gain (log) and phase (linear) versus frequency (log); see Figure 4.97. The *stability criterion* is that the difference of slopes between the open-loop gain curve and the ideal closed-loop gain curve, at their intersection, should ideally be 6 dB/octave, but in no case as much as 12 dB/octave.

4x.1 From Philbrick to SMT

High-gain dc-coupled feedback amplifiers date back to Bell Labs in the early 1940s, at least, when they were built with vacuum tubes and used in WWII artillery "fire control" (gun aiming) in conjunction with the early radars (famously the SCR584) of that era. The first op-amps with differential inputs (designed by L. Julie at Columbia University) appeared soon after, in 1947 (when they also acquired the name of "operational amplifier"). And the first commercially available op-amps came from the company founded by George Philbrick (who had worked at Bell Labs), in the form of the GAP/R K2-W (Figure 4x.1). It was the jellybean of its day, costing a mere $24 according to the 1958 price list.[4]

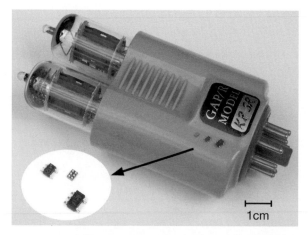

Figure 4x.1. From our collection of dinosaurs, a specimen of the Philbrick K2 vacuum-tube op-amp (introduced in 1952). Sitting on its hip, like fleas on a dog, are three contemporary op-amps in diminutive surface-mount packages. With the exception of available output voltage swing, these little beasts can run metaphorical circles around their forebear.

These things sold by the gazillion, and launched the general use of op-amps in analog circuit design. There were improved versions of the Philbrick vacuum-tube op-amp, followed by discrete transistor op-amps (for example, the P65, see Figs. 4x.3 and 4x.4), and then monolithic integrated circuit op-amps. There was a brief overlap era

when discrete op-amps outperformed ICs, but, as the saying goes, "that was then." Some high points in the contemporary history of op-amps include the μA709 and LM301 (the first easy-to-use op-amps, both designed by the legendary Bob Widlar), the μA741 (internally compensated, and perhaps the most popular op-amp of all time), the OP-07 (precision), the LF355–7, LF411, and TL081-series (JFET input, low-bias), the OPA655-6 (JFET, fast!), the OPA627 (JFET, quiet!), and finally the contemporary proliferation of low-voltage rail-to-rail I/O CMOS op-amps (too many part numbers to mention; but we're fond of the LMC6xxx series, e.g., the LMC6482). Contemporary op-amps are monolithic, and getting smaller and smaller (see the photograph!). They are available in a vast variety: the spectacularly useful DigiKey website, when queried with "operational amplifier," responds with "You have selected 27,589 items, spanning 1,104 pages." Whew![5]

A trip down memory lane.
Figure 4x.2 is the circuit of the famous K2-W. The unfamiliar components are the 12AX7A dual triode vacuum tubes, and the NE-2 neon bulbs (here used as zener-like dc level shifters). The front-end is the fa-

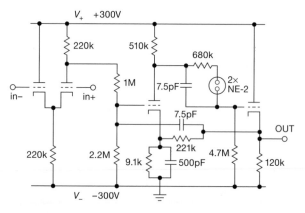

Figure 4x.2. Circuit diagram of Philbrick's original K2-W op-amp.

miliar "long-tailed pair" differential amplifier, which is followed by a single-ended stage of inverting gain, and finally a cathode follower to create the relatively low-impedance ($1/g_m \approx 1\,\mathrm{k}\Omega$) output. There's a 7.5 pF capacitor for feedback frequency compensation, and there's also a bit of gain-boosting *positive* feedback (the 221k resistor). A datasheet for this pioneering op-amp lists $G_{OL}=15{,}000$, BW$>100\,\mathrm{kHz}$ ($G_{CL}=-1$), $I_B<100\,\mathrm{nA}$, $Z_{in}>100\,\mathrm{M\Omega}$, $V_{os}=-1.5\,\mathrm{V}\pm0.5\,\mathrm{V}$, V_{CM}(operating)$=\pm50\,\mathrm{V}$, and $V_{out}=\pm50\,\mathrm{V}$.

---

[4] This and other fascinating documents can be found in abundance at www.philbrickarchive.org.

[5] To be fair, we should point out that they count separately different packaging of the same op-amp.

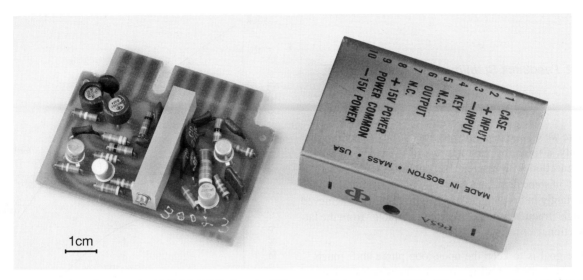

Figure 4x.3. Another dinosaur, the Philbrick P65A discrete BJT op-amp (introduced in 1962): "All silicon...low in cost" ($95 then, equivalent to $795 in 2019!). Quite a contrast, with some contemporary op-amps (of far greater performance) selling for less than ten cents apiece.

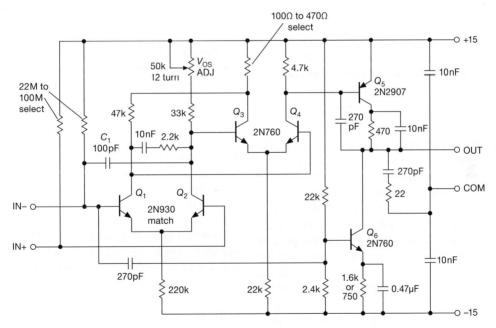

Figure 4x.4. Circuit diagram of Philbrick's first silicon BJT op-amp, the P65. This puppy had an f_T of 1.6 Mc (that's MHz), an open-loop gain of 20,000, and an output swing to ±10 V (with ±1.1 mA of drive).

Figure 4x.4 is the circuit of their first silicon discrete BJT op-amp (pictured in Figure 4x.3). Here they squandered two additional active components (hey, transistors cost only a few dollars apiece), to make a cascaded pair of long-tailed pairs, followed by a single *pnp* common-emitter stage (Q_5) with 1.5 mA or 3 mA current-sink pull-down (Q_6). They selected matched transistors for the input stage (with a thermal-equalizing clip), and they provided a 12-turn offset trimmer (notice the hole in the metal case). Even with these measures, they still had to select resistor values for input-stage biasing and second-stage pullup.

4x.2 Feedback Stability and Phase Margins

Here we elaborate on the business of feedback amplifier stability, a topic that first appears in Chapter 2 (§2.5.4B) in connection with discrete BJT amplifiers, and which is treated in some detail in Chapter 4 (§4.9.1); there it figures importantly in op-amp circuits, where stability is achieved through "frequency compensation." To quote[6] from the latter section,

> The goal is to keep the open-loop phase shift much less than $180°$ at all frequencies for which the loop gain is greater than 1. Assuming that the op-amp may be used as a follower, the words "loop gain" in the last sentence can be replaced by "open-loop gain." The easiest way to do this is to add enough *additional* capacitance at the point in the circuit that produces the initial 6 dB/octave rolloff, so that the open-loop gain drops to unity at about the 3 dB frequency of the next "natural" RC filter. In this way the open-loop phase shift is held at a constant $90°$ over most of the passband, increasing toward $180°$ only as the gain approaches unity.

That deliberate "additional capacitance" creates the dominant pole; and if chosen as described, the amplifier will be stable when configured as a unity-gain follower (which is the least stable configuration – go back and read the whole of §4.9.1 if you're confused about this point).

Of course, you can always pile on additional capacitance, ensuring greater stability; but you pay the price in terms of bandwidth. You can go the other way, reducing the compensation capacitance to gain some speed. In the latter case the amplifier's open-loop phase shift ϕ will be closer to $180°$ at unity gain; there will be less *phase margin* (the value of $180° - \phi$ at unity gain).[7] Hence the question: How does phase margin depend upon the location of the next RC characteristic frequency? And, more broadly,

Figure 4x.5. Bode plot (open-loop gain and phase versus frequency) for an amplifier like that in Figure 4x.6, illustrating bandwidth and phase margin for closed-loop gains of 1 and 100. Here f_c is the open-loop unity-gain crossover frequency; absent the second pole (at f_2), $f_c = $ GBW, the gain–bandwidth product.

how do performance characteristics such as overshoot, settling time, and frequency peaking depend on phase margin? These questions are relevant not only to op-amp circuits, but also to discrete amplifiers with feedback (such as the high-voltage amplifier of §4x.23).

Look at Figure 4x.5, a classic Bode plot showing an amplifier's open-loop gain G_{OL} and phase shift ϕ as a function of frequency. This amplifier has dominant pole compensation that by itself would roll off to unity gain at frequency f_1, also known as the gain–bandwidth product GBW. We've cascaded an RC-style roll-off with characteristic frequency f_2 (the "second pole"), causing the overall amplifier's actual gain to reach unity at frequency f_c (subscript indicating *crossover* frequency). The corresponding phase shift plot is a lagging $-90°$ over most of the amplifier's bandwidth, increasing to $-135°$ at f_2 as it heads toward $-180°$ well beyond f_2. If this amplifier is configured as a unity-gain follower, i.e., $G_{CL} = 1$ (don't worry about the $G_{CL} = 100$ case, we'll get to that later), the loop gain is the same as the amplifier's open-loop gain, falling to unity at f_c, with phase shift (and phase margin) as shown.

---

[6] Quoting from one's own book?! Well, if it was good enough for J.S. Bach (whose Christmas Oratorio, for example, lifted music from at least three of his secular cantatas, BWV 213–215, as well as from his St. Mark Passion, the latter itself hijacking major sections from cantatas BWV 54 and 198), then it's surely good enough for us!

[7] Which can be mitigated by introducing a zero to cancel the second pole; we won't deal with pole-zero compensation (§4.9.2C) here.

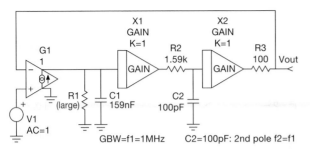

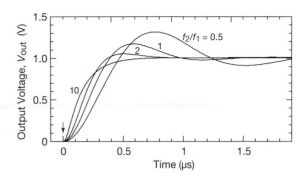

Figure 4x.6. SPICE model of a feedback amplifier composed of an op-amp with single dominant pole compensation (transconductance gain stage G1, integrating capacitor C1, buffer stage X1) followed by a single-pole RC to model the op-amp's second pole (just past its f_T).

Figure 4x.7. Closed-loop response to a 1 V input step at $t=0$ (arrow), from a SPICE simulation of the circuit of Fig. 4x.6, plotted for several values of f_2/f_1.

4x.2.1 Sliding f_2: phase margin and circuit performance

To find out how the phase margin and circuit performance changes with pole frequency f_2, we set up the SPICE simulation circuit shown in Figure 4x.6. A simulation like this is easily created, and it lets you explore circuit performance as you adjust various parametric "knobs." Here gain stage G1 is a transconductance amplifier, with a gain of 1000 mS, i.e., $g_m = 1$ A/V. With its capacitive C_1 load[8] the stage has a voltage gain of magnitude $G_V = g_m X_{C_1} = g_m/\omega C_1 = 1/2\pi f C_1$, that is, a gain falling at 6 dB/octave, crossing unity gain at $f_1 = 1$ MHz. The voltage buffer X1 drives the second-pole roll-off network $R_2 C_2$, whose buffered output is also the feedback for an overall unity gain follower. For the component values shown, $f_2 = 1$ MHz; that is, the ratio $f_2/f_1 = 1$.

We ran simulations, varying C_2 to produce f_2/f_1 ratios going from 0.5 to 10. For each value we ran both a time-domain step response, and a frequency-domain sweep of closed-loop gain and phase shift. Figure 4x.7 shows a subset of the SPICE output waveforms for a unit input step, and Figure 4x.8 shows the analogous closed-loop Bode plots.

From a set of such simulations we tabulated for each value of f_2/f_1 the phase margin, the step response (delay time, overshoot, and settling time), and the amount of gain peaking (i.e., magnitude of gain at the frequency where the gain was highest). The results are plotted in Figures 4x.9–4x.12. It is noteworthy from Figure 4x.9 that the phase margin does not decline as fast as one would at first expect, when the pole frequency f_2 is reduced well below f_1. For example, you might expect (and you often hear people say) that the phase margin for $f_2 = f_1$ would be 45°, whereas it

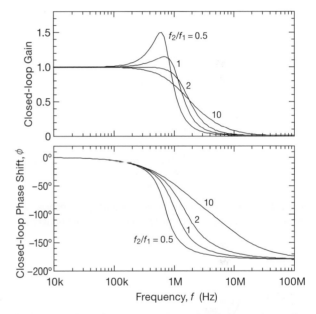

Figure 4x.8. Closed-loop gain and phase shift, from a SPICE simulation of the circuit of Fig. 4x.6, plotted for several values of f_2/f_1.

is in fact 52°. How can that be? The explanation is simply that the presence of the second pole pushes down the unity-gain frequency (from f_1 to f_c in Fig. 4x.5; for $f_2 = f_1$ we find $f_c \approx 0.78 f_1$, see the minitable in Fig. 4x.10), at which there's more phase margin; and the lower f_2, the lower f_c. It's also evident, from Figure 4x.11, that a modest reduction in phase margin causes some overshoot but little peaking, whereas further reductions lead quickly to large peaks in the frequency response.

In Figure 4x.9 we've plotted also the phase margin when two RC-style roll-offs (two poles) are placed at f_2; with this configuration the phase margin deteriorates rapidly, as expected, as f_2 approaches f_1. A pole-zero compensation

---

[8] The parallel resistor R_1 is needed so the simulation can figure out the quiescent bias condition; it does not affect the dynamic results.

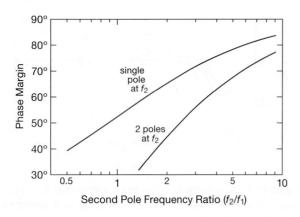

Figure 4x.9. Phase margin versus location of the second pole frequency f_2.

f_2/f_1	f_c/f_1
10	0.99
5	0.98
2.5	0.93
2	0.91
1.5	0.86
1.25	0.83
1	0.78
0.75	0.72
0.63	0.68
0.5	0.61

Figure 4x.10. Reduction of unity-gain frequency f_c, relative to its single-pole frequency f_1, caused by a second pole at frequency f_2.

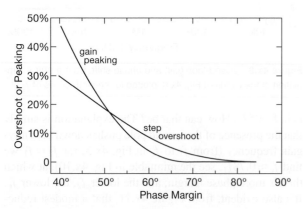

Figure 4x.11. Overshoot and gain peaking versus phase margin.

scheme is well suited to a situation like this. In an intermediate case where there's a pole at f_2 and another at a somewhat higher frequency f_3, the phase margin plot would lie between the two curves of Figure 4x.9.

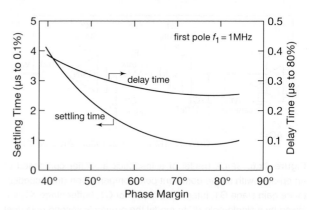

Figure 4x.12. Step response (delay time and settling time) versus phase margin.

4x.2.2 What about amplifiers with $G_{CL} > 1$?

For clarity of exposition we've thus far treated only the case of a unity gain follower. It's no more complicated, though, to deal with the situation where the target closed-loop gain is greater than unity, i.e., where the feedback signal is attenuated by some factor $B = 1/G_{CL}$. In the Bode plot of open-loop gain and phase versus frequency (Fig. 4x.5) we simply draw a line across the gain plot at $G = 100$, finding the intersection with the open-loop gain at which the loop gain has fallen to unity; here that's about 10 kHz. The corresponding point on the phase plot gives us the phase shift, and the (considerably better) phase margin. You can understand this in another way: In this example the second pole at f_2 is now *above* this unity-gain frequency; so 10 kHz is the "new f_1," which is now comfortably below the second pole frequency f_2 (i.e., we have a larger value of f_2/f_1), with correspondingly better phase margin.

4x.2.3 Applying Bode plots to amplifier design

In practice you want to optimize speed and stability trade-offs when designing a feedback amplifier. This usually means driving up the frequency of the unavoidable second pole frequency f_2 (by reducing capacitances, raising operating currents, using a cascode stage, etc.). Then you can increase the GBW (by reducing the dominant pole compensation capacitance), bounded by a decreasing phase margin and consequent performance degradation. This process is iterative, with additional measures to drive up f_2, perhaps until you reach a third pole (call it f_3).

In a complex design (for example the high-voltage amplifier in §4x.23) it may be difficult to identify pole frequencies f_2, f_3, etc., but if you can construct a Bode plot (by measurement, or by simulation), you can use its values of phase margin, in conjunction with the plots in Figures 4x.11–4x.12, to get a reasonable first estimate of amplifier performance. Once you have settled on a baseline design, you can model it in detail with SPICE if you wish, iterate again, and ultimately measure and validate its performance on the bench.

4x.2.4 Afterword: High-speed op-amps

Later in this chapter (§§4x.5 and 4x.6) we discuss in detail high-speed op-amps, an increasingly important topic in the contemporary world of wideband analog sensors and signals. A "high-speed" op-amp may mean one with high gain–bandwidth product f_T, or one with a high slew rate, or one with a fast settling time; or, depending on application, it may mean an op-amp with a very high frequency at which the closed-loop gain has fallen by 3 dB. Interestingly, these goals can be (and often are) in conflict.

In this section we revisit the SPICE model of Figure 4x.6, decorating it with a third pole (which becomes important in fast op-amps) and watching its performance as we play games with the location of the dominant pole (which is what sets f_T). There's some nonintuitive stuff going on that a good circuit designer needs to know about, because the goals of the *op-amp's* designer may not align with those of your application.

Here's how to look at it: for a fast op-amp, the frequencies of the inevitable second and third poles are set by the circuit design and the fabrication process, and are presumably aimed as high as possible, as constrained by power dissipation and manufacturing cost goals. Then the op-amp designer creates a stable and "clean" op-amp by reducing the amplifier's low-frequency gain; that is, by choosing a (dominant pole) compensation capacitor large enough to put the amplifier's GBW (f_T) well below the higher pole frequencies. That way you get an op-amp with a healthy phase margin, and a Bode plot exhibiting a nice straight line of G_{OL} all the way down to unity gain.

But the designer may want to offer an op-amp of greater GBW, while stuck with those pesky second and third poles. This can be done, of course, at the expense of phase margin, but not without consequences – such as response peaking, overshoot, and extended settling time in the closed-loop configuration. These may be acceptable, depending on the application. For instance, a curious and nonintuitive result of such "twisting the dragon's tail" is an extended −3 dB frequency with low closed-loop gains: that roll-off frequency can be significantly *higher* than f_T! But you may not wish to use such an op-amp in a fast servo loop.

A. SPICEing the 3-pole op-amp

To learn how this all goes, we built a 3-pole SPICE model, initially setting the compensation pole for a GBW of 25 MHz (100 kΩ and 6.37 nF for R_1 and C_1 in the extended version of Fig. 4x.6, respectively). That would be a safe choice for an op-amp whose higher poles are at 75 MHz and 150 MHz (i.e., three and six times f_T). When closed with unity gain (i.e., a follower) the response is smooth, with no peaking and a −3 dB frequency of about 36 MHz (44% higher than f_T). Figure 4x.13 shows the follower response for both this choice of GBW, and also for two more aggressive choices of GBW: 50 MHz and 100 MHz; and Figures 4x.14 and 4x.15 show the corresponding step response and settling behavior.

The bottom line

An op-amp that has been tailored for higher GBW may not settle any faster than a slower op-amp, and it may exhibit unacceptable levels of response peaking and step-waveform overshoot. And its reduced phase margin may render it unusable in a fast loop. See additional discussion about slewing and settling in §4x.9, and further advice about the behavior and selection of high-speed op-amps in §§4x.5 and 4x.6.

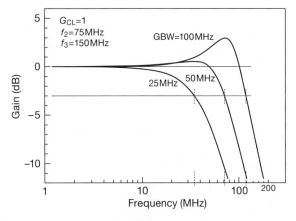

Figure 4x.13. Unity-gain frequency response for an op-amp SPICE model with three choices of GBW, given the pair of higher poles listed.

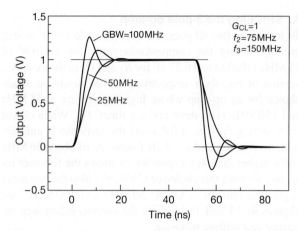

Figure 4x.14. Output voltage waveform for the same op-amp model as in Fig. 4x.13, when driven with a 50 ns positive pulse of 1 V amplitude.

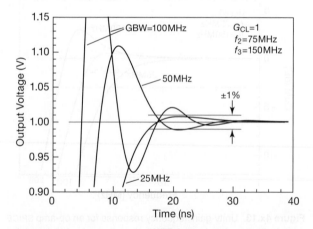

Figure 4x.15. Detail of the settling behavior of Fig. 4x.14. Surprisingly, the "slowest" amplifier settles fastest to the ±1% limits shown.

4x.3 Transresistance Amplifiers

We introduced the basic current-to-voltage, or *transresistance*, amplifier configuration in §4.3.1: an op-amp with feedback resistor R_f converts an input current I_{in} at the summing junction to an output voltage $V_{out} = -R_f I_{in}$. It is called transresistance because its "gain" (output/input) has units of resistance: Gain $= V_{out}/I_{in} = R_f$. (You often see the term "transimpedance" and "TIA" used instead, perhaps suggesting that you ought to be worrying about more general feedback circuits, and phase shifts; but in the real world people design these things as simple current-to-voltage amplifiers, and so we often say "transresistance amplifier," or "current-to-voltage amplifier.")

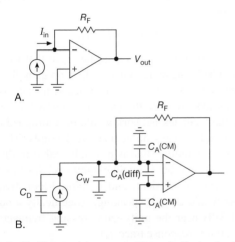

Figure 4x.16. Transresistance amplifiers. A. Basic circuit. B. Including real-world parasitic input capacitances.

4x.3.1 Stability problem

The basic transresistance amplifier is shown again in Figure 4x.16A, and with real-world complications in Figure 4x.16B. The problem with the simple circuit is, simply, that it will probably oscillate! That's because photodiodes (and other detectors, or current-output devices in general) have some intrinsic capacitance C_D, and this capacitance at the input forms a lowpass filter with R_f (with $-3\,\mathrm{dB}$ "breakpoint" $f_{RC_{in}} = 1/2\pi R_f C_D$), hence a lagging phase shift that approaches $-90°$ well beyond $f_{RC_{in}}$. That's often well below the op-amp's gain–bandwidth product f_T,

so the effect is to add nearly $90°$ of lagging phase shift to the feedback path, augmenting the op-amp's $90°$ (or greater) lagging internal phase shift. The situation is shown in the Bode plot of Figure 4x.17, where the feedback network contributes a second pole that increases the roll-off to $12\,\mathrm{dB/octave}$, and the phase shift to $-180°$, at a frequency where the loop gain is still greater than unity; and it crosses the unity gain axis with still greater slope. That's the prescription for oscillation.

Note, by the way, the additional capacitances from the summing junction to ground shown in Figure 4x.16B: the op-amp's input capacitance (both differential and common-mode), and wiring capacitance. For the purposes of circuit behavior, they're all in parallel: $C_{in} = C_D + C_A + C_W$. Which capacitance dominates the sum depends on the size of the detector, the op-amp's internal input circuit, and the wiring. With a fast, small-geometry detector, the op-amp's input capacitance may well dominate; whereas a large-area detector's capacitance is likely to dominate (unless significant lengths of shielded cable are used with a remote detector). As we'll see shortly, the more capacitance you have, the poorer the performance (in terms of speed and noise). So it's always best to avoid adding significant capacitance, where possible. For example, if your detector is some distance from the rest of your circuit, it is often a good idea to put the transresistance preamp right at the detector, bringing the amplified voltage output back through shielded cable; this also has the advantage of minimizing noise pickup on the low-level, high-impedance input signal via capacitive and inductive pickup, ground loops, and the like. Also, detector capacitance decreases markedly with increasing applied back-bias, so speed is improved (but leakage current is introduced) by returning the detector common terminal to a quiet (i.e., well bypassed) bias supply instead of ground (Figure 4x.18).

4x.3.2 Stability solution

The simplest solution to this problem is to put a small parallel compensation capacitance C_f across the feedback resistor, as in Figure 4x.19. It's easiest to understand what's going on with a Bode plot (Figure 4x.17). C_f stops the $6\,\mathrm{dB/octave}$ roll-off of the feedback network at frequency $f_c = 1/2\pi R_f C_f$ (that's where the magnitude of C_f's reactance equals R_f, and is roughly the roll-off response frequency of the amplifier), which makes the overall roll-off of loop gain revert to its original $6\,\mathrm{dB/octave}$ slope (and corresponding $90°$ lagging phase shift). (In official jargon this is known as putting a "zero" into the feedback network.) The trick is to choose C_f so that the resulting closed-

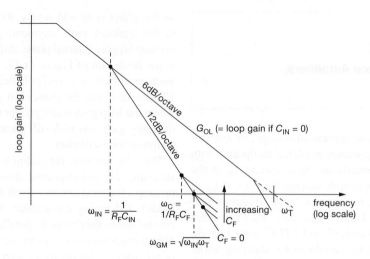

Figure 4x.17. Bode plot (log magnitude of gain vs log frequency for the transresistance amplifier). For stability the closed-loop gain curve must intercept the unity-gain axis at a 6 dB/octave slope.

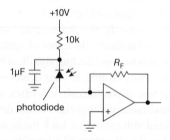

Figure 4x.18. Reverse biasing a photodiode decreases capacitance and increases speed (but at the expense of "dark current"). Be sure to use a clean, bypassed bias supply; choose the series resistor so that the drop across it is small compared with the bias voltage, at the maximum anticipated photodiode current.

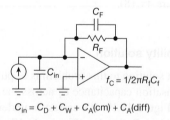

Figure 4x.19. Transresistance amplifier with stabilizing feedback capacitor C_f.

loop gain plot (Figure 4x.17) has reverted safely back to 6 dB/octave somewhat before reaching the unity gain axis.

Here's how you do it: First note that the unstabilized amplifier has the loop gain crossing the axis halfway (logarith-

mically) between $f_{RC\text{in}}$ and f_T; that is, at a frequency[9] that is the geometric mean:

$$f_{GM} = \sqrt{f_{RC_{in}} f_T}. \qquad (4x.1)$$

If we were to choose C_f so as to put f_c at that frequency, we would be living dangerously – the loop gain plot would be in the midst of reverting to 6 dB/octave as it crossed the unity gain axis; to state things more accurately, the phase shift of the feedback network would have dropped to 45°, as RC circuits always do at their 3 dB points. The result would be an amplifier that is probably stable in the sense of not oscillating, but it might exhibit overshoot and ringing following a transient; and its closed-loop frequency response would exhibit "peaking," specifically a bump of about 1.3 dB near the unity-gain crossing frequency f_{GM} (we'll call this response trace f_a).

So, we choose C_f a bit larger. A common procedure is to choose C_f so that $f_c = 1/2\pi R_f C_f = \sqrt{f_{RC\text{in}} f_T/2}$, i.e., at about 70% of the geometric mean. This generally ensures good stability, and produces a closed-loop response that is maximally flat (actually, a second order Butterworth, see Chapter 6), without any peaking, and is down 3 dB at f_{GM} (Figure 4x.20). You'll often see a parameter called the "damping ratio," with the symbol ζ (zeta). The choice

$$f_c = 0.7 f_{GM} \qquad (4x.2)$$

_____

[9] An op-amp's f_T is the extrapolated frequency at which the log–log curve of open-loop gain versus frequency crosses the unity gain axis, extended from a lower frequency where there is plenty of gain and where the slope is 6 dB/octave (i.e., $\propto 1/f$). On datasheets this is usually called the gain–bandwidth product, GBW.

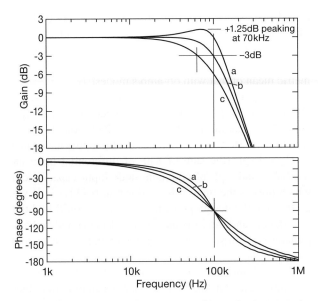

Figure 4x.20. TIA frequency response (normalized transresistance gain and phase vs frequency) for three choices of C_f, namely those corresponding to characteristic frequencies $f_c = 0.7 f_{GM}/\zeta$, with the damping ratio ζ equal to (a) 0.7, underdamped, (b) 1.0, critically damped, and (c) 1.4, overdamped. This SPICE-modeled transimpedance amplifier's bandwidth is 100 kHz; the convergence in gain at high frequencies is controlled by the op-amp, with a slope of 12 dB/octave.

corresponds to a damping ratio $\zeta = 1$.

If C_f is chosen still larger, it produces an overall amplifier bandwidth at a lower frequency (which we'll call f_b) equal to the roll-off frequency of the RC, i.e., $f_b = f_c = 1/2\pi R_f C_f$; the corresponding frequency response exhibits the usual slow roll-off characteristic of a single RC lowpass (a single "on-axis pole," Figure 1.104), with the familiar RC step response (Figure 1.34). But if C_f is chosen for maximally flat response (i.e., $f_c = 0.7 f_{GM}$), the result is to introduce just the right amount of "peaking" to extend the amplifier's response to $f_b = f_{GM}$, and to speed the step response so it smartly moves to the new output voltage with minimal overshoot (Curve b of the SPICE results in Figures 4x.20 and 4x.21, and seen in the measured $C_f = 2.4$ pF waveform of Figure 4x.26).[10]

It's customary to define the bandwidth of an amplifier by measuring the frequency at which the gain has fallen by 3 dB. But this simple measure does not take into account the possibility of overdamping, gain peaking due to underdamping, etc. The figure shows that an underdamped

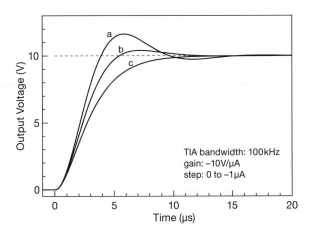

Figure 4x.21. TIA step response for the same three choices of C_f as in Fig. 4x.20.

amplifier appears to have more bandwidth. One useful approach to determining bandwidth is to measure the frequency at which the phase shift reaches $-45°$. This is particularly relevant if the amplifier is used inside a feedback loop, e.g., in a scanning tunneling microscope preamp (see §8.11.12). Measured in this way,[11] an underdamped amplifier has more bandwidth, and an overdamped amplifier has less. As an example, here are SPICE results for the -3 dB and $-45°$ frequencies of a sample TIA with $f_c = 100$ kHz:

	damping, ζ	$f_{-3\,dB}$	$f_{-45°}$
a	0.7	129 kHz	62.7 kHz
b	1.0	100 kHz	51.5 kHz
c	1.4	66 kHz	40.8 kHz

4x.3.3 An example: PIN diode amplifier

As an example, let's design a photodiode amplifier for use with a typical silicon PIN diode of 5 mm² active area. These popular devices come in a TO-5 transistor package with glass window, and are meant to be operated with a back bias of 10 V to 20 V. Examples are the S1223 from Hamamatsu and the PIN-5D from UDT, with closely similar characteristics: terminal capacitance $C_D = 10$ pF at 20 V back bias, cutoff frequency (3 dB down) $f_c = 30$ MHz (corresponding to a rise time $t_r \approx 0.35/f_c = 12$ ns), and a red-weighted visible-light response rising to a maximum close to 1 μm wavelength.

A comment on detector speed: The rise time specification tells you the "datasheet speed" of the detector (usually specified for some wavelength of incident light, and with

---

[10] In the language of "root locus" in the *s*-plane, one would say that a single on-axis dominant pole has morphed into a pair of off-axis poles.

[11] That is, ignoring other aspects such as settling time, ringing, and the like.

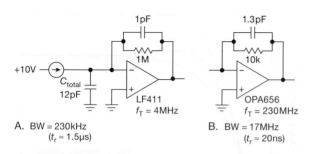

A. BW = 230 kHz
    ($t_r \approx 1.5\,\mu s$)

B. BW = 17 MHz
    ($t_r \approx 20\,ns$)

Figure 4x.22. Photodiode amplifier examples. A. Using jellybean LF411 (f_T=4 MHz), with G=1 V/μA. B. Using wideband OPA656 (f_T=230 MHz), with G=10 mV/μA.

some standard load resistance, usually 50 Ω); it depends on the detector's capacitance (which forms an RC time constant with the load resistance), and also upon the physics of charge carrier transit time in the detector itself (which in turn depends on semiconductor properties, junction geometry, and applied bias voltage). Depending upon your circuit, you may or may not achieve that "50 Ω–load" detector speed.[12] As we'll see, achieving adequate bandwidth may be harder than you think!

Let's arbitrarily choose the transresistance gain (V_{out}/I_{in}) to be 1 MΩ; that becomes the value of the feedback resistor: $R_f = 1$ MΩ. We'll see shortly that this is not a wise choice, if we care about speed. For the op-amp let's start with our standard jellybean LF411, with a gain–bandwidth product $f_T = 4$ MHz (typ). The amplifier datasheet gives no information about input capacitance, but it's probably safe to guess a value of about $C_A = 2$ pF, giving a total input capacitance $C_{in} = C_D + C_A = 12$ pF. In combination with the 1 MΩ feedback resistor, this produces a roll-off beginning at $f_{RCin} = 13$ kHz.

Next we calculate the value of feedback capacitor C_f to ensure stable operation. The geometric mean of f_{RCin} and f_T is $f_{GM} = 2.3 \times 10^5$ Hz. For optimum transient performance and good stability, we now choose C_f so that its characteristic frequency, in combination with the existing R_f, is 70% of that value: $1/2\pi R_f C_f = 0.7 f_{GM}$, giving $C_f = 1.0$ pF. The resulting amplifier has a 3 dB bandwidth of $f_b = 230$ kHz, (equal to f_{GM}), and a rise time of approximately $t_r \approx 0.35/f_b = 1.5\,\mu s$. (Figure 4x.22A)

A. Gaining speed

Our amplifier's bandwidth is only 1% of the detector's datasheet speed! And we got only 230 kHz response, even

though we used a 4 MHz op-amp. What's going on here? There are two problems, actually: The large feedback resistor formed a very low frequency roll-off (at 13 kHz) with the input capacitance; and the final bandwidth is the geometric mean of that with op-amp's modest f_T.

Let's try a faster amplifier: The low-noise JFET-input OPA627 (a JFET version of the popular low-noise bipolar OP-27) has an f_T of 16 MHz, which sounds like it should help. However, it also has a total input capacitance of $C_A = 15$ pF (the sum of 8 pF of differential input capacitance and 7 pF of common-mode input capacitance), which pushes the input roll-off down to 6.4 kHz.[13] If you go through the design procedure as above, you'll find that $C_f = 0.7$ pF, and the 3 dB bandwidth of the completed amplifier is $f_b = f_{GM} = 320$ kHz, a minor improvement over our first design.

We can improve things by using an OPA637, which is a decompensated OPA627 (G_{min}=5). Note an important fact: it's not necessary to use unity-gain compensated op-amps in a transresistance configuration if the op-amp's second breakpoint (its "second pole," the frequency at which its open-loop gain begins dropping at 12 dB/octave) is well above f_c. The OPA637's f_T of 80 MHz allows us to extend f_c from 320 kHz to a more respectable 715 kHz. But we're still suffering from the penalty of the op-amp's high input capacitance.

B. "Pedal to the metal"

OK, let's really step on the gas: The high-speed JFET-input OPA656 has an f_T of 230 MHz, and total input capacitance of $C_A = 3.5$ pF. Going through the same design procedure, you'll find that f_{RCin} is still 13 kHz (as with the LF411), but the completed amplifier's bandwidth is now $f_b = 1.7$ MHz (with a smaller value of feedback capacitance $C_f = 0.13$ pF – see comments below). This is almost an order of magnitude better speed than our first pathetic attempt (because the op-amp's f_T is nearly 2 orders of magnitude higher); but it's still more than an order of magnitude slower than the detector itself (recall $f_c = 30$ MHz). We can do somewhat better here by using a decompensated op-amp (an OPA657, f_T=1.6 GHz, G_{min}=7), which pushes f_b up to a more respectable 4 MHz. We can't go much further down the path of increasing f_T, certainly not the factor of several hundred that we evidently still need.

One reasonable solution, if full speed is needed, is to trade off noise performance for speed, by reducing the gain

---

[12] However (looking on the bright side) you may in fact be able to do better (e.g., when the detector is loaded into the low-impedance presented by a good TIA, or when it is bootstrapped).

[13] The higher capacitance is related to the op-amp's lower e_n specification, 4.5 nV/$\sqrt{Hz}$ versus 25 nV/$\sqrt{Hz}$; we'll see the important significance of that later.

of the transresistance stage, then follow it with a wideband voltage amplifier. For example, if we reduce the feedback resistor to $R_f = 10\,\mathrm{k\Omega}$ (gain of 10 V/mA), we drive the input pole up by a factor of 100, to $f_{RCin} = 1.3\,\mathrm{MHz}$. The completed amplifier's resulting bandwidth goes up by the square root of that factor, or a factor of 10, to $f_b = 17\,\mathrm{MHz}$; the corresponding value for C_f is 1.3 pF. With this design we are getting most of the detector's speed (and we could do still better with the decompensated OPA657). Whether we can accept the lower gain depends on issues of noise, which are discussed in Chapter 8 (§8.11).

Another interesting solution is to *bootstrap* the detector, greatly reducing the effective input capacitance seen at the TIA's input; we illustrate this important technique in §4x.3.4.

C. Sub-picofarad capacitors
The calculated value of feedback capacitance in our last iteration – $C_f = 0.13\,\mathrm{pF}$ – sounds awfully small; can you actually get such capacitors? That's an interesting question, but you might ask first how much "parasitic" capacitance there is between the leads of the feedback resistor itself. We treat this and similar topics in Chapter 1x (properties of components); we have found, by actual measurement, that a standard metal-film resistor ("RN55D-type") has something like $0.07\,\mathrm{pF}$ – $0.15\,\mathrm{pF}$ of parasitic parallel capacitance, the exact value depending on manufacturer and resistance value. So, you might need to add a tiny bit of capacitance across the resistor, perhaps using a "gimmick," the official term for a pair of short insulated wires that you twist up until there's enough capacitance. When dealing with circuits like this, in which a fraction of a picofarad has important effects, be careful about component placement and lead dress; for example, the feedback resistor (and perhaps the inverting input pin of the op-amp) should be raised up from the circuit board to minimize capacitance to ground and to other signals. You often see similar advice when dealing with ultra-low input currents (femtoamps), namely to float the input leads, or support them on a Teflon standoff insulator.

What should you do if the calculated feedback capacitance comes out *less* than the parasitic capacitance of the feedback resistor? One solution is to reduce the feedback resistor value until the calculated capacitance is about equal to the parasitic capacitance. This reduces the transresistance gain of the amplifier, of course, perhaps requiring additional gain downstream.[14] A clever alternative is

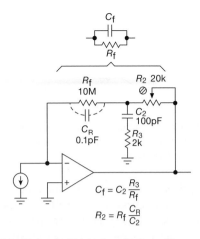

Figure 4x.23. "Pole-zero" trick when the parasitic capacitance C_R of the gain-setting feedback resistor R_f is itself larger than the calculated shunt feedback capacitance.

shown in Figure 4x.23: Here the unavoidable time constant $R_f C_R$ of the feedback resistor with its own parasitic capacitance C_R (a "zero") is canceled by a deliberate lagging time constant $R_2 C_2$ (a "pole"); then the addition of a resistor R_3 reintroduces a zero with time constant $C_2 R_3$. For example, if a certain circuit needed $C_f = 0.02\,\mathrm{pF}$ across a 10 MΩ feedback resistor, we would be in trouble because of the $\sim 0.1\,\mathrm{pF}$ of parasitic capacitance. We cancel this by choosing $R_2 = 10\mathrm{k}$ and $C_2 = 100\,\mathrm{pF}$; then we choose $R_3 = 2\mathrm{k}$, as shown.[15] We haven't seen this trick described elsewhere, but we've used it successfully in several wideband photodiode amplifiers.

Figure 4x.24 shows another trick, helpful when dealing with these small-value compensation capacitors. Here the effective feedback capacitance is the fraction of C_f set by trimmer R_1.

To recapitulate the major results: The "bandwidth" of this properly stabilized current-to-voltage amplifier is far less than the op-amp's gain–bandwidth product f_T. It is, rather, at the geometric mean of that frequency and the (much lower) characteristic frequency set by the time constant of the total input capacitance and the feedback resistor. This shows why input capacitance compromises speed,

---

[14] It also increases R_f's Johnson noise contribution – see the extensive discussion in Chapter 8 (§8.11).

[15] That is, $R_3 C_2 = R_f C_f$, where C_f was the desired feedback capacitance (0.2 pF) appropriate to the gain-setting feedback resistance R_f (10 MΩ). Note how the new effective C_f is well predicted, set by low-tolerance parts. R_2 has to be adjustable, because we don't know the value of the stray capacitance C_R; it should be adjusted for a flat gain response in the crossover region, $f = 1/2\pi R_f C_R$, about 160 kHz in this case, well below the circuit's f_{3dB} bandwidth. Choosing a 20k trimmer for R_2 allows canceling 0 pF to 0.2 pF of stray capacitance C_R.

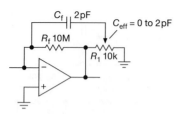

Figure 4x.24. Creating a "tunable" low-value feedback capacitor. The total effective feedback capacitance includes ~ 0.1 pF of stray capacitance of R_f (not shown).

and the surprising need for op-amps that are much faster than you might have guessed. And we'll see presently how input capacitance also degrades noise performance (we saw it initially in Chapter 8 of AoE3 (§8.11.3)).

4x.3.4 A complete photodiode amplifier design

The circuit of Figure 4x.25 should help tie these ideas together.[16] It's the basic design of RIS-617, a photodiode amplifier that has been in wide use in the laboratories at our Rowland Institute. For the transimpedance stage we chose the JFET-input OPA637 (or equivalent ADA4637) for its combination of low input current and low noise voltage (e_n=4.5 nV/$\sqrt{\text{Hz}}$), combined with wide bandwidth (f_T=80 MHz). This op-amp is the decompensated ($G >$ 5) version of the unity-gain-stable OPA627 (which has f_T=16 MHz); it's suitable for a transimpedance application like this, owing to the aggressive compensation provided by the external compensation capacitor C_f. And of course that extra bandwidth translates into improved amplifier speed.

Less obvious is the desirability of low op-amp input noise voltage, e_n. At first sight it might seem to be of little concern, perhaps contributing only that quantity of noise voltage at the output. That would be wrong. In fact, as we saw in Chapter 8, the op-amp's e_n grinds up against the input capacitance C_{in} to create an effective input noise current $i_n = e_n \omega C_{in}$ (which we like to call "$e_n C$" noise). This can easily dominate over all other sources of noise, particularly when you're striving for substantial bandwidth.

In this circuit we chose R_f for a modest first-stage transimpedance gain (0.1 V/μA), with a second stage of selectable voltage gain to set the overall instrument gain. The trade-off is speed (smaller R_f, thus greater f_{RCin}) versus noise (larger R_f, thus less Johnson noise current $i_n = \sqrt{4kT/R_f}$; see §8.11).

In a transimpedance amplifier input capacitance is the

villain: it drives down the bandwidth (via necessarily larger C_f), and it drives up the noise (via the input i_n produced by the op-amp's e_n imposed across C_{in}). Because high-performance transimpedance amplifiers are generally expected to work properly with rather large input capacitances[17] (up to 1000 pF), we added a 2-stage bootstrap follower ($Q_1 Q_2$), which reduces the effective input capacitance at signal frequencies by roughly a factor of ten (thus 100 pF maximum). Q_2 is a very low noise (0.8 nV/$\sqrt{\text{Hz}}$) JFET[18] of high transconductance (~ 25 mS, thus $\sim 40\,\Omega$ output impedance), here buffered by Q_1 for plenty of drive muscle. See §8.11 for more detail.

To figure the compensation capacitor C_f, we take the maximum effective C_{in}=100 pF, for which f_{RCin}=16 kHz, f_c=1.1 MHz, and critical damping ($\zeta = 1$) would require C_f=2.1 pF. The 4 pF shown in the diagram is quite conservative, and produces an overdamped response, as shown in the measured response traces of Figure 4x.26.

The second stage is a wideband voltage amplifier, here implemented with a current feedback (CFB) op-amp. The LT1217 maintains 5 MHz bandwidth at G=10, with decent accuracy (V_{os}=3 mV max) and noise (e_n=6.5 nV/$\sqrt{\text{Hz}}$). The offset trim network is worthwhile, given the 0.5 mV maximum offset of the input stage; you can think of it as a trim of the combined 2-stage offset, if that makes you happier. Both amplifier stages use "high-voltage" op-amps (i.e., ± 15 V supplies), permitting output swings to ± 10 V.[19]

The lowpass filter $R_5 C_2$ between the stages, with its 300 kHz breakpoint, is important in reducing out-of-band noise. However, as discussed in §8.11.3, we struggle with an ugly current noise whose spectrum rises with frequency. The single lowpass pole introduced at f_c cancels the rising noise density, but still leaves a noise spectrum flat with frequency (although we do benefit from an additional pole at f_{GM}).

Sharper lowpass filter.

Better, though, to add a steeper lowpass cutoff to limit the rms noise degradation, which we can do by turning U_2 into a second-order filter, with the addition of two parts, see

---

[16] "That rug really tied the room together..."

[17] It's unusual to find a case where the bootstrap isn't a critically important part, the exception being tiny sensors (of the type used for fiber optic receivers).

[18] Sadly *discontinued!* But the CPH3910 from ONSemi is just as good, and available as a dual (CPH6904).

[19] With the photodiode configured to sink current, as shown here, the output will only go positive; but the circuit happily accepts input currents of either polarity.

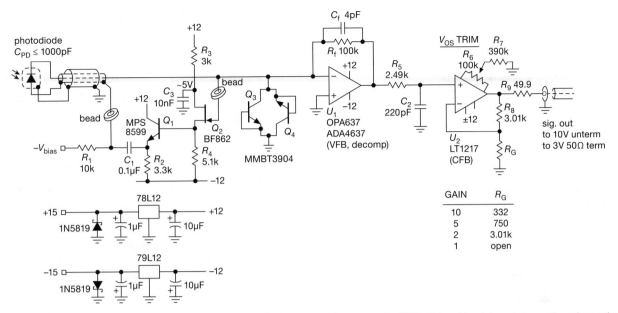

Figure 4x.25. A complete photodiode amplifier, suitable for input capacitances up to 1000 pF. Input bootstrapping greatly reduces the effective photodiode and cable capacitance, for enhanced speed and reduced noise.

Figures 4x.27 and 4x.28.[20] The filter's second pole is less effective for gains <10, where however you're dealing with larger signals, thus less sensitivity to noise filtering. From the response curves (Fig. 4x.28) you might initially choose $C_1 = 140$ pF for its pretty curve; but sometimes it's nice to exploit a peaky response (e.g., with $C_1 = 160$ pF) to extend somewhat the response of an amplifier in its rolloff region.

Expensive amplifiers need protection; U_1 will set you back $30 (!), so we added diode clamps Q_3Q_4 (an *npn* base–collector junction is an inexpensive diode of low capacitance and very low leakage; see for example Figure 5.2).

4x.3.5 Gain-switching

The TIA stage in Figure 4x.25 has been set at a fixed gain ($G=v_{out}/i_{in}=-100$kΩ), with a selection of higher gain steps provided by the 1–2–5–10× voltage-amplifying stage U_2. That's OK – but we could do better if instead we increased R_f to go to higher gain: that's because the (fixed,

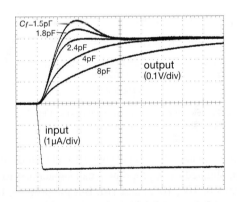

Figure 4x.26. Measured response of the transresistance stage (U_1) of Figure 4x.25 to a current step input, with an input (summing-junction) capacitance of 100 pF, for several values of feedback capacitor C_f. Horizontal: 200 ns/div.

100k) low value of R_f introduces more current noise than the alternative of using higher values of R_f when we want more gain.[21]

Figure 4x.29 shows two approaches to finessing this problem. We've used the same $f_T = 80$ MHz op-amp, and we've assumed there's some 25 pF of input capacitance. We should be able to achieve 500 kHz of bandwidth with a gain of 1 MΩ. In circuit A we took the simple solution of a

---

[20] This is a modification of the VCVS lowpass filter (Fig. 6.28A), in which we've set $G=10$ and, keeping $R_1=R_2$, we've chosen the capacitor values to produce a smooth low-Q roll-off to -12 dB/octave. Put another way, we've broken the rules for a VCVS Butterworth lowpass filter: for $G=10$ choose C_1 and C_2 to be 0.4 and 2.5 times the respective canonical value of $C=1/2\pi Rf_c$. This generalization of the Sallen-and-Key filter was discussed in AoE3's §6.3.2D.

[21] Recall §8.1.1, where resistor current noise is seen to go as $i_n=\sqrt{4kT/R}$.

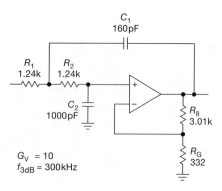

$G_V = 10$
$f_{3dB} = 300\,kHz$

Figure 4x.27. Two-pole lowpass filter with $G=10$ to replace U_2 in Fig. 4x.25.

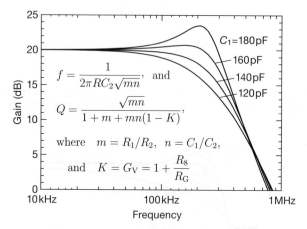

$$f = \frac{1}{2\pi RC_2\sqrt{mn}}, \text{ and}$$

$$Q = \frac{\sqrt{mn}}{1+m+mn(1-K)},$$

where $m = R_1/R_2$, $n = C_1/C_2$,

and $K = G_V = 1 + \dfrac{R_8}{R_G}$

Figure 4x.28. Response of the lowpass filter of Fig. 4x.27, as simulated in SPICE. Note the sensitivity to C_1.

3-position (center-OFF) toggle switch to select the feedback resistor, shunted by appropriate small capacitors, to select TIA gains of 10k, 100k, and 1M. But the problem here is that the switch, in the middle (open) position, introduces some 0.5 pF to 1.5 pF of capacitance that effectively shunts R_f, overriding the 0.3 pF that's supposed to be across the 1M feedback resistor (there's about 0.1 pF parasitic capacitance of R_f itself, thus the 0.2 pF explicit capacitor shown). That kills our target 500 kHz bandwidth![22] We might try replacing the toggle switch with a CMOS SPDT switch; but the parasitic capacitance in the OFF state is even worse than the mechanical switch – about 5 pF.

Circuit B is the better solution: here we've used a CMOS switch, wired "backwards" (in current-steering mode),

---

[22] There is a workaround, of sorts, by adding 10 to 20 pF capacitors to ground at each switch terminal – but this is ugly, loading the 80 MHz op-amp, and it's only a partial solution because there's some capacitive signal leaking through, lowering the gain.

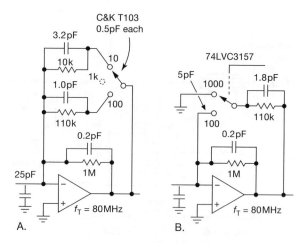

Figure 4x.29. Minimizing bandwidth-killing capacitance while gain-switching the TIA with several values of feedback resistor. Circuit B nicely circumvents the degrading effects of parasitic switch capacitance in circuit A.

with the two signal terminals always at ground potential. They switch *currents* (rather than voltages), a scheme we first encountered in Figure 13.47 in Keysight's "Multislope III" converter. The ground potential means we can use a low-voltage switch, even with high-voltage signals. We chose a '3157-type switch (see §13.8.5 and Table 13.7) for its low capacitance (accepting its moderately high R_{ON}).

In the high-gain position (switch open) we're adding an acceptable 5 pF to the summing junction, while the damaging 110k and 1.8 pF currents are safely shunted to ground. In the low-gain 100k position the switch capacitance increases to 17 pF, forcing us to increase C_f to 1.8 pF, slightly reducing the bandwidth. The 74LVC1G3157 is available in SOT23-6 and SC70-6 packages, the latter small enough to squeeze in at the summing junction. If you want three gain choices with R_f switching, you can add a second switch.

4x.3.6 Some loose ends

• Note that you generally can assume that the op-amp is behaving like a classical 6 dB/octave, 90° lagging phase amplifier; you can take a "single dominant pole" model, in engineer's lingo, and you don't have to worry about additional phase shifts that usually creep in as you approach the op-amp's f_T. That's because the large feedback resistors used in these amplifiers usually put the input pole f_{RCin} at a very low frequency, such that the trouble region at $f_{GM} = \sqrt{f_{RCin}f_T}$ is far below the frequencies at which the op-amp departs from the single-pole model.

• For the same reason – namely that the loop gain is

brought to unity far below the op-amp's f_T – you can safely use *decompensated* op-amps in this circuit, as we did with the OPA637 in our second design attempt. That more than doubled the speed, because it has 5 times greater bandwidth ($f_T = 80\,\text{MHz}$); but it is only stable at closed-loop gains of 5 or greater. A warning, however: if you use decompensated op-amps in this circuit, make sure that the ratio of C_{in} to C_f is greater than the minimum specified closed-loop gain (this will usually be true), because that ratio sets the high-frequency closed-loop gain of the transresistance amplifier. Also note that circuit stability depends upon a minimum input capacitance, so the circuit may oscillate with the input unplugged.

- Following similar reasoning, it is possible to raise the gain–bandwidth product inside the feedback loop by cascading two op-amp stages, properly configured. The stability of this arrangement depends on careful placement of the unity-gain crossing of the loop gain within this transresistance configuration – don't try this trick with a conventional *voltage* amplifier!

- We have only briefly discussed here the important issue of *noise* in transresistance amplifiers (which are often called upon to amplify very small signals). That is treated in Chapter 8 of AoE (devoted to low-noise design, including both discrete and op-amp voltage amplifiers), where there's a discussion of the unfortunate property of transresistance amplifiers of converting internal op-amp voltage noise into an effective input current noise, which rises proportional to frequency and to total C_{in}.

- As in Figure 4x.25, it's good hygiene to put low-capacitance protective clamps at the input of TIAs in which there's risk of high-voltage transients (from a biased detector, etc.). A nice (but unrelated) trick we've seen[23] is the deliberate use of the CMOS op-amp's input protection diodes to reset the integrator capacitor – by simply pulsing the supply rails with a momentary (and current-limited) polarity reversal.

- To trim the compensation it's a good idea to use a test fixture that can provide a clean nanoamp-scale square wave. This is discussed in §8.11.13 in AoE3, with a suggested circuit (Fig. 8.91).

4x.3.7 Designs by the masters: A wide-range linear transimpedance amplifier

The dynamic range (i.e., ratio of maximum to minimum input current) of a resistance-feedback transimpedance amplifier like that in Figure 4x.16A is limited by several factors. First, the op-amp's input bias current sets an approximate lower bound on measurable currents, typically somewhere around a picoamp for op-amps of reasonable precision (but as low as 10 fA for less accurate CMOS parts[24]).

Second, for a given feedback resistance R_f, the current range is bounded at the high end by the op-amp's supply voltage, i.e., $I_{max} \lesssim V_S/R_f$; and at the low end it becomes inaccurate when the output approaches the op-amp's offset voltage, i.e., $I_{min} \gtrsim V_{os}/R_f$. These constraints limit the dynamic range to $I_{max}/I_{min} \lesssim V_S/V_{os}$. In practical terms, that's roughly a dynamic range of 10^5 ($\sim 10\text{V}/100\mu\text{V}$) for low-offset op-amps.

One way to get a larger dynamic range is to use nonlinear diode-like feedback to create a logarithmic response (see §4x.20). For some applications this is just what you want. But this method suffers from several drawbacks: (a) you can't average or lowpass-filter the output voltage to get the average input current (because the average of the log is not the log of the average); (b) it's difficult to get significant precision, say at the part-per-thousand level, owing to drift and calibration uncertainties; and (c) you often want good linearity with signals of both polarities (and which cross through zero).

What you want, then, is a linear precision TIA that somehow spans multiple sensitivity ranges simultaneously. Figure 4x.30 shows an elegant implementation, devised by Stephen Eckel and his team at Yale University.[25] The basic topology is a standard resistive-feedback TIA, here implemented with op-amp A_1 and a series string of resistors (R_1–R_3, with successive ratios of 100:1) that individually would span four decades of full-scale sensitivity. To this basic circuit JFETs Q_1–Q_4 have been added, which go into conduction progressively as each range reaches full-scale output; this prevents op-amp A_1 from saturating, as explained below.

Here's how it works: for the lowest input currents ($I_0 < 100\,\text{nA}$) the voltage developed at A_1's output is just $I_0(R_1+R_2+R_3)$, and output amplifier A_2's gain (approximately $\times 2$) is chosen to produce 10 V output (called V_1) for 100 nA input. That input current is full-scale for the most sensitive range, and at that current the op-amp's output is approximately 5 V. The other output amplifiers A_3

---

23 Thanks to Bernie Gottschalk for this elegant suggestion.

24 With the stunning exception of the ADA4530-1, with its 1 fA typ (20 fA max) bias current at 25°C, remarkably combined with a precise 9 μV typ (50 μV max) offset voltage. Its noise performance is, uh, underwhelming – some 80 nV/$\sqrt{\text{Hz}}$ at 10 Hz.

25 S. Eckel, A.O. Sushkov, and S.K. Lamoreaux, "A high dynamic range, linear response transimpedance amplifier," *Rev. Sci. Instrum.*, **83**, 026106 (2012).

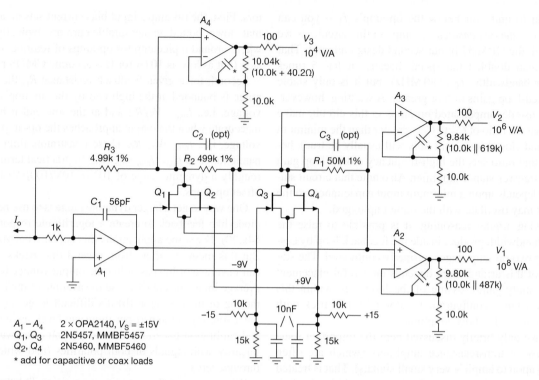

Figure 4x.30. Wide dynamic-range linear transimpedance amplifier, with three simultaneous outputs with sensitivities in ratios of 100:1. The JFETs Q_1–Q_4 shunt current around each gain-setting feedback resistor after the respective stage passes its full-scale output.

and A_4 generate the outputs for the higher-current ranges: A_3 picks off the voltage developed by I_0 flowing through R_2+R_3, and A_4 picks off the voltage developed by I_0 flowing through R_3 alone.

Now for the trick: for input currents significantly greater than 100 nA, A_1's output would saturate, but that is prevented by Q_3 or Q_4. For example, an input sinking current of 200 nA, which would bring A_1's output to +10 V, instead causes p-channel JFET Q_4 to conduct (its gate is biased at +9 V), effectively shunting R_1 and preventing saturation. So the TIA loop remains in the active region, and A_3's output (V_2, with its sensitivity of 1 V/μA) will be at the correct +0.1 V. Similarly, input currents great enough to bring V_2 beyond full-scale cause Q_2 to shunt current around R_2, again preventing A_1's saturation.

Figure 4x.31 shows measured performance, plotted on log–log axes, illustrating nicely the stacking of simultaneous linear outputs, each of which saturates just beyond full-scale. These data were taken for positive output polarity only (i.e., input sinking current, shown as I_0 in the circuit diagram), but the circuit as drawn works properly for both polarities, notionally illustrated (on linear axes) in Figure 4x.32 and seen accurately in the measured data of

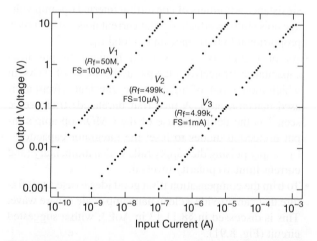

Figure 4x.31. Measured simultaneous outputs of the wide-range transimpedance amplifier of Figure 4x.30, demonstrating a dynamic range of 10^7:1. For V_2 and V_3, corrections (i.e., zero-current offsets) of 0.12 mV and 0.045 mV were applied; they barely nudged the lowest few points.

Figure 4x.33. If only one polarity is required, the complementary JFETs in Figure 4x.30 can be omitted.

A few comments on the circuit: (a) C_1 provides fre-

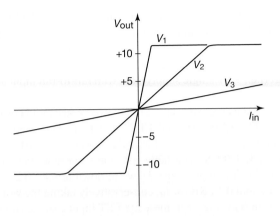

Figure 4x.32. The wide-range TIA, implemented with split supplies and with JFETs of both polarities (Fig. 4x.30), generates simultaneous linear outputs that transition smoothly through zero, as indicated in this sketch (with gain ratios of 5:1) and in the measured data in the next figure.

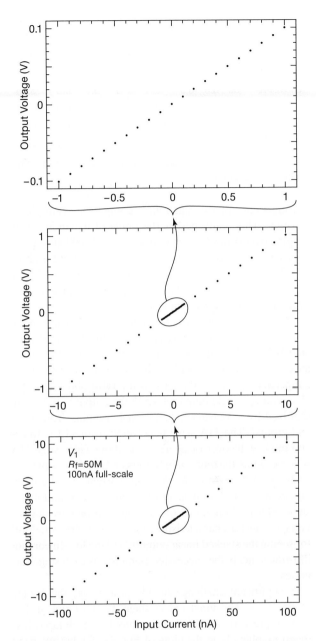

quency compensation; the 56 pF value we used would need to be increased for a capacitive input source.

(b) Additional compensation capacitors C_2 and C_3 can be used to tailor frequency response of the individual ranges; we omitted them for our measurements.

(c) The series 1k input resistor protects A_1 if the input is overdriven; it has no effect on performance.

(d) We chose OPA2140 op-amps for their low V_{os}, low I_B, and good value (see the mini-table below); the low-current performance could be improved somewhat with OPA627B's (expensive), or considerably more with the ADA4530-1 (but with poorer noise performance).

(e) To obtain the full dynamic range, it's necessary that the leakage currents of JFETs Q_1–Q_4 be no more than a few picoamps. The guidance provided by the datasheets will leave you sleepless: worst-case gate reverse currents of 1000 pA and 5000 pA (for *n*- and *p*-channel, respectively, at room temperature and 15 V–20 V reverse bias). And, if you worry about *channel* leakage when cutoff, the datasheet is particularly unhelpful, specifying only a range of gate voltages to produce "cutoff" drain currents of 10,000 pA and 1,000,000 pA (for *n* and *p* types, respectively). Happily, the manufacturers are highly conservative, and the actual situation is far better: we measured total leakage currents (at room temperature, and at gate-to-drain voltages to 15 V) ranging from 0.1 pA to 3 pA for 17 JFET samples; see the discussion in §2x.1.

Figure 4x.33. These linear plots of output voltage versus input current for the most sensitive scale (i.e., V_1: 0.1V/nA) show the TIA's linearity straight through zero input current. No corrections have been applied to these measured data.

4x.3.8 A "starlight-to-sunlight" linear photometer

Here's a nice application for a wide-range linear transimpedance amplifier of the kind described in §4x.3.7 (which you should read first): a photometer that measures

Type[a]	V_{os} (μV)		I_b (pA)		e_n (nV/◊)		i_n (fA/◊)	Price qty 10 (US$)
	typ	max	typ	max	10Hz	1kHz	typ	
OPA627B	40	100	1	5	15	5.2	1.6	30.81
OPA2140[b]	30	120	0.5	10	8	5.1	0.8	5.38
ADA4530-1[c]	9	50	0.001	0.02	80	16	0.07	21.72

Notes: ◊ = nV/√Hz. (a) except as noted, V_S=36V max total supply. (b) dual. (c) V_S=16V max total.

Figure 4x.34. Op-amp choices for wide-range linear transimpedance amplifier.

a wide range of illumination,[26] going from sunlight to starlight. This really pushes the limits of dynamic range, ranging over some 8 orders of magnitude: bright sunlight is approximately 110,000 lux, and dark-sky starlight is about 0.002 lux (100 times darker than the full moon's 0.25 lux). And we'd really like 9 orders of magnitude, to give us 10% accuracy at the lowest light levels.

Because the photocurrent is unipolarity, we can run the circuit from a single supply polarity (almost – see discussion below). We also need only p-channel JFETs for the successive resistor clamping circuits. We chose the transimpedance gains of the taps in the ratio 300:1 (R_1, R_2, R_3), with full-scale output of +5 V; the output voltages are digitized by a 3-channel, 12-bit ADC running from +5 V and ground. The TIA stage is powered from +12 V, to accommodate the over-range behavior of the wide-range TIA scheme, while the unity-gain buffers and the ADC run from +5 V; the 10k resistors R_4–R_6 limit current into the input clamp diodes to less than 1 mA. Figures 4x.36 and 4x.37 show SPICE simulations of the input-current to output-voltage transfer characteristics: the log–log plot is helpful for seeing the stacked linear outputs, and the log–linear plot illustrates nicely the successive clamping as each stage saturates.

A significant challenge in this design was the choice of op-amps. The input (TIA) stage A_1 must run at 10–12 V to preserve a 5 V full-scale output on each tap (illustrated graphically in the plots of Fig. 4x.37), but we need more: its input current must be down in the 0.5LSB range (i.e., <6 pA), and its offset voltage should be less than the

---

[26] Officially, *illuminance*, which is the flux of light per unit area (lumens/meter) upon a surface, weighted to take account of the eye's spectral sensitivity. Photometric units can drive you nuts, with candela, lumen, lux, and a host of confusing names like radiant flux, radiant intensity, radiance, irradiance, radiosity, radiant exitance, radiant exposure, luminous flux, luminous intensity, luminance, illuminance; and most of these can host the modifier "spectral," meaning the same photometric quantity per unit frequency (or per unit wavelength). See also §9x.22.

voltage step corresponding to 0.5LSB of the ADC (i.e., <0.6 mV). It also must operate with the inputs at ground, and its output voltage range must extend to ground (RRO). For the buffer op-amps A_3 and A_4 we can use a low-voltage part (i.e., +5 V single supply), but with rail-to-rail input and output.

With these constraints we found two good candidates for A_1 – TI's LMP7701 and LTC's LTC6240HV – and two candidate dual op-amps for A_3A_4 – ADI's AD8616 and LTC's LTC6078. The table lists their relevant specifications. For the input stage A_1 we're down around 0.2 LSB for V_{os} and 0.1 LSB for I_B, conservatively taking the worst-case (max) values. But these are FET inputs, so we should expect significantly higher bias currents at elevated temperatures; even so, we're OK up to 60°C or so. For the followers A_3A_4 the dual op-amps AD8616 and LTC6078 meet our requirements.

The photodiode was chosen to match the amplifier's gain and dynamic range. The S1133-14 is an inexpensive silicon photodiode in a ceramic package, producing 3 mA in full sunlight; it's being operated here in photovoltaic mode (zero bias), but it does see the op-amp's offset voltage across its terminals. No worry, though – the S1133-14 photodiode spec shows dark current (with an extravagant 10 mV bias) as 0.2 pA (typ), and sloping down to an extrapolated current less than 0.1 pA at 1 mV bias. The S2387-33R photodiode does even better, in fact specifying a maximum leakage of 5 pA at 10 mV bias (and a typical value of 0.1 pA).

"Getting to ground"

And now for the troublesome business of RRO op-amps not delivering on their promise: the cruel fact is that most RRO op-amps operating with a single positive supply cannot bring their output fully to ground, even when unloaded. The V_{OL} specs in the table (taken from the datasheets) show this problem, which we discuss in further detail in §§4x.11.3 and 4x.11.4; but, put simply, the quiescent current through the op-amp's push–pull output stage produces a drop across the pulldown transistor's R_{ON}. In §4x.11.4 we show two ways of dealing with this (sinking current from the op-amp's output, or bringing the op-amp's negative supply terminal a hundred millivolts or so below ground). We prefer the latter, and we've indicated such a connection on the schematic.

Some additional circuit details: (a) Because full-scale current for the least-sensitive range is 5 mA (compared with 1 mA for the circuit of Fig. 4x.30), we chose a larger JFET (type J175), which specifies I_{DSS}=7 mA (min). You may worry about JFET leakage current compromising the

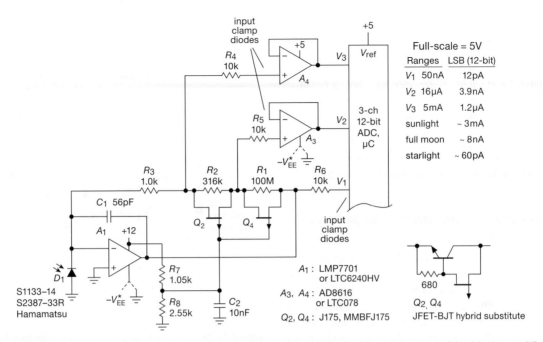

Figure 4x.35. Starlight-to-sunlight linear photometer, based on the method of §4x.3.7. The V_{SS} terminal of the "single-supply" op-amps should be powered from -100 mV or so, to ensure operation all the way to ground; see §4x.11.3.

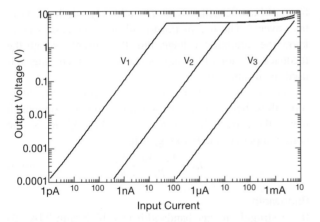

Figure 4x.36. SPICE simulation of the three output voltages versus (sinking) input current, for the circuit of Fig. 4x.35.

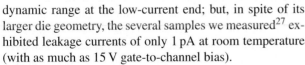

Figure 4x.37. Same data as Fig. 4x.36, plotted on log-linear scales.

dynamic range at the low-current end; but, in spite of its larger die geometry, the several samples we measured[27] exhibited leakage currents of only 1 pA at room temperature (with as much as 15 V gate-to-channel bias).

(b) Another way to extend the operating current is to add a small *npn* BJT to Q_2 and Q_4 as shown in the inset in Figure 4x.35. If you go this route, be warned that BJT leakage

is highly unpredictable, as we found by actual measurement, see §2x.1.

(c) The loose V_P specification makes it difficult to set Q_2's and Q_4's gate bias: 3 V (min), 6 V (max). We've complained about this JFET ugliness before, you know the drill. It's probably best to trim the bias manually (and while you're at it, measure the output with zero input current, to ensure that leakage effects do not prevent proper operation down to $\sim$10 pA or so).

---

[27] See the discussion of transistor leakage in §2x.1.

Type	V_{supply} (V)	$I_b{}^a$ typ (pA)	$I_b{}^a$ max (pA)	V_{os} typ (µV)	V_{os} max (µV)	V_{OL} typ (mV)	V_{OL} max (mV)	I_S typ (mA)	GBW typ (MHz)
LMP7701	2.7-12	0.2[b]	1[b]	37	200	40[c]	50[c]	0.8	2.5
LTC6240HV	2.8-11	0.5[b]	1[b]	60	250	15	30	2.7	18
AD8616[d]	2.7-5.5	0.2	1	80	500	7.5[e]	15[e]	1.7[f]	24
LTC6078[d]	2.7-5.5	0.2	1	25	100	1	–	0.06[f]	0.75

Notes: (a) at T=25°C. (b) at V_S=+10V. (c) when sinking 0.5mA.
(d) dual. (e) when sinking 1mA. (f) per amplifier.

Figure 4x.38. Op-amp choices for Fig. 4x.35.

4x.3.9 Autoranging wideband transimpedance amplifier

In §4x.3.7 we presented a linear TIA with a seven-decade dynamic range. The trick was to generate three simultaneous linear outputs, each with its own gain factor (going by factors of 100); so the output with the highest gain pins at full scale, while the others continue working. The input current is then read from the most sensitive non-saturated range. We used this same trick in §4x.3.8 to create a "starlight-to-sunlight photometer" with linear outputs and with sensitivity to photocurrents from 1 pA to 5 mA (a factor of 5×10^9).

As nice as those circuits are, they do have the drawback of rather limited bandwidth. For example, in Figure 4x.30 the "bypass" JFETs Q_3 and Q_4 add bandwidth-robbing capacitance to ground at the downstream side of the highest-value (50MΩ) feedback resistor R_1. Ordinary (single-gain) transimpedance amplifiers, lacking bypass JFETs, do not suffer from this problem.

But there's another way to achieve wide dynamic range in a linear transimpedance amplifier, based on a suggestion by the ever-creative John Larkin. Look at Figure 4x.39, which illustrates this novel technique in the context of a wideband photodiode amplifier (where we've chosen OSI's PIN-13D silicon photodiode: 13 mm² area in a TO-5 hermetic package).

In this unusual circuit the lower-sensitivity ranges start operating after each more-sensitive range saturates, and (unlike the earlier schemes) the input current is derived as a weighted sum of the outputs of all the individual gain stages, as we'll see presently.

In this single-polarity circuit, TIA amplifier A_1 (a wideband RRIO with 3 pA bias current) operates normally until enough input current causes its output to saturate at the positive supply rail. Further input current causes A_1 to lose control of its summing-junction node, whereupon the voltage at its inverting input drops below ground enough to

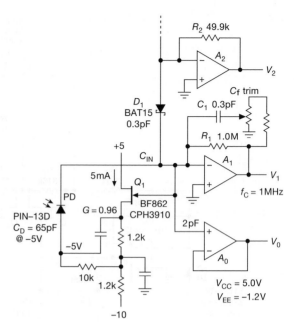

Figure 4x.39. Linear autoranging TIA. When high-gain TIA A_1 saturates, lower-gain A_2 absorbs the additional input current, while A_0 corrects for the offset at A_1's summing junction.

forward bias Schottky diode D_1, allowing amplifier A_2 to take over.[28] The use of rail-to-rail op-amps allows us to know the saturation voltage, thus the current through the feedback resistor when stage A_1 (and subsequent stages, if used) is in saturation.

Once A_2 takes over, A_1's summing junction (buffered by A_0) will be below ground by a Schottky diode drop, which means the current through R_1 is now $(V_1 - V_0)/R_1$, so the correct input current (sinking) is just

$$I_{in} = \frac{V_1 - V_0}{R_1} + \frac{V_2}{R_2} \tag{4x.3}$$

Bandwidth

It's a struggle to get bandwidth in a high-gain TIA. To achieve 1 MHz in a TIA with 1MΩ feedback resistance we needed a feedback capacitance C_f of 0.16 pF, implemented here with a 0.3 pF capacitor "tuned" with a trimmer.

The total capacitance at the summing-junction node C_{in} is the sum of Q_1's feedback capacitance C_{rss}, plus the common-mode and differential capacitances of A_0 and A_1, plus D_1's capacitance at zero bias, plus the (bootstrapped) capacitance of the photodiode.[29] The latter is a bit less than 5% of the photodiode's 65 pF capacitance at −5 V, thanks

---

[28] The photodiode's bias will drop slightly, but this is of no consequence.
[29] See §8.11.9.

to the bootstrap from JFET follower Q_1 (whose voltage gain is >0.95). These terms add up to about 11.2 pF, for an f_{RC} of 14.2 kHz – the penalty over a conventional single-gain TIA is only 2.3 pF, great!. With the 100 MHz rail-to-rail OPA357 for A_1 and a CPH3910[30] for Q_1, we get an f_{GM} of 1.2 MHz, and the damping factor for $f_c = 1.0$ MHz is about 0.8 (see §4x.3.2). Overall we're getting a bandwidth improvement (for equivalent transimpedance gain) of some $10\times–20\times$, as compared with the JFET-bypass scheme of §4x.3.8.

Some details.

(1) For D_1 we chose a small-die Schottky diode, BAT15-03W, for its low 0.3 pF capacitance. At 100 μA (where A_2 goes into saturation) the diode's drop is about 180 mV, which adds about 4% current through R_1; that is why follower A_0 is needed to get 1% precision in this circuit.

(2) Because the summing junction is pulled a diode drop below ground when A_1 saturates, it's necessary[31] to power its negative rail a volt or so below ground; here we chose -1.2 V (two diode drops, easily generated).

(3) We're powering our op-amps with a precision $+5.0$ V supply. The OPA357 has a 7.5 V absolute maximum rating, enough for us to use more than a -1.2 V negative rail. If you wanted a higher saturated output voltage (say $+10.0$ V), this would be a good place to use a composite amplifier configuration (see Figure 8.78 for inspiration).

(4) This scheme can be extended to additional stages, by adding another Schottky diode and (lower gain) TIA stages at the top. Each stage (except the topmost) needs a follower to track its summing-junction offset at saturation; and each stage increases the negative offset of lower stages by a diode drop. So it may be better to use the "JFET-bypass" method (Figure 4x.35) for the lower-gain stages, where higher capacitances wouldn't matter. But in the example here, A_1 is the special high-gain stage where it's hard to achieve a 1 MHz bandwidth in a TIA stage with 1MΩ feedback resistance.

4x.3.10 Multiple-range cascode-bootstrap wideband TIA

Continuing the theme of wideband transimpedance amplifiers, the critical bandwidth-robbing (and e_nC noise-degrading) issue of summing-junction capacitance in photodiode amplifiers can be nicely addressed by a combination of an isolating cascode transistor and a bootstrap of the low side of the photodiode. We've seen these before

(JFET bootstrap: Figure 4x.39 and §8.11.9; and cascode with bootstrap: §8.11.10), but not in the context of gain-switching.

Here we combine these bandwidth-enhancing techniques with the ability to switch the transimpedance gain. Figure 4x.29 showed the best way to connect an SPDT switch to change the gain of a wideband TIA, but it did not include circuitry to reduce summing-junction capacitance, which is almost always needed in a wideband TIA.

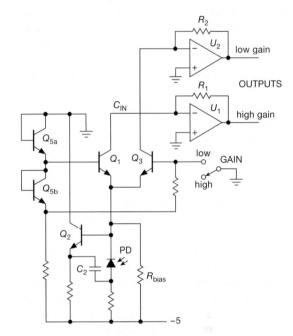

Figure 4x.40. Cascode-bootstrap wideband transimpedance amplifier with multiple switch-selected gain outputs.

Figure 4x.40, based on suggestions by Phil Hobbs,[32] fills this gap. In this circuit Q_2 is the bootstrap follower that dramatically lowers the effective photodiode capacitance, which is then isolated from the summing junction by cascode transistor Q_1. This enables both greater bandwidth and lower e_nC noise.

Before explaining the gain-switching, some important comments on the circuit so far:

(1) Q_2's operating current must be somewhat higher than the maximum photodiode current, making its base current

---

[30] To replace the popular BF862, inexplicably discontinued by NXP.

[31] But see §4x.11.3.

[32] For a good first reading assignment, try his "Photodiode Front Ends – The REAL Story," *Optics and Photonics News*, April 2001, pp 42–45. And follow that with his book *Building Electro-Optical Systems, Making It All Work*, 2nd ed., Wiley (2009), a fine collection of tricks for designing cascode photodiode amplifiers, including series peaking inductors and T-coils to extend the bandwidth, noise cancellers, and more.

a possible issue at low photodiode currents. Hobbs suggests using a high-beta MPSA18 (similar to 2N5089); the MMBT6429 and MMBT5962 are candidate surface-mount alternatives.

(2) The photodiode capacitance C_{PD} is in parallel with Q_1's base-emitter capacitance C_π (see §2x.11). This lowers Q_1's effective f_T, which we want to keep well above f_c. For example, a 2N5089 has an f_T of about 2 MHz at 10 μA (see Figure 2x.80 and note how f_T increases roughly proportional to I_C at low currents, due to fixed $C_\pi = C_{je}$ values below 30 μA). To make the cascode work at low photodiode currents we exploit Q_2's bootstrapping to reduce C_{PD} to values well below C_π of Q_1. We may also need to add R_{bias} to maintain a minimum current through Q_1 (Hobbs added 8 μA).[33]

Now for the gain-switching: We start by adding dual transistor[34] Q_5 to generate two stacked voltages just below ground, and we use the first voltage to bias our primary cascode transistor Q_1. Generally you'd dedicate Q_1 to the highest-gain stage. The second lower voltage is used to back-bias the unused secondary cascode-transistor candidates Q_3, Q_4, etc. These transistors can be turned on individually to take over the photodiode current, by switching their base to ground.[35] This scheme can be extended to additional range stages, each with its separate output. Unlike the two autoranging approaches discussed earlier, in this scheme you must select the single active range and output. However, you could use one of the cascode transistors to feed a JFET-bypass or diode-stacked autoranging circuit, as described in §§4x.3.7, 4x.3.8, and 4x.3.9.

---

[33] You can cancel most of this extra current with resistors from the summing junction to the positive supply, for example as shown in Fig. 8.87. But if you do that in the gain-switched circuit here (Fig. 4x.40), the "off" op-amp will saturate at the negative rail). These considerations reduce the attractiveness of this circuit for photocurrents below 100 nA.

[34] Some dual *npn* transistor choices are the DMMT3904W or MMDT3904. We like Diodes Inc.'s selection of dual transistors in SOT-23-6 (SOT-26) or SC-70-6 (SOT-363) and smaller(!) packages.

[35] Increasing the photodiode's back-bias by half a volt, of no consequence.

4x.4 Unity-Gain Buffers

This is a class of wideband unity-gain (usually) amplifiers, intended for substantial output currents (~ 100 mA or more), to be used within a feedback loop to supplement the output current of an ordinary op-amp; by itself such a buffer is neither precision (offset voltages typically of 10 mV or more) nor necessarily gain-accurate (for some parts the voltage gain can be $G_V = 0.95$ or even lower when loaded). But they are useful in many situations where an op-amp alone won't do the job. Table 4x.1 on page 303 lists all the unity-gain buffers we could find.

We made use of these handy ICs for several applications in AoE3:

Signal buffer into cable impedance

See, for example, the photomultiplier amplifier of Figure 12.83 in §12.6.2, where we used a BUF634 wideband unity-gain buffer to launch clean pulses into a $50\,\Omega$ cable.

High output current

See, for example, the current source circuit of Figure 5.69B in §5.14.2D, where a high-current buffer completes a precision bipolarity current source.

Preserving precision

See, for example, the precision dc amplifier of Figure 4.87 in §4.8.1, where an LT1010 unity-gain buffer (IC_5) isolates high-current loads, preventing thermal-gradient degradation of the precision op-amp at the output.

4x.4.1 Stability of the composite amplifier

In these applications a feedback path was closed around a composite amplifier (op-amp plus buffer), with the op-amp providing the loop gain for accuracy, and the buffer providing muscle (and load isolation). As we explained there (e.g., in §4.8.1), a cascade of two amplifiers, each of which is by itself stable in a feedback circuit, may become unstable according to the usual criterion – a phase shift approaching 180° at the unity-gain frequency of the composite amplifier. So, when the loop is closed around a unity-gain buffer (Fig. 4x.41), you need to include a compensation capacitor C_C if both amplifiers have comparable

open-loop bandwidths. Put another way, you need C_C if the buffer A_2 adds significant lagging phase shift (say $\geq 20°$) at the unity-gain frequency f_T of the op-amp A_1.

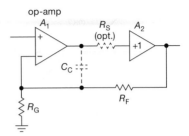

Figure 4x.41. If the roll-off frequencies of the op-amp and buffer are comparable, the accumulated phase shifts cause instability. Use a compensation capacitor C_C to bring the closed-loop gain safely down to unity at a lower frequency, here $f \approx 1/2\pi(R_F \| R_G)C_C$; or use a faster buffer. The optional series resistor R_S is often desirable to prevent peaking and overshoot, see Fig. 4x.47.

You might think that the composite amplifier (op-amp plus buffer) would be stable without the stabilizing compensation loop (C_C and R_F in Fig. 4x.41) if either the buffer's 3 dB bandwidth is significantly greater than the op-amp's unity-gain frequency f_T, or vice-versa. But it's not that simple: indeed, in the former situation the combination is stable, but not the reverse. To see why, look at Figure 4x.42, where we've drawn Bode plots for both cases, assuming the combination is configured for unity gain. The top plot represents $f_T(\text{buffer}) > f_T(\text{op} - \text{amp})$, where the buffer only adds phase lag somewhat beyond the unity-gain frequency of the pair, a stable situation. By contrast, with the roles reversed (middle plot), the op-amp indeed contributes additional phase lag only beyond the buffer's f_T; but there's lots of overall gain there, dropping to unity only many decades of frequency later. That loop is unstable, and requires the $C_C R_F$ compensation of Fig. 4x.41.

Well, you say, let's just push the op-amp's f_T out so far that the buffer's phase shift doesn't kick in until the loop gain has fallen to unity (bottom plot of Fig. 4x.42). Nice thought – but doomed to instability, because the buffer's phase shift is sure to increase beyond 90° before the op-amp's f_T; that is, neither op-amps nor buffers obey a pure single-pole response.[36] Anyway, what are you *thinking!* – why in heaven's name would you throw away all that op-amp bandwidth with a sluggish buffer?

---

[36] And why should they? An op-amp designer puts the dominant pole as high in frequency as possible, while still retaining a decent phase margin at the resultant f_T. That's guaranteed to slide up close to the next pole, causing additional phase shift beyond f_T. Ditto for the design of the buffer.

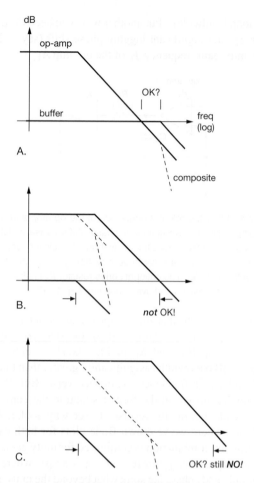

Figure 4x.42. Exploring op-amp+buffer stability (unity-gain follower): A. The composite is stable if the buffer's bandwidth is significantly greater than the unity-gain bandwidth (f_T) of the op-amp (but don't cut it too close – beware the op-amp's second pole, likely to kick in near its f_T). B. It does not work the other way around, because the combined phase shift reaches 180° well before the composite's gain falls to unity. C. How about a *much* faster op-amp? In the ideal case shown, it looks OK; but it isn't, because the buffer is likely to have a second pole long before the op-amp's f_T. Besides, why would you waste all that fast op-amp's bandwidth with a slow buffer?

To test these conclusions on the bench, we wired up a buffered op-amp, configured for unity-gain, and drove it with a 1Vpp 1MHz square wave (Fig. 4x.43). Trace A shows the clean (but not terribly fast) response with a slow op-amp (LF412: 4MHz) driving a somewhat faster buffer (BUF634 at minimum bandwidth setting: 22MHz). For trace B we substituted an LM4562 (55MHz), a situation like the middle Bode plot in Figure 4x.42, with the expected instability (oscillation at ∼20MHz); we tamed it by

adding the $C_C R_F$ compensation network (a bit undercompensated in trace C, critically damped in trace D). Finally, we removed the compensation network and tied the "bandwidth" pin on the BUF634 to V_-, raising its f_T to 160MHz, which once again restores stability, this time with improved speed (the little jiggles are layout and probing artifacts).

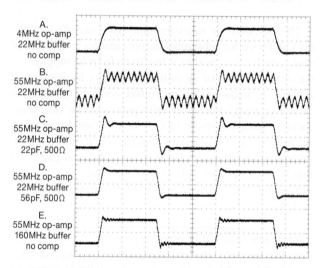

Figure 4x.43. Exploring op-amp+buffer unity-gain stability on the bench. $R_S=100\,\Omega$, $R_L=75\,\Omega$; see text for other component choices. Horizontal: 200 ns/div; vertical: 0.5 V/div.

4x.4.2 Some more applications

In addition to the examples we reviewed above, here are some other situations where you should consider a unity-gain buffer:

Isolating capacitive loads

Most op-amp circuits are unstable with capacitive loads, sometimes even as little as a few hundred picofarads. Once again, buffers come to the rescue, for example the LT1010 happily drives loads up to 1 µF or more. Also, there are substantial reactive currents ($I=CdV/dt$) when driving capacitive loads, so you may need the output drive capability, quite apart from issues of stability. But note: unity-gain buffers prefer a small resistor in series with the output when driving capacitive loads, see the caution in §4x.4.3 below.

Supply-rail splitter

A bare buffer makes a fine supply-rail splitter, of the sort shown back in Figure 4.72B (and repeated here in

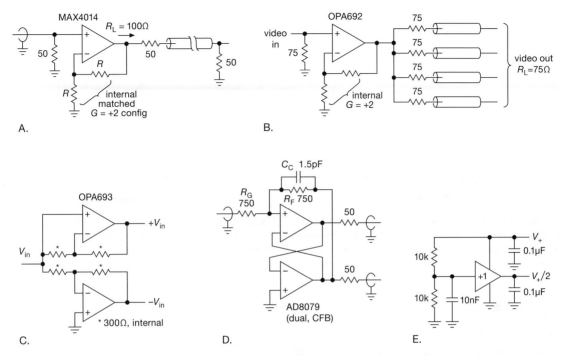

Figure 4x.44. Some buffer applications. A. Back-terminated cable driver, with $G_V=2$; because of the series termination it sees a load of $2R_0$, or $100\,\Omega$. B. Video distribution amplifier. C and D. Single-ended to differential amplifier. E. Rail splitter; a small series resistor ($\sim 25\,\Omega$) is helpful to isolate the buffer from the capacitive load.

Fig. 4x.44E). No great accuracy is needed here – don't bother with an op-amp.

Fault protection

Some op-amps are easily damaged under fault conditions, for example a persistent short circuit. By contrast, buffers expect such mistreatment, and are generally protected against overcurrent and overtemperature.

Video/RF buffer

Nearly half of the buffers in Table 4x.1 include an internal pair of matched resistors, to allow gain choices of -1, $+1$, or $+2$ (Fig. 4x.45). The latter choice is ideal for video or RF cable-driving applications (Fig. 4x.44A,B), where a series-terminated ("back-terminated") driver matches the cable's characteristic impedance (thus suppressing reflections, if any, from a mismatch at the far end); this arrangement also relaxes the requirements on the driver, which sees a load resistance equal to twice the cable's impedance.

Figures 4x.44C,D show two ways to generate differential RF or video outputs from a single-ended input.

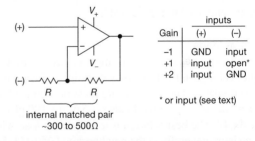

Figure 4x.45. Many wideband buffers include an internal matched resistor pair, allowing a pin-selectable choice of three gain settings.

L272A

And then there's venerable L272A "power op-amp," which runs on 32 V supplies and can deliver 700 mA output. Some people use it as a buffer, because it's cheap ($0.32) and comes as a dual, in a mini-DIP package.[37] It does include thermal limiting, but it has a limited 350 kHz bandwidth, and special care would be necessary if you place it inside a feedback loop. Instead you could simply insert it at the output, and live with its 7 mV typical (30 mV max!) offset voltage.

---

[37] The L272A is available from Fairchild (ON Semi) and the L272 ($V_{os}=$ 15 mV typ) from ST, along with an SOIC-16 version.

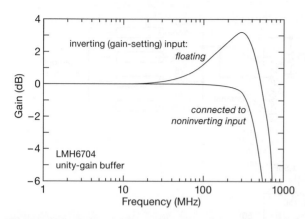

Figure 4x.46. Buffers with internal gain-setting resistors let you configure for gains of −1, +1, or +2. When configured for unity gain, however, there may be peaking if the inverting (gain-setting) input is left floating, as shown in this graph from the LMH6704 datasheet; connecting both inputs together cures the problem.

4x.4.3 Some cautions

Unity-gain configuration
For buffers that allow gain choices (-1, $+1$, $+2$), the default configuration for unity ($+1$) gain is to leave the "inverting input" open. However, parasitic capacitance tends to produce some response peaking; this is cured by connecting the inverting input to the noninverting input, as seen in Figure 4x.46 (and used in Fig. 4x.44C).

Capacitive loads
When driving a capacitive load, most buffers exhibit significant peaking (in the frequency domain) and ringing (in the time domain). These effects are most pronounced with low-to moderate capacitive loads ($\sim 5\,\text{pF}$ to $100\,\text{pF}$), see Figure 4x.47. The best solution is to isolate the load with a small resistor, generally in the neighborhood of $10\,\Omega$–$50\,\Omega$ (Fig. 4x.48).

Input resistor
When used within a feedback loop (the usual situation), where the buffer is driven by an op-amp's very low output impedance, it is often a good idea to include a small series resistor, say $100\,\Omega$ (see Fig. 4x.49); the legendary Bob Widler (who designed the LT1010) used them, and (or *so*) we use them. You can, too.

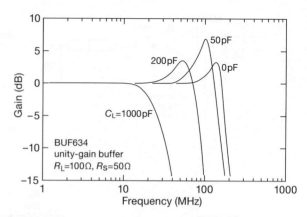

Figure 4x.47. Unity-gain buffers don't take kindly to capacitive loads, even as small as a few picofarads. Fix the problem with a series isolation resistor (Fig. 4x.48).

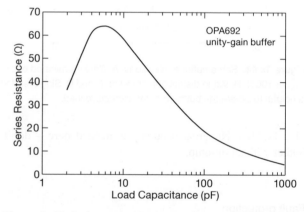

Figure 4x.48. Isolate capacitive loads with a small series resistor at the output, as suggested in this datasheet graph of "Recommended R_S vs Capacitive Load."

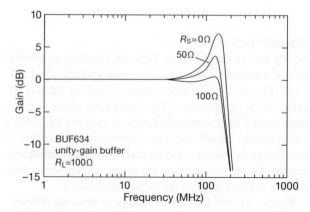

Figure 4x.49. Unity-gain buffers prefer a non-zero source resistance, as seen in these datasheet response curves for the versatile BUF634.

Table 4x.1: Unity-Gain Buffers[a]

Part #	Qty/pkg	V_{supply}[d] min (V)	V_{supply}[d] max (V)	I_Q[e] typ (mA)	V_{os} max (mV)	Z_{in} typ (MΩ/pF)	max I_{out} typ (mA)	BW[b] typ (MHz)	SR typ (V/µs)	PSRR V+,V- typ (dB)	To rails? in −	in +	out −	out +	Packages[z]	Price[f] qty10 ($US)	Comments
LT1010	1	4.5	±20	9	100	100µA	150	20	75[m]	-	-	-	-	-	DIP8, DFN8, T220-5	2.92	1
OPA832	1	±1.4	±5.5	3.9	7	0.4/1.2	85	100	350	68	●	-	-	-	SO8, SOT23-5	1.19	2
HA-5002	1	±12	±15	8.3	20	3/-	220	110	1300	54	-	-	-	-	SO8, PLCC20, T8.C-8	5.63	3
LMH6321	1	5	±15	11	35	0.25/3.5	300	110	2900	66,64	-	-	-	-	SO8ppad, DD7	5.45	4
BUF634	1	±2.3	±18	15[c]	100	8/8[c]	250	180[c]	2000	-	-	-	-	-	DIP8, SO8, T220, DD5	8.90	5
MAX4014-22	1-4	±1.6	±5.5	5.5	20	3/1	120	200	600	66	●	-	●	●	SO8, SOT23-5, µM8	4.54	6
MAX4214-22	1-4	±1.6	±5.5	5.5	10	3/1	120	230	600	66	●	-	●	●	SO8, SOT23-5, µM8	6.88	7
MAX4285	1,2	±1.4	±3.3	20	8	0.6/2	80	250	350	50	-	-	-	-	SO8, SOT23-6, µM8	2.04	8
AD8079	2	±3	±6	5	15	10/1.5	70	260	800	69,50	-	-	-	-	SO8	7.82	9
OPA633	1	±5	±16	21	15	1.5/1.6	100	260	2500	72	-	-	-	-	DIP8	13.02	10
OPA692	1	±2.5	±6	5.1	2.5	0.1/2	190	280	2000	58	-	-	-	-	SO8, SOT23-6	2.70	11
HA4600	1	±4.5	±5.5	11	10	0.4/1	20	480	1700	-	-	-	-	-	SO8, SOT23-6	2.90	12
LMH6704	1	±4.5	±6	12	7	1/1	90	650	3000	52	-	-	-	-	SO8, SOT23-6	3.12	13
MAX4200	1,2	±4.5	±5.5	2.2	15	0.5/2	90	660	4200	72	-	-	-	-	SO8, SOT23-5	1.39	14
HFA1112	1	±4.5	±5.5	21	25	0.05/2	60	850	2400	45	-	-	-	-	DIP8, SO8	7.46	15
BUF602	1	2.8	12.6	5.8	30	1/2.1	60	1000	8000	54	-	-	-	-	SO8, SOT23-5	2.36	16
OPA693	1	±2.5	±6	13	2	0.3/1.2	120	1400	2500	58	-	-	-	-	SO8, SOT23-6	3.20	17
LTC6417	1	4.75	5.3	123	3.2	0.018/1	100[m]	1600	10000	72	-	-	-	-	QFN20	5.56	18
OPA695	1	±2.5	±6	13	3	0.28/1.2	120	1700	4300	55	-	-	-	-	SO8, SOT23-6, VS8	3.88	19
LMH6559	1	3	±5	10	20	0.2/1.7	75	1750	4600	63	-	-	-	-	SO8, SOT23-5	1.02	20
LTC6416	1	2.7	3.9	42	5	0.012/1	20[m]	2000	3400	80	-	-	-	-	DFN10	4.92	21

Notes: (a) listed by increasing 3dB bandwidth. (b) 3dB BW, $G=1$. (c) with max BW Rext. (d) in most cases a dual-supply part can operate with a single supply. (e) per amplifier. (f) price for first p/n. (m) min or max. (z) DD=DDpak, SO=SOIC, T220=TO-220, µM=µMAX, VS=VSSOP.

Comments: (1) price is for DIP8; US$5.53 in TO220-5. (2) $G=-1, +1, +2$; video buffer. (3) "replaces LH0002." (4) adj curr lim 10-300mA; "equiv to HA-5002." (5) adj BW via ext res; I_Q=1.5mA, BW=30MHz, Z_{in}=80MΩ at low I_Q setting. (6) $G=-1, +2$; disable pin for IQ=400µA; also SO14, QSOP16; 4019 has disable for I_Q=400µA. (7) $G=-1, +2$; disable pin for I_Q=400µA; also SO14, QSOP16. (8) MAX4287,88 are dual; disable pin for I_Q=1mA. (9) $G=-1, +1, +2$. (10) replaces HA-5033; "improved LH0033, LT1010, H0S200." (11) $G=-1, +1, +2$; video buffer; disable pin for I_Q=150µA. (12) disable pin for I_Q=100µA. (13) $G=-1, +1, +2$; 2.3nV/√Hz; disable pin for I_Q=250µA. (14) MAX4203 is dual. (15) $G=-1, +1, +2$. (16) mid-point ref, for ac coupling. (17) $G=-1, +1, +2$; video buffer; disable pin for I_Q=170µA. (18) ADC buffer, differential IO; output clamping; 1.5nV/√Hz; disable for I_Q=24mA(!). (19) CFB op-amp; disable pin for I_Q=170µA. (20) gain flat to 0.1dB to 200MHz. (21) ADC buffer, diff'l IO; output clamping; 1.8nV/√Hz.

4x.5 High-Speed Op-amps I: Voltage Feedback

In this section[38] and the next we take up the increasingly relevant use of op-amps in high-speed and high-bandwidth applications. In the early days of IC op-amps we were lucky to get more than a few megahertz of gain–bandwidth product and a few volts per microsecond of slew rate,[39] whereas the miracles wrought by the semiconductor industry now bring us unity-gain-stable op-amps with GBWs above 500 MHz, or slew rates up to 3000 V/μs. To get a flavor of these devices, take a look at Table 4x.2 beginning on page 312, which we'll soon visit in some detail.

When looking for fast op-amps, you'll often wind up with a *current-feedback* ("CFB," or sometimes "CFA") topology, which differs from the "standard" (VFB) op-amps we're most accustomed to. We've seen CFB op-amps before: If you look at Tables 4.2a (Representative Op-amps), 5.4 (Representative High-speed Op-amps), and 8.3c (High-speed Low-noise Op-amps), you'll find some labeled CFB. Looking more closely, you'll find that they generally excel in slew rate (this trend is prominent in Table 8.3c, where the group labeled "HV bipolar CFB" beats the pants off the "HV bipolar VFB," and similarly for the low-voltage CFB versus VFB groups of op-amps), and (though not evident from the table listings) they deliver wide bandwidth even when configured for substantial closed-loop gain. CFB op-amps have drawbacks, however, including asymmetrical input characteristics (the inverting input is low impedance), lower loop gain, larger offset voltage, and somewhat demanding requirements on the feedback network. In the next section (4x.6) we'll take a look at these unusual op-amp specimens.

In this section, however, we'll look at fast voltage-feedback op-amps, after clarifying a few sometimes-confusing amplifier terms: voltage-feedback, current-feedback, transconductance amplifier, and transimpedance amplifier.

---

[38] References to §4x.5 from within AoE3 (referring to composite amplifiers) are redirected to §4x.4.

[39] For example, the vintage LM301 and μA741 op-amps had f_T=1 MHz and SR=0.5 V/μs in unity-gain-stable configurations.

4x.5.1 Voltage feedback and current feedback

A *voltage-feedback* amplifier ("VFB" or "VFA") is a configuration in which the error signal is a voltage (or voltage difference), as for example with a conventional op-amp, where the voltage difference at the (high-impedance) input terminals controls the output voltage. By contrast, in a *current-feedback* configuration ("CFB" or "CFA") it is an error *current* that drives the amplifier's output in a direction to eliminate the error; Figure 4x.50's block diagrams illustrate what's going on in the two topologies. The VFB op-amp is familiar, whereas the CFB can be confusing at first. But we've seen discrete-component examples of the latter, for instance the "shunt-feedback" voltage amplifier of Figure 2.92, in which the error current applied to the (low-impedance) terminal of the input stage (Q_1's emitter) acts to stabilize the output quiescent point and the circuit gain of the 2-stage amplifier, and in which the external voltage input is applied to the (high-impedance) noninverting input (Q_1's base), producing an amplified voltage signal at the output. And we'll see plenty more of the CFB op-amp in §4x.6.

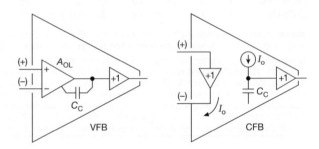

Figure 4x.50. Voltage-feedback and current-feedback op-amp architectures. The VFB's gain block has symmetrical high-impedance inputs, whereas in the CFB two unity-gain voltage amplifiers plus a current mirror creates a high-gain wideband amplifier, with a low-impedance inverting input.

A. Some confusing terms

Before going further, it's worth clarifying some confusing terms. A *transimpedance amplifier* (Fig. 4x.51A) produces a voltage output in response to a current input; its gain has units of V/I, thus impedance. We've seen such amplifiers before (see for example Figs. 4.22 or 8.69), created with ordinary VFB op-amps; in that configuration the transimpedance gain equals R_f, the feedback resistor. In other words, a VFB op-amp, with appropriate external circuitry, can be configured as an overall transimpedance amplifier *circuit*. By contrast, a CFB op-amp, by its very nature, is itself a transimpedance amplifier with respect to its inverting input. The gain of a CFB op-amp with respect to its inverting input, therefore, is an impedance. For example, the AD844 specifies a typical transresistance gain ($\Delta V_{OUT}/\Delta I_{IN-}$) of $R_{OL}=3\,M\Omega$. With respect to the noninverting input, however, it's a voltage amplifier, and so the gain is unitless, $A_V=6\times10^4$ typ, according to the datasheet. The input impedance of a voltage amplifier should be high, and that of a transimpedance amplifier should be low. The AD844 obliges – the datasheet specifies $10\,M\Omega$ and $50\,\Omega$ (typ) for the noninverting and inverting inputs, respectively.

The inverse of a transimpedance amplifier is a *transconductance amplifier*, that is, a circuit that converts a voltage input to an output current; its gain has units of I/V, thus conductance. Both BJTs (Ebers–Moll view) and FETs are transconductance amplifiers. Another description is simply a voltage-controlled current source, of which we've seen many, built with discrete components or with op-amps (e.g., Figs. 2.31 or 4.14). There are some op-amps that, analogous to CFB op-amps, do the opposite – voltage input to current output – though such *operational transconductance amplifiers* (OTA) are relatively minor players in the op-amp landscape. Some examples are the classic (now obsolete) CA3080, and a handful of available types, such as the LM13700 (Fig. 4x.51B), and the OPA860–61 (Fig. 4x.51C). The former has a standard op-amp input pair (i.e., high-impedance differential), whereas the OPA860-types use an asymmetrical input structure similar to the CFB: they label the inputs B, E, and the output C, analogous to a BJT; but it behaves like a zero-offset BJT that operates in both polarities, with output current proportional to input voltage, and with a transconductance that is settable via an external pin. This is discussed in §2.3.6, where it is called a "perfect transistor" (with some aliases such as "current conveyor," "diamond transistor," "transconductor," and "voltage-controlled current source").

4x.5.2 Overview of the table

This is one impressive table! As is always the case with circuit design, there is no one "best" choice – you must trade among desirable characteristics. For high-speed circuits the most relevant parameters are likely to be bandwidth, slew rate, settling time, supply current, and output current, combined with the usual suspects (input current and offset voltage, supply voltage and output swing, noise voltage and current, package choices, price, and the like). Herewith some commentary on the business of choosing an op-amp from the splendid array represented in the table.

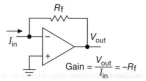

A. Transimpedance amplifier

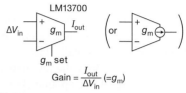

B. Transconductance amplifier (OTA)

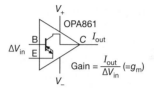

C. "Perfect transistor"

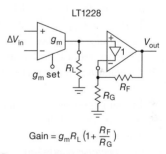

$$\text{Gain} = g_m R_L \left(1 + \frac{R_F}{R_G}\right)$$

D. LT1228: OTA–CFB hybrid

Figure 4x.51. Amplifiers not to be confused with current-feedback op-amps. A. Transimpedance amplifier (TIA). B. Operational transconductance amplifier (OTA) with symmetrical voltage inputs (standard symbol on left; more revealing symbol on right, favored by one of the authors). C. Operational transconductance amplifier with CFB-style inputs; add a voltage buffer and you have a CFB. D. The elegant LT1228 is a hybrid, combining an OTA with a CFB output stage to exploit the latter's wideband capability.

Table arrangement
Within each category (BJT, JFET, and CMOS, each subdivided into low-voltage and high-voltage parts) we've sorted the op-amps by increasing gain–bandwidth product, GBW. For parts specified over a range of supply voltages we show the parameters for the highest voltage.

Note that the column headings are not identical for the different pages. For example, JFET and CMOS have a "shdn pin" column (power disable), whereas the bipolar parts have a "comp pin" column. High-voltage BJT op-amps also have a bias-current cancellation column, see below. What's more, the units in some columns change dramatically, with pA input currents for JFETs and CMOS, compared with μA for BJTs.

What's not in the table

This table includes op-amps that use what we call "VFB+CFB" architecture: these look externally like an ordinary VFB op-amp, but internally they use a matched inverting-input follower to drive the low-impedance input of an internal current-feedback op-amp (Fig. 4x.63C,D). True CFB op-amps are listed instead in Table 4x.3 on page 322; and wideband fully differential op-amps are listed in Table 5.10 on page 375 of AoE3.

Gain–bandwidth product, GBW

Note that GBW (f_T) is the unity-gain extrapolation of the open-loop gain versus frequency curve; it's what matters when dealing with op-amp frequency compensation, transimpedance amplifier calculations, and the like.[40] For example, a decompensated op-amp like the LT1794 (G=10, min), whose open-loop gain G_{OL} is 10 at 20 MHz, has GBW=200 MHz, even though G_{OL} has fallen to unity at 60 MHz (owing to the more rapid falloff from the second pole, which is what disqualifies it for service as a unity-gain follower).[41]

Another bandwidth parameter you'll sometimes see is the −0.1 dB bandwidth, or the −3 dB bandwidth. These are good things to know. But you should also know that they can be misleading, because bandwidth "peaking" can extend the low-gain (G_{CL}=1 or 2) closed-loop bandwidth, in some cases well beyond the GBW(!) – see, for example, the datasheet plot of the AD8038's closed-loop gain versus frequency (Fig. 4x.52). Thus a caution: closed-loop bandwidth numbers are not a reliable guide to an op-amp's true (extrapolated) GBW.[42]

Full-power bandwidth

The full-swing bandwidth is not listed, but is given approximately by $f_{max}=SR/2\pi V_{peak}$; put another way, the mini-

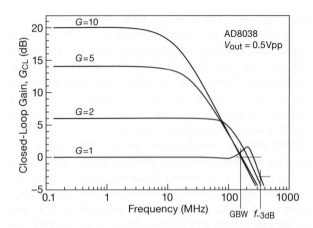

Figure 4x.52. The AD8038's −3 dB bandwidth, when configured for unity gain (i.e., as a follower) is about 340 MHz, significantly greater than its gain–bandwidth product (GBW, here ∼170 MHz), owing to "peaking."

mum slew rate for a sinewave of amplitude A and frequency f is $SR(min)=2\pi f A$.

Parts with "comp" pins

Most of the op-amps with *comp* pins are decompensated types, and we may show an "o" in the *min gain* column to indicate that you can add compensation for optimal bandwidth, slewing, and overshoot characteristics. Be careful to avoid adding stray capacitance to those pins.

Clamping

There's a nice capability that op-amps with *comp* pins may offer: adding output-voltage clamping, with a pair of low-capacitance diodes to the upper and lower clamp voltages. For example, the 500 MHz LT1222 suggests using a 1N5711 Schottky type to clamp node currents. That classic diode has less than 2 pF at 0 V. But you can do even better, for example the BAT15 has less than 0.3 pF at 0 V. Its cousin, the BAT-15-04W, has two properly-arranged diodes in a 3-pin SOT-323 package.

Harder-to-find variants

We've listed the package multiples (singles, duals, quads), with the part number and tabulated characteristics corresponding to the bold-face number/package; but beware that the single (as opposed to dual and quad) versions are frequently poorly stocked. Likewise, the most accurate versions of an op-amp can be hard to find (particularly for JFET types). Usually we list the part number for the single-amplifier type, even though the dual might be more popular. In many case the modified part numbers may be obvi-

---

[40] See, for example, §§4.9, 8.11, and 4x.3.1.

[41] Some manufacturers don't tabulate GBW, providing instead a unity-gain frequency; for those op-amps we've used their graphs of G_{OL} versus frequency to deduce their GBW.

[42] As an extreme example, the AD817's datasheet lists a −1 dB flatness bandwidth of 70 MHz (typ), measured with G_{CL}=1, while the op-amp's typical GBW is 50 MHz.

ous, for example adding a 2 or 4 before the single-version's number, or changing the last digit.

Settling time

The tabulated settling times include slewing time; the slewing amplitude is not standardized, but is usually mentioned on the datasheet.

Input bias-current cancellation

Input bias-current cancellation is a useful option, available only for BJT op-amps. It is particularly relevant for high-speed op-amps, with their higher operating currents. The design aims to cancel the input transistor pair's base current, ideally reducing the input bias current by one to two orders of magnitude. In Figures 5.6 and 5.38 we showed how BJT amplifiers can have lower input bias currents than a JFET op-amp, because the latter suffers from exponentially increasing input leakage at high temperatures.

The Table's high-voltage BJT op-amp section has a column indicating if this feature is included. High-speed BJT op-amps would have rather high input base currents, were it not for this feature. For example, in §5.10.8 we show how the base currents can be reduced by factors of 10 to 50. The 18 MHz OPA209 has a typical input bias current spec of only 1 nA, which is astonishing.

Low-voltage BJT op-amps generally don't include bias-current cancellation, because they have to work close to the supply rails, leaving no operating room for the cancellation transistors; so we omitted this column. However, low-voltage op-amps in the table that do include cancellation are the LT1803, LT6220, LT1800, and the 890 MHz LTC6228. The latter part lets you turn the cancellation on and off;[43] note also that the cancellation is ineffective for input common-mode voltages below 500 mV (see, for example, Fig. 4x.53).

Later we discuss bias-current cancellation and show how it's achieved, in §4x.10; and in §5.10.8 we explored its impact on bias-current noise.

Input noise, e_n and i_n

Wideband op-amps often have a high $1/f$ noise corner frequency, as high as 100 kHz or even 1 MHz. So they can be quite noisy at "low" frequencies like 1 kHz or 10 kHz. Also (as described in AoE3's §5.11.3), chopper-type op-amps can be afflicted by substantial high-frequency noise around the chopping frequency. Be sure to check the datasheets for noise parameters in your frequency range of interest.

---

[43] The price you pay is that the input current noise and offset current are degraded when cancellation is enabled.

Quiescent current, I_Q

Values shown are for the package multiple indicated in boldface. Decompensated op-amps ($G_{min}>1$) typically have quiescent currents significantly lower than unity-gain-stable parts of similar bandwidth; for example, the quiescent current of a $G_{min}=10$ op-amp might be one tenth that of a unity-gain-stable part of comparable GBW.

Input current, I_b

JFET- and MOSFET-input op-amps have far lower dc input (leakage) current than BJT-input types – note that the units for the latter are *micro*amps, compared with *pico*amps for the FET types. Some BJT types use input-bias cancellation, to reduce the input current (but not the current *noise*!); compare, for example, the LT6220 and LT6233, which have comparable speed but typical input currents differing by a factor of 100. The input *noise* current, by contrast, is somewhat higher in the bias-canceled part (LT6220), as expected. Note also that the bias cancellation becomes less effective for inputs near the supply rails, especially for rail-to-rail input op-amps; see, for example, Figure 4x.53. Finally, with matched input resistances you can greatly reduce the voltage offset of high-I_b parts (be sure to check the offset current specifications).

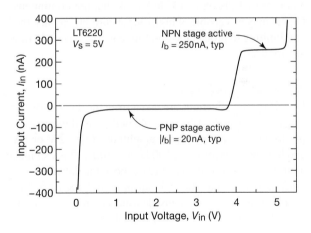

Figure 4x.53. Input bias cancellation may become ineffective near the supply rails, particularly for a RRI op-amp like the LT6220.

Maximum differential input voltage

The table doesn't have a column listing the maximum allowed differential input voltage (although it probably should!). Many, if not most, low- and moderate-speed op-amps allow you to present differential input voltages up to the supply rails. That's also true for many parts in this table, especially high-voltage JFET op-amps. But very high-speed op-amps, which we cover here, by necessity have very small low-capacitance input transistors. These are fragile and need protection, which often is implemented with a circuit involving back-to-back input diodes. We recommend checking the *Absolute Maximum Ratings* in the manufacturer's datasheet.

Some parts with input protection (e.g., the LM7171) allow up to ± 10 V. But many others (e.g., the THS4031) allow only ± 4 V. The LM7121 allows only ± 2 V. CMOS-input parts often have a pair of input diodes and are spec'd at ± 0.5 V, for example the TLV172 and OPA1688. Some BJT-input op-amps also use back-to-back protection diodes, for example the OPA1611 and LTC6228. Excess differential input voltages can cause input currents to flow, and there are maximum specs for that as well, typically 10 mA. Be sure to check the datasheet's *Absolute Maximum Ratings – they mean it!*

Output current, I_{out}
High-speed op-amps are often called upon to drive coaxial lines or other forms of terminated loads, so you need plenty of drive capability; after all, 5 V into 50 Ω is 100 mA. Most of the op-amps in the table struggle to get there. There are some that can drive high current for a short time, for example the LTC6268 (>135 mA peak, but not continuous). Others may be capable of high output current, but lack thermal limiting to protect the op-amp (e.g., the AD8009, rated at 175 mA).

Distortion and overload
Distortion specifications are taken from the datasheets. Because there is no universal set of conditions, these may apply to widely varying signal amplitudes (e.g., 2 V or 8 V) and load resistances (e.g., 100 Ω or 500 Ω). This can make a dramatic difference – be sure to check the datasheet. Another important parameter is an op-amp's recovery from overload; it appears in many datasheets, but we've not included it in these already-crowded tables.

Fully differential op-amps.
When seeking high-performance low-distortion op-amps, don't overlook fully differential operational amplifiers. These amplifiers have two outputs, and the output signal is best taken as the difference between them. Ideally, if the two output stages have identical error flaws, the differential error will cancel. We cover these op-amps extensively in AoE3's §5.17. Consider, for example, TI's LME49724 50 MHz op-amp (discussed extensively in those pages). It can produce a 50 Vpp output swing, with ± 15 V sup-

plies, it claims distortion as low as 0.3 ppm distortion (see Fig. 5.43), and it sells for as little as \$1.70 (qty 100).

RF susceptibility
BJT-input op-amps can be highly susceptible to radiofrequency signals (EMI – electromagnetic interference), because the input BJT's junctions act as demodulating rectifiers. JFET and CMOS op-amps are far less affected, but the price you pay is poorer noise performance. Larkin reports good results with the precision low-noise "EMI-enhanced" ADA4522.

4x.5.3 Scatterplots: Seeking trends

It's both interesting and useful to see how a parameter like GBW correlates with an op-amp's quiescent current I_Q, or with its slew rate; or how a higher I_Q is required to achieve a lower noise voltage e_n in a bipolar op-amp.

A nice way to see these trends is to make a *scatterplot* of a pair of such parameters, which is easily done when you've got the data in a spreadsheet; happily, that's what we do as the first step in making the component tables.[44]

In this section we've made nine such scatterplots, using the full set of data in Table 4x.2, from which we can see general trends. Perhaps more usefully, they can show which op-amps excel in delivering better than average performance. And they illustrate the trade-offs in selecting an op-amp (e.g., power dissipation versus speed); understanding how to make good trade-offs is an important design skill. Let's start our survey with GBW versus quiescent current.

GBW versus I_Q scatterplots
The first two plots (Figs. 4x.54 and 4x.55) relate to the idea that a high-speed amplifier's basic task is to deliver bandwidth in exchange for power dissipation. This is particularly true for battery-powered applications. Looked at this way, some "slow" amplifiers should be considered fast – for example the LT1226 delivers 1000 MHz with just 7 mA (the highest point in Fig. 4x.54); how did they do that? hah – it's a decompensated op-amp ($G_{min}=25$)! Another part that excels is the MIC920 (standing tall at 67 MHz, 0.55 mA), and they did that without cheating[45] (it's unity-gain stable); plus, it has an astonishing 2500 V/μs slew rate. And this puppy sells for just \$0.25 – our hats are off to this amazing op-amp!

Moving to some faster op-amps, we find some that are

---

[44] See "A note on the tools" at the end of the *Preface to the Third Edition*.
[45] Perhaps they used the VFB+CFB trick (§4x.6.3) – they aren't talking.

outliers in relatively low quiescent current, for example the LT6233-10 (375 MHz, 1.15 mA) and the LT6230-10 (1450 MHz, 3.2 mA). These do well partly because they are decompensated op-amps ($G \geq 10$): comparing them with their unity-gain-stable brethren, we see they gain substantial speed (LT6233: 60 MHz, 17 V/μs; LT6230: 215 MHz, 70 V/μs), with GBWs and slew rates about $6\times$ greater.

These parts gain speed by running their input stages at high current, so they suffer from high input currents (1.2 μA and 5 μA, respectively, which can be partly alleviated with matched input source resistances). They also have unimpressive slew rates, at least when compared with CFB op-amps, and with the impressive VFB+CFB types, which excel in both GBW and slew rate, for example the unity-gain compensated LT1815 (220 MHz, 6.3 mA, 1500 V/μs) or the $G \geq 2$ LM7171 (200 MHz, 6.5 mA, 4100 V/μs).

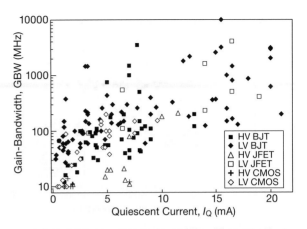

Figure 4x.55. VFB op-amps: GBW versus quiescent current (log–linear). These axes emphasize the faster op-amps.

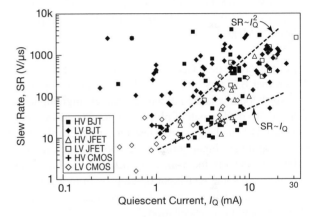

Figure 4x.56. VFB op-amps: Slew rate versus quiescent current.

Slew rate versus GBW scatterplot

This plot (Fig. 4x.57) illustrates nicely the minimum slew rate ($SR = 0.32f_T$) we deduced in §4x.9 for a BJT differential amplifier stage with no emitter degeneration; the plot shows many BJT op-amps clustering around that minimum. But, as we discussed in §4x.9.1, you can increase the slew rate (for a given GBW) by lowering the input-stage transconductance, either by adding emitter degeneration, or by using devices of lower transconductance (JFETs, MOSFETs); the latter effect is seen in the FET devices in the scatterplot. Section 4x.9.1 describes some tricks to boost the slew rate in a BJT op-amp (decompensation; cross-coupled current-boosting input transistors; Butler current-boosting stage; or current feedback). And indeed, the op-amps plotted whose slew rate is farthest above the line are in fact BJT types, some of which are $100\times$ higher than the line. However, enhanced slew-

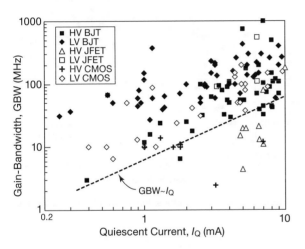

Figure 4x.54. VFB op-amps: GBW versus quiescent current (log–log). These axes expand the low-current (and generally slower) parts. Roughly speaking, bandwidth is proportional to quiescent current. There are, of course, many low-I_Q op-amps with GBW<10 MHz; but we've somewhat arbitrarily ruled that a "fast" op-amp shouldn't be that slow.

Slew rate versus I_Q scatterplot

In the next plot (Fig. 4x.56) you might expect slew rate to be proportional to supply current; but the data shows $SR \sim I_Q^2$ to better fit the trend. You can think of it this way: faster op-amps (GBW proportional to I_Q) have smaller compensation capacitors, so slew rate ($\sim I/C$) improves as the square of quiescent current.

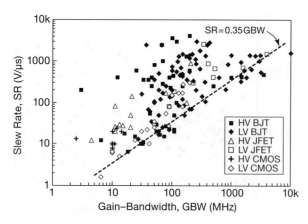

Figure 4x.57. VFB op-amps: Slew rate versus GBW.

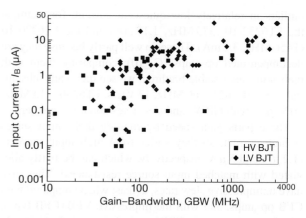

Figure 4x.58. VFB op-amps: Input current versus GBW.

ing can incur a steep price, for example in voltage noise: the LT1363 (70 MHz, 1000 V/μs) has 9 nV/$\sqrt{\text{Hz}}$, ten times higher than the 0.85 nV/$\sqrt{\text{Hz}}$ of the LT1028 of comparable GBW (75 MHz, 15 V/μs).

Input current versus GBW and SR scatterplots

Given that BJT types have the advantage in fast op-amps, it's worth looking at their downside, namely high input current. We've made a pair of plots (Figs. 4x.58 and 4x.59), I_B versus GBW and I_B versus slew rate, respectively. Most of the op-amps' input currents are above 0.1μA (and we arbitrarily cut off the y-axis at 40μA at the high end), but there are some BJT-input op-amps with input currents down around 10 nA or less: those are high-voltage types that use input bias-cancellation.[46]

The general trend in the I_B versus GBW plot is higher bias current for greater bandwidth, a consequence of higher collector current. The trend is less evident in the I_B versus slew rate plot, mostly owing to slew-rate-enhancing tricks (§4x.9.1).

Noise versus quiescent current scatterplot

The next scatterplot (Fig. 4x.60) shows e_n versus I_Q, with an arbitrarily placed[47] inverse square-root line that characterizes voltage noise in a BJT (see §8.3.1 in AoE3). What this plot shows is that higher currents are *required* to achieve low noise.

---

[46] Bias cancellation is generally impractical with low-voltage op-amps, whose inputs need to work to the supply rails, though we do see three exceptions in the plots.

[47] It's high by an order of magnitude, if you're considering noise in a single-ended stage with a perfect BJT ($r_{bb}=0$). But op-amps are differential, and, besides, there's plenty more going on in there.

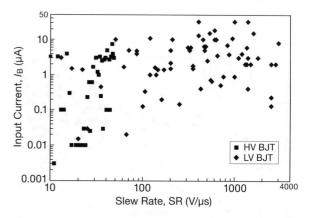

Figure 4x.59. VFB op-amps: Input current versus slew rate.

Settling time versus slew rate or GBW scatterplots

The final scatterplots (Figs. 4x.61 and 4x.62) largely show the expected trend: faster op-amps generally settle faster. The correlation with slew rate (Fig. 4x.61) is particularly constrained as you get to the faster parts (recall that settling time includes slewing time). However, the correlation with GBW is more interesting: while it's clear you *need* high GBW to get fast settling, a high GBW does not *ensure* fast settling. That's because op-amp designers, striving for maximum GBW, will sometimes increase the internal gain, moving f_T up closer to the second pole: they get a "faster" op-amp, and with even greater bandwidth at $G=1$ or 2, owing to peaking (see Fig. 4x.52). But the reduced phase margin lengthens the settling time, compared with a less aggressive op-amp whose second pole is kept well beyond f_T.

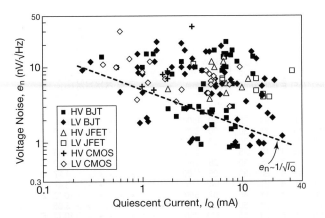

Figure 4x.60. VFB op-amps: Voltage noise versus quiescent current.

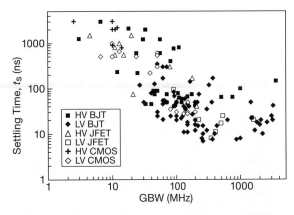

Figure 4x.62. VFB op-amps: Settling time versus GBW.

Low-voltage op-amps usually excel

Looking over these scatterplots, you see a general tendency for the low-voltage op-amps to do somewhat better than their high-voltage brethren (one exception is in input current, where the HV bipolar parts have the luxury of input bias-compensation). And the trend in contemporary circuit design is toward lower supply voltages. But a dose of reality: when you need large swings or large dynamic range, you've got no choice.

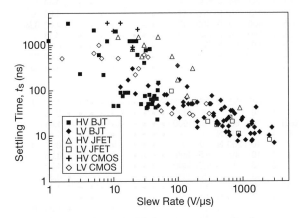

Figure 4x.61. VFB op-amps: Settling time versus slew rate.

Table 4x.2: High-Speed Op-amps I: VFB[a] (page 1 of 4)

Part #	# per pkg	Range (V)	Iq[t] (mA)	Ib[b] typ (pA)	Ib max (pA)	Vos typ (mV)	Vos max (mV)	ΔVos typ (µV/°C)	CMRR min (dB)	en 1kHz (nV/√Hz)	in[k] 1kHz (fA/√Hz)	Cin (pF)	min gain	GBW typ (MHz)	Iout typ (mA)	thermal pad	Slew typ (V/µs)	Settle[h] (ns)	Swing IN+	Swing IN−	Swing OUT+	Swing OUT−	null pins	shdn pin	DIP avail	SOIC etc	SOT-23	SC70 or smaller	Cost ($US)	Notes
JFET, high-voltage																														
LF357	1	10-36	5	30	200	3	10	5	80	12	10	3	5	20	20		50	1500	−	−	−	−	−	−	•	−	−	−	obs	A
OPA140	1,2,4	4.5-40	1.8	0.5	10	0.03	0.12	0.35	126	5.1	0.8	10	1	11	30		20	800	−	−	•	•	−	−	−	−	•	−	2.66	
OPA1641	1,2,4	4.5-40	1.8	2	20	1	3.5	−	120	5.1	0.8	8	1	11	30		26	900[e]	−	−	•	•	−	−	−	•	•	•	1.98	B
AD744	1,2	9-36	3.5	30	100	0.3	1	5	78	16	10	5.5	1	13	25		75	200	−	−	−	−	−	−	•	•	−	−	4.68	
OPA827	1	8-40	4.8	15	50	0.08	0.15	1.5	104	3.8	2.2	9	1	15	30		28	550	−	−	•	−	−	−	−	•	−	−	9.00	C
OPA2604A	1,2	9-50	5.3	50	−	1	5	8	80	10	4	10	1	20	35		25	1500	−	−	•	−	−	−	•	•	−	−	4.27	D
AD825	1	10-36	6.5	10	30	1	2	10	69	12	10	6	1	21	50		140	75	−	−	−	−	−	−	−	•	−	−	3.47	
ADA4627	1	10-36	7	1	5	0.12	0.3	1	100	4.5	1.6	4	1	11	45		82	550	−	−	−	−	−	−	−	•	−	−	10.12	E
OPA2810	1,2	4.7-28	3.5	2	10	0.1	0.5	2	90	5.7	0.7	2.5	1	70	50		170	40	•	•	•	•	−	−	−	•	−	•	3.35	
ADA4637	1	10-36	7	1	5	0.12	0.3	1	100	6.1	1.6	8	5	79	45		170	300	−	−	−	−	−	−	−	•	−	−	10.12	F
THS4601	1	10-33	10	30	100	1	4	10	100	5.4	5.5	6.5	1	180	50		100	170	−	−	−	−	−	−	−	•	−	−	5.63	
THS4631	1	10-33	11.5	50	100	0.26	0.5	2.5	95	7	20	3.9	1	210	95		900	40	−	−	−	−	−	−	−	•	−	−	5.21	
CMOS, high-voltage																														
OPA454	1	10-120	3.2	1.4	100	0.2	4	1.6	100	35	40	10	1	2.5	50	P	13	3000	−	−	−	−	−	−	•	−	−	−	7.50	HV
OPA552	1	8-60	7	20	100	1	3	7	92	14	3.5	6	5	12	380	P	24	2200	−	−	−	−	−	−	−	•	−	−	3.36	H
TLV172	1,2,4	4.5-40	8	10	−	0.5	1.7	1	94	9	1.8	4	1	10	75		10	2000	•	•	•	•	−	−	−	•	−	•	0.72	
OPA1688	2,4	4.5-40	1.6	10	20	0.25	1.5	0.5	104	8	1.8	7	1	10	60		8	3000	−	−	•	•	−	−	−	•	−	•	1.22	
OPA192	1,2,4	4.5-36	1	5	20	5µV	25µV	0.1	120	5.5	1.5	6	1	10	55		20	900	•	•	•	•	−	−	−	•	−	•	2.08	
OPA172	1,2,4	4.5-40	1.8	5	15	0.2	1	0.3	110	7	1.6	4	1	11	30		10	2000	−	−	•	•	−	−	−	•	−	•	2.66	
OPA189	1,2	4.5-40	1.3	70	300	0.4µV	3µV	0.01	146	5	165	6	1	14	50		20	800	•	•	•	•	−	−	−	•	−	•	1.70	J
OPA1678	2,4	4.5-40	2	10	−	0.5	2		100	4.5	3	6	1	16	40		9	−	−	−	•	•	−	−	−	•	−	•	s 0.44	
JFET, low-voltage																														
AD8033	1,2	5-24	3.3	1.5	11	1	2	4	89	11	0.7	2.3	1	90	35		80	95	•	•	•	•	−	−	−	•	•	•	1.85	
AD8065	1	5-24	6.4	2	6	0.4	1.5	1	85	7	0.6	4.5	1	110	35		180	55	•	•	•	•	−	−	−	•	•	•	3.45	
OPA656	1	9-13	14	2	20	0.25	1.8	2	80	7	1.3	3.5	1	230	50	N	290	20	•	•	•	•	−	−	−	•	−	•	5.59	K
OPA659	1	8-13	32	10	50	1	5	10	68	8.9	1.8	2.5	1	350	70	P	2550	8	•	•	•	•	−	−	−	•	−	•	7.48	
ADA4817	1,2	5-10.6	19	2	20	0.4	2	7	77	4	2.5	1.5	1	410	70	P	870	11	•	•	•	•	−	−	−	•	−	•	4.93	L
LTC6268	1,2	3.1-5.5	17	3fA	20fA	0.2	0.7	4	72	4.3	7	0.5	1	500	80	B	400	17	•	•	•	•	−	−	−	•	−	•	4.02	G,M
AD8067	1	5-24	6.4	0.5	5	0.2	1	1	89	6.6	0.6	2.5	8	540	30	N	640	25	•	•	•	•	−	−	−	•	−	•	4.18	
OPA657	1	9-13	14	2	20	0.25	1.8	2	80	4.8	1.3	5.2	7	1600	50	N	700	20	•	•	•	•	−	−	−	•	−	•	10.01	
OPA818	1	6-13	28	4	20	0.35	1.25	3	70	2.2	2.5	2.4	7	2700	50	P	1400	5.7	•	•	•	•	−	−	−	•	−	•	8.00	SD
LTC6268-10	1,2	3.1-5.5	17	3fA	20fA	0.2	0.7	4	72	4	7	0.5	10	4000	80	B	1500	−	•	•	•	•	−	−	−	•	−	•	4.02	G
CMOS, low-voltage																														
MAX4239	1	2.7-6	0.6	1	2	0.1µV	2µV	0	115	30	−	−	10	6.5	30	N	1.6	500	−	•	•	•	−	•	−	•	•	•	1.86	J
OPA316	1,2,4	1.8-7	0.4	5	15	0.5	2.5	2	70	11	1.3	4	1	10		6	1000	•	•	•	•	−	−	−	•	•	•	0.75	
AD8605	1,2,4	2.7-6	1.0	0.2	1	0.02	0.07	1	85	6.5	0.01	0.01	1	10	80		5	1000	•	•	•	•	−	−	−	•	•	•	1.44	
MCP6293	1,2,4	2.4-7	1.0	1	−		3	1.7	70	8.7	3	6	1	10	25	N	7	−	•	•	•	•	−	−	−	•	•	•	0.43	
MAX4231	1,2,4	2.7-6	1.2	−	17	−	8	3	46	12	−	8	1	10	200	N	10	−	•	•	•	•	−	−	−	•	•	•	0.71	Q
TLV9062	1,2,4	1.8-6	0.54	0.5	−	0.3	1.6	0.5	80	10	23	4	1	10	40		6.5	500	•	•	•	•	−	−	−	•	•	•	0.33	
OPA338	1,2	2.5-7.5	0.53	0.2	10	0.5	3	2	74	6	0.6	4	5	13	7		4.6	1400	•	•	•	•	−	−	−	•	•	•	1.36	
TLV6741	1	2.2-6	0.89	10	−	0.15	1	0.35	95	3.7	26	6	1	13	−		5	650	•	•	•	•	−	−	−	•	•	•	0.36	
OPA320	1,2	1.8-6	1.45	0.2	0.9	0.04	0.15	1.5	100	7	0.6	9	1	20	65		10	250	•	•	•	•	−	−	−	•	•	•	s 1.41	
AD8616	1,2,4	2.7-6	1.7	0.2	1	0.02	0.06[g]	1	80	7	50	7	1	24	150	N	12	500	•	•	•	•	−	−	−	•	•	•	2.51	
LMV861	1,2	2.7-6	2.5	1	10	0.27	1	0.7	78	5	15	20	1	31	−	N	20	−	•	•	•	•	−	−	−	•	•	•	1.30	R
MAX4488	1,2,4	2.7-6	2.2	1	150	0.07	0.35	0.3	90	3.5	0.5	−	5	42	40		10	−	•	•	•	•	−	−	−	•	•	•	0.84	
MAX44280	1	1.7-6	0.75	10	500	0.01	0.05	0.8	75	13	1.2	0.4	5	50	30		30	600	•	•	•	•	−	−	−	•	•	•	1.30	G,S,T
OPA350	1,2,4	2.5-7	5.2	0.5	10	0.15	0.5	4	74	7	4	6.5	1	38	40	N	22	220	•	•	•	•	−	−	−	•	•	•	1.67	
OPA365	1,2	2.2-5.5	4.6	0.2	10	0.1	0.2	1	100	4.5	4	6	1	50	50	N	25	300	•	•	•	•	−	−	−	•	•	•	1.75	S
LTC6244	2	2.8-6	4.8	1	75	0.04	0.1	0.7	74	8	0.6	3.5	1	50	20		35	535	•	•	•	•	−	−	−	•	•	•	1.98	V
OPA358	1	2.7-3.6	5.2	0.3	50	2	6	5	60	6.4		1.5	1	80	50	N	55	35	•	•	•	•	−	−	−	•	−	•	s 0.93	W
LMP7717	1,2	1.8-6	1.15	50	1000	10µV	0.15	1	85	6.2	10	15	10	88	36		28	−	•	•	•	•	−	−	−	•	•	•	1.72	
OPA380	1,2	2.7-7	7.5	3	50	4µV	25µV	0.03	100	5.8	10	3	1	90	75	N	80	−	•	•	•	•	−	−	−	•	−	•	5.39	X,Y
OPA300	1,2	2.7-7	7.5	0.1	5	1	5	2.5	66	3	1.5	6	1	150	50	N	80	30	•	•	•	•	−	−	−	•	•	•	1.72	
TLV3541	1,2,4	2.7-7.5	5.2	3	−	2	10	4.5	66	7.5	50	2	1	100	100	T	150	30	•	•	•	•	−	−	−	•	•	•	0.91	
ADA4891	1,2,3,4	2.7-6	4.4	2	50	2.5	10	6	88 t	9	−	3.2	1	130	125	N	170	28	•	•	•	•	−	−	−	•	•	•	1.08	W
OPA354	1,2	2.7-7.5	4.9	3	50	2	8	4	66	6.5	50	2	1	100	100	B	150	30	•	•	•	•	−	−	−	•	•	•	1.51	
OPA357	1,2	2.7-7.5	4.9	3	50	2	8	4	66	6.5	50	2	1	100	100	B	150	30	•	•	•	•	−	−	−	•	•	•	1.34	
LMH6601	1	2.4-6	9.6	5	50	1	2.4	5	56	10	50	1.3	1	155	100	N	275	50	•	•	•	•	−	−	−	•	•	•	s 1.43	
OPA355	1,2,4	2.7-7.5	4.9	3	50	2	9	7	66	5.8	50	1.5	1	200	60	T	300	30	•	•	•	•	−	−	−	•	•	•	1.38	
OPA859	1	3.5-5.5	20.5	0.5	5	0.9	5	2	70	3.3	−	0.6	1	140	64		1150	8	−	−	•	•	−	−	−	•	−	•	3.10	
OPA858	1	3.3-5.5	20.5	0.5	5	0.7	5	2	70	2.5	−	0.6	7	130	64		2000	8	−	−	•	•	−	−	−	•	−	•	4.18	

Table 4x.2: High-Speed Op-amps I: VFB[a] (page 2 of 4)

Part #	# per pkg	Supply[p] Range (V)	I_q[t] (mA)	Input Current[b] @25°C typ (µA)	max (µA)	bias cancel	Offset Voltage V_{os} typ (mV)	max (mV)	ΔV_{os} typ (µV/°C)	CMRR min (dB)	Noise[t] e_n 1kHz (nV/√Hz)	i_n[k] 1kHz (fA/√Hz)	C_{in} (pF)	min gain	GBW typ (MHz)	I_{out} typ (mA)	thermal pad	Slew typ (V/µs)	Settle[h] typ (ns)	Swing to Supply IN +/- OUT +/-	null pins / comp pin / shdn pin / DIP avail / SOIC etc / SOT-23 / SC70 or smaller	Cost ($US)	Notes
bipolar, high-voltage																							
LT1351	1,2,4	5-36	0.39	0.02	0.05	-	0.2	0.6	3	80	14	0.5	3	1	3	12		200	1250	- - - -	• - • • • s -	2.65	Z
ADA4075	2	5-36	1.8	0.03	0.1	•	0.2	1	0.3	110	2.8	1.2	2.4	1	6.5	30		12	3000	- - - -	- - - - • -	1.68	AA
LT1354	1,2,4	5-36	1.0	0.08	0.3	•	0.3	0.8	5	80	10	0.6	3	1	12	30		400	230	- - - -	• • • • • • -	2.40	Z
LM7321	1,2	2.5-35	1.1	1.2	2	-	0.7	5	2	80	15	1.3	-	1	16	65	N	18	-	• • • • •	- - • •	1.88	AB
OPA209	1,2,4	4.5-36	2.2	1nA	5nA	-	0.04	0.15	1	120	2.2	0.5	4	1	18	65	N	6.4	2100	- - - -	• • • •	2.25	AJ
LM8261	1,2	2.5-32	1.3	1	2	-	0.7	7	2	84	15	1	-	1	24	35		15	600	• • • •	• - • •	2.45	AD
LT1357	1	5-36	2	0.12	0.5	-	0.2	0.6	5	80	8	0.8	3	1	25	30		600	220	- - - -	• - • • -	2.60	Z,AE
LT1630	2,4	2.7-36	4.1	0.55	1.1	•	0.22	1	4.5	89	6	0.9	3	1	30	35		10	1200	• • • •	• - • •	3.45	AF
OPA1632	2	4-40	2.6	1.2	2	-	0.1	0.5	0.5	110	2.8	0.8	0.9	1	32	100		10	75	- - - -	- - - •	3.58	AG
AD817	1	5-36	7	3.300	6.6	-	0.5	2	10	86	15	1.5	1.5	1	33	50		350	45	- - - -	• - • • -	3.33	AB
OPA228	2	5-36	4	0.003	0.01	•	0.01	0.08	0.1	120	3	0.4	12	5	33	35		11	2000	- - - -	• - • • -	2.54	AH
LT6274	1,2	5-36	1.6	0.1	0.5	-	0.15	0.4	4	90	10	0.5	3	1	40	35		2200	40	- - - -	s -	3.15	Z
ADA4898	1,2	9-36	8.1	0.1	0.4	•	0.02	0.12	1	103	0.9	2.4	3.2	1	43	40	P	55	90	- - - -	- • - • -	5.37	
LT1220	1	8-36	8	0.1	0.3	•	0.5	1	20	92	17	2	2	1	45	25		250	75	- - - -	• - • •	3.57	
LT1224	1,2,4	5-36	7	4	8	-	0.5	2	25	86	22	1.5	2	1	45	40		400	90	- - - -	• - • •	3.30	BY
LT1360	1,2,4	5-36	4	0.3	1	-	0.3	1	9	86	9	0.9	3	1	50	26		800	90	- - - -	• - • •	2.29	Z,AB
LMP8671	1,2,4	5-46	5	0.01	0.08	•	0.10	0.40	0.1	105	2.5	1.6	-	1	55	50	N	20	1200	- - - -	• - - • -	2.89	AC
LME49710	1	5-36	4.8	7nA	0.07	•	0.05	0.7	0.2	110	2.5	1.6	-	1	55	26		20	1200	- - - -	• - • • -	1.97	AC
LM4562	2	5-36	5	0.01	0.07	•	0.1	0.7	0.2	110	2.7	1.6	-	1	55	23		20	1200	- - - -	• - • • -	2.45	AC,AK
LME49720	2	5-36	5	0.01	0.07	•	0.1	0.7	0.2	110	2.7	1.6	-	1	55	26		20	1200	- - - -	• - • • -	2.48	AL,AK
LME49870	1	5-44	5	0.01	0.07	•	0.14	0.7	0.1	110	2.5	1.6	-	1	55	31		20	1200	- - - -	• - • • -	2.00	AC
LME49860	2	5-44	5.1	0.01	0.07	•	0.1	0.7	0.2	110	2.7	1.6	-	1	55	30		20	1200	- - - -	• - • • -	3.49	AM
MAX9632	1	4.6-40	3.9	0.03	0.18	•	0.03	0.13	0.15	120	0.94	3.8	-	1	55	50	P	30	600	- - • -	- • - - •	8.07	AN
OP-37A	4	5-44	3	0.01	0.04	•	0.01	0.03	0.2	114	3	1	-	5	63	-		17	-	- - - -	• - • • -	2.10	AO
MIC920	1	5-20	0.55	0.23	0.6	-	0.43	5	1	75	10	-	-	1	67	45		2500	-	- - - -	s • -	0.25	
LT1363	1,2,4	5-36	6.3	0.6	2	-	0.5	1.5	10	84	9	1	3	1	70	50	N	1000	50	- - - -	• - • • -	2.75	
LT1028A	1	8-44	7.4	0.025	0.09	•	0.01	0.04	0.2	108	0.85	4.7	5	o	75	-		15	-	- - - -	• - • • -	5.05	
OPA1611	1,2	4.5-40	3.6	0.06	0.25	-	0.1	0.5	1	110	1.1	1.7	8	1	80	40		27	500[e]	- • - • -	• - • • -	3.60	AP
OPA211	1,2	4.5-40	3.6	0.06	0.2	-	0.03	0.13	0.35	114	1.1	1.7	8	1	80	40		27	400	- - - -	• - • •	6.40	
ISL55001	1	5-33	9	1.7	3.5	-	0.06	3.0	18	70	12	1.5	1	1	70	120	N	280	75	- - - -	- - • -	3.71	
LT1468	1	7-36	3.9	3nA	0.01	•	0.03	0.08	0.7	96	5	0.6	4	1	90	22		23	770	- - - -	• - • • -	4.26	AQ
THS4061	1,2	9-33	7.8	3	6	-	2.5	8	15	70	14.5	1.6	2	1	90	80	P	400	40	- - - -	• - • • -	4.62	AR
THS4081	1,2	9-33	3.4	1.2	6	-	1	7	15	78	10	0.7	1.5	1	95	65	P	230	43	- - - -	• - • • -	4.06	AR
LM6171	1,2	5-36	2.5	1	3	-	1.5	3	6	80	12	1	-	1	100	90	N	3600	50	- - - -	• - • • -	2.53	Z,AS
AD818A	1,2	5-36	7	3.3	6.6	-	0.5	2	10	86	10	1.5	1.5	2	105	50	N	450	80	- - - -	• - • • -	3.61	
AD797B	1	10-36	8.2	0.25	0.9	•	0.01	0.04	-	114	0.9	2	20	o	110	50	N	20	800	- - - -	• - • • -	8.50	AT
AD8022	2	4.5-26	4	2.5	5	-	1.5	6	-	98	2.5	1.2	0.7	1	110	55	N	50	62	- - - -	- • •	4.14	
LME49990	1	10-38	9	0.03	0.5	•	0.13	1	2	110	0.88	2.8	-		110	50	nc	22	0.6	- - - -	- • - • -	4.16	AU
LM7372	2	10-36	13	2.7	10	-	2	8	12	75	14	1.5	-	2	120	250	P	3000	50	- - - -	• - • • -	4.23	Z,AV
THS4031	1,2	9-33	8.5	3	6	-	0.5	2	2	85	1.6	1.2	2	1	120	60	P	100	60	- - - -	• - • • -	5.60	AR
LT1221	1	8-36	8	0.1	0.3	•	0.2	0.6	15	92	6	2	2	4	150	25		250	66	- - - -	• - • • -	3.42	
LM7121	1	9-36	5.3	5	10	-	0.9	8	-	73	17	1.9	2.3	1	175	40		1300	74	- - - -	• v -	2.61	AB
LT1794	2	8-36	13	0.1	4	•	1	5	10	74	8	0.8	3	10	200	500	B	600	-	- - - -	• - -	6.70	Z,AW
LM7171	1	9-36	6.5	2.7	10	-	0.20	3	35	85	14	1.5	-	2	200	90		4100	42	- - - -	• - • • -	2.27	Z
LT1222	1	8-36	8	0.1	0.3	•	0.1	0.3	5	100	3	2	2	10	500	25		250	65	- - - -	• - • • -	3.47	AX
AD829	1	9-36	5	3.3	7	-	0.2	1	0.3	100	1.7	1.5	5	o	750	o 25		150[r]	65	- - - -	• - • • -	4.82	AY
AD8021	1	4.5-26	7	7.5	11	-	0.4	1	0.5	86	2.1	2.1	1	o	1500	60		420	23	- - - -	- • • -	2.42	AZ
LT1226	1	7-36	7	4	8	-	0.3	1	6	94	2.6	1.5	2	25	1000	40		400	100	- - - -	• - • • -	3.40	
THS4021	1,2	9-33	7.8	3	6	-	0.5	2	15	-	1.5	2	1.5	10	3500	80		470	145	- - - -	• - - • -	8.20	P

313

Table 4x.2: High-Speed Op-amps I: VFB[a] (page 3 of 4)

Part	# per pkg	Supply[p] Range (V)	Supply[p] I_q[t] (mA)	Input Current[b] @25°C typ (μA)	Input Current[b] @25°C max (μA)	V_{os} typ (mV)	V_{os} max (mV)	ΔV_{os} typ (μV/°C)	CMRR min (dB)	e_n 10kHz (nV/√Hz)	i_n[k] 10kHz (pA/√Hz)	C_{in} (pF)	min gain	GBW typ (MHz)	I_{out} typ (mA)	thermal pad	Slew typ (V/μs)	Settle[h] (ns)	Swing to Supply IN +	Swing to Supply IN −	Swing to Supply OUT +	Swing to Supply OUT −	null pins	comp pin	shdn pin	DIP avail	SOIC, etc	SOT-23	SC70 or smaller	Cost ($US)	Notes
bipolar, low-voltage																															
ADA4841-1	1,2	3-12.6	1.1	3	5.3	0.04	0.3	1	86	2.1	1.4	3	1	27	30		13	120	-	-	-	-	-	-	-		•	-	•	3.49	
OPA835	1,2	2.5-5.5	0.25	0.2	0.4	0.12	0.5	1.4	91	9.3	0.45	1.2	1	31	40		160	50	-	•	•	•	-	-	-		•	d	•	2.45	
EL5151	1	6-13.2	1.35	0.02	0.1	0.5	1	2	85	12	1	1	1	40	70		67	80	-	-	-	•	-	-	-		•	-	•	2.45	
AD8029	1,2,4	2.7-12.6	1.4	1.7	2.8	1.6	5	30	80	17	1.4	2	1	43	150		62	80	•	•	•	•	-	-	-		•	-	•	1.54	
AD8031B	1,2	2.7-12	0.9	0.45	1.2	0.5	1.5	5	60	15	1	1.6	1	45	35		35	125	•	•	•	•	-	-	-		•	-	•	4.83	
OPA837	1,2	2.7-5.4	0.6	0.34	0.52	0.03	0.13	0.4	95	4.7	0.5	2.0	1	50	70		105	25	-	•	•	•	-	-	-		•	-	•	1.79	
LT1803	1,2,4	3-12.6	2.5	0.125	0.75	0.35	2.5	10	78	21	2.5	2	1	50	45		100	350	•	•	•	•	-	-	-		•	-	•	2.85	BA
OPA838	1	2.7-5.4	0.95	1.5	2.8	0.02	0.13	0.4	95	1.8	1	2.1	6	50	40		350	30	-	•	•	•	-	-	-		•	-	•	2.47	
ADA4851-1	1,2,4	3-12.6	2.5	2.2	3.9	0.6	3.4	4	86	10	2.5	1.2	1	60	85		200	55	-	•	•	•	-	-	-		•	-	•	1.40	
ISL28190	1,2	2.5-5.5	8.5	10	16	0.24	0.7	1.9	78	1	2.1	-	1	60	80		50	45	-	-	•	•	-	-	-		•	-	•	1.37	
LT6220	1,2,4	2.2-13	0.9	0.015	0.15	0.07	0.35	1.5	85	10	0.8	2	1	60	45		20	300	•	•	•	•	-	-	-		•	-	•	1.75	BA
LT6233	1,2,4	3-12.6	1.15	1.5	3	0.1	0.5	6	90	1.9	0.43	3.7	1	60	55		17	170	-	•	•	•	-	-	-		•	-	•	2.15	
MAX4389	1,2,4	4.5-12	3.2	2.5	15	5	18	15	70	13	3.1	1	1	60	50		500	21	-	•	•	•	-	-	-		•	-	•	1.50	
ADA4853-1	1,2,3	2.7-5.5	1.4	1	1.5	2	3.3	1.6	79	22	2.2	0.6	1	64	95		120	54	-	•	•	•	-	-	-		•	s	•	1.55	
MIC920	1	5-20	0.55	0.23	0.6	0.43	5	1	75	10	-	-	1	67	45		2500	-	-	-	-	-	-	-	-		•	s	•	0.25	
LT1800	1,2,4	2.3-12.6	1.8	0.03	0.4	0.45	1.5	1.5	85	8.5	1	2	1	70	40		25	300	•	•	•	•	-	-	-		•	-	•	1.45	BA
AD8091	1,2	3-12.6	4.8	1.4	2.6	1.8	11	10	72	16	0.9	1.4	1	70	45		170	50	-	•	•	•	-	-	-		•	-	•	1.46	
OPA2889	2	2.6-13	0.9	0.15	0.75	1.5	5	-	70	8.4	0.7	0.8	1	80	50		250	25	-	-	-	-	-	-	-		•	-	•	2.56	BB
LT6202	1,2,4	3-12.6	2.8	1	7	1.6	3	7.5	83	2.8	1.1	1.8	1	90	30		22	85	•	•	•	•	-	-	-		•	-	•	1.90	
LMH6642	1,2,4	2.7-13	2.7	1.6	2.6	1	5	5	74	17	0.9	2	1	92	75		135	68	-	•	•	•	-	-	-		•	-	•	1.11	
AD8051	1,2,4	3-12.6	4.4	1.4	2.5	1.7	10	10	72	16	0.85	1.4	1	100	45		145	50	-	•	•	•	-	-	-		•	-	•	1.75	
LT6205	1,2,4	3-12.6	4	18	30	1	4.5	10	78	9	4	2.0	1	100	50		600	25	-	•	•	•	-	-	-		•	-	•	1.31	
LT1812	1,2,4	3-12.6	2.7	0.9	4	0.4	1.5	-	75	8	1	2.0	1	100	60		750	30	-	•	•	•	-	-	-		•	-	•	1.59	Z
EL8101	1	4-5.5	2	1.5	2.1	0.8	6	3	70	10	1	0.5	1	100	65		200	20	-	•	•	•	-	-	-		•	-	•	2.11	
OPA836	1,2	2.5-5.5	1	0.65	1	0.07	0.4	1	94	4.6	0.75	1.2	1	118	40		560	22	-	•	•	•	-	-	-		•	d	•	2.57	
THS4221	1,2,3	3-16.5	14	0.9	3	3	10	10	74	13	0.8	1	1	125	100		990	25	-	•	•	•	-	-	-		•	-	•	2.01	BC
AD8027	1,2	2.5-5.5	6	4	6	0.2	0.8	2	90	4.3	1.6	2	1	130	95	N	85	40	•	•	•	•	-	-	-		•	-	•	2.16	BD
LMH6611	1,2	2.7-11	3.3	6.5	10	0.07	1.5	4	81	10	2	2.5	1	135	120	N	460	60	-	•	•	•	-	-	-		•	-	•	1.34	
LMH6657	1,2	3-13	6.2	5	20	1.1	5	2	72	19	7.5	1.8	1	140	100		470	37	-	•	•	•	-	-	-		•	-	•	1.59	
AD8057	1,2	3-12.6	5.4	0.5	2.5	1	2.5	3	48	7	0.7	2	1	140	-		700	35	-	•	•	•	-	-	-		•	-	•	1.54	
AD9631	1	6-12.6	17	2	7	3	10	10	70	7	2.5	1.2	1	130	50		1300	11	-	-	-	-	-	-	•		•	-	•	8.30	
AD9632	1	6-12.6	16	2	7	2	5	10	70	4.3	2	1.2	2	150	50		1500	11	-	-	-	-	-	-	•		•	-	•	7.99	
AD8038	1,2	3-12.6	1.0	0.4	0.75	0.5	3	4.5	61	8	0.6	2	1	150	-	N	425	18	-	-	-	-	-	-	-		•	-	•	1.54	
MAX4451	1,2	4.5-12	6.5	6.5	20	4	26	8	70	10	1.8	1	1	150			485	16	-	•	•	•	-	-	-		•	-	•	1.79	
ADA4897	1,2	2.7-11	3	11	17	0.03	0.5	0.2	92	1.0	2.8	11	1	150	50		120	45	-	n	•	•	•	•	-		•	-	•	3.64	BE
LMH6622	2	4-13	4.3	4.7	10	0.2	1.2	2.5	80	1.6	1.5	1	1	160	100		85	40	-	-	•	•	-	-	-		•	-	•	3.20	
LT6200	1,2	3-12.6	16.5	10	40	0.2	1.2	8	75	0.95	2.2	4	1	165	65		50	140	•	•	•	•	-	-	-		•	-	•	2.99	BF
LT1809	1,2	2.5-12	12.5	1.8	8	0.6	2.5	-	66	16	5	2	1	180	85		350	27	•	•	•	•	-	-	-		•	-	•	1.94	
LT6236	1,2	3-12.6	3.2	5	10	0.1	0.5	0.5	95	1.1	1	6.5	1	200	30		70	50	-	•	•	•	-	-	-		•	-	•	2.35	BG
AD8055	1,2	8-13.2	5.4	0.4	1.2	3	5	6	75	6	1	3	1	200	60		1400	20	-	•	•	•	-	-	-		•	-	•	1.54	
LMH6628	2	5-13	9	0.7	20	0.5	2	5	57	2	2	1.5	1	200	50		550	12	-	-	-	-	-	-	-		•	-	•	4.39	
AD8037	1	6-12.6	21	4	10	2	7	10	66	6.7	2.2	1.2	2	200	70	N	1200	16	-	-	c	c	-	-	•		•	-	•	7.10	BH
LT6230	1,2,4	3-12.6	3.2	5	10	0.1	0.5	0.5	95	1.1	1	6.5	1	215	30		70	50	-	•	•	•	-	-	-		•	-	•	1.80	BJ
LT1815	1,2,4	4-12.6	6.3	2	8	0.2	1.5	10	75	6	1.3	2	1	220	80		1500	15	-	•	•	•	-	-	p		•	-	•	1.22	Z,BK
LMH6611	1,2	2.7-12	3	6	10	0.02	1.5	4	79	10	2	2.5	1	230	70		330	74	-	•	•	•	-	-	-		•	-	•	1.34	
OPA698	1	5-13	16	3	10	2	5	15	55	5.6	2.2	1.0	1	250	90	N	1100	8	-	-	c	c	-	-	•		•	-	•	4.14	BL
LMH6654	1,2	4-13	4.5	5	12	1	3	6	70	4.5	1.7	1.8	1	260	30		200	15	-	•	•	•	-	-	-		•	-	•	1.60	
ADA4855	3	3-6	7.8	3.8	-	1.3	3	-	94 t	6.8	2	0.5	1	260	57	P	870	10	-	•	•	•	-	-	-		•	-	•	2.50	BM
EL5104	1,2,3	4-13	9.5	8	30	3	10	50	56	10	54	1	1	264	160	N	3000	7	-	•	•	•	-	-	-		•	-	•	2.50	Z
AD8061	1,2	2.7-8	6.8	3.5	9	1	6	3.5	62	8.5	1.2	1	1	300	50		650	35	-	•	•	•	-	-	-		•	-	•	1.93	
OPA690	1,2,3	5-13	5.5	3	10	1	4	10	60	5.5	3.1	0.9	1	300	190	N	1800	12	-	•	•	•	-	-	-		•	-	•	2.69	BN
LT1807	1,2	2.5-13	11	1	5	0.1	0.7	1.5	83	3.5	1.5	2	1	325	70		140	120	•	•	•	•	-	-	-		•	-	•	3.90	BO
THS4211	1	5-16.5	19	-	-	3	12	-	-	-	-	-	1	350	170	N	850	22					-	-	-		•	-	•	5.45	
ADA4899	1	4.5-12	14.7	6	12	0.01	0.21	5	90	1	2.6	4.4	1	360	130	P	310	50	-	-	-	-	•	•	•		•	-	•	3.53	
AD8045	1	3-12.6	15	2	6.3	0.2	1	8	83	3	3	1.3	1	370	55	P	1350	7.5	-	-	-	-	-	-	•		•	-	•	2.59	BP

	# per pkg[f]	Supply[p] Range (V)	I_q[t] (mA)	Input Current[b] @25°C typ (µA)	max (µA)	Offset Voltage V_{os} typ (mV)	max (mV)	ΔV_{os} typ (µV/°C)	CMRR min (dB)	Noise[t] e_n 1kHz (nV/√Hz)	i_n[k] 1kHz (fA/√Hz)	C_{in} (pF)	min gain	GBW typ (MHz)	I_{out} typ (mA)	thermal pad	Slew typ (V/µs)	Settle[h] typ (ns)	Swing to Supply IN/OUT + - + - / pkgs	Cost ($US)	Notes
bipolar, low-voltage (cont'd)																					
LT6233-10	1	3-12.6	1.15	-	4	-	0.6	0.5	90	1.9	0.43	3.7	10	375	55		105	170	- - • - - - - - • - - - •	2.79	
LT1818	1,2	4-12.6	9.0	2	8	0.2	1.5	10	75	6	1.2	1.5	1	400	70	N	2500	10	- - - - - - - - • - • - •	2.12	Z,BQ
THS4275	1	5-16.5	22	-	-	5	10	-	-	3	3	-	1	425	80	N	950	25	- - - - - - - - • - - - •	2.74	
ADA4857	1,2	5-11	5	2	3.3	2	4.5	2.3	78	4.4	1.5	2	1	430	50	P	2800	15	- - - - - - - - • - • - •	3.51	
MAX4104	1	7-12	20	32	70	1	6	2.5	80	2.1	3.1	-	1	625	55	N	400	20	- - - - - - - - - - • - •	2.74	
LT6200-5	1,2	3-12.6	16.5	10	40	0.2	1.2	8	75	0.95	2.2	4	5	800	65		210	-	• • • • - - - - • - - - •	2.99	BR
THS4304	1	2.7-6	18	7	12	0.5	4	5	80	2.4	2.1	1.5	1	870	90	N	830	7.5	- • - - - - - - • - - - •	4.63	
LTC6228	1	2.8-12	16	0.6	2.5	0.02	0.1	0.4	100	0.88	6	3.5	1	890	110		500	26	- • - - - - - - • - - - s	2.76	BA
OPA699	1	5-13	16	3	10	1.5	5	15	55	4.1	2	1.0	4	1000	90		1400	8	- - c c - - - - • - - - •	4.23	BS
ADA4895	2	2.7-11	3	11	17	0.03	0.5	0.2	92	1.0	2.8	11	10	1460	50		943	22	- n • • - - - - • - - - •	4.48	
LT6200-10	1,2	3-12.6	16.5	10	40	0.2	1.2	8	75	0.95	2.2	4	10	1450	65		450	-	• • • • - - - - • - - - •	2.99	BR
LT6230-10	1	3-12.6	3.2	5	10	0.1	0.5	0.5	95	1.1	1	6.5	10	1450	30		320	50	- - - - - - - - • - - - •	2.74	
OPA2846	2	8-13	25	10	20	0.15	0.65	2.8	95	1.2	2.8	2	7	1650	80	N	600	12	- - - - - - - - • - - - •	7.15	BN
LMH6624	1,2	5-13	12	13	20	0.1	0.5	0.2	90	0.9	2.3	2	10	1800	100		400	18	- - - - - - - - • - - - •	3.20	BT
MAX4105	1	3.5-12	20	32	70	1	6	2.5	80	2.1	3.1	-	5	2050	70	N	1400	20	- - - - - - - - • - - - •	3.49	
OPA846	1	5-13	12.6	10	19	0.15	0.6	0.4	95	1.2	2.8	1.8	7	2200	80		625	10	- - - - - - - - • - - - •	3.74	
AD8099	1	5-12.6	15	6	13	0.1	0.5	2.3	98	0.95	5.2	2	o	2550	100	N	1350	18	- - - - - - - - • • - - -	3.69	BU
MAX4304	1	7-12	20	32	70	1	6	2.5	80	2.1	3.1	-	2	3000	55	N	1000	25	- - - - - - - - • - - - •	3.20	
MAX4305	1	3.5-12	20	32	70	1	6	2.5	80	2.1	3.1	-	10	3100	70	N	1400	20	- - - - - - - - • - - - •	3.49	
LMH6629	1	2.7-6	15.5	15	23	0.15	0.78	0.5	82	0.69	2.6	1.7	4	3200	200	P	530	42	- • - - - - - - • - - q -	4.00	BV
LMH6629	1	2.7-6	15.5	15	23	0.15	0.78	0.5	82	0.69	2.6	1.7	10	4000	200	P	1600	42	- • - - - - - - • - - q -	4.00	BW
OPA855	1	3.2 5.5	17.8	12	19	0.2	1.5	0.5	90	0.98	2.5	0.6	7	8000	85		2750	2.3	- - - - - - - - - - - - •	4.00	BX

NOTES: (a) within each group, listed in order of increasing bandwidth. (b) some op-amps (e.g., the OPA1688) have back-to-back protection diodes across their inputs; be sure to check the respective datasheets. (c) output-voltage clamping pins, set pins. (d) dual, multiple package types. (e) estimate. (f) bold indicates the number of amplifiers in the listed part number. (g) SOT-23 package may have larger V_{os} errors. (h) wildly varying measurement conditions and rules. (k) datasheet value is calculated, in most cases. (n) near to rail. (o) optimize compensation pin, for low or high gains. (p) programmable I_Q, range shown. (q) and other small packages, such as WSON. (r) tested at $G=20$ with no compensation. (s) and other smaller packages. (t) typical. (v) SOT-23 version has reduced supply-voltage limit.

COMMENTS: A. a part for manufacturers to recreate. B. audio. C. cheaper than '627. D. 3ppm; single = OPA604. E. OPA627 DIP. F. OPA637 DIP. G. guarded ESR input-protection diodes for very low capacitance.. H. HV; DDPak-7. J. chopper. K. 500MHz at G=1; '657 for 1.6GHz at G>5. L. pad must be grounded to avoid pin-6 feedback. M. excellent TIA response. Q. pulsed output driver. R. RF EMI hardened. S. charge pump. T. self-cal on power-up. V. LTC6244HV for 12V supply. W. video, G=2. X. transimpedance. Y. auto-zero. Z. VFB+CFB. AA. cross-coupled + diff. AB. stable with unlim cap-load. AC. 0.3ppm dist. AD. SOT-23, 50mA out; dual='8272. AE. 1% dist at 10MHz. AF. fastest HV RRIO? AG. audio dual, PowerPAD. AH. improved OP-37; decomp '227; dual '228. AJ. 0.25ppm dist. AK. TO-99 pkg also. AL. same as LM4562. AM. ±20V. AN. fastest RRO. AO. classic industrial. AP. audio, 0.15ppm dist. AQ. 0.7ppm dist. AR. THS4031 family, note e_n. AS. favorite; 6172 dual. AT. spec'd at ±5V. AU. 0.1ppm dist. AV. PowerPAD. AW. DSL driver, metal pad. AX. comp/clamp pin. AY. VFB, decomp node pin. AZ. drive 16-bit ADC, choose C_C. BA. input-stage bias-canceling. BB. MSOP10 part can do multiplexing. BC. 4-member family. BD. RRIO with settable input crossover. BE. ADA4897-1 has disable. BF. -10 decomp version; 1% dist at 50MHz. BG. replaces LT6230, lower noise above 100kHz. BH. input clamp pins; '8036 for G=1. BJ. '6230-10 decomp. BK. programmable I_Q, GBW. BL. V_{out} clamp, 400mV min, ±4.3V max for ±5V. BM. 16LFCSP only! BN. VFB. BO. LT1806 SOT-23. BP. improved SOIC pinout. BQ. JW (?) favorite. BR. 1% dist at 50MHz. BS. clipping limiter. BT. 0.03% dist at 10MHz. BU. external C_{comp}. BV. with prog G_{min} set to 4. BW. with prog G_{min} set to 10. BX. for TIA amps. BY. LT1224 is single; dual and quad versions are LT1208 and LT1209. HV. very high voltage. SD. has shutdown.

4x.6 High-speed Op-amps II: Current Feedback

As explained in the previous section, a current-feedback op-amp consists of a high-impedance noninverting *voltage* input, and a low-impedance inverting *current* input. The op-amp's output is a voltage. It's a strange beast, acting as a transimpedance amplifier (current in, voltage out) with respect to its inverting input, and simultaneously a voltage amplifier with respect to its noninverting input. Figure 4x.50 showed it in its most basic form, with Figure 4x.63A revealing the most common implementation: the input stage is a zero-offset unity-gain follower whose "output" is the op-amp's inverting *input* terminal!

4x.6.1 Properties of CFBs

Current-feedback op-amps are most often used as closed-loop noninverting voltage amplifiers, for example as seen in the inset of Figure 4x.64. In this configuration the closed-loop bandwidth is primarily set by R_f (as the graphs in Figs. 4x.65 and 4x.66 show), with R_g setting the voltage gain.

A. Closed-loop bandwidth

Before going on, it's worth pausing to develop some understanding why the closed-loop bandwidth of a CFB voltage amplifier is relatively constant, unlike the situation with a VFB (with its fixed gain–bandwidth product), where the closed-loop bandwidth decreases inversely as the gain is increased (i.e., BW $\approx f_T/G_{CL}$).

Here's a way to think about it: (a) From Figure 4x.50, you see that the current into the inverting terminal, set by the feedback resistor, is mirrored to the output buffer stage, where it sees a fixed load capacitance C_C (the compensation capacitance); so the amplifier's closed-loop bandwidth (and slew rate) is determined by R_f. (b) Now, adding the gain-setting resistor R_g sets the voltage gain ($G_{CL}=1+R_f/R_g$, just as in a VFB amplifier); but it also causes the source impedance seen at the inverting input (which is what creates the input current) to decrease, increasing the gain–bandwidth product comparably, thus preserving the closed-loop bandwidth. (c) This holds only to the extent that the CFB's inverting-terminal input resistance is small compared with the Thévenin resistance ($R_f \parallel R_g$); this fails for large gains, nicely seen in the graphs

of Figures 4x.64–4x.66. Of course, this kind of qualitative intuition cannot replace an honest analysis; for that, take a look at the postscript section 4x.6.5 ("Bandwidth and gain in CFBs").

B. Slew rate and output current

As noted earlier, CFB op-amps deliver impressive slew rates, when compared with VFB op-amps of comparable supply current or bandwidth. That's because the slew rate is not limited by the rate at which the input-stage's quiescent current can charge the compensation capacitor, as it is in a conventional VFB op-amp (see for example Fig. 4.43); instead, in the standard CFB op-amp the feedback current driven into the inverting input (which can be much larger than the stage's quiescent current) is mirrored to slew the compensation capacitance.

This can be seen nicely in the tabulated specifications in Table 8.3c: comparing the groups labeled "HV bipolar VFB" and "HV bipolar CFB," the average ratio of slew rate (in units of V/μs) to supply current (in mA) is 21 and 570 (V/mA·μs), respectively; i.e., the CFBs beat the VFBs by a factor of 25 in that figure of merit. In those same two groups of op-amps, the slowest-slewing CFB is more than triple the fastest-slewing VFB. These puppies can really slew!

Interestingly, CFB op-amps tend to have significantly higher output current ratings. Comparing the same groups in Table 8.3c, the I_{out}(min) of the CFBs average 190 mA, versus 48 mA for the VFBs. Current-feedback op-amps are good when you need lots of bandwidth, slewing, and output current.

C. The feedback network and stability

The closed-loop bandwidth increases as you decrease the feedback resistor R_f, until you reach some minimum value beyond which the amplifier becomes unstable. Most CFB datasheets give plenty of guidance on this, including graphs of response peaking (see for example Fig. 4x.67). Although there are no issues of stability if you choose to use a far larger value of feedback resistance (as we've hinted at in Figures 4x.65 and 4x.66), in most cases the datasheet will tell you what R_f should be, with graphs like Figure 4x.67; evidently chip designers want their customers to get maximum performance. Some datasheets include a table of recommended R_f, R_g pair values for each of a list of closed-loop gain values, including corresponding bandwidths; see Figure 4x.68 for a nice example, where the optimum resistances are given for each of the three available packages. The closed-loop gain of a CFB op-amp is ultimately limited by the impedance R_o at its inverting input: for R_g well below R_o, the op-amp acts like a VFB part with fixed f_T. This affects particularly low-power CFBs.

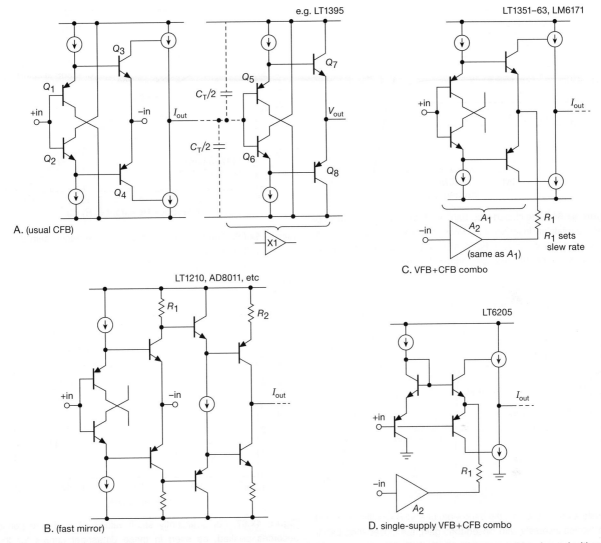

Figure 4x.63. Current-feedback op-amp topologies. Configuration A is the basic form; in B the input-stage current is mirrored with gain R_1/R_2; configuration C buffers the inverting input, creating a high-impedance voltage-input pair that looks like a normal VFB; the variant D is similar, with single-supply operation down to the negative rail (the latter two are marked with **Z** in the VFB table). For all configurations the intermediate current I_{out}, loaded by internal capacitance C_T, is converted to an output *voltage* by the follower stage shown explicitly in configuration A.

Unlike VFBs, where stability is enhanced with a small capacitor across the feedback resistor (see for example Fig. 4.104), CFBs react strongly (and unhappily) to such medicine: it takes only a small capacitance across R_f (or, similarly, from the inverting terminal to ground) to push them into instability and oscillation. See the impressive results in Figures 4x.69 and 4x.70 produced by just a few picofarads. A personal anecdote: in making the measurements for Figure 4x.66, the unity-gain datapoints showed anomalously large bandwidths – even with $R_f = 30\,\mathrm{k\Omega}$ the

bandwidth was suspiciously "too good." Turns out capacitance at pin 2 of the DIP was the culprit, tamed by bending it up and away from the board, then connecting R_f to the flying lead.

D. Input current and precision
Current-feedback op-amps, with their asymmetric input structure, are generally mediocre in precision (V_{os}), and of course the inverting input presents a low impedance and significant input current. Looking again at the LT1223, we

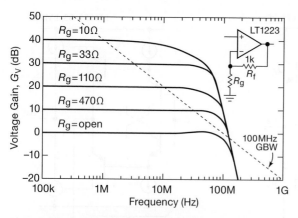

Figure 4x.64. The closed-loop bandwidth of current-feedback op-amps is relatively insensitive to closed-loop voltage gain, as shown in the LT1223's datasheet. For comparison the dashed line represents a constant 100 MHz gain–bandwidth product.

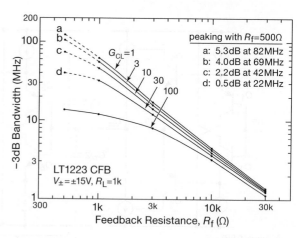

Figure 4x.66. Another way to show the closed-loop gain dependence on R_f and G_{CL} is to plot a family of bandwidth curves versus feedback resistance, shown here with data measured on our bench.

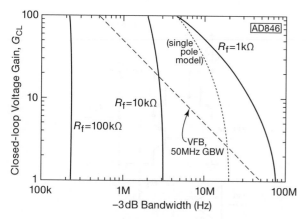

Figure 4x.65. Whereas the bandwidth of a VFB amplifier (dashed line) varies inversely with closed-loop gain, the closed-loop bandwidth of a CFB is determined primarily by the feedback resistor R_f. Additional poles in the AD846 cause peakiness in the 1 kΩ curve at low gains; the dotted curve illustrates ideal behavior, see §4x.6.5. (Adapted from the AD846 datasheet.)

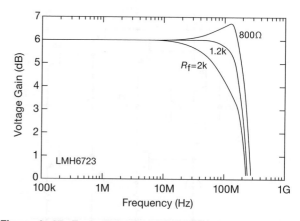

Figure 4x.67. For sufficiently small values of R_f the response becomes peaked, as seen in these datasheet curves for the LMH6723. A small amount of peaking extends the bandwidth, but don't overdo it!

see $V_{os} = \pm 1$ mV (typ), ± 3 mV (max); and both inputs have quiescent bias currents of $\pm 1\,\mu$A (typ), $\pm 3\,\mu$A (max). The input impedance of the noninverting input is 10 MΩ (typ), but the inverting input looks like some tens of ohms (not specified on the datasheet).

Quite apart from low offset voltage, a good precision op-amp should have large open-loop voltage gain. The exemplary OPA277 VFB precision op-amp ($V_{os}=20\,\mu$V max) has a typical open-loop voltage gain of 140 dB (10^7), compared with 89 dB (3×10^4) for the LT1223 CFB op-amp. Owing to the low gain of the latter, it requires 300 μV of

input difference to swing its output 10 V, compared with just 1 μV for the high-gain VFB.

4x.6.2 Care and feeding of CFBs

Current-feedback op-amps can be fussy things, primarily perhaps because their wide bandwidth makes them susceptible to instability if they are not treated with respect. That respect begins with following the datasheet's recommendations for resistor values, as just described. But you have to be particularly careful (*obsessive* might be a better description) about bypassing; we mistakenly soldered 10 nF chip bypass caps (plus larger tantalums) at the supply pins of an

Recommended Component Values

Package	AD8001AN (PDIP)					AD8001AR (SOIC)					AD8001ART (SOT23-5)				
Gain	−1	+1	+2	+10	+100	−1	+1	+2	+10	+100	−1	+1	+2	+10	+100
R_F (Ω)	649	1050	750	470	1000	604	953	681	470	1000	845	1000	768	470	1000
R_G (Ω)	649		750	51	10	604		681	51	10	845		768	51	10
small-signal BW (MHz)	340	880	460	260	20	370	710	440	260	20	240	795	380	260	20
0.1dB flatness (MHz)	105	70	105			130	100	120			110	300	300		

Figure 4x.68. Op-amp manufacturers are not shy when it comes to telling you what resistor values to use with their CFBs, even specifying it separately for each package style (data from AD8001 datasheet).

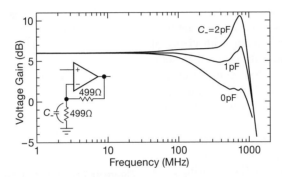

Figure 4x.69. Current-feedback op-amps are sensitive to capacitance at the inverting input, as shown in the AD8009 datasheet. If you're careful, this can be used to extend amplifier bandwidth (see the LT1228 datasheet, for example). But don't put a capacitor across R_f in CFB circuits.

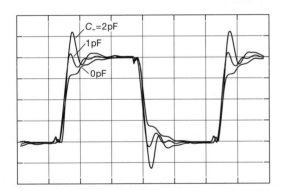

Figure 4x.70. An amplifier with peaking in its frequency response will exhibit transient overshoot, as seen here for the same AD8009 circuit as in Fig. 4x.69. Vertical: 50 mV/div; Horizontal: 2 ns/div.

AD846 (the datasheet[48] said to use 100 nF, but we forgot), which proceeded to oscillate around 100 MHz. The problem is widespread enough to merit bragging rights about

forgiving CFBs, for example "The LT1223 is very stable even with minimal supply bypassing, however, the transient response will suffer if the supply rings."

Another fussiness is the sensitivity to capacitance at the inverting input, as seen in Figures 4x.69 and 4x.70. Just a few picofarads can kill you, and you'll find advice like this on some datasheets: "The [PCB] ground plane should be removed from the area near the input pins to reduce stray capacitance."

4x.6.3 "Hybrid" VFB+CFB op-amps

The drawbacks of the conventional CFB op-amp (Fig. 4x.63A) – namely its unfriendly inverting input and its mediocre input offset voltage – can be nicely circumvented by buffering the inverting input and including the current-feedback resistor R_f within the chip, as in Figure 4x.63C. The resultant hybrid, which we might christen a "CFB in VFB's clothing," combines the best properties of CFBs and VFBs: high slew rate and bandwidth, with symmetrical high-impedance voltage inputs. The 100 MHz LM6171, for instance, appeals to designers with a banner proclaiming "Easy-To-Use Voltage Feedback Topology" followed by "Very High Slew Rate: 3600 V/μs." They

---

[48] "A 0.1 μF ceramic and a 2.2 μF electrolytic capacitor as shown in Figure 35 placed as close as possible to the amplifier (with short lead lengths to power supply common) will assure adequate high frequency bypassing, in most applications." They do not further expound on the meaning of "most."

don't brag about precision, however, probably because it's, uh, underwhelming (V_{os} of 3 mV typ, 6 mV max). LTC's LT1351–63 series of successively wider bandwidth op-amps (also VFB+CFB hybrid) does considerably better in this regard, with typical offset voltages ranging from 0.2 mV to 0.5 mV for parts of bandwidths from 3 MHz to 70 MHz. Their LT6205–07 series uses the topology of Figure 4x.63D to create an analogous single-supply VFB+CFB part. See also the discussion in §4x.9.1.

4x.6.4 When to use CFBs

Current-feedback op-amps are good when you want high slew-rate, or plenty of voltage gain along with plenty of bandwidth. They tend to have lower distortion than comparable VFBs at high frequencies. And you can change gains without significantly changing the bandwidth, a useful property in a programmable-gain amplifier (PGA) that might precede an ADC system.

Voltage-feedback op-amps are more familiar, and they come in a bewildering variety (precision, low-bias, high-voltage, rail-to-rail, etc.). They're versatile, allowing great variety in the feedback network (think of active filters). Our general advice is to choose VFBs for pretty much everything other than the special applications for which CFBs excel.

4x.6.5 Mathematical postscript: bandwidth and gain in CFBs

It's instructive to see how the closed-loop voltage gain G_{CL} of a given CFB depends primarily on the value of the feedback resistor R_f, and only weakly on the target closed-loop gain (as determined by the gain-setting resistor R_g). Although one may have some confidence in an intuitive understanding, there's no substitute for a proper analytic derivation, which we'll do here. Plus, some folks just love to see the math.

The familiar case: voltage-feedback op-amp
An "ordinary" (aka *voltage-feedback*, or VFB) op-amp, configured as a noninverting voltage amplifier with feedback resistor R_f from the output to the inverting input, and gain-setting resistor R_g from the inverting input to ground (Fig. 4x.71A), has an ideal voltage gain (in the limit of infinite open-loop voltage gain $A_{OL}=\infty$), which we'll call G_∞, of

$$G_\infty \equiv 1 + R_f/R_g.$$

As we saw back in Chapter 2 (eq'n 2.16), in the realistic case of *finite* open-loop gain A_{OL}, the closed-loop voltage

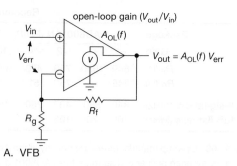

A. VFB

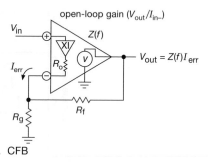

B. CFB

Figure 4x.71. VFB and CFB op-amps compared. A VFB amplifies the voltage error at its high-impedance input pair to produce a voltage output, with unitless gain $A_{OL}(f)$; a CFB converts a current (I_{err}) at its low-impedance inverting input to a voltage output, with transimpedance gain $Z(f)$.

gain becomes

$$G_{CL} = \frac{A_{OL}}{1 + A_{OL}B},$$

which (because the ideal closed-loop gain $G_\infty=1/B$) we can rewrite as

$$G_{CL} = G_\infty \frac{1}{1 + \dfrac{G_\infty}{A_{OL}(f)}}, \qquad (4x.4)$$

a form that shows in plain sight how the ideal gain is degraded, according to the ratio of target gain (G_∞) to (finite) open-loop gain.[49] We've here written the open-loop gain as $A_{OL}(f)$, emphasizing that the latter depends on frequency, extrapolating to unit gain at the unity-gain bandwidth f_T. So, for a VFB voltage-amplifier circuit, the closed-loop bandwidth is approximately[50] the frequency at which the open-loop gain $A_{OL}(f)$ has fallen to the target gain G_∞;

---

[49] Note that the term G_∞/A_{OL} in the denominator is just the inverse of the loop gain AB. A large loop gain ensures a closed-loop gain close to the ideal G_∞.

[50] We say *approximately* because the gain $A_{OL}(f)$ is complex, and accompanied by a 90° phase shift over most of the interesting range of frequency, see §4.4.2A.

another way to say it is that the closed-loop bandwidth is approximately f_T/G_∞.

The unfamiliar case: current-feedback op-amp

Now let's look at the scene where we substitute a CFB op-amp, again in the noninverting voltage amplifier configuration (Fig. 4x.71B). We'll cast the closed-loop gain in a form similar to eq'n 4x.4, to make the differences evident.

Initially let's assume that the internal impedance R_o of the amplifier is zero. Then the CFB op-amp produces an output voltage

$$V_{out} = I_{err}Z(f)$$

with

$$I_{err} = \frac{V_{in}}{R_g} - \frac{V_{out} - V_{in}}{R_f}.$$

Substituting and rearranging, we get

$$G_{CL} = \frac{V_{out}}{V_{in}} = G_\infty \frac{1}{1 + \dfrac{R_f}{Z(f)}}, \qquad (4x.5)$$

where the target gain $G_\infty = 1 + R_f/R_g$, as before.

Unlike the case with the VFB, the bandwidth-limiting term in the denominator depends only on the amplifier's native bandwidth (set by the roll-off of its transimpedance gain Z_f) and on the value of the feedback resistor R_f. It does not depend on the gain-setting resistor R_g. This behavior is seen, approximately, in Figure 4x.64.

Notice, though, that the bandwidth *does* decrease at the highest gains in that figure, and in others (e.g., Fig. 4x.66). To finish the discussion we have to take into account the non-zero impedance R_o seen at the CFB's inverting input. When that is included, the voltage at the inverting input becomes $V_- = V_{in} - I_{err}R_o$, so the input error current is

$$I_{err} = \frac{V_{in} - I_{err}R_o}{R_g} - \frac{V_{out} - V_{in} + I_{err}R_o}{R_f}.$$

With a bit of fiddling you arrive at the grand finale

$$G_{CL} = \frac{V_{out}}{V_{in}} = G_\infty \frac{1}{1 + \dfrac{R_f + R_o G_\infty}{Z(f)}}. \qquad (4x.6)$$

With R_o included, the closed-loop gain includes an additional bandwidth-limiting term $R_o G_\infty$ in the denominator; for a given feedback resistor R_f, the bandwidth will decrease with increasing target gain when the product of the op-amp's R_o and the target gain G_∞ is comparable to or greater than R_f. That is the cause of the dropoff in Figures 4x.64, 4x.65, and 4x.66. (It is not hard to show that the extra term $R_o G_\infty$ becomes important when R_o is not small compared with R_g, as we expect from the qualitative argument in §4x.6.1.)

Let's try this out, for some CFB op-amp. It's not so easy, though because many CFB op-amp datasheets omit a value for R_o, and those that do include it (Analog Devices is exemplary in this regard) do not usually list bandwidth versus gain for a given R_f. Happily the AD846 (no longer in production ... but datasheets live forever!) lets us test eq'n 4x.6. The datasheet specifies $R_o = 50\,\Omega$; so voltage gains of 1 and 100 correspond to values of $R_o G_\infty$ of $50\,\Omega$ and $5\,k\Omega$. With a feedback resistor of $1\,k\Omega$ the closed-loop bandwidths should be approximately in the ratio of 5:1, which is in good agreement with the dotted curve[51] in Figure 4x.65. And for a $10\,k\Omega$ feedback resistor, eq'n 4x.6 predicts a ratio of 1.5:1, in good agreement with the $R_f = 10\,k\Omega$ curve.

4x.6.6 Remarks on the table

We've scoured manufacturers' CFB offerings, collected together in Table 4x.3 on the following page, analogous to the four-pages of fast VFB op-amps listed in the previous section. Some general comments follow:

Bandwidth

As discussed earlier (see Fig. 4x.65 if you've forgotten), CFB op-amps (unlike VFB) cannot be characterized by a gain–bandwidth product. In the table we list more useful bandwidth specs: the $-3\,dB$ frequency for unity gain (or the lowest usable gain), and for $G \geq 5$, which gives some idea of the high-gain capability. We also list (where available) the frequency at which the low-gain configuration rolls off by $0.1\,dB$; this is useful for precision applications, such as an ADC driver. Be aware that small changes in op-amp compensation can cause large changes in the bandwidth; see, for example, the MAX4223 and 4224. Note, also, that op-amps with bandwidths greater than about $100\,MHz$ often have greatly reduced $-3\,dB$ frequencies for large output swings (such as $1\,V$); be sure to read the datasheet carefully! Another speed-related parameter is overload recovery time, which we've not listed, but which can usually be found on datasheets.

Limited specifications

Many CFB op-amps provide no specifications for gains other than $G=1$ or 2. These are aimed at the cable-driver market, where you need to drive a back-terminated and

---

[51] A complication: with small R_f, CFB amplifiers exhibit peaked response at high frequencies, as seen for example in Fig. 4x.67; that's generally not a good thing, but it does extend the official $-3\,dB$ bandwidth. The AD846 acknowledges this artifact, and adds the dotted curve with the notation "single-pole model, AD846."

Table 4x.3: High-Speed Op-amps II: CFB[a]

Part #	qty per pkg[f]	Supply[p] Range (V)	Iq[t] (mA)	Bias Curr @25°C Iin+ typ (µA)	Iin− typ (µA)	Vos max (mV)	CMRR min (dB)	Noise[t] @10kHz en (nV/√Hz)	in[x] (pA/√Hz)	min gain[n]	Bandwidth −3dB G=1 (MHz)	G>5[g] (MHz)	−0.1dB[w] (MHz)	Iout typ (mA)	Slew typ (V/µs)	Settle 0.1% (ns)	Cin[c] (pF)	null pins	comp pin	shdn pin	DIP avail	SOIC-8	SOT-23	Power pkg	Cost ($US)	Comments
BJT, high-voltage																										
LT1217	1	9-36	1	0.1	0.1	3	60	6.5	0.7	1	10	8		50	500	280	1.5	•	-	•	•	-	-		4.42	A
LT1210	1	10-36	9-35	2	10	15	55	3	2	1	53	48		1100	900	-	2	-	•	•	-	-	-	•	8.93	B,D
AD844	1	9-36	6.5	0.15	0.2	3	90	2	10	2	60	33		60	2000	100	2	•	•	-	-	u	-		5.23	
LT1206	1,2	10-36	20	2	10	10	55	3.6	2	1	60	40		250	900	-	2	-	•	•	•	u	-		5.70	C,D
LM6181	1	9-36	7.5	0.5	2	5	50	4	16	1	100	80		90	1400	50	-	-	•	•	•	-	-		2.67	
THS3120	1	9-33	7	1	3	8	60	2.5	1	1	130	105	90	200	620	7	0.4	-	•	•	-	•	-	•	5.24	E
AD812	2	3-36	4.5	0.3	7	5	55	3.5	1.5	1	145	40	30	30	1600	40	1.7	-	-	-	•	-	-		5.22	F
THS3122	2	9-33	8.4	0.33	6	20	63	2.2	2.9	1	160	120	30	440	1550	53	2	-	•	•	-	•	-	•	9.24	F
LT1223	1	9-36	6	1	1	3	56	3.3	2.2	1	200	10		60	1000	75	1.5	•	•	-	•	-	-		5.17	
THS3091	1	10-33	9.5	3.5	4	3	62	2	14	1	235	190	95	175	7300	42	0.1	-	•	•	-	•	-	•	8.21	G
THS3095	1	10-33	9.5	3.5	4	3	62	2	14	1	235	190	95	175	7300	42	0.1	•	•	•	-	•	-	•	8.65	H
LT1227	1,2,4	4-36	10	0.3	10	10	55	3.2	1.7	1	280[k]	80	60	60	1100	50	3	•	•	•	•	•	-	•	3.57	J
THS3061	1,2	10-33	8.3	6	2	3.5	72	2.6	20	1	300	260	120	145	5700	30	1	-	-	-	-	•	-	•	7.28	K
THS6012	2	9-33	11.5	4	3	5	100[b]	1.7	12	1	315		40	500	1300	70	1.4	-	-	-	-	•	-	•	6.98	L
THS3001	1	9-33	6.6	2	1	3	65	1.6	13	1	420	350	115	100	6500	25	7.5	-	-	-	•	-	-	•	6.93	
THS3491	1	14-33	16.7	2	7	2	69	1.7	15	2	900	450	350	420	8000	7	1.2	-	-	•	-	•	-	•	11.49	M
BJT, low-voltage																										
AD8017	2	4.4-13	14[h]	16	1	3	59	1.9	23	2	160		70	270	1600	35	2.4	-	-	-	-	•	-	•	4.74	N
AD8010	2	9-12.6	15.5	6	10	12	50	2	3	1	230	100	60	200	800	25	2.8	-	-	-	-	•	-	•	6.26	N
AD8023	2	4.2-15	6.2	5	15	5	50	2	14	3	125	38	7	100	1200	30	2	-	-	-	-	u	-	-	9.19	P
ADA4310	2	5-12	0.7-8	2	6	1[t]	62[t]	2.9	22	5	190	190		120	820	-	-	-	-	-	-	-	v	-	2.10	F,N
LT6211	1,2	3-13	0.3-6	3.5	14	6	46	6.5	4.5	1	200[e]		75	700	20	2		-	-	p	-	•	-	•	2.29	B,Q
OPA2677	2	5-13	9	10	10	4.5	51	2	16	1	220	250	80	500	2000	-	2	-	-	•	-	•	•	•	3.48	S
OPA2674	2	5-13	2-9	10	10	4.5	51	2	16	1	250	220	100	500	2000	-	2	-	-	•	-	•	•	•	2.62	B,T
AD8004	4	4-12.6	3.5	40	35	3.5	52	1.5	38	1	250		30	50	3000	21	1.5	-	-	-	-	u	-	-	8.66	
OPA691	1,2,3	5-13	5.1	15	5	2.5	52	1.7	3.1	1	280	210	90	190	2100	12	2	-	-	•	-	•	•	•	2.41	
LT6559	3	4-12.6	4.6	10	10	10	42	4.5	6	1	300		150	100	800	25	2	-	-	-	•	-	q	-	1.39	P
LT1399HV	2,3	4-15.5	4.6	10	10	10	42	4.5	6	1	300		150	100	800	25	2	-	-	•	-	u	-	-	4.25	P
AD8011	1	4-12.6	1	5	5	5	52	2	5	1	400	57	25	30	2000	25	2.3	-	-	•	-	•	•	-	5.32	
LT1395	1,2,4	4-12.6	4.6	10	10	10	42	4.5	6	1	400		100	80	800	25	2.0	-	-	•	-	•	•	-	2.28	
LMH6723	1,2,4	4.5-13	1	2	0.4	3	57	4.3	6	1	370	150	100	110	600	30	1.5	-	-	•	-	•	•	-	2.03	
LMH6720	1,4	5-13.5	5.6	1	4.0	6	48	3.4	1.2	1	400		120	70	1800	12	1.0	-	-	•	-	•	•	-	2.47	
EL5162	1,2,3	5-13.2	1.5	0.5	2.0	5	50	3	6.5	1	500	110	30	100	4000	25	1	-	-	•	-	•	•	•	2.39	
MAX4223	1,2	5.7-12	6	2	4	4	55	2	3	1	1000	230	300	80	800	8	0.8	-	-	•	-	•	•	-	8.36	
EL5164	1,2,3	5-13.2	3.5	2	2	5	50	2.1	13	1	500	230	30	140	4000	25	1	-	-	-	•	-	-	•	2.82	
TSH350	1	4.5-6	4.1	12	1	4	56	1.5	20	1	550	125	120	205	940	-	-	-	-	•	-	•	•	•	2.61	F
MAX4224	1,2	5.7-12	6	2	4	5	55	2	3	2	600	230	200	80	1400	5	0.8	-	-	•	-	•	•	-	8.36	
HFA1135	1	9-11	6.9	6	0.4	5	40	3.5	2.5	1	660		25	60	1200	23	2	-	-	•	-	•	-	-	3.82	V
AD8007	1,2	5-12	9	4	0.4	4	56	2.7	2	1	650		90	70	1000	18	1	-	-	-	-	•	s	-	3.05	
LMH6702	1	5-13.5	12.5	6	8	4.5	45	1.8	3.4	1	720	140	140	100	3100	13	1.6	-	-	-	-	•	-	-	3.04	
ADA4860-1	1	5-12.6	5.2	1	1	13	55	4	1.5	1	800	320	125	30	790	8	1.5	-	-	-	-	-	•	-	1.31	W
HFA1130	1	9-12	21	25	12	6	40	4	18	1	850	200	80	60	2300	11	2	-	-	-	-	-	-	-	4.36	V
AD8001	1,2	6-12.6	5	3	5	5.5	50	2	2	1	880	200	145	85	1200	10	1.5	-	-	-	-	•	•	-	2.72	
AD8009	1	5-12.6	14	50	50	5	50	1.9	46	1	1000	350	75	175	5500	10	2.6	-	-	-	-	•	•	-	3.19	
EL5166	1	5-12.6	8.5	0.7	0.7	5	52	1.7	13	1	1400	260	100	110	6000	8	1.5	-	-	•	-	•	•	-	2.95	
OPA694	1,2	7-13	5.8	5	2	3	55	2.1	22	1	1500	250	90	80	1700	20	1.2	-	-	•	-	•	•	-	3.38	
AD8000	1,3	4.5-13	13.5	−5	−3	10	52	1.6	3.4	1	1580	330	190	100	4100	12	3.6	-	-	•	-	•	s	•	3.00	
OPA695	1,2,3	5-13	13	13	20	3	51	1.8	18	1	1700	450	320	120	4300	10	1.2	-	-	•	-	•	•	•	3.42	
THS3201	1,2	9-16.5	14	14	13	3	60	1.7	13	1	1800	565	380	100	10500	20	1.0	-	-	-	-	•	•	•	5.74	

NOTES: (a) ordered by increasing unity-gain −3dB frequency. (b) balanced differential. (c) input capacitance from noninverting input to gnd. (d) dual. (e) 10MHz and 170V/µs at I_Q=0.3mA. (f) bold is number of amplifiers in listed part number. (g) for G=5 to 10. (h) both amplifiers. (k) for G=2. (m) maximum. (n) min gain = 2 means G≥2 or −1. (p) programmable I_Q, range shown. (q) QFN pkg. (s) SC-70 available. (t) typ. (u) SOIC-14 or SOIC-16. (v) many-pin SOP or MSOP pkg. (w) at G=1 or 2. (x) noninverting input.

COMMENTS: **A.** very low I_Q. **B.** programmable I_Q. **C.** Ccomp pin for optimum compensation to >1nF C-load. **D.** power pkgs: DD and TO220-7, 5°C/W. **E.** Winfield's fave. **F.** output swings to within 1 volt of the rails. **G.** Larkin uses this one, in DDA pkg. **H.** THS3091 with shutdown. **J.** video amp. **K.** Larkin shuns this one. **L.** balanced line driver, SOIC-20. **M.** upgrade for THS3091, -95. **N.** line driver. **P.** video driver. **Q.** rail-to-rail output! LT6210 is SOT23-6. **S.** OPA2674 has current lim, adjustable I_Q. **T.** no powerpad, use OPA2677. **V.**. Vout clamps! **W.** lowest cost.

end-terminated coax line (that attenuates the signal by a factor of two, thus the need for a $G=2$ buffer). Some of these parts are low cost, and may be suitable at higher gains, but the datasheets provide little guidance.

Power dissipation
High-voltage parts with high quiescent current may need to be operated at lower voltage (say ± 5 V to ± 8 V) to prevent overheating. And supply current may increase with large output swings at high frequencies. Be generous with PCB copper, to carry off heat.

Output swing
Most CFB op-amps require significant supply-rail headroom, which means that "low-voltage" op-amps (which can in principle run from a 5 V total supply) will often need more (e.g., ± 5 V). Op-amps that can swing to within a volt of the rails are noted in the table (note F); the LT6211 is the unique exception with its rail-to-rail output.

Packages
Smaller is usually better! The SOT-23 versions of wideband low-voltage parts, with their reduced lead inductance, often have less peaking and overshoot than the SOIC-8 versions. When selecting CFB (or other!) op-amps, be aware that the single (versus dual) versions are often not stocked at distributors. Duals also have different part numbers, not listed.

4x.7 Power Supply Rejection Ratio

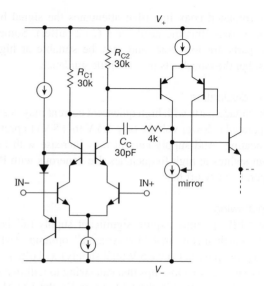

Figure 4x.73. Simplified schematic of the LT1097 op-amp, showing unbalanced paths of positive-rail fluctuations, owing to shunting of one side of the (otherwise symmetrical) input-stage differential output (collectors of the cascode) by the compensation capacitor at higher frequencies.

We encountered PSRR in Chapter 5 (§5.7.5), when dealing with precision design, where we pointed out several important features (uh, maybe *warnings* is more accurate): (a) PSRR decreases with increasing frequency, in a manner similar to the open-loop gain; (b) PSRR is usually specified for a unity closed-loop gain configuration; (c) the PSRR is generally different with respect to the positive and negative supply rails.

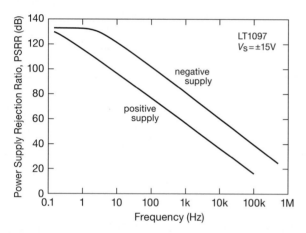

Figure 4x.72. LT1097 PSRR versus frequency, as shown in the datasheet plot.

Here we look a bit deeper into the latter effect. Figure 4x.72 shows the datasheet plot of PSRR versus frequency for the LT1097 precision BJT op-amp. At low frequencies the PSRR is comparable to the op-amp's open-loop gain (128 dB, typ), with the PSRR falling roughly at 20 dB/decade. Notice, however, that the PSRR with respect to the positive supply is some 24 dB worse than that with respect to the negative supply, for frequencies above about 10 Hz. (Interestingly, the tabulated data states only that the PSRR is 130 dB typ, 114 dB min, with no commentary about this disparity, nor about its behavior for closed-loop gains greater than unity.)

Why the difference? Look at the simplified circuit in Figure 4x.73. The input-stage *npn* BJT differential amplifier (with bootstrapped cascode) drives a second-stage *pnp* difference amplifier, with single-ended output to the unity-gain push–pull output follower. What spoils the symme-

try of this neat picture is the compensation network, which couples back to one side of the input stage's collector pair; so voltage fluctuations on the positive rail see an unsymmetrical path to the first stage's output, being shunted (at high frequencies) by the C_c path. So signals on the V_+ rail are coupled more strongly than those on the V_- rail.

Depending on the details of the op-amp's internal guts, this effect can afflict either supply rail. Figure 4x.74 shows the datasheet plot of PSRR versus frequency for the LT1055 precision JFET op-amp.[52] Here it's signals on the *negative* rail that are more strongly coupled. Once again the simplified circuit (Fig. 4x.76) reveals the reason; this time the compensation network couples to one side of the *p*-channel JFET differential amplifier's drain pulldown. As with the LT1097, the tabulated PSRR for this op-amp is silent on the disparity, stating only that the PSRR is 106 dB typ, 90 dB min.

The poorer negative-rail PSRR of the LT1055 is typical of many op-amps, particularly those with the "Widlar architecture" (see Fig. 5.13), which use a feedback compensation capacitor to a negative-rail common-emitter stage; Figure 4x.75, a simplified circuit of our canonical

---

[52] Both examples are from Linear Technology (now part of Analog Devices); we like their products, but the reason we chose these examples is that LTC includes (simplified) schematics, a rarity in contemporary datasheets.

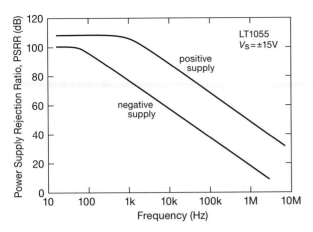

Figure 4x.74. LT1055 PSRR versus frequency, as shown in the datasheet plot.

LF411, is typical. Over frequencies from 100 Hz to 1 MHz its negative-rail PSRR is 30–40 dB worse than the positive-rail PSRR.

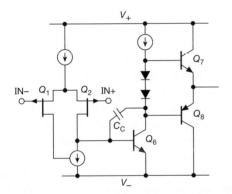

Figure 4x.75. Simplified schematic of the LF411 op-amp, a "Widlar" configuration whose unbalanced compensation path degrades negative-rail PSRR.

In fairness we should point out that the tabulated data is a dc value (though not so stated). But the lesson is clear: spend some time with the pretty graphs in the datasheet for any component you care about (and while you're at it, read carefully the application notes – there can be some pretty scary stuff there).

An obvious question is whether there are op-amp topologies that exhibit identical PSRR with respect to both supply rails. The answer is yes – take a look, for example, at the datasheet for the LT1351, a highly symmetrical op-amp[53]

whose plot of PSRR versus frequency is a single curve with the notation "$-$PSRR$=+$PSRR."

Wrapup

Be careful when looking at power-supply rejection specifications in op-amp datasheets. Quite often the tabulated value is accurate only for PSRR with respect to one of the supply rails (and they may not tell you which!), at dc and low frequencies only, and for the follower ($G=1$) configuration. The good news, though, is that it's easy enough to filter the rails to keep them clean at frequencies that matter (the powerline frequency and harmonics, and signal frequencies present elsewhere in the system)

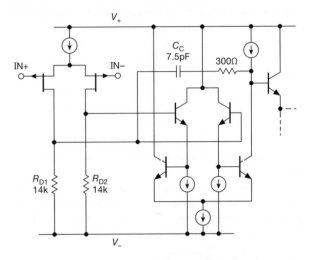

Figure 4x.76. Simplified schematic of the LT1055 op-amp, showing unbalanced paths of negative-rail fluctuations, owing to shunting of one side of the (otherwise symmetrical) input-stage differential output (JFET drains) by the compensation capacitor at higher frequencies.

---

[53] Internally it is a current feedback architecture, with the inverting input buffered so that both inputs are symmetrically high impedance; this is the "CFB+VFB" architecture (Fig. 4x.63C,D).

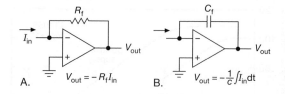

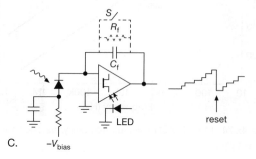

Figure 4x.77. Transimpedance amplifiers. A. Traditional resistive-feedback circuit. B. Capacitive-feedback TIA (bias not shown). C. Bias/reset schemes used in preamp TIA for germanium X-ray detectors.

4x.8 Capacitive-Feedback Transimpedance Amplifiers

The conventional transimpedance amplifier (current-in, voltage-out) consists of an inverting gain stage with resistive feedback R_f, with gain (i.e., V_{out}/I_{in}, units of impedance) of $G = -R_f$, see Figure 4x.77A.[54] As explained in §8.11, however, the feedback resistor is itself a source of (Johnson) noise current, specifically a noise current density $i_{nR} = \sqrt{4kT/R_f}$. To reduce the noise source you can go to large values of R_f, but that may result in unusably large gain, and also stability problems (treated in detail in §4x.3).

But there's another solution, usable in some special cases. The idea is to replace the feedback resistor with a *capacitor* (Fig 4x.77B), which creates an integrator of the input current, then differentiate (or take discrete differences of) the output to recover the correct transimpedance signal.[55]

This pretty scheme gets a bit complicated, though, when you realize that, just as with any integrator, you've got to provide some way to prevent the output from wandering off to saturation. There are two ways to do this:

"Dynamic charge restoration"
This is a fancy name for putting a large-value resistor R_f across the integrating capacitor, as in Figure 4x.77C, choosing R_f so large that its noise contribution (which goes inversely as the square root of R_f) is insignificant. This works OK as long as the average dc input current (producing an average output voltage of $-I_{in}R_f$) does not bring the amplifier into saturation (and similarly for low-frequency components of the input signal current). As with any integrator, this circuit behaves like a good integrator for input signal frequencies large compared with the $R_f C_f$ breakpoint, so you have to differentiate the output to flatten the response, i.e., to recover a true transimpedance output signal.

Pulsed reset
As with any integrator, you can reset the output to zero by shorting the capacitor, with a mechanical or FET switch (see Fig. 4.18A,B). Doing so interrupts the output, of course, but that's OK if you don't mind a bit of dead time (e.g., in a situation where the input is inactive at known times, or if you're detecting individual events at a low rate); additional circuitry disables output during the reset cycle. To prevent charge injection, the reset FET can be turned on with an LED flash; this is sometimes called "pulsed-optical reset," and it's often used in cryogenic semiconductor particle detectors (germanium, or lithium-drifted silicon).

4x.8.1 Capacitive-feedback TIA for gigabit optical receivers

A nice example of a low-noise capacitive-feedback TIA is the design by Shahdoost et al.,[56] delivering excellent performance at low operating current. Figure 4x.78 shows a clean eye diagram[57] of a 2.5 Gb/s pseudorandom bitstream, with dc operating power of just 14 mW. At a considerably lower data rate (but for the same noise-reduction goal) capacitive-feedback TIAs were used in the supermar-

---

[54] We've seen these earlier, for example Figure 4x.16, or (in AoE3) §4.3.1C (Fig. 4.22) or Figure 8.69.

[55] Why would any sane person choose to "integrate, then differentiate"? Because the first stage dominates the amplifier's noise contribution, so eliminating the Johnson noise source R_f creates a much quieter transimpedance amplifier.

[56] S. Shahdoost, A. Medi, and N. Saniei, "Design of low-noise transimpedance amplifiers with capacitive feedback," *Analog Integr. Circ. Sig. Process*, **86**, 233–240 (2016).

[57] See, for example, Figs. 12.130–12.132 in AoE3.

ket check-out barcode readers devised by IBM in the early 1980s.[58]

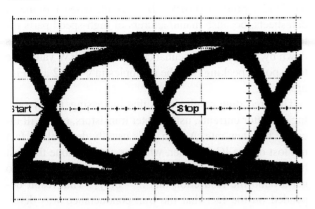

Figure 4x.78. Measured eye diagram of capacitive-feedback TIA, with pseudorandom bitrate of 2.5 Gb/s. (Shahdoost et al., used with permission.)

---

[58] R. L. Garwin, private communication.

4x.9 Slew Rate: A Detailed Look

Here we take a closer look at op-amp slew rate, its relation to bandwidth, and techniques for its enhancement. The reader should look first at the basic treatments in §4.4 and §5.8, where a differential transconductance stage drives current into the compensation capacitor.

It's easy enough to get a *quantitative* expression for op-amp slew rate. First, let us write an expression for the open-loop small-signal ac voltage gain, ignoring phase shifts:

$$A_V = g_m X_C = g_m/2\pi f C$$

from which the unity-gain bandwidth product (the frequency at which $A_V = 1$) is

$$f_T = \frac{1}{2\pi}\frac{g_m}{C}.$$

Now, the slew rate is determined by a current I_E charging a capacitance C:

$$S = \frac{dV}{dt} = \frac{I_E}{C}.$$

For the case of a BJT differential amplifier with no emitter resistors, g_m is related to I_E by

$$g_m = \frac{1}{r_e} = \frac{I_E}{2V_T} = \frac{I_E}{50\,\text{mV}}.$$

By substituting this into the slew-rate formula, we find

$$S = 2V_T\frac{g_m}{C}$$

i.e., the slew rate is proportional to g_m/C, just the same as the unity-gain bandwidth! In fact,

$$S = 4\pi V_T f_T = 0.32 f_T$$

with f_T in MHz and S in V/μs.

This is independent of the particular values of C, g_m, I_E, etc., and it gives a good estimate of slew rate (e.g., the classic 741, with $f_T \approx 1.5$ MHz, has a slew rate of 0.5 V/μs). It shows that an op-amp with greater gain–bandwidth product f_T will have a higher slew rate. You can't improve matters in a slow op-amp by merely increasing input-stage current I_E, because the increased gain (from increased g_m) then requires a correspondingly increased value of C for compensation. Adding gain anywhere else in the op-amp doesn't help either.

Important note: In this discussion and the accompanying equations we use the symbol "f_T" to mean the gain–bandwidth product (GBW). Be careful, when looking at datasheets, because manufacturers of high-speed parts often use the frequency at which the gain of a follower is down by 3 dB, which is enhanced over the GBW value due to response peaking; see the examples in §4x.5.2 and Figure 4x.52.

The preceding result shows that increasing f_T (by raising collector currents, using faster transistors, etc.) will increase the slew rate. A high f_T is usually desirable, a fact not lost on the IC designer, who has already done the best he can with what's on the chip. However, there is a way to get around the restriction that $S = 0.3 f_T$. That result depended on the fact that the transconductance was determined by I_E (through $g_m = I_E/2V_T$). You can use simple tricks to raise I_E (and therefore the slew rate) while keeping f_T (and therefore compensation) fixed. The easiest is to add some emitter resistance to the input differential amplifier. Let's imagine we do something like that, causing I_E to increase by a factor m while holding g_m constant. Then, by going through the preceding derivation, you would find

$$S = 0.3 m f_T \quad \text{and} \quad m = 3.1 S/f_T$$

Exercise 4x.1. Prove that such a trick does what we claim.

Likewise, the use of JFETs or MOSFETs in the differential stage has a similar effect, owing to their lower inherent transconductance. For example, compare the precision LT1007 (BJT input stage, R_E=0) with the LF411 (JFET input stage, R_s=0):

	LT1007	LF411
f_T	8 MHz	4 MHz
Slew rate	2.5 V/μs	15 V/μs
SR/f_T	0.31	3.75

The LT1007 with its BJT input pair has $m = 3.1 S/f_T = 0.97$, while the JFET pair has an order of magnitude lower transconductance at comparable operating current (and $m = 11.6$). This effect is seen nicely also in the measured data of Figure 4x.79.

4x.9.1 Increasing slew rate

Here, then, are some ways to obtain a high slew rate:

• Use an op-amp with high f_T.
• Increase f_T by using a smaller compensation capacitor; of course, this is possible only in applications where the closed-loop gain is greater than unity.
• Reduce the input-stage transconductance g_m by adding

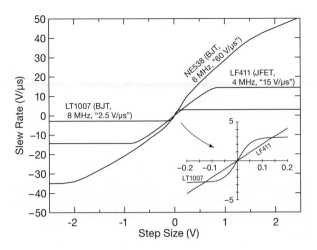

Figure 4x.79. Op-amp slew rate versus input error, from measurements of slew rate versus input differential step size. The lower input-stage transconductance of the JFET-input LF411 results in greater ultimate slew rate, but only with significantly larger differential input voltage. The NE538 uses a cross-coupled input stage for still greater ultimate slew rate.

emitter resistors (if BJTs) or by using FETs (perhaps with additional source resistors); then reduce C or raise I_E (or I_S) proportionately.

- Decompensate the op-amp, adding a minimum gain requirement, or allowing it to ring when configured for low gains.
- Use a different input-stage circuit, for example a cross-coupled current-boosting configuration, or a pairing of high- and low-transconductance differential pairs.
- use a current-feedback (CFB) op-amp, or one of the many CFB variants that buffer the inverting input in order to present a VFB-like external appearance.

The third technique (reduced g_m) is used in many op-amps. As an example, the HA2605 and HA2505 op-amps[59] are nearly identical, except for the inclusion of emitter resistors in the input stage of the HA2505. The emitter resistors increase the slew rate, at the expense of open-loop gain. The following data demonstrate this trade-off.

	HA2605	HA2505
f_T	12 MHz	12 MHz
Slew rate	7 V/μs	30 V/μs
Open-loop gain	150,000	25,000
m	1.8	7.8

Note that FET-input op-amps benefit naturally in this re-

<hr/>

[59] No longer available for use in circuits, but excellent for a textbook example.

gard from reduced input-stage transconductance. Here, for example, are the specifications of the op-amp pair plotted in Figure 5.12:

	LT1007	LF411
f_T	8 MHz	4 MHz
Slew rate	2.5 V/μs	15 V/μs
Open-loop gain	20,000,000	200,000
Enhancement (m)	1.0	12.5

In one implementation the fifth technique exploits "cross-coupled transconductance reduction," which involves having a second set of transistors available at the input stage, biding their time during small signal swings, but ready to help out with some extra current when needed. Figure 4x.80 shows the basic structure: the input differential pair Q_1Q_2 behaves conventionally for small inputs; but a large input differential (imagine IN− is brought $\gg 60$ mV below IN+) raises Q_2's collector current well beyond its usual limit of twice its quiescent current, via R_E. This scheme has the advantage of improved noise and offset performance, at the expense of some complexity, as compared with the simple emitter resistor scheme. This technique seems to have originated in parts from Signetics and Harris (the NE535/8, HA5141/51, etc.), and is used in contemporary parts like the ADA4075 and TLE2141 families (with enhancement factors m=6 and 25, respectively). The measured data of Figure 4x.79 includes one such specimen (NE538).

Using reduced-transconductance input transistors and adding degeneration emitter resistors with their Johnson noise can be expected to degrade an op-amp's noise performance. Indeed, reviewing 300 op-amps in our working table, the JFETs have e_n from 10 nV/$\sqrt{\text{Hz}}$ to 40 nV/$\sqrt{\text{Hz}}$ (although a few fast ones are in the 5 nV/$\sqrt{\text{Hz}}$ territory, e.g., the OPA627 and OPA827); degenerated JFETs have 15 nV/$\sqrt{\text{Hz}}$ to 50 nV/$\sqrt{\text{Hz}}$ noise; degenerated BJTs are not much better; and cross-coupled BJTs have 10 nV/$\sqrt{\text{Hz}}$ to 30 nV/$\sqrt{\text{Hz}}$ noise. All this is to highlight a notable exception, the ADA4075-2 cross-coupled op-amp with m=6.1, which has just 2.8 nV/$\sqrt{\text{Hz}}$ of noise, yeah! Because it has 1 mV max offset voltage, however, it didn't gain entry to our precision op-amp table (Table 5.2).

Another input-stage variant is the Butler circuit,[60] where a conventional high-g_m BJT pair is paralleled by a low-g_m JFET pair (operating at substantially higher current: $I_F \gg I_B$), as shown in Figure 4x.81. For small input differentials the circuit behaves like an ordinary BJT op-amp, but larger differentials cause greatly increased differential

<hr/>

[60] Well described in his US patent 5,101,126.

output current as the JFETs become unbalanced. This technique is used in the OPA275/285, with a slew-rate enhancement of $m=8$, producing precision low-noise op-amp with high slew-rates. This technique can be generalized: the low-g_m pair can be implemented as emitter-degenerated BJTs (i.e., $R_E > 0$), and the scheme can be extended to include several paralleled stages.

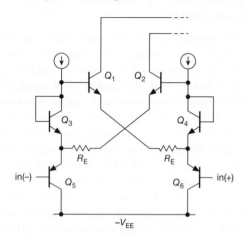

Figure 4x.80. A cross-coupled input stage enhances the slew rate.

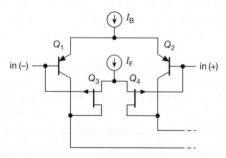

Figure 4x.81. Butler input stage: paralleled BJT and JFET differential amplifiers, with $I_F \gg I_B$.

Finally, a current feedback op-amp can provide very high slew rates, set by the feedback resistor to the (low-impedance) inverting input. For example, the LT1210 CFB op-amp with datasheet-specified $G=2$ feedback network ($R_f = R_g = 1.5\,\mathrm{k}$) achieves an impressive slew rate of $900\,\mathrm{V}/\mu\mathrm{s}$. Given its f_T value of 55 MHz, that's an enhancement factor $m=55$. Taking this a step further, there are numerous examples of what might colorfully be described as "a CFB in VFB's clothing": these add a unity-gain buffer plus internal feedback resistor to the inverting input of a CFB (Fig. 4x.63C,D and §4x.6.3), so the combination looks externally like a VFB. So you get high input impedance at both terminals, combined with the dy-

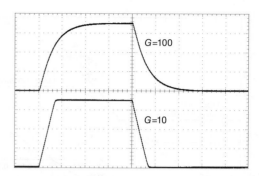

Figure 4x.82. Generating a 70 V pulse with an OPA454 high-voltage op-amp, configured as a noninverting amplifier with two choices of voltage gain. Vertical: 20 V/div; Horizontal: 10 μs/div.

namic performance of a CFB. An example is the LT1351–63 family (with increasing bandwidths): the slowest part (the LT1351) runs at 0.25 mA and has an f_T of 3 MHz, yet achieves an astonishing slew rate of $200\,\mathrm{V}/\mu\mathrm{s}$; that's an enhancement factor $m=220$! Intersil's EL5100 family of "Slew-enhanced VFAs" work similarly; the EL5100 (200 MHz, 2.5 mA) has SR=$2200\,\mathrm{V}/\mu\mathrm{s}$ ($m=37$), while the EL5104 (700 MHz, 9.5 mA) has SR=7000 ($m=33$). TI's OPA690 (500 MHz, 5.5 mA) has SR=$1800\,\mathrm{V}/\mu\mathrm{s}$ ($m=12$); and National has an extensive series of these, for example the LM7171 (200 MHz, 6.5 mA) has SR=$4100\,\mathrm{V}/\mu\mathrm{s}$ ($m=68$). This is clearly a favored architecture for enhanced slew rate amplifiers.

4x.9.2 Case study: high-voltage pulse generator

In a recent posting on an electronics Internet forum ("Improving High Voltage Op-Amp's Slew Rate") the following question came up:

Q. I'm trying to make a 70-volt pulse with 10 μs rise and fall times, using an OPA454 high-voltage op-amp (which accepts a total supply voltage to 100 V). The datasheet says its slew rate is 13V/μs, yet I'm getting sloppy 20 μs pulses instead of the expected $\sim$5 μs (70 V divided by 13 V/μs). I'm driving my $G=100$ amplifier with a pulse from a 3.3-volt FPGA. What's wrong?

A. We hooked up your suggested circuit, and measured the output pulse shape (Figure 4x.82, top trace). The inexpensive ($7.50) OPA454 high-voltage op-amp looks like a candidate for the job, with its 13 V/μs slew rate. But slew rate isn't the only issue affecting pulse response; what about *bandwidth*? The OPA454's f_T spec is 2.5 MHz, so a $G=100$ amplifier will have a bandwidth of only 25 kHz. And (taking an approximate model) a 25 kHz *RC* filter

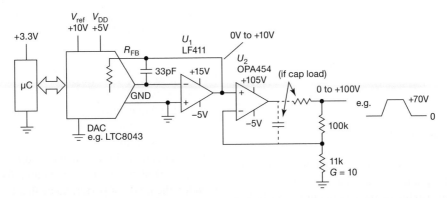

Figure 4x.83. Big signals from tiny beginnings: The high-voltage OPA454 output stage is configured for low gain, to maximize bandwidth; the correspondingly large 0 V–10 V input signal range ensures that the high-voltage OPA454 is driven to its maximum slew rate. See §13.2 for teachings on DACs, and Chapter 15 for microcontrollers.

would have a time constant of $\tau = 1/(2\pi \cdot \mathrm{BW}) = 6.4\,\mu\mathrm{s}$ – just what we see.

The OP (original poster) was advised to use two amplifiers, and to reduce the HV output stage's gain. The bottom trace shows the result, with the OPA454 now configured for $G=10$, creating textbook pulses with $7\,\mu\mathrm{s}$ rising and falling linear ramps. Two things are going on here: first the bandwidth is increased to 250 kHz, with corresponding $0.6\,\mu\mathrm{s}$ small-signal rise time. Second, in a large-signal analysis the amplifier is presented with a 7 V (rather than 0.7 V) pulse, which ensures that the input error can remain high enough to drive the OPA454 at its full slew rate for the entire rise time. The OP could use a circuit like that in Figure 4x.83, where a 15 V op-amp creates a large enough input signal.

It's instructive to compare the OPA454 JFET op-amp with our jellybean LF411 JFET op-amp, whose slew rate versus input error is shown in Figure 4x.79. The OPA454 has $m=16.1$ compared with $m=11.6$ for the '411, so we would guess that its input stage is further degenerated, and thus a higher input error would be required for full slewing, e.g., ~1.1 V for the '454 versus 0.8 V for the '411. We were curious and decided to measure slew rate versus input error for the OPA454. Figure 4x.84 shows the somewhat unexpected result: ~600 mV produces full slew rate. This tells us the op-amp must be slightly decompensated to achieve its high slew rate (relative to its bandwidth), which would make sense for a high-voltage part that would rarely be used with $G=1$. This is confirmed by examining the datasheet plots of open-loop gain (2nd-pole rolloff near 0 dB) and small-signal step-response (35% overshoot and ringing).

One more observation. The top trace with $G=100$ shows

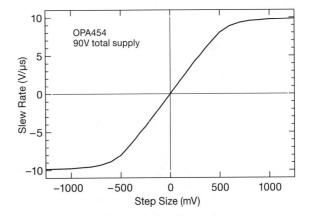

Figure 4x.84. Measured slew rate versus input error for the OPA454 high-voltage JFET-input op-amp.

a full slew rate at the very beginning of the step, when a 700 mV error is presented across the op-amp's inputs. Because we'd expect the slew rate to slow as the input error drops, perhaps that can explain the slow rise time. Perhaps, but from a look at Figures 4x.82 and 4x.84 it's apparent that, for a 70 V output step with $G=100$, this op-amp is bandwidth limited (rather than slew-rate limited).

4x.10 Bias-Current Cancellation

Input current matters, when you've got a signal of high source impedance, or when you're trying to measure small currents. As we've seen in Chapter 4, the input "bias" current of FET-input op-amps is just leakage current, and it can range from low (in the picoamps) to outrageously low (a few femtoamps). The price you pay is poorer accuracy – larger offset voltage, and offset voltage drift (tempco) – and higher noise voltage. An example of a very low input current op-amp is the JFET-input OPA129A, whose typical values are listed in this comparison table:

	OPA129A (JFET)	LT1028 (BJT)	units
I_{in}	**0.03**	25000	pA
V_{os}	500	**10**	μV
ΔV_{os}	3	**0.2**	μV/°C
e_n	17	**0.85**	nV/$\sqrt{\text{Hz}}$

By contrast, op-amps with BJT input stages (particularly the "precision" types, see Chapter 5) excel in accuracy and noise, with offset voltages measured in the 10's of microvolts, tempcos in the tenths of microvolts per °C, and voltage noise in the single-digit nanovolt per root-hertz. An example is the LT1028, whose typical values are listed alongside those of the JFET OPA129A. (In this table **boldface** indicates the winners in each parameter. As in golf, you want a lower score.) It beats the low-bias OPA129A by factors of 15 to 50 in the "accuracy" parameters, but at what a cost – a million times as much input current! (And, with its large dc input current, its corresponding current *noise* is also much larger.)

4x.10.1 The best of both worlds?

Is there a way to assemble the best of each? Sort of. With some serious offset trimming the accuracy of FET-input op-amps can be greatly improved; the JFET OPA627 achieves $40\,\mu$V and $0.4\,\mu$V/°C – only four times poorer than the LT1028 – but at a cost (literally!) of $30. And its large-area input transistors keep the noise down to a pretty impressive $4.5\,$nV/$\sqrt{\text{Hz}}$ (still five times poorer than the LT1028), but with a compromise in input current (1 pA, thirty times poorer than the low-bias OPA129A – and of

course rising rapidly with temperature, a further factor of 1000 to 1 nA at 60°C).

The other approach is to reduce the input current of the inherently accurate and quiet BJT-input op-amps. This turns out to be quite straightforward, and successful. It's used in many BJT-input op-amps, recognizable by their unusually low input current (often down in the picoamp territory, rather than the expected nanoamps), and (sometimes) by the inclusion of a "$\pm$" sign in front of the tabulated I_B value. And, unlike FETs, the input current does not exhibit an exponential rise with temperature. To steal some later thunder, the kinds of performance that the wizards of Silicon Valley achieve this way is exemplified by the LT6010 (or its predecessor, the LT1012): I_B (typical) of $20\,$pA (a thousand times better than the LT1028), while preserving the latter's accuracy (the same $10\,\mu$V and $0.2\,\mu$V/°C).[61]

4x.10.2 Bias cancellation: the circuits

The basic approach is (a) begin with transistors of very high current gain ("superbeta"), then (b) reduce the (already small) input current by subtracting a dc current of the same magnitude. Let's take it in several steps.

A. Simplest: Mirroring the base current of a cascode twin

Figure 4x.85 shows the idea, simplified to reveal its essence. The collector currents of the superbeta matched differential pair Q_1Q_2 are intercepted on their way up to the usual collector load (most simply, resistors to $+V_{CC}$; more realistically, a current mirror), by Q_3Q_4. The latter are β-matched to Q_1Q_2, and their bases are held at a fixed dc voltage that is one diode drop below the indicated "V_{bias}." (In this notional circuit the bias voltage would be chosen a few volts above the anticipated input level. We'll do this a better way in subsequent circuits; work with us, for now.)

So, the base currents I_b' in the cascode transistors Q_3Q_4 are approximately the same as I_b, the base currents in Q_1Q_2. Hence the nice trick: mirror I_b' to create sourcing replicas I_b'', which we then connect at the inputs to cancel the (sinking) input currents I_b. Voilá – zero input current!

That's the basic idea. Of course, the cancellation is not perfect. It can be only as good as the beta matching: a 10% mismatch, for example, would result in a net input current that is reduced to 10% of its original current; not bad, but ideally we'd like a factor of 100 (reducing an un-

---

[61] It is only average in noise, however ($14\,$nV/$\sqrt{\text{Hz}}$), a design trade-off forced by the desire to run the input stage at lower currents, and the use of superbeta transistors, which have higher $r_{bb'}$ noise.

canceled base current of a few nanoamps down to the tens-of-picoamps territory).

Even with perfect beta matching, there's another problem, namely the dependence of β on V_{CE}. And in this circuit the V_{CE} of input pair Q_1Q_2 depends on input common-mode voltage, while the cascode pair Q_3Q_4 operates at constant V_{CE}. Which brings up another flaw: the fixed bias on the cascode pair severely limits the input common-mode range. This circuit needs some serious fixing.

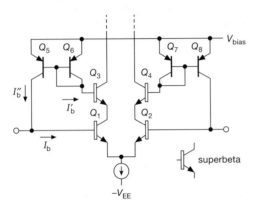

Figure 4x.85. Notional bias-cancellation scheme. The base current of a β-matched cascode pair is mirrored and summed.

B. Better: Bootstrapping the cascode bias

Figure 4x.86 shows how to fix both the severe common-mode restriction, and also the V_{CE} mismatch problem. In Figure 4x.86A the base bias voltage for the cascode is rigged to be two diode drops (that is, $2V_{BE}$) above the emitter voltage of the input pair. It thus follows the input common-mode signal, taking care of the V_{CM} problem, and simultaneously eliminating the V_{CE} variation in the input pair. But now we've simply shifted the problem of V_{CE} variation to the cascode pair! That's fixed in Figure 4x.86B, where a second level of cascode ($Q_{10}Q_{11}$), also with bootstrapped bias, ensures that both Q_1Q_2 and also Q_3Q_4 operate with fixed V_{CE}. If you count your diode drops correctly, you'll find that those four transistors all have V_{CE} equal to one diode drop. Now Q_{10} and Q_{11} are stuck with the V_{CE} variation, but that doesn't matter: it's only the Q_3Q_4 pair (from which the accurate bias-current replica is extracted) that have to be treated with care.

C. Another way: replicating the emitter current

Instead of extracting the bias-current sample from the signal-current-carrying collectors, we can take advantage of precision matching to create a second current sink that is matched to that of the input pair. Running that through

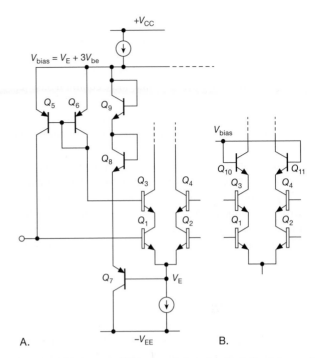

Figure 4x.86. Improved bias cancellation circuits. A. Bootstrapped bias for the cascode pair greatly extends the common-mode input range; this scheme dates back at least to 1979, when it was used in the OP-07. B. A second level of cascode, with bootstrapped bias, prevents V_{CE} variation in the Q_3Q_4 pair that creates the input current replica. This was used in Raytheon's version of the LT1012.

a transistor twin of the input pair then creates the matching base current needed to cancel the input bias. This technique has several advantages: (a) better tracking at elevated temperatures; (b) better tracking over common-mode input voltages; (c) improved signal-frequency performance; and (d) possibility of fine trimming of the cancellation during wafer testing.

Figure 4x.87 shows a simplified diagram of the somewhat convoluted circuit.[62] Some serious head-scratching will get you to an understanding. Spoiler: don't read further, if you like puzzles!

OK, if you're reading this, you don't like puzzles. Here's how the circuit works, assuming the operating current and beta indicated: (a) A matched current sink (in the ratio 1:3) runs Q_5 at $20\,\mu A$ (via current mirror Q_9Q_{10}), thus a base current that is beta times smaller (here 10 nA). (b) The input pair Q_1Q_2 each run at the same $20\,\mu A$ (the $60\,\mu A$ emitter sink is reduced by Q_5's current, then split two ways), so

---

[62] See Dobkin, Erdi, and Nelson's 1986 patent #4575685, "Arrangement for cancelling the input bias current, at picoampere levels, in an integrated circuit."

their base currents are also 10 nA, assuming good matching. (c) Q_8 is a triple matched current source that replicates Q_5's base current, canceling the input currents.

The Q_8 mirror (think of it as three matched transistors sharing a common emitter terminal and a common base terminal) is a bit unusual, because the programming current (Q_5's base) is connected only to a collector, rather than to both base and one collector, as in the conventional mirror. Here the loop is closed instead by Q_5's collector, in a tight feedback loop that pulls the mirror's emitters down until the collector sources a base current into Q_5 appropriate to its $20\,\mu A$ collector current. The other two outputs from the mirror are compelled to source the same current, of course, which is just what's needed to cancel the input currents of Q_1Q_2. You might be concerned about the current that the mirror diverts from Q_5's collector; but cast aside your worries – that current is down by a factor of nearly 1000, because the mirror transistors are running only at the superbetas' 10 nA base current.

You can count diode drops, going up and going down, and find that both the input pair and the Q_5 replicant run at V_{CE} equal to one diode drop. You'd also find that Q_8 runs at $V_{CE}=0$! That would be cause for worry; in practice you fix this by running Q_6 and Q_7 at a higher current density, leaving about 100 mV for Q_8's V_{CE}. This and other subtle circuit mismatches are accommodated during chip layout by adjustment of transistor area scalings.

4x.10.3 Bias cancellation: how well does it work?

Pretty well! Most manufacturers don't tell you the operating currents, or transistor betas, of their products. But you can find some exceptions, such as the oft-mentioned LT1028: it is a BJT-input high-voltage part with best-in-class noise voltage ($e_n=0.85\,\mathrm{nV}/\sqrt{\mathrm{Hz}}$ typical at 1 kHz), which requires relatively high collector current to keep the transistors' noise-generating r_e low enough (see the discussion in Chapter 8). Linear Technology is unusually forthcoming about their products, and the schematic shown in the LT1028A datasheet includes input-stage collector currents ($900\,\mu A$), and even the (uncanceled) base currents ($4.5\,\mu A$). So we know that the nominal beta is 200 (thus not superbeta). The tabular data lists a (canceled) bias current spec of $I_B = \pm25\,\mathrm{nA}$ typ, and ±90 nA max – so their bias current cancellation is typically accurate to $\pm0.6\%$, and to $\pm2\%$ worst-case.

Note that op-amps with bias cancellation will list an input *offset* current spec (i.e., mismatch of input currents) that is comparable to the bias current spec itself, in contrast to uncanceled op-amps for which I_{os} is ordinarily $10\times$–

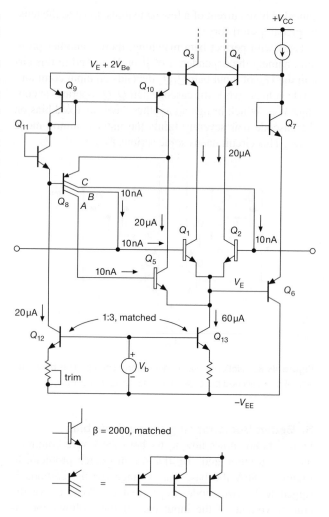

Figure 4x.87. Alternative bias-cancellation circuit, exploiting a matched emitter current sink to bias a matched superbeta transistor. The values shown assume $\beta=2000$ input transistors running at $I_C=20\,\mu A$. This circuit is shown in the LT6010 datasheet (for which $I_{in}=\pm20$ pA typ), and variations appear in the LT1012/97 datasheets and in the Dobkin et al. patent #4575685.

$100\times$ smaller than I_B): for the LT1028A the typical values are ±12 nA and ±25 nA, respectively, and for many other bias-canceled parts (e.g., LT6010, OPA277P, OP97A) the listed I_{os} magnitude is the same as the listed I_B. Compare that with the (uncanceled) LT1013, where $I_{os} = \pm0.15$ nA and $I_B = 12$ nA.

Finally, it's important to realize that the input *noise* current of a bias-canceled op-amp will be considerably larger than you would expect if you were to calculate the shot noise arising from the net (i.e., canceled) input bias current. Rather, you need to calculate the shot noise from the uncanceled base currents (and then apply a factor of $\sqrt{2}$ to account for the additional noise in the cancellation current). For example, using the LT6010's value of $I_B = \pm 20\,\text{pA}$ (typ), you'd incorrectly estimate a shot noise current of $i_n \approx 2.5\,\text{fA}/\sqrt{\text{Hz}}$, whereas the datasheet lists a typical value (at 1 kHz, well above the $1/f$ corner) of $100\,\text{fA}/\sqrt{\text{Hz}}$; similarly for the LT1028 (which specifies $1000\,\text{fA}/\sqrt{\text{Hz}}$, versus the $90\,\text{fA}/\sqrt{\text{Hz}}$ you would incorrectly estimate from the net I_B).

A caution: Some datasheets list greatly optimistic values for i_n, evidently making exactly this error. For example, the bias-canceled LT1012's datasheet shows a typical i_n leveling off to $6\,\text{fA}/\sqrt{\text{Hz}}$ (beyond the $1/f$ corner), which is what you'd calculate from the specified net (i.e., canceled) input current of $\pm 100\,\text{pA}$ max, whereas you would expect a value about ten times larger (assuming the uncanceled base current is about $100\times$ larger). We were skeptical of the datasheet's claimed i_n, so we measured it (along with others that appeared to be similarly in error), and found $i_n \approx 55\,\text{fA}/\sqrt{\text{Hz}}$.[63] This error shares some of the characteristics of an *epidemic*, having infected also the datasheets of auto-zero op-amps. For example, the exemplary AD8628A (listed in Table 5.6) specifies an input noise current density of $5\,\text{fA}/\sqrt{\text{Hz}}$; imagine our surprise when we measured a value 30 times larger. Not to be outdone, the MCP6V06 auto-zero op-amp's specification of $0.6\,\text{fA}/\sqrt{\text{Hz}}$ is rather at odds with its measured $170\,\text{fA}/\sqrt{\text{Hz}}$. See the discussion in §8.9.

---

[63] Datasheets for the closely similar OP-97 and LT1097 make the same error, evidently corrected in that of the later LT6010 (the recommended successor to the LT1012).

4x.11 Rail-to-Rail Op-amps

We introduced RRIO op-amps in AoE's Chapter 4, and dealt with them in some detail in Chapter 5 (§5.9), where we discussed, among other things, their input and output properties. RRIO come not without some drawbacks, which we'll summarize here, for review, before exploring the very interesting topic of achieving "true" rail-to-rail output swing.

4x.11.1 Rail-to-rail inputs

Most rail-to-rail-input (RRI) op-amps use complementary pairs of differential input stages, which solves the problem of operation all the way to (and a bit beyond) each supply rail. But, as described in §5.9, this structure has some undesirable side effects, such as an abrupt change in input current (if BJT input transistors), an abrupt change in offset voltage, and a rise in distortion when input signals move through the crossover region. Figure 5.29 (duplicated here as Figure 4x.88) shows the disastrous degradation of offset voltage in two typical RRI op-amps, and a nice on-chip solution (internal charge-pump creation of a beyond-the-rail supply voltage, circumventing the need for paired complementary input stages).

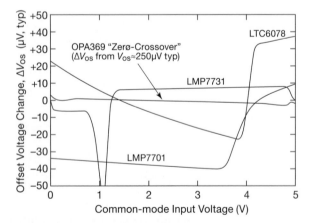

Figure 4x.88. Op-amps with rail-to-rail inputs usually exhibit a shift of V_{OS} as the input voltage passes control from one input pair to the other. The OPA369 circumvents this by using a single input pair, powered beyond the rail by an on-chip charge pump.

4x.11.2 Rail-to-rail outputs

Op-amps with rail-to-rail outputs suffer several drawbacks, mostly related to the common-source (or common-emitter) structure of the push–pull output stage, necessary to allow swings to the respective rails. The consequence is a high native output impedance (drain or collector), which means that the open-loop gain depends strongly on load impedance, and that a capacitive load creates undesirable phase shifts. The output stage also has inherently higher distortion, and, as we'll see next, in many cases it cannot swing *completely* to the rails (owing to quiescent current flowing through the output transistors' saturated R_{on}).

4x.11.3 Output near ground: when "RRO" *isn't*

In §4x.3.8 we encountered a peculiarity (perhaps that's too diplomatic a term) of op-amps with rail-to-rail output stages, namely their inability to swing all the way to the negative supply voltage (i.e., ground, when powered from a single positive supply), even when the op-amp's output is unloaded. The deficit isn't large – typically a few millivolts – but it's enough to seriously compromise the dynamic range. Taking the example of the wide-range transimpedance amplifier in that discussion (Fig. 4x.35), where the op-amp's output is digitized by a modest 12-bit ADC with input span of 0 V to 4.095 V (thus an LSB of 1 mV), an op-amp that swings only to within 10 mV of ground is throwing away 3 bits (10 LSBs) of dynamic range. For a seriously accurate 16-bit ADC (a plausible choice since the LSB of 62 μV is comparable to worst-case offset errors in the suggested op-amp types), the inability of the op-amp outputs to go below 10 mV, say, corresponds to an ADC span of the bottom 160 LSBs.

It may seem puzzling that the output-stage nMOS pulldown switch (see Fig. 5.32B) does not go all the way to ground with no load attached. But it's not puzzling when you realize that output-stage quiescent current flowing through the on-resistance of the nMOS pulldown limits the negative swing to $V_{min}=I_Q R_{ON}$. You can circumvent this problem with a current-sinking load, with current set to I_Q or greater.

We got interested in this problem, and rigged up some tests to explore the behavior of several interesting RRO op-amps. Figure 4x.89 shows what you get when you wire these as unity-gain followers, input grounded, and output connected to a current source varying from −0.5 mA (sinking) to +1 mA (sourcing). For example, our favorite jellybean LMC6482 sits at 7 mV (unloaded), going linearly to ground when 140 μA is sunk at its output. Evidently that's the quiescent current when the output is near ground. You can learn more: the slope of V_{out} versus I_{sink} reveals the nMOS output resistance, here about 57 Ω.

So, one way to make these puppies behave is to attach a current sink to the output; this requires a negative voltage source somewhere in the circuit.[64] If a negative voltage is not available, you can use a charge-pump voltage inverter to create it – see the short discussion in §4x.11.4, where we discuss also the (perhaps better) alternative of running the op-amp's negative supply terminal a few tenths of a volt below ground.

Another possibility is to choose an RRO op-amp whose output is well behaved, as some in Figure 4x.89 appear to be. This eliminates the need for a negative supply, but it has the drawback of limiting your choice of op-amp types. We explored this scene for three of these RRO op-amps, wired as followers and driving each with a 25 mVpp triangle wave at 1 kHz. Figure 4x.90 shows the output waveforms under various load conditions (unloaded, resistor to ground, or current sink load), and with the input waveform just touching ground or with it offset so the negative peaks go 10 mV below ground.[65]

The top waveform is the input signal, when set for $V_{min}=0$ V, used for all three parts shown in the lower panels. The LMC6482, whose curves in Figure 4x.89 predict failure near ground, bottoms out around 7 mV, even with a 1k load to ground (waveform A), but delivers clean waveforms to ground with a 200 μA sinking load (waveform B). In fact, it works well under this load (waveform C) even when the input triangle wave is offset to bring its negative peaks down to −10 mV.

Based on Figure 4x.89 we had hoped for better with the LTC6078; and its performance is pretty good (middle panel of Fig. 4x.90), but it flames out at around 1 mV (waveform A) unless a current-sinking load is attached (waveforms B and C). The AD8616, by contrast, heroically swings clear to zero on this scale (bottom panel, waveform A). Heroically, but not magically, as demonstrated by our attempt to have it do the impossible with a negative input (waveform B) but without a current-sinking load; but the clean saturation without evidence of recovery transients is admirable. As with the other op-amps, a current sink at

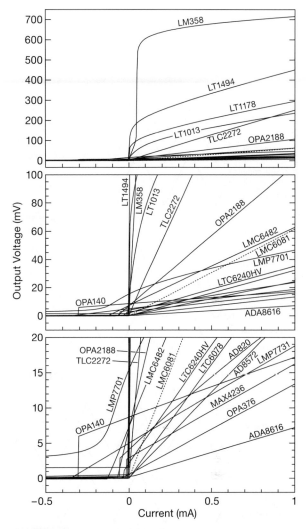

Figure 4x.89. Not all op-amps that specify output swing to the negative rail (including RRIO and RRO types) will reach that voltage, as seen in these plots of V_{out} versus I_{load} for 18 representative op-amp types. For these plots the op-amps were wired as followers with grounded input, with $V_+=5$ V and $V_-=0$ V.

---

[64] A simpler solution you might invent is a simple resistor to ground, but it does not work: even a small value like 1 kΩ sinks only 10 μA when the output is 10 mV above ground, far less than the quiescent current you're trying to cancel.

[65] We were interested in the latter because we worried about the dynamic behavior of a feedback circuit where the input voltage may flirt with ground, causing the output to saturate at zero volts. Such a nonlinear situation allows for no overshoot, and could exhibit unwholesome behavior.

the output (here requiring nearly a milliamp) allows it to swing negative.

Instead of a current sink at the op-amp's output, you can simply supply a slight negative voltage (100 mV, say) to the op-amp's negative supply terminal. This method works for any single-supply components (comparators, ADCs, etc.) that are struggling in those last few millivolts near ground; we discuss it next.

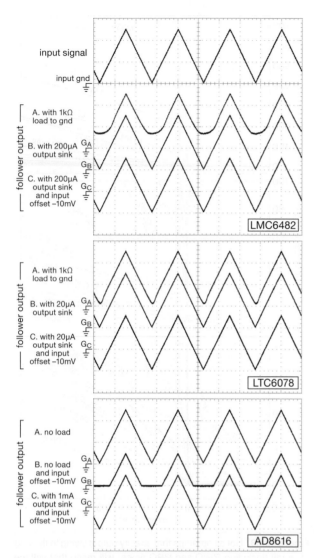

Figure 4x.90. Three RRIO op-amp followers, powered from +5 V and ground, and driven with a 25 mV triangle wave whose negative peaks are at ground or offset to −10 mV. The output waveforms correspond to open, resistive, or current-sinking loads, as indicated. Vertical: 10 mV/div; horizontal: 400 μs/div.

4x.11.4 Offsetting the negative supply terminal

In the good ol' days of ±15 V supply rails we never worried about behavior near ground; couldn't care less. But it's a real issue with single-supply designs, and particularly low-voltage parts; there RRO and RRIO op-amps will get you to ground... almost. As we discussed in the previous section, most RRO op-amps flame out a few millivolts above the negative (i.e., ground) rail. Not bad, but a serious drawback if you're needing the full dynamic range,

as we did in the wide-range TIA circuits of §§4x.3.7 and 4x.3.8. As we saw, one solution is to sink some current from the output. A more general solution, which deals also with close-to-ground problems in other single-supply parts (like comparators and single-supply ADCs) is to supply a small negative voltage to the part's negative supply pin.

If you've got a negative voltage available, you can generate a $\sim$250 mV negative supply by simply biasing a Schottky diode into forward conduction: an SD103 diode, for example, has a forward drop of 250–300 mV at 5 mA (a 1N5817 drops 175–225 mV at 10 mA). We used this idea in Figure 4x.91B, where we show also a flying-capacitor voltage inverter to generate the necessary negative rail. That figure shows the very nice LTC1550 part,[66] which includes a linear post-regulator for low ripple and noise; its 900 kHz oscillator frequency lets you use pleasantly small-value (0.1 μF) flying capacitors.

Evidently the need for low-voltage negative supplies has not escaped the semiconductor industry, who offer the LM7705 voltage inverter illustrated in Figure 4x.91A. It also includes a linear post-regulator, but set to −230 mV. This part is small and inexpensive ($0.66 in unit quantities), but does require large-value capacitors because of its lower operating frequency (92 kHz). It was clearly designed for exactly this purpose; the datasheet says it all:

> The LM7705 device is a switched capacitor voltage inverter with a low noise, −0.23 V fixed negative voltage regulator. This device is designed to be used with low voltage amplifiers to enable the amplifier's output to swing to zero volts. The −0.23 V is used to supply the negative supply pin of an amplifier while maintaining less than 5.5 V across the amplifier. Rail-to-Rail output amplifiers cannot output zero volts when operating from a single-supply voltage and can result in error accumulation due to amplifier output saturation voltage being amplified by following gain stages. A small negative supply voltage... will help maintain an accurate zero through a signal processing chain. Additionally, when an amplifier is used to drive an input of the ADC, the amplifier can output a zero voltage signal and the full input range of an ADC can be used.

Note that the increment to the "total supply voltage" is so small that you don't have to decrease the positive supply voltage: you can still use a +5 V positive rail for a low-voltage part that specifies supply voltages from 2.7 V to 5.5 V, for example.

---

[66] We show the −2.5 V part, but you can get other pre-set voltages, or an adjustable part with 1.22 V internal reference.

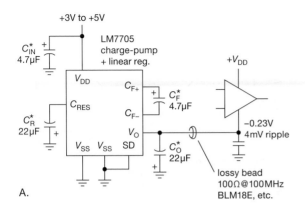

A.

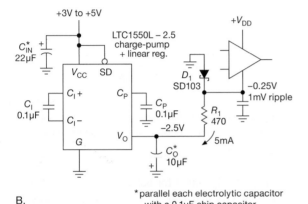

B.

Figure 4x.91. Charge-pump voltage inverters are a nice way to generate a ~0.25 V negative rail, so RRO op-amps (and other parts) can operate cleanly to zero volts.

4x.11.5 Designs by the masters: the Monticelli output stage

The output stage of a conventional (not RRO) op-amp is ordinarily a complementary push–pull follower (or some variation thereupon), biased with some conduction overlap to prevent crossover distortion at mid-supply (see §5.8.3). By contrast, the output complementary pair in an RRO op-amp is configured as a push–pull common-source *amplifier*, see Figure 5.32. That's necessary for the output to reach the rails (absent a second set of beyond-the-rails supply voltages). But it creates problems, owing to its inherently high output impedance. Some of these (described in §5.8.3) are high open-loop output impedance, especially at low frequencies; instability with capacitive loads; and dependence of open-loop gain on load resistance.

The rail-to-rail output stage also presents challenges to the chip designer when it comes to quiescent biasing and reduction of crossover distortion. An elegant solution is

the output stage designed by Dennis Monticelli, described briefly (in the context of low-distortion) in §5.9.2.[67] At the broad brush-stroke level, it is a clever circuit that biases the push–pull output transistors to maintain current overlap at crossover, and, better still, with continuing current through both transistors *throughout the output swing*. Read the basic description in §5.8.3 first, then join us for an exploration of its detailed workings.

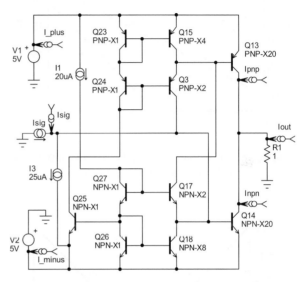

Figure 4x.92. SPICEing the Monticelli rail-to-rail output stage. Both input and output signals are *currents*, effortlessly probed by plopping down a "current probe" (if only real life were that simple!).

This is one cool circuit! We ran a SPICE simulation, to check out its linearity, and to see the overlap of source and sink currents over the full swing. Figure 4x.92 shows the circuit, as entered in IntuSoft's ICAP/4 SPICE software.

First we explored the transfer function, by sweeping the input current and watching the output current and the individual source and sink currents. In Figure 4x.93 the output current looks quite linear, to the eye. Also, you can see (maybe) that both transistors stay in conduction (just barely) over the full cycle. To explore this latter point, we plot in Figure 4x.94 the source ($I+$, from Q_{13}) and sink ($I-$, from Q_{14}) currents versus output current I_{out},

---

[67] See his patent US4570128, and his IEEE *JSSC* paper (SC-21, #6, 1986), in which he says "The output stage (Figure 8) must solve a level shifting problem that has plagued rail-to-rail designs for some time. Elaborate solutions have been proposed that combine multiple embedded feedback loops that are in effect op amps within op amps. To succeed as a general-purpose quad, a simpler solution had to be found." Although originally developed at NSC, this circuit (or close variations) is popular with op-amp designers at Analog Devices and at TI (even before it swallowed NSC).

on both coarse and expanded scales; the latter shows the $100\,\mu A$–$200\,\mu A$ current in the "wrong" transistor. By comparison, these currents drop fully to zero in a normal class-AB push–pull output stage, as shown in the corresponding SPICE simulation of Figure 4x.95.

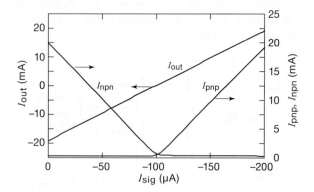

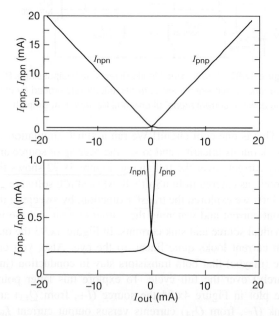

Figure 4x.93. Source, sink, and output currents for a BJT implementation of the Monticelli rail-to-rail output circuit, as modeled in SPICE.

Figure 4x.94. Source and sink currents versus output current for the Monticelli circuit simulation. Note that both transistors remain in conduction over the full cycle.

Seeing this circuit being adopted by many designers of high-performance op-amps in the last 25 years, we continue to be impressed by Monticelli's creation. He originally developed the circuit to work with silicon-gate MOSFETs, but it works especially well with BJTs. His

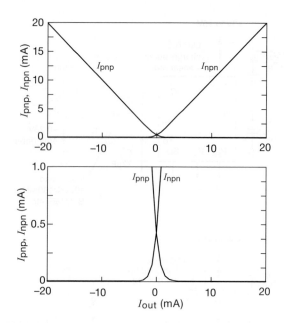

Figure 4x.95. By contrast, the currents in a conventional class-AB push–pull follower output stage go completely to zero beyond the narrow crossover region, as seen in the SPICE simulation of the circuit of Figure 5.32.

circuit is perfectly suited for balanced push–pull drive (to the bases of Q_{13} and Q_{14}, in Fig. 4x.92). It works well at high frequencies – for example, the OPA1611 has an 80 MHz GBW, and it can deliver lots of output drive (the OPA209 is happy delivering 65 mA, which is plenty for a 40 V op-amp).

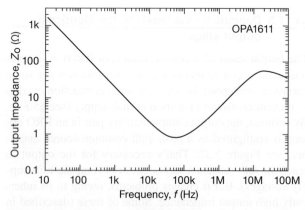

Figure 4x.96. Open-loop output impedance of the OPA1611 op-amp, which incorporates a Monticelli output stage (with capacitive feedback to lower the output impedance).

Many op-amp datasheets don't reveal the inner workings of their output stage, but sometimes you can recognize a

Monticelli rail-to-rail output stage from a distinctive open-loop output impedance versus frequency plot. Many of the TI op-amps with a Monticelli output stage add an output feedback capacitor to an input-stage gain node, which causes Z_{out} to drop to an amazingly low value, e.g., $1\,\Omega$ at $100\,\mathrm{kHz}$ for the OPA1611, before rising and then falling again; see for example Figure 5.34 in AoE3, or Figure 28 from the OPA1611 datasheet (plotted here as Fig. 4x.96).

4x.12 Slewing and Settling

Here we expand on topics treated earlier, in §§5.8 and 4x.9. If you apply an input step to an amplifier, the time required to get (and stay) within a specified accuracy (say 0.1%) of the final output is the *settling time*, see Figure 5.14. This involves two distinct phases: (a) slew-rate-limited ramping brings the output close to the asymptotic level, followed by (b) a bandwidth-limited approach (often exponential) to the final output voltage; see Figure 4x.97 for some typical behavior.

In this section we'll see that it's easy to estimate the approximate times of both phases, which is both intellectually satisfying and also useful when settling time is not specified on an op-amp's datasheet.

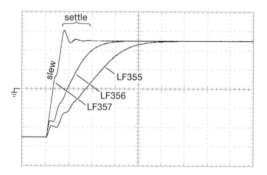

Figure 4x.97. Measured 10 V output-step response for three JFET op-amps, configured for $G_V = -5$. Vertical: 2 V/div; Horizontal: 400 ns/div.

4x.12.1 Dependence on f_T

The *slewing* phase takes approximately $t_{slew} \approx V_{step}/S$, thus linear in the magnitude of output voltage step. And for a well-behaved op-amp (clean single-pole roll-off to unity loop-gain) the *settling* phase takes approximately 5τ (to 1%) or 7τ (to 0.1%), etc., where the time constant $\tau \approx G_{CL}/2\pi f_T$. We'll see presently that the slewing time dominates for the classic BJT op-amp configuration,[68] but for op-amps with slew-rate enhancing configurations (see §4x.9) the final settling portion of the waveform can dominate.

---

[68] Sometimes called the "Widlar circuit," see Fig. 5.13.

Let's look at some examples. In Figure 4x.97 we compare three JFET-input op-amp relatives (LF355–357), with specified slew rates of 5 V/μs, 12 V/μs, and 50 V/μs, respectively. From these we estimate slewing times for the 10 V output step of 2 μs, 0.83 μs, and 0.2 μs, in reasonable agreement with the observed traces. And from the specified f_T values (2.5 MHz, 5 MHz, and 20 MHz) we estimate settling times (to 1%) of approximately 1.3 μs, 0.64 μs, and 0.16 μs. The slower op-amps are in rough agreement, but the LF357 exhibits overshoot and ringing (a consequence of insufficient phase margin) that extends its final portion of settling by a factor of four. For these op-amps the final settling times are not insignificant compared with the slewing times, a consequence of the low transconductance of the JFET input stages, as we'll see below.

Figure 4x.98 represents a pair of classic BJT op-amps, where the slewing phase is clearly dominant. As before, we estimate (from specified $S=0.2$ V/μs) a slewing time of 50 μs, and from the bandwidth ($f_T=0.7$ MHz) a settling time to 1% of 5 μs; the latter is not visible at this scale (but see Fig. 4x.102), with the slewing clearly dominating.

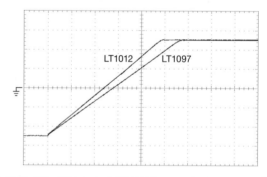

Figure 4x.98. 10 V output-step response for two conventional (Widlar circuit) BJT op-amps. Vertical: 2 V/div; Horizontal: 10 μs/div.

Figure 4x.99 shows an expanded view of the final settling of an interesting pair of op-amps, ones in which the slewing and final settling times are comparable. Here the vertical scale has been expanded 20-fold, so one major division represents 1% of the 10 V output step. As remarked earlier, the LF357 decompensated op-amp exhibits ringing, which can be seen here to extend the total settling time to roughly twice the slewing time (note position of input step); in fact, the datasheet's settling time specification (1.5 μs to 0.01%) for the faster-slewing LF357 is the same as for the 4× slower-slewing LF356. The faster LT1357, by contrast, is well-behaved in its approach to the asymptotic voltage. What is special is its extraordinarily high slew rate (600 V/μs), given its modest f_T (15 MHz); read on . . .

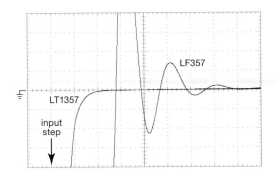

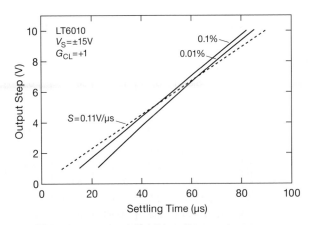

Figure 4x.99. Expanded vertical scale, showing the final approach of a pair of op-amps, with vertical major divisions representing 1% of the 10 V output step. Vertical: 100 mV/div; Horizontal: 100 ns/div.

Figure 4x.100. Datasheet graph showing typical settling times versus output voltage step. The dashed line shows slewing time for the specified slew rate; the latter dominates total settling time.

A. Slew-rate enhanced op-amps

To understand the impressive settling-time performance of the modest-bandwidth LT1357, recall that back in §§5.8.1 and 4x.9 we saw that a simple op-amp model (BJT differential input stage with single-pole capacitive-load roll-off) predicts a slew rate proportional to bandwidth, $S=0.32f_T$. For such op-amps (like the LT1012 and LT1097 in Fig. 4x.98) the ratio of slewing time ($\Delta V/S$) to final settling time constant ($\tau \propto 1/f_T$) is constant (for a given step size), and the slewing time always dominates for large output step size. But op-amps with enhanced slew rate (like the internal current-feedback LT1357) speed up the slewing portion, as seen in Figure 4x.99. It's easy to show that an enhancement factor m ($S=0.3mf_T$) of roughly 25–50 (depending on step size and settling precision) creates comparable slew and final settling times; the LT1357 has $m=125$. See the tabulated parameters of these op-amps relevant to slewing and settling (Table 4x.4 on the next page).

Some datasheets conveniently show graphs of settling time versus output step size; see for example Figures 4x.100 and 4x.101, which show settling times for the conventional BJT LT6010 (analogous to the LT1012) and for the slew-rate enhanced (emitter-degenerated) TLE2141. We've overlaid a dashed line showing slewing time, from which you can see that the total settling time of the LT6010 ($m=1$) is dominated by slewing, whereas for the TLE2141 ($m=24$) the slewing accounts for about half of the total settling time.

4x.12.2 A caution: 'scope overdrive artifacts

Figure 4x.102 shows an expanded view of the final approach of the (slow, conventional) LT1012 and LT1097 op-amps (whose full slew is shown in Fig. 4x.98). There's

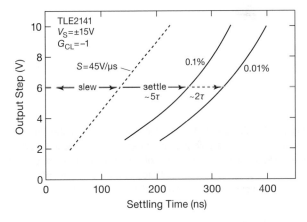

Figure 4x.101. Datasheet graph showing typical settling times versus output voltage step for an op-amp with enhanced slew rate. The dashed line shows slewing time for this op-amp, which uses a cross-coupled current-boosting input stage (see §4x.9).

some op-amp misbehavior evident, with an initial 0.5% overshoot followed by a fast correction before the final smooth landing. What's more interesting here, though, is the potential for grossly malformed waveforms when you overdrive a 'scope input. Here we've expanded the vertical scale so the initial signal voltage (-10 V) is below the centerline of the screen by 12 screen heights. Scopes don't specify what happens when you overdrive them, and the results can be bizarre. We encountered this first when making Figure 4x.99, where the LT1357's approach to the final voltage was a bit wiggly. To investigate whether this was a 'scope artifact, we created a clamped output (2.5 kΩ series resistor, SD101 Schottky diode to ground, measured

Table 4x.4: Slew Rate and Settling Time of Selected Op-amps

| | | JFET | | | | CFB/VFB | emitter degen | | | |
	LF355	LF356	LF357	LT1012	LT1097	LT1357	LT1224	TLE2141	LT6010	units
specs Slew	5	12	50	0.2	0.2	600	400	45	0.11	V/µs
Settle 0.10%	–	–	–	–	–	0.12	0.09	0.34	–	µs
Settle 0.01%	4	1.5	1.5	–	–	0.22	–	0.4	85	µs
f_T	2.5	5	20	0.7	0.7	15	45	5.9	0.35	MHz
calculated $S = 0.32 f_T$	0.8	1.6	6.4	0.22	0.22	4.8	14.4	1.9	0.11	V/µs
m	6.3	7.5	7.8	0.9	0.9	125	28	24	1.0	–
$\tau = 1/2\pi f_T$	64	32	8.0	230	230	11	3.5	27	460	ns
T_{slew} (10V)	2.0	0.83	0.20	50	50	0.017	0.025	0.22	91	µs
$T_{slew} + 5\tau$	2.3	0.99	0.24	51	51	0.072	0.043	0.36	93	µs

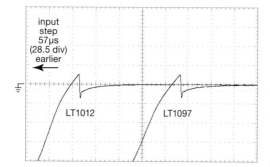

Figure 4x.102. Expanded vertical scale, showing the final approach of the same pair of op-amps as in Fig. 4x.98, with vertical major divisions representing 1% of the 10 V output step. On this faster horizontal scale the input step is 28.5 divisions off the left side of the figure. Vertical: 100 mV/div; Horizontal: 2 µs/div.

trace of Figure 4x.104. Aha! The 'scope distorted the trace by some 6 screenfuls (46 vertical divisions), recovering over the next thousand microseconds. This unhelpful behavior occurred even with a modest overdrive of just 1.5 screenfuls below the bottom horizontal level (Fig. 4x.105). This behavior is an extreme example of "The scope is lying to you" (Appendix O, The Oscilloscope).

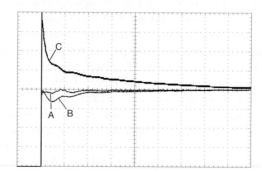

Figure 4x.103. Examining 'scope overdrive artifacts: the LT1224's output step from −10 V settles to ground, with a vertical scale factor such that the starting voltage is 24 screen heights below the bottom. A. small overdrive (clamped to −0.5 V, 6 divisions below bottom); B. large overdrive (24 screen heights below bottom); C. same as (B), but as seen on inexpensive 'scope (see text). Vertical: 50 mV/div; Horizontal: 4 µs/div (A & B), 5 µs/div (C).

with <1 pF FET probe) and looked at both the clamped and overdriven inputs.

Figure 4x.103 shows an example of what you see when you seriously overdrive the 'scope: Here the expanded vertical sensitivity (50 mV/div) corresponds to an off-scale signal (initially at −10 V) coming up from 24 screenfuls below the visible screen. Trace A is the clamped (accurate) signal, and trace B is what the overdriven 'scope (Tek TDS3044B) shows. For purposes of settling time, a vertical division is 0.5%, so the artifact is significant, but not devastating.

For comparison we measured the same signal with an inexpensive 'scope, with curious results: as we expanded the vertical scale, with ground centered, the trace bottomed, and we had to add in a large vertical offset to get it back on screen. Trace C shows its (overdriven) trace, rather more seriously in error. But, we wondered, what was that large offset? To find out we slowed down the sweep, getting the

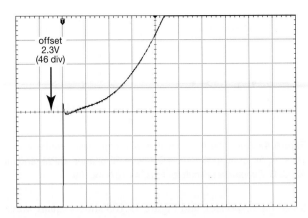

Figure 4x.104. Slowing down the sweep shows the extended over-drive recovery of trace C of Fig. 4x.103. Vertical: 50 mV/div; Horizontal: 200 μs/div.

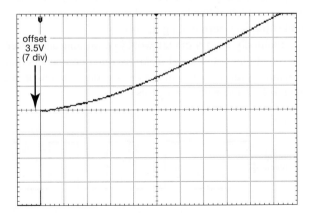

Figure 4x.105. Same as Fig. 4x.104, but with vertical scale reduced so that initial overdrive is 1.5 screen heights below bottom. Vertical: 500 mV/div; Horizontal: 200 μs/div.

4x.13 Resistorless Op-amp Gain Stage

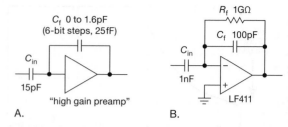

Figure 4x.106. Inverting amplifier with capacitive feedback network. A. As found in Weeroc's Citiroc1A 32-channel SiPM readout ASIC. B. Simple test circuit for Figs. 4x.107–4x.109.

From time to time you see tentative discussions of using capacitors (instead of resistors) in an op-amp voltage amplifier. Usually these are dismissed as impractical, for several reasons: varying input impedance (if inverting), requirement that the op-amp drive a capacitive load (the feedback network), absence of dc feedback, and the like.

These are all good points. So, imagine our surprise when we saw just such a gain stage used in a linear preamp ASIC intended for use with silicon photomultipliers (SiPM). The ASIC is the Citiroc 1A (Cerenkov Imaging Telescope Integrated Read-Out Chip), a product of the French company Weeroc.[69] One nice feature of this elegant ASIC is the ability to individually trim the voltage gain of the pair of amplifiers in each of its 32 channels,[70] which is done by selecting one of 64 values of feedback capacitor C_f (Fig. 4x.106A), going from 0 to 1.6 pF in steps of 25 fF. The voltage gain is then $G_V = -C_{in}/C_f$. Evidently this capacitive feedback scheme was chosen[71] for several reasons: (a) it incorporates dc blocking inherently, allowing easy channel-by-channel dc trimming of detector bias; (b) the designers already had a nice integrating "charge amplifier" design with switchable time constant, so all they needed to do was to replace the input resistor with a capacitor; (c) it is easy to integrate switchable capacitors in silicon (see, for example, the discussion of charge-redistribution ADCs in §13.7); and (d) the signals of interest are fast, and do not require gain at low frequencies. The simplified circuit in the figure omits the essential dc bias resistor (a large-value R_b across C_f), whose resistance should be large compared with the reactance of C_f at signal frequencies, not a serious constraint given the ns-scale waveforms produced by SiPMs.

Out of curiosity we rigged up an LF411 with a capacitive divider network, with C_{in}=1 nF and C_f=100 pF (Fig. 4x.106B, target gain of 20 dB); for dc bias we put a 1 GΩ resistor across C_f. Figure 4x.107 shows input and output when driven with a 1 kHz square wave (a good indica-tor of flat response), and Figure 4x.108 shows square wave response over 3 decades (composite figure, each trace is at a different horizontal scale). When swept with a sinewave you get the response of Figure 4x.109.

Given the reactive (and frequency-dependent) input impedance, it's important that the input signal looks like a voltage source to C_{in} for the highest signal frequencies; that is, its Thévenin impedance should be small compared with $1/2\pi f_{max}C_{in}$. For our test circuit we used a signal generator with 50 Ω output impedance, small compared with the reactance of 1 nF even at 100 kHz (where X_C=1.6 kΩ); and for the SiPMs for which the Weeroc chip was designed, the pulse of charge (from the $\sim10^6\times$ amplified photoelectron) is converted to a low-impedance voltage pulse across a 50 Ω (typical) terminating resistor.

Admittedly such a circuit would be an unusual choice for something constructed from discrete components; but evidently it makes sense in the context of a silicon circuit intended for use with fast signals.

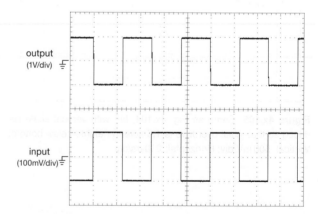

Figure 4x.107. Inverting op-amp stage (Fig. 4x.106B) with capacitive gain-setting divider: C_{in}=1 nF, C_f=100 pF, R_f=1GΩ. Horizontal: 400μs/div.

[69] www.weeroc.com.

[70] A related feature is a SiPM bias trim on each channel, for detector gain leveling.

[71] Thanks to Jean-Baptiste Cizel and Julien Fleury at Weeroc for clarification.

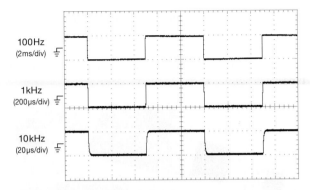

Figure 4x.108. Output of same circuit as Fig. 4x.107, for three decades of square wave frequency; note horizontal scale factors. Vertical: 2 V/div.

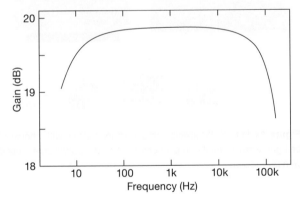

Figure 4x.109. Sinewave gain versus frequency for the circuit of Fig. 4x.106B.

4x.14 Silicon Photomultipliers

This topic properly belongs in the optoelectronic detectors section of Chapter 12, but, having mentioned them in the last section, we cannot resist a short description of these elegant devices that are in many ways superior to the conventional (vacuum) photomultiplier tubes (PMTs) described in §12.6.2.

4x.14.1 SiPM characteristics

Like a PMT, a silicon photomultiplier[72] (SiPM) responds to incident visible-light (or near infrared) photons, producing a fast (ns-scale) output pulse of some 10^6 electrons per detected photoelectron (p.e.). The latter is created by photoelectric absorption of the incident photon, with a photon detection efficiency[73] (PDE) typically around 20% (for a PMT), and 25–50% for a SiPM. And, like a PMT, a SiPM responds linearly to the photon number detected within its resolving time; that is, a "pileup" of n photo-electrons produces an output pulse n times larger than that produced by a single photoelectron.

Better still, as we'll see presently, the SiPM is superior in several ways: (a) unlike a PMT, which ages over its lifetime, losing much of its gain after some 100 coulombs of integrated output charge, the SiPM appears to have no wearout mechanism, and it is not damaged by exposure (while powered) to daylight; (b) the SiPM runs on pleasantly low voltage, typically in the range of 25 V to 60 V, compared with the $\sim$1 kV (or more) HV required for PMTs; (c) SiPMs are far less expensive than PMTs, costing typically less than US$10 in quantity, a hundred times less than typical PMTs; (d) SiPMs and SiPM arrays are highly uniform in sensitivity (spatial variation across the sensitive area) and gain, compared with PMTs; and (e) the single-p.e. output of SiPMs is far above the noise level, unlike PMTs where it is buried in the exponential noise tail.

---

[72] Hamamatsu, a leader in optical detectors, calls theirs Multi-Pixel Photon Counters, MPPC.

[73] The *photon detection efficiency* is the product of quantum efficiency (the probability that a photon produces a electron–hole pair) times avalanche probability (the probability that the photoelectron produces an electron avalanche) times the fill factor (the fraction of total detector area that is sensitive); that is, PDE=QE$\times P_{av}\times$FF.

SiPMs are available in standard photodetector packages (through-hole and surface-mount), and you can get square arrays of matched SiPMs in sizes up to 12×12; Figures 4x.110 and 4x.111 show some currently available types.

Figure 4x.110. Silicon photomultipliers come in through-hole and surface-mount packages, as well as multi-element arrays. (Hamamatsu Photonics, used with permission.)

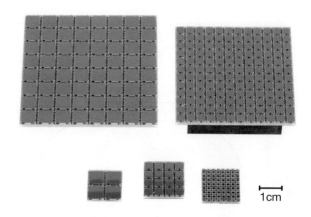

Figure 4x.111. SiPM arrays come in various sizes, with choice of ball-grid array or multi-pin connectors. (SensL Technologies, used with permission.)

4x.14.2 SiPM construction

What's not to like about SiPMs? Not much![74] Let's see how these devices do their magic. The basic structure is a rectangular array of thousands of individual avalanche photodiodes (APDs), see Figure 4x.112. An APD is a type of photodiode that is operated with a reverse bias sufficient to generate additional electron–hole pairs from the single

---

[74] To be fair, we need to admit that they have relatively high single-p.e. dark-count rates, and available types are no larger than 6 mm$\times$6 mm (compared with flat-faced photomultipliers up to 133 mm diameter, and some specialized types that are twice the diameter of a basketball).

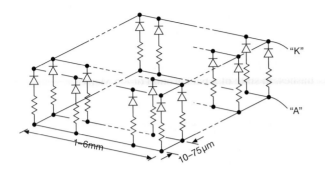

Figure 4x.112. A solid-state photomultiplier (SiPM) is an array of thousands of individual Geiger-mode avalanche photodiodes (APDs), each with a quenching resistor, connected in parallel to form a single-pixel two-terminal device.

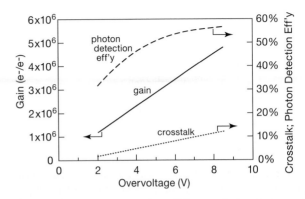

Figure 4x.113. Photon detection efficiency, electron gain, and crosstalk, as a function of applied overvoltage $V_{op} - V_{br}$, from the datasheet for Hamamatsu's S13360-series 50 μm-pitch silicon photomultiplier.

electron–hole pair created by absorption of a photon. At modest reverse bias voltages APDs operate in a linear gain region, producing some 10 to 100 or so additional electron–hole pairs per photoelectron. If the bias is increased sufficiently, however, the device operates in nonlinear "Geiger mode" (like a Geiger counter), generating a full-sized output pulse in response to one or more photoelectrons.

In the SiPM the thousands of individual micro-APDs (of size 10 μm–75 μm) are operated in Geiger mode, with individual resistors that quench the avalanche; they are all connected in parallel, producing a single large-area pixel (a square, typically of size 1 mm to 6 mm). Although the response of each APD is nonlinear (producing a full-size Geiger pulse of some 10^6 electrons), the whole array behaves like a very linear device when detecting multiple photons arriving within the resolving time; that is because multiple photons land on different micro-APDs, producing a linear summed response.

4x.14.3 SiPM characteristics, electronics, and waveforms

Figure 4x.113 shows how the gain (avalanche electrons per photoelectron) varies with overvoltage (the amount by which the operating voltage V_{op} exceeds the Geiger threshold breakdown voltage V_{br}, the latter typically[75] around 25–60 V). At higher voltages you get better photon detection efficiency and more gain, but at the cost of higher crosstalk and afterpulsing. SiPMs operate satisfactorily at room temperature, but they typically exhibit a single-p.e. "dark count" rate of about 50 kcps per square millimeter of sensitive area at 25°C; at −20°C the rate

drops to ∼1 kcps/mm². Typical SiPM spectral response extends from 300 nm to 750 nm (at 10% PDE), peaking at 450 nm (∼40–50% PDE) which is well matched to applications such as Cerenkov atmospheric gamma-ray astronomy, or scintillator-based PET (positron-emission tomography) medical scanners.[76] For applications that require extended red response there are *n*-on-*p* SiPMs with response out to 1 μm (Hamamatsu S13720 series; SensL R-series). To extend further into the near infrared you can get devices constructed of InGaAs, with response from about 900–1700 nm; for low light-level applications (e.g., astronomy) these are best operated with Peltier coolers to reduce the dark-count rate. Currently these devices are small (0.2 mm square) and expensive; with the potential market of automotive anti-collision lidar, however, there is hope for reduced cost.

Because of the high gain and low noise level, it doesn't take much to amplify the single-p.e. output pulses. A wideband amplifier IC works just fine, as in Figure 4x.114, where we've used an inexpensive 750 MHz bandwidth IC with 31 dB gain; it's a nice part – doesn't even require an output pullup inductor or resistor. Figure 4x.115 shows a single-shot capture, triggered on the occasional 3-p.e. pileup, showing the linear and uniform response to single- and multiple-p.e. events. With uniform illumination you don't get significant pileup at rates of 20 Mcps or more, as seen in Figure 4x.116 where an LED provided dc illumination sufficient to produce 25 Mcps.

We are engaged in a project[77] to detect intentional

---

[75] Hamamatsu SiPMs operate around 50–65 V, whereas SiPMs from SensL ("sense light") operate around 25 V.

[76] Where the SiPM speed allows time-of-flight accuracies of order 0.2 ns, enabling some degree of precision in range along the line-of-emission.

[77] https://oirlab.ucsd.edu/PANOSETI.html.

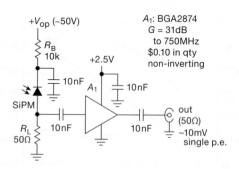

Figure 4x.114. Simple SiPM readout electronics. In our deployed electronics we reduced the input blocking capacitor to 100 pF, to suppress the relatively long recovery tail (see, e.g., in Fig. 4x.117).

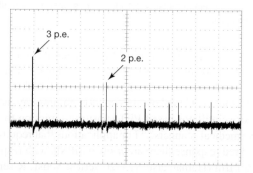

Figure 4x.115. Single-shot capture from a Hamamatsu S13360-3025CS SiPM, triggered at 2.5 p.e., with steady background of 0.5 Mcps. Vertical: 10 mV/div; Horizontal: 2 μs/div.

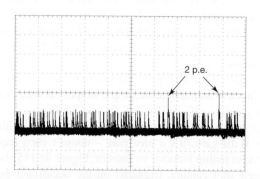

Figure 4x.116. Single-shot capture, background rate ∼25 Mcps. Vertical: 12 mV/div; Horizontal: 400 ns/div.

laser flashes from other technological civilizations (optical SETI), for which we are planning an array of some 10^5 SiPMs at the focal planes of some 10^2 0.5m-class Fresnel lenses staring at the night sky. The dark night sky produces a background flux of ∼5–20 Mcps per SiPM, against which we wish to detect a solitary pulse producing >5 photoelectrons (the laser flash). We built a test setup, illuminating

the SiPM at those background rates, with a pulsed LED generating occasional 5 p.e. pulses. Figure 4x.117 shows a typical detected pulse, with 'scope trigger set at 4.5 p.e. At that trigger threshold the 'scope triggered only during LED flashes, in spite of a background of ∼5 Mcps.

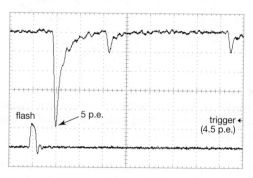

Figure 4x.117. Single-shot triggered at 4.5 p.e. (response to LED flash), with background ∼5 Mcps. Here we used an inverting amplifier. The 40 ns delay is caused by a long fiber path. Vertical: 5 V (LED pulse) and 200 mV/div (SiPM); Horizontal: 40 ns/div.

4x.15 External Current Limiting

All op-amps include internal current-limiting, a necessary measure to prevent destruction if the output is shorted, or subject to other punitive mishaps. This is nicely illustrated in the circuit of the LF411 (Fig. 4.43), where $\gtrsim 25$ mA through sense resistor R_5 turns on Q_9, preventing further base drive to output pull-up Q_7 (and with similar circuitry for pull-down). There's no real precision in this simple circuit – you'll find rather loose specs (if any); for example, the LT1354 lists a short-circuit current I_{SC} of 30 mA (min), 42 mA (typ), and ...*no* spec for maximum. And, as you might guess from the variation of base–emitter voltage with temperature, the current limit also varies considerably over temperature; the graphs in the LF411 datasheet, for example, show "typical" current limits of 20 mA, 25 mA, and 35 mA at 125°C, 25°C, and −55°C, respectively. While we're busy complaining, let's add that an op-amp's recovery from bumping into its current limit can be, well, *ugly* – see for example Figure 4x.118.

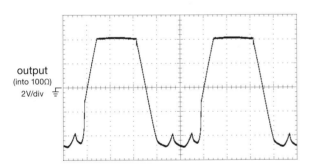

Figure 4x.118. This is what you get when you (over-) drive an OPA277 follower with a 5 kHz sinewave of 8 V amplitude into 100 Ω. Recovery from current limit is a sight to behold. Horizontal: 40 μs/div.

Good enough, if your objective is merely to prevent self immolation. But, depending on what you're driving, you may want a predictable current limit, settable to some current less than the op-amp's native I_{SC}. And you can improve greatly on the op-amp's recovery behavior. Figure 4x.119 shows a nice circuit solution. Output current comes through Schottky diodes D_2 (sourcing) or D_4 (sinking), strictly limited by the respective current sources I_{lim}^+ and I_{lim}^-. When in a current-limit condition, the feedback

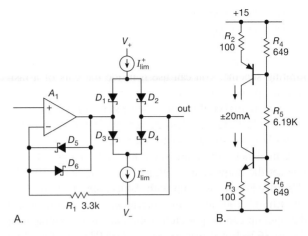

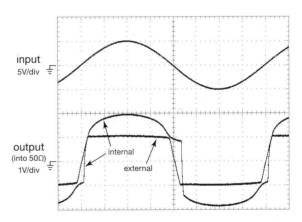

Figure 4x.119. External current limiting circuit, applicable to any op-amp. The current sources can be replaced by resistors if you don't care about constancy of I_{limit} versus output voltage. The current limits I_{lim}^+ and I_{lim}^- can be set independently (they do not need to be equal).

Figure 4x.120. We drove an LT1354 follower, loaded with 50 Ω, with a 5 V amplitude sinewave at 313 kHz, and compared output waveforms with internal limiting and with external limiting (the circuit of Fig. 4x.119). This op-amp emerges from current limit with some nasty twists and turns, happily tamed with external limiting. Horizontal: 400 ns/div.

through R_1 is inoperative, so diodes D_5 and D_6 take over, limiting the op-amp's output swing to a few tenths of a volt beyond the input signal; this prevents op-amp saturation, and speeds up recovery (the op-amp need slew through only a diode drop).

The values shown set a current limit of ±20 mA, and produce good limiting and recovery, as seen in the measured traces of Figure 4x.120. The particular part we chose is one of LTC's C-Load™ op-amps, with the interesting property of being stable into any capacitive load. We wondered if their scheme of deriving compensation from the output node might improve recovery from current-limit, but, alas, it was as ill-behaved as any other op-amp.

As an example of a good application for this current-limiting scheme, you can use it inside the guts of a fast-slewing power amplifier: the latter often use a current-feedback (CFB) configuration (see §4x.6), which can benefit from a fast and well-defined current limit. Figure 4x.121 shows how to assemble such a circuit.

A final note: there are some rarefied op-amps that let you set the output current limit, for example the LT1970 "500 mA Power Op Amp with Adjustable Precision Current Limit." You can set the current with 1% accuracy, and the op-amp can be powered from total supply voltages of 5 V to 36 V. But it was too slow (3.6 MHz and 1.6 V/μs) for our application, so we used an LT1220 (45 MHz and 250 V/μs) with the external current limit circuit.

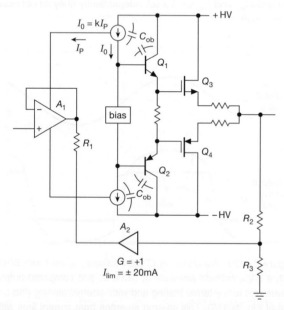

Figure 4x.121. Block diagram of a fast-slewing power amplifier, with current-feedback (CFB) configuration. The return path exploits current-limited unity-gain amplifier A_2, as in Fig. 4x.119. The maximum slew rate is $S = i/C$, where i is A_2's current limit, multiplied by the gain of the current mirrors, and C is the sum of the input capacitances of the four BJTs; i.e., $S \approx kI_{\text{lim}}/4C_{\text{ob}}$. For operation at higher voltages you can put a cascode transistor in each of A_1's supply rails, clamped to, say, ± 15 V.

4x.16 Designs by the Masters: Bulletproof Input Protection

Imagine you are a designer at a world-class instrument company in 1991, about to introduce a breakthrough product – an advanced $6\frac{1}{2}$-digit DMM with consistent 0.002% accuracy, with dc voltage ranges to 1000 V, resistance ranges to 100 MΩ, true RMS voltage and current, and a few other goodies. And all that for a thousand dollars, an unheard-of combination of features and price. That was the legendary HP 34401A multimeter, whose praises we sang in §5.12 (its precision analog front-end) and §13.8.6 (its precision conversion circuitry).

Here's the rub: by allowing inputs to a kilovolt (unusual in DMMs of that era), you've got to protect the precision circuitry from such voltages, *even with the instrument set for a sensitive range, or set for a different function such as ohms.* That's a tall order – when measuring resistance, for example, the instrument applies a current and measures the resulting small dc voltage. So the current source has to withstand abuses such as a ±1 kV applied across the "resistance" input terminals. And, if that seems like a manageable task, consider the challenge of multi-kilovolt transients, for example a carpet-shock jolt of, say, 10 kV into the meter's high-Z input (selectable 10 MΩ or, get this, >10 GΩ on the dc voltage ranges to 10 V).

Well, they did it. Figure 4x.122 shows how. Let's learn from the masters.

Input clamp

Taking it "from the top" (top-left), the signal chain begins with a 1.5 kV gas discharge tube (GDT) surge arrestor, here placed in series with a 1.1 kV metal-oxide varistor (MOV) (we'll see why, presently). Why not just a MOV? Because MOVs have lots of capacitance (a few hundred picofarads, for kilovolt-rated types), and their leakage current below their clamping threshold (microamps to milliamps) would seriously degrade the instrument's otherwise high input impedance. GDTs, by contrast, have very low loading: around 1 pF, and leakage currents less than 1 pA ($R>10$ GΩ). OK, you say, then why not just the GDT? Good question.

The reason is that the GDT acts not like a *clamp*, but rather like a *crowbar* – that is, once triggered by an over-

voltage, it goes into heavy conduction, bringing the terminal voltage down to ~15 V or so. And, just like an SCR crowbar, it won't let go until the offending input stops providing current. So, for example, if a transient spike were present on an otherwise perfectly satisfactory input of, say, 500 Vdc, a GDT would crowbar the input to a low voltage, and continue to hold the input near ground. By adding a series MOV, the GDT clamp current ceases after the transient overvoltage has passed, and the instrument continues to function normally. The 10 nF bypass across the MOV ensures that the GDT alone sets the trip voltage, and the 2 MΩ shunt discharges the capacitance afterwards. (Why a pair of 1 M resistors in series, rather than a single 2 M resistor? Because a 2 M resistor would have to dissipate 0.5 W and withstand a kilovolt, requiring a larger high-voltage resistor. A series string of small SMT resistors takes up less space, and costs less – would you believe, $0.002 in 10k-piece quantity![78])

GDTs are rugged brutes – they come in ratings from 75 V to 7.5 kV, can handle many cycles of 10 A or so, and a few impulses of hundreds to tens of thousands of amps; their ceramic packages typically measure about 1 cm in diameter and length, they're available in through-hole or SMT, and they cost around a dollar in small quantities.[79]

These components deal with *over*voltages (>1 kV), but the instrument must not be damaged by applied voltages (of either polarity) less than a kilovolt. Let's see how the HP designers contended with such harassment, both in resistance- and voltage-measuring modes.

Resistance ranges

Moving along to the right, there's relay K_1, which disconnects the input during mode and range changes, then the node where the resistance-measuring current source connects. How can the current source protective circuitry contend with user abuses such as inadvertent application of a

---

[78] Curiously (and uncharacteristically for HP engineers), the specified SMT 1206-size 1M resistors are pushed past their limits here: an input over 1.5 kV would subject them to at least 550 V, but they are rated only to 200 V "working," 400 V "overload," and 500 V "dielectric withstanding" voltages. (To be fair, though, they would see a full 550 V only during a rare – and likely transient – fault event that triggers the GDT.) By contrast, the series string of 13 kΩ resistors (discussed below) are operated conservatively: they are rated to 500 V and 1.5 W each, but in this circuit they see at most 125 V and 1.2 W at the maximum permitted 1 kV input.

[79] Littelfuse has a nice "Product Catalog & Design Guide" (ECN141N, current version 11/15); see also the "Bourns Gas Discharge Tubes Short-Form Brochure," and TDK/Epcos's "Surge Arresters and Switching Spark Gaps."

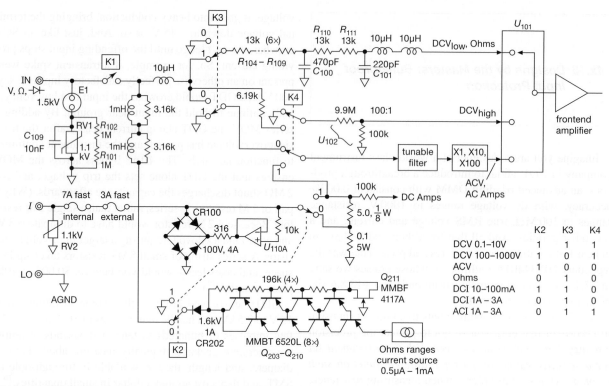

Figure 4x.122. Input protection circuitry for the Agilent 34401A DMM.

kilovolt (of either polarity) to the input terminals? And the circuitry must not interfere with normal operation, where the instrument applies precise currents in the range of 1 mA (low resistance ranges) down to 0.5 μA (high ranges).

The designers came up with an elegant solution. The precision current source itself (shown in Figure 4x.123) is straightforward, though with output compliance to about +12 V. The secret sauce is in the protection network: (a) series inductors prevent fast transients, followed by an HV series diode to deal with a possible applied positive voltage; (b) in the event of a negative applied voltage, the potential is spread equally along the string of eight *pnp* transistors, biased into conduction by the low-leakage JFET diode; and (c) in normal operation (i.e., measuring an unknown resistance) that diode is reverse-biased and the test current flows through the right-hand pair of base–emitter diodes, the four base resistors, and the HV diode. During resistance measurements the voltage drop through those components is no more than 3 V (why?), well within the compliance range.

During resistance measurements, of course, the (small) voltage across the unknown resistor must be measured precisely; that is, the DMM must both apply a test current

and measure the resulting voltage drop. So the dc voltage-measuring circuits must tolerate ±1 kV input, whether measuring resistance or voltage. Let's see how.

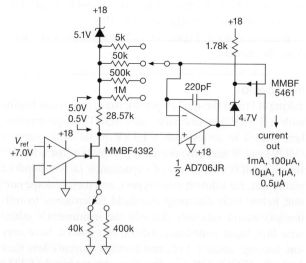

Figure 4x.123. Precision current source for resistance measurements with the Agilent 34401A DMM.

Voltage ranges

For the high-voltage ranges (100 V, 1000 V full-scale) the input is attenuated 100:1 with a resistive network (U102, given only a proprietary part number of 1NB4-5035; we trust it is designed to handle a continuously applied kilovolt), which prevents downstream damage. But in the low ranges (100 mV, 1V, 10V full-scale) relay K_3 steers the input through a series string of eight 13 kΩ resistors, along with a pair of small shunt capacitors C100 and C101 (to kill transients – think of them as forming capacitive dividers with the parasitic shunt capacitance of the resistors) and a pair of series inductors. As we remarked in footnote 78, by using a string of eight surface-mount resistors the designers kept them within their ratings, even when 1 kV is applied continuously at the input with the DMM set to a low range. In such an overvoltage situation the downstream end of the resistor string is clamped to the supply rails by the input protection diodes of the analog switch IC that routes the various inputs to the front-end amplifier (shown in Fig. 5.59), which happily withstand a continuous fault current of 9 mA (with transients filtered by the upstream capacitors and inductors). How about the resistors themselves, when thus maltreated? In the worst case the resistor string R104–R111, which are 2512-size thick-film resistors, rated at 1 W and 500 V, would dissipate 1.2 W per resistor; that's 20% over the specified continuous power dissipation (rated at 70°C ambient temperature), but the datasheet for this part (Vishay CRCW2512-133J) also allows "short time overload" to 250% rated power. Evidently that was good enough for the master designers.[80]

Current ranges

For current measurements a separate input terminal is used, and the DMM measures the voltage across a small current-sensing resistor (0.1 Ω for 1A and 3A ranges, 5.1 Ω for 10 mA and 100 mA ranges). Here there's no hope of preserving instrument functioning if a stiff voltage is applied across the terminals, so the designers threw in the towel and simply wired a 4 A power diode-bridge clamp (CR100) to ground, cleverly bootstrapping the intermediate node (opamp U110A) to eliminate leakage current during normal operation. During a fault condition the clamp will conduct, so they interposed a pair of fast-acting fuses – 3A externally accessible, and a 7A internal backup.[81] The designers added a varistor clamp to limit voltages if a fuse has been annihilated.

Does it work?

Yes! Despite some serious abuse, we've not managed to blow the thing out . . . yet.

---

[80] Or perhaps they used the high-power variant, p/n CRCW2512-133J-HP, rated at 1.5 W continuous at 70°C, to forestall a bad grade from the teacher.

[81] Probably to protect things if a desperate user puts an oversize fuse in the external holder.

4x.17 Canceling Base-Current Error in the Current Source

Here's an elegant method[82] of eliminating base-current error in the op-amp-assisted current source. Base-current error can become significant in low-beta BJTs, notably in high-current or high-voltage devices – see, for example, Table 2.2 (Bipolar Power Transistors, many of which have minimum betas of 50 or less), or datasheets for the convenient MPSA42 and MPSA92 (300 V low-capacitance small-signal BJTs, *npn* and *pnp* respectively, with minimum betas of 25 at currents of 1 mA or less).

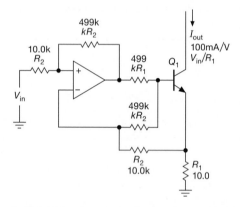

Figure 4x.124. This elegant circuit cleverly cancels the error caused by finite beta in current-sink transistor Q_1, so that I_{out} is accurately equal to V_{in}/R_1.

In Figure 4x.124 the base current is sensed by the series base resistor and, by way of the wickedly clever feedback network (a difference amplifier in disguise), the op-amp creates appropriate base drive to make I_{out} independent of Q_1's beta. You may enjoy proving this for yourself.[83]

---

[82] Nicely explained by Christian de Grodinsky, in his publication "Error compensation improves bipolar current sinks," *EDN*, 6 July 2006. There is evidence that it was independently devised earlier (though not published) by Carl T. Nelson, working at National Semiconductor.

[83] We did, but we found it necessary to exploit the two inequalities. If anyone out there can demonstrate that $I_{out}=V_{in}/R_1$ is strictly true, for any ratios of R_2/R_1 and arbitrary k, please let us know!

And Ali Mehmed did just that: in five pages of algebraic manipulations he proved that the result is exact for any resistor ratio and for any value of k, ignoring the effects of op-amp input current and offset voltage.

Then Emanuele della Valle raised the ante by sending a one-page proof. Not to be outdone, Jake Thomas came up with *two* simpler proofs, one by superposition, the other by recognizing a difference amplifier; the latter is less than a half page: nice!

4x.18 Analog "Function" Circuits

We've seen op-amps used to make amplifiers, summing circuits, current sources, integrators, and so on. These functions can be combined, to powerful effect. One important case is the use of op-amps within an electromechanical *control loop*, for example to control liquid levels or temperatures in a chemical plant, or pen position in a plotter. In those applications you often see "PID" (proportional, integrator, differentiator) controllers, in which the error amplifier consists of a parallel combination of integrator, differentiator, and proportional gain blocks (see §15.6).

Another example is the use of op-amp integrators and summing amplifiers to model a system of linear differential equations. With the addition of analog multipliers you can model also nonlinear equations. A lovely example comes from the theory of so-called *dynamical systems*, sometimes known as "chaotic systems," in which a set of simple rules, or equations, leads to extremely complex behavior. The results are astonishing, and often gorgeous – for example the intriguing Mandelbrot set of fractals.

4x.18.1 The Lorenz attractor

In 1963 Edmund Lorenz published[84] the following set of coupled nonlinear first-order ordinary differential equations:

$$dx/dt = \sigma(y-x)$$
$$dy/dt = \rho x - y - xz$$
$$dz/dt = xy - \beta z$$

with suggested parameters $\sigma=10$, $\rho=28$, and $\beta=8/3$. The solution executes a trajectory, plotted in three dimensions, that winds around and around, neither predictable nor random, occupying a region known as its *attractor*.

With lots of computing power you can approximate the equations numerically, and many handsome plots can be found on the Internet. More interesting, from our point of view, is the ease with which the system of equations can be implemented with just three op-amps (each does both an integration and a sum) and two analog multipliers (to form the products xy and xz).

---

[84] E. N. Lorenz, *J. Atmos. Sci.*, **20**, 130 (1963).

Figure 4x.125 shows such an implementation, using LF412 dual op-amps and MPY634 4-quadrant analog multipliers. The op-amps are wired as integrators, with the various terms that make up each derivative summed at the inputs. The resistor values are scaled to $1\,\text{M}\Omega$, thus for example R_3 weights the variable x with a factor of 28 ($1\,\text{M}\Omega/35.7\,\text{k}\Omega$); this is combined with $-y$ (unit weight) and $-xz$ (also unit weight, given the scale factor of the multiplier and our choice of 0.1 V equal to 1 unit – the latter to accommodate excursions of ±100 units within the circuit's range of ±10 volts).

This circuit just sits there and produces the three voltages $x(t)$, $y(t)$, and $z(t)$; if you hook x and z into a 'scope, you get a pattern like that shown in Figure 4x.126, the characteristic "owl's face" of the Lorenz attractor. The curve plays out in time, sometimes appearing to hesitate as it scales the boundary and decides which basin to drop back into. The value of C, the three integrator capacitors, sets the time scale: at $0.47\,\mu\text{F}$ it does a leisurely wander, at $0.1\,\mu\text{F}$ it winds around like someone on a mission, and at $0.002\,\mu\text{F}$ it is fiercely busy solving its equations and delighting its audience.

4x.18.2 Summing amplifiers

Back in §4.3.1D we showed the simplest example of a linear summing circuit, in which each input voltage generates a current into the summing junction of an inverting amplifier. It does what it is supposed to, but the output is inverted. Sometimes you don't want that. Or maybe you're operating with a single supply so you can't have negative output voltages.

A. Non-inverting Adder

Happily you can hook up a noninverting summing circuit (Fig. 4x.127B) by simply averaging the input voltages, then amplifying with appropriate gain. As with the inverting summing circuit, you can have a weighted average if you use unequal input resistors. There's a minor drawback to this circuit – the inputs see a current that depends on the other inputs (the inverting circuit's virtual ground circumvents this); but we're assuming the inputs are stiff voltage sources in either circuit, so this ordinarily is not a problem (if it is, buffer the inputs).

B. Adder–subtractor

Going a step further, you can combine inverting and non-inverting summing circuits to create an "adder–subtractor," a useful analog function circuit. Figure 4x.128 shows how. We've drawn it in full generality, with arbitrary weight-

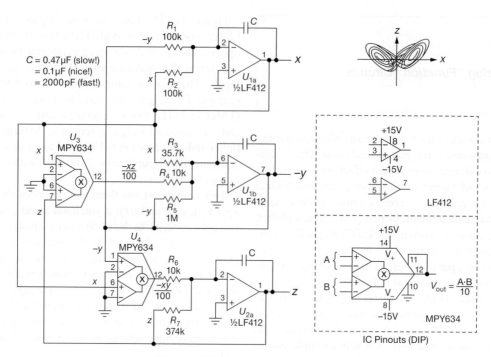

Figure 4x.125. Lorenz equation solver.

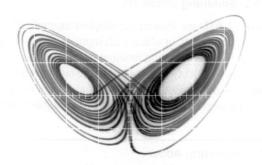

Figure 4x.126. Output projected into XZ plane.

ing coefficients. This circuit has the additional pleasant property of presenting equal driving impedances to the op-amp inputs (to minimize bias-current error). Following the exemplary presentation of Dan Sheingold,[85] here is the design procedure:

1. Write the adder–subtractor's desired behavior as
$$V_{out} = p_1 V_{1p} + p_2 V_{2p} + \cdots - (n_1 V_{1n} + n_2 V_{2n} + \cdots),$$
where the p_i and n_i are the weighting coefficients of the adding and subtracting inputs, respectively.
2. Choose a desired R_0.

---

[85] D. Sheingold, "Simple rules for choosing resistor values in adder-subtractor circuits," *Analog Dialogue*, **10**, 1, p. 14 (1976).

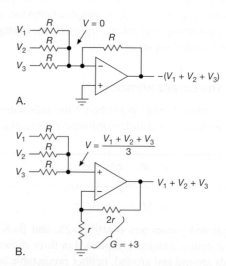

Figure 4x.127. Simple analog summing circuits. A. inverting; B. noninverting. The virtual ground in circuit A isolates the signal inputs, unlike circuit B.

3. Calculate two sums: $S_p = \sum p_i$ and $S_n = 1 + \sum n_i$.
4. Calculate R_F: it is R_0 times the larger of the two sums.
5. Calculate a quantity $\Delta = S_n - S_p$.
6. If Δ is positive, the circuit needs the resistor R_+; if Δ is negative, the circuit needs the resistor R_-.
7. In either case the value of that resistor is R_F/Δ.

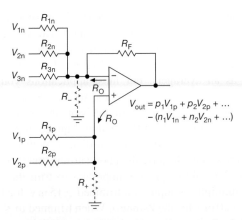

Figure 4x.128. Weighted analog adder–subtractor circuit; the design procedure accommodates arbitrary weighting factors, and presents matched impedances to the op-amp inputs. See text for the design procedure, which calculates the resistor values from the weighting coefficients p_i and n_i, and the desired impedance R_0 seen at the op-amp's inputs. Only one of the resistors R_- or R_+ is used.

8. Finally, calculate the weighting resistors R_{ip} and R_{in} as $R_{ip} = R_F/p_i$ and $R_{in} = R_F/n_i$.

Sounds complicated... but it's not. Let's do an example. Suppose we want to generate an output $V_{out} = V_{1p} + 3V_{2p} - 5V_{1n}$, and we want the op-amp's input to see a driving impedance of $R_0=10\,k\Omega$ at both inputs. Following Sheingold's prescription, we find $S_p=4$ and $S_n=6$; so $R_F=60k$. Next, $\Delta=2$, so we need a resistor R_+, and its value is 30k. Finally, the weighting resistors are $R_{1p}=60k$, $R_{2p}=20k$, and $R_{1n}=12k$. You can easily check that the op-amp sees driving impedances of 10k at both inputs, and that the inverting gain is -5 (the non-inverting gains are slightly trickier, but still easy enough to check; they're correct).

As with the non-inverting adder, there is no virtual ground, so each input sees a current that depends on the other inputs.

4x.19 Normalizing Transimpedance Amplifier

Here's a design problem that has come up several times among our experimentalist colleagues: they need to steer a laser beam onto a pair (or quad) of PIN diode detectors, with a feedback system that can control the deflection of the beam.[86] So, you'd like to create a signal proportional to the difference in photocurrents (let's call that $\Delta I_{BA} \equiv I_B - I_A$). Easy enough. But now comes the challenge: the laser *intensity* is also being controlled, perhaps through as much as two to three orders of magnitude. So, for a given offset of beam position, the error signal itself (i.e., difference of photocurrents in the adjacent PIN diodes) will vary by several orders of magnitude. This creates a real problem for a controller feedback loop, because the loop gain will vary in accordance with the laser intensity.

What's needed is an output error voltage that is proportional to the beam position error, but independent of laser beam intensity. The solution: divide the photocurrent difference by the photocurrent sum, to generate a fractional error signal: $\Delta I_\% = \Delta I_{BA} / (I_B + I_A)$. Since the inputs are currents, a solution is to convert them to voltages with transimpedance amplifiers, then divide the difference voltage by the sum voltage in an analog multiplier/divider. Figure 4x.129 shows a circuit of such a "normalizing TIA."

Circuit description
The front-end is a pair of TIAs, with electronically switchable gain via analog switch U_3 (a 4000-series logic part[87]) The inputs, protected with clamp diodes (the MMBD1403

is a handy 10-cent diode pair in SOT-23), drive an LT1792 precision low-noise JFET-input op-amp. The voltage outputs are differenced in ×10 differential amplifier U_4, and summed in U_5 (a unity-gain differential amplifier, here wired as an averager). For most applications you want some smoothing of these signals, so we added a pair of simple RC lowpass filters with precision unity-gain buffers U_6, whose outputs drive the analog 4-quadrant multiplier U_7. The latter is configured as a divider, with scale factor as indicated. Because the photocurrents may vary over a wide dynamic range, it's important to trim the offsets at the multiplier's inputs (untrimmed ±15 mV for the X-input, ±20 mV for the Z-input). When trimmed to ≤ 1 mV, say, the circuit should provide good normalizing accuracy over some 70 dB or more. The optional ± full-scale output offset would be useful as a fine trim on beam position, if needed.

Some additional comments: (a) You could configure the gain switch U_3 with its common terminal at the op-amp output (we did this, in an earlier circuit revision), but that would limit the signal swing (20 Vpp absolute max); by putting the common at virtual ground you do not limit the full ±12 V swing (note the ±1 V clamp on the unused inputs, to suppress cross-channel leakage). (b) The input op-amps are powered from filtered ±15 V ("±15Q"), derived from the 15 V split supplies with 2-stage RC filtering (10 Ω, 47 µF). (c) The summing amplifier U_5 is configured to produce the average, (A+B)/2 (rather than the sum, A+B); this preserves ±10 V full-scale without saturation. (d) If you want to stabilize with an offset position (in which one photocurrent is always larger than the other), you could take a different approach to the problem, as taught by Phil Hobbs,[88] namely by exploiting the logarithmic (Ebers–Moll) properties of a BJT, in a circuit he calls a "noise canceler."

---

[86] Interesting physics here: for beam steering you can launch an acoustic wave in a diffracting material (an acousto-optic deflector, or AOD); the acoustic frequency sets the wavelength of the Bragg-diffracting standing waves, hence the angle of laser-beam reflection (and, helpfully, the acoustic amplitude controls the intensity of the beam). Alternatively, you can use a piezo- or galvo-steered mirror. Another interesting laser challenge: you might want to stabilize the *polarization* (rather than position) of a laser beam; for that task you'd use a polarizing beam-splitter and photodiode pair for sensing, and an electro-optic modulator (EOM) to control the polarization.

[87] Why not use a 5 V 74HC4052? Because they have surprisingly large capacitance – 80 pF for some manufacturers – compared with 18 pF for the high-voltage CD4052. The price you pay is higher R_{ON} (which we address by running the part on a full 15 V, where its R_{ON}=125 Ω typ, compared with 470 Ω at 5 V).

[88] *Building Electro-Optical Systems: Making It All Work*, Wiley (2009), pp. 723ff.

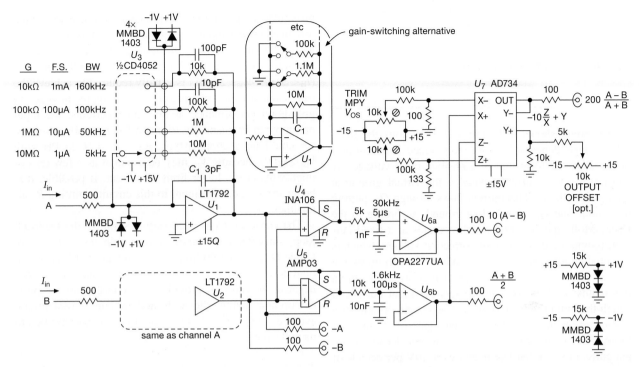

Figure 4x.129. By taking the ratio of sum and difference signals, this "normalizing TIA" outputs a voltage proportional to the fractional unbalance of input currents. The gain-switching alternative, though notionally less elegant, introduces less capacitance at the input node, and additionally allows the use of low-voltage ($\sim 5\,\text{V}$) switches

4x.20 Logarithmic Amplifier

The circuit shown in Figure 4x.130 exploits the logarithmic dependence of V_{BE} on I_C to produce an output proportional to the logarithm of a positive input voltage. R_1 converts V_{in} to a current, owing to the virtual ground at the inverting input. That current flows through Q_1, putting its emitter one V_{BE} drop below ground, according to the Ebers–Moll equation. Q_2, which operates at a fixed current, provides a diode drop of correction voltage, which is essential for temperature compensation. The current source (which can be a resistor, since point B is always within a few tenths of a volt of ground) sets the input current at which the output voltage is zero. The second op-amp is a non-inverting amplifier with a voltage gain of 16, in order to give an output voltage of -1.0 volt per decade of input current (recall that V_{BE} increases 60 mV per decade of collector current).

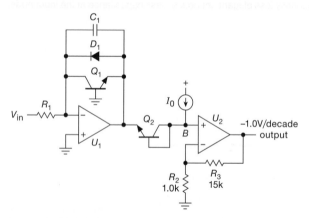

Figure 4x.130. Logarithmic converter. Q_1 and Q_2 comprise a monolithic matched pair.

Some further details: Q_1's base could have been connected to its collector, but the base current would then have caused an error (remember that I_C is an accurate exponential function of V_{BE}). In this circuit the base is at the same voltage as the collector because of the virtual ground, but there is no base-current error. Q_1 and Q_2 should be a matched pair, thermally coupled (a matched monolithic pair like the SSM2212 is ideal, with a typical V_{os} of $10\,\mu V$). This circuit will give accurate logarithmic output over seven decades of current or more (1 nA to 10 mA, ap-

proximately), providing that low-leakage transistors and a low-bias-current input op-amp are used. A BJT-input op-amp, with input currents in the nanoamp range, is generally unsuitable, and a FET-input op-amp like the OPA192 ($I_B<20$ pA, $V_{os}<25\,\mu A$) is usually required to achieve the full seven decades of linearity. Furthermore, in order to give good performance at low input currents, the input op-amp can be manually trimmed for zero offset voltage (or could be a chopper-type, e.g., the LTC2050H with $V_{os}<3\,\mu V$), because V_{in} may be as small as a few tens of microvolts at the lower limit of current. If possible, it is better to use a current input to this circuit, omitting R_1 altogether.

The capacitor C_1 is necessary to stabilize the feedback loop, since Q_1 contributes voltage gain inside the loop. Diode D_1 is necessary to prevent base–emitter breakdown (and destruction) of Q_1 in the event the input voltage goes negative, because Q_1 provides no feedback path for positive op-amp output voltage. Both these minor problems are avoided if Q_1 is wired as a diode, i.e., with its base tied to its collector.

4x.20.1 Temperature compensation of gain

Q_2 compensates changes in Q_1's V_{BE} drop as the ambient temperature changes, but the changes in the slope of the curve of V_{BE} versus I_C are not compensated. In §2.3.2 we saw that the "60 mV/decade" is proportional to absolute temperature. The output voltage of this circuit will look as shown in Figure 4x.131. Compensation is perfect at an input current equal to I_0, Q_2's collector current. A change in temperature of 30°C causes a 10% change in slope, with corresponding error in output voltage. The usual solution to

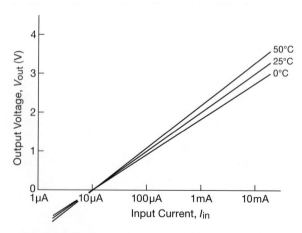

Figure 4x.131. Log converter's temperature dependence.

this problem is to replace R_2 with a series combination of an ordinary resistor and a resistor of positive temperature coefficient (PTC). Knowing the temperature coefficient of the resistor (e.g., the NXP KTY82 or Littelfuse 202PS1J positive tempco resistors, $2\,k\Omega$ at $25°C$ with a coefficient of $+0.4\%/°C$) allows you to calculate the value of the ordinary resistor to put in series in order to effect perfect compensation. For instance, with one of these 2k PTCs, you would add a $383\,\Omega$ series resistor.

There are several logarithmic converter modules available as complete integrated circuits. These offer very good performance, including internal temperature compensation. Some manufacturers are Analog Devices, Burr–Brown, Philbrick, Intersil, and National Semiconductor.

Exercise 4x.2. Finish up the log converter circuit by (a) drawing the current source explicitly and (b) using a 2k $+0.4\%/°C$ PTC resistor for thermal slope compensation. Choose values so that $V_{\text{out}} = +1$ volt per decade, and provide an output offset control so that V_{out} can be set to zero for any desired input current (do this with an inverting amplifier offset circuit, not by adjusting I_0).

4x.21 A Circuit Cure for Diode Leakage

Quite often a clever circuit configuration can provide a solution to problems caused by non-ideal behavior of circuit components. Such solutions are aesthetically pleasing as well as economical.

Suppose we want the best possible performance in a peak detector, i.e., highest ratio of output slew rate to droop. If the lowest-input-current op-amps are used in a peak-detector circuit (some are available with bias currents as low as 0.003 pA), the droop will be dominated by diode leakage; i.e., the best available diodes have higher leakage currents than the op-amps' bias currents. Figure 4x.132 shows a clever circuit solution. As before, the voltage on the capacitor follows a rising input waveform: U_{1a} charges the capacitor through both diodes and is unaffected by U_{1b}'s output. When the input drops below the peak value, U_{1a} goes into negative saturation, but U_{1b} holds point X at the capacitor voltage, eliminating leakage altogether in D_2. D_1's small leakage current flows through R_1, with negligible drop across the resistor. Of course, both op-amps must have low bias current. The OPA2192 is a reasonable choice here, with its combination of precision ($V_{os} = 25\,\mu V$, max) and low input current (20 pA, max). Because U_{1a} sees a capacitive load, it may be necessary to include a small series resistor (shown in brackets) to ensure stability.[89] This circuit is analogous to the so-called guard circuits used for high-impedance or small-signal measurements.

Note that the input op-amps in both peak-detector circuits spend most of their time in negative saturation, only popping up when the input level exceeds the peak voltage previously stored on the capacitor. However, as we saw in the active rectifier circuit (Section 4.10), the journey from negative saturation can take a while (e.g., 1 μs–2 μs for the OPA2192). This may restrict your choice to high-slew-rate op-amps.

A few additional points: (a) The JFET gate-channel diode of a small-geometry part like the MMBF4117 (or PN4117) has considerably lower leakage and capacitance than a "low-leakage" diode like the 1N3595 shown; we've often used JFETs for this purpose. (b) For really low input current, you should consider a low-voltage CMOS op-amp,

for example the 6 V LMP7721 (20 fA max!), or the 16 V LMC6001A (25 fA max).

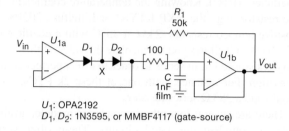

U_1: OPA2192
D_1, D_2: 1N3595, or MMBF4117 (gate-source)

Figure 4x.132. Bootstrapping the peak detector to eliminate diode leakage.

---

[89] The OPA2192's datasheet claims that it is stable into 1 nF, but the impedance of the added diodes degrades unity-gain stability.

4x.22 Capacitive Loads: Another View

We've seen the stability problems you can have when an op-amp (or other feedback circuit) drives a capacitive load, as for example in a sample-and-hold circuit (§4.5.2) or a supply splitter, or with voltage regulators (§9.1.1C and 9.3.12F). The easy way to describe this is to say that the load capacitance, in combination with the amplifier's output resistance, introduces an additional lagging phase shift within a feedback loop that already has 90° (or more) of lagging phase shift within the op-amp itself. The latter is due to the op-amp's deliberate frequency compensation (§4.9); and the combination can be toxic – 180° of phase shift (which turns negative feedback into positive feedback) at a frequency at which there's enough loop gain to sustain an oscillation.

There's another way to view the situation. Look at Figure 4x.133, which shows the characteristic Bode plot of open-loop gain (falling as $1/f$ at most frequencies), and also a simplified op-amp model that assumes some value of (open-loop) output-stage resistance r_{out}. Imagine for now that we've wired it as a follower (B=1). Then, as we saw in §2.5.3, the magic of feedback produces a greatly reduced output impedance, whose magnitude is

$$|Z_{out}| = r_{out} \frac{1}{1 + G_{OL}B}.$$

Now, here's the new insight: G_{OL} is falling as $1/f$, so r_{out} is rising approximately as f (as long as the loop gain $G_{OL}B \gg 1$). Hmmm ... that's the same characteristic as an *inductor!* So the follower's output looks inductive, which suggests that it can form a resonant circuit with a capacitive load.[90] Note that this doesn't require there to be any actual inductor anywhere.

Can this be correct? An easy test is to apply a sinusoidal current to the follower's output terminal (with the input grounded), and see how the output voltage responds (Fig. 4x.134). If it's inductive the voltage should *lead* the current by 90°, with a magnitude that increases proportional to frequency. Figure 4x.135 shows what we saw, with

our favorite LF411 jellybean, when driven with a 1 mA(pp) sinewave current at 250 kHz and at 500 kHz.

We repeated this at a bunch of frequencies, producing the plot of $|Z_{out}|$ versus frequency in Figure 4x.136. It's approximately linear at low frequencies, flattening out asymptotically to r_{out} as the loop gain becomes comparable to unity, in accordance with the equation above.

4x.22.1 Frequency of oscillation

Attach a capacitive load, and you've got a resonant circuit with $f = 1/2\pi\sqrt{L_{out}C_{load}}$. Easy to test: we put 10 nF across the follower's output, and got a robust oscillation at 650 kHz. That's a bit higher than the 585 kHz you'd predict for L_{out}=7.4 μH, but comfortably close to the 690 kHz resonant frequency set by the 5.3 μH effective incremental inductance at that frequency (see Fig. 4x.136).

That's good agreement.[91] But there's more to learn: What happens, for example, if the op-amp loop is configured for a closed-loop gain of ×10? This simple inductive-output model predicts that the output inductance increases by a factor of 10 (because the loop gain goes to unity at about $f_T/10$), so the resonant frequency should decrease by the square root of 10.

We tried it, with the same 10 nF load capacitor, and ... *no oscillation!* Here's one plausible explanation: as we raise G_{CL}, f_T drops proportionally, but the resonant frequency falls only as the square root. So the loop gain at the resonant frequency drops to less than unity. This is consistent with the observation of plenty of ringing and overshoot, but no sustained oscillation. So we increased C_{load} to 100 nF, at which point the circuit oscillated nicely, at about 100 kHz.

There's an additional effect in play here. Look back at the Bode plot in Figure 4x.133. When configured as a follower (B=1), the op-amp's phase shift at the unity loop-gain frequency is somewhat more than 90° (the datasheet shows a phase margin of ∼55°, thus a lagging phase shift of 125°). So for a sustaining oscillation the LC resonant circuit needs to provide an additional 55°, which it's happy to do: recall that the phase shift of an LC goes abruptly from +90° to −90° through resonance (the impedance of a series LC goes from negative imaginary to positive imaginary, and vice-versa for a parallel LC). However, when configured for higher closed-loop gain, say ×10, the loop-gain is considerably better behaved, with phase margin approaching 90°. So it's more stable with a capacitive load,

---

[90] A similar "impedance-rising-proportional-to-frequency" effect came up earlier, in connection with emitter followers. See "If it quacks like an inDUCKtor ..." (§2x.13).

[91] "The most dangerous situation," according to analog guru Jim Williams. That's because you don't look deeper, once you think you've got the right answer.

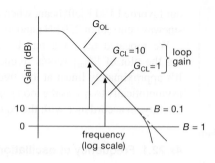
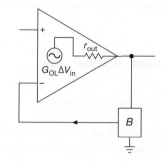

Figure 4x.133. A simple model for calculating the closed-loop output impedance of an op-amp: an output source resistance r_{out}, and an open-loop gain falling as $1/f$.

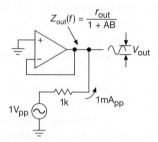

Figure 4x.134. Measuring the output impedance Z_{out} of an op-amp follower.

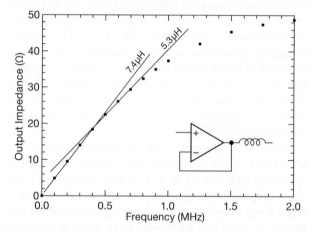

Figure 4x.136. Measured Z_{out} versus frequency for an LF411. The slope at low frequencies approximates that of a 7.4 μH inductor.

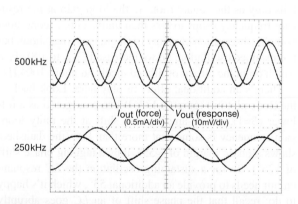

Figure 4x.135. Measured waveforms for an LF411 op-amp in the test setup of Fig. 4x.134, at two frequencies approaching the unity-gain frequency F_T. Note that the voltage *leads* the current by ~90° – it's *inductive!*

and more easily damped (preventing the load from introducing the required nearly 90° lagging phase shift within the loop).

4x.22.2 So, how about a few equations?

We've deliberately kept this discussion non-mathematical, to encourage intuition. Now that you've got it, consider this:

We can express the op-amp's gain, in the region where it is falling as $1/f$, as

$$G_{OL} = -j\frac{f_T}{f},$$

where f_T is the unity-gain frequency (more properly, the frequency at which the extrapolation of the straight portion of G_{OL} versus f reaches unit gain), and the $-j$ represents the 90° lagging phase shift (see §1.7.3). So we can write the output impedance, in all its complex glory, as

$$\mathbf{Z}_{out} = r_{out}\frac{1}{1+(-j\frac{f_T}{f})B}.$$

For loop gain significantly greater than unity we can ignore the 1 in the denominator, and therefore (for a follower)

$$\mathbf{Z}_{\mathrm{out}} = jr_{\mathrm{out}}\frac{f}{f_{\mathrm{T}}}.$$

It's pure imaginary, and proportional to frequency. And that's what inductors do for a living. So, equating this $\mathbf{Z}_{\mathrm{out}}$ to the quantity $j2\pi fL$, the follower's output impedance is equivalent to an inductor of value

$$L = r_{\mathrm{out}}/2\pi f_{\mathrm{T}}.$$

For equation-lovers, this is the only way to understand what's going on. The rest of us, however, are satisfied with the idea that something whose impedance rises proportional to frequency might as well be called an inductor. It quacks correctly.

4x.23 Precision High-Voltage Amplifier

Ordinary op-amps are fine for driving loads that require swings up to ± 15 V (or even ± 20 V) – these are commonly called "high-voltage" op-amps (see, for example, Table 4.2), to be distinguished from the much larger selection of op-amps that tolerate total supply voltage spans only to 15 V or less. And there is a limited selection of specialty op-amps intended for higher voltages still, for example the monolithic OPA454 (100 V maximum total supply voltage) or LTC6090 (140 V total supply); these op-amps cost about \$4 and \$6 (qty 100), respectively, and come in small packages with exposed pads so that they can dissipate as much as a few watts when attached to a few square inches of PCB foil.

You can get op-amps with much higher voltage ratings, but they are hybrid (not monolithic), and generally rather expensive. For example, the Apex PA95 is rated to 900 V total supply (e.g., ± 450 V); it can source or sink up to 100 mA (but limited by 30 W maximum power dissipation), and its output stage has class-AB biasing with provision for current limiting. It has a slew rate of 30 V/μs, and it costs about \$190 in single quantities.[92] Apex's highest voltage op-amp (the PA99) takes you all the way to 2500 V total supply, with output current to 50 mA (but limited to 35 W dissipation), and with 30 V/μs slew rate; it costs about \$850 in single quantities.

But you can roll your own high-voltage amplifier (to total supply voltages of a kilovolt or more) from discrete parts, and at a fraction of the cost of a commercial hybrid part, particularly if you are not looking for the utmost in speed and output current. Many applications (such as piezo drivers, deflection electrodes, scanning coils, and semiconductor test equipment) are well served by such an amplifier; and sometimes you need dozens of them in a complex setup.[93] And, you can optimize the design for what you really need – in this case we've aimed for *precision*, with a low-offset op-amp ($V_{os(max)} = 25\,\mu$V for the best grade; that's $200\times$ better than the PA95) and plenty of loop gain at low frequencies. We've worked also to ensure stability into the highly capacitive loads that you see in these applications: this circuit works well into any load capacitance up to 25 nF or so (albeit with reduced speed for $C_{load} > 500$ pF).

We've evolved many such designs over the years, targeted at research applications in the physical sciences. Here we present, in tutorial fashion, a fully evolved (yet relatively simple) design with good performance: ± 500 V output swing (or single-ended to +1 kV), ± 5 mA output current over the full voltage range, and 7.5 V/μs slew rate. The output stage is class-AB biased for good linearity, and it incorporates internal current limiting. The amplifier has a fully differential input with good accuracy (or a single-ended input with excellent accuracy), and the design can be extended to higher voltages. The total parts cost is about \$25 (unit quantity), or \$20 (qty 25).[94]

4x.23.1 Overview

The full schematic is shown in Figure 4x.137, and in the most high-level block diagram in Figure 4x.138. It consists of a precision control op-amp driving a high-voltage output stage, working together as a composite amplifier (see §4x.4, and examples such as Figure 5.47 in Chapter 5). It is an elaboration of the simple high-voltage regulator circuit of Figure 9.110 (where the goal is to generate a settable high-voltage unipolarity dc output). There are some nice challenges here, notably in configuring feedback to ensure stability around the composite amplifier, and in biasing and protecting the high-voltage transistors. Let's walk through this amplifier, to see how it can be designed and understood with simple back-of-the-envelope calculations.

4x.23.2 High-voltage output stage

The output stage is easy enough to understand: Q_5 is a common-source inverting amplifier with very high voltage gain at low frequencies (because its drain load is a current source, Q_8–Q_{10}). You could, of course, ask what the voltage gain is *at dc*; here it is limited only by the parallel resistance of Q_5's drain and that of the current

---

[92] Its bargain-priced PA97 cousin (\$150) may tempt you; but it's good to only 10 mA and 8 V/μs, it lacks current-limiting, and its output stage is biased class-C. Plenty of downside, not much savings.

[93] In our labs we commonly use 12 to 32 amplifiers to drive sets of electrodes, typically four that act as beam benders, plus 2–4 more for electric field shims; a charged-particle beam system may have multiple such guiding locations. This is in addition to accelerating and gating elec-

trodes. In another experiment nearly a hundred high-voltage amplifiers are used in guiding and trapping charged particles.

[94] The authors can supply bare boards to individuals wishing to replicate this circuit; write to winfieldhill@yahoo.com, referencing the "AMP-62" design.

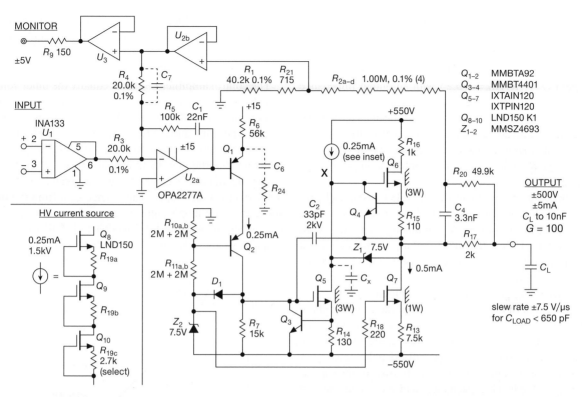

Figure 4x.137. Precision high-voltage amplifier (±500 V, ±5 mA, 7.5 V/μs), implemented with MOSFET totem-pole output stage with class-AB pull-down current sink on the output source follower. See text for modifications for 0 to +1000 V unipolarity output.

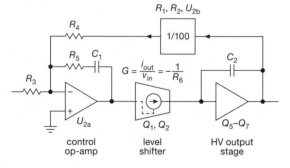

Figure 4x.138. Block diagram of the amplifier of Figure 4x.137.

source, $G_{dc} = g_{m5}R_\parallel$. With that load resistance (we estimate ~ 100 MΩ, see Fig. 9.41) and an estimated $g_{m5} \approx 2.2$ mS (based upon measured values[95]), the voltage gain of the

[95] As discussed in §3x.2, the transconductance of an FET operating in the subthreshold region of drain current is roughly proportional to drain current (analogous to a BJT's transconductance $g_m = I_C/V_T$), but with a scale factor n in the denominator: $g_m = I_D/nV_T$. The measured transconductances of an IXTP1N120 MOSFET correspond to values of n of 4.24, 4.55, and 4.77 at drain currents 1 μA, 30 μA, and 700 μA, respectively, thus a transconductance of $g_m \approx 2.2$ mS at $I_D = 0.5$ mA.

output stage at dc is of order 2×10^5 (106 dB). But of course Q_5's voltage gain at any frequency depends on its total load *impedance*, which includes plenty of capacitance – both that contributed by the transistors themselves and indirectly through load capacitance C_L connected to the amplifier's output. Finally, the output stage's frequency response is deliberately tailored by feedback capacitor C_2 to ensure stability, as we'll see presently.

MOSFET Q_6 is a voltage follower, biased with a 0.5 mA pull-down current sink Q_7; Q_6 can source output currents up to ~ 5.5 mA (set by current limit transistor Q_4). When *sinking* current from a load, Q_7 takes care of currents less than 0.5 mA, above which the gain transistor Q_5 takes over, through otherwise non-conducting diode Z_1 (a low-capacitance zener). Analogous to Q_4, transistor Q_3 limits Q_5's current to ~ 4.5 mA. These current-limit values keep the short-circuit power dissipation in Q_5 and Q_6 safely less than 3 W, which is quite conservative, given their low thermal resistance of $R_{\Theta JC} = 2^\circ$C/W – a small heatsink (or even a few square inches of PCB foil) is adequate to keep the junction temperature below its 150°C rating; see §9.4 for lots of heatsinking know-how.

Because Q_5's input is referenced to the negative rail, it's necessary to level-shift the op-amp's output (which is referenced to ground). That's done by common-emitter amplifier Q_1, whose collector resistor R_7 sets the unloaded voltage gain $G = -R_7/R_6 = -0.25$ with an output impedance equal to R_7. However, feedback through C_2 forces the input impedance at Q_5's gate to a low impedance at signal frequencies, so you can think of the level shifter's output as a signal *current* into a summing junction at Q_5's gate: $i_{sig} = -v_{2a}/R_6$, where v_{2a} is the frontend's output voltage at op-amp U_{2a}.

In the level-shifter circuit the cascode transistor Q_2, biased halfway to the negative rail, allows these 300 V transistors to span the full 550 V. An analogous trick is used to extend to 1.5 kV the 500 V rating of the individual depletion-mode current source transistors Q_8–Q_{10}; see the LND150 datasheet, Figure 3.9, and the measured current of a series string of five LND150's in Figure 3x.59.

It's essential in this design to control the gain and phase shift, both of this output stage and of the whole feedback loop. For this process of "frequency compensation" we need to take account of internal roll-off poles (e.g., Q_5's high-impedance output loaded by its own drain capacitance, Q_6's feedback capacitance, and the capacitance of the Q_8–Q_{10} current source), as well as additional poles contributed by the load capacitance C_L. Compensation capacitor C_2 is part of this process, which we'll describe below, after an explanation of the front-end amplifier and gain-setting feedback.

4x.23.3 Front-end amplifier stage

The unity-gain difference amplifier U_1 can be omitted if a good ground-referenced input is available. It's nice to have a differential input, but you pay a price in precision: The INA133's maximum offset voltage and drift ($\pm450\,\mu$V and $\pm5\,\mu$V/°C for the best grade) is $20\times$ worse (!) than that of the OPA2277 ($\pm25\,\mu$V and $\pm0.25\,\mu$V/°C for the best grade).

Comparing its specifications to that of the OPA2277, you might worry also about the INA133's *noise* contribution (60 nV/$\sqrt{\text{Hz}}$, compared with 8 nV/$\sqrt{\text{Hz}}$ for the OPA2277). That's not worth worrying about, though, for several reasons: (a) The noise introduced by the feedback divider resistor R_1 is 25 nV/$\sqrt{\text{Hz}}$, or 0.8 μVrms in the amplifier's 1kHz bandwidth, and we cannot much reduce this without dissipating too much power in R_2; (b) an amplifier like this is usually driven from a 16-bit digital-to-analog converter, for which an LSB corresponds to 150 μV for ±5 V programming range; (c) the MOSFET driver Q_5 is itself noisy,

and its input noise looks $\times4$ worse still when reflected back (by the level shifter's gain of 1/4) to the op-amp's output; and (d) in a typical system it's likely that other things (e.g., power supply noise) will add more noise anyway.

The control amplifier U_{2a} sums equally the input voltage and the inverted $V_{out}/100$ feedback signal, for an overall voltage gain of -100; i.e., a ±5 V input produces a ∓500 V output. The level shifter Q_1Q_2 (dc gain of -0.25) converts an op-amp output near ground to a ~4 V bias at Q_5's gate, in the neighborhood of its threshold voltage. Overall feedback at dc, of course, closes the loop accurately to produce the correct HV dc output. The voltage gain from the input signal to the op-amp's output is enormous at dc ($G_{dc} \approx 10^7$, typ), where there is no local feedback; it falls to $G = 5$ at frequencies above about 75 Hz (the R_5C_1 characteristic fre-

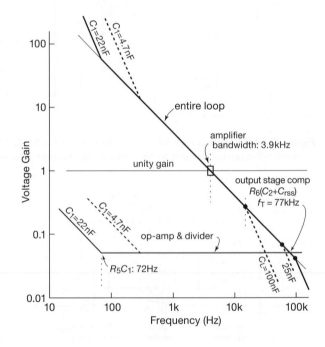

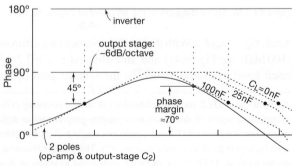

Figure 4x.139. Bode plot (gain and phase shift) of the amplifier of Figure 4x.137.

quency), where R_5 introduces a zero to prevent further roll-off. This deliberate roll-off is a portion of the overall frequency compensation, discussed next.

4x.23.4 Feedback stability

A multistage amplifier like this presents stability challenges, especially because we want it to drive loads whose capacitance may range from near zero, to 1000 pF (e.g., 10 m of coax cable), and perhaps even to 10 nF or more (e.g., a piezo element). And notice that the combined open-loop voltage gain *at dc* of the two cascaded stages is enormous, well over 200 dB ($10^7 \times 10^5$)!

This sounds positively scary, and one's first (and cowardly!) instinct might be to reduce greatly the open-loop gain. But, happily, a stable composite amplifier configuration can be rigged up (while maintaining plenty of accuracy-enhancing low-frequency feedback) by careful application of a Bode plot; here we can do it with just three compensation components. Let's see how.

Look at the well-decorated Bode plot in Figure 4x.139, which shows the amplifier's loop gain and phase margin versus frequency. As usual, we do not care about the gain at dc, the roll-off frequency of the dominant pole, or even the gain at any particular frequency. What we *do* care about is the frequency at which the closed-loop gain passes through unity (the amplifier's "bandwidth," or *transition frequency* f_T), and particularly the slope of the gain in the neighborhood of that frequency. We strive for a well-behaved gain curve falling at 6 dB/octave at f_T, with troublesome poles (which cause a steeper slope) kept at arm's length (ideally distant at a factor of ten – a decade – in frequency). A -6 dB/octave slope corresponds to a phase margin of 90°; you lose 6° if there's a pole a decade away, 18° for a pole a half-decade away, and 52° for a pole right at f_T (Fig. 4x.9). Even with the latter the amplifier would be stable, but it would exhibit some peaking in the frequency domain, and some overshoot in the time domain.

Look first at the curve marked "op-amp & divider," which represents U_{2a}'s closed-loop gain ($-R_5/R_4$, or ×5 at high frequencies) multiplied by the "gain" (×1/100) of the feedback divider R_1R_2. To retain lots of low-frequency gain we put C_1 in series with gain-setting R_5, and we choose this time constant large enough to keep its low-frequency 6 dB/octave slope below ~ 70 Hz, safely to the left of the unity-gain crossing of the composite amplifier, which we'll see is around 4 kHz.[96]

Now we multiply that curve by the output stage's gain to get the overall amplifier's closed-loop gain (the curve marked "entire loop"), whose unity gain crossing is the amplifier's bandwidth f_T. We tailor the roll-off of the output stage with compensation capacitor C_2, chosen large enough to put the overall f_T (i.e., output stage $G_{out} = 20$) safely below the frequency of the next pole. The latter (as we'll see presently) is around 100 kHz when the output is unloaded, but drops to ~ 60 kHz for 25 nF capacitive load, and down to ~ 15 kHz for a giant 100 nF load. With our choice of $C_2 = 33$ pF the amplifier bandwidth f_{-3dB} is 3.9 kHz, as shown, more than a decade below the pole produced by a 25 nF load (and a factor of four even for a highly capacitive 100 nF load).

This choice of C_2 produces conservative phase margins. Look at the aligned "Phase" plots of the feedback signal, which are keyed to the "entire loop" gain plot. The top line represents the overall inverting feedback connection (the op-amp, level shifter, and output stage are each inverting), from which the 6 dB/octave slope from C_2 over most of the plotted frequency range introduces a 90° lagging phase shift. That leaves a 90° baseline phase margin, reduced at low frequencies (where we don't care) by the rising gain of the op-amp stage, and reduced at high frequencies (where we care very much!) by additional output-stage poles (from its own self-capacitance, aided and abetted by load capacitance).

You can plot the phase margin, approximately, by drawing dashed-line asymptotes through 45° at the corresponding pole frequencies, and extending to 0° and 90° a decade on either side. This we've done in the Bode phase plot, for three choices of load capacitance (see below). Then you can draw a smoothed curve (done here for $C_L = 100$ nF), from which you can read off the phase margin at the overall amplifier's f_T; here that's about 70°, a highly conservative phase margin. That is reflected in the small overshoot of measured step response (Figure 4x.143). To reduce clutter we plotted the smoothed curve only for a 100 nF load capacitance; but analogous curves for $C_L = 0$ and 25 nF would show estimated phase margins of ~ 85° and ~ 75°, respectively.

---

[96] A value of $C_1 = 4.7$ nF would put this R_5C_1 "zero" approximately a decade below the composite amplifier's f_T, for fully adequate phase margin. We conservatively chose an additional factor of five to prevent entirely any overshoot with a step input; overshoot is bad news in high-voltage amplifiers. Note also that (as is usual in op-amp circuits) the op-amp's *open-loop* 6 dB/octave gain curve (and 90° lagging phase shift), which begins at frequencies of just a few hertz, is not seen in the *closed*-loop response – feedback creates a flat response, up to a frequency of the op-amp's GBW/G_{CL}, or ~ 0.2 MHz here. That is, this op-amp stage exhibits lagging phase shifts only below 70 Hz and above 0.2 MHz.

4x.23.5 Circuit capacitances and capacitive loads

In this discussion we've simply asserted the output-stage roll-off frequencies caused by internal capacitances and by capacitive loading, sidestepping the actual calculation. We asserted, but did not calculate, because we wanted to display the Bode plot and nail down the stability criteria and thus the required compensation networks.

Now it's time to circle back and do the calculations. Figure 4x.140 shows the capacitances that determine the output stage bandwidth, and Figure 4x.141 is a busy plot, Bode-style, of the output-stage voltage gain (i.e., from Q_1's base to the amplifier's output terminal) versus frequency. For the latter plot we've drawn curves for several combinations of internal and load capacitances, to promote an intuitive understanding of their combined bandwidth-limiting effects. In this figure the descending bold line of output-stage gain (with all capacitances included) corresponds to the similarly bold line of overall amplifier gain presented in Figure 4x.139.

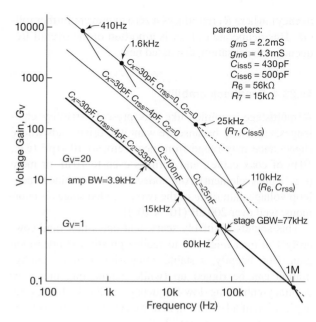

Figure 4x.141. Bode plot of output-stage voltage gain, showing effects of circuit and load capacitances.

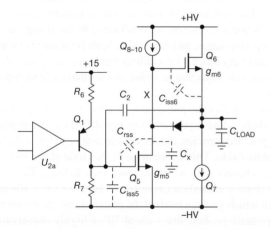

Figure 4x.140. Bandwidth-limiting capacitances affecting the output stage. Capacitance C_x represents C_{rss} of Q_6 plus the combined C_{oss}'s of Q_5 and Q_8–Q_{10}.

Taking it in stages, we'll first calculate the voltage gain with no load, ignoring for the moment both the compensation capacitor C_2 and the feedback capacitance C_{rss} of Q_5. After that we'll include these capacitances. And finally we'll pile on load capacitance, a situation discussed in isolation in §3x.8.2. We'll call the signal at Q_1's base v_{in}, and the amplifier's output v_{out}; the overall voltage gain is $G_V = v_{out}/v_{in}$. In the following we'll write gain magnitudes, i.e., we won't keep track of minus signs for inversion.

A. No load, no feedback capacitance

The voltage gain from Q_5's gate to the output (call it G_5) is simply $G_5 = g_{m5}/\omega C_x$, where C_x is the total capacitance seen at Q_5's drain. We multiply that by the gain from Q_1's base to Q_5's gate (call it G_1). At low frequencies (where the reactance of C_{iss5} is large compared with R_7) $G_1 = R_7/R_6$; at high frequencies the decreasing reactance of C_{iss5} swamps R_7, and $G_1 = X_{Ciss5}/R_6 = 1/R_6\omega C_{iss5}$. The crossover is at the frequency where the reactance of C_{iss5} equals R_7, i.e., $f = 1/2\pi R_7 C_{iss5}$; for our $C_{iss5} = 430$ pF the crossover is at 25 kHz.

Multiplying the two stages of gain, we have

$$G_V = \frac{R_7}{R_6}\frac{g_{m5}}{\omega C_x} \quad (f < 25\,\text{kHz})$$

$$G_V = \frac{1}{R_6}\frac{g_{m5}}{\omega^2 C_x C_{iss5}} \quad (f > 25\,\text{kHz})$$

which evaluates to $G_V = 3100$ at 1 kHz, for example, and slopes down at 6 dB/octave to 25 kHz, after which it slopes at 12 dB/octave. This is the top curve in Figure 4x.141.

B. Add feedback capacitance

The addition of feedback capacitances C_{rss5} and C_2 makes Q_5's gate a summing junction, with the Miller feedback current combining with the signal input current from Q_1's collector. The impedance at that node is far smaller than both R_7 and the reactance[97] of C_{iss5}, so we're in the high-frequency regime where Q_1 contributes a current v_{in}/R_6. If you work it out, you'll find that the overall gain is given by

$$G_V = \frac{g_{m5}}{\omega C_x} \frac{1}{R_6} \frac{1}{\omega(C_{iss5} + \frac{g_{m5}}{\omega C_x} C_{rss})}.$$

At low frequencies the gain falls as $1/f$ (i.e., 6 dB/octave) because of the coefficient of C_{rss}, then as $1/f^2$ beginning at a frequency where the two denominator terms are equal: $f = (C_{rss}/C_{iss5})(g_{m5}/2\pi C_x)$. That corresponds to the break-points seen at 110 kHz (middle gain curve) or at 1 MHz (bottom curve).

C. Add load capacitance

We saw in §3x.8.2 that a capacitively loaded MOSFET follower with current drive exhibits a breakpoint (transition from 6 dB/octave to 12 dB/octave) at a frequency

$$f_1 = \frac{g_{m6}}{2\pi C_{iss6}} \frac{C_x}{C_{LOAD}}, \qquad (4x.7)$$

where we've substituted the corresponding parameters and circuit designators appropriate to our amplifier. For the case of no feedback capacitance ($C_2 = C_{rss} = 0$) and a load capacitance of 25 nF, for example, we find $f_1 = 1.6$ kHz, as indicated in Figure 4x.141. Following the capacitively loaded curves down to their intersection with the bold line, we find the breakpoints of the fully implemented output stage (i.e., including both C_{rss} and compensation capacitor C_2). The effect of the feedback capacitances is to lower the output-stage gain by a factor of $R_7 g_{m5}(C_2 + C_{iss5})/C_x$, about ×40, and to raise the frequency of the breakpoints by the same factor. The resulting bold gain curve, with these breakpoints, was used for the complete amplifier Bode plot of Figure 4x.139.

D. Output series resistor

Sharp-eyed readers are probably wondering about R_{17} and associated components in the amplifier circuit. There are several things going on here. (a) We like to add a series output resistor to help protect the MOSFETs in the event of a sudden output short, with its high dV/dt; (b) It's sometimes helpful to filter the output, for example to limit wideband noise; and, perhaps most important, (c) The output network provides additional stability when driving capacitive loads, by adding a zero to combat the capacitive load's

pole.[98] Put quantitatively, if we equate f_1 in eq'n 4x.7 to $1/2\pi R_{17}C_{LOAD}$, we find that a value $R_{17} = C_{iss6}/C_x g_{m6} = 3.8$k cancels the C_{LOAD} pole. This works over the full range of load capacitance in our amplifier, and the resulting output routed by C_4 to the feedback network has a simple 6 dB/octave roll-off, with corresponding lack of excess phase shift. However, the calculated resistor value may be higher than you would like, slowing the output response into C_{LOAD} and degrading output swing at high load currents. This would be especially true for fast-slewing high-frequency versions of the amplifier.

E. SPICE analysis

Often you don't need to wrestle with detailed stability calculations when designing a low-speed composite amplifier, because (as in this example) you've got highly conservative phase margins. If you want greater bandwidth, though, you may find yourself with deteriorating phase margins, and you'll have to balance delicate trade-offs. If you lack the benefit of a bench prototype, you may want to use a SPICE model to analyze the circuit as it evolves. A simplified model can work fine, but as we pointed out in §3x.5.5, you cannot use the manufacturer's MOSFET model for small-signal analysis, because we're operating in the subthreshold region (see Figure 3x.36), where its g_m values peak and then collapse as the drain current improperly drops to zero.

A solution, for small-signal analysis at fixed quiescent currents, is to substitute a VCCS (voltage-controlled current-source) model for the MOSFET's g_m, as shown in Figure 4x.142. The MOSFETs are replaced by their capacitances and by a VCCS gain element G, with the gain to g_m at the MOSFET's operating current, programmed by the voltage across C_{iss}.

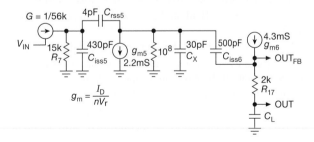

Figure 4x.142. SPICE model of the output stage, with parameter values listed in Figure 4x.141.

---

[97] We can write $X_{node} = X_{Crss}/A$, where the gain $A = g_m X_{Cx}$; in that way we find that the feedback capacitance produces a flat (and resistive) node impedance of $C_x/g_m C_{rss} \approx 450\,\Omega$. This dominates over the effect of C_{iss5} up to 850 kHz, if the effect on C_x of additional output load capacitance is ignored.

[98] This is a widely used technique when driving capacitive loads, for example with an op-amp; see §4.6.2 and Figure 4.78.

4x.23.6 Output slew rate

Look back at the amplifier circuit: ignoring for the moment any attached capacitive loads, the slew rate of the output stage is limited by Q_1's full-scale current signal of $\pm 0.25\,\mathrm{mA}$ driving the summing junction at Q_5's gate, which must balance the current coming back through C_2 that is produced by that slew rate. That is, $S \equiv dV/dt = i/C_2 = 7.5\,\mathrm{V}/\mu\mathrm{s}$. Of course, the circuit must have sufficient drive current to slew its own capacitances (the sum of C_{oss5}, C_{oss7}, C_{rss6}, and $C_{\mathrm{oss10}}/3$, where C_{oss5} designates the output capacitance C_{oss} of Q_5, etc.), as well as any attached load capacitance. The circuit capacitances add up to $\sim 30\,\mathrm{pF}$, requiring only $i = C\,dV/dt = 0.2\,\mathrm{mA}$ for the maximum calculated slew rate of $7.5\,\mathrm{V}/\mu\mathrm{s}$. And, by an analogous calculation, load capacitances up to $650\,\mathrm{pF}$ (requiring $\pm 5\,\mathrm{mA}$ for $\pm 7.5\,\mathrm{V}/\mu\mathrm{s}$) do not degrade the maximum slew rate. For greater load capacitance the slew rate is current-limited to $S = 5\mathrm{mA}/C_\mathrm{L}$, thus for example $1.5\,\mathrm{V}/\mu\mathrm{s}$ into $3.3\,\mathrm{nF}$ (as seen in Figure 4x.144). It's important to remember that our small-signal analysis does not apply during slewing, but it does bear directly on the settling response following recovery from integrator-induced error.

4x.23.7 Measured performance

We put together the circuit of Figure 4x.137 and ran it through its paces (and hey, we also got to tune up the compensation networks – to paraphrase Colonel Rosa Klebb in *From Russia with Love*, "Theory iz useful, but iz no substitute for experience."[99]). When driven with a low-level sinewave,[100] the measured bandwidth of 3.8 kHz was in good agreement with the 3.9 kHz estimated value, and there was no evidence of peaking in the frequency domain with load capacitances up to 3.3 nF; there was about 0.2 dB peaking with $C_\mathrm{L} = 10\,\mathrm{nF}$ and 2 dB with 25 nF, both around 900 Hz.

Figure 4x.143 shows the response to a small (40 mV) input square wave. For load capacitances up to a few nanofarads the output displays clean steps with no visible overshoot; it's quite happy driving 10 nF ($\sim 6\%$ overshoot, settling in $\sim 0.5\,\mathrm{ms}$), and even with a heavy $0.1\,\mu\mathrm{F}$ load the overshoot is reasonably well controlled, with adequate damping.

This is, of course, a *high-voltage* amplifier, so we juiced

---

[99] To which SPECTRE operative Morzheny replies "I agree: ve use live targets as vell."

[100] Small-signal measurements are a good way to investigate the desirable linearity and uniform closed-loop frequency response with a smooth roll-off (absence of peaking), thus quick settling after slewing.

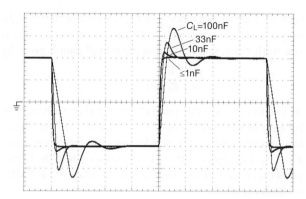

Figure 4x.143. Small-signal step response of the amplifier of Figure 4x.137. Even with a load capacitance of $0.1\,\mu\mathrm{F}$ the amplifier is stable, though with significant overshoot and ringing. Horizontal: 1 ms/div.; Vertical: 1 V/div.

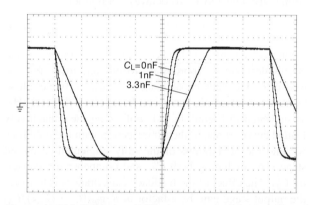

Figure 4x.144. Full-swing (1 kVpp) response of the amplifier of Figure 4x.137. The circuit's output current limit of $\pm 5\,\mathrm{mA}$ limits the slew rate of into 3.3 nF: $dV/dt = I/C = 1.5\,\mathrm{V}/\mu\mathrm{s}$. Horizontal: $400\,\mu\mathrm{s}$/div.; Vertical: 200 V/div.

it up (and only got shocked once, but *ouch!*), and drove it full-swing into some capacitive loads. Figure 4x.144 shows the output (note change of scales) when driven with a 10 Vpp square wave, i.e., a $-500\,\mathrm{V}$ to $+500\,\mathrm{V}$ square wave output. Here the current limit of $\pm 5\,\mathrm{mA}$ sets the slew rate into the most capacitive load; once again there is no visible overshoot for load capacitances to 1 nF, with just a hint into 3.3 nF.

Figure 4x.145 shows what happens when driving large steps into very high load capacitances. Here we use a 6 Vpp square wave, to prevent clipping during the substantial overshoot. The good news is that the amplifier can drive 25 nF; the bad news is that you can expect 25% overshoot in addition to current-limited slewing.[101] If you don't

---

[101] The overshoot comes from "integrator windup," a phenomenon in two-

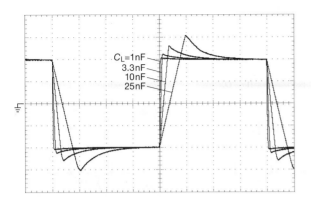

Figure 4x.145. Large-signal (600 Vpp) response into load capacitances to 25 nF. The slew rate into $C_L \geq 3.3$ nF is set by the ± 5 mA current limit. The overshoot can be eliminated (at the expense of speed) with an input RC slowdown of time constant 2.5 ms. Horizontal: 4 ms/div.; Vertical: 150 V/div.

care about optimizing speed, you can address the overshoot nicely by putting an RC slowdown at the amplifier's input (split R_3 into a series pair of 10k resistors, with a capacitor C to ground from the midpoint). We found that $C = 0.47\,\mu\text{F}$ (a time constant of 2.5 ms) prevented any overshoot for load capacitances up to 25 nF; use $C = 2\,\mu\text{F}$ (10 ms) if you want to prevent overshoot for loads to 100 nF.

4x.23.8 Variations: unipolarity, higher voltages, greater speed

You can work variations on this amplifier, for example to run it with positive polarity output only, or with higher total supply voltages. Here are some ideas:

Unipolarity: −5 V to +1000 V operation
The basic design can be easily modified to create a +1 kV amplifier, with −15 V and +1100 V supplies. Omit Q_2 and replace $R_{10}R_{11}$ with a single 47k resistor. Because U_2 is a ± 15 V op-amp, its output could saturate Q_1, inverting the feedback phase and thus locking up the amplifier output at the rail; this can be addressed with a zener diode in series with Q_1's base and a pull-up resistor to +15 V, to provide at least 6 V of operating room for the MOSFET. Be sure to reduce R_6 accordingly.

Higher voltages
For operation with total supply voltages well above 1000 V you can extend the cascode-connected level shifter Q_1Q_2 with additional *pnp* cascodes, biased with an extended resistor string. And you can increase the voltage rating of Q_5's drain current source Q_8–Q_{10} with a longer string of depletion-mode MOSFETs. Although you can get higher voltage MOSFETs to replace Q_5–Q_7, it's usually more practical to add less expensive and more readily available MOSFETs in cascode series, see the extensive discussion in §3x.6.5. The resistor ladder approach, shown in Figure 9.111, has worked well for us. As illustrated there, it's a good idea to use 10 V gate-protection zener diodes and $100\,\Omega$–$220\,\Omega$ gate-isolating resistors.

Greater bandwidth and slew rate
The design we've described is quite conservative, tolerant of high output capacitance and requiring only a minimum of compensation components. But it's also quite slow. To speed it up you can increase the operating currents and reduce C_2. To deal with highly capacitive loads, you can increase the current-limit set by R_{14} and R_{15}. These changes may require larger heatsinks for Q_5 and Q_6. You can increase the loop gain with R_{24} effectively in parallel with R_6, coupled at high frequencies with C_6. If the resulting loop stability is too degraded, you can add a compensation zero with C_7; you can also add a zero with a resistor in series with C_2.

Figure 4x.146 shows three additional speed-enhancing techniques. In circuit A a common-emitter BJT ($Q_{11}Q_{12}$) increases the transconductance of the output gain stage (typically by a factor of $4\times$–$10\times$), with MOSFET Q_5 now serving as a high-voltage cascode (and eliminating its C_{rss} Miller feedback capacitance). In circuit B the emitter follower Q_{13} drives Q_5's input and Miller-feedback capacitances, dramatically increasing the slew rate. An elaboration of this latter idea is to drive Q_5's input capacitance with a high-current op-amp. For example, we used a BUF634 (unity gain, 250 mA output current capability, 180 MHz, 2000 V/μs) with a 6.8 Ω series gate resistor to drive a 1.5 kV MOSFET (STP4N150) to make a linear −1500 V ramp with a 100 V/μs–3 kV/μs controllable slew rate. A transconductance op-amp (OPA660, 15 mA max) provided the programmable ramp current into a 3 pF integrating feedback capacitor. Draw the circuit yourself![102]

---

stage servo systems where the control compensation is in the input stage, whose error-integrating capacitor charges while waiting for the output stage to recover from a nonlinear "fault" situation such as slewing or current limiting.

[102] Then look at §3x.20.

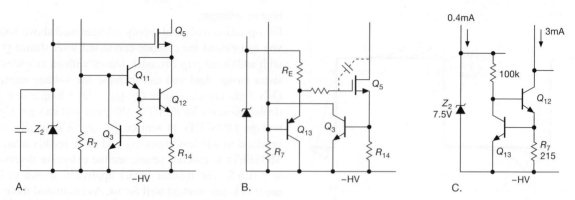

Figure 4x.146. Speeding up the output stage. A. Raising g_m with a BJT driver and HV MOSFET cascode; B. Driving output-stage capacitance with an emitter follower; C. Replacing R_7 with a 3 mA current sink (§4x.23.9).

A. MOSFET transistor choices

Back in Chapter 3x (§3x.5) we discussed in detail the use of MOSFETs in linear circuits, including the properties of high-voltage MOSFETs, subjects relevant to amplifiers like this. That section pairs nicely with the precision amplifier here, as does Table 3x.3, where we've collected a selection of MOSFET candidates for high-voltage amplifier applications, with corresponding measured transconductance data (g_m).

4x.23.9 Faster HV amplifier: 1MHz and 1200V

We've built versions of this amplifier with surprisingly high small-signal −3 dB bandwidths. To do this it was necessary to dramatically improve the performance of the output stage. Here we discuss a 1.2 kV output stage with a −3 dB bandwidth of 1 MHz. Taking a realistic application scenario, we assume the amplifier is driving a 20 pF electrode, plus one meter of 50 Ω coax (97 pF/m), for a total load capacitance of ~120 pF. As will become evident, we'll need to search for unusual components to improve the amplifier's response, and make circuit changes as well.

We're trying to increase the bandwidth of the high-voltage amplifier of Figure 4x.137 by a factor of ten or more. Since the bandwidth is limited by the combination of drive currents and the capacitances being driven, we need to seek components and configurations that *minimize capacitance*, and at the same time we need to *boost operating currents*. Both factors involve a delicate dance, particularly in combination with high supply voltages – for example, small-die transistors tend to come in small packages, so their higher thermal resistance compromises heat removal. And, as we'll see, the circuit changes needed to improve speed will have ripple effects throughout the design.

As shown in Figure 4x.138, the output stage's integrating feedback capacitor C_2 creates the classic −6 dB/octave gain slope, with an ultimate −90° phase shift. But the output stage is just one portion of the entire feedback loop, so we want to avoid adding much additional phase shift up to (and well past) the complete amplifier's f_T bandwidth. So we need to worry about Q_6's (source follower) 2nd gain pole, f_1, which combines to create a −12 dB/octave ultimate slope; we'll assume a 120 pF load at the output. It's good if the f_1 pole is well above f_T, but we'll use its −12 dB/octave intercept with an output-stage gain of 50 to discover at what frequency its effect on the C_X node will be felt. We want this to be at least five times f_T.

A. Transistor choices

Turning first to component choices, below is a mini-table of MOSFET choices (some of which are not available from regular distributors, but which can be found on Alibaba); see also Table 3x.3 for more choices. The best choice will depend on output current requirements and the heatsinking arrangement.

For our design we chose Sanyo's 2SK1412, a 1.5 kV 20 W TO-220F (isolated) part, with $C_{iss} = 40$ pF, $C_{oss} = 12$ pF, and $C_{rss} = 3$ pF. If we bias Q_6 at 1 mA, the 2SK1412's transconductance is 10 mS, from which we estimate (using eq'n 3x.6) an f_T of 40 MHz. Next we estimate Q_5's C_X node capacitance: C_{oss5} plus C_{rss6}, plus another 5 pF for Q_{10} and stray, thus C_X is about 20 pF. Now we can estimate f_1, by applying eq'n 3x.15, to find $f_1 \approx 40\text{MHz} \times (20\text{pF}/120\text{pF}) = 6.4$ MHz.

That looks good, but we've conveniently ignored a critical part, the zener diode Z_1 across Q_6's gate-to-source. *Awwkk*, the capacitance of a small 7.5 V zener is a gigantic 55 pF, which reduces f_T and f_1 by almost 60%! (This is the

Table 4x.5: HV MOSFET Choices.

Type	V_{DSS} (V)	Packages TO-220	DPak	$R_{\Theta JC}$ (°C/W)	C_{iss} (pF)	C_{oss} (pF)	C_{rss} (pF)	FOM (W/pF)
RJK6024DPE	600	-	•	6.4	37.5	7.5	0.9	2.6
2SK1412	1500	•	-	6.3	40	12	3.0	1.7
IXTY01N100	1000	-	•	5.0	54	6.9	2.0	3.6
IXTP02N120	1200	•	•	3.8	104	8.6	1.9	3.8
STD1HN60K3	600	-	•	4.6	140	13	2.0	2.1
FQD1N60C	600	-	•	4.5	130	19	3.5	1.5
FQP2N60C	600	•	-	2.3	180	20	4.3	2.7
FQD1N80	800	-	•	2.8	150	20	2.7	2.2
STP2N105K5	1050	•	•	2.1	115	15	1.5	4.0
IXTP06N120	1200	•	-	3.0	236	15	3.2	2.8

value for 20mA-rated '5236 types, but it's also the same for the 50 μA-rated '4693 zeners.)

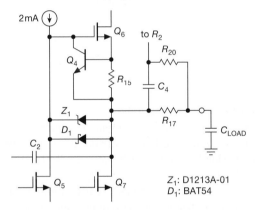

Figure 4x.147. Signal-protection zener Z_1's capacitance (0.85 pF) is far less than ordinary zeners. Shunt Schottky diode D_1 assists during Q_5's current-sinking activity.

B. Circuit changes

Zener capacitance.

We made extensive changes to Figure 4x.137's circuit, both to solve this "zener problem," and to address other issues that arose when we increased operating currents. First we replaced the protection zener Z_1: the D1213A-01 is a TVS (transient suppression) device, intended for signal lines, with just 0.85 pF capacitance, and a roughly 10 V breakdown. That's nice, but it's a small-die part with poor forward current-handling capability, so we put a BAT54 Schottky diode in parallel, adding another 4 pF to the C_{iss} value (Fig. 4x.147). The BAT54 drops 1.0 V max at 100 mA, so it's about as small as we dared go. This zener substitution reduced the C_X node capacitance

by some 50 pF, increasing the f_T and f_1 values to 35 MHz and 5.7 MHz.[103]

Current source.

The next major circuit change was to the Q_8–Q_{10} current-source stack. We used the configuration of Figure 3x.57B, which allowed us to dissipate power equally in three transistors, with a portion of the heat dissipated in 1210-size resistors. With the current set to 2.0 mA,[104] the stack must dissipate up to 2.5 W at 1.2 kV (see §3x.6.5). But, looking on the bright side, it can slew the C_X node at $dV/dt = I/C = 100$ V/μs, a further 8× improvement.

Gain and stability.

Next we calculate the transconductance of Q_5 at 2 mA to be 20 mS, and its stage gain into the 20 pF C_X node, $G = g_m X_C$, to be $G = 158$ at 1 MHz. Assuming we're counting on a gain of 50 for the Q_5 stage, which is 3 times lower, the impact of our $f_1 = 5.7$ MHz pole is extended by $\sqrt{3}$, to about 10 MHz. So we're happy to see that the Q_6 source follower's second pole is more than fast enough for our 1 MHz goal.

An ancillary goal was maximizing the amplifier's loop gain at low-to-moderate frequencies, so we replaced R_7 with a current sink, see Figure 4x.146C. We set the current sink to 3 mA, so to achieve symmetrical slewing capability into C_2 the output from Q_1 needs to range from 0 to 6 mA.

This is a good point to analyze the C_2 scene. We'd like the output stage gain to be greater than 50 at 1 MHz. We started by checking distributor stock and choosing a

---

[103] If Q_7 were permitted a larger heatsink, Q_6's bias could be raised to 2 mA, doubling these bandwidth numbers.

[104] At lower supply voltages and/or with additional heat-sinking, this current can be as high as 10mA.

low-value high-voltage capacitor. The selection of stocked high-voltage capacitors, with values below 10 pF, isn't very good, but we were happy to settle on a 3.3 pF 1 kV part. Next we chose $R_6 = 415\,\Omega$, for which we estimate $G = 115$ at 1 MHz. But there's also Q_5's C_{rss}, which adds an additional 3 pF. When added to C_2, we get $G = 58$, which is still a fine result. We also added a 27k resistor, R_8, in series with C_2, to produce a stabilizing zero in the loop at 1.8 MHz.

The ± 3 mA current into the 6 pF $C_2 + C_{rss}$ summing-junction node should make the output slew at 500 V/μs. However, we saw above that the anemic 2 mA pull-up current into C_X limits the positive slew rate to 100 V/μs (though we could always match the negative slewing by increasing the current to 10 mA). To calculate the full-power bandwidth, use $S = \omega A$ to find the maximum frequency for which the amplifier can make a 1200 Vpp sine wave: $f_{max} = S/2\pi A = 26.5$ kHz. For an example of an amplifier with $f_{max} = 3$ MHz, see §2x.14.4.

Output crossover.

This may be a good place to discuss an aspect of this circuit that may be bothering some, the output-transistor deadband. Unlike some of our previous designs, we have chosen a super-simple output circuit, without any class-AB properties. Most of the time the Q_6 follower is operating in class-A mode, due to the Q_7 current sink. But there's a roughly 4 V difference between Q_5's drain voltage (to Q_6's gate) when sourcing output current, and its voltage when it is sinking output current via the BAT54. Rapid negative-going events have Q_5 taking over in less than 8 ns (4V/[500V/μs]), given the fast negative slewing. For positive-going events, from Q_5 negative control (unusual), slewing takes 40ns (4V/[100V/μs]). (Add a little extra time for slewing to get established.) At any rate, after slewing in either direction, class-A control of the Q_6 follower is quickly re-established. That's why we can get away with analyzing the Q_6 follower circuit as a class-A circuit.

Integrator windup.

A second consideration is "integrator windup." This is a condition where the U_{2a} error integrator creates excessive error signal while the output stage is slewing. As a result, there can be an output overshoot after the output reaches its intended voltage. You can see evidence of this in Figure 4x.145. We minimized this effect with a high output-stage gain, forcing a lower R_5, and a long C_1 time constant.

Output isolation.

Finally, we chose $460\,\Omega$ for R_{17}, to isolate the amplifier from the 120 pF capacitive load above 3 MHz. In a common split-feedback arrangement, R_{20} provides low-frequency (and dc) feedback while C_4 provides the isolated version of the output stage at high frequencies.

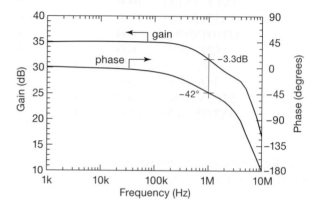

Figure 4x.148. Bode plot (gain and phase versus frequency) for the 1200 V output-stage when driving a 120 pF load. The gentle increase in phase shift beyond the -3 dB point is a desirable attribute for use within a servo loop.

Current-feedback configuration.

Another way to increase the output-stage gain at high frequencies is to add the $C_6 R_{24}$ network in Q_1's emitter. Instead we added another resistor (R_{26}, 22.7k) from the output back to Q_1's emitter, effectively converting the output stage into a current-feedback (CFB) amplifier, suitable for stand-alone testing of the output stage alone.[105] Figure 4x.148 shows the measured gain and phase with the amplifier set for a gain of 55.6. It has a measured gain of 56 from 3 to 100 kHz, down 3 dB at 905 kHz, where its phase shift is $-40°$. We're comfortable calling that a "1 MHz amplifier."

Wideband voltage divider.

We made other changes to complete a wideband version of the amplifier,[106] but we'll just mention one more notable issue. The $R_1 R_2$ divider, with U_{2b}, needs to have a 5 to 10 MHz bandwidth, but we don't want to make the high-voltage resistor R_2 too small. At first we struggled with U_{2b}'s input capacitance, substituting a separate op-amp with low input capacitance; and we were also forced

---

[105] We drove the CFB amplifier from the base of Q_1, with appropriate dc offset. We used a 2 W resistor, so we had to be careful to keep the average dc output voltage less than ± 200 V.

[106] To learn more about this project, inquire about the AMP-62B.

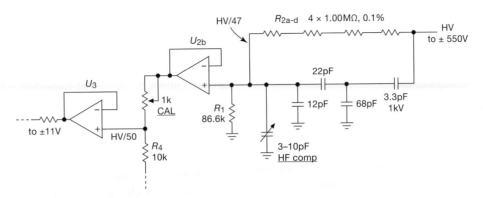

Figure 4x.149. Wide-bandwidth high-voltage divider.

to reduce the values of R_1 and R_2. We also struggled with oscillations due to the wide bandwidth of the U_{2b} op-amp, and the high RF impedance of its input node. Finally we realized that a multi-stage capacitive divider made more sense, see Figure 4x.149. This produces good wideband accuracy, lets us choose op-amps with higher input capacitance, and presents the op-amp with a low-impedance input node at high frequencies (15 to 20 pF).

Low-value capacitors tend to have poor accuracy, so the second stage of the divider includes a trimming capacitor, to set the capacitive gain to match that of the resistive divider. The first attenuation is 1/47, and the op-amp is budgeted for 3 pF of input capacitance. We added a 1k CAL trimpot (plenty of bandwidth) at U_{2b}'s output, to bring the total attenuation to a calibrated 1/50.[107]

---

[107] In practice, you adjust the HF COMP capacitor trimmer on the edges of a square wave (as you do with a scope probe), and adjust the CAL pot at dc.

4x.24 High-Voltage Bipolarity Current Source

If you want to know how to make an accurate voltage-controlled current source capable of wide compliance ($\pm 350\,\text{V}$), output currents to $\pm 50\,\text{mA}$, and with low output capacitance, you've come to the right place.

Top-level view

Op-amp-based current sources, like the Howland circuit and its relatives (see Figures 4.14 and 4.15) require the op-amp's output to slew to follow changes in load voltage. That makes them slow, as if they had a high apparent parallel output capacitance. By contrast, current sources whose output comes from the collector of a bipolar transistor are fast, and have intrinsically low capacitances, see §4.2.5. The challenge here is to create a circuit configuration that exploits this potential in a split supply configuration.

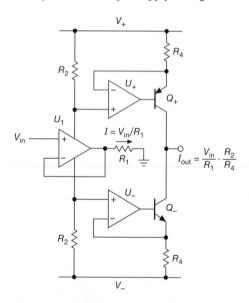

Figure 4x.150. Bipolarity current source block diagram.

Block diagram

The basic idea (Figure 4x.150) is to use a pair of accurate current mirrors (one at the positive rail, the other at the negative rail) to combine the unbalanced current produced by an op-amp transconductance input stage U_1. As we'll see presently, U_1 generates an unbalanced current on

its supply rails of $I = V_{\text{in}}/R_1$, which is mirrored (with gain $G_I = R_2/R_4$) by op-amps U_+ and U_- and associated transistors. Their output currents are combined to produce I_{out}. The quiescent current of U_1, gained up by G_I, forms the standing current of the output transistors – but these currents cancel and do not appear at the output. The resulting summed output current is therefore

$$I_{\text{out}} = \frac{R_2}{R_4}\frac{V_{\text{in}}}{R_1}$$

which is 5 mA/V for the resistor values in the final circuit.

This circuit is OK, as far as it goes. But it's limited to normal op-amp voltages. That limitation can be overcome by interposing a high-voltage cascode transistor between U_1's current outputs (its supply rails) and the rail-referenced current mirrors. First, though, an explanation of the weird input stage configuration.

The transconductance stage

The input stage creates an unbalanced current on its *supply rails*, which we exploit in an unusual arrangement first suggested by the legendary Bob Widlar,[108] see Figure 4x.151. An input voltage V_{in} forces a current V_{in}/R_1 through R_1 to ground; this must equal exactly the unbalanced supply currents, because the op-amp's input draws no current. (The op-amp's balanced quiescent current is also present at both supply pins, the supply pin having extra current depending on the polarity of V_{in}.) The op-amp's operating current cancels at the output, including any transient currents it may have as it slews, etc.

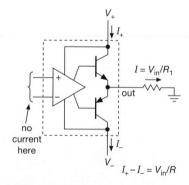

Figure 4x.151. The difference in currents at an op-amp's supply rails must equal its output current (apart from negligible input current). The dashed outline encloses the op-amp's innards, where the push–pull output is shown explicitly.

---

[108] Figure 30 in National Semiconductor Corporation's Application Note AN-29, available as document snoa624b at www.ti.com.

The whole enchilada

The full circuit is shown in Figure 4x.152. Input stage U_1 (an AD548) is a JFET op-amp with low quiescent current (0.2 mA max), to minimize the standing current in output transistors Q_5–Q_8; that current is $10I_{Q(U1)}$, thus a maximum of 2 mA. That would amount to a quiescent power dissipation of 0.35 W in each transistor, but the dissipation is less, as we'll see, owing to voltage-equalizing resistors R_6–R_9. Cascode transistors Q_1 and Q_2 level-shift the op-amp's rail currents to high voltages for the output stage; they also act as emitter followers, to provide low voltages to run the op-amp.

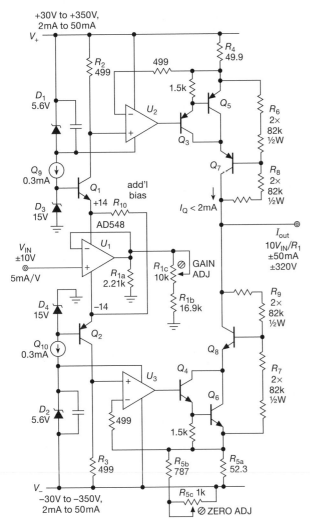

Figure 4x.152. Programmable high-voltage bipolarity current source.

The output stage consists of a push–pull pair of complementary op-amp-controlled current mirrors. It appears

complicated, but it's not. Looking at the top half, and ignoring Q_7 for the moment, Q_5 is the controlled current source, wired in a Darlington configuration with Q_3 for greatly reduced base current. These transistors are rated at 400 V, so we stack Q_7 atop Q_5's collector, with R_6 and R_8 dividing the drop equally (and you could stack additional transistors, as needed).

Here's something that may worry you: the equalizing resistors R_6, R_8 appear to shunt the current source's output, defeating the desired high impedance. But not to worry, the combination of transistors and equalizing resistors behaves like a 2-terminal device (apart from the insignificant base current), as is evident from Figure 4x.153. That is, its "current gain" in this approximation is $G_I = 1.00$, regardless of current flowing through R_6, R_8. It is only necessary that the latter resistances be large enough to ensure that the transistors remain in conduction.

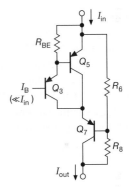

Figure 4x.153. Perhaps defying intuition, the voltage-equalizing resistors R_6 and R_8 do not degrade the high impedance of the current source, because the configuration behaves like a 2-terminal component (apart from base current).

4x.24.1 Performance issues

As with any circuit design, there are lots of devils in the details. Here are some things to think about.

Power dissipation

Good starting choices for the (high-voltage) transistors are MPSA42 and MPSA92 (or SMT-packaged MMBTA42 and MMBTA92) for Q_1–Q_4, and FZT458 and FZT558 (in 2 W SOT-223) for Q_5–Q_8. If the circuit is operated at lower voltages, the current-setting resistors R_4 and R_5 could be reduced to increase the maximum output current. More capable output transistors could be used for Q_5–Q_8, though likely at the expense of output capacitance.

Current errors

Several factors contribute to a small output-current off-set. The AD548B op-amp's offset voltage is 0.5 mV max, enough to create a 2.5 μA output offset. A 1% net error in the R_2/R_3 and R_4/R_5 resistor ratio matching could create an additional 20 μA output offset, and a 2% beta mismatch in Q_1 and Q_2 could add 40 μA of output offset. Zero-adjust pot R_{5c} lets you zero out potential offset errors. There are also sources of possible *gain* errors, due to resistor and beta mismatches; gain-adjust pot R_{1c} lets you trim those out.

Slew rate: current

The low power AD548 (0.2 mA) is relatively slow, 1.8 V/μs, which corresponds to an output-current slewing rate of 9 mA/μs. A faster op-amp would increase the standing quiescent current in the output-transistor stack, perhaps dictating the use of higher-power transistors.

Slew rate: voltage

Output-voltage slewing is closely related to the output-node capacitance. The suggested FZT458 and FZT558 for Q_5–Q_8 have less than 10 pF of capacitance. With 25 pF of load capacitance, voltage slewing could be as fast as $S = I/C = 300$ V/μs for a 10 mA output. Op-amps U_2 and U_3 must handle the C_{ob} dynamic currents of Q_3 and Q_4, but C_{ob} is less than 2 pF for the suggested transistors.

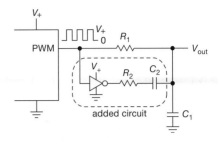

Figure 4x.154. PWM ripple canceler: adding an ac-coupled inverted pulse current to the traditional R_1C_1 greatly reduces the ripple amplitude. The resistors R_1 and R_2 should be of equal value.

4x.25 Ripple Reduction in PWM

From the ever-inventive Stephen Woodward comes this astonishingly effective (and remarkably simple!) trick to reduce greatly the residual ripple in a pulse-width-modulated DAC. Recall (from §§13.2.8 and 15.6) that you can use PWM as a simple DAC, smoothing the rectangular-wave output with an RC lowpass averager (or relying on the load's inherent averaging, as in a thermal regulator). But in choosing the averaging time constant you're caught in an unfavorable trade-off: a short time constant responds more quickly, but produces greater residual ripple at the switching frequency. As summarized by Woodward,[109] a traditional RC-filtered PWM produces a worst-case ripple amplitude (at 50% switching, i.e., mid-scale) of

$$v_{\text{pp–ripple}} \approx \frac{T_{\text{PWM}}}{4RC} V_{\text{fs}} \qquad (4\text{x}.8)$$

where V_{fs} is the full-scale output range, T_{PWM} is the switching period $(1/f)$, and the ripple is assumed small compared with V_{fs} (i.e., $RC \gg T_{\text{PWM}}$).

Now here's the rub: you'd probably choose RC large enough to keep the ripple down to an equivalent LSB of your converter. But a long time constant means a slow response. It's not hard to show that the time for an RC-filtered PWM to settle to within its ripple voltage of its final value is

$$t_{\text{settle}} = RC \log_e \frac{V_{\text{fs}}}{v_{\text{ripple}}} \approx \frac{T_{\text{PWM}}}{4} \frac{V_{\text{fs}}}{v_{\text{ripple}}} \log_e \frac{V_{\text{fs}}}{v_{\text{ripple}}}. \qquad (4\text{x}.9)$$

As Woodward helpfully points out, the settling time really gets long when you're trying for more than a few bits of resolution. He illustrates with a modest 8-bit example, where you'd choose RC so that the peak-to-peak ripple equals an LSB (i.e., $v_{\text{ripple}} = V_{\text{fs}}/256$), thus (via eq'n 4x.8) $RC = 64 T_{\text{pwm}}$. The resulting settling time (to an LSB) is then (via eq'n 4x.9) $t_{\text{settle}} = 64 T_{\text{PWM}} \log_e 256 = 355 T_{\text{PWM}}$. This is a *long time*: a PWM clocking at 10 kHz would take all of 35 milliseconds to settle (you can think of that as a meager 30 conversions/sec from a 10 kHz clock).

What can be done? Woodward to the rescue – kill the ripple by ac-coupling an inverted PWM to the smoothing capacitor (Fig. 4x.154). This is easily done: if you're using logic levels (the usual case), just use an inverting gate, as indicated.[110] With $R_1 = R_2$ and $C_1 = C_2$, this reduces the worst-case ripple and settling time to[111]

$$v_{\text{pp–ripple}} \approx \frac{1}{2} \left(\frac{T_{\text{PWM}}}{4RC} \right)^2 V_{\text{fs}} \qquad (4\text{x}.10)$$

$$t_{\text{settle}} \approx \frac{T_{\text{PWM}}}{2} \left(\frac{V_{\text{fs}}}{v_{\text{ripple}}} \right)^{\frac{1}{2}} \log_e \frac{V_{\text{fs}}}{v_{\text{ripple}}}. \qquad (4\text{x}.11)$$

Revisiting the 8-bit example, we get to reduce RC to just $2.8 T_{\text{PWM}}$ for an LSB ripple, with a resultant settling time $t_{\text{settle}} = 44 T_{\text{PWM}} = 4.4$ ms, or 8 times faster. In the general case of n-bit conversion, this trick reduces the settling time by a factor of $2^{\frac{n}{2}-1}$, assuming as above that the RC time constant is chosen for an LSB worst-case ripple.

Can it really be this good? To find out, we rigged up the circuit of Figure 4x.154 on the bench, and measured some waveforms. Figure 4x.155 compares the steady-state ripple with and without the canceler, for a half-scale output (i.e., a square wave, or 50% duty cycle). We chose RC values (267 kΩ, 0.24 μF) to produce a 1LSB ripple (80 mV for 5 V full-scale) for 6-bit conversion with the conventional circuit, then added the canceling circuit with the same RC values; the comparison is stunning.

To illustrate better the benefits of the canceler circuit, we ran the circuits with a shorter clock cycle (2 kHz) and an RC time constant (13 kΩ and 0.24 μF, thus ~3 ms) corresponding to about 1/6 of a clock cycle. The measured traces in Figure 4x.156 show the situation nicely: Compared with the traditional RC filter (trace A), the canceler has far lower ripple, but with slower response and settling time (trace B).

---

[109] S. Woodward, "Cancel PWM DAC ripple with analog subtraction," *EDN*, 28 Nov 2017.

[110] If, striving for greater precision, you're using a precision voltage reference and an SPDT analog switch, just use a second switch, with the inputs reversed, see Fig. 4x.158.

[111] These differ by factors of 1/2 and 2, respectively, from the corresponding equations in Woodward's article; they have been validated with SPICE simulations.

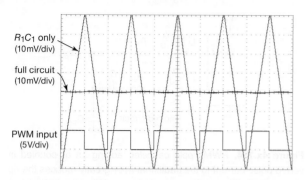

Figure 4x.155. Testing the PWM ripple canceler of Fig. 4x.154. The 250 Hz PWM square wave has 5 Vpp amplitude. Full-scale of the ac-coupled waveforms corresponds to 1LSB for 6-bit conversion. Horizontal: 2 ms/div.

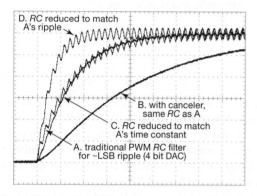

Figure 4x.156. Ripple canceler versus traditional *RC* filter, with 2 kHz 6 Vpp square wave input: A. simple *RC*, time constant of ~3 ms (6 PWM cycles). B. ripple canceler, same *RC* values. C. ripple canceler, time constant shortened to match speed of simple *RC* filter. D. ripple canceler, time constant shortened further to match ripple of simple *RC* filter. Vertical: 0.5 V/div; Horizontal: 2 ms/div.

By shortening the canceler's time constant by a factor of 3 ($R=4.4$k) the response (trace C) approximately matches that of trace A, but still with significantly reduced ripple. If instead you want to match the *ripple* of the traditional *RC* filter, you can shorten further the time constant of the canceler (trace D), to optimize settling time.

Finally, while happily seated at our favorite test bench, we again compared slewing/settling performance of the two circuits, this time selecting *RC* values that produced ripple amplitudes equal to 1LSB at a realistically high resolution (8 bits this time) for each configuration separately. This is what you would do, of course, to take full advantage of the circuit. We used a 1 kHz logic-level square wave ($V_s=4.8$ V), with $R=267$k and $C=283$ nF (traditional *RC* circuit) or 16 nF (canceler circuit). Figure 4x.157 shows

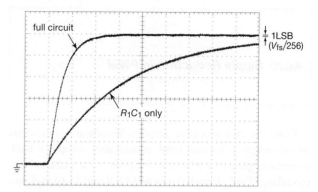

Figure 4x.157. Comparing slewing and settling for the ripple canceler, with *RC* values chosen for equal ripple (1LSB/8-bit). The input step is from zero to 50% full-scale. Vertical: 400 mV/div; Horizontal: 20 ms/div.

the measured waveforms, nicely displaying the $\sim 8\times$ faster settling when the canceler circuit is deployed.

The Rules

In the spirit of an engineer's rulebook, let's simplify these results for easy design of the Woodward circuit.

1. To match the *dynamics* (slew rate) of the traditional *RC* PWM filter, simply reduce the *RC* product by a factor of three. The resulting ripple will be reduced by a factor of $T_{\mathrm{PWM}}/2RC$.

2. To produce ripple equal to an LSB of an equivalent *n*-bit DAC, choose $RC=0.7\cdot 2^{\frac{n}{2}-2}\,T_{\mathrm{PWM}}$. The settling time to 1LSB will be approximately $t_{\mathrm{settle}}=0.35\,n\,2^{\frac{n}{2}}\,T_{\mathrm{PWM}}$, a reduction (relative to the traditional *RC* filter) by a factor of $2^{\frac{n}{2}-1}$.

Some comments

(1) Logic power supplies are notoriously inaccurate, so if you want to exploit the PWM's potential accuracy (it's digitally timed, after all), use a stable reference and a pair of SPDT analog switches,[112] as in Figure 4x.158.

(2) How sensitive is the cancellation to mismatch of resistors R_1 and R_2? To find out, we ran a SPICE simulation, an ideal tool for exploring this kind of question. Figure 4x.159 shows a blowup of the waveform as it approaches the final voltage, for matched resistors and for mismatches of $\pm 10\%$. The latter roughly double the ripple amplitude, whereas a simulation with $\pm 2\%$ mismatched resistors showed negligible increase in ripple. The bottom line: standard 1% resistors are adequate for this circuit.

---

[112] Or you can use a single DPDT analog switch, for example the TS5A23157.

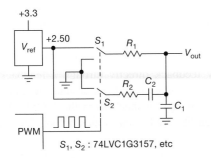

Figure 4x.158. Use a stable reference and analog switches for better precision and stability in a PWM DAC. With this scheme no logic inverter is needed.

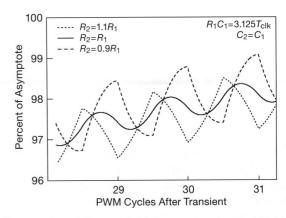

Figure 4x.159. This SPICE simulation shows the effect of a 10% mismatch in R_1 and R_2; for adequate performance use 1% resistors.

(3) How about capacitor mismatch? Further simulations revealed that the performance is insensitive to capacitor matching; the ripple was unchanged with a mismatch of $\pm 5\%$. You can think of C_2 as an ac coupling capacitor, with R_2 providing the well-matched pulses of current. Although the circuit works well enough with significantly mismatched capacitors, it seems best to choose $C_2 = C_1$, with 10% tolerance capacitors altogether adequate.

(4) But hey, why bother with all this when you can get inexpensive and tiny DACs to do the same job? For example, the Microchip MCP4921 is a 12-bit DAC with SPI input that costs just $2 in unit quantities, and it comes in DIP, SOIC, MSOP, and DFN packages. No need for any R's and C's, and in MSOP-8 it takes up less board space than the inverter and RC pairs needed for Woodward's circuit.

(5) We can answer that! Sometimes you've got a microcontroller with PWM and $\overline{\text{PWM}}$ outputs already available, so by adding an extra R and C you get significantly improved performance over the conventional RC filter.

(6) And, as long as we're interested in PWM as a way to generate clean analog outputs, we cannot resist sending you to the remarkable LTC2645 family of quad PWM-to-voltage-output DACs: they come in 8-, 10-, and 12-bit versions, and they feature *zero* latency, because they use the timing of successive edges to set the output voltage. To quote from the datasheet, "The DAC outputs update and settle to 12-bit accuracy within $8\mu s$ typically, ... eliminating voltage ripple and replacing slow analog filters and buffer amplifiers." What's not to like about these parts, which include a precision 10 ppm/°C internal reference and rail-to-rail output buffers, are guaranteed monotonic, and can source and sink 5 mA when operated from 3 V (or 10 mA from a 5 V supply)?

For additional amusements on PWM, be sure to check out §9x.5.

4x.26 Nodal Loop Analysis: MOSFET Current Source

We favor a largely non-mathematical approach to circuit design; as we said back in 1980's first edition "the treatment is largely nonmathematical, with strong encouragement of circuit brainstorming and mental (or, at most, back-of-the-envelope) calculation of circuit values and performance." But sometimes, when you have a thorny circuit problem, you just need to run the mathematical machinery. This can be done in several ways: a contemporary favorite technique is to create a SPICE model, and let a computer do the heavy lifting, as illustrated in the exploration of BJT amplifier distortion (§2x.4). But SPICE has its limitations, for example poorly modeling MOSFETs at low currents (see Fig. 3x.36), or poorly modeling capacitively loaded op-amps at high frequencies.

Classically trained circuit designers (and those who like mathematical models) favor a second technique, *nodal analysis*, in which circuit components (both passive and active) are modeled by their transfer characteristics. This loop analysis is analogous to the process of analyzing a multiply-connected dc circuit (by going around loops, applying Kirchhoff's voltage and current laws), but with the complication that dc voltages and currents are replaced by small-signal voltages and currents, resistances are replaced by complex impedances, and active components are modeled by their terminal impedances and transfer functions.

As we'll see below, you wind up with a set of linear equations, after which some algebraic manipulations get you to measures of performance like gain, bandwidth, phase margin, and the like. We saw an example of this earlier, in the treatment of current-source noise (Fig. 8.21) in the main volume, and in Chapter 3x's treatment of the capacitively loaded source follower (Figs. 3x.72 and 3x.74). Here we illustrate its use in a hybrid design – the op-amp-controlled MOSFET current source – a circuit in which both SPICE limitations above are in play.

4x.26.1 Example: MOSFET current source

Let's say we want to make a fast MOSFET-based current sink, with op-amp control to ensure accuracy. The problem that we worry about is the MOSFET's large gate capaci-

tance (which can be as high as 5000 pF), a capacitive load that drives many op-amps to instability. The usual solution, (Fig. 4x.160) is to isolate the capacitive load with a series resistor (R_3), and then split the feedback path to route high frequency feedback directly through a feedback capacitor (C_2), thus doing an end-run around the lagging-phase (stability-robbing) feedback from the MOSFET's source. We've seen this trick many times before, for example in

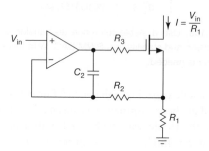

Figure 4x.160. Conventional MOSFET current source with op-amp control.

Figure 4.105 (phase mangled by transformer-coupled feedback), Figures 9.2 and 13.38 (capacitively loaded linear voltage regulator), Figure 4.76 (op-amp rail splitter), or Figure 3x.114 (high-Z, high-voltage pulse monitor).

Our usual approach has been to choose component values based rationally on experience: current-setting resistor R_1 is chosen for a volt or so at maximum current; load-isolating R_3 is ordinarily a few hundred ohms or so; R_2 is chosen much larger than R_1 (for negligible diversion of current); and C_2 is chosen so the time constant R_2C_2 of the "slow" feedback (i.e., through R_2) is much longer than the op-amp's inverse bandwidth (that is, $R_2C_2f_T \gg 1$).

This approach works well enough; but we may wish to know more. Dealing with op-amp feedback-loop performance in the presence of capacitive loading is a well-known problem. Most of the analysis we've seen concentrates on the extra pole created by capacitive load, but we'd like a more comprehensive analytic solution. Some designers turn to SPICE to analyze the circuit, but be aware that the MOSFET's SPICE model will likely be seriously defective at low currents (see a typical example in Fig. 3x.36, where the drain current plummets to zero at $V_{GS} = 4.2$ V), with possible problems also in the op-amp's capacitive-load model at high frequencies.[113]

---

[113] If you do wish to run SPICE on a circuit with a power MOSFET, be sure to create a vetted subthreshold MOSFET model, as we describe in §3x.5.5. Even if you are operating mostly at high currents, the ON/OFF transitions will go through the subthreshold region. The analytic approach we describe here must be evaluated at both the high- and low-

A. Nodal model

The first step is to draw (by hand!) a model circuit (Fig. 4x.161) showing components and nodes.[114] We've replaced the schematic's MOSFET with its operating elements: the gate–source capacitance C_{iss} (here labeled C_1), and its transconductance g_m, neither of which are shown in the earlier schematic.

Figure 4x.161. Hand-drawn nodal-analysis circuit for the MOSFET current source.

B. KCL equations

Next we write loop equations (as with a dc circuit), using small-signal voltages and currents (v_n and i_n), and Laplace s-plane notation (see §1x.5), $s = j\omega$. These will let us see node voltages, currents, and phase as functions of frequency. To write a KCL equation for a node, we set the sum of currents leaving the node to zero; doing this for v_g and v_s gives:

$$v_g: \quad \frac{v_g - v_3}{R_3} + (v_g - v_s)sC_1 = 0,$$

$$v_s: \quad \frac{v_s}{R_1} - (v_g - v_s)sC_1 - g_m(v_g - v_s) = 0.$$

We can't evaluate v_3 with KCL, because we don't know the op-amp's output current; but we already know the equation for the op-amp stage, where the output v_3 is given by the R_2C_2 integration of the input error. It's a non-inverting stage, so it includes an extra v_1 term:

$$v_3 = v_1 + (v_1 - v_s)\frac{1}{s\tau_2}.$$

We assume the op-amp's gain–bandwidth product GBW is much greater than the signal bandwidth $1/2\pi\tau_2$ (where $\tau_2 = R_2C_2$), justifying the simple op-amp equation for v_3. We also welcome the lower op-amp output impedance Z_{out}, its open-loop R_{out} being reduced by the ratio of the two bandwidths to be small compared with R_3.

C. Node equations

We have a nodal equation for each of the three unknown voltages. In principle this allows us to calculate important aspects of the circuit, such as the gain to the R_1 current-sense resistor v_s/v_1, the overall current–source gain i_d/v_1, and the impedance Z_g of the node v_g.

When doing the math to find solutions to a set of simultaneous equations,[115] it's easy to get bogged down in a mass of expanding expressions, so we were happy to turn the task over to Alan Stern, the Institute's mathematician; he came up with a compact set of equations[116] for v_s, v_3, and v_1, further simplified by defining intermediate variables

$$\alpha = 1 + sR_1C_1 + g_mR_1, \quad \beta = sR_3C_1$$

to get

$$v_s = v_g\frac{\alpha - 1}{\alpha},$$

$$v_3 = v_g\frac{\alpha + \beta}{\alpha}, \qquad \text{and}$$

$$v_1 = \frac{v_g}{\alpha}\left(\alpha - 1 + \frac{(1+\beta)s\tau_2}{1 + s\tau_2}\right).$$

---

current operating extremes of your circuit; but once the two sets of formulas are in hand, you are in a good position to see what the trade-offs are, and optimize the circuit. If you want a quick look at frequency- and transient-response curves for the nodal-analysis circuit, you can make a SPICE model of Fig. 4x.161 with trial component values; then make a family of curves, stepping Q_1's g_m values corresponding to different output currents.

[114] We made some simplifications. The op-amp's second pole is omitted, as discussed later. The MOSFET's feedback capacitance C_{rss} is omitted, for two reasons: first, we assume the output load for the current source has a low ac impedance (e.g., a piezo actuator), but, for high-Z loads, Miller capacitance effects can be large; and second, because the MOSFETs we chose for designs like this have a very low feedback capacitance, 25 to 100 times lower than the input capacitance C_{iss} (e.g., 7.5 pF versus 550 pF for our high-current example part and 3.2 pF versus 270 pF for our low-current example). If either of these conditions fails to hold, you may need to include C_{rss} in the analysis circuit. We've also assumed that the MOSFET's transconductance does not much depend on V_{DS}, which is generally the case for high-voltage MOSFETs operated above about 15 V.

[115] The node equations can be organized in a matrix form, for which standard tools are available to find numerical solutions. If an analytical solution is desired, and the matrix is simple enough, you can manually invert the matrix. Or you can use a tool like Mathematica.

[116] Alan started by separating the v_g and v_s terms in the first equation, and solving for v_s as a function of v_g and v_s. Similarly he solved the second equation for v_s as a function of v_g. Substituting these into the third equation gave three voltages (v_s, v_3, and v_1), all as functions of v_g, making it straightforward to derive various circuit parameters such as $G = v_s/v_1$ and $Z_g = v_g/i_3$.

D. Results

First we'll get Z_g, the impedance looking into the MOS-FET's gate node v_g; this determines the gate load on the op-amp plus its resistor R_3. For this we need $i_3 = (v_3 - v_g)/R_3$, giving

$$Z_g = \frac{v_g}{i_3} = R_3 \frac{\alpha}{\beta}.$$

Substituting α and β, we find the gate impedance to be $Z_g = R_1(1 + g_m/sC_1) + 1/sC_1$. Recognizing that the MOS-FET's $f_T = g_m/2\pi C_{iss}$ (see eq'n 3x.6), we can rewrite the equation in terms of f_T and the signal test frequency f:

$$Z_g = R_1 \left(1 - \frac{jf_T}{f}\right) + \frac{1}{sC_1}. \tag{4x.12}$$

From eq'n 4x.12 we see that the MOSFET isolates the gate node from R_1 at low frequencies, by the ratio f_T/f. This can be a huge benefit when dealing with a small R_1, given the high values we may see for f_T (hundreds of mega-hertz, or even over 1 GHz for our example part below at 2.5 A). Eq'n 4x.12 also tells us that Z_g effectively looks capacitive, roughly $C_{eff} = C_{iss}/g_m R_1$ for $g_m R_1 \gg 1$. This in-sight helps us choose the value for R_3, and we can calculate $f_p = 1/2\pi R_3 C_{eff}$ to see the location of the additional pole. At high frequencies, or low $g_m R_1$, the bootstrap action be-comes ineffective, and hence Z_g reduces to R_1 and C_{iss} in series, so we may be forced to add a buffer to the op-amp (similar to that in Figure 3x.129).

The MOSFET's **source-follower gain into a resistive load** (eq'n 4x.13 below) is interesting (see eq'n 3x.10 for gain into a *capacitive* load). This is the gate-to-source voltage gain, which we'll call G_s. At low frequencies the equation reduces to $G_s = g_m R_1/(1 + g_m R_1)$. The gain-degradation can be significant if $g_m R_1$ is small enough. For example, in our 2.5 A current source example below, with $R_1 = 2\,\Omega$ and with the MOSFET operating at 50 mA with $g_m = 600\,\text{mS}$, the dc gain is only 0.65, and the gate looks like $C_{iss} = 550\,\text{pF}$ in series with $2\,\Omega$, a real problem for the op-amp and R_3. But if we could use $R_1 = 100\,\Omega$ for opera-tion at 50 mA, then C_{eff} would be reduced to just 10 pF:

$$G_s = \frac{v_s}{v_g} = \frac{\alpha - 1}{\alpha} = \frac{1 + jf/f_T}{1 + jf/f_T + 1/(g_m R_1)}. \tag{4x.13}$$

Voltage gain to sense resistor.

Next we can find the gain v_s/v_1 from the input to the current-setting resistor R_1 (with its current i_1). We know $v_s = i_1 R_1 = (i_d + i_3)R_1$ (ignoring R_2's small contribution), which is the signal presented to the op-amp feedback loop (there's a direct correspondence between i_1 and i_d, as we'll see presently). Combining this equation with the MOS-FET's f_T, we get

$$G_v = \frac{v_s}{v_1} = \frac{\alpha - 1}{\alpha - 1 + \frac{(1 + \beta)s\tau_2}{1 + s\tau_2}}. \tag{4x.14}$$

We derived a number of fully-expanded and then simpli-fied versions of this equation, each with various features. Here's one of them without any omissions:

$$G_v = \frac{1}{1 + \frac{s\tau_2}{(1 + s\tau_2)} \left(\frac{1 + s\tau_3 + s\tau_5}{g_m R_1 + s\tau_1} + s\tau_5 \right)}, \tag{4x.15}$$

where $\tau_1 = R_1 C_1$, $\tau_3 = R_3 C_1$, and $\tau_5 = R_3 C_{rss}$, giving us the effect of Q_1's feedback capacitance, assuming that the MOSFET's drain is at ac ground (otherwise boost C_{rss} by a Miller-effect ratio). The τ_5 term can be ignored if $C_{rss} \ll C_{eff}$, where $C_{eff} = C_{iss}/g_m R_1$. Omitting τ_5, eq'n 4x.15 can be manipulated to take the form

$$G_v = \frac{1}{1 + \frac{(1/s\tau_1) + (R_3/R_1)}{(1 - jf_c/f)(1 - jf_T/f)}}. \tag{4x.16}$$

If $f_c = 1/2\pi R_2 C_2$ is low enough, the equation predicts a rolloff well above f_c, at $f > f_T R_1/R_3$. Keep in mind, though, that f_T degrades rapidly at low currents, dramat-ically slowing the current-source response. Recall that f_c is the op-amp's error-correction bandwidth, and for signal frequencies f well below f_c that term drops out. At those frequencies what matters is the ratio f/f_T compared with R_1/R_3 in the denominator, which determines if G_v is atten-uated. At low operating currents, f_T will be dramatically reduced, and the overall response dropoff can fall far below f_c and f_T, with an $(f/f_T)(R_3/R_1)$ term in the denominator.

With some more math[117] you can evaluate whether the G_v frequency-response equation will develop some peak-ing before rolling off. It will, if this expression is less than zero:

$$\frac{2\tau_1 + \tau_2}{2g_m R_1} + \tau_2 + \tau_4 - \tau_3,$$

where $\tau_4 = R_4 C_2$ (the R_4 is in series with C_2, and is in-troduced below). It's easy for this to happen if R_3 is too large, or if f_c is too high (i.e., $R_2 C_2$ is too small). For pro-grammable designs you'll want to do the test at both high currents, and also at low MOSFET currents where g_m is smallest. At low currents you're more likely to see a greatly slowed response, rather than response peaking.

---

[117] Write the full equation for the magnitude of G_v, then take its derivative with respect to frequency.

Current-source gain.

Ultimately we want to know the output current as a function of the input voltage, i.e., $G_i = i_d/v_1$. We can get an expression for G_i by noting that $i_d = i_1 - i_3 = v_s/R_1 - i_3$ (ignoring R_2's small current), and that $i_3 = (v_3 - v_g)/R_3$. It gets seriously messy when this is all assembled; but it turns out that G_i is simply related to G_v: $G_i = G_v g_m/(\alpha - 1)$, which can be further manipulated into

$$G_i = \frac{G_v}{R_1}\frac{1}{1 + jf/f_T}. \tag{4x.17}$$

This tells us the MOSFET routes most of the R_1 current to its drain, at frequencies up to its f_T. Normally f_T will be much greater than the bandwidth f_c of the op-amp stage, and indeed of the entire MOSFET circuit, so $G_i \approx G_v/R_1$. This also means the R_1 node voltage v_s is a useful proxy for the output current i_d, and we can take scope measurements directly on R_1.

4x.26.2 Example: fast 2.5 A pulsed current

In §3x.22 we described a pulsed current source used to drive a piezo actuator. For that task we used the circuit of Figure 3x.134, where a MOSFET gate driver (§3x.14) with a trimmed drive-voltage supply was used to drive the MOSFET to just the right level to produce 2.5 A pulses through R_1. We had considered using an op-amp circuit like Figure 4x.160, but the (admittedly kludgey!) gate-driver approach seemed to be the best way to get < 30 ns switching speeds. But now, with our new-found (and hard-fought!) wisdom, we can use the equations derived above to evaluate and optimize the op-amp approach. We rigged up the test circuit of Figure 4x.162, with a beefy FDP7N60NZ MOSFET (with a substantial C_{iss} of 550 pF), which, when run at 2.5 A, has a transconductance of 6.4 S and an f_T of 1850 MHz. Keeping the same $2\,\Omega$ value for R_1[118] that we used for the earlier approach of Figure 3x.134, eq'n 4x.12 predicts $Z_g = -j150\,\Omega$ at 25 MHz, equivalent to a modest 43 pF capacitance to ground. If we choose $R_3 = 50\,\Omega$, the $R_3 C_{eff}$ pole frequency is $f_p = 75$ MHz, and we should have no problem driving this gate load... things are looking very promising.

We picked a modest op-amp $f_c = 10$ MHz rolloff for our calculations; with the component values above, G_v in gain eq'n 4x.15 has identical terms in the numerator and denominator, plus an additional f/f_T term, which predicts operation of the full circuit to beyond 45 MHz. This is well beyond the op-amp's 10 MHz rolloff, where the op-amp simply becomes a voltage follower; we're then depending on

the MOSFET driving R_1, without further help from the op-amp's feedback loop.[119] The deadline for final book submission was now only hours away(!), and we were ready for bench measurements.

We used the circuit of Figure 4x.162 to produce 200 ns 1.2 A pulses at 100 kHz into a $2\,\Omega$ load returned to +50 V. For these tests we pulsed from a 25 mA base current to 1.2 A (fast pulse-startup from zero will require further development). We were pleased to see a fast 25 ns risetime, with minimal overshoot (Fig. 4x.163). Sometimes it's helpful to add a zero in the op-amp feedback path, as here with added 750 Ω resistor R_4 in series with C_2. But we have to confess that Q_1 failed to be as fast as we'd calculated; and the predicted speeds above about 20 MHz failed to materialize.

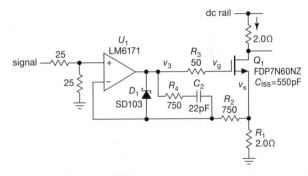

Figure 4x.162. High-voltage 2.5 A programmable current source.

Running a high-current circuit at low current.

In our tests above we set the zero-current level to 25 mA, or 1% of the circuit's design level. At that current the MOSFET is operating in the subthreshold region, where its transconductance could drop from 6400 mS to as low as 300 mS (see, e.g., Fig. 3x.38), and where f_T drops to just 87 MHz and $g_m R_1$ is just 0.6. That's too low to apply eq'n 4x.12 for Z_g and C_{eff} calculations, so the gate will look simply like C_{iss} in series with R_3. This means the V_g node is working with an $R_3 C_{iss} = 28$ ns time constant. That's not as bad as we feared (we'd prefer not having to increase

---

[118] See §1x.2 for details about low-inductance sense resistors.

[119] This was confirmed by SPICE simulating the nodal circuit of Fig. 4x.161, and by GNUPLOT running eq'n 4x.14. For a dose of reality, we added in the effects of the MOSFET's (previously neglected) feedback capacitance of 45 pF (its value at $V_{DS} = 10$ V), whereupon a SPICE simulation showed operation was limited to 25 MHz. This is overly conservative, however, since the MOSFET's feedback capacitance drops to just 7 pF above 25 V. For these simulations we tied C_{rss} from node v_g to ground, given that the MOSFET's drain is clamped by the highly-capacitive piezo load.

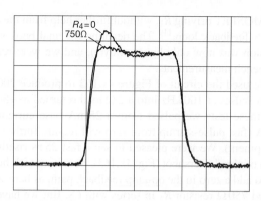

Figure 4x.163. Measured pulse response of the current source of Fig. 4x.162, showing 25 ns rise and fall times, and the damping effect of R_4. Horizontal: 50 ns/div; Vertical: 250 mA/div.

R_1 above $2\,\Omega$). But paradoxically it's sometimes harder to run one of these circuits at low currents rather than at high currents! However, we found we could run the circuit of Figure 4x.162 at 50 mA and still get 25 ns risetimes, but only if we increased R_1 to $50\,\Omega$ (i.e., 100 mA full-scale).

Low-current, programmable over a wide range.

There are special considerations for fast current sources that are programmable over a wide range of operating currents. Take for example a typical small power MOSFET, the IXTA06N120, chosen from Table 3x.3. We'll run it at 10 mA,[120] where its transconductance will be on the order of 12% that of a BJT, or about 50 mS, see Figures 3x.38–3x.40. The '06N120 has $C_{iss} = 270\,\text{pF}$, and its $f_T = g_m/2\pi C_{iss} = 30\,\text{MHz}$ at 10 mA. The gate impedance Z_g is high at low frequencies (eq'n 4x.12), say $f \ll f_T/30$. Let's take $R_1 = 500\,\Omega$. Then at 5 MHz the R_1 term equals $j2550\,\Omega$, which looks like a 12.5 pF capacitance. Very good – the MOSFET's gain, in collaboration with R_1, has reduced its huge 270 pF input capacitance to about 12 pF. The MOSFET's C_{rss} is still relatively insignificant – 3 pF at 25 V. So, after adding R_3, we see the op-amp is not significantly loaded. The 9 pF total load capacitance forms a pole with R_3, so the latter must not be made too large; for $R_3 = 470\,\Omega$, for example, the pole f_p will be at 37 MHz.

The MOSFET's transconductance varies with current, so with a voltage-programmable current source it's necessary to evaluate two sets of formulas at the extremes of its operating range. At currents above 10 to 100 mA the transconductance trends toward $g_m \propto \sqrt{I_d}$, but lower down (in the subthreshold region) g_m is proportional to drain current I_d; see the graph in Figure 3x.40 of measured

values for typical high-voltage devices. You may need to make some transconductance measurements on the particular part you're using.

Now look at the situation at the low-current end, say $10\,\mu\text{A}$. Here $g_m = I_d/nV_T = 0.1\,\text{mS}$, so the product $g_m R_1$ is less than unity. That means the MOSFET is failing to bootstrap its C_{iss}, and Z_g looks like 270 pF in series with R_1. Happily, the $500\,\Omega$ value of R_1 is a load the op-amp can handle, but the series gate resistor R_3's value of $470\,\Omega$ is a bit too large – a better choice would be more like $220\,\Omega$ or even $100\,\Omega$. Operating at $10\,\mu\text{A}$, the current-source response degrades above 50 kHz, dropping 3 dB at 5 MHz for $R_3 = 220\,\Omega$ (but staying better than $-2\,\text{dB}$ with R_3 reduced to $100\,\Omega$).

Simplify with low-cost op-amps.

We ignored the op-amp's second pole above, by choosing one with a GBW higher than our cutoff frequency, f_c, set by $R_2 C_2$. But you can do an analysis with our equations for a bare-bones circuit with a low-cost (low-performance) op-amp like an LM358 or LM321. You'd omit R_2 and C_2 from the circuit, so the cutoff frequency f_c is simply the op-amp's GBW, and $\tau_2 = 1/(2\pi \cdot \text{GBW})$. Then use eq'n 4x.12, and make sure $R_3 + Z_g$ is much larger than the op-amp's open-loop output impedance (the latter typically a few hundred ohms). If Z_g is high enough well past the GBW frequency, you may be able also to eliminate R_3. Simple circuits like this are common with slow op-amps, small MOSFETs, and large R_1 values.

---

[120] Values that might be used to make a version of Fig. 3x.136.

ADVANCED TOPICS IN POWER CONTROL

CHAPTER **9x**

In this chapter supplement we treat a number of more advanced topics dealing with power: ac–dc and dc–dc converters, power switching, and the like. In keeping with the objective of these "x-chapters," we have intentionally kept the discussions terse. The topics are presented in no particular order, and with no attempt to connect them logically. Think of them as a collection of short stories.

Among the collection of diverse topics, you'll find

(a) reverse polarity protection;
(b) lithium-ion cells, monitoring, and protection;
(c) PWM for dc motors;
(d) power-factor correction;
(e) high-side sensing;
(f) ultra-isolated dc–dc converter;
(g) low-power non-isolated ac–dc converters;
(h) the bus converter;
(i) negative regulated supplies;
(j) thermometry; and
(k) transient voltage protection and transient thermal response.

Review of Chapter 9 of AoE3

To bring the reader up to speed, we start this chapter with the end-of-chapter review from AoE3's Chapter 9:

A. Voltage Regulator Taxonomy.

Voltage regulators provide the stable dc voltages needed to power all manner of electronic circuits. The simplest (and least noisy) type is the *linear* regulator (Figure 9.2), in which the output error signal, suitably amplified and compensated, is used to control a linear "pass transistor" (BJT or MOSFET) that is in series with a higher (and perhaps unregulated) dc input voltage. Linear regulators are not power efficient, with dissipation $P_{diss} = I_{out}(V_{in} - V_{out})$, and they are not able to a produce dc output that is larger than the input, nor a dc output of reversed polarity.

The *switching* regulator (or switching *converter, switch-mode power supply*, SMPS, or just "switcher," §9.6) addresses these shortcomings, at the cost of some induced switching noise and greater complexity. Most switching power supplies use one or more inductors (or transformers), and one or more saturated switches (usually MOSFETs) operating at relatively high switching frequencies (50 kHz–5 MHz), to convert a dc input (which may be unregulated) to one or more stable and regulated output voltages; the latter can be lower or higher than the input voltage, or they can be of opposite polarity. The inductor(s) store and then transfer energy, in discrete switching cycles, from the input to the output, with the switch(es) controlling the conduction paths; with ideal components there would be no dissipation, and the conversion would be 100% efficient. The output error signal, suitably amplified and compensated, is used to vary either the pulse width ("PWM") or the pulse frequency ("PFM"). Switching converters can be *non-isolated* (i.e., input and output sharing a common ground, Figure 9.61), or *isolated* (e.g., when powered from the ac powerline, Figure 9.73); for each class there are dozens of topologies, see ¶ D below.

A minor subclass of switching converter is the *inductorless converter* (or "charge-pump" converter; see §9.6.3), where a combination of several switches and one or more "flying" capacitors is used to create a dc output that can be a multiple of the dc input, or an output of opposite polarity (or a combination of both). For many of these, the output(s)

track the dc input (i.e., unregulated), but there are also variants that regulate the output by controlling the switching cycle. See Table 9.4 and Figures 9.56 and 9.57.

¶B. The DC Input.

Regardless of the kind of converter or regulator circuit, you need to provide some form of dc input. This may be poorly regulated, as from a battery (portable equipment) or from rectified ac (line-powered equipment, Figures 9.25 and 9.48); or it may be an existing stable dc voltage already present within a circuit (e.g., Figure 9.64). For a line-powered instrument that uses a linear regulator, the "unregulated" dc input (with some ac ripple) consists of a transformer (for both galvanic isolation and voltage transformation) plus rectifier (for conversion to dc) plus bulk storage capacitor(s) (to smooth the ripple from the rectified ac). By contrast, in a line-powered (confusingly called "off-line") switcher the powerline transformer is omitted, because a transformer in the isolated switcher circuit provides galvanic isolation and it is far smaller and lighter since it operates at the much higher switching frequency; see Figure 9.48.

A diode bridge and storage capacitor converts an ac input to full-wave unregulated dc, whether from a powerline transformer or directly from the ac line. Ignoring winding resistance and inductances, the dc output voltage is approximately $V_{dc} = 1.4V_{rms} - 2V_{diode}$, and the peak-to-peak ripple voltage is approximately[1] $\Delta V_{ripple(pp)} \approx I_{load}/2fC$, where C is the capacitance of the output dc storage capacitor and f is the ac input frequency (60 Hz or 50 Hz, depending on geographic and political boundaries). The ac input current is confined to relatively short pulses during the part of the waveform leading up to the positive and negative peaks (see Figures 9.51 and 9.78). This low "power factor" waveform is undesirable, because it produces excessive I^2R heating and more stressful peak currents. For this reason all but the smallest switching converters use a power-factor correction (PFC) input stage (Figure 9.77) to spread out the current waveform and thus create an input current approximately proportional to the instantaneous ac input voltage.

Line-powered instruments need a few additional components, both for safety and convenience. These include a fuse, switch, and optional line filter and transient suppressor; these are often combined in an IEC "power entry module," see Figure 9.49.

---

[1] From $I = C\,dV/dt$, assuming approximately constant discharge current I.

¶C. Linear Voltage Regulators.

The basic linear voltage regulator compares a sample of the dc output voltage with an internal voltage reference (see ¶ G, below) in an *error amplifier* that provides negative feedback to a *pass transistor*; see Figure 9.2. The dc output voltage can be greater or less than the reference (Figures 9.4 and 9.5), but it is always less than the dc input. You can think of this as a feedback power amplifier, which is prone to instability with capacitive loads, thus the compensation capacitor C_c in Figure 9.2D,E. The final circuit in that figure shows a current limiting circuit (R_{CL} and Q_2), and also an *overvoltage crowbar* (D_1 and Q_3, see ¶ J below) to protect the load in the event of a regulator fault.

All the components of the original 723-type linear voltage regulator can be integrated onto a single IC (Figure 9.6), forming a "3-terminal" fixed regulator, e.g., the classic 78xx-style (where "xx" is its output voltage). These require only external bypass capacitors (Figure 9.8) to make a complete regulator. There are only a few standard voltages available, however (e.g., +3.3 V, +5 V, +15 V); so a popular variant is the 3-terminal *adjustable* regulator (e.g., the classic 317-type, see Figure 9.9), which lets you adjust the output voltage with an external resistive divider (Figure 9.10). Figures 9.14, 9.16, and 9.18 show some application hints for this very flexible regulator. Both fixed and adjustable 3-terminal regulators are also available in negative polarities (79xx and LM337, respectively), as well as in low-current variants (78Lxx and LM317L, respectively).

One drawback of these classic linear regulators is their need for an input voltage that is at least ~2 V greater than the output (its *dropout voltage*); that is needed because their pass transistor operates as an emitter follower (thus at least a V_{BE} drop) and the current-limit circuit can drop up to another V_{BE}. Two volts may not sound like a lot, but it looms large in a low-voltage regulator circuit, e.g., one with a +2.5 V dc output. To circumvent this problem, you can use a *low-dropout* (LDO) regulator, in which the pass transistor (BJT or MOSFET) is configured as a common-emitter (or common-source) amplifier, see Figure 9.20; the resulting dropout voltages are down in the tenths of a volt (Figure 9.24). LDOs are nice, but they cost more, and they are more prone to instability because their high output impedance (collector or drain) causes a lagging phase shift into the substantial load capacitance. LDOs may require significant minimum output bypass capacitance (as much as $10\,\mu F$ or $47\,\mu F$), often constrained with both a minimum and maximum equivalent series resistance (ESR), e.g., $0.1\,\Omega$ min, $1\,\Omega$ max; see Table 9.3.

¶D. Switching Converter Topologies.

The basic *non-isolated* switcher topologies are the *buck* (or "step-down"), the *boost* (or "step-up"), and the *invert* (or "inverting buck–boost"); see §9.6.4 and Figure 9.61. The power train of these each uses one inductor, one switch, and one diode (or a second switch acting as an active rectifier), in addition to input and output storage capacitors. A complete converter requires additional components: an oscillator, comparator, error amplifier, drive circuits, and provisions for compensation and fault protection; see for example Figure 9.65. As with linear regulators, the semiconductor manufacturers have stepped in to provide most of the necessary components as packaged ICs, see Tables 9.5a,b and 9.6.

For the buck converter $V_{out} < V_{in}$, and for the boost converter $V_{out} > V_{in}$. The inverting converter produces an output of opposite polarity to the input, and whose voltage magnitude can be larger or smaller than the input voltage (this is true also of the remarkable Ćuk converter, §9.6.8H). The respective dc output voltages are $V_{out(buck)}=DV_{in}$, $V_{out(boost)}=V_{in}/(1-D)$, and $V_{out(invert)}=-V_{in}D/(1-D)$, where D is the switch-ON duty cycle $D=t_{on}/T$. There are also *non*-inverting buck-boost topologies that permit the output voltage range to bracket the input (i.e., able to go above or below the input). Examples are the 2-switch buck–boost (two switches, two diodes, one inductor), and the SEPIC (one switch, one diode, two inductors), see Figure 9.70. Of course, a switching converter with a transformer (whether isolated or not) provides flexibility in output polarity, as well as improved performance for large ratio voltage conversion.

Isolated switching converters use a transformer (for isolation), in addition to one or more inductors (for energy storage), see Figure 9.73. In the *flyback* converter (Figure 9.73A) the transformer acts also as the energy-storage inductor (thus no additional inductor), whereas in the *forward* converter and *bridge* converters (Figures 9.73B–D) the transformer is "just a transformer," and the diodes and inductor complete the energy storage and transfer. The respective dc output voltages are $V_{out(flyback)}=V_{in}[N_{sec}/N_{pri}][D/(1-D)]$ and $V_{out(forward)}=DV_{in}(N_{sec}/N_{pri})$. Speaking generally, flyback converters are used in low-power applications ($\lesssim 200\,W$), forward converters in medium-power applications (to ~500 W), and bridge converters for real power applications.

¶E. Switcher Regulation: Hysteretic, Voltage Mode, and Current Mode.

There are several ways to regulate a switching converter's dc output voltage. Simplest is *hysteretic* feedback, in which the error signal simply enables or disables successive switching cycles. It's a form of simple "bang–bang" control, with no stability issues that require a compensation network; see Figure 9.64 for a buck converter design with the popular MC34063. Proportional PWM control is better, and comes in two flavors: voltage mode and current mode. Both methods compare the output voltage with a fixed reference to regulate the output voltage, but they do it in different ways. In *voltage-mode* PWM, the output voltage error signal is compared with the internal oscillator's sawtooth waveform to control the primary switch's ON duration, whereas in *current-mode* PWM the comparison ramp is generated by the rising inductor current, with the internal oscillator used only to initiate each conduction cycle. See §9.6.9, and particularly Figures 9.71 and 9.72. In either case the controller terminates a conduction cycle if the switch exceeds a peak current, the input drops below an "undervoltage lockout" threshold, or the chip exceeds a maximum temperature. Figure 9.65 shows a simple voltage-mode PWM buck converter.

Voltage-mode and current-mode control loops both require compensation for stability, and each has its advantages and disadvantages. Current-mode controllers appear to be winning the popularity contest, owing to their better transient response, inherent switch protection (owing to pulse-by-pulse current limiting), improved outer-loop phase margin, and ability to be paralleled.

¶F. Switching Converter Miscellany.

Switching conversion is a rich subject, many details of which are well beyond the scope of this chapter (or this book). Some topics – ripple current and inductor design, core saturation and reset, magnetizing inductance and snubbing, soft start, diode recovery, CCM and DCM conduction modes, switching losses, loop compensation, burst mode, inrush current, isolation barriers, PFC, switching *amplifiers* – were treated lightly here, and in more detail in this chapter (Chapter 9x). Consider this chapter's treatment of switching converters as a lengthy introduction to a specialty field that can easily consume a professional lifetime.

¶G. Voltage References.

A stable voltage reference is needed in any voltage regulator, as well as in accurate applications such as precision current sources, A/D and D/A conversion, and voltage- and current-measurement circuits. Often a good voltage reference is included in a regulator or converter IC (see for example Table 13.1), but you may want the improved performance you can get with a high-quality external reference. And, often, you need a stand-alone voltage reference for other uses in a circuit.

The simplest voltage reference is the discrete *zener diode* (§9.10.1), but most voltage references are multicomponent integrated circuits that behave externally either like a high-performance zener ("2-terminal," or *shunt*; Table 9.7), or like an extremely good linear regulator ("3-terminal," or *series*; Table 9.8). Shunt references must be biased into conduction (just like a zener) by supplying current from a higher-voltage rail (use a resistor or a current source), while series references are powered by connecting their supply pin directly to the dc supply. References of either type are available in a small set of standard voltages, typically in the range of 1.25–10.0 V.

The discrete 2-terminal zener is fine for non-critical applications, but its typical accuracy of ±5% is inadequate for precision circuits. Integrated references of either kind are far better, with worst-case accuracies in the range of 0.02% to 1%, and tempcos ranging from 1 ppm/°C to 100 ppm/°C, as seen in the tables. Most integrated references are based on a circuit that temperature compensates the V_{BE} of a BJT (a so-called "bandgap reference"), generating a stable voltage of approximately 1.24 V; but others use a buried zener diode with $V_Z \approx 7$ V. The latter are generally quieter, but bandgap references can operate from low supply voltages and are widely available in voltages of 1.24 V, 2.50 V, etc. Two newer technologies with surprisingly good performance are the *JFET pinchoff reference* from ADI (the ADR400 "XFET" references), and the *floating-gate* reference from Intersil. Both exhibit excellent tempco and low noise. Other important characteristics of voltage references are *regulation* (R_{out} for shunt types, PSRR for series types), minimum *load capacitance* and stability into capacitive loads, *trim* and *filter* pins, and *package* style.

¶H. Heat and Power Dissipation.

Along with power electronics comes… *heat* You remove it with a combination of convection (air flow) and conduction (thermal contact with a heat-dissipating *heatsink*). Conductive heat flow is proportional to the temperature difference between the hot and cold sides (Newton's law of cooling), $\Delta T = P_{diss} R_\Theta$, where R_Θ is known as the "thermal resistance." For a succession of conductive joints the thermal resistances add; thus, for example, the junction temperature T_J of a power semiconductor dissipating P_{diss} watts is $T_J = T_A + P_{diss}(R_{\Theta JC} + R_{\Theta CS} + R_{\Theta SA})$, where T_A is the ambient temperature, and the successive R_Θ's represent the thermal resistances from junction to case, case to heatsink, and heatsink to ambient. Printed circuit foil patterns are often adequate for dissipation of a few watts or less (Figure 9.45); finned heatsinks or metallic chassis surfaces are used for greater heat removal, with forced airflow generally needed when the power dissipation reaches levels of 50 W or more (Figure 9.43). Semiconductor devices can withstand considerably greater power dissipation during short pulses; this is sometimes specified as a graph of *transient thermal resistance* (i.e., R_Θ versus pulse duration and duty cycle), or as elevated contours on a plot of Safe Operating Area (see ¶I).

¶I. Safe Operating Area.

A power transistor (whether BJT or MOSFET) has specified maximum values of voltage and current, and also (because of maximum allowed junction temperature) a maximum product $V_{DS}I_D$ (i.e., power dissipation) for a given case temperature; the latter is just $V_{DS}I_D \leq (T_{J(max)} - T_C)/R_{\Theta JC}$. These limits define a *safe operating area* (SOA, §9.4.2), usually shown as contours on log–log axes of current versus voltage; see for example Figure 3.95. That plot shows two further features: (a) greater dissipation is allowed for short pulses; (b) the SOA of BJTs (but not MOSFETs) is further constrained by a phenomenon known as "second breakdown."

¶J. Overvoltage Crowbars.

Some failure modes of power converters cause output overvoltage, for example a shorted pass transistor in a linear regulator, or loss of feedback control in a switcher. This is likely to damage or destroy powered circuitry. An *overvoltage crowbar* (§9.13.1) senses overvoltage and triggers an SCR to short the output. A less brute-force technique shuts down conversion when an overvoltage is sensed; these are indicated in the "OVP" column in Table 9.5b.

¶K. Current Sources.

"Regulator" usually means a stable *voltage* source; but there are many uses for a controllable *current* source (§9.3.14). Three-terminal linear regulators are easily coaxed into current-source service (§9.3.14A). There are also dedicated current-source ICs like the LM334 and REF200. JFETs make convenient 2-terminal current sources, and depletion-mode MOSFETs make excellent current sources that can operate up to voltages as high as 1 kV, see §9.3.14C. And don't forget about the humble discrete BJT current source (§§2.2.6 and 2.3.7B), or the op-amp current-source circuits (§4.2.5: Howland; op-amp + transistor).

9x.1 Reverse Polarity Protection

Circuits that connect to a source of dc power are usually damaged by reversal of polarity; this can happen easily, for example when attaching the connector to a small 9 V battery, or (more seriously) when replacing a 12 V lead–acid "gel cell" type battery (which unwisely have identical lugs for + and −, and which can supply a short circuit current of a hundred amps or so without feeling much pain).

Standard solutions (Fig. 9x.1) include (i) a series diode, preferably Schottky if low voltage; (ii) a shunt high-current diode; or (iii) a shunt high-current zener diode. The series diode must be able to carry the full load current, and withstand the full supply voltage in reverse; and it introduces a diode drop, which is serious in low-voltage applications. The shunt diode methods fix the diode drop problem, but they crudely crowbar the reversed voltage source, during which they must withstand the full short-circuit supply current.

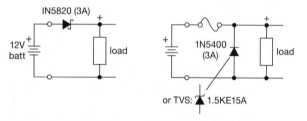

Figure 9x.1. Reverse polarity protection with series or clamp diodes. 12 V gel cell batteries have unpolarized terminal lugs. A fuse is needed if supply current is not adequately limited.

Fig. 9x.2 is an elegant alternative, for supply voltages greater than 9 V, and particularly for high-current loads. It is simply a series MOSFET of low R_{ON}, acting as an active (switchable) series element. You can think of this as a perfect series diode, implemented with a MOSFET active rectifier.

Here's how it works: When the input is connected correctly, the p-channel MOSFET's gate is forward biased (clamped below $V_{GS}(max)$, if necessary, by the small zener), so it acts like a series resistance of R_{ON}, in this case 20 mΩ or less. Note that the p-channel MOSFET is running "backwards" (drain more positive than source). There's no harm in this, and in fact the substrate diode

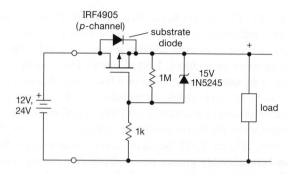

Figure 9x.2. Active polarity protection with series p-channel MOS-FET. The relatively small gate drive resistor ensures rapid turn-on into the large gate capacitance; the 1 MΩ resistor discharges the gate voltage to prevent residual conduction if the battery is quickly (and incorrectly) reconnected.

would merely go into conduction if V_{DS} became greater than a diode drop. But the low R_{ON} keeps the drop low – 0.1 V (max) at 5 A – so there's little power dissipated (0.5 W at 5 A). If the input is connected backward, the MOSFET is OFF (and V_{DS} is in the "normal" direction). So we get the best of both worlds: small forward drop (and low dissipation), and no crowbarring of the supply.

We used a modest *p-channel* MOSFET (IRF4905: 55 V, 50 A, 20 mΩ max at $V_{gs} = -10$ V; about \$2) in order to have a common negative rail. You can get p-channel MOS-FETs with less ON-resistance, for example the FDB9503L-F085 (40 V, 110 A, 2.6 mΩ max; about \$3.90), or the SQM40031EL (40 V, 120 A, 3 mΩ max; about \$3.35). But you can do even better with an *n*-channel MOSFET (lower R_{ON}, higher voltage) if you don't mind having the series element on the negative side, or for negative supplies (e.g., the FDBL86361-F085 *n*-channel MOSFET, 80 V, 300 A, 1.4 mΩ max, about \$4; or the IPT007N06, 60 V, 300 A, 0.75 mΩ max, about \$7.50).

<table>
<tr><td>

9x.2 Lithium-Ion Single-Cell Power Subsystem

</td></tr>
</table>

Lithium-ion ("Li-ion") cells and batteries have earned a fine reputation for performance and reliability over the last few decades, in cell phones and other handheld electronics devices. They have a high energy density and retain their charge far better than the NiCd and NiMH batteries they replaced. They also have a conveniently high cell voltage, about 3.6 V to 3.9 V during operation, and 4 V to 4.3 V just after full charge, voltages that are well-suited to low-voltage microcontrollers and electronics.

But their improved performance (and energetic chemistry) comes with new risks, requiring precautionary safety measures. This is especially true for multi-cell batteries, which we'll discuss later. Thankfully, IC Manufacturers have created special circuits for the care and feeding of Li-ion cells. We'll explore a few of these ICs in the Li-ion single-cell charger subsystem of Figure 9x.3.

9x.2.1 Charger features

We've chosen components to support 1.8 V to 7 V loads (set by a resistor ratio), at currents to 1 A or more. We like TI's BQ24070 because, among its other pleasant properties, it independently provides the regulated +4.4 V output when V_{in} is present, even with a dead or absent Li-ion cell.[2] It accepts a dc charging input V_{in} that includes the useful +4.4 V to +6 V range[3] provided by a USB host or ac wall adapter, which it regulates to a nominal output voltage of +4.4 V (via Q_1) when the dc input is present; otherwise it connects the Li-ion cell to the output via Q_2.

It's increasingly common to see battery operated devices getting their dc and charging power from a USB host computer, or from ac adapters that provide the same +5 V. USB hosts provide up to 100 mA initially, which can be negotiated to 500 mA (or 900 mA for USB3) via communication with the slave; ac wall adapters with a USB slave connector can provide more current, often 1 A. And, of course,

wall adapters that are not pretending to be a USB source are available with regulated dc at many amperes.

The '070's operation is customized with three resistors and a few logic-level inputs, which we've connected to a microcontroller (Chapter 15). These inputs let you instruct the chip to limit the input current to either 100 mA or 500 mA, to satisfy the power limits of a USB host; otherwise it can handle up to 2 A. It also senses severe drops in V_{in} and scales back the current draw allowed through Q_1. The '070 is a linear regulator in a small package with a thermal pad, and if necessary it scales back current setpoints to keep its die temperature below 125°C.[4]

The BQ24070's most important task is taking care of the Li-ion cell. It regulates the charging current, via Q_2, and changes to a float-charge mode when the 4.2 V full voltage is reached. Charging is terminated after the current drops to 10% of its maximum level. As a safety precaution, it's also terminated after a preset time. The '070 constantly observes the Li-ion cell's temperature by means of a 10k thermistor mounted in contact with the cell, and it suspends charging if the cell overheats ($> 45°C$) or is too cold ($< 0°C$). Some charger ICs (for example the LTC4070) also lower the float voltage at high cell temperatures. When no external power is present, the BQ24070 connects the cell to the load, via Q_2, but with current limit and short-circuit protection.

9x.2.2 Monitor and Protect

The precautions mentioned above, and others discussed in the '070 datasheet, would seem to be sufficient for taking care of the Li-ion cell. But we're just getting started. The cell's health is monitored by the microcontroller shown. It's customary to add a second chip, which continually monitors the cell's voltage, current, and temperature; it also keeps track of its actual charge level, via a charge integrator, which you can think of as a "gas-gauge" measure of the cell's remaining capacity in ampere-hours. In our case, the DS2762 IC has a "1-wire" serial data link (see §14.7.3) to provide this detailed information, so the microcontroller's software can keep track of the condition of the cell.[5] Fre-

---

[2] The LTC4160 provides similar capabilities.

[3] Its specifications allow operation to +16 V input, but at higher input voltages the charging and output currents are greatly limited by dissipation in its (linear) regulator.

[4] Don't confuse this with conventional thermal limit circuitry, which normally causes hysteretic cycling (the BQ24070 also includes such a 155°C thermal trip). Here the chip linearly servos its maximum currents, as needed, to maintain stable and safe operation at its 125°C setpoint.

[5] We've discussed ways to avoid external abuse of Li-Ion cells, but not how to discover internal cell faults that might signal incipient battery failure. Symptoms of trouble include slow current taper after reach-

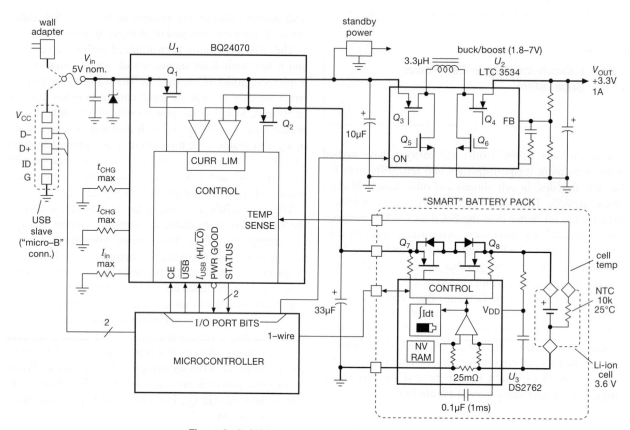

Figure 9x.3. Lithium-ion charger and power supervisor.

quently gathered gas-gauge data can be continuously analyzed to look for these signs of trouble, and act on them.

But we're *still* not done! It's important to include a protective device to disconnect the cell completely from all other circuitry under potentially dangerous conditions. These are if the cell's voltage discharges too far (below 2.6 V), or rises too high (above 4.275 V), or if its current exceeds a safe limit (1.9 A), or if a load short is detected (above 8 A, requiring response within $200\mu s$), or if the cell overheats. This functionality is often provided by an additional third IC. Here, however, those functions are happily included in the multifunction DS2762 ("High-Precision Li+ Battery Monitor With Alerts"). As shown in the figure, this IC is meant to be bundled with the Li-ion cell, making up a four-terminal power module (two cell wires,

a thermistor, and a 1-wire serial digital link). This IC has several other useful features, including a unique 64-bit serial number to help you keep track of the cell's history and verify its provenance, an EEPROM (shadowed by RAM), a locked EEPROM to store manufacturer information, and a 16-byte SRAM, powered by the cell. And it has multiple programmable alarms that signal through the 1-wire pin.

The DS2762 keeps track of cell current with a low-side sensing precision differential amplifier ($V_{os} \approx 15\,\mu A$) and integrating ADC, and an optional internal 25 mΩ resistor. With this resistor the state-of-charge resolution is 0.25 mA-hr, which is accumulated to 15-bits (plus sign) for a full scale of 8.2 A-hr. Two external *p*-channel MOSFETs, wired drain-to-drain, are used to disconnect an endangered Li-ion cell from external influences, as shown.

ing full charge voltage, loss of charge voltage during long rest periods, noisy voltage profiles during charge or discharge, charge capacity greatly exceeding discharge capacity, and excessive temperature rise during charging; see, for example, *Power Electronics Technology*, "Detecting Internal Li-ion Cell Fault Development in Real Time," **36**, 3, 37 (March, 2010).

9x.2.3 Output voltage regulator

Li-ion cell voltages range from 3 to 4 volts, reaching 4.1 V at full charge. A switching converter creates the regulated output voltage V_{out}. This could be something like +3.3 V, in which case the cell voltage can be above or below the output voltage. So we need a buck-boost converter, for example the LTC3534 that we've chosen. This part features four internal MOSFET switches, allowing buck or boost modes with a single small 3.3 μH inductor. It accepts input voltages in the range 2.4 V–7 V, with regulated output (set by a resistive divider) in the range 1.8 V–7 V.[6] It includes a fixed 1 A current limit, which works by sensing the voltage drop across the switching MOSFETs. For a low-current application, say below 1 mA, you can enable a "burst mode" for high efficiency.

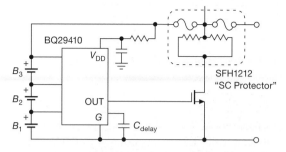

Figure 9x.4. Multi-cell battery protector.

9x.2.4 Multiple cells: a "battery"

These are some of the protective measures used to provide safe operation of a Li-ion cell. But what if you've got a *battery*, consisting of several cells in series? You can't be sure all of the cells are OK, based just upon the total voltage; for example, one of the cells may become completely discharged. There are a number of steps that are commonly used in the industry. First, battery manufacturers provide a means of bypassing charging current around a fully charged cell. Second, in a more complex monitoring function, the single-cell protective measures above are implemented individually for each cell in the stack. And, lastly, it is thought that another layer of protection is required, namely a way to disconnect completely (and permanently) the battery from both charger *and load*. A circuit to do this is shown in Figure 9x.4. TI's BQ29410 is typical of the parts designed for this second-level protection. It handles battery stacks of 2, 3, or 4 cells, continuously checking the health of every cell. If any cell's voltage exceeds 4.35 V, it turns on a MOSFET, blowing a special 3-terminal thermal fuse within seconds.[7] At that point the entire battery is disabled, and permanently disconnected from the outside world.

Figure 9x.5 shows a good example of a multi-cell protection IC that requires no microcontroller or software. It monitors current, temperature, and the individual cell voltages, and disables charging or discharging, as appropriate, according to the health of the stack. Some other pleasant features of this chip are its low quiescent current (6μA), separate overtemperature thresholds for charge and discharge, and the ability to cascade multiple chips to accommodate stacks comprising up to 20 cells.

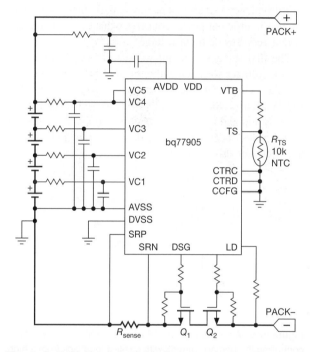

Figure 9x.5. Multi-cell stackable battery protector with current, temperature, and cell voltage monitors.

The care and feeding of huge batteries made up of thousands of cells is a serious matter – every cell needs careful attention. Semiconductor manufacturers have risen to the challenge, e.g., NXP's 33771 lithium-ion battery-cell controller IC, in a 64-pin package, intended for electric-vehicle applications. Each chip cares for a 14-cell cluster, up to 75 V, and communicates with an array of sister chips and a central processor, using a differential two-wire transformer-isolated databus. It takes voltage, coulomb, and temperature measurements, and controls balancing load resistors.

---

[6] It would be nice if it could go lower; but the minimum output voltage has to be high enough to ensure turn-on of *p*-channel MOSFET Q_4.

[7] For example, the "Self Control Protector" (SCP) series from Dexerials (formerly from Sony Chemical).

9x.3 Low-Voltage Boost Converters

It's handy to be able to run portable electronics on a single alkaline cell (1.35 V fresh, down to 0.9 V at end-of-life). It has also become fashionable to use ambient energy sources ("energy harvesting"), such as solar, vibration, or radiofrequency signals that are present. So there is growing interest in efficient power converters that can start up, and run, at very low dc input voltage.[8]

The first thing you might wonder is how a circuit can run on voltages well below the threshold V_{GS} of MOSFET switches. Good teachings on this come from the legendary Jim Williams, whose article[9] "J-FET-Based DC/DC Converter Starts and Runs from 300 mV Supply" explains one useful trick. Look at Figure 9x.6, a circuit we put together to check out this idea. The way to get around the gate

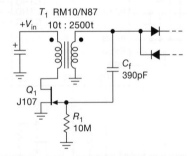

Figure 9x.6. A depletion-mode FET (here the high-g_m JFET J107) controlling the primary current, with positive feedback from a high-ratio secondary, lets this low-voltage dc–dc converter start up at a mere 20 mV. This trick is used in several of LTC's energy-harvesting converters, for example the LTC3108 (Fig. 9x.8).

threshold problem is to use a *depletion-mode* FET, so the "threshold voltage" is zero! For our circuit we used a JFET, in a simple self-excited oscillator circuit. This converter, though not terribly efficient, does manage to start up at 20 mV (Fig. 9x.7). And once you've got a converter started, you can use a higher-voltage auxiliary output to bootstrap parts of the converter that want a higher operating voltage.

Linear Technology (now part of Analog Devices) offers several energy-harvesting ICs, and you can get a nice

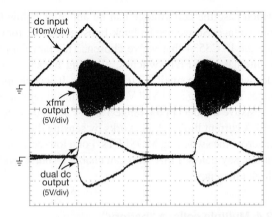

Figure 9x.7. The low-voltage converter of Fig. 9x.6 starts up at 20 mV of dc input, and will continue to run until the input voltage drops below 10 mV.

evaluation kit (e.g., the DC2042A) that lets you play with several of them. Some part numbers to start with are the LTC3105 and 3106 (solar power, 200–330 mV min), and LTC3107, 3108, and 3109 (thermal, 20 mV min). This is an active field, with offerings from the likes of Analog Devices, Advanced Linear Devices, Cypress, Silicon Laboratories, ST Microelectronics, TI, and others.

---

[8] We mentioned this challenge in §15.8.4, where we promised this followup.

[9] *Linear Technology Magazine*, Sept 2006, pp 34–35.

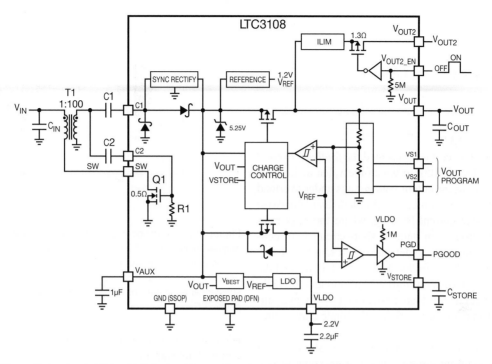

Figure 9x.8. Linear Technology's LTC3108 "Ultralow Voltage Step-Up Converter and Power Manager" uses the nice trick of Fig. 9x.6 to enable energy harvesting from low-voltage sources such as thermoelectric generators, thermopiles, and solar cells. (Adapted from LTC3108 datasheet.)

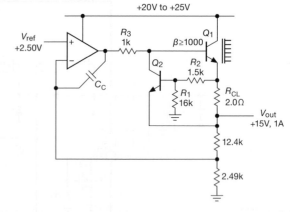

Figure 9x.9. We've added current-limiting resistor R_3 to the AoE3's original foldback circuit (Fig. 9.105), so transistor Q_2 doesn't force the op-amp's output into current limit. However, this forces us to use a Darlington pass transistor (or a MOSFET), because of the substantial base current (10–20 mA) needed to drive an ordinary power BJT.

9x.4 Foldback Current Limiting

Addendum to §9.13.3

Back in AoE3 Chapter 9 we presented the *foldback current limiter* circuit (Fig. 9.105), in which the value of limiting current *decreases* as the output voltage sags under overload conditions. It addresses nicely a problem that otherwise arises with simple current limiting, where the pass transistor's power dissipation under a short-circuit fault condition is far greater than under maximum rated load current. In that example the foldback circuit safely limited the pass transistor's dissipation (during short circuit) to 7.5 W, which is less than its full-load dissipation (10 W), and, importantly, less than the dissipation with simple current limiting (23 W).

In response to a reader's comment,[10] we should point out that such a current-limiting circuit needs some way to limit the control amplifier's output current when the limiter is active. In Figure 9.105 we relied on the op-amp's internal current limit (typically ~20 mA), but in practice it's a good idea to use a current-limiting resistor, as shown here in Figure 9x.9.[11]

Modified topology

A simple circuit modification improves the foldback circuit. Look at Figure 9x.10, where we compare the two versions, this time using a MOSFET pass transistor. Both circuits are configured for a maximum output of 15 V and 10 A, and both fold back to 2 A when the output is shorted to ground. What's different is that R_2 of the foldback divider returns to the input rail in the modified circuit. And this reduces greatly the size of the current-sense resistor R_s,[12] with much less dissipation. For the circuit values shown, the sense resistor in circuit A dissipates 25 W (and drops 2.5 V) at full load, compared with just 4 W (and 0.4 V drop) for circuit B.

As with any foldback current limiter, you must ensure

adequate current at zero output voltage, otherwise the circuit will not start up.

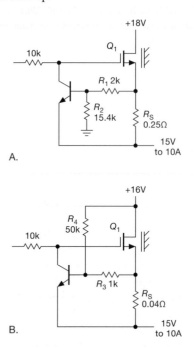

Figure 9x.10. Returning resistor R_2 to the positive input (circuit B) reduces the size of sense resistor R_S in a foldback current limiter, compared with the conventional circuit A; the reduction in power dissipation can be substantial (here 4 W versus 25 W).

---

[10] A welcome email from Marco Sartore, around St. Valentine's Day, 2019.

[11] We couldn't do that in Fig. 9.105, given the $\gtrsim$10 mA base drive needed there.

[12] A way to think about it is that, in circuit A the sense resistor has to be larger than it would be in a simple current-limit circuit to overcome the downward drop of divider $R_1 R_2$, whereas for circuit B the sense resistor has to be smaller.

9x.5 PWM for DC Motors

Those of us who grew up with electric train sets remember well the difficulty in controlling their speed: the train is stationary; you gradually advance the control knob or level, but the train stubbornly sits there, then a moment later it rushes off. In those days the same recalcitrance afflicted sewing machines, with their pedal-operated speed control. It was a wonder anyone could control those things. But now you can get tools with speed controls that actually work – battery-powered drills, for example, where the speed is well-controlled as you advance the trigger, in spite of large variations of mechanical load as you drill your way through tough material.

9x.5.1 The myth: PWM as secret sauce

These latter tools use permanent-magnet (PM) dc motors, with pulse-width modulation of the battery's dc voltage. They deliver plenty of torque at low speed, in contrast to the sewing machines and train sets of yesteryear. So it's easy to conclude (incorrectly, as we'll see) that it's the magic of PWM that delivers the desirable feature of well-controlled speed with adequate torque. To elaborate on this thinking, you might imagine that the pulsed current delivers high peak torque, with the motor gliding between pulses, and thus that PWM is more effective than the alternative of applying a steady dc voltage equal to the average PWM voltage (i.e., peak voltage times duty-cycle). This has caused considerable confusion, with statements like this: "Controlling the speed of something like a dc motor by varying the voltage (even in an efficient way) is a poor technique. It limits the motor to low torque at lower speeds because of limited power. That is why it is more common to see PWM (Pulse-Width Modulation) used to control motors (and lights/LEDs as well). PWM allows motors to have more torque at low speeds."

Such thinking would suggest that PWM is superior to applying a variable dc voltage, and that the experience with trains and sewing machines provides confirmation.

A. An experiment

With characteristic arrogance, we decided to get to the bottom of this. First step, echoing eevblog's Dave Jones ("Don't turn it on, *Take it apart!*"), was a look at the innards of a typical battery-powered drill (Fig. 9x.11). Not much there – in particular there is no tachometer or other rotational-speed feedback, just PWM to a 2-terminal dc motor. Figure 9x.12 shows the voltage waveforms at increasing degrees of trigger depression.

Now for the fun: we intercepted the motor leads and applied dc from an external power supply. We did this to test the proposition that PWM is "better" than steady dc in terms of torque at low speed. We used a big drill bit to bore into hunks of plywood (Fig. 9x.13) at low speed, comparing the drill's muscle with both PWM and dc drive. And we found...(drumroll)...that there was no discernible difference!

B. Toy trains and sewing machines

But this disagrees with our childhood "knowledge." What's going on? It turns out that those controllers of yesteryear did not provide a variable (and stiff) voltage source – instead they used a fixed voltage (either dc or ac) in series with a variable impedance (a *rheostat*, fancy name for a wirewound potentiometer that can dissipate at least a few watts). If you think about it, this is a *terrible* way to control a motor: Imagine you've set the rheostat so the motor's turning slowly, and then you increase the mechanical load on the motor; the latter then slows down, presenting a lower impedance (because of lower "back-EMF"), which causes its voltage (and therefore rotational speed) to drop. Similarly, if the motor is turning at some speed under load and you decrease the load, the motor will turn faster, whereupon the increased back-EMF causes the voltage (and therefore motor speed) to rise still further.

C. Another experiment

To compare the merits of voltage-drive with what we might call rheostat-drive, we rigged up two experiments, each of which let us see the ramp-up of a dc motor when driven either with a fixed low voltage (i.e., less than the motor's rated voltage), or with a series resistor from a higher voltage, the resistor chosen to produce the same lower motor voltage once things had stabilized. The ramp-up from a cold start is a simple proxy for torque, and it's easier to measure.

1. dc fan with tachometer

For this experiment we used a 120 mm 12 V brushless cooling fan (Delta AFB1212SH); it has a logic-level tachometer output that generates two pulses per revolution. We ran it at half voltage (6 V), using either a 6 V dc supply, or a 12 V supply with a 24 Ω series resistor (the value

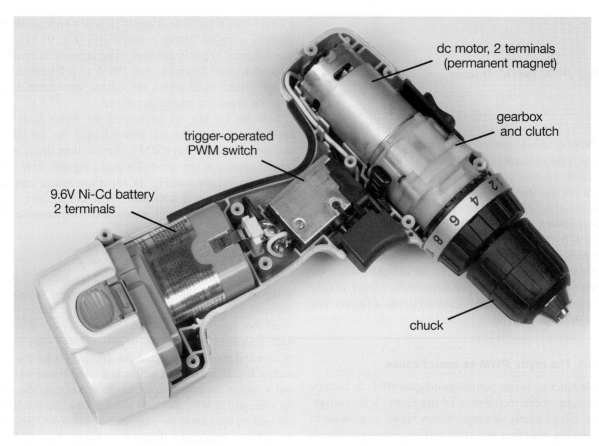

Figure 9x.11. Half-clamshell view of a variable-speed electric drill (Makita 6260D). The shiny module behind the trigger generates full-voltage PWM to the permanent-magnet dc gearhead motor.

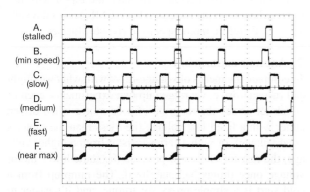

Figure 9x.12. PWM voltage waveforms applied to the dc motor in Fig. 9x.11 as the trigger is advanced. When it is fully squeezed the waveform transitions to steady 10 V dc. Horizontal: 40 μs/div; Vertical: 20 V/div.

Figure 9x.13. Victimized plywood, following drill-torque tests.

that produced 6 V across the fan after settling to its final speed). Figure 9x.14 shows the results (both fan voltage and tachometer pulses) when ramping up from a cold start.

It's clear that a series resistor causes considerably reduced torque (slower ramp-up of both voltage and speed), as just explained in §9x.5.1B.

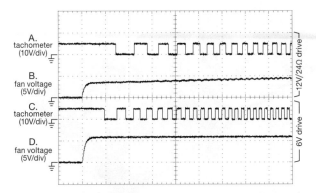

Figure 9x.14. Tachometer output (A, C) and fan voltage (B, D) for a 12 V dc fan, as measured from a cold start for two configurations that run the fan at half voltage (6 V): The top pair used a 12 V supply with 24 Ω series resistor; the bottom pair used a 6 V supply and no resistor. Horizontal: 100 ms/div.

2. dc motor-generator

For this experiment we used a pair of permanent-magnet dc motors[13] (marked Pittman 3140-0665) that we found in our vast collection of weird flea-market junk. They specify a strange operating voltage of 19.1 V (bringing to mind the famous outburst of the legendary I.I. Rabi, "who ordered *that*?"[14]) We used some heat-shrink tubing to couple the shafts, with the passive motor acting as a dc generator whose unloaded output voltage is proportional to rotational speed. Figure 9x.15 shows the results (motor drive voltage, and speed-proportional generator output voltage), demonstrating once again that the worst way to control motor speed is with a rheostat. Of perhaps greater interest, if you like myth debunking, is the demonstration that the thing behaves exactly the same with PWM (20 V amplitude, 50% duty cycle, 5 kHz) as with dc drive (10 V).

9x.5.2 Wrapup: PWM versus dc for motor drive

Evidently the poor performance of those rheostat-controlled sewing machines and toy trains was a result of using a variable impedance in series with the motor's power supply, nicely seen in the data above. In those heady

---

[13] These included a nice quadrature optical encoder on the back end; evidently these were used to drive the pen assembly in HP 7470A *x-y* plotters.

[14] Referring to the just-discovered muon interloper, not to some motor's odd choice of voltage.

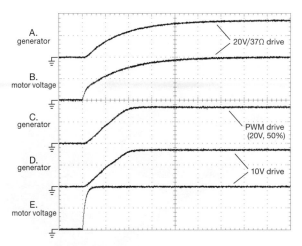

Figure 9x.15. This time we tricked a pair of PM dc motors into acting as a motor-generator, with the passive motor's generated dc output serving as a measure of shaft rotation rate. As in Fig. 9x.14, the top pair (A,B) shows startup to half-voltage (10 V) with a series resistor, whereas the bottom pair (D,E) is powered with a 10 V voltage source. Trace C demonstrates the equivalence of PWM (versus dc) drive. Horizontal: 100 ms/div; Vertical: 5 V/div.

years (of the 1950s) a rheostat was far less expensive than a variable transformer ("Variac"), and before the era of silicon power semiconductors it was difficult to implement a high-current variable dc supply.

Figure 9x.16. A vintage sewing-machine pedal-operated rheostat-type control. This thing gets *hot* and runs on powerline voltage, so the power resistor is kept safely inside a well-ventilated cage.

It's no longer difficult or expensive to make the latter; so, looping back to the confusion about PWM versus dc drive, and given the prevalence of PWM motor control, are

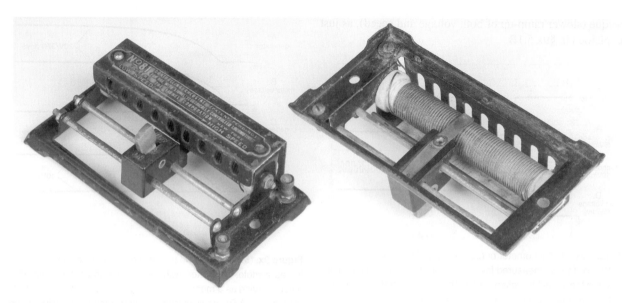

Figure 9x.17. Vintage Lionel model 81 train-set rheostat. The handle moves a sliding contact along the wirewound power resistor inside the ventilated enclosure.

we to conclude that PWM provides superior control of PM dc motors?

The answer is no: PWM is simply a convenient way to achieve the benefits of variable dc drive, without having to build a dc–dc converter, with its inductive energy storage, capacitive smoothing components, and feedback regulation. In effect, with PWM the motor's inductance and mechanical inertia[15] substitutes for the *LC* components in a dc–dc converter. You can think of PWM+motor as comprising a buck converter, equivalent to a dc voltage equal to the switched voltage times the switching duty cycle.[16] So all that's needed is a PWM switching signal to drive a MOSFET, with a catch diode (or MOSFET) to complete the motor drive. Easy peasy.

Looking slightly deeper, there are some disadvantages to PWM (compared with dc), namely the high switching frequency (and its harmonics) generates additional losses in the magnetic materials, and there are dynamic switching losses associated with switch and parasitic capacitances ("hard switching"). Overall, however, PWM is easy, and efficient enough to make it the technique of choice in dc motor control.[17]

One might add, as an afterthought, that if your goal is

to tightly control a motor's rpm over varying mechanical load, you might consider actively closing a PID loop, adjusting PWM duty cycle with tachometer feedback.[18] Active control beats the pants off passive control; the latter depends on motor current changing according to the difference between applied voltage (via PWM, or whatever) and the speed-proportional back-EMF, a control mechanism that is compromised by non-zero resistive losses in the motor windings.[19] Of course, active control requires a speed sensor (tachometer), which may not be practical in many situations. An interesting alternative is to drive the motor from a dc source that presents a negative output resistance approximately equal to the motor's resistance R_m; if this appeals to you, checkout Burr-Brown's Application Bulletin AB-152 ("DC Motor Speed Controller: Control a DC Motor without Tachometer Feedback"), available from

---

[15] Typical switching frequencies are upward of a few kHz (to put them above audibility), well above the response of the motor.

[16] And, as with a buck converter, the current out of the motor's return lead is dc, with little ripple.

[17] And in some cases the pulsating torque caused by PWM can be bene-

ficial, particularly at low speeds where it can overcome "stiction" and "cogging."

[18] Jim Roberge made a nice video in 1985 demonstrating just such a system – go to YouTube.com and enter this string: uHtKGf4AymM. We are indebted to Prof. Steven Leeb at MIT for this link, as well as for teachings on the secrets of dc motors.

[19] Going into this a bit further, the voltage difference ($\Delta V = V_{PWM} - V_{back-emf}$) appears across the winding's total impedance ($R + jX$), but what creates torque is the *current* through the winding. So, for example, if inductance dominates the winding's impedance, then the torque will increase only according to the time integral of the speed error. The differential term in a PID loop could do wonders here.

the TI website as document SBOA043; it shows you how to rig up such a linear driver. We tried it, and it works!

9x.5.3 Afterword: DC motor model

We've been a bit casual about what's actually going on inside the dc motor, having arrogantly asserted that there's a "back-EMF" that's proportional to rotation speed, and that the torque depends on the current. It's worth looking into this a bit more. The treatment that follows[20] was suggested by Steven Leeb, to whom we are indebted.

Figure 9x.18 is the basic model for a permanent-magnet dc motor (with brushes or equivalent commutation). *Mechanically* it's a machine whose shaft is turning at an angular velocity ω radians/s, while providing a torque into some mechanical load of τ newton-meters (abbreviated N·m). *Electrically* it looks like a voltage source ("back-EMF") V_{bemf}, proportional to angular velocity, in series with the motor's winding resistance R_m and inductance L_m. The torque is strictly proportional to current, $\tau = k_i I$ (Lorenz force on a current-carrying wire), and the back-EMF is strictly proportional to rotational velocity, $V_{bemf} = k_v \omega$ (Faraday's law of induction, with the motor acting as a generator).

$$V_{bemf} = k_v \omega \qquad \tau = k_i I$$

Figure 9x.18. Simplified electrical and mechanical model of a permanent-magnet dc motor. R_m and L_m represent the resistance and inductance of the motor windings. The motor generates a torque proportional to the current flowing through the windings; and their motion through the magnetic field generates a "back-EMF" proportional to rotational speed. The proportionality constants are equal: $k_v = k_i$.

Now here's something totally cool: the constants k_v and k_i are equal! This is easy to see, by equating the electrical power converted to motion, $P_{elec} = V_{bemf}I = k_v \omega I$, to the mechanical power delivered $P_{mech} = \tau \omega = k_i I \omega$.[21]

Now to the business of dc (or PWM) drive versus rheostat drive. Looking first at the limiting cases, for a perfect *current source* drive (i.e., of unlimited voltage compliance) the motor simply produces constant torque. With no mechanical load it will accelerate without limit, and the terminal voltage will track the angular speed. With a mechanical load whose torque is proportional to speed (say $\tau = \beta \omega$, a viscous drag) the motor will accelerate until the load's torque matches the motor's current-determined torque: $\omega_{final} = kI/\beta$. And if the "coefficient of viscosity" β changes, so will the equilibrium speed. It's rubbery – it behaves like the hard-to-control sewing machine.

At the other limit (a *voltage source*), the motor's steady-state velocity is simply $\omega_{final} = V/k$, the speed at which the back-EMF equals the applied voltage. For a lossless motor ($R_m = 0$) that speed is independent of load torque – any attempt to lower the speed by increasing the load causes a mismatch of drive voltage and back-EMF, and is thus met with an enormous current (the voltage mismatch across $R_m - 0$) that maintains the speed.[22]

For the realistic case of finite (but small) R_m, the acceleration to final speed is finite; and loading the motor with mechanical torque causes some degree of slowing, because the current required to produce the torque creates some voltage drop across R_m, thus leaving somewhat less voltage to balance the back-EMF. But the motor fights the slowdown: the reduced back-EMF puts more voltage across R_m, thus more current (and more torque). The lower R_m, the more effective the motor is in running at a load-independent constant speed.

For the in-between case (driving the motor with a rheostat in series with a dc voltage), the situation is also in-between. Rather than hand-waving, let's put this more quantitatively. Figure 9x.19 shows the arrangement, where we've ignored the (small) winding impedance; i.e., an ideal lossless motor. For a given supply voltage V_s and rheostat setting R_s, the motor current (for rotation speed ω) is just

$$I = \frac{V_s - V_{bemf}}{R_s} = \frac{V_s - k\omega}{R_s}, \qquad (9x.1)$$

---

[20] In which we ignore the transient effects due to the (often ignorable) motor inductance. Strictly speaking, the motor's (electrical, energy-storing) inductance and its (mechanical, energy-storing) moment of inertia form a second-order coupled system, with some natural frequency and damping factor. But the electrical time scale is ordinarily far faster than the mechanical timescale, justifying the simplified (electrically quasi-static) first-order treatment below. Of course, for the steady-state behavior one can always ignore inductance (and inertia as well).

[21] Note that it's the motor's internal voltage V_{bemf} that figures into this proof. Because of the motor's resistance R_m, the voltage you supply to run the motor (call it V_s) is greater than V_{bemf}, unless you have an "ideal motor" ($R_m = 0$). But $k_v = k_i$, regardless.

[22] And, in that unrealistic case of $R_m = 0$, the motor draws infinite current initially, producing infinite acceleration until reaching final speed.

producing a motor torque

$$\tau = kI = \frac{k}{R_s}(V_s - k\omega) = \frac{kV_s}{R_s} - \frac{k^2\omega}{R_s}. \qquad (9x.2)$$

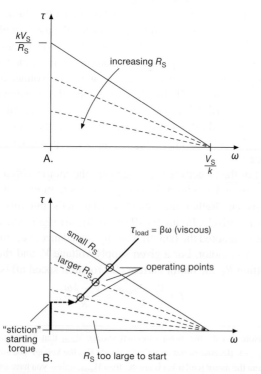

Figure 9x.19. A fixed dc supply powers an ideal dc motor, with (undesirable) speed control through a rheostat (variable resistor).

Equation 9x.2 lets us plot a family of curves of torque versus angular speed (Fig. 9x.20A). These show that the unloaded speed is just $\omega = V_s/k$, the speed at which the back-EMF plus resistive drop equals the supply voltage. Since there's no load, it's insensitive to series resistance (though in reality the motor itself has some bearing and air friction). These curves also show that the starting torque ($\omega=0$) increases inversely with series resistance.

From these operating curves (analogous to *load lines*, see Appendix F in AoE3) it's easy to figure out how a loaded motor will behave, by overlaying a curve representing the mechanical load's torque versus angular speed (Fig. 9x.20B). Here we've drawn a viscous load ($\tau \propto \omega$) that is sticky at rest and requires a minimum torque to overcome its "stiction" (static friction). For the motor to start with that load you must have $R_s < kV_s/\tau_{\text{stiction}}$: a graphical representation of the annoying "deadband" affliction of those rheostat controllers of yesteryear.

The bottom line:
for good control of motor speed in situations of varying mechanical loading, power your dc motors from a low-impedance voltage source, adjusting the voltage to set the rotation speed. Equivalently, you can use pulse-width modulation to control motor speed, starting from a voltage source that runs the motor at maximum rated speed for 100% duty cycle. And for the ultimate control, use tachometer feedback.

A. Series resistance: Op-amp analogy
Here's another way[23] to think about this business of series resistance. Look at Figure 9x.21. We apply a supply voltage V_s to the (non-ideal) motor, creating a current I that produces a torque $\tau=kI$. But the current is proportional to $V_s - V_{\text{bemf}}$ and inversely proportional to R_m: $I=(V_s-k\omega)/R_m$. That is, you can think of the motor as a differential amplifier that converts the difference voltage $\Delta V = V_s - V_{\text{bemf}}$ to torque, with gain ($\tau/\Delta V$) equal to k/R_m.

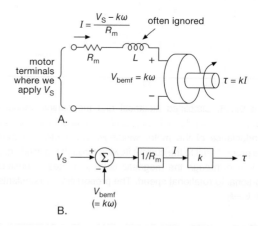

Figure 9x.21. Motor model, viewed as a gain block that converts the voltage difference $V_s - k\omega$ to a torque τ, with a "gain" of k/R_m.

Viewed this way, a good motor has high gain (k/R_m), just as a good op-amp has high open-loop gain G_{OL}. And it doesn't matter if the resistance comes from the motor or

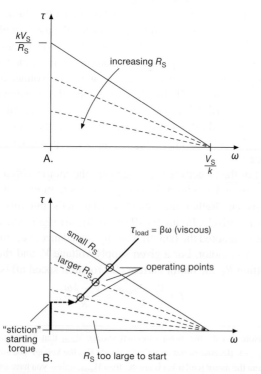

Figure 9x.20. A. Operating characteristic (load line) for the circuit of Fig. 9x.19, according to eq'n 9x.2. B. Determining operating points for a linear load cursed with some stiction.

---
[23] Steven Leeb, again!

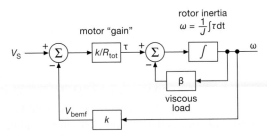

Figure 9x.22. The motor as an amplifier – losing the loop. The inner loop is the mechanical system (viscous load torque proportional to speed, plus rotor inertia) that converts torque to angular speed; the outer loop takes it from there.

an external series resistance R_s, which simply adds to the motor's internal resistance for a final (even lower) gain of k/R_{total}.

Finally, closing the loop around the "amplifier" gives us the circuit of Figure 9x.22. The inner (mechanical) loop includes a proportional ("viscous") torque drag, plus rotor inertia (which integrates net torque to spin up the angular speed); the outer loop is the by-now familiar voltage-to-torque characteristic of the motor, including the effect of its internal resistance and any external series resistance that is foolishly added. Don't add the latter!

9x.6 Transformer + Rectifier + Capacitor = Giant Spikes!

Most designers don't realize that a simple linear unregulated dc supply can (and often does) generate large microsecond-scale spikes, at the powerline frequency; these generate a buzzing sound in low-level audio circuits, and generally create havoc. They're easily tamed, once you know about them.

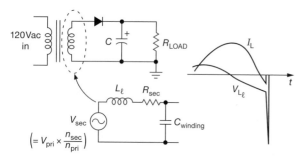

Figure 9x.23. Half-wave rectifier with transformer leakage inductance.

9x.6.1 The effect

We can understand this under-appreciated effect by looking at a half-wave supply (Fig. 9x.23). The problem is caused by the combination of the transformer's leakage inductance (see §§1.5.2 and 9.5.2 and the L_L values listed in Table 9x.1) and the rectifier's reverse recovery time (§1x.7.2): first, the (series) leakage inductance causes the current through the rectifier to lag the voltage across the inductor, so by the time the rectifier current reaches zero there is a significant reversed voltage across the inductor; that means that the current is decreasing at a significant rate, according to $V = L\,dI/dt$. Now, if the diode had no stored charge (zero reverse recovery time) there would be no problem – diode conduction would cease at zero current; but the diode, with its stored charge, continues to conduct as the current through it goes negative, until the stored charge is removed, whereupon the current stops abruptly.

This is the essential combination: the diode abruptly stops conducting with reversed current flowing through the inductor. Inductors don't like to have their current stopped

abruptly ($V = L\,dI/dt$ again), and respond by increasing the voltage to maintain continuity of current. In this case that means a negative voltage spike at the transformer secondary.

The effect can be quite large. Fig. 9x.24 shows measured waveforms for a 10 V unregulated half-wave rectifier circuit. We used a Signal Transformer Co. "split bobbin" type, designed for good isolation (low inter-winding capacitance, high breakdown voltage), whose reduced winding coupling results in significant leakage inductance. The diode was a 1N4001 (1 A conventional rectifier), the filter capacitor was 3,300 μF electrolytic, and the load was a 20 Ω resistor. You can see the flattening of transformer output (with its series L_L and winding resistance), and the diode current causing capacitor charging; the mischievous spike occurs when the current goes to zero. The expanded trace (Fig. 9x.24B) shows that the diode current goes negative for about 8 μs before abruptly ceasing (or trying to, anyway!), at which point the spike erupts.

9x.6.2 Calculations and cures

For low-voltage unregulated supplies you can use Schottky rectifiers to prevent this effect. In addition (and more generally) you can suppress the spike by putting a capacitor (or series RC combination) across the transformer secondary; this provides a conduction path for the leakage inductance's suddenly orphaned current, as seen in Fig. 9x.24C.[24]

To choose values for a damping RC circuit, we can use the measured leakage inductance L_L to estimate spike size, slew rate, and resonant Q; we'll illustrate with the transformer used for the waveforms above (Signal 241-4-10: 10 Vrms, 0.5 A), which has a measured secondary $L_L = 2$ mH. The energy stored in the leakage inductance at the moment of snapoff ($\frac{1}{2}L_L I_{snap}^2$) is transferred to the capacitor ($\frac{1}{2}CV_{pk}^2$), so the peak voltage is

$$V_{pk} = I\sqrt{L/C}.$$

For our circuit, where the measured $I_{snap} = 13$ mA (Fig. 9x.24B), a 1 μF damping capacitor results in $V_{pk} = 0.6$ V (compared with the 35 V undamped spike, whose size is set by the effective shunt capacitance of the transformer secondary winding and the rectifier, here roughly 300 pF). With that damping capacitor[25] the slew rate is

---

[24] You sometimes see a small ($\sim 0.1\mu$F) capacitor across each diode of a transformer-powered bridge rectifier.

[25] A 1μF capacitor might seem large, but the (reactive) current through

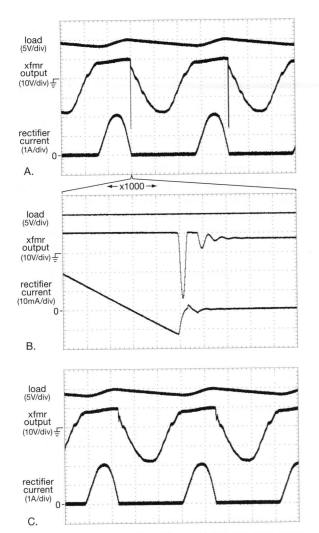

Figure 9x.24. Measured waveforms for the circuit of Fig. 9x.23, showing transformer output voltage, dc output voltage, and diode current. A. two full cycles (4 ms/div); note large negative spike at transformer output. B. expanded view (4 μs/div and 10 mA/div), showing detail of current reversal and of 35 V, 1 μs spike. C. addition of snubber (10 Ω and 1 μF in series) across secondary.

Table 9x.1: Small Transformer Parasitics[a]

V_{sec} (Vrms)	I_{sec} (Arms)	R_{pri} (Ω)	R_{sec} (Ω)	R_{sec}^{eff} (Ω)	$L_{L(pri)}$ (mH)	$L_{L(sec)}$ (mH)	Part #
24	0.1	372	23.8	47.8	377	25.1	241-3-24
24	0.2	138	8.33	17.0	178	11.3	241-4-24
24	1.25	21.5	1.31	2.59	59	3.42	241-6-24
24	2.4	9.6	0.55	1.09	40	2.18	241-7-24
24	4.0	4.2	0.24	0.46	25	1.34	241-8-24

Notes: (a) All are Signal Transformer 241-series "split bobbin" power transformers, 120Vac primary, 24V secondary.

given by $Q = \omega L/R$: our value of $10\,\Omega$ gives $Q \approx 3$, somewhat underdamped (even including the secondary winding resistance, which was measured to be $1.3\,\Omega$); an external resistor of $R = 39\,\Omega$ would produce critical damping, though the added impedance would nearly double the voltage spike, by adding a $\sim 0.4\,\text{V}$ step ($I_{snap}R$); perhaps of greater concern, that step is characterized by high slew rate.

Table 9x.1 lists measured parasitic parameters ($R_{winding}$, L_L) for a series of typical small ac power transformers of successively larger frame sizes (thus power ratings). The leakage inductances were measured with an impedance meter, in each case with the other winding shorted. For resistive losses it's often convenient to assign a single "effective winding resistance" to the secondary, by combining a reflected primary resistance: $R_{eff}(sec) = R_{sec} + R_{pri}/N^2$, where N is the turns ratio n_{pri}/n_{sec}; that way, you have just one resistance value for calculations. Here, where the open-circuit output voltage for the transformers went from 30 Vrms down to 28 Vrms (going down the table), the turns ratio N went from 4.0 to 4.3. Normal transformer design typically chooses wire sizes to equalize resistive losses in primary and secondary; and that is the case with this series of transformers, where R_{sec}^{eff}/N^2 is very closely equal to twice R_{sec}.

given by

$$\text{SR} = dV/dt = I_{snap}/C_{total}$$

which evaluates to 13 V/ms; this is harmless, being only a factor of 4 greater than the secondary ac waveform's $\text{SR} = 2\pi f A = 3.8$ V/ms. Finally, the series resistor R is chosen for adequate damping of the series *RLC* circuit, whose Q is

it is 4 mA, negligible compared with the power transformer's 500 mA current rating.

9x.7 Low-Voltage Clamp/Crowbar

Contemporary digital circuits run on low voltage, for example 1.8 V or 2.5 V, and are intolerant of overvoltage spikes. At historical voltages of 5 V one could use a passive zener clamp, or for higher currents a conventional voltage crowbar with zener and SCR. But at these very low voltages, with correspondingly narrow overvoltage tolerance, the situation becomes more difficult. And because a large digital system may run at prodigious currents – 50 A or more – a clamp or crowbar may have to withstand 100 A or more.

While worrying about this problem, we happened upon an article[26] that described a combination clamp/crowbar in which a power MOSFET provided shunt clamping, with an SCR crowbar triggered only for large currents (> 6 A), and then with an RC time delay; for much larger currents, however, (> 12 A) the SCR is triggered immediately. (We could not find the full article, only the text; but the description was pretty clear.)

This is a good idea! That is because you often see minor transients for which you do not wish to trigger an SCR, which crowbars the supply and requires shutdown and restarting. However, the implementation was unsuitable for the low-voltage supplies we are talking about (because it used a 2.5 V reference trigger, and because it used a MOSFET clamp, requiring a few volts of gate drive); it also had the disadvantage of crowbarring the supply indefinitely, should the SCR trigger.

9x.7.1 New clamp/crowbar

A. Circuit operation

Refer to the circuit diagram (Fig. 9x.25). The trigger U_1 is a low-voltage version of the venerable TL431, called a TLV431; it is a 3-terminal shunt regulator with a 1.24 V internal reference. The divider $R_1 R_3$ samples the input voltage across $V_+ V_-$, so when the divided portion across R_3 exceeds 1.24 V, U_1 goes into conduction, turning on Q_3, which turns on power npn Q_5. The latter is the clamp transistor, and is happy to sink a few amps just about forever (it's a 15 A device, on a heatsink). The clamping action

[26] Keith Billings, "Crowbar Methods Protect Overvoltage Transients," *Power Electronics Technology*, June 2002

takes place quickly – a microsecond or two – thus as long as the power rail slew rate is adequately limited (with bypassing and perhaps a bit of inductance) the voltage will not spike beyond the trip voltage.

For currents greater than 4 A the emitter voltage exceeds a diode drop, and will ultimately trigger the SCR, with a delay set by $R_8 C_2$, whose time constant is $150\,\mu$s. The triggering delay depends on the voltage across R_9, and can be somewhat more or less than that time constant. However, for clamp currents greater than about 8 A the delay circuit is bypassed by diode D_1, and the SCR fires promptly; that's what you want, of course, when the clamp cannot sink the full current behind the overvoltage fault.

The circuitry to the right of Q_1 is a third stage, which could be omitted for relatively low-current systems, but which is essential for the high currents we're interested in. It cuts ac power to the supply after the SCR is triggered. This is done by using the SCR's ON-state forward gate voltage to switch on Q_2, which turns on Q_4, which powers the "open" relay coil of a series latching relay K_1 in the ac line. Power can be restored with a manual pushbutton or a logic level input to switch Q_6, which powers the "close" relay coil.

B. Additional details

Because of the very high clamp/crowbar currents that can flow, we've provided a pair of sense inputs; we've also provided a jumper option for setting the trip voltage with a fixed resistor or a trimmer (1.5 V–6.5 V with a 5k pot at R_{14}). We put no current limit atop U_1, for maximum speed; the clamp action protects U_1. The relay driver circuit, powered by auxiliary +28 Vdc (from T_1, U_2, and C_1) uses cascaded switches ($Q_2 Q_4$) so that Q_2 can run at low V_{BE}, ensuring reliable conduction when the SCR is in the crowbar state.

C. Performance

We tested the breadboard crowbar by setting the trip point to +3.0 V, then charging a large capacitor to an overvoltage and abruptly connecting it directly across the +2.5 V regulated output of a high-current dc bench supply. A 47 μF capacitor, charged to +10 V, caused the clamp to work, but the SCR did not fire; the clamp current peaked at 5 A, with a total duration of 25 μs, and the resulting power supply voltage transient, slewing at 0.1 V/μs, reached +3.3 V before clamping quickly back to +3.0 V. After 50 μs the voltage had returned to its nominal +2.5 V. With 30,000 μF charged to +15 V, however, the clamp activated, reaching about 12 A, then (5 μs later) the SCR triggered; the peak transient was held to +3.4 V.

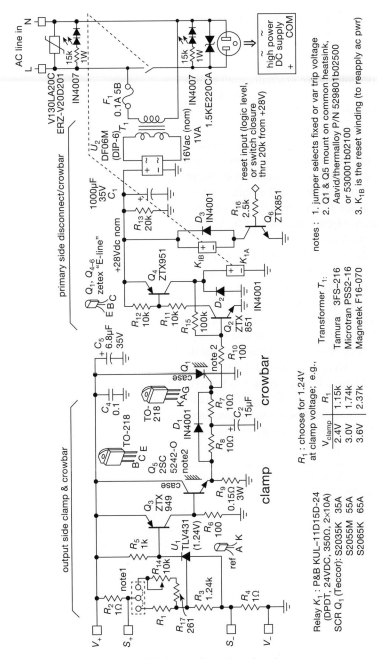

Figure 9x.25. Low-voltage clamp/crowbar.

This kind of testing is probably a bit unrealistic, because large and energetic charged capacitors don't that often fall across your low-voltage rails. So we did some tests with high-current fixed-voltage switchers, simply bringing the clamp/crowbar's trip point down gradually until something exciting happened. First we used a +3.3 V 80 A Vicor "MegaPAC" ac–dc switcher: the crowbar fired, of course, then the ac shut down. The current waveform was a flat 100 A for the 125 ms it took for the switcher to exhaust its internal storage, after which the current went to zero abruptly. At the leading edge there was a current transient to 220 A, about 50 μs wide, with low-Q ringing transition over ~1 ms to the steady 100 A. Paralleling 47,000 μF across the dc rails increased the transient to 250 A[a], and extended its duration somewhat.

Finally, we performed the same torture with a +5 V 120 A Vicor ac–dc switcher, with unsurprising results: the current waveform is a slightly downsloped 150 A, of 75 ms duration, with an initial transient to 200 A for roughly 50 μs.

We performed these tests a dozen times in quick succession, after which the SCR and heatsink were just slightly warm to the touch. Inspection of the non-repetitive peak surge current curves for the 65 A SCR showed that we were within the safe operating region by a factor of 3 to 5; there is an additional safety factor because those surge ratings assume an operating and fully stressed SCR (maximum rated current at maximum case temperature) prior to the transient.

---

[a] Probably ESR and wiring limited; the latter consisted of a pair of 10 gauge copper wires, 5′ long, thus 10 mΩ.

9x.8 High-Efficiency ("Green") Switching Power Supplies

Many ac-powered switching power supplies have to operate continuously, but spend much of their time operating at very light loads, or at no load. The authors' homes and laboratories are filled with examples, each consuming 0.5 W to as much as 10 W while in standby or performing light duty; altogether too much wasted power. (A dozen electronic devices, idling in "standby" at 5 W apiece, adds up to nearly a thousand dollars after ten years.)

Some products do not require their dc power supplies to be isolated from the ac line, and this allows the use of simple circuit tricks to achieve low power. We explore this idea in §9x.15. Turning to isolated offline supplies, we find a multitude of simple designs with limited features, meant for modest power levels like 5 watts, such as Figure 9.74. This blocking-oscillator circuit does fine in the no-load contest, at about 100 mW.

It's more of a challenge when designing an offline supply to operate on "world power" (from 85 V to 265 V ac), with additional important features (regulation, ripple, transient performance, overcurrent and overtemperature limit, and so on). Our real-world example in §9.8 (Astrodyne OFM-1501, with its Power Integrations TOP201 switch-

ing IC) is typical; it consumes 0.8 W under no load. Ideally we'd like a sophisticated line-powered switcher to consume less than 0.1 W when unloaded (which is well below current "Energy Star" requirements).

The semiconductor industry has responded, and IC manufacturers offer a number of controllers that can meet our goal. Here we'll examine one, TI's UCC28610 "Green-Mode Flyback Controller." Converters made with this part can deliver dc in the range of 12 W to 65 W, and yet consume less than 95 mW from the ac line when unloaded, or when providing up to 25 mW of standby dc. The 8-pin '28610 is both more complex than the 3-pin TOP201, and more capable. For example, the TOP201 obtains its operating power from a rectified auxiliary transformer winding (as with most offline designs); but the '28610 devotes an extra pin to monitor this winding's waveform for zero crossings, and uses that information to ensure that it always operates in discontinuous-current mode (DCM), thereby simplifying design and operation. This extra connection also lets TI add an over-voltage protection feature, using transformer read-back to limit excess output voltage if the normal feedback pathway fails. The '28610 also adds valuable features in the green department, as we'll see.

Here's a short list of power-saving design hints, tagged with numerals on the simplified schematic of Figure 9x.26. We'll see that what matters is more than simply the controller chip's power consumption; the clever features incorporated into the IC make a big difference.

(1) input snubber The capacitor discharge resistor R_x in parallel with filter capacitor C_x can consume as much as

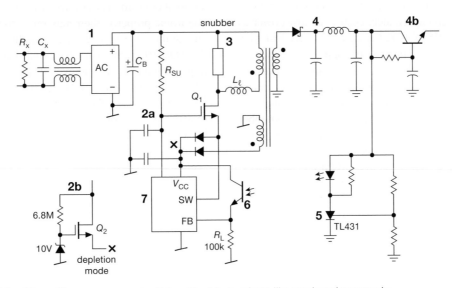

Figure 9x.26. Adding "Green" to an ac-powered switcher. The labels refer to like-numbered paragraphs.

0.1 W. (R_x is sometimes omitted in higher-power units, because the switcher's normal input current is adequate to discharge C_x rapidly.) But a clever solution is provided by Power Integrations' CapZero (§9.5.1E), which disconnects the discharge resistor when it senses that ac input is present. Alternately we could rig up the functional equivalent: a depletion-mode MOSFET and series resistor to discharge the high voltage, with negative gate drive to turn off the MOSFET when ac is present.

(2) startup Offline switching circuits generally get their operating power from a rectified auxiliary transformer winding, which is fine *after* the controller IC starts running. But they need a startup current, which is usually provided by a resistor, R_{su} coming down from the rectified ac line voltage. For example, the blocking oscillator in Figure 9.74 uses a 470k resistor, delivering about $350\,\mu A$ and consuming about 60 mW. One oft-used improvement: increase the resistor value by $15\times$, to run a zener diode at $\sim 25\,\mu A$, say, thus biasing the gate of a depletion-mode MOSFET, Q_2, as shown in item **(2b)**. Q_2's source terminal provides the startup current, but it is switched off once the transformer is running (by the $\sim 15\,V$ at the source terminal). TI plays an even cleverer trick in their '28610: they use the power switching MOSFET Q_1 to do the job, thus eliminating a high-voltage part. They start the IC with current from the Q_1's source; once things are up and running, they switch the source to ground to drive the transformer during each conduction cycle (a "cascode," see §2.4.5B).

(3) switching losses Now that we have MOSFET switch Q_1 running, let's consider switching losses. First, avoid large-die MOSFETs: their low R_{DS}(on) is good, but that comes with high capacitance, costing us lost power as the internal $\frac{1}{2}CV^2$ of stored energy is largely lost each switching cycle.[27] For the '28610 controller, the switching transistor Q_1 is external, so we can use an external MOSFET of our choosing. Second, use good transformer design to minimize the leakage inductance L_L, and avoid overly large capacitor values in the snubber network across the transformer primary. Or, for low-power applications, use a zener clamp instead. TI's advice is to use a higher-voltage MOSFET, say 800 V, and just let the transformer ring. Finally, many switching controller ICs revert to "burst mode" (see next item) at low loading. The '28610's approach, when lightly loaded, is instructive: first it reduces the switch-

ing frequency (FM mode) to 33 kHz (compare with the fixed 100 kHz of the TOP201), then it reduces the peak cycle-by-cycle primary current (AM mode) to 1/3 of its normal level (compare with a fixed current for the TOP201); the result is reduced switching and snubber losses.

(4) output ripple Most switchers operate in burst mode when lightly loaded. That is, the converter inhibits conduction cycles when the output voltage rises above the target output, and it waits until it drops below a lower threshold before resuming operation. During burst operation the output ripple is greater than during normal operation. This may require additional output filtering. One nice solution is a capacitance multiplier (**(4b)** in the figure). But note that almost any filter you choose will reduce efficiency at full power – in this case a V_{BE} of $\sim 700\,mV$ or more is lost in the series transistor.

(5) output reference and feedback The opto-isolator voltage-feedback path runs continuously, so its power is an issue. The jellybean TL431 reference used in the Astrodyne OFM-1501 needs 1 mA just to operate, thus 15 mW for a 15 V supply. We could use one of the 1.24 V $80\,\mu A$ low-power equivalents: TLV431, LMV43, or TS431. These parts also have less than $0.5\,\mu A$ reference current, about 1/10th that of a TL431, which means we can use $10\times$ higher-value divider resistors. Curiously, in their '28610 design example TI uses a zener diode and resistor in series with the optocoupler's LED, running about 1 mA zener current. Evidently the designers of a sophisticated green controller missed an obvious power-saving opportunity!

(6) control current The controller feedback circuits share one ironic property: their maximum control current occurs at the lowest power-usage condition. For example, the TOP201 wants 6 mA to signal that it's in the low-power standby state (and even if the opto-coupler has a current transfer ratio (CTR) of 200%, we'd need another 3 mA on the secondary side). So, with a typical value of 10 V to 12 V for the auxiliary supply, and comparable output voltages, we're consuming $\sim 100\,mW$ right there! By comparison, the corresponding current in the '28610 is $200\,\mu A$.

(7) controller features Quite apart from the operating current of the controller IC itself, it should be evident from our review that the design features of the controller greatly affects the power consumed by other elements of the switching converter. Review the datasheet carefully, and choose wisely.

---

[27] In some configurations it is possible to approach "zero-voltage switching," by playing your cards right; but it's better to minimize the stored energy to begin with.

A bit of historical perspective

Keep in mind that elegant all-in-one ICs like the TOP201 harken back to a different set of design compromises that characterized an era in which green "EnergyStar" issues were not in the forefront.

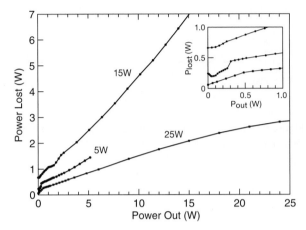

Figure 9x.27. Measured power loss and conversion efficiency of three ac-powered switching converters.

One last important issue: there are trade-offs to consider. Low-power flyback converters are generally intrinsically inefficient, first due to flyback copper and core losses, and second due to lack of a CCM mode (see §9.6.4). TI's UCC28610 circuits, with their very low standby power of about 0.1 W, can deliver up to 65 watts at full load, with a typical 85% efficiency (see Figure 9x.27, which shows the losses of two typical conventional converters, compared with the '28610 running at 25 W; note also the expansion in the inset graph). So, for example, a 20 W converter would waste about 3 watts. If operated at full power 1/3 of the time, you'd be wasting a bit over 1 W on average. For such applications you would do better by choosing a converter with higher full-power efficiency (say 95%), even if it had greater standby losses (say 0.5 W).

9x.9 Power Factor Correction (PFC)

We introduced this topic, in its two guises, in AoE3. (a) In §1.7.6 we talked about what we might call *linear* power factor, the ratio of actual power to the volt–ampere product in a linear (but reactive) circuit with sinusoidal waveforms:

$$\text{PF(linear)} = \frac{\text{power}}{|V||I|} = |\cos\theta|, \qquad (9x.3)$$

where θ is the phase angle between voltage and current. (b) In §9.7.1 we illustrated what we might call *distortion* power factor, the ratio of sinusoidal in-phase rms current to total rms current in a nonlinear load (e.g., rectifier + storage capacitor) where the current is not proportional to the voltage:

$$\text{PF(distortion)} = \frac{I_{\text{rms}}(\text{fundamental})}{I_{\text{rms}}(\text{total})}. \qquad (9x.4)$$

In either case a power factor less than unity causes excess I^2R power loss in the wiring. The linear situation is easily addressed, because reactive loads simply produce a phase shift of the still-sinusoidal current: thus in industrial plants with lots of rotating machinery, it is common to use banks of large capacitors to correct the lagging inductive current.

For the nonlinear case (think of a switchmode power converter with the powerline connected through a bridge rectifier to a bulk storage capacitor) the problem is not *phase*, it's the large rms-to-average power ratio caused by the short charging pulses.[28] What is needed is an input circuit that spreads out the current waveform to match the (sinusoidal) voltage waveform. As we showed in Figure 9.77 (reproduced here in Fig. 9x.30), the trick is to rig up an input boost converter, cleverly configured to control its switch conduction duty cycle to keep the input current proportional to input voltage *within each ac line cycle*, while simultaneously maintaining output voltage regulation. The result – the powerline is tricked into thinking it's driving a proper resistive load, as nicely seen in Figure 9x.28 (Fig. 9.78 in AoE3).

---

[28] This is often described as *harmonic currents*, the undesirable portions of the load current that are at frequencies other than the powerline sinusoid – that is, at multiples (harmonics) of the fundamental powerline frequency.

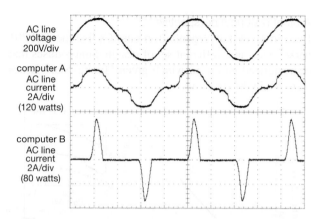

Figure 9x.28. A tale of two computers. Computer A has a PFC-input power supply, causing its input current to track the input voltage. The power supply in computer B, built ten years earlier, lacks PFC; its input bridge rectifier charges the storage capacitor with short-duration current surges. Horizontal: 4 ms/div.

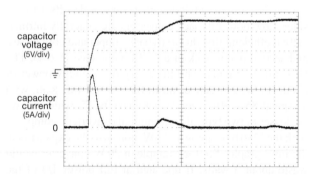

Figure 9x.29. This unregulated linear supply, good for about 5 W, powered the controls of a small refrigerator. Even with its 19 Vct input stepdown transformer, it caused nearly 15 A of 120 Vac inrush current (1800 W peak power!). Horizontal: 2 ms/div.

It's worth noting that, quite apart from the near-sinusoidal input-current waveforms in a PFC supply, another benefit of PFC is the great reduction in peak inrush current during initial powering of these supplies. In that spirit we conclude this short section with an annotated zoo of inrush waveforms, measured with a selection of power supplies we happened to have on hand.

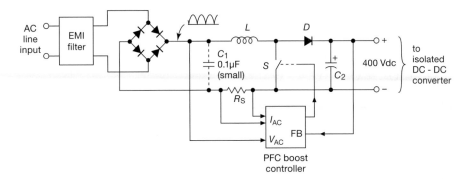

Figure 9x.30. A power factor correction (PFC) front-end consists of a boost converter running from the (unfiltered) full-wave rectified line-voltage waveform, controlled by a special PFC chip that operates the switch to maintain the input current approximately proportional to the input voltage.

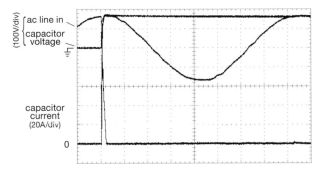

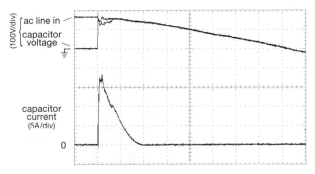

Figure 9x.31. Here we charged a $100\,\mu$F capacitor through a half-wave rectifier, closing the switch at the peak of the input powerline waveform. The peak inrush current is nearly 100 A (10 kW!). Horizontal: 2 ms/div. For a better view of the gory details, see the next figure.

Figure 9x.33. This 15 W switchmode converter (shown at top right in Fig. 9.1) generated an 18 A (2 kW!) inrush current when switched on at the peak of the ac powerline cycle. It evidently lacks input PFC, confirmed by its circuit (Fig. 9.83 in AoE3). Horizontal: $400\,\mu$s/div.

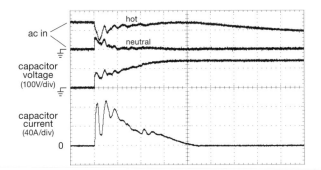

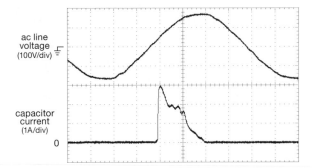

Figure 9x.32. With the same setup as in Fig. 9x.31, we speeded up the horizontal sweep ($100\,\mu$s/div) to reveal the impressive crow-barring of the ac powerline by this hardswitched $100\,\mu$F storage capacitor.

Figure 9x.34. The inrush current is greatly reduced if you switch on the power near a zero-crossing of the input ac waveform. Here we powered up the converter of Fig. 9x.33 with a "ZVS" solid-state relay, which switches on a bit late (about 10° past zero-crossing). The peak current is reduced by a factor of six, while the charging time is extended by about the same factor. Horizontal: 2 ms/div.

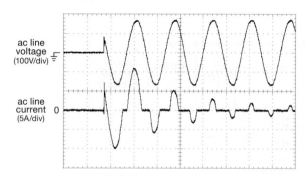

Figure 9x.35. Now for some *real* power: this is a computer 250 W switchmode converter with PFC (ATX-style, FSP brand). It's rated at 15 times the power output of the non-PFC supply of Fig. 9x.33, but with half the peak inrush current – not bad! Horizontal: 10 ms/div.

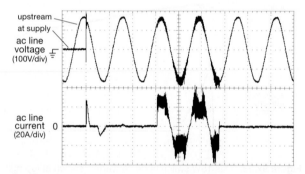

Figure 9x.36. This "no-name" 460 W ATX switchmode supply also has PFC, but it's plagued by oscillations, excessive peak current, and poor sinusoidal tracking. Horizontal: 10 ms/div.

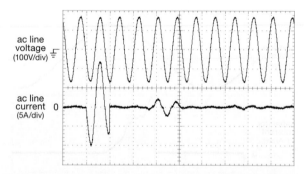

Figure 9x.37. The best of the batch of ATX switchmode supplies: this 550 W Corsair power supply exhibits excellent sinusoidal tracking, with peak inrush current of just 10 A in spite of its significantly greater power capability compared with the supplies of Figs. 9x.35 and 9x.36. Horizontal: 20 ms/div.

9x.10 High-Side High-Voltage Switching

We dipped our toe into this topic in Chapter 12, where we showed some ways to use *p*-channel MOSFETs as high-side switches; and we showed the treacherous waters that there reside, largely related to the difficulty of protecting the series switch against fault conditions (most notably a shorted output).

Much of the time a better solution is the use of a protected high-side switch, basically a smart IC that incorporates fault sensing (overcurrent, overvoltage, over temperature) and shuts down the drive (see also §12.4.4). These devices (see Table 9x.2) all have built-in level-shifting circuitry so you can drive them from ground-referenced logic inputs (Fig. 9x.38). Devices with I_S in the comment column provide a current-monitor output.

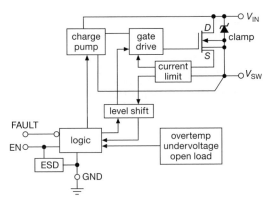

Figure 9x.38. High-side intelligent switches have a charge pump and level-shifting circuits to drive the *n*-channel gate beyond the drain input supply; and they protectively monitor fault conditions such as overvoltage, overcurrent, and overtemp.

The protected devices include a switching power MOSFET. Most are *n*-channel types, with their gates driven by charge pumps, but some use *p*-channel MOSFETs, as noted in the comments. These have much faster switching times, because they directly drive the gates.

Most of the parts allow excess turn-off voltages for inductive flyback load dumping, and many include an active-clamping function for this purpose. We have not listed the maximum allowed voltages for these, but generally the maximum V_{in} value is also the maximum input-to-output pin voltage.

To escape the modest ratings in the table (highest voltage: 65 V), you can make your own high-side switches, with current limiting, as illustrated in Figure 12.45. There Q_1 switches a current, so with a single MPSA42 you could switch 250 V, or you could cascode a few for much higher voltages.

If you are willing to give up junction temperature protection, there are times you may prefer a driver that works with an external discrete MOSFET or IGBT switch. Some of these are intelligent, such as the ACPL-332J, with a "desat" function exploiting switch desaturation to detect short circuits, and produce a "fault" output.

There are situations where you don't need an intelligent high-side switch, you just need a switch. Simple drivers like the FAN7371 come in handy for driving an *n*-channel MOSFET, switching loads anywhere from 10 V to 600 V.

See also the discussion of switch driver ICs in §3x.15. These can work up to 600 or 1200 V, but the high side also works well down to zero. Most include a matching low-side driver (see Figure 3x.108 and Table 3x.5) so they're well suited for use in half-bridge or H-bridge dc–dc converter circuits. Most of the IC choices do not include any protection features, but some, like the IR22141 in Table 3x.5, include desaturation detection on both the high and low sides. We discuss high-side current sensing techniques in the next section, which can be used at high voltages, but switches like those discussed here may need a faster response. The desaturation shutdown speed is typically set to 1 μs–2 μs, slow enough to avoid high $L dI/dt$ transients, but fast enough that the switch's thermal mass can prevent damage.

Table 9x.2: Selected High-Side Switches[a]

Type	Switches	V_{in} min (V)	V_{in} max (V)	I_o max (A)	R_{DS} typ (mΩ)	I_s typ (mA)	V_L min (V)	t_{ON} typ (ms)	Cost qty25 ($US)	Logic input?	Fault output	Active clamp?	Package	Comments
FDG6323L	1	2.5	8	0.6	550	b	1.5	0.01	0.35	•	•	-	SC-70-6	nMOS drvr + pMOS hi-side[b]
TPS22960	2	1.8	6	0.5	435	0.00	1.6	0.08	0.95	•	-	-	SOT-23-8	p-channel switch
FPF2110	1	1.8	8	0.4	160	0.08	1.8	0.03	0.99	•	•	-	SOT-23-5	p-channel switch
FPF2123[g]	1	1.8	8	1.5[c]	160	0.08	1.8	0.03	1.05	•	-	-	SOT-23-5	p-channel switch
MIC2514	1	3	14	1.5	900[d]	0.08	2.3	0.01	1.90	•	-	-	SOT-23-5	p-channel switch
STMPS2151	1	2.7	6	0.5	90	0.04	2.2	1	0.81	•	•	-	SOT-23-5	
AP2156	2	2.7	5.5	0.8	100	0.09	2.2	0.6	0.85	•	2	•	SOP-8	USB-port power
TPS2041	1	2.7	5.5	0.7	80	0.08	2.2	2.5	2.59	•	•	-	DIP-8	
BTS452	1	6	62	1.8	150	0.8	2.5	0.08	1.97	•	•	•	TO-252-4	
BTS410	1	4.7	65	2.7	190	1.0	2.5	0.10	3.15	•	•	•	TO-220-5	see also BTS462T
BTS611	2	5	43	2.3	200	4	4	0.20	3.17	•	1	•	TO-220-7	
IPS511	1	6	32	5	135	0.7	3.3	0.05	1.76	•	•	•	TO-220-5	
FPF2702	1	2.8	36	2[h]	88	0.09	0.8	2.7	2.10	•	•	-	SO-8	adjustable current limit
IPS6031	1	6	32	16	60	2.2	3.6	0.04	2.52	•	•	•	TO-220-5	
BUK202-50Y	1	5	50	20	28	2.2	3.3	0.14	4.16	•	•	•	TO-220-5	
BTS432	1	4.5	63	35	30	1.1	2.7	0.16	4.85	•	•	•	TO-220-5	
BTS6142	1	5.5	45	25	12	1.4	n	0.25	2.65	n	e	•	TO-252-5	$I_S = I_L / 10k$ ($\pm$20% at 30A)
BTS6133	1	5.5	38	33	10	1.4	n	0.25	3.97	n	e	•	TO-252-5	$I_S = I_L / 9.7k$ ($\pm$10% at 30A)
VN920	1	5.5	36	30	16	5[k,m]	3.6	0.10	3.10	•	•	•	TO-220-5	
BTS442	1	4.5	63	70	15	1.1	2.7	0.35[f]	5.37	•	•	•	TO-220-5	
IPS6011	1	6	35	60	14	2.2	3.3	0.07	4.40	•	•	•	TO-220-5	
BTS6144	1	5.5	30	37	9	2.2	n	0.30	4.77	n	e	•	TO-220-7	$I_S = I_L / 12.5k$
BTS555	1	5.0	44	165	1.9	0.8	n	0.6[f]	5.80	n	e	•	TO-218-5	to 480A surge, $I_S = I_L / 30.2k$

Notes: (a) all are n-channel switches, with charge pumps, unless marked; all have overcurrent and overtemperature protection. (b) transistor pair, not intelligent, add your own source and gate resistors. (c) adjustable 0.15–1.5A. (d) at 12V. (e) one pin signals load current and faults. (f) max. (g) shuts down after 10ms, retries every 160ms thereafter. (h) adjustable 0.4–2A. (k) 10μA when off. (m) max. (n) Requires MOSFET closure to GND, sinking I_S.

9x.11 High-Side Current Sensing

There are lots of circuit situations where you need to sense current flowing on the dc power supply's "high side." Most often that's because you need to keep a low impedance in the common (ground) path, which you don't want to interrupt with a current-sensing resistor. What you do, then, is to sense the voltage drop across a small-value resistor in series with the positive supply. So far, so good. The challenge comes when you need to convert the measured high-side current to a voltage signal that is referenced to ground, especially if the dc rail is at a high voltage.

In this section we'll look at a few examples, enriched with a table of high-side current-sensing ICs.

9x.11.1 Pulse generator overcurrent limit

Here's a nice example (Fig. 9x.39) of high-side current monitoring at work, applied to a high-energy pulse generator – the Rowland Institute's RIS-764, a very fast high-voltage high-power pulse generator (up to 20 A and 850 V; see §3x.15.2). That's dangerous territory, and the power components are at risk of damage if it's operated at too high a repetition rate. But that was exactly our goal, running at pulse rates up to 10 MHz or so. A weak point is its super-fast MOSFET gate driver circuit.

The output stage consists of a push–pull pair of SiC (silicon carbide) *n*-channel MOSFETs, whose gates are driven to +16 V (ON) and −4 V (OFF) by fast gate drivers (Fig. 3x.111). Because the upper MOSFET is a "flying switch," we needed a low-capacitance isolated dc–dc converter. We also needed fast drivers, in this case finally settling on the UCC27538:[29] the good news is that it's fast (17 ns into 1800 pF); the bad news is that it comes only in an itty-bitty package (SOT23-6, would you believe?). When you're switching fast, the driver's dynamic current (switching the sum of its own capacitance and that of the MOSFET gate) can approach hundreds of milliamps or more, thus several watts or more of dissipation.

For example, imagine driving the 260 pF of gate capacitance (C_{iss}) of our chosen SiC MOSFET through 25V[30] at 10 MHz: that results in $P=CV^2f=1.6$ W of driver dissipation. That's too much for a small SOT23 package (whose typical junction-to-board thermal resistances are of order $\Theta_{JB} \approx 70°C/W$), so it's important to ensure that the drivers stay within allowed dissipation – a lesson we learned the hard way when a failed driver took out about a dozen other parts! Since the driver rails are fixed, supply current is a good proxy for power dissipation, and it suffices to disable pulsing when excessive current is sensed on the +12 V rail that powers the dc–dc isolated converters.

We used the elegant LTC6101, intended for high-side current monitoring. It is powered by the same high-side rail that it is sensing, which can range from +4 V to +60 V (+100 V for the LTC6101HV version), and it converts the sensed voltage drop across R_S to an output current, with the scale factor set by the external resistor R_G: $I_{out}=V_{sense}/R_G$. That current, flowing through R_1, generates a voltage (with respect to ground) of $V_{out}=I_S R_S R_1/R_G$, or, for our chosen component values, 1.0 V per amp. The LTC6101's good accuracy (offset voltage of 85 μV typ, 300 μV max) allowed us to use a low-value sense resistor (0.025 Ω), minimizing dissipation (56 mW at the 1.5 A set point) while preserving adequate accuracy ($\pm$12 mA maximum error). One thing to watch out for, when you send a current down from a high-side circuit, is excessive power dissipation in the high-side current source; here that's not a problem – the LTC6101's dissipation at the maximum 1.5 A supply current is just 4 mW (do the math to check our result).

The rest of this circuit fragment is straightforward: Hysteretic comparator U_{2b} disables gate U_3 and blocks pulsing when the sensed current reaches 1 A, but allows for self-reset; C_1 allows for momentary power transients, permissible owing to thermal mass. Undervoltage comparator U_{2a} locks out pulsing when the +12 V rail is below 11 volts. The two comparators are wired-ORed, and light a FAULT indicator (with 40 ms stretch) when either comparator triggers. For the simple overcurrent task at hand, the LTC6101 has adequate precision; for applications requiring greater accuracy there are chopper-type ("auto-zero") parts available, for example the voltage-output INA210–15 series (0.55 μV typ, 35 μV max), see Table 9x.3 on page 426.

---

[29] The circuit evolved through nine successive versions as we addressed (and conquered) various demons; variations included using paralleled gate drivers, or slower drivers in TO-220 power packages attached to the board's large fan-cooled heatsink, or fast but sensitive drivers (with frequency-measuring cutoff circuitry).

[30] Silicon carbide MOSFETs excel in low gate capacitance, but they require a larger gate swing than silicon MOSFETS; the C2M0280120D devices we used recommend +20 V and −5 V.

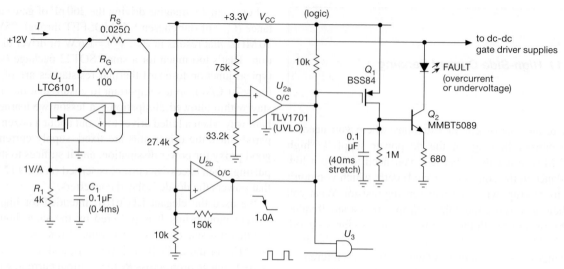

Figure 9x.39. Pulse generator overcurrent inhibit with high-side current sensing.

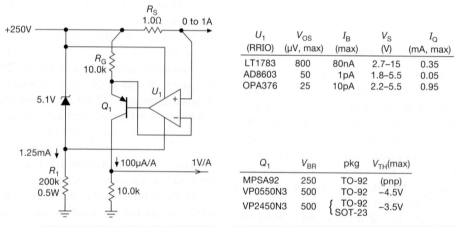

Figure 9x.40. High-side current monitor implemented with discrete current-source transistor and op-amp.

U_1 (RRIO)	V_{OS} (μV, max)	I_B (max)	V_S (V)	I_Q (mA, max)
LT1783	800	80nA	2.7–15	0.35
AD8603	50	1pA	1.8–5.5	0.05
OPA376	25	10pA	2.2–5.5	0.95

Q_1	V_{BR}	pkg	V_{TH}(max)
MPSA92	250	TO-92	(pnp)
VP0550N3	500	TO-92	−4.5V
VP2450N3	500	{ TO-92 / SOT-23	−3.5V

Epilogue:

For operation up to moderate frequencies (say 4 MHz) the SOT-23 drivers work fine. But ultimately we wanted more, and used instead TI's excellent UCC21520 dual gate driver. It's almost as fast (19 ns), comes in a larger package (SOIC16), and, best of all, it includes the 1.5 kV isolator for the high-side driver, so for input you need only ground-referenced logic levels; it even includes programmable dead-time (to prevent MOSFET shoot-through): it's one-stop shopping! To reduce power dissipation in the driver we buffered the outputs with push–pull BJT followers, followed by robust (1812-size) series gate resistors, unloading about 75% of the total driver dissipation.

9x.11.2 Current monitor for high-voltage amplifier

Another high-voltage application, this time for a linear amplifier with large (and, later, bipolarity) output swings. Here we put the monitor on the "high side," rather than meddling with the ground return, or (would you try this?) on the signal-bearing output line. Figure 9x.40 shows the circuit fragment, where the current-sensing circuitry is similar to that of Figure 9x.39, but implemented with a discrete high-voltage BJT or MOSFET (to send the current proxy down), driven by an op-amp that is powered between the HV rail and a zener-drop downward.

Because of the large voltage span, you've got to keep currents low to circumvent the need for heatsinking. So we chose candidate op-amps that draw less than a milliamp, and we pick sensing resistor R_S and gain-setting resistor R_G for a proxy current of $100\,\mu$A (thus 25 mW) at the full-scale 1 amp load current.

The small-signal MPSA92 *pnp* transistor is a good choice for this application; it's good for 300 V, and it has lower capacitance (thus better speed) than the *p*-channel MOSFETs listed in the figure. But even the latter top out at 500 V, so for higher voltages you need to stack a series chain, as in Figure 9x.41. Note in that circuit that both the resistive divider's current and the base currents of the lower members of the string (Q_2, Q_3, etc.) are included within the current source loop, and delivered to the output, so no error is incurred. There *is* a (small) error, however, from Q_1's base current, in both circuits (Figs. 9x.40 and 9x.41). That error is absent if you use MOSFETs, of course. But there's a nice circuit you can use for *canceling* the base-current error with BJTs – see the elegant trick in §4x.17.

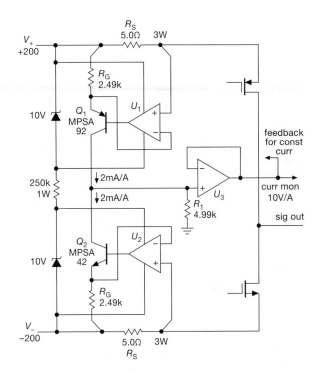

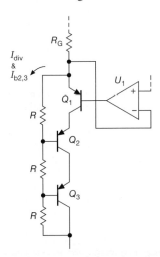

Figure 9x.41. Extending voltage range with series-connected BJTs. Note that the base divider string with its associated currents does not introduce error, because those currents are both measured and delivered to the output.

A. Current monitor for HV bipolarity amplifier

Now things get interesting! Elaborating on this idea, we designed a *bipolarity* linear amplifier, with output swings to ±200 V rails. How to monitor the output current? You could imagine a "flying" current monitor in series with the output; but you'd have to worry about common-mode rejection, and figure out how to send the (bipolarity) output-current signal down to ground.

But there's a better way: just replicate the rail-monitoring technique of Figure 9x.40 on both rails, and combine their output currents at a node near ground. That way the current monitors operate at stable (though high-voltage) potentials. Figure 9x.42 shows how.

Figure 9x.42. Output current monitor for a high-voltage bipolarity linear amplifier, implemented as summed currents from high-side current monitors on both rails. The push–pull MOSFETs are the output stage of the overall amplifier, the ghostly hints of a rather complex underlying circuit.

In this circuit we replicated the current monitor of Figure 9x.40 on *both* supply rails: U_1Q_1 sources a current of 2 mA/A of current sensed on the positive rail, and U_2Q_2 sinks an analogous current for the negative rail. The (signed) sum of these currents, converted to a voltage by R_1, is a measure of the load current, with a scale factor of 10 V/A. Perhaps most interesting, that monitor voltage, representing output current, can be used as a feedback signal: you've got a high-voltage current-output linear amplifier! In our design we included a switch, so the user can select between feedback signals representing output voltage or output current.

A subtle (or maybe not-so-subtle) point: the push–pull output stage of our amplifier was operated class-AB, i.e., with substantial quiescent current flowing from V_+ to V_-, for enhanced speed and linearity. Because those supply-rail currents are monitored, you might at first think that would put a fly (actually, two flies) in our nice ointment. But no, the standing currents cancel out in the monitoring system, so only the actual delivered output current appears in the summed monitor output (and in the feedback signal, when configured for *current* output) – cute!

Table 9x.3: High-Side Current-Sensing ICs[a]

Part #	V_{supply} min (V)	V_{supply} max (V)	I_Q^x typ (µA)	V_{high} range (V)	CMRR typ (dB)	V_{os} max (µV)	BW[b] typ (MHz)	Output Format[f]	Bidirectional?	Pkgs[c]	Price[z] qty10 ($US)	Comments
AD7294	4.5	5.5	70mW	5 to 60[d]	80	8000[e]	3µs	V	•	TQ64, CS56	16.69	1
ADM1075	−35	−80	5500[p]	−35 to −80	-	170	200µs	D	no	TS28, CS28	9.23	2
INA133U	4.5	36	950	$2V_- + 2$ to $2V_+ - 2$	90	450	1.5[g]	V	•	SO8	3.25	3
INA138	2.7	36	25	2.7 to 36[i]	120	1000	$4/R_{load}(k\Omega)$	I	no	SOT23-5	1.68	4
INA139	2.7	40	60	2.7 to 40[i]	115	1500	$4.4/R_{load}(k\Omega)$	I	no	SOT23-5	1.48	5
INA170	2.7	40	75	2.7 to 60[i]	120	1000	$4/R_{load}(k\Omega)$	I	h	MS8	2.18	6
INA193–95	2.7	18	700	−16 to +80[i]	94[j]	2000	0.5	V	no	SOT23-5	2.02	7
INA199	2.7	26	65	−0.3 to +26	120	150	$3/G_V$	V	h	SC70-6, UQ10	0.63	8
INA200–02	2.7	18	1400	−16 to +80[i]	100[j]	2500	0.5	V	no	SO8, VS8	2.44	9
INA206–08	2.7	18	1800	−16 to +80[i]	100	2500	0.5	V	no	SO14, TS14, VS10	1.86	10
INA210–15	2.7	26	65	−0.3 to +26[i]	140	35	0.014	V	h	SC70, UQ10	1.27	11
INA220	3	5.5	700	0 to 26[i]	120	50[u]	600µs[y]	D	•	VS10	2.59	12
INA223	2.7	5.5	200[n]	0 to 26[i]	120	150	$3/G_V$	V	no	SON10	2.89	13
INA225	2.7	36	300	0 to 36[i]	105	150	0.25	V	h	MS8	2.43	14
INA226	2.7	5.5	330	0 to 36[i]	140	10	8ms[v]	D	•	VS10	3.09	15
INA240	2.7	5.5	1800	−4 to +80[i]	132	25	0.4	V	h	TS8	2.17	16
INA250	2.7	36	200	−0.1 to +36[i]	102–118	100–20mA	0.05	V	h	TS16	2.85	17
INA270–71	2.7	18	700	−16 to +80[i]	120	2500	0.13	V	no	SO8, VS8	2.21	18
INA282–86	2.7	18	600	−14 to +80[i]	140	70	0.01	V	h	SO8, VS8	1.41	19
LMP8480–81	4.5	76	88	4 to 76[i]	124	265	0.27	V	s	VS8	2.26	20
LMP8601–03	3	5.5	1000	−22 to +60[v]	105	1000	0.06	V	h	SO8, VS8	3.02	21
LMP8640	2.7	12	720	−2 to +42[i]	103[m]	900	0.95	V	no	SOT23-6	2.33	22
LMP8645	2.7	12	450	−2 to +42[i]	95[m]	1000	0.85 $(G_V=2)$	V, I	no	SOT23-6	2.33	23
LT1638	2.7	44	170	V_- to $V_- + 44$	76[m]	600	1.1	V	•	DIP8, SO8	2.60	24
LT2940	6	80	3500	4 to 80[i]	-	7000	1	I	•	MS12, DFN12	2.93	25
LTC2945	4	80	800	0 to 80[i]	-	75	100ms	I2C	no	MS12, QFN12	5.73	26
LTC2946	2.7	100	900	0 to 100[i]	-	50	65ms	I2C	no	MS16, DFN16	4.80	27
LTC2947	4.5	15	9000	−0.1 to 15[i]	120[m]	9mA	100ms	I2C/SPI	no	QFN32	8.50	28
LTC4151	7	80	1200	near V_S	-	100	130ms	I2C	no	MS10, DFN10	5.55	29
LTC6101	80	60	250	V_S−1.5 to V_S+0.5	140	300	0.14	I	no	SOT23-5, MS8	2.50	30

Notes: (a) most have full-scale voltage range of 80–500mV. (b) or conversion time; for parts with multiple G_V choices, the BW of the lowest gain or lowest-numbered part is listed. (c) CS=LFCSP; MS=MSOP; SO=SOIC; TQ=TQFP; TS=TSSOP; UQ=UQFN; VS=VSSOP. (d) requires high-side V_P supply; logic and analog circuits referenced to low side. (e) can be chopped to remove offset. (f) D=PMBus/I2C; must isolate to create ground-referenced logic signals. (g) −3dB BW. (h) offsetting input pin. (i) for any supply voltage. (j) 120dB for V_{cm}>12V. (k) by suffix. (m) min or max. (n) 0.1µA disabled. (o) pin settable. (p) internal shunt regulator, must supply 5.5mA min via ext resistor. (q) bus programmable. (r) by part number. (s) LMP8481 only. (t) typical. (u) at maximum PGA gain. (v) down to 140µs, with degraded noise and accy. (w) g_m=200µA/V, ext resistor sets voltage gain, follower output. (x) per amplifier. (y) down to 90µs with degraded accy. (z) of first p/n.

COMMENTS: (1) multifunction, 2 x hi-side, 12-bit 3µs SAR adc x9 MUX, plus temp and dacs; I2C config. **(2)** hot-swap controller, gate control output for soft-start, overcurrent. **(3)** difference amplifier, G_V=1. **(4)** g_m=200µA/V: G_V=1 to 100, set by R_L: G_V = 0.2R_L(kΩ); INA168: to 60V. **(5)** g_m=1000µA/V: G_V=1 to 100, set by R_L: G_V = R_L(kΩ); INA169: to 60V. **(6)** g_m=1000µA/V; buffered offset input. **(7)** G_V=20; INA194,195: G=50,100; alternate pinouts: INA196–198. **(8)** "zero-drift"; 1% accy; G_V^k=50,100,200. **(9)** + uncommitted comparator; G_V=20; INA201,202: G=50,100. **(10)** +2 uncommitted comparators; G_V=20; INA207,208: G=50,100. **(11)** G_V=200; "zero-drift," 0.02% accy; INA211-215: G=500, 1k, 50, 100, 75. **(12)** current, voltage, power; PGA; 12-bit ADC; 0.2% accy; INA219: SOT23-8, SOIC8. **(13)** G_V=20,128,300[q]; "zero-drift"; 0.2% accy; voltage, current, power monitor; PMBus/I2C config. **(14)** G_V=25,50,100,200[o]; "zero-drift"; 0.05% accy. **(15)** current, voltage, power; 16-bit ADC; 0.2% accy. **(16)** G_V=20,50,100,200[k]; "zero-drift"; enhanced PWM rejection. **(17)** G=0.2, 0.5, 0.8, 2V/A[k]; "zero-drift"; internal 2mohm 0.1% R_S; 0.3% accy. **(18)** G_V=14, 20[r]; external filter pin. **(19)** G_V=50,100,200,500,1000; "zero-drift"; 0.4% accy. **(21)** G_V=20,50,100[k]; 0.5% max gain error. **(20)** G_V=20,60,100[k]; 0.6% max gain error. **(22)** G_V=20,50,100[k]; 0.25% max gain error; LMP8640HV: −2 to +76V. **(23)** G_V=1-100[w]; LMP8645HV: −2 to +76V. **(24)** "over-the-top" micropower RRIO op-amp; LT1639=quad. **(25)** g_m=1000µA/V. **(26)** current, voltage, power; 12-bit SAR adc; -1 version has inverted data. **(27)** current, voltage, power, charge, energy; 12-bit SAR; 1% accy; -1 version has inverted data. **(28)** I, V, P, Q, energy; internal 300µΩ R_S, 1% current accy; 15-bit ΔΣ ADC. **(29)** current, voltage; 12-bit ΔΣ ADC; -1 has separate SDAI and SDAO. **(30)** R_1 sets g_m: g_m = 1/R_1(MΩ); Vout=IR$_S$R$_L$/R_1; LTC6101HV: 5 to 100V.

9x.12 High-Voltage Discharge Circuit

When you've got high voltages rattling around in a circuit, it's a good idea to ensure that storage capacitors get themselves discharged when the thing is turned off. We

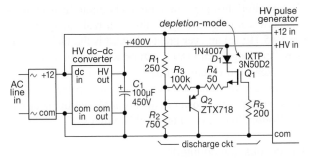

Figure 9x.43. Depletion-mode MOSFET Q_1 discharges the 100 μF HV capacitor C_1 when ac power is removed, but is inactive when the circuit is powered.

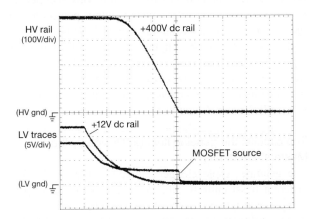

Figure 9x.44. Measured waveforms for the high-voltage active-discharge circuit of Fig. 9x.43. Horizontal: 400 ms/div.

mentioned this in AoE3 §3.6.2B, where we showed an example circuit.[31] This sort of "active bleeder" is a good thing – it draws no current from the high-voltage rail during powered operation, unlike the classic fixed bleeder resistor – but it springs into action, with relatively high discharge currents available when the low-voltage supply decays after ac power removal. The idea (Fig. 9x.43) is to use an

n-channel depletion-mode MOSFET to discharge the HV rail: its gate is at ground potential, but its source terminal is biased $\sim$10 V positive by any available low-voltage supply that's part of the overall HV circuit.

We rigged up this circuit, and Figure 9x.44 shows measured waveforms, commencing with the removal of ac power. The MOSFET starts conducting when the source terminal has dropped to about +4 V, with the 50 Ω resistor providing source biasing to limit the discharge current to about 50 mA (estimated by noticing that $\sim$2.4 V is across R_4, and confirmed by the observed discharge rate $I = C\,dV/dt$).

---

[31] But see the corrected circuit here and at artofelectronics.net/errata.

When we bought a hand-me-down (i.e., used) iPhone recently, it was listed as

> **Condition: New** – This Certified Refurbished product is factory refurbished, shows limited or no wear, and includes all original accessories plus a 90-day warranty.

Although one could take exception to this use of the word "new," the phone itself was fine – but the "original accessories" were highly suspicious. Most interesting was the little white 5 W power brick, outwardly a pretty good imitation of the real thing (Fig. 9x.45), but on closer inspection an obvious fake. It put up a good fight, but a misspelling clearly brands it counterfeit. And although it copied the model number, manufacturer, and most other aspects of the design, it had the decency to omit the top line, and to replace the official Underwriters Laboratories (U$_L$) symbol with a meaningless "M." And, perhaps revealing of a guilty conscience,[32] the author(s) downgraded the country of origin to a lower-case "china."

On the bench

We powered up these puppies and looked at the dc output under various loads. The Apple charger delivered on its promise of "5 V, 1 A" (Fig. 9x.46), with a worst-case ripple of 20 mVpp, and 25 mV drop going from no-load to full-load. By contrast, the fake's performance was, well, *terrible* (Fig. 9x.47), with 1 V spikes (at about 20 Hz) when unloaded, 1 Vpp ripple and high-frequency hash when loaded, and a maximum output current of just half its rated 1 A. As The Donald might say, "you're fired!"

The inside story

As the entertaining Dave Jones (of eevblog.com) likes to say, "don't turn it on, *take it apart!*" We did both. The powerline side of the fake seemed loose, and it popped out easily, held in place only by a pair of diminutive 0.5 mm plastic protrusions. The real Apple charger was a different story – no amount of pulling or prying did anything – we

had to cut all the way around with a Dremel rotary tool to separate the base. Figure 9x.48 compares the innards, revealing a complex high-density PCB board pair in the genuine charger, each packed with SMT parts, compared with a simple single-sided phenolic PCB with just a few through-hole parts in the fake. Obviously there's a lot less circuitry in the fake, and it shows in the measured performance.

We'll get to the circuits presently, but a few first impressions upon examining the fake: (a) it has no filter inductors, neither at input nor at output; (b) it has no line-rated "Y-capacitor" bridging the input and output, instead using a generic ceramic disc type; this is a real safety hazard, see §9.5.1E; (c) an ever greater hazard is the lack of sufficient "creepage" path (§9.7.2D) between input and output – we measured extensive trace pairs separated by a scant 0.6 mm; and (d) it has no overtemperature protection, and no fuse to interrupt power in the event of a component failure. Bottom line: this thing should be illegal.

The circuits: the counterfeit

We traced out the circuit of the counterfeit charger, an easy task because it has only 15 parts, all of which are discrete components (no ICs); it is shown in Figure 9x.49. The powerline ac is applied directly to a full-wave rectifier (no fuse, no interference filter), whose output powers a simple *blocking oscillator* running at about 100 kHz. The secondary is configured as a flyback (i.e., no conduction during the primary power cycle), again without RFI filtration. Voltage regulation (if you can dignify it with that name) is provided by the optocoupler, which shunts oscillator drive when 3.9 V zener Z_1 conducts.

The circuits: the genuine item (a "Design by the Masters" candidate)

The genuine Apple charger's circuit[33] (Fig. 9x.50) is a different beast entirely. The designers, while packing 68 components into the same 1 cubic inch, adhered to industry standards: overtemperature shutdown, input fusing, RFI filtration at both input and output, ac-rated Y-capacitor (Y_1), and 5 mm minimum creepage paths. The assembly is exemplary: all components are well anchored, with insulating tape and elastomeric material applied liberally, and with a metallic shield covering the SMPS control circuitry (seen in the lower left image of Fig. 9x.48.) Their circuit exploits a "quasi-resonant" current-mode controller (ST's L6565, evidently, according to Shirriff), which minimizes

---

[32] Or maybe thinking of this item as highly breakable (like dishware), which, as we'll see presently, is close to the truth.

[33] Rearranged and annotated from the very fine work of Ken Shirriff, see his teardown and tutorial at righto.com/charger.

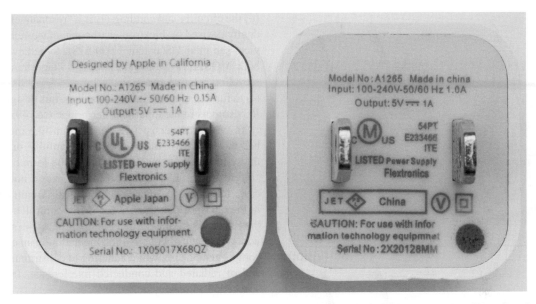

Figure 9x.45. Real Apple chargers don't misspell "equipment," nor do they forget that the powerline input current is way less than the 5 V dc output current.

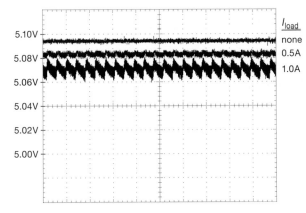

Figure 9x.46. Apple A1265 5 W charger output, for full-load, half-load, and no-load conditions; note expanded vertical scale, with offset zero. Horizontal: 20 ms/div.

switching losses through zero-voltage switching (ZVS, see §9.6.8D) by sensing transformer demagnetization (the ZCD "zero-current detect" input), and which implements low-power features such as very low quiescent current and frequency foldback at light loads. The result is a safe and electrically quiet power supply with good efficiency: 77% at 50% load, compared with 72% for the simpler counterfeit charger.[34]

Some comments on the circuit:

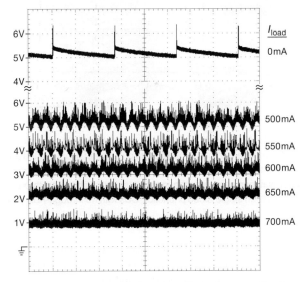

Figure 9x.47. Imitation 5 W charger output under several load conditions (the no-load trace is plotted separately because it overlaps the half-load trace). Note coarse vertical scale, compared with Fig. 9x.46 where the vertical axis was expanded 50×. Horizontal: 20 ms/div.

---

[34] Perhaps surprisingly, the sophisticated Apple charger consumes more *standby* power than the cheap imitation: 200 mW versus 30 mW, respec-

tively, by actual measurement. To put these numbers into perspective, running 200 mW for a year consumes about $0.35 worth of electricity (at a typical domestic electricity cost of $0.20/kWh). That's not much – but then again if you have a dozen little gadgets idling at 200 mW, the

dc (output)
ac (input)

1cm

Figure 9x.48. A world of difference inside: the Apple charger (left side) is replete with quality SMT parts on a pair of double-sided and plated-through fiberglass FR-4 PCBs, with careful attention to mandated safety clearances; the imitation is built with cheap through-hole parts on single-sided phenolic PCBs with no through-hole plating.

(a) voltage feedback compares the dc output with the 2.5 V threshold of TL431 shunt regulator U_2, optically coupled via OC_1 to the controller's error amplifier;

(b) a separate feedback path is used to trigger shutdown (via latch Q_3Q_4) on either of overvoltage or overtemperature, again using a TL431 (U_3) to set thresholds, but without the need for linear compensation;

(c) because the switching cycle depends on zero-current timing, the frequency changes with load, going from 135 kHz (full load) to 320 kHz (no load);

(d) the circuitry surrounding Q_2 is a "resonant clamp," evidently devised by Flextronics (the manufacturer of this device, see their US patent 7,924,578), to achieve ZVS for both MOSFETs while minimizing voltage transients and recycling reactive energy;

(e) to further suppress inductive spikes, this design includes *RC* dampers in the input rectifier bridge (see §9x.6).

To put some perspective on a design like this, remember that much of the circuitry is helpfully provided by the semiconductor manufacturers, in the form of so-called "reference designs," and also in application notes. In this case, for example, much of the critical circuitry is found in the datasheet for the L6565 SMPS controller IC and in the Flextronics patent. The larger part of Apple's challenge was to package a reliable implementation in an elegant and compact enclosure, while adhering to constraints of safety, and of radiated and conducted interference. In this they succeeded admirably.

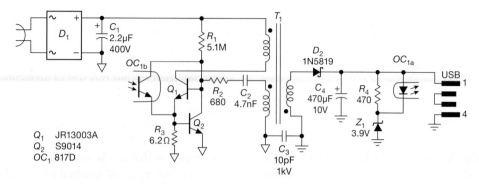

Figure 9x.49. Circuit of the counterfeit Apple "5 W" charger: 15 components in all, in a simple blocking oscillator configuration that provides at most 2.5 W. Regulatory and safety violations include lack of fuse or temperature sensing, lack of input or output filters, non-ac rated C_3, and multiple dangerous 0.6 mm creepage paths.

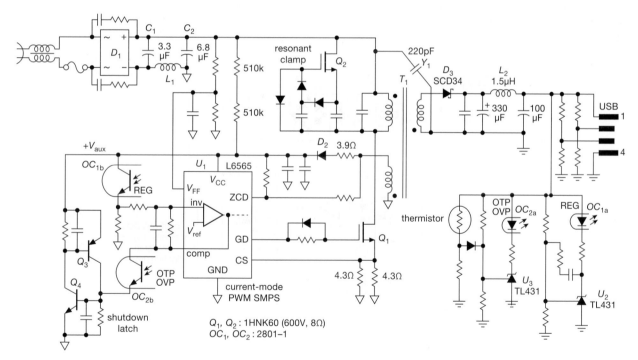

Figure 9x.50. Circuit of a genuine Apple 5 W charger, adapted from the reverse-engineered schematic by Ken Shirriff: 68 components in a sophisticated current-mode configuration, with a full complement of safety and interference measures – fuse, filters, overtemperature sensing, zero-voltage switching, spike suppression, resonant clamping, and fully compliant creepage and gap dimensions.

9x.14 Low-Noise Isolated Power

Let's say you want to make sensitive measurements of low-level analog signals with a high-resolution ADC, and convey the results to a microcontroller; or perhaps going in the opposite direction, you want to generate stable dc voltages from a DAC, with low noise and microvolt stability. For these kinds of instrumentation challenges you'll want to isolate the low-level analog front-end from the noisy digital controller, using digital signal isolators (e.g., optocouplers, see §12.7) to separate the grounds. You'll need isolated dc supplies, also, especially because some of these applications may run the analog circuitry with its "ground" floating at a substantial voltage difference (tens or hundreds of volts offset) from the controller's ground.

Figure 9x.51 shows the basic idea, with some typical part numbers that permit data rates to 25 Mbps or more (we've listed both opto- and capacitively-coupled isolators, for variety). The figure shows a 2 W isolated power converter, intended for just this sort of application; it converts +5 Vdc at the driver side to an unregulated (and floating) 7 Vdc nominal output,[35] which is easily regulated down to 5 V with a low-dropout linear regulator, as shown.

So far, so good. The problem we've encountered in this tidy scheme is that the power isolator (a switching converter clocked at several hundred kilohertz) couples some of its clocking signal onto the isolated output. This appears as a spiky common-mode signal whose amplitude is typically 100 mV or more. Figures 9x.52 and 9x.53 show measured waveforms of the voltage impressed on the isolated ground (triangle symbol in Fig. 9x.51) with respect to the input-side ground (normal ground symbol), for two representative isolated power converters. These measurements

---

[35] Many other input and output voltage options are available, including models with dual ±5 V or ±15 V outputs.

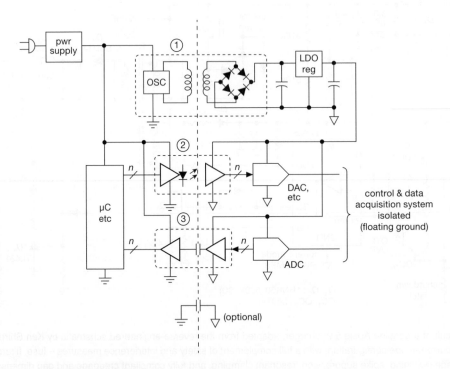

(1) ISO-PWR: DCP0207P (+5 Vin, 7 V(unreg) out)
(2) opto: ACPL-772L, PC412S, etc
(3) capacitor: ISO7220, etc

Figure 9x.51. An isolated system with its own floating "ground" (▽) needs a thoroughly isolated dc supply. It's important that no signals associated with the power conversion (either on the isolated dc power or its ground) are coupled through to interfere with the low-level signals.

were made across a 50 Ω resistor bridging the two grounds, and with a 25 mA resistive load at the 12 Vdc isolated output; in each case we took a waveform with and without a bypass capacitor.

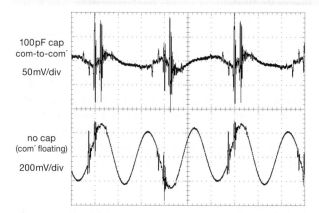

Figure 9x.52. Noise coupled to the isolated ground (com′) with respect to driver-side ground (com), measured across a 50 Ω bridging resistor, both with and without an optional common-to-common bypass capacitor. Dc–dc converter: DCP021515P. Horizontal: 400 ns/div.

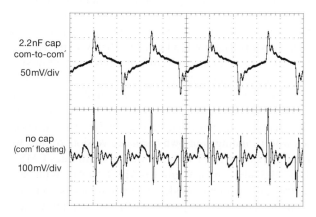

Figure 9x.53. Same setup as Fig. 9x.52, for the LT3999 "Low Noise Isolated Power Supply." Horizontal: 400 ns/div.

These switching-noise signals are *common-mode* – they appear on both the isolated common and its corresponding regulated dc voltage. If shielding and grounding in the isolated system were perfect, so that the power common and all signal commons joined in a single point, well-bonded to the shield (if any), they might not be of serious concern. But in real life there is inductance and resistance in the grounding paths, allowing a portion of common-mode noise to appear at sensitive inputs; and so, as one user of research instrumentation put it, "common-mode noise, if left

to its own devices, will find a way to become normal-mode noise."

In practice these converter-noise voltages coupled onto the sensitive analog circuitry have proven quite troublesome, amounting to some 1000 LSBs of a 16-bit conversion. And this induced noise extends to many megahertz, as can be seen in the spectra in Figure 9x.54. What is needed is a dc–dc converter whose output acts like an isolated battery.

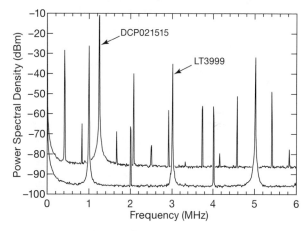

Figure 9x.54. Power spectra for the two dc–dc converters of Fig. 9x.52 and 9x.53, measured with the same configuration as used for their lower traces.

Taking our cue from the "ultra-isolated" powerline transformers that come from manufacturers like Topaz,[36] we developed a converter design with far better isolation. The tricks are (a) using sinusoidal drive, to eliminate those pesky edge transients, (b) using balanced (differential) drive, to eliminate common-mode primary signal, (c) surrounding the output winding of the transformer with an electrostatic shield, to suppress capacitive coupling, and (d) using a common-mode choke at the isolated output, to add common-mode series impedance at signal frequencies.

Figure 9x.55A shows the scheme, with a custom transformer wound on an RM10-size ferrite core. For shielding we used copper foil tape with an insulating layer of Kapton: both primary and secondary windings are 22 turns each (2 mH), with a layer of foil on both sides of the secondary. The foil layers are each a bit more than one turn, overlapped but insulated so they do not form shorted turns. The transformer is "inside-out": the primary surrounds the shielded secondary. We wound the output choke on another

---

[36] Whose transformers specify an inter-winding mutual capacitance of 0.003 pF: the capacitive coupling is blocked by the grounded shield.

RM10 core, 161 bifilar turns of #28 magnet wire (110 mH). The inductance was chosen so that its self-resonant frequency matched the 100 kHz fundamental drive frequency (100 kHz), to further suppress coupled energy.

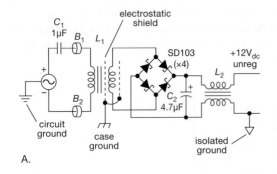

A.

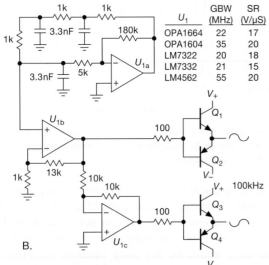

U_1	GBW (MHz)	SR (V/μS)
OPA1664	22	17
OPA1604	35	20
LM7322	20	18
LM7332	21	15
LM4562	55	20

B.

B_1, B_2: Toshiba AB4X2X8W
L_1: EPCOS (TDK) RM10/N87 core, A_L = 4200, p/n B65813JR87
   $N_P = N_S$ = 22t #30, primary over shielded secondary
L_2: same core, 161t #28, bifilar
Q_1, Q_3: BD139-16
Q_2, Q_4: BD140-16

Figure 9x.55. Suppressing converter noise with sinewave drive through a shielded transformer. A. Circuit. B. Buffered phase-shift oscillator adequate for the job. See Fig. 9x.59 for a triangle-wave or trapezoidal-wave alternative, and Fig. 9x.61 for a clocked square-to-sine converter. Fig. 9x.60 shows what the cores look like.

For the sinusoidal drive (Fig. 9x.55B) we used a phase-shift oscillator (see §7.1.5C), whose paired outputs of opposite phase are buffered by simple push–pull BJT followers. The op-amps listed are appropriate: their supply voltage range extends to ±16 V, and they have enough bandwidth (note U_{1a}'s inverting gain of ×36 requires

$f_T \gg 4$ MHz) and slew rate (100 kHz at 10 V peak amplitude requires at least SR=6.3 V/μs) to do the job. However, by actual measurement (see below) a triangle-wave drive[37] is equally effective in reducing coupled ground signals. Figure 9x.59 shows a simple implementation, which, owing to ratiometric design, exhibits an oscillation frequency independent of supply voltage: $f=R_3/4R_1R_2C_1$. By clamping the output you could generate a trapezoidal waveform of the same frequency.

Figure 9x.56 shows the coupled ground noise, for three choices of bypass capacitor, measured with the same setup and loading as for the lower traces in Figures 9x.52 and 9x.53. The shielded sinewave converter improves on the latter by about 50 dB in peak amplitude; the reduction in *peak spectral power* is comparable or greater – some 50–70 dB, as seen in the spectra of Figure 9x.57.

A few additional notes: (a) The push–pull driver stages are outside the op-amps' feedback loops, which avoids problems of feedback stability; we did not bias them into crossover-free class-AB conduction, having found no improvement with the latter. (b) The Toshiba "Amobeads" (B_1 and B_2; see also §1x.4.3E) were effective in suppressing some minor ringing at a few MHz, which was seen at portions of the waveform (e.g., at onset or termination of diode conduction). (c) With power supplies (V_+, V_-) of ±12 V, the circuit of Figure 9x.55 delivered an isolated 12 Vdc (unregulated) output when loaded to 50 mA, rising another volt when loaded only to 25 mA. For other output or supply voltages, change the turns ratio of L_1 accordingly to set the nominal unregulated V_{out}; the gain of U_{1b} can be altered for minor trimming. (d) We tried substituting some commercial common-mode chokes for L_2, to see whether it's worth winding your own resonated version; for this we tried an Eaton CMT3-1-R (5.4 mH), a Coiltronics CMS3-14-R (1.3 mH), and a Würth #74429 (6.5 mH). With each of these the common-mode suppression was poorer than with our optimized L_2 by some 20 dB, i.e., no improvement over the performance with no choke at all.

Which suppression tricks matter the most?

We threw the kitchen sink at this problem, combining suppression measures of (a) sinusoidal drive, (b) symmetrical push–pull drive, (c) electrostatic shielding of transformer windings, and (d) resonant common-mode choke. To see how effective each of these are, we measured coupled signal amplitudes and spectra for various reduced configurations. The data is summarized in Figure 9x.58, which

---

[37] Probably also a trapezoidal waveform, i.e., a slew-rate-limited square wave.

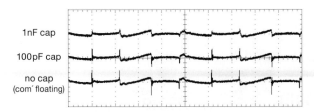

Figure 9x.56. Coupled ground noise for the shielded-transformer converter of Fig. 9x.55, measured across a 50 Ω bridging resistor (with three values of bypass capacitor), with its 12 Vdc output loaded to 25 mA. Note greatly expanded vertical sensitivity (50× to 200×), compared with Figs. 9x.52 and 9x.53. Horizontal: 2 μs/div; Vertical: 1 mV/div.

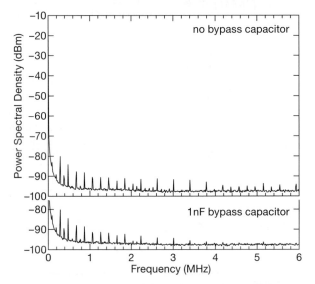

Figure 9x.57. Power spectrum for the shielded-transformer converter running at 100 kHz, measured with the same configuration as used for Fig. 9x.56 (and plotted on the same scale as Fig. 9x.54), showing a reduction in peak spectral power by some 50–70 dB.

SQUARE	SINE	TRIANGLE	SINGLE-ENDED	PUSH-PULL	SHIELDED	COMMON-MODE CHOKE	Vpp across 50 Ω (mV)	Peak Spectral Line (dBm)
–	●	–	–	●	●	●	1.0	–80
–	–	●	–	●	●	●	1.0	–80
–	●	–	–	●	●	–	16	–60
–	●	–	–	●	–	–	26*	–34
–	●	–	●	–	–	–	62*	–29
●	–	–	–	●	●	●	24	–72
●	–	–	–	●	–	–	1400	–34
●	–	–	●	–	–	–	2200	–33

Figure 9x.58. "What matters?" Approximate ground-coupled clocking noise: signal level at the isolated common, measured across 50 Ω with respect to driver-side ground. Those marked with an asterisk exhibited wide pulse-like waveforms, all the rest were narrow spikes.

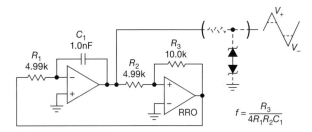

$$f = \frac{R_3}{4R_1R_2C_1}$$

Figure 9x.59. Triangle-wave oscillator circuit, simpler than the sinewave version of Fig. 9x.55B yet retaining equivalent performance. This replaces U_{1a} and associated components in that circuit. The optional clamp creates a more efficient trapezoidal waveform without compromising performance (adjust subsequent gain appropriately).

demonstrates that, well, almost *everything* matters, if you want optimum suppression of clock signal. However, the drive waveform need not be sinusoidal, as long as it does not have discontinuities – it's OK to use a triangle wave (or a slew-rate-limited square wave), as in Figure 9x.59. Parts like LTC's LT1533 exploit this property (though their literature[38] does not address coupled common-mode clock noise, only switching noise as seen across the dc output (i.e., normal-mode noise).

For some applications it's important to synchronize the

various clocking signals in a system, so you don't get beat notes and mixing products. For such applications you can use the circuit in Figure 9x.61 to create a very clean sinewave from a logic-level square wave of 50% duty cycle (use a toggling flip-flop, if needed, to ensure equality of HIGH and LOW durations). Here we used an LVC gate to drive a series-resonant *LC*, the latter forming a bandpass filter at 100 kHz for killing the harmonics; the additional shunt capacitance and *RC* section removed narrow spikes coupled through the inductor's parasitic shunt capacitance. With this sinewave source the isolated power supply re-

---

[38] See especially Jim Williams' LTC App Note 70, "A Monolithic Switching Regulator with 100 μV Output Noise," October 1997.

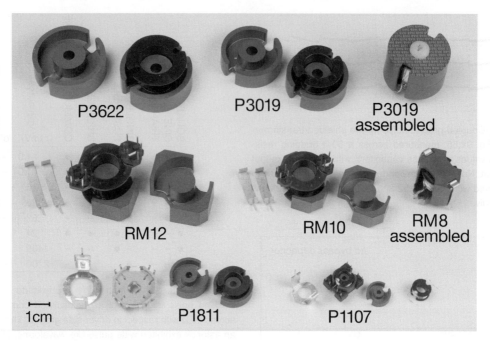

Figure 9x.60. Several sizes and styles of ferrite cores. Pot cores are sized by outer and inner diameter; e.g., a P3019 is 3 cm OD by 1.9 cm assembled height. The core halves are pictured with their matching "coil former" bobbins; note also the clamps and (for the two smallest pot-cores) the "terminal carriers." We used an RM10 core for L_1 and L_2 of Fig. 9x.55.

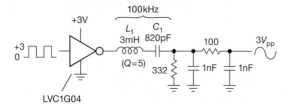

Figure 9x.61. A series resonant LC converts a logic-level square wave into a clean sinewave, so good you can't tell it's not a genuine sinewave when viewed on a 'scope (Fig. 9x.62). Trim C_1 to resonance (maximum sinewave amplitude) at the switching frequency.

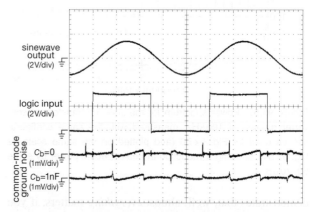

Figure 9x.62. Measured waveforms when driving the converter of Fig. 9x.55 with sinewaves from the square-to-sine circuit of Fig. 9x.61. For L_1 we used a TDK 1433507C 3.3 mH inductor resonated with a 820 pF series capacitor (a Pulse Engineering PE-50502NL with 680 pF works identically). Horizontal: 2 μs/div.

tained its excellent common-mode isolation, as can be seen in the traces of Figure 9x.62.

This circuit design, combining symmetrical push–pull sinewave drive, electrostatically shielded transformer, and resonated common-mode choke, is really effective in suppressing common-mode switching noise – a 60 dB improvement over commercial modules is nothing to sneeze at. But for perspective keep in mind that many applications are rather insensitive to common-mode coupled noise on their power rails, particularly with robust grounding paths and shielding. You may need to resort to the kind of measures we've described only in unusual circumstances (for example, the delicate physics experiment that provided in-

spiration, in which programmable voltages are required, stable and quiet at the microvolt level, and they must ride on top of another programmable voltage, all this at the end of a 10 m length of cable).

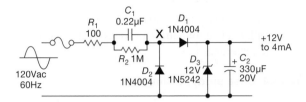

9x.15 Low-Current Non-isolated DC Supplies

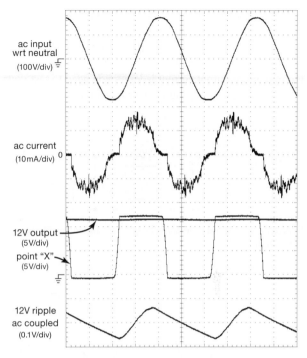

Figure 9x.63. An easy and inexpensive way to get a few milliamps of non-isolated dc from the powerline. C_1 is a 275 Vac line-rated ("X2") film capacitor, and R_1 is a 1 W surge-rated wirewound resistor. Always include a fuse on line-powered circuits.

In §9x.8 we hinted that there are tricks for generating non-isolated low-current dc directly from the ac powerline. Here are some ideas.

9x.15.1 Simplest circuit: reactance-limited zener bias

Figure 9x.63 shows a simple way to generate low-voltage non-isolated dc, to power some little gadget that can safely float at line voltage. The basic idea is to use a lossless (i.e., reactive) current-limiting impedance (the 0.22 μF capacitor) to bias a zener (this is sometimes called a "capacitive-dropper"). With the components shown this circuit can supply four milliamps at 12 Vdc output. In this circuit D_2 prevents C_1 from charging up (in the good ol' days of analog television, such a voltage clamp on the analog video signal performed "dc restoration"), and bleeder resistor R_2 discharges C_1 after power is removed. The 100 Ω series resistor (e.g., a TE EP1W 1 W "anti-surge" wirewound, only $0.10!) has little effect on the output capability, but greatly reduces the inrush current, to a worst-case[39] maximum of 3 A; the resistor's peak dissipation (in that worst-case situation) is about 1 kW, but the duration is less than a millisecond, so we're saved by the resistor's "transient power" rating, see §1x.2.6, and particularly Figure 1x.38. The input capacitor C_1 should be an ac-line rated "X2" capacitor (see §§1x.3.11E and 9.5.1E), for example a Panasonic ECQUA metalized polypropylene film type ($0.25 in qty 100). This circuit delivers 12 Vdc with about 120 mVpp

[39] Worst-case would involve applying power at the negative peak of the ac cycle, with C_1 fully charged from prior disconnection at the positive peak of the ac cycle.

Figure 9x.64. Measured waveforms for the circuit of Fig. 9x.63. Because of the proliferation of switchmode converters, the ac line voltage is afflicted with small switching spikes, seen dramatically here in the input current waveform, owing to the differentiator action of input capacitor C_1. *Safety warning:* Be sure to use an ac isolation transformer if you want to make measurements like this.

ripple at 4 mA load (Fig. 9x.64),[40] with, uh, *modest* efficiency (about 10% at full load). It's a "full-range" supply – it can run from 90 Vac to 240 Vac, with higher output current capability at higher powerline voltages.

9x.15.2 Improved circuit: full-wave rectifier

You can double the performance by simply substituting a full-wave bridge for D_1D_2; and this lets you halve the value of the output storage capacitor for the same ripple. With the same input capacitor C_1 it delivers twice the output current (here 8 mA). Figures 9x.65 and 9x.66 show the circuit and measured waveforms.

[40] A note about the dramatic noise seen in the ac current waveform. C_1 forms a differentiator, $I=C\,dV/dt$, so its impedance drops (i.e., "current gain" increases) with increasing frequency, up to an R_1C_1 cutoff around 7.2 kHz for this circuit. The gain at 7 kHz is 120× that at the ac sinewave signal at 60 Hz, exaggerating the high-frequency powerline noise. But this noise doesn't interfere with the function of the capacitive-dropper supply, and if we were to increase R_1 to reduce it, we'd simply be wasting more power. Interestingly, the jaggedy current waveform was relatively stationary in our lab, evidence of a proliferation of switching-mode loads.

For both of these circuits the input capacitor C_1 delivers an ac *current*, $I_{ac} \approx 2\pi V f C_1$; the output voltage falls rapidly for greater load currents. For the values in Figure 9x.65 that's about 10 mA for 60 Hz 120 Vac. At low line voltage the maximum output current is less – you need to test your circuit at lowest anticipated powerline voltage.

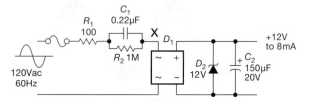

Figure 9x.65. Fullwave version of Fig. 9x.63 delivers twice the output current.

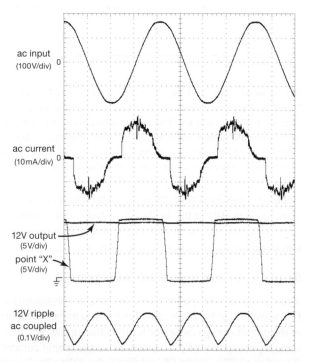

Figure 9x.66. Measured waveforms for the circuit of Fig. 9x.65. Characteristic of full-wave rectification, the ripple frequency is doubled, to 120 Hz. There's no common connection between ac input and dc output, so we used a differential probe for the top trace.

9x.15.3 Why hasn't Silicon Valley responded?

It has! Check out the TPS7A78 "Smart AC/DC Linear Voltage Regulator" from Texas Instruments, which they modestly describe as a "non-isolated power solution for 70 Vac to 270 Vac" that supports output voltages from

1.25 V to 5 V, with up to 75% efficiency and just 15 mW of standby power consumption. It's intended for applications such as smoke detectors, thermostats, RF-transponding electric meters, and building/factory automation.

It's an elegant device (Fig. 9x.67): it includes on-chip LDO regulator, power-fail and power-good indication, and a four-stage flying-capacitor charge pump to reduce the rectified ac voltage by a factor of four.[41] To implement all of its features it requires a dozen external support components; still, not bad for a tiny (5×6 mm) IC that costs $2.50 in small quantities.

9x.15.4 Case study: ceiling fan

After some six years of faithful service, our elegant (and expensive) Casablanca "Brescia" ceiling fan started acting erratically. It would occasionally work, suggesting that the motor and microprocessor were not at fault; that was a good thing, given the fan's complexity (see US Patent 4,716,409!).

Time for a "teardown"! A quick voltmeter measurement showed -3.5 V on the -5 V supply, caused by apparent failure of storage capacitor C_1 in the non-isolated dc supply of Figure 9x.68. This supply[42] is capable of delivering up to 50 mA (to power the microcontroller and associated electronics). It differs from the simple supplies described above: The idea is to use an emitter follower with a zener reference on its base to charge a storage capacitor to $V_Z - V_{BE}$. This way the transistor will rapidly charge the capacitor during the early part of each cycle, while the ac line voltage is still low, minimizing the voltage drop across the transistor and collector resistor, and thus minimizing the power dissipation during the charging time. Nice!

Why did it fail? We ran a SPICE simulation of the capacitor's voltage and current waveforms, suspecting that excessive ripple current was the culprit. The capacitor reaches full charge after about 7 cycles of ac input voltage (during which the peak capacitor current reaches 400 mA), after which its rms ripple current (under full 50 mA output load) settles down to 76 mA.

How does that compare with typical 1000 μF 25 V electrolytic capacitors? Looking at the Nichicon UVZ series of small low-cost capacitors, for example, that capacitor value is rated at a maximum 60 Hz ripple current of 510 mA. That maximum rated ripple works out to 109 mW dissipation in the capacitor's 60 Hz ESR of 0.42 Ω, whereas our considerably lower ripple would dissipate only 2.4 mW in the ca-

---

[41] See the next subsection for a related technique.

[42] Shown here for clarity as a *positive* 11 V unregulated supply with +5 V linear regulator. In the actual device they used a *negative* 11 V supply, with LM7905 −5 V post regulator; see following text.

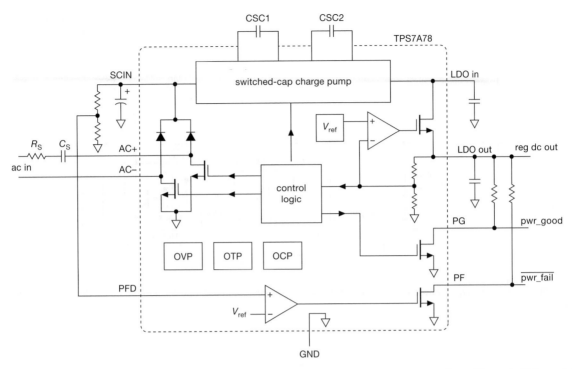

Figure 9x.67. The TPS7A78 generates up to 120 mA of non-isolated regulated dc from the ac powerline, with good efficiency, internal protection, and indicator outputs.

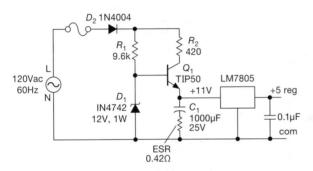

Figure 9x.68. Non-isolated 5 Vdc supply used in ceiling fan.

pacitor's ESR (just 2% of rated maximum); and even at a dc load current of 100 mA the dissipation is only 9.4 mW (8.6% of rating).

So? We trust that the designers rated their components conservatively, and the capacitor they chose had no right to fail, at least from ripple current. But, nothing's perfect; the Nichicon UVZ series has a rated endurance of 1000 hours at rated voltage and 105°C (i.e., it should return to nominal tolerance after that punishment). This sample was not subjected to such treatment... but it did live considerably longer (>50,000 hours). Anyway, we replaced it, in fact with a heftier capacitor (Panasonic FC-series, rated at

1.6 A ripple current), and our fan continues to twirl, happily, in its ceiling abode.

A note on polarity: The actual circuit uses a *negative* 11 V non-isolated ac-to-dc stage, with a negative LM7905 linear regulator to generate the 5 V needed for the microprocessor. We all instinctively dislike circuits that run on a single negative rail, so why did the designers do that? For a very good reason: that allows the triacs to operate with negative gate pulses ("quadrants II and III"), and it allows gate drive with a low-cost Darlington (ULN2003); see Figure 9x.69. Triacs require greater drive in quadrant IV (and some don't permit it at all) – see the footnote in §9.13.1B.

9x.15.5 Inverse Marx generator

The last circuit[43] (Fig. 9x.70) is best described as an "inverse Marx generator" or "inverse Cockcroft–Walton generator," since it does the opposite of the classic boost circuits (which charge their capacitors in parallel, then discharge them in series). Here the capacitor string is charged in series (during the positive half-cycle), then discharged in parallel during the negative half-cycle (in this respect it resembles the TI device above, with its 4:1 charge-pump voltage divider).

---

[43] Suggested by the ever-creative Steve Cerwin.

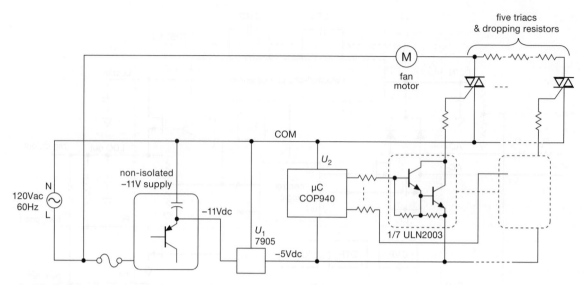

Figure 9x.69. Running the logic between −5 V and common allows quadrant II and III operation of the triacs, driven with a convenient *npn* Darlington array.

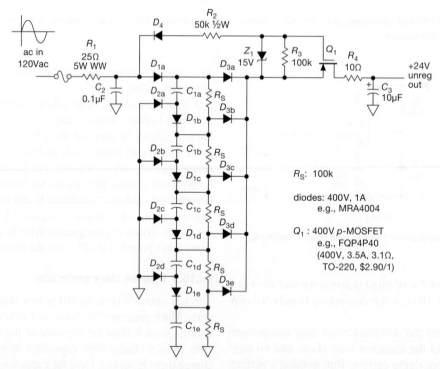

Figure 9x.70. Inverse Marx generator: charge capacitors in series, discharge in parallel. You can use five MMBD5004S (series diode pair) and two MMBD5004C (dual common cathode), which come in SOT-23 and cost $0.12 (qty 100), instead of fourteen '4004 single diodes.

During the positive half-cycle the series string of capacitors C_{1a}–C_{1e} are charged through diodes D_{1a}–D_{1e}, with equalized voltages from the resistor string R_S, while back-biased diodes D_{2a}–D_{2d} stand idly by. The source terminal of p-channel MOSFET Q_1 swings to the full positive line voltage, but the transistor is non-conducting because its gate is held at the source voltage, via R_3. When the input swings negative, Q_1 is brought into conduction via D_4, with V_{gs} limited to 15 V by zener Z_1. What it conducts is the positive voltage presented to its source by diodes D_{3a}–D_{3e}, each of which conducts (approximately) the voltage of a single C_1. The lowpass filter R_4C_3 blocks any spikes at the input. In this circuit the peak inrush current is limited by R_1 to 7 A. As with Figure 9x.63, that is well within the transient peak power rating of a 2 W wirewound resistor. And, as in that earlier circuit, you should always fuse the incoming ac line.

9x.16 Bus Converter: the "DC Transformer"

It would be nice to have a "dc transformer," to convert one dc voltage rail to another; and, like the usual ac transformer, it should work in both directions. We've seen plenty of dc–dc converters in Chapter 9 – these convert an input dc supply rail to an output dc at a different voltage (boost, buck, invert, etc., recall §9.6). These *switchmode* devices act like a one-way (input to output) dc transformer, but they do their magic by converting to ac internally, with magnetics and rectification (or flying capacitors) to create the output dc, and with feedback duty-cycle control to set the regulated output voltage.

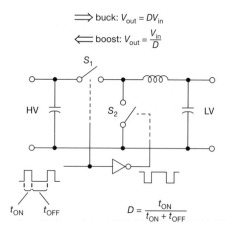

Figure 9x.71. A buck converter with active switches is also a boost converter running the other way. Here the duty cycle D refers to switch S_1's ON fraction.

Perhaps surprisingly, you *can* create a bidirectional switchmode dc–dc converter: look at Figure 9x.71, where we've drawn a conventional non-isolated buck converter, using an active switch (S_2) instead of a passive catch-diode. With input from the left, it's a buck converter. But if the input dc comes from the right, it's a boost! You can think of this as a bidirectional dc transformer, working both ways with a "turns ratio" $N = n_{pri}/n_{sec}$: it steps the dc voltage up or down by a factor N (depending on which side is the "input"), and (if lossless) it conserves energy. For example, if V_{out} is the lower voltage, $I_{out} = NI_{in}$.

The "transformer turns-ratio N" is set by the switching duty cycle; for example, a 25% duty cycle (fraction of time

S_1 in ON) is a turns ratio $N = 4$. It's the same equation as for the classical buck converter (eq'n 9.3d in Chapter 9),

$$N = \frac{V_{out}}{V_{in}} = D, \qquad (9x.5)$$

where D is the fraction of time switch S_1 is conducting.

9x.16.1 Differences from classic switch-mode converter

The dc bus converter, though similar to the standard non-isolated buck or boost, differs in two important ways:

- a bus converter has a fixed duty cycle, rather than a feedback loop to generate a target output voltage; so a bus converter keeps the voltage *ratio* (rather than the output voltage) approximately constant.
- a bus converter must use a synchronous switch (MOSFET S_2) instead of a freewheeling catch diode, because it must allow current to flow in either direction when ON.

9x.16.2 Bus converter applications

When might you want a bidirectional dc transformer, anyway? A situation where you have several dc buses: for example, a contemporary automobile with a 48 V bus, but encumbered with many legacy 12 V accessories. It's convenient to have power moving between the buses; and it's essential if one side suffers a "load dump" (the sudden disconnection of load on the 12 V side, which would cause a voltage surge unless transferred to the 48 V side).

Another example is generating a high-power split supply. Say we start with a 72 V offline switching supply; use that to drive a 2:1 bus converter (50% duty cycle), which creates a 36 V center-tap. Now ground that output, and you've got a $\pm$36 V high-current supply, ideal for a high-power audio system (for example). It works because the bus converter's output (the new "ground") will happily source or sink current, making it a stiff voltage midpoint.

9x.16.3 Bus converter example

Commercial dc bus converters that are aimed at high-power applications (e.g., the automobile example above) take advantage of multiphase switching, with transformers to take care of the voltage ratio. In the modest example here, we've taken a simpler approach, using a single inductor with a settable duty cycle D to determine the voltage ratio. The basic scheme is shown in Figure 9x.72, where a half-bridge driver IC provides the gate drive to the switches. The IR2085 includes an internal oscillator, but with a fixed

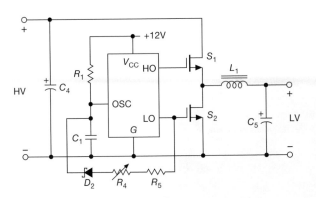

Figure 9x.72. Block diagram of dc bus converter. R_4 adjusts the duty cycle, thus the voltage conversion ratio.

bootstrap components D_1 and C_B to generate the $V_{CC}+12$ V needed to drive the high-side n-channel MOSFET's gate.

In this circuit we've added D_2 and R_4+R_5 to alter the duty cycle around the IR2085's fixed 50% internal setting. That IC has a bus-voltage limit of 100 V, but the robust (50 A) 55 V MOSFETs limit the high-side voltage. This thing is capable of plenty of power – for example, it can convert between 48 V@5 A and 24 V@10 A (50% duty cycle), a very respectable 250 W.[44] We tested it at somewhat lower voltages (24 V in, 12 V out) and power, where it delivered a measured conversion efficiency of 95% or better from 10 W to 100 W.

9x.16.4 A few comments

Effective series resistance

The bus converter ideally transforms voltage perfectly in either direction (by the ratio N), but with an added series loss resistance, analogous to the copper winding resistance in an ac transformer. The effective series resistance is partially due to the expected MOSFET's $R_{DS(on)}$, but mostly due to switching dead-time, when neither the high-side nor the low-side MOSFETs are turned ON (by design, the dead-time helps to ensure that both MOSFETs won't be on at

duty cycle of D=50%; but a little trickery lets you alter D (refer back to Fig. 7.10B, where we played a similar trick on the venerable 555). Here the left side is the higher voltage input (say 48 V) and the right side is the lower voltage rail (say 12 V).

Figure 9x.73 shows the full circuit, with our PCB version pictured in Figure 9x.74. Much of it should be familiar: the usual gate coupling networks driving the switches S_1 and S_2 (Q_1 and Q_2, respectively), the energy-storage inductor L_1, the large storage capacitors C_4 and C_5, a 4-terminal current-sense resistor R_S for overcurrent protection, and the

---

[44] It uses the impressive inductor seen in the center of the last row in the photograph of Fig. 1x.72.

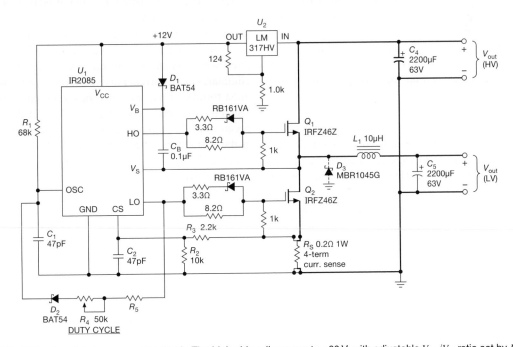

Figure 9x.73. Bidirectional dc bus converter circuit. The high-side rail can run to +60 V, with adjustable V_{out}/V_{in} ratio set by R_4.

Figure 9x.74. The bus converter of Fig. 9x.73 fits nicely on a 6×8 cm 2-layer PCB. Here we added an input fuse (automotive type), and we doubled the output bypass capacitance.

the same time, to prevent high supply-rail shoot-through current spikes). In this circuit we're running the IR2085 around 150 kHz with a dead-time of about 100 ns, thus a 3% time during which the inductor node is not driven.

The relatively large dead-time is a consequence of our having used a simple half-bridge driver IC, taking advantage of its built-in dead-time conduction gap. But its 100 ns gap is rather conservative, and you would do better to separate the waveform generation from the gate driving circuitry.

Tracking ratio

Because the converter generates a fixed voltage ratio, and does not use a feedback loop, slow variations in input voltage are faithfully transformed. Here "slow" might range up to a few kilohertz, that is, a smallish fraction of the switching frequency. But that's not an "ac signal," because the converter can't work with bipolarity signals: it's ac riding on dc.

Current pulsation and multiphase

We claimed our converter conserved power, thus the claim that the high-voltage input current was lower than the output current, in proportion to the voltage step-down ratio. But it's not that simple. Take, for example, our 48 V to 12 V converter: input switch S_1 is ON only 25% of the time, but

during that time the input current is the *same* as the average output current; it's only in the *average* that it's reduced by a factor of four. The problem is that our simple converter used a single inductor, with pulsed conduction at the input. In a more sophisticated configuration you would create an interleaved multiphase converter with four inductors (each with its pair of switches[45]), and with the switches interleaved so that each operates in sequence. This largely eliminates input current pulsation. And, in a further burst of good fortune, it happens that the *output* currents from the four phases overlap considerably (33% in this example), because the S_2 output switches are ON three times as long as the S_1 input switches, during which their output currents combine.[46]

This is why you see a nest of converter inductors surrounding the processor on computer motherboards, performing multiphase (8 or even 16 phases) buck conversion from a 12 V bus to a 1 V CPU core voltage. And, moving back to our favorite dual-bus automotive theme, Figure 9x.75 shows a reference design for a 2 kW bidirec-

---

[45] As in the photograph of Fig. 9x.75.

[46] The basic timing and monitoring tasks are increasingly being done with inexpensive and fast microcontrollers; this makes a lot of sense, especially because there's no feedback or stability issues with a fixed-ratio bus controller.

Figure 9x.75. A *serious* dc bus converter: 4-phase bidirectional 12V/48V automotive module, good for efficient 2 kW conversion. Courtesy of Texas Instruments Incorporated.

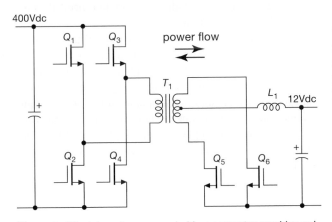

Figure 9x.76. A transformer-coupled bus converter provides galvanic isolation between input and output (i.e., separate grounds), and it permits large ratios of input/output via the transformer's turns ratio.

Isolated bus converter

The bus converters above, which we've called dc transformers, lack one feature of real ac transformers: they are non-isolated, sharing a common ground. But that's not necessary – they can be (ac-) transformer coupled, as in Figure 9x.76. Besides allowing independent grounds, the transformer's turns ratio lets you maintain high efficiency (reasonable switching duty-cycle) even when transforming large voltage ratios (as in the 400V/12V converter shown).

tional bus converter from Texas Instruments, intended for 12V/48V automotive systems.

9x.17 Negative-Input Switching Converters

The world seems to favor positive polarities (in spite of the electron's birthright to the contrary), as evidenced by a real paucity of negative-input voltage regulators and switching converters (try it yourself – do a search at DigiKey or Mouser or Newark). Happily, you can corral a positive-polarity switching regulator into negative-polarity service; read on.

9x.17.1 Negative buck from positive boost

The basic idea is to trick the switcher into thinking it's doing the usual positive output, but, unbeknownst to it, we tie its positive input to ground, and connect its ground terminal to the unregulated negative input. There's two complications, though: (a) because a positive boost converter has its switch returned to its ground pin, the flipped configuration now has the switch tied to $-V_{in}$, the connection you'd use for a *buck* converter; and (b) the usual voltage-regulating feedback connection would become *positive* feedback (the controller would interpret a larger negative output, in its positive-thinking, as requiring more output), so it's necessary to invert the sign of feedback. Figure 9x.77 shows the scheme.[47]

The converter's SW pin now serves as the usual series switch in a buck converter. Transistor Q_1 inverts the swing of the V_{out} pin, with gain $-R_2/R_1$, thus generating a voltage across R_2 (i.e., from FB to G) with the correct sign of feedback; the error amplifier equates that to V_{ref} to set the output voltage $V_{out} = -V_{ref}R_1/R_2$. In this circuit Q_2 compensates Q_1's V_{BE} drop; it could be omitted (with approximate V_{BE} accounted for) if precision and temperature stability is not critical.

A few comments: (a) we have not shown external compensation components, which will differ from those for the conventional positive-boost circuit;[48] (b) the converter is

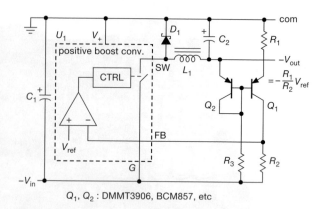

Figure 9x.77. A positive-polarity boost converter configured as a negative-polarity buck converter.

subjected to the full input voltage, which is higher than the output voltage, contrary to the usual situation with a boost converter IC; (c) similarly, the switch is carrying higher voltage but lower current than in its conventional configuration.

9x.17.2 Negative boost from positive buck

Similarly, a positive buck converter can be wired as a negative-input, negative-output *boost* converter. The feedback is simpler here, with no inversion needed; Figure 9x.78 shows the circuit.[49]

As with the previous circuit, some comments are in order: (a) once again, you need to worry about compensation, which will be different in this configuration; (b) opposite to its normal use, the output voltage is higher than the input voltage, but at start-up it must operate at the (lower) input voltage and continue to operate at the (higher) output voltage; (c) similarly, the input current is higher than the output current, so you need to choose the converter's current rating accordingly.

---

[47] Adapted from Figure 2 of Ajmal Godil's "Turn Positive Buck/Boost Circuits Negative," *Electronic Design*, 6 July 2006: http://electronicdesign.com/boards/turn-positive-buckboost-circuits-negative.

[48] The transfer function is different, as well as the operating mode (e.g., a boost converter has continuous output current and discontinuous input current, the opposite of the buck converter).

[49] This circuit is shown also (Fig. 4) in the Godil reference, but note that the GND pin there should be connected to $-V_{out}$, not to $-V_{in}$. See also Mark Pieper's information-rich TI 2013 publication "Designing a negative boost converter from a standard positive buck converter," SLYT516.

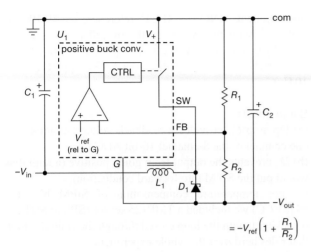

$$= -V_{\text{ref}}\left(1 + \frac{R_1}{R_2}\right)$$

Figure 9x.78. A positive-polarity buck converter configured as a negative-polarity boost converter.

9x.18 Precision Negative Bias Supply for Silicon Photomultipliers

Here's a straightforward circuit problem we encountered during the design of a large array (200,000) of silicon photomultipliers (SiPM, see §4x.14) for optical SETI. These detectors run with a reverse bias in the range of 25–60 V, for which some SiPM manufacturers provide stable and low-noise dc supplies that are programmable and temperature compensated. An example is the C11204 from Hamamatsu, a tiny module (2 mm thick and 11.5 mm square) that runs from +5 V, is serial programmable from 40–90 V with 1.8 mV resolution, 10 ppm/°C stability, and 100 μVpp typical ripple, and which includes voltage and current monitoring along with programmable temperature compensation specified by two coefficients.

The problem, for us, was that the 64-channel ASIC front-end amplifier we chose (Maroc3A), originally designed for multi-anode PMTs, expects negative-going input pulses, whereas the "normal" SiPM configuration produces positive-going pulses, biased with a positive bias supply; see, for example, Figure 4x.114. As it happens, SiPM manufacturers like Hamamatsu make excellent positive bias supplies for SiPMs (including appropriate temperature compensation, serial programmability, etc.), but no one makes a low-noise SiPM bias supply with *negative* output polarity. So we had to design our own.

You might start by looking for a quiet negative regulator that works in this range of voltage; but you'd be disappointed. However, the low SiPM current ($\leq$2 mA for an 8×8 SiPM array with common bias, each pixel of which is cranking along at a maximum count rate of $\sim$50 Mcps) lets us do the deed simply with an HV op-amp. Figure 9x.79 shows what we came up with (only one channel of four is shown). We looked at available op-amps that can operate up to 100 V total supply (Table 4.2b), and chose the LTC6090[50] because it is stable, ($\pm$5 μV/°C max), decently quiet (14 nV/$\sqrt{\text{Hz}}$ at 1 kHz), and has both an overtemperature output and an output-disable pin.

We biased the op-amp from +5 V and -70 V, and set the gain to -25. Its input comes from a filtered output of one channel of the 16-bit DAC, with a particularly quiet (1.2 μVpp, 0.1–10 Hz) 2.50 V external reference

(the excellent ADR441 "XFET" reference, see §9.10.3 in AoE3).[51] With this reference the overall noise is dominated by that of the DAC itself (6 μVpp, 0.1–10 Hz), which is somewhat noisier than the op-amp (3 μVpp RTI, 0.1–10 Hz).

Some details
(a) Op-amp U_4 provides a readback voltage monitor, via one channel of the 8-channel 16-bit ADC U_5.

(b) D_4 prevents the output from going positive (a fault that would put the SiPM into forward conduction).

(c) For temperature compensation of SiPM bias ($\sim$ 54 mV/°C) we included a TMP125 sensor (SPI, in SOT23-6 package), with the loop closed through the central server that rides herd over the whole experiment.

(d) Perhaps most interesting in this otherwise-conventional circuit, Q_1–Q_3 and associated circuitry disables the op-amp's output if its supply rails are not present. Here's how it works: You don't want this function to depend on other supply voltages, which may not be present during startup; so the trick is to use a depletion-mode MOSFET (Q_3) to hold the op-amp's DIS' pin at ground until both the +5 V and -70 V supplies have turned on. You can think of this circuit as an "analog AND gate" – for Q_3's gate to be pulled negative (thus releasing the internally pulled-up DIS' pin) you need both a negative voltage at Q_2's emitter, and positive base current being sourced by Q_1. Both transistors are operated as saturated switches, with zeners D_1 and D_3 preventing conduction until their respective supplies have gotten most of the way to their final voltages. We leave the reader to take it from there.

(e) Finally, we needed a clean -70 V supply, and we didn't want to squander a regulated bench supply that goes to that high a voltage. So we cooked up a simple circuit, described in the next section.

---

[50] Other candidates: ADA4700, OPA445, and OPA454.

[51] There are plenty of DACs with internal bandgap references, but none of them is as quiet as what you get with this configuration (e.g., 12 μVpp, 0.1–10 Hz, for the DAC8564's internal 2.50 V reference; that's 20 dB worse than the ADR441 external reference we chose).

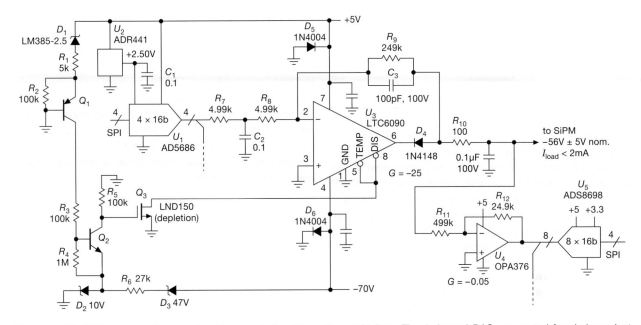

Figure 9x.79. Digitally controlled negative bias supply for silicon photomultipliers. The 4-channel DAC can control four independent outputs, and the 8-channel ADC is for voltage readback (and other analog voltages). The circuitry associated with Q_3 disables the output until the op-amp's supply voltages are established.

9x.19 High-Voltage Negative Regulator

We needed a -70 V regulated supply for the SiPM precision bias supply of the previous section. We wanted a quiet supply, with just a few milliamps of output current, for which a linear regulator is ideal. But there are no $\sim$100 V negative regulators (see, for example, Tables 9.2 and 9.3, where negative linear regulators top out at -40 V). Happily, it's easy to use a positive regulator to make a negative supply, if the unregulated dc input is floating – see Figure 9x.80.

Here we've used the very useful TL783 MOSFET positive regulator (Table 9.2), which is like the venerable 317 but with 125 V $V_{in}-V_{out}$ rating (and, owing to its n-MOS pass transistor, a larger dropout voltage: up to 15 V at its full 0.7 A current, but no more than 5 V at our 20 mA maximum load). The circuit is the standard 3-terminal positive regulator, but with its positive "output" grounded.[52] Diode D_1 protects the regulator against input faults (see Fig. 9.14F in §9.3.4 of AoE3), a healthy precaution when using higher voltage regulators.

Just as with the 317, the TL783 requires a minimum load current; it's specified as 15 mA at full 125 V drop, but here we're dropping at most 30 V, for which the minimum load current of 6 mA is conservatively satisfied by the 10 mA through $R_1 R_2$. A comment on the choice of the line transformer: the combination of leakage inductance and winding resistance makes it difficult to predict the input voltage to the regulator with reasonable accuracy (see §9.5.2); we addressed this annoying detail by buying three samples (48 V, 56 V, and 60 V rms ratings) and testing them in-circuit. Turns out only the 60 V choice provided adequate margin under worst-case combination of full load current and -10% line voltage.

---

[52] Serendipitously, the mounting tab of the TO-220 regulator package is the output terminal, so we just screwed it to the case, no insulator!

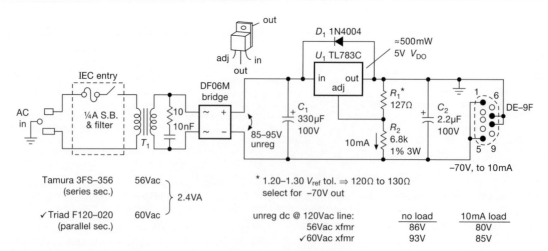

Figure 9x.80. Positive regulator with grounded "output" makes a regulated negative supply. At 10 mA load the measured 120 Hz ripple across C_1 was 300 mVpp and the output ripple was 0.5 mVpp.

9x.20 The Capacitance Multiplier, Revisited

For our actual fielded system of some 2×10^5 SiPMs (see §§4x.14, 9x.18, and 9x.19) we needed two dozen $-70\,V$ quiet supplies, each capable of output currents to 2.5 A. Although this can be done with linear regulators, we chose instead Vicor's QPAC low-noise switcher modules, given the $\sim 4\,kW$ total power requirement. These things are pretty quiet, with noise and ripple down in the tens of millivolts; but for added noise suppression we decided to use capacitance multiplier stages (discussed in AoE3 at §8.15.1, and used in several low-noise circuits, for example Fig. 8.92).

We started with the basic capacitance multiplier (lower half of Fig. 9x.82), here implemented with a *pnp* Darlington pass transistor. Given the voltage and current levels, it is tempting to consider using a MOSFET (to circumvent SOA second-breakdown problems); but a MOSFET operated with low drain-source voltage drop (as it would be in a capacitance multiplier, with the drain and gate voltages nearly the same) is still largely in the linear (resistive) region – see Figure 3.13 – thus poor ripple rejection. The advantage here goes to the humble BJT, operating nicely in the high collector-impedance (constant current) regime.

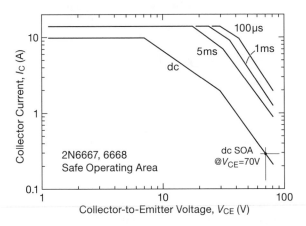

Figure 9x.81. The 2N6667-68 "10A" power Darlington BJT is limited to far lower currents when operated at the high-voltage end of its V_{CE} range. In our application an output short would put 70 V across it, where the steady-state current must not exceed 0.3 A.

OK – but we need to worry about violating the safe operating area (Fig. 9x.81), for example in a fault condition where the output is shorted, thus putting the full input volt-

age across the pass transistor, limited only by the upstream supply's current limit. So we added a voltage clamp (top portion of Figure 9x.82) to create a reasonable first trial circuit.

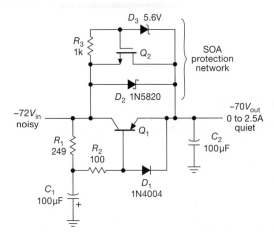

Figure 9x.82. Capacitance multiplier for a $-70\,V$, 175 W supply. Q_1 is a *pnp* small power Darlington, e.g., 2N6667 or TIP147T. The MOSFET barnacle clinging to its C-E terminals keeps Q_1 from seeing more than 10 V in the event of an output short to ground; even a modest power MOSFET will do, for example an IRFZ14 (60 V, 10 A, TO-220, \$0.78 in unit quantity).

Here the *pnp* Darlington Q_1 pass transistor maintains about 1.5 V drop; R_2 suppresses oscillation, and D_1 protects the base–emitter diode if the *input* is brought suddenly to ground. Because Q_1's dc SOA at 70 V is a meager 0.3 A, we've bridged its collector–emitter terminals with *n*-channel MOSFET Q_2, which acts as a $\sim 10\,V$ clamp in case the output is shorted; Schottky rectifier D_2 protects Q_2 in some unspecified fault condition (although Q_2's intrinsic body diode would probably do the job without assistance). Note that an inexpensive *n*-channel device is fine here, even though it's a negative supply, because the Q_2 clamp is a 2-terminal circuit. In the shorted-output fault condition there is just $\sim 10\,V$ across both transistors, thus no danger of SOA violation; the transistors then dissipate $P_{\text{diss}} \approx 10 I_{\text{lim}}$, where I_{lim} is the upstream supply's current limit (about 3 A in this case).

Performance

How well does the circuit perform? Figure 9x.83 shows input and output ripple for an input source with $\sim 120\,mVpp$ ripple at 200 kHz. The output amplitude is down by a factor of 60 (35 dB). Evidently the capacitance multiplier is doing its job.

We subjected the circuit to some torture tests, by shorting the output (nice sparks!), and looked at the resulting voltage drop across the pass transistor. Figure 9x.84 shows

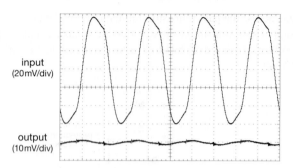

Figure 9x.83. Measured ripple reduction for the circuit of Fig. 9x.82. The upper trace is the −72 V dc input, and the lower trace is the quiet output. Both are ac-coupled. Horizontal: 2μs/div.

single-shot captures of the voltage across Q_1 (and Q_2). The bottom trace corresponds to the circuit of Figure 9x.82, clamping at about 12 V; replacing the gate zener with a signal diode reduces the clamp voltage to 8 V (top trace). The initial spike may be a probing artifact, given the voltage slew rates caused by such violence.

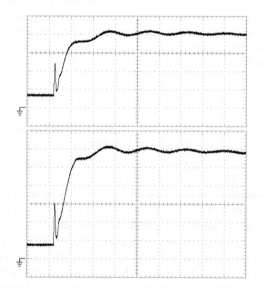

Figure 9x.84. Voltage across Q_1 when the input is at −72 V and the output is shorted. Bottom: circuit of Fig. 9x.82; Top: similar, but with zener D_3 replaced by a 1N914 diode. Vertical: 2 V/div; Horizontal: 4 μs/div.

Variations

There's room for plenty of creativity with circuits like this. Figure 9x.85 shows a few. Circuit A increases the drop across the pass transistor (by charging C_1 to the lowest point of the ripple, via D_4) to accommodate a dc input plagued with lots of ripple or spikes; alternatively you can

connect a large-value resistor across C_1, to increase the drop across Q_1. Circuit B addresses the issue of power dissipation under extended fault conditions (shorted output) by substituting a BJT clamp to lower the clamp voltage; this would be unsuitable, of course, if the input has large ripple amplitude. Circuit C addresses a ubiquitous problem with switchmode power supplies, namely their *common-mode* switching noise. The latter can be troublesome in the extreme: for example, when we were measuring output noise in these circuits, we would clip the scope ground to the output common, and then probe the same point (i.e., the scope's input tied to its own ground), and, as often as not, we'd see substantial switching spikes. That's due to the spikey ground current forced through the scope probe's ground. We complained about this in AoE3 – see for example the grouchy caption to Figure 9.53.

The capacitance multiplier can be used in linear circuits (such as active filters) to increase the effective value of a capacitor, or to "tune" the value of a capacitor. See §1x.8.2 for a discussion and example circuits.

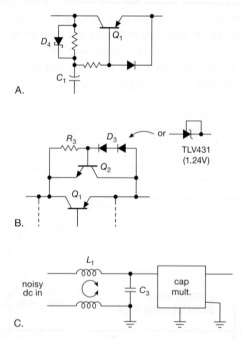

Figure 9x.85. Decorating the capacitance multiplier for the negative supply of Fig. 9x.82. A. Schottky diode D_4 charges C_1 to the minimum of the peak input ripple, to increase Q_1's drop if the input ripple is large. B. Substituting a BJT for clamp transistor Q_2 reduces the clamp voltage. C. A common-mode input choke suppresses switchmode ground currents; the choke can also be placed at the output of the capacitance multiplier.

9x.21 Precision Low-Noise Laboratory Power Supply

9x.21.1 Overview

Figure 9x.86 shows the block diagram, and Figure 9x.87 shows the full circuit in detail. This design is an elaboration (and modernization) of circuits used in the 1970s-vintage HP6205 bench supplies.

The block diagram reveals an unusual topology, where the error amplifiers (U_6 and U_7 for constant current and constant voltage operation, respectively) sit at the positive rail, rather than the usual low-side placement. This is a better arrangement because they control the high-side series-pass transistor Q_2; if you put the error amplifiers at the low side, you'd have to relay the control signals up to the high potential.[53]

In detail, voltage-mode error amplifier U_7 compares the output voltage (at the sense terminals) with the drop across the $10\,k\Omega$ voltage-setting control, the latter biased with a clean 5 mA current. Feedback passes through the diode OR gate pair to driver Q_5 and pass transistor Q_2, the latter equipped with fast current limit via R_{CL} and Q_{CL}. However, if the drop across current-sensing resistor R_S reaches the

We wanted to try our hand at an ambitious goal: a general-purpose bench-type dc power supply with
(a) high stability,
(b) exceptionally low noise,
(c) constant voltage or (precision) constant current modes, and
(d) full protection against overload (including continuous short circuit).
Commercially available power supplies, though endowed with nice digital programming features, tend to fall short in noise performance and in providing seriously stable constant current operation.

---

[53] Why not put the pass transistor on the negative rail? you ask ... because then the pass transistor would be *p*-type, available in far fewer choices, and with generally poorer performance.

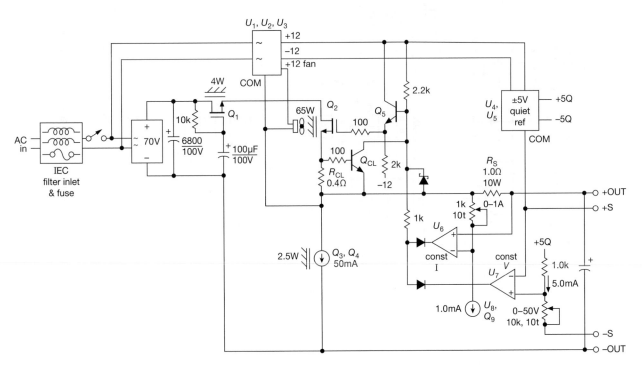

Figure 9x.86. Block diagram of a precision low-noise dc power supply. The boxes with ac inputs and dc outputs include transformers and bridge rectifiers, (and linear voltage regulators for the +12 V and −12 V outputs).

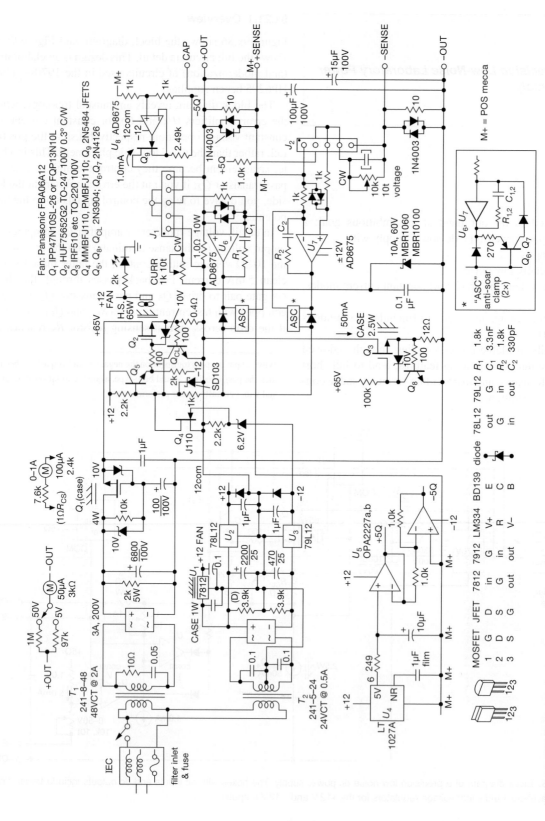

Figure 9x.87. Plenty of nice circuit tricks contribute to excellent stability and low noise in this "laboratory" dc power supply.

454

voltage across the 1 kΩ current-setting control, the current-mode error amplifier U_6 takes over, through the same feedback path.

A few additional comments at this block-diagram level: (a) transistor Q_1 and associated components comprise a capacitance multiplier (see §8.15.1), to suppress unregulated ripple and thus give the regulator a head start in producing clean dc;[54] (b) the current sink $Q_3 Q_4$ provides a minimum load for MOSFET Q_2, maintaining reasonable transconductance for ease of feedback compensation; (c) things get hot – this is, after all, a *power* supply – so Q_1–Q_3 live on a fan-cooled heatsink.

9x.21.2 Circuit details

Now let's drill down into the circuit details (Fig. 9x.87), perhaps with an admiring glance at the creature's handsome portrait (Fig. 9x.88).

---

[54] This is a trick that's often easy to retrofit to all kinds of commercial power supplies, dramatically lowering their output noise. We have rescued more than one sensitive experiment this way.

Unregulated dc supplies.

Beginning at the ac power entry, we use a line filter (combined with fusing in an IEC entry module), then a pair of transformer-bridge-storage capacitors for the dc power (T_1) and for the low-level circuits (T_2). Both transformers include transient-snubbing capacitors (see §9x.6), and the +12 Vdc fan supply uses a separate regulator to isolate any fan-commutating noise (see §9x.21.3) from the control circuits. At full current there's about a volt of ripple across the 6,800 μF storage capacitor, so we added the MOSFET capacitance multiplier Q_1, with zener gate protection (for both turn-on and turn-off) as shown: at full 1 A load current the residual ripple on the +65 V unregulated supply is only 2.3 mVpp, not bad. We put the usual reverse protection diodes across the ±12 V split supply outputs, to prevent destructive reverse biasing of powered stages during power cycling.

Pass transistor

Power MOSFET Q_2 does the heavy lifting: it comes in a hefty TO-247 power package ($R_{\Theta JC}$=0.29°C/W), thus just 19°C rise of junction temperature (with respect to case) at maximum dissipation (65 W). We bolted it onto a CPU-style heatsink (intended for ~100 W continuous dissipation) with its supplied 12 V brushless fan (later replaced, see below). Driver transistor Q_5 is biased (pull-down to

Figure 9x.88. Portrait of a precision laboratory power supply, here set to 30 V and loaded with 40 Ω.

−12 V) at about 10 mA, for good response into the MOS-FET's gate capacitance (the latter, about 7.5 nF, will slew at 1.3 V/μs with that current, thus response times of just a couple of microseconds to bring the MOSFET's current down dramatically). Note that the collector of current-limit transistor Q_{CL} is connected to the driver's base, not to the MOSFET's gate; the latter connection would require Q_{CL} to fight the driver's ∼100 mA emitter current, not a happy situation.

The MOSFET's quiescent bias of ∼50 mA is provided by Q_3, with a simple V_{BE}-moderated current sink; note the zener gate protection, given the +65 V gate bias (perhaps one could get away without the zener, if you put your trust in Q_8; but we feel better with such "feel-good" measures – a gate insulator, once punctured, never recovers).

There's an interesting bit of circuit trickery in JFET Q_4: its function is to prevent nasty output voltage transients during power supply turn-on. During those milliseconds the op-amp supplies are ramping up, perhaps unsymmetrically, and the loop is not yet in control; so, until the −12 V supply has reached at least −8 V or so, the JFET is in heavy conduction, holding the driver and pass transistor in a non-conducting state – cute! The circuit works effectively also when the power supply is switched off: the storage capacitor for the −12 V supply is deliberately chosen smaller than the corresponding capacitor for the +12 V supply, so it discharges faster (note further that the −12 V supply has an additional 10 mA load from Q_5's pull-down).

Voltage references

For superior stability and noise you want a good voltage reference. Here we chose the excellent LT1027A (2 ppm/°C, max), with a noise reduction pin that brings the noise density down to 20 nV/√Hz all the way down to 100 Hz. We further filtered its output with an RC low-pass filter (f_{3dB}=64 Hz), then buffered and inverted its +5 V output with the even-quieter OPA2227 (3 nV/√Hz down to 10 Hz), keeping the resistor values small (1 kΩ: e_n=4 nV/√Hz). The resulting "quiet" output voltage pair +5Q and −5Q are the references for the constant-voltage and constant-current modes, respectively.

Control loop

The error amplifier for both control loops is the excellent AD8675, a BJT op-amp characterized by low noise (2.8 nV/√Hz typ, corner frequency ∼10 Hz), low bias current (±2 nA max), and precision (V_{os}=10 μV typ, 75 μV max). Its integrated low-frequency (0.1–10 Hz) noise voltage is just 0.1 μVpp (see Fig. 8.64). They are stabilized with a pole-zero network, trimmed for optimum transient

response. Measured transients when switching a 0.5 A load (with several values of output capacitance) are 0.2 V and 5 μs recovery (15 μF), 0.07 V and 20 μs recovery (100 μF), and 0.015 V and 100 μs recovery (1000 μF).

As with the voltage reference circuitry, we kept resistor values small to minimize Johnson noise. We also added an "anti-soar clamp" circuit, because the op-amp output of the control loop that is inactive (e.g., the constant-current loop when running with constant voltage) would rise to positive saturation (about 10 V above +V_{out}), and have to slew back down if conditions required it to spring into action. The clamp circuit prevents saturation, so the inactive op-amp's output need slew through only two diode drops when summoned.

Other circuitry

Back-to-back power diodes limit the permitted drop from sense terminals to output terminals, and a larger power diode prevents reverse output polarity. A pair of terminal blocks (J_1 and J_2) permit external programming, either with an external resistor or an external voltage (or current), for example from a DAC. (See if you can figure out the connections to do this.) Protective clamp diodes at the op-amp inputs recognize the propensity of users to assert unwelcome voltages on such terminals. Analog panel meters display output voltage (2 ranges) and load current; these could (and probably should) be replaced by digital displays.

9x.21.3 Performance

Noise and ripple

Confident in our design, we built the power supply and expected to see vanishingly small ripple and noise. That confidence was misplaced: initial testing revealed significant 120 Hz ripple, as seen in Figure 9x.89A, in spite of careful layout to minimize wiring loops that are susceptible to magnetic pickup from transformer fringing fields. Although 0.1 mVrms of ripple would ordinarily be considered pretty good, we wanted better. Much better. So we tried substituting a toroidal power transformer (a favorite choice among audiophiles), and we tried Schottky rectifiers to eliminate stored-charge spikes (§9x.6). No joy.

In desperation we repackaged the ac power components in a separate enclosure, connected with an umbilical cord. That did the job (Fig. 9x.89B), pushing the line-frequency ripple down to just a few microvolts. Now the dominant contamination was some sort of positive-going spikey stuff...what can that be? Triggering on the spikes (Fig. 9x.89C) instead of the powerline (as in A and B), we found a periodicity of about 77 Hz (13 ms period), with

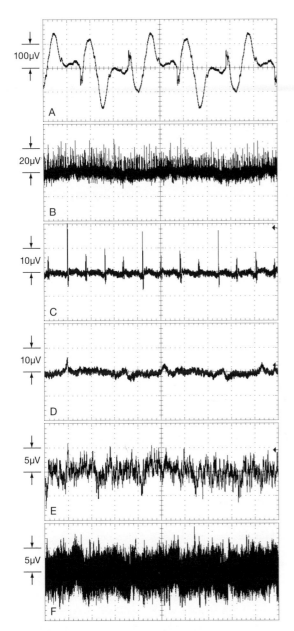

Figure 9x.89. Measured dc output noise of the lab supply of Fig. 9x.87, 10 Vdc output into 100 mA load; all traces are 4 ms/div, with ac line trigger unless trigger point is shown. A. First build, transformer internal, 8-trace avg. B. Same as A, but transformer remote; note scale change. C. Same as B, but triggered as shown, with scale change; spikes are from dc fan. D. Same as C, but replaced noisy StarTech fan with quiet Panasonic. E. Same as D, but single-shot capture; note scale change. F. Same as E, but peak single-shot capture.

the pulse rate changes!). We tried some filtering, but ultimately the cure was to replace the electrically noisy fan (StarTech FANDURONTB, 12 V 220 mA) with one of our favorites (Panasonic/NMB FBA06A12), with an added bypass capacitor at the fan terminals (and a clamp-on ferrite core on the lead wires for good measure). That did the job (Fig. 9x.89D), pushing the residual ripple and noise down to very respectable $10\,\mu$Vpp levels.

It's probably not a fair comparison, but, what the heck, we measured three commercial dc bench supplies that were around the lab; the ripple and noise of these supplies were worse than ours by factors of 35 dB, 20 dB, and 85 dB, respectively (Figs. 9x.90A–C, respectively).

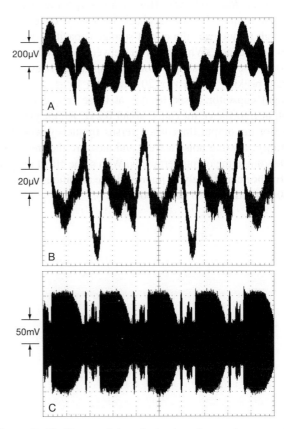

Figure 9x.90. Measured dc output noise of several commercial bench supplies with same 100 mA load, for comparison; all traces are peak capture, 4 ms/div, with ac line trigger. A. BK Precision 1786B (32 V @ 3 A), 10 Vdc output; note vertical scale. B. Leader LPS151 (±25V @ 0.5 A), 10 Vdc output; note scale change. C. Xantrex XFR7.5-130 (7.5 V @ 130 A switcher), 7.5 V output; note scale change.

second and fourth harmonics. A bit of sleuthing uncovered the culprit – the brushless dc fan (block the flow and

Stability

Low noise and ripple is important, but so is stability. A good dc power supply should deliver constant voltage under variations of either load current or ac line voltage. We measured the lab supply, delivering +10 V into a load that switched between no load and full load (1 A); the measured voltage change was less than 1 mV. Likewise, varying the ac line voltage by ±10% produced less than 1 mV change in dc output voltage.

What about signal-frequency variations of load current? We measured the output voltage variations while injecting 20 mApp signal currents at a range of frequencies. At both low frequencies (<1 kHz) and high frequencies (>100 kHz) the output voltage varied less than a millivolt (i.e., $R_{out}<0.05\,\Omega$). This is not surprising, because at low frequencies the feedback loop is in control, and at high frequencies the output storage capacitors produce low impedance. What about in between? Indeed, around 10 kHz the power supply's output impedance rises somewhat, peaking at 8 kHz or 20 kHz (with 15 μF or 100 μF output capacitor, respectively); at those frequencies the output impedance measured about 0.2 Ω. An elegant way to plot the output impedance versus frequency would be to inject a white noise current and measure the frequency spectrum at the output terminals. With some effort the performance could likely be improved; the ambitious reader is encouraged to do so (and report back!).

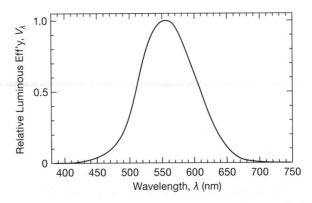

Figure 9x.91. Sensitivity of the "standard observer" at normal light levels (photopic vision), normalized to peak sensitivity at 555 nm. Luminous flux (in lumens) is a measure of light output, taking account of the eye's sensitivity versus wavelength.

9x.22 Lumens to Watts (Optical)

Photometric quantities (lumens, lux, candelas, etc.) have got to be the most confusing measurements on the planet. They deal, variously, with the amount of light cast onto a surface or emitted from a surface ("illuminance," units are *lux*), the areal brightness of the source itself ("luminance," units are *candelas/m²*, also known as *nits*), the total emitted optical power ("luminous flux," units are *lumens*); and, if that's not enough complexity, these quantities themselves have angular dependence ("luminous intensity," units are *candelas*), and wavelength dependence ("spectral intensity," "spectral flux," and a dozen more.[55]) To make a bad situation even worse, they also come in both metric and English units.[56] And, perhaps for historical reasons, they come in various aliases.[57]

Here we address only a single question, one that has come up often in our work with LEDs: how much optical power (in watts[58]) is emitted by an LED whose optical output is specified in lumens? Put another way, what is the efficiency of the emitter – optical watts output divided by dc watts input?

This should be simple, since lumens (abbreviated "lm") measure total optical output power (luminous flux), right? Not quite, because luminous flux (units of lumens) is normalized by the spectral response of the human eye (the term *radiant flux* is used for total optical power, units of watts). So what you need to know is the conversion factor from lumens to watts *as a function of wavelength*.

So, here's how it goes: the eye's maximum photopic[59]

response peaks at 555 nm (approximately pure green), see Figure 9x.91. At that wavelength 1 watt of emitted optical power equals 683 lumens. So, for example, take a 555 nm green LED[60] rated at 165 lm (typ) at 350 mA forward current (at which the forward voltage is 2.97 V). From the conversion factor that's 0.24 W of optical power; the "wall-plug" efficiency (here dc-to-light) is therefore $P_{opt}/P_{dc}=23\%$. That's about ten times better than the 2.2% figure for standard incandescent light bulbs – not bad.

For the 555 nm wavelength of this example the photopic normalizing factor (V_λ, known as *luminous efficiency*) is unity. For other wavelengths you've got to look up the corresponding photopic factor. We'll save you the trouble – Table 9x.4 lists values going from bluer-than-blue to redder-than-red. To illustrate the calculation, consider a Luminus CBT-120-R red LED, with peak emission at 628 nm, where its luminous flux is specified as 825 lm (typ) at 18 A forward current. At that wavelength, interpolating from Table 9x.4, $V_\lambda=0.288$, so $1 W=683 \cdot V_\lambda=197$ lm, and the emitted optical power is 4.19 W. Since the dc forward voltage is 2.3 V, the dc-to-optical efficiency is 10.1%.

A caution: Don't get confused by the term "luminous efficiency," (the quantity V_λ in Table 9x.4), which sounds suspiciously like what we've just calculated; it is simply the (dimensionless) normalized response curve of the human eye. And, don't get (even more) confused[61] by the term *luminous efficacy*, which is simply (why can't optical

---

[55] Would you believe: spectral intensity, spectral radiance, spectral irradiance, spectral flux density, spectral radiosity, spectral exitance, spectral exposure, spectral hemispherical emissivity, spectral directional emissivity, spectral hemispherical absorptance, spectral directional absorptance, spectral hemispherical reflectance, spectral directional reflectance, spectral hemispherical transmittance, spectral directional transmittance, spectral hemispherical attenuation coefficient, and spectral directional attenuation coefficient.

[56] For example, illuminance has the metric unit *lux* (1 lux=1 lumen/m²), and the English unit foot-candle (1 ft-candle=1 lumen/ft²).

[57] Need an example? Illuminance can also be stated in *phot* or *nox*, which are 10^4 lux and 10^{-3} lux, respectively.

[58] See §9x.23 for doing something interesting with those watts.

[59] Photopic refers to the receptors called *cones*, which are densest in the central vision, and responsible for color vision. The receptors called *rods* are more sensitive; their response is called "scotopic."

[60] These data are taken from the datasheet for the Osram LCG H9RM-LXLZ-1 green LED intended for projector applications.

[61] Evidently plenty of people get confused, enough to earn a warning on Wikipedia.

things be simple?) the quantity $683V_\lambda$, that is, the lumens per (optical) watt.[62]

These examples were simple because the emitted light was concentrated at a single wavelength. For emitters with a broad spectrum you've got to integrate luminous flux spectral density (lumens per nanometer) over wavelength, weighted by the photopic response of Table 9x.5. This has been done for some common emitters; for example, a 100 W incandescent lamp (an endangered species, and with good reason) has a luminous efficacy of approximately 20 lumens/W, the low figure owing to its infrared-weighted spectrum (which peaks at an IR wavelength of 1000 nm).

Table 9x.4: Photopic Luminous Efficiency[a]

λ (nm)	V_λ	λ (nm)	V_λ	λ (nm)	V_λ
380	0	520	0.7100	650	0.1070
390	0.0001	530	0.8620	660	0.0610
400	0.0004	540	0.9540	670	0.0320
410	0.0012	550	0.9950	680	0.0170
420	0.0040	555	1.0000	690	0.0082
430	0.0116	560	0.9950	700	0.0041
440	0.0230	570	0.9520	710	0.0021
450	0.0380	580	0.8700	720	0.0010
460	0.0600	590	0.7570	730	0.0005
470	0.0910	600	0.6310	740	0.0003
480	0.1390	610	0.5030	750	0.0001
490	0.2080	620	0.3810	760	0.0001
500	0.3230	630	0.2650	770	0
510	0.5030	640	0.1750		

Notes: (a) or "relative spectral luminous efficiency of radiant energy," the sensitivity of a standard observer to wavelengths of light, normalized to unity at the peak sensitivity at 555nm.

We conclude with a few useful conversions:
1 W (optical) = $5.036 \times 10^{15} \lambda_{nm}$ photons/sec
1 lumen = 7.373×10^{12} photons/sec at $\lambda = 555$ nm
1 lux = 1 lumen/m^2
1 candela = 1 lumen/steradian.

---

[62] Continuing the theme of confusion, you'll find that the term *efficacy* is sometimes used to mean what we've called dc-to-optical efficiency.

9x.23 Sending Power on a Beam of Light

Radiation of light conveys energy, sometimes illustrated in physics courses with a charming antique *Crookes radiometer* (Fig. 9x.92), with a set of vanes that are blackened on one side, shiny on the other, and in which the vanes spin around happily when illuminated.[63] In a more practical vein (pun, get it?), you can exploit light's energy flux to power a circuit with no wire connections – thus achieving complete galvanic isolation.

In one common application diode emitter(s) illuminate a series-connected stack of photovoltaics to generate an isolated dc source for high-side gate drive. We saw this back in §3.5.6B, where a PVI5033 (dual-channel LED to PV-stack) was used to generate MOSFET gate drive (Fig. 3.107A). Parts like that can create enough bias to switch a MOSFET, but they are feeble, and slow: the PVI5033's photovoltaic stack generates 5–10 V, with a short-circuit current of just a few microamps, and with a time constant of a few milliseconds. That's not enough to power much of anything, nor can it switch a MOSFET's input capacitance quickly (e.g., the PVI5033's 5 μA output would take some 5 ms to switch a classic power MOSFET like the IRF520N, with its total gate charge of Q_g=25 nC); recall that's why hefty gate driver ICs like the TC4420-series are rated to source or sink many *amperes* of peak current).

But it's possible to generate and transmit plenty of optical power, most easily with power LEDs or, better, diode lasers. The latter couple nicely to low-loss optical fiber, a convenient medium for safe transport to the receiving photovoltaic. The common 62.5/125 μm graded-index glass fiber, for example, has an attenuation of less than 0.7 dB/km at minimum-dispersion infrared (λ=1300 nm), and 3 dB/km at commonly used "redder-than-red" (800–850 nm) wavelengths for power transmission. Power-over-fiber has many attractive features: it's non-conducting, therefore immune to transients (e.g., lightning) and magnetic fields, and does not create sparks; it has perfect gal-

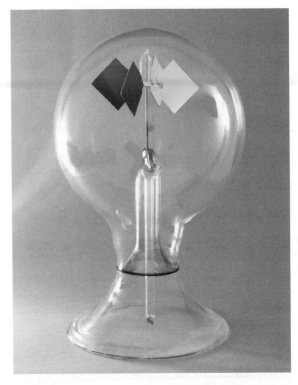

Figure 9x.92. Crookes Radiometer, invented in 1873, and responsible for a decade's worth of argument over the cause of its "backwards" rotation. These things are the very devil to photograph: low light (so it doesn't twirl), wait for it to stop jiggling (and in a favorable orientation), and (worst of all) its shiny surface reflects everything it sees back into the camera; spent a couple of frustrating hours before getting this portrait.

vanic isolation, and thus total common-mode rejection; it's lightweight, and can be used to send high-bandwidth data as well as power. Some disadvantages are cost, efficiency, and the need for careful design to ensure eye safety.

You can buy components for ready-to-go fiber power systems (Fig. 9x.93). For example, the Lumentum L4-2486-005 fiber-coupled diode laser puts out up to 2 W at 830 nm into a 60/125 μm fiber; when coupled to a Broadcom AFBR-POC406L photovoltaic "optical power converter" it generates 6 V at 120 mA at the receiving end.[64] Figure 9x.94 shows the datasheet's photovoltaic *I–V* characteristic (their lower-voltage AFBR-POC404L puts out comparable power – 650 mW – into 3 V loads).

These power levels are altogether adequate for operat-

---

[63] Oddly enough, they spin *backward*: You'd think light bouncing off the shiny side would impart twice as much momentum as light absorbed by the dark side, so the shiny side should be in retreat. But it goes the other way. The reason is far more complicated than the simplistic explanation seen in some textbooks (which at least acknowledge the paradox; others just get it completely wrong).

[64] The Lumentum products originated with JDS Uniphase, who also manufactured photovoltaic converters. You can read about these products in the Lumentum Technical Note 30175827-900-0314 (poweroverfiber-tn-pv-ae).

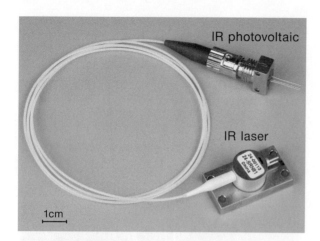

Figure 9x.93. JDSU's 2W power-over-fiber laser (now available from Lumentum) delivers 500 mW at 3 Vdc or 5 Vdc when coupled to a power-conversion photovoltaic.

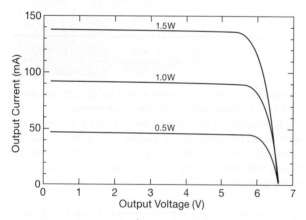

Figure 9x.94. Output current versus voltage for the Broadcom AFBR-POCx06L 6V photovoltaic power converter, from their datasheet.

ing small analog or digital systems. But if the reader finds them, uh, anemic, take a look at the web page of Powerlight Technologies (http://powerlighttech.com/power-over-fiber-2), where you'll learn that "PowerLight Power Over Fiber technology uses high-intensity light to transfer up to kilowatts of power via fiber optic cable. Whether tethered above ground or undersea, unmanned aerial vehicles (UAVs) and unmanned undersea vehicles (UUVs) using our technology can last longer, go farther and maneuver more."

Returning to laboratory-instrumentation power levels, Tektronix uses power-over-fiber (and a return signal-over-fiber path) in their remarkable IsoVu™ probe system (Fig. 9x.95): one fiber sends power to the remote probe,

two more exchange digital control signals, and an analog pair is used to send an optical "carrier" which is returned bearing wideband (to 1 GHz) analog signals back from the probe head (Fig. 9x.96). The result is complete galvanic isolation, with the ability to probe signals with extraordinary common-mode rejection (160 dB at low frequencies, falling to 110 dB at 800 MHz). That's necessary if, for example, you want to see high-speed gate-source voltage differences on a flying MOSFET. And the complete isolation means you can probe signals sitting atop high-voltage electrodes, or otherwise floating at nasty potentials.[65]

Figure 9x.95. Tektronix IsoVu probing system: completely isolated, with power and signals over fiber. (Copyright© Tektronix. Reprinted with permission. All Rights Reserved.)

Just for the fun of it, we decided to cobble together a fiber power link. First stop: eBay, where we bought[66] a JDSU 2486-L3 830 nm 2 W laser with 3′ 60/125 μm ST-terminated fiber.[67] While at eBay we also got some 40 mm square (25×35 mm active area) monocrystalline[68] 2 V photovoltaic stacks ($17.60 for ten of them, cheap!). We chose

---

[65] As is often the case, every silver lining has a cloud: these things are not inexpensive – they start at +41 dB$.

[66] Starlight Photonics Inc., aka "Junktronix."

[67] JDSU's products are now part of Lumentum, who sell these laser sources as their L4-2486-xxx; however, they do not sell matching photovoltaics.

[68] Interestingly, *amorphous* silicon photovoltaics do not work with infrared, they cut off completely above 800 nm.

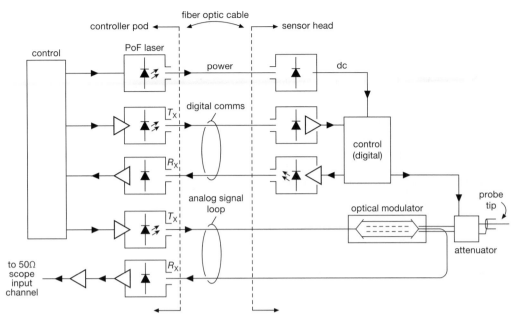

Figure 9x.96. Five fibers carry digital signals, analog signals, and power in the IsoVu isolated probing system. Most of the secret sauce resides in the wideband, high-impedance, low-power, linearized optical modulator (a Mach–Zehnder electro-optic interferometer), which elegantly eliminates the need for a power-hungry wideband front-end amplifier.

the latter because they were a hundred times cheaper than those nifty Broadcom photovoltaics. Not as compact or pretty – we put the PV cell at one end of a 6″ long 2.5″ mailing tube, with an ST feedthrough at the other end – but it worked well enough, putting out 150 mA at 2 V.[69]

This got us thinking about power transfer with *LED* sources (instead of lasers), for applications where the compactness of fiber is not needed. So we fished through our junkbox(es), where we found a collection of high-power LEDs in the form of "chip on board" (COB) arrays, some of which are shown in Figure 9x.97. These are widely available in many colors, and they put out an impressive amount of light. The red ones (625 nm) we had on hand are rated at 100 W (dc input), with roughly 25% conversion efficiency to optical power output. When we placed our 4-cell 25×35 mm photovoltaic about 25 mm away from the LED (the latter running at its maximum rated 4 A) we measured an open-circuit voltage of 2.6 V and a short-circuit current of 400 mA; the corresponding figures for a larger 10-cell photovoltaic (65 mm square) were 6.2 V and 200 mA. Figure 9x.98 shows curves of the measured photovoltaic output voltage versus load current, for a range

of LED drive currents; and Figure 9x.99 demonstrates that this crazy stuff really works.

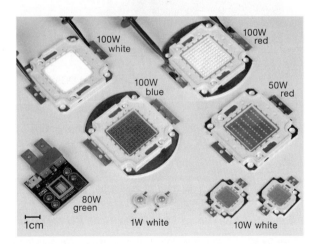

Figure 9x.97. "Chip-on-board" (COB) LED arrays are available in many colors as well as white (yellow phosphor over blue LEDs); they are inexpensive ($15 or so for the 100 W variety) and are used for area lighting. The green "photonic lattice" emitter at lower left (available in matching red and blue) is used in place of halogen or arc lamps in digital projectors; it puts out 2000 lumens from a 0.12 cm² emitting area. To photograph the powered COBs at top we had to run the current *way* down – at full blast these things are painful to look at.

---

[69] Photovoltaics intended for conversion of ∼800 nm laser light use GaAs rather than silicon, because the bandgap is better matched, producing conversion efficiencies in the neighborhood of 40–50%.

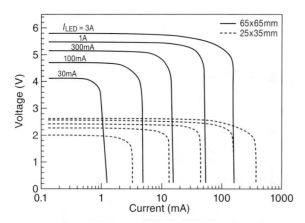

Figure 9x.98. Measured output voltage versus load current for two monocrystalline silicon photovoltaics ("solar cells"), measured at half-decade steps of LED drive current. For variety we plotted V versus I, rather than the other way around as in Fig. 9x.94.

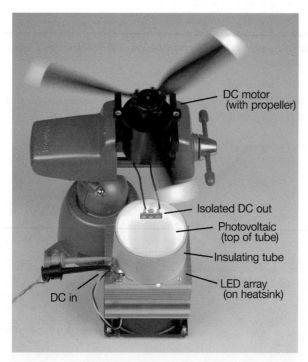

Figure 9x.99. "Photomotive force!" (just kidding). We coupled a red LED array to a silicon photovoltaic, which produced far more dc output than needed to spin this 3-bladed propeller.

9x.24 "It's Too Hot" Redux

Back in §9.4.1 we introduced the sizzle test, a sure indication that your components are, well, *too hot*. But it would be useful to add a bit of nuance to the subject. We were reminded of this the other day, when we ran a power-hungry circuit (4.5 kW peak) for a few hours to see if the components were happy – they didn't sizzle, but they weren't exactly cool, either, and we wanted to know about how hot they were. Looked around for the Tempilstik calibrated crayons (couldn't find them), then grabbed the Fluke IR temperature probe (darn, it doesn't work any more), finally looked for a thermocouple (couldn't find one, either). What to do? There must be a "rule of finger," based on how long you can stand to hold your finger on the heatsink, right?

9x.24.1 The finger test

Yup, turns out there is, according to the all-knowing Google, which found what we wanted in Geoff Phillips' *Newnes Electronics Toolkit* (but not without a struggle... had to go through a few screenfuls). To spare readers comparable hassle, we here reproduce the "Engineer's Finger Test" from Geoff's book:

skin sticks	$\leq -20°C$
cool can of beer	$4°C$
barely getting warm	$30°C$
finger temperature	$37°C$
getting warm	$40°C$
can hold palm on indefinitely	$50°C$
can hold finger on for seconds	$60°C$
can hold finger on for 1 sec*	$70°C$
can hold finger on <1 sec	$80°C$
finger is withdrawn quickly	$\geq 90°C$
blister (not recommended)	$\geq 120°C$
* Mnemonic: 1 SEcond is SEventy degrees	

We applied the test to our circuit (more than about ten seconds – ouch), which evidently was running around $60°C$. Nice!

Crickets

While we're on the subject of unusual methods of thermometry, we can't resist a bit of backwoods lore about The Cricket As A Thermometer (the title of a charming one-page article by A.E. Dolbear in the November 1897 issue of *The American Naturalist*). Crickets are cold-blooded, so their metabolism (thus rate of chirping) depends strongly on temperature. Dolbear evidently enjoyed the evening symphony of a field of crickets "keeping time as if led by the wand of a conductor" (how quaint was scientific writing in those calmer days – and how brief the articles!). Here's the rule, as formulated by Dolbear: "Let T stand for temperature and N the rate per minute. $T = 50 + (N - 40)/4$. For example. What is the temperature when the concert of crickets is 100 per minute? $T = 50 + (100 - 40)/4 = 65°$." (Fahrenheit, one assumes)

Devotees of cricket thermometry may enjoy additional discussion found in Miller and Friedman's handy reference *Photonics Rules of Thumb* (chapter title: Crickets as Thermometers). There you can learn[70] alternative formulations, for example:

• count the clicks for 14 sec, then add 40.
• count for 24 sec, then add 35.
• count for 15 sec, then add 40.

Astute readers will no doubt have noticed that these are all formulated in the King's units, reflecting the provincialism of English-speaking colonies. Those who traffic in Celsius may wish to adapt these formulae appropriately.

9x.24.2 Better thermometry

Let's get serious – what techniques are available to get reliable temperature measurements? Turns out there are plenty – the major players (in alphabetical order) are (a) indicating wax or paint, (b) infrared thermal-imaging camera ("FLIR"), (c) infrared thermometer (pyrometer), (d) resistance temperature detector ("RTD"), (e) silicon sensor, (f) thermistor, and (g) thermocouple. Here follows a brief rundown on their characteristics, and (at the end) advice on when to use each.

Change-of-state indicators

You can get a set of color-coded crayons, with calibrated melting points, that provide a simple and quick way to check if a surface has reached some temperature. They are made by companies like Tempil ("Tempilstik®"), who offer more than a hundred selected temperatures (from $38°C$

---

[70] You can also learn that some authorities assert that the best results come from the *snowy tree cricket*, an insect related to the katydid.

to 2000°C), as well as kits of ten or twenty round-number crayons (going in 25°F increments from 125°F).[71] You can also get analogous products as pellets, paints, or stick-on labels. They are generally irreversible (except for liquid-crystal types), which is helpful if you want to know if the temperature has exceeded some limit (in transit, for example). These things are used by brazers, welders, and heat-treaters; they are of limited utility in electronics.

Thermal cutouts

These thermostatic switches undergo a (reversible) change of state, but unlike the previous category (which merely indicate when a temperature has been reached), these are 2-terminal mechanical switches that are commonly used to protect a circuit from overheating (e.g., from loss of air-flow, or excessive heatsink dissipation). They reset when the temperature drops back down. You can get them in normally open or normally closed configurations (close on heat or open on heat, respectively), with current ratings (resistive) from about 100 mA to 10 A or more. They come in useful form-factors: surface-mount, DIP, power transistor style (TO-220), and flanged (0.5″ round), and in a full range of temperature thresholds (from −30°C to +300°C or more). They're priced in the range of $5–$10 in small quantities; not the cheapest component in the toolbox (many microcontrollers cost less!), but well worth it if it protects an instrument that's worth much more. We first encountered these as TI "Klixon" devices (a brand now owned by Sensata, along with their Airpax brand of thermostatic switches); other manufacturers include Bourns, Cantherm, Honeywell, Kemet, Senasys, and White-Rodgers.

Another type of thermal circuit protector is the fusible thermal cutoff (TCO). These are basically thermal fuses, and they do not reset. The most common types come in small axial-lead cylindrical packages, 2–3 mm in diameter and 5–10 mm long. You see them in consumer items like hair dryers and motors; they're inexpensive – $0.25–0.50 in small quantities – and made by companies like Cantherm, Littelfuse, Panasonic, and TDK.

IR camera (FLIR)

Warm objects emit infrared radiation, peaking around $\lambda=10\,\mu$m ("long-wavelength infrared," LWIR) at room temperature, shifting to shorter wavelength[72] (and higher

intensity[73]) at higher temperatures. So you can image objects in their IR emission with a forward-looking infrared camera ("FLIR").

Traditionally this required expensive cryogenically cooled detector arrays, but recent developments in un-cooled (i.e., ambient temperature) LWIR bolometric array imagers[74] have made such cameras widely available at reasonable cost. For example, you can get a 160×120 real-time imager that connects to a smartphone for $400 (FLIR One Pro, from FLIR Systems), or competing units from Seek Thermal (CompactXR, 206×156, $300; or Compact-Pro, 320×240, $500); Figure 9x.100, taken with the latter imager, reveals what's hottest when you grind some foil

Figure 9x.100. The action at the tip of Jim's Dremel tool tops the heat map in this monochrome FLIR image.

---

[73] "Stefan–Boltzmann Law": the power radiated per unit area from an ideal blackbody is $M = \sigma T^4$, where T is in Kelvin, the Stefan–Boltzmann constant $\sigma=5.67\times10^{-8}$ Wm^{-2}K^{-4}, and M has units W/m^2 of radiating area. In practice one includes on the right-hand side a multiplier ε, the *emissivity* (a unitless factor between 0 and 1, which in general depends on wavelength) by which the actual emitted power is reduced. Most ordinary objects have $0.5 < \varepsilon < 1$, but reflective objects (shiny things) can have significantly lower emissivities. An obscure but interesting fact: gas lamps exploit the emissivity variation of thorium oxide to create a brilliant white light – their fragile "mantle" is a fabric of the oxide, whose low IR emissivity suppresses radiation at long wavelengths, whereas at visible wavelength (where its emissivity is high) it belts out the photons. Thorium is mildly radioactive (1.4×10^{10} yr half-life), but in this application it is thorium's configuration of *electrons* that does the magic; i.e., it has nothing to do with its nuclear properties (which make it radioactive).

[74] The individual elements are thin resistive pixels, typically $10\,\mu$m–$25\,\mu$m in size, made of IR absorbing materials (amorphous silicon, or vanadium oxide), suspended on small struts a few microns above a silicon IC that biases and reads out the resistance changes.

---

[71] You can buy them from distributors like Grainger or McMaster-Carr.

[72] "Wien's displacement law": $\lambda_{max}(\mu\text{m})=2900/T_{\text{Kelvin}}$.

off a PC board; and Figure 9x.101 shows "thermal scars," the ghostly images of residual warmth. And Figure 9x.102 shows how we used it to locate a stuck power rail on an 8-layer high-density PC board with twelve separate power supplies. Standalone imagers with better sensitivity, speed, and resolution (up to 1280×1024) run upwards of $10k (topping out around $100k), and are available from companies like FLIR Systems, Fluke, Sofradir, and others.

Even the inexpensive imagers can resolve temperature differences of less than 1°C (the *thermal sensitivity*). That's altogether adequate to locate hot spots on a circuit board or within an instrument. Figure 9x.103 shows a beautiful example, albeit from a high-resolution (and definitely not inexpensive!) cooled camera.

An important caution: as described in footnote 73, objects that are reflective in the infrared (i.e., of low emissivity ε) will not render accurately with an IR camera. In the extreme case of shiny metal, what the camera will see is the IR of the surrounding objects, reflected as in a mirror. In such cases you need to coat the object with a material of high IR emissivity – for example black electrical tape (Scotch 88) or (paradoxically) "white-out" correction fluid; both have $\varepsilon \approx 0.96$.[75]

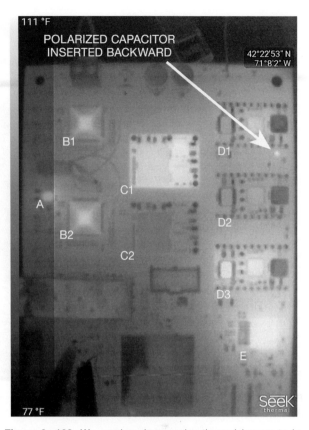

Figure 9x.102. We used an inexpensive thermal imager to locate the culprit (a reversed polarity capacitor) that was loading the power rail (and warming itself). A is a linear LDO regulator; B1,B2 are controllers for switching converters C1–2 and D1–3; and E is a high-power switching converter.

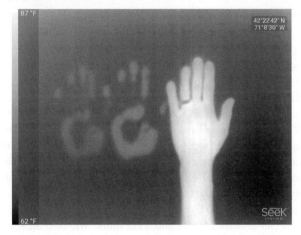

Figure 9x.101. It's easy to spot "thermal scars," as in this shot of an author's hand's history of three placements on a wooden door.

IR thermometer (pyrometer)

These are inexpensive single-point LWIR instruments, typically configured as a handheld (pistol-shaped) trigger-operated device with LCD readout. They come with various angular cone-of-acceptance sizes (usually stated as a

ratio, e.g., 20:1 means that it accepts the radiation from a 1 cm area at a standoff distance of 20 cm. Typical accuracies are ±2°C, with repeatability of 0.5°C. Additional features include min/max memory, display of difference and average, USB interface, and built-in visible laser pointer. Suppliers include Amprobe, Anaheim Scientific, Extech, FLIR Systems, Fluke, Omega, and Testo. Most models range in price from $50 to $150.

Until recently you could get the guts of an IR-sensing thermometer in TI's elegant TMP006 chip: It has a tiny thermopile (stack of thermocouples) that responds to incident IR, and an on-chip silicon temperature sensor (±0.5°C accuracy, typ) to correct for local ambient temperature. It isn't done yet – it includes an on-chip 16-bit delta-sigma ADC (with on-chip voltage reference), and it sends the result (accurate to ±0.5°C, typ) out on an SPI interface. The on-chip ambient temperature sensor operates from −40°C to +125°C, and its IR sensor works over

---

[75] Some others recommended by FLIR include Kapton tape, masking tape, flat (matte surface) paints, and Dr. Scholl's foot powder spray.

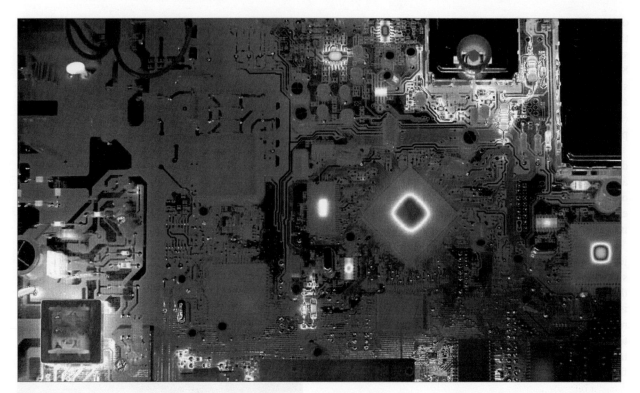

Figure 9x.103. Thermal image of a powered PCB, taken by a high-end cooled InSb high-resolution MWIR camera (FLIR Systems A8000sc-Series; used with permission).

a narrower range, sensing object temperatures that differ from ambient by −20°C to +40°C. How's that for a tiny 1.6×1.6 mm chip that costs $5 (unit qty)?

And if you don't need tiny, check out the Melexis MLX90614: it comes in a metal TO-39 can that includes the same stuff as the TI part, but with a far wider IR sensor temperature range (−70°C to +380°C, albeit poorer accuracy at the extremes), and with choice of angular field-of-view. It incorporates a 17-bit ADC and considerable DSP, along with configuration EEPROM, and it outputs PWM or 2-wire SMBus (a variant of I^2C); it costs $14 in unit quantity.

RTD

The resistance of metallic conductors increases with increasing temperature, an effect that is exploited in the *resistance temperature detector*[76] (RTD). The coefficient is

not large – for the noble metal platinum it is +0.385%/°C at 0°C – but it is highly stable and repeatable, and Pt RTDs are the most accurate temperature sensors over the temperature range of about 14 K to 1235 K (−259°C to +962°C).[77] RTDs can also be made from nickel or copper, but platinum is superior, both because of its extended temperature range (nickel becomes nonlinear above 300°C, and copper oxidizes above 150°C), and because of its excellent linearity, repeatability, and chemical inertness. Most Pt-RTDs are supplied with 100 Ω resistance at 0°C ("Pt100"), though you can also find 500 Ω and 1 kΩ units. We used an RTD in the thermal controller in §15.6.

For optimum performance, the wire in a Pt RTD must be kept free of strain; this can be arranged by allowing the coil of wire to sit loosely in a hollow cavity of a ceramic or glass capsule, perhaps surrounded by powdered insulating material. Such "coil-element" RTDs are widely available

---

[76] Proposed in 1871 at the annual Bakerian Lecture, given by the awardee of the Bakerian Prize (established in 1775, and continuing to this day), described by the Royal Society (UK) as "the premier lecture in physical sciences." The lecture that year, entitled "on Electrical Resistance," was given by Dr. Charles William Siemens (birth name Wilhelm), brother

of Werner Siemens, the latter the co-founder of Siemens AG, currently the largest manufacturing and electronics company in Europe.

[77] In fact, over this range they constitute the equipment calibration standard set by the International Temperature Scale of 1990 (ITS-90). Pure platinum has a slightly higher tempco (+0.3925%/°C at 0°C), but the wire in most Pt RTDs is lightly doped to conform to the standard.

commercially for \$10–50 (check out Omega Engineering, or Adafruit); typically these units are intended for temperatures from about $-200°C$ to $+500°C$. If you want a smaller sensor, and don't need such a wide temperature range, the recent thin-film types are quite adequate, and inexpensive. You can get them in standard surface-mount sizes (e.g., 0603, 0805, 1206), at prices around \$2 in unit quantity. The types that are soldered onto a PCB typically specify operation from $-50°C$ to $+150°C$; some are available with leads attached, with operation from $-70°C$ to $+500°C$. These inexpensive RTDs are quite stable – typically they specify long-term stability of $\pm0.1°C$ – and they come with initial accuracies of $\pm0.3°C$ or $\pm0.6°C$ ("F0.3" and "F0.6" respectively). Check out offerings from Heraeus, Honeywell, US Sensor, and Vishay Beyschlag. You can also get Pt-RTDs in flat stick-on configurations, both in wire-wound and thin-film construction (e.g., Omega Engineering type SRDT and SA1, respectively).

A couple of cautions:

(a) RTDs require an applied voltage, which produces some self-heating; so it's best to keep the current less than a milliamp or so – for an 0805-size RTD the self-heating in still air is about $0.8°C/mW$, so 1 mA through $100\,\Omega$ would produce a bit less than $0.1°$ of self-heating.

(b) for leaded RTDs, the wire leads contribute to the total resistance; it's not a large effect – a foot of 26-gauge copper wire pair adds about $80\,m\Omega$ (0.08% of $100\,\Omega$, equivalent to an error of $+0.2°C$). But most leaded RTDs are available in a 3-wire configuration to eliminate the error[78] (which is small if the RTD is located close to the electronics, but which can be substantial when it's many meters away).

To achieve the full benefit of an RTD you need a precision amplifier. Happily there are ICs that take aim at the problem, for example the MAX31865. It includes everything you need: bias supply, 2-, 3-, or 4-wire configurations, ability to handle $100\,\Omega$ to $1\,k\Omega$ RTDs, 15-bit delta-sigma ADC with 50/60 Hz nulling, total accuracy (over all operating conditions) of $0.5°C$, fast conversion time (21 ms), and SPI digital interface. What's not to like about this \$3.50 (unit qty) part?[79]

Silicon sensor

As we discussed in §9.10.2, the predictable variation of V_{BE} with temperature (when compared with a stable bandgap reference) lets you make an accurate on-chip temperature sensor. These come in many varieties. There are plenty of analog-out types with convenient scale factors, for example the 3-terminal LM35/TMP35 ($10\,mV/°C$, 0 V at $0°C$) or LM61 ($10\,mV/°C$, offset 0.6 V at $0°C$, so it works to $-30°C$ with a single supply); the LM35 can operate from $-55°C$ to $+150°C$ if you provide a $50\,\mu A$ resistive pull-down to a negative rail.[80] These voltage-output sensors are easy to use, and accurate enough for most purposes (typically $\pm2°C$); they cost about a dollar (qty 25).[81]

You can use these things with a microcontroller that has on-chip ADC (that describes pretty much all of them), but for ease of use in digital systems (and generally with improved accuracy) there are many V_{BE}-based "smart" temperature sensors with SPI or I^{2}C local buses.[82] A good example is Microchip's MCP9808, which works from $-40°C$ to $+125°C$, accurate to $\pm1°C$ (max, full temperature range), and $\pm0.5°C$ max from $-20°C$ to $+100°C$; it supports I^{2}C/SMBus, and costs about a dollar (in unit qty) – not bad for a part that can assert an ALERT pin upon a user-programmable limit (or window), uses 0.2 mA, and comes in both MSOP-8 (nice) and DFN-8 (annoying) packages. Other buses are supported – for example Maxim's MAX31723 (SPI, accurate to $\pm0.5°C$ max, \$2 in unit qty), and their DS18B20 (1-Wire interface, see §14.7.3). Digital temperature sensors are available from many semiconductor manufacturers, including Analog Devices, Honeywell, Maxim, Melexis, Microchip, Sensiron, Silicon Labs, and TI.

At their best, these devices rival the accuracy of the best RTD devices (albeit within their narrower temperature range); an example is TI's TMP117, whose maximum error bands are shown in Figure 9x.104, compared with the corresponding error bands of the four classes of RTDs. The TMP117 has nice "creature features," including 16-bit resolution ($0.0078°C$ LSB), operation from 1.8 V to 5.5 V, programmable limit alerts, offset correction, se-

---

[78] See if you can figure out why three leads (i.e., only one sensing lead) are sufficient, when the usual Kelvin-sensing arrangement requires four leads.

Hints:

(a) the other two leads must be matched in resistance, and

(b) it may help to draw a bridge arrangement to see what happens. And a footnote to this footnote: you *can* get 4-lead Pt-RTDs, preferable for the highest precision.

[79] Which you can get in a convenient \$15 break-out board from Adafruit, who also provide a helpful tutorial.

[80] If you are a devotee of the King's units, the LM34 is your friend: $10\,mV/°F$.

[81] In contrast to these analog voltage-output sensors, the LM334 (Fig. 2x.13) provides a *current*-output that is proportional to absolute temperature (PTAT); it has a tempco of about $+0.34\%/°C$ at $25°C$.

[82] These likely use a popular technique that exploits the *change* in V_{BE} when two different currents are applied to a diode-connected BJT: $\Delta V_{BE} = (kT/q)\log_e(I_2/I_1)$. You can add your own diode-connected BJT to some ICs using this technique; e.g., Microchip's MCP9904 handles three such sensors.

lectable averaging, and true micropower operation (3.5 μA operating, 150 nA typ when shutdown) for negligible self-heating (<0.1°C). For this performance you pay about $3.50 (unit qty). For some reason TI chose to support only I^{2}C and SMBus, and to supply this fine part only in a tiny (2mm×2mm) WSON package.

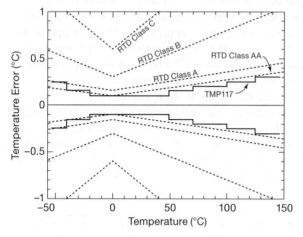

Figure 9x.104. TI's TMP117 silicon-based sensor matches the performance of the best RTDs, over its −55°C to +150°C operating range.

As long as you're putting down a chip to measure temperature, why not have it tell you humidity (and maybe barometric pressure)? Check out devices from Sensiron (e.g., their SHT30-DIS-B, I^{2}C, humidity accuracy to ±2% and temperature accuracy to ±0.3°C, $4 in unit qty), or comparable units from Silabs (e.g., Si7021-A20) or TI (e.g., their HDC1010). For barometric pressure combined with temperature check out the ST LPS25 (±2°C typ, absolute pressure[83] 260–1260 hPa, ±1 hPa typ uncalibrated, ±0.2 hPa typ after 1-point calibration, both I^{2}C and SPI interfaces, $3 in unit qty). Similar sensors are made by many manufacturers (likely owing to the automotive market), check out listings at your favorite electronics distributor.

Thermistor

A thermistor (thermal resistor) is a 2-terminal passive resistive device with a (respectably large) temperature coefficient; the most common types are NTC (negative tempco, typically about −4%/°C at room temperature), in

the form of tiny beads or pills of semiconductor material. They are inexpensive (as little at $0.06 in small quantities), easy to use (because of the large tempco), and available in interchangeable types with good precision (±1%) and conformance to a standard curve. You can get thermistors in many forms: standard SMT chip sizes, discs, beads (with leads), stud-mounting, TO-220, immersion-sealed (in metal, glass, or plastic), etc. Thermistors come in several standard resistances (specified at 25°C), the most common being 10 kΩ.[84] They also come in a narrow range of temperature coefficients, usually specified by the so-called beta, or B-value.[85] Here are the relevant expressions for resistance as a function of temperature, and vice versa, according to the 2-parameter (R_0, B) model:

$$R = R_0 e^{-B\left(\frac{1}{T_0} - \frac{1}{T}\right)}$$
$$= r_\infty e^{B/T} \quad (\text{where} \quad r_\infty = R_0 e^{-B/T_0}) \quad (9x.6)$$
$$T = \frac{B}{\ln(R/r_\infty)} \quad (9x.7)$$

where the temperature T is in kelvins, R_0 is the resistance at 25°C, and T_0 is 298.15 K (+25°C).

For example, taking the most common thermistor, which has R_0=10k and B=3950 K (that's 3950 kelvins), you can calculate that it will have a resistance of 3588 Ω at 50°C; this is in excellent agreement with a table of values (calculated with a more accurate approximation to the true resistance versus temperature), which lists $R(50)$=3592 Ω (i.e., an error of just 0.03°C). This good agreement is no accident: that B value has a subscript ("$B_{B25/50}$=3950K") indicating that it is calculated between that pair of temperatures (and is a good approximation for temperatures in that general vicinity).[86] If we're more daring and try using the R_0, B model with $B_{B25/50}$ to predict the resistance at 150°C, for example, we get R_{150C}=199.7 Ω, a resistance error of 13% (and a temperature error of 5°C, based on the table value $T(R)$=145°C). Thermistors sometimes list several B values for other spans (e.g., $B_{0/50}$, $B_{25/75}$, $B_{25/85}$, or $B_{25/100}$).[87]

A handy trick when using a thermistor sensor is to

---

[83] What's an *hPa*? A *hecto*pascal, of course, which equals 10^2 pascals (1 Pa=1 newton/m^2); 1 hPa equals 1 millibar, or approximately 10^{-3} of average atmospheric pressure (a "standard atmosphere" is 1013.25 hPa). For us old-timers 1 hPa is roughly 1 mmHg.

[84] Some other popular values are 2.252k, 5k, 30k, and 100k.

[85] A more accurate model of R versus T is the 3-parameter Steinhart–Hart equation, to which the 2-parameter formulation in terms of R_0 and B approximates and is "good enough" for most purposes. Best is interpolation from a lookup table, which is no sweat in these days of microcontrollers and smart ASICs.

[86] That is, $B_{T1/T2}$ is calculated as $B_{T1/T2}=[\ln(R_2/R_1)]/(1/T_2 - 1/T_1)$.

[87] For example, a tiny Murata thermistor bead with a giant part number (NXFT15XV103FA2B150) specifies $B_{25/50}$=3936 K, $B_{25/80}$=3971 K, $B_{25/85}$=3977 K, and $B_{25/100}$=3988 K.

linearize the resistance-to-temperature relationship over a narrow range, by simply adding a series resistor. Choose a value equal to the thermistor's value in the middle of the important part of the temperature range, e.g., 10k for a thermistor that's 10k at 25°C (a common value). If you use 0.1% resistors with 0.2°C thermistors, any calibration you make will track well away from the calibration temperature.

Thermistors are inexpensive, quite stable[88] and accurate, and (being passive) amenable to remote thermometry in unpleasant environments (liquids, etc.). They span a modest range of temperatures (−80°C to +150°C or more), but they do require some signal conditioning (conversion of resistance to temperature). Happily, the latter requirement is handled nicely by accommodating ICs, for example the SPI-interface MAX6682, which performs a simple linearization of NTC thermistors that's good enough over a limited range (e.g., ±0.2°C over 10°C–40°C using the series-resistor trick mentioned earlier); it avoids self-heating by biasing the thermistor only during the measurement. Maxim's RTD chip (MAX31865) can also be used for precision readout of thermistor resistance, but it does no temperature conversion or linearization, so you're on your own. Thermistors are handy for simple applications such as a high/low alarm or a thermal stabilizer – you can roll your own with a resistive divider and comparator, or you can use something like Analog Devices' ADN8834 "Ultra-compact 1.5 A Thermoelectric Cooler (TEC) Controller," which incorporates a pair of precision chopper amplifiers, provision for PID compensation, and even includes a power H-bridge so that the TEC can seamlessly morph from cooling to heating, as required.

Thermocouple

This method goes back 50 years earlier than the platinum RTD, dating to 1821, when Thomas Johann Seebeck, a German physicist, observed that if a pair of wires of dissimilar metals are joined at both ends, and if the ends are heated to different temperatures, a magnetic field is created; he called it thermo-magnetism,[89] but a better way to view it is that the thermal gradient creates an open-circuit voltage.

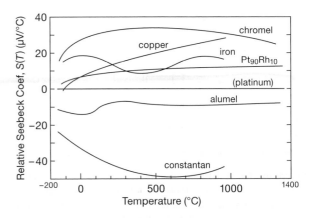

Figure 9x.105. Seebeck coefficient versus temperature, relative to platinum, for some common thermocouple metals. (After Dike, P. H., *Thermoelectric Thermometry*, Leeds and Northrup, 1954.)

This *Seebeck Effect* is the basis of the thermocouple, a method of thermometry used for more than a century, particularly useful for high temperatures encountered in ovens, etc. The voltage developed along a conductor of uniform composition depends only on the temperature difference at the endpoints; to form a *thermocouple*, a second conductor with a different Seebeck coefficient is joined at the temperature-sensing end, and terminated with a distinctive (color-coded) 2-terminal connector at the measuring end. Then the voltage between the two terminals depends on (a) the *difference* of Seebeck coefficients, and (b) the temperature difference between the sensing end (where the conductors are joined) and the measuring end. The voltages are not large, generally in the tens of microvolts per degree Celsius (Fig. 9x.105), but they are stable and can easily enough be measured for results accurate to a degree or so.

Because the voltage you measure depends on the temperature difference between the sensing end and the connector end, you need to know the latter's temperature (call it T_{ref}). This is called "cold-junction compensation" (based on the traditional ice water reference), and it's usually done by attaching some form of temperature sensor (Silicon IC sensor, RTD, or thermistor) to the junction block (Fig. 9x.106). This local measurement, combined with the measured voltage difference, allows you to calculate the temperature at the (remote) sensing end, given the characteristics of the thermocouple alloys (their Seebeck coefficient).

You can get nice ICs that take care of all this, for example the AD8495 (analog, voltage output), or the MAX31856 (digital). The former is designed for type-K thermocouples, with an output of 5 mV/°C (plus settable

---

[88] The best are extraordinarily stable: for example, the document "Stability and Long Term Aging of NTC Thermistors" from U.S. Sensor Corp. asserts that "thermistors can be produced with typical drift of only 0.001°C to 0.002°C per year." This rivals the performance of platinum RTDs.

[89] *Magnetische Polarisation der Metalle und Erze durch Temperatur-Differenz.*

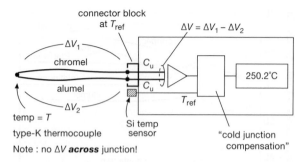

Figure 9x.106. A thermocouple measures an unknown temperature T by exploiting the difference in "thermal EMFs" developed along two conductors of different material (here the nickel alloys chromel and alumel, which form a type-K thermocouple) that span a temperature difference (here T at the sensing end, T_{ref} at the meter/display end). Do not believe anyone (even the President) who tells you that the voltage is developed across the junction.

offset), and achieves an accuracy of $\pm 2°C$ (max) over temperatures of $-25°C$ to $+400°C$; it costs $6 in unit quantities. The MAX31856 goes quite a bit further: it can handle eight different types of thermocouples (B, E, J, K, N, R, S, and T), and it uses lookup tables to correct for the nonlinear relationship of Seebeck voltage versus temperature (seen in Figs. 9x.105 and 9x.107). With those corrections it achieves accuracies of about $1°C$ over the full thermocouple temperature range (the specification is stated as $\pm 0.15\%$ max error); so, for example, you could get accurate results from a type-K thermocouple from $-200°C$ to $+1372°C$. This IC outputs digital temperature, in Celsius, via an SPI interface. It includes nice features like 50/60 Hz filtering and thermocouple fault detection. Not bad for an IC that costs less than $5 in unit quantities!

But why not just buy a thermocouple meter? Why not, indeed. An example is the Amprobe TMD-52, which looks like a handheld DMM, and which accepts four kinds of thermocouples (E, J, K, T), and is accurate to 0.1%. It has a backlit LCD display (°C or °F, your choice), and it has two input connectors so you can read T_1, T_2, or $T_1 - T_2$; it comes with two type-K thermocouples ($-200°C$ to $+1372°C$), and it costs about $80.

These temperature ranges ($1372°C$ is "white hot") illustrate an important benefit of thermocouples: they can measure really hot things, in fact the widest range of any contacting temperature measurement. See "Wrapup" below for a summary of the pros and cons of thermometry techniques.

Thermocouple potpourri

(a) The thermal voltage is not created "at the junction," or "across the junction" – it's created along the wire, by the temperature gradient; the junction only connects the two wires, putting that end of the two wires at the same voltage. (b) Some thermocouple alloys are magnetic (e.g., type-K), and exhibit some wiggles in their thermal EMF at the Curie point (the magnetic transition temperature). (c) A series string of thermocouples can generate a large enough voltage to operate a gas valve (that's the *thermopile* that holds open the gas valve in a water heater or furnace); our house has a gas boiler with a "millivolt" thermostat system – so no external electrical power is needed. (d) A thermopile heated by a radioactive thermal generator (RTG) is used to power spacecraft visiting the outer solar system (where there's not much sunlight). (e) In spite of the small thermal voltages, thermocouples can be used over large distances, up to a few hundred feet or so; in such cases it's a good idea to use a shielded thermocouple cable, with the shield acting as a guard; and good powerline rejection and transient protection is essential. (f) The properties of thermocouple wires can change with time, for example if they oxidize at high temperatures, or are exposed to liquids or other corrosive influences. (g) And, as with any contacting temperature sensor, you've got to ensure that the thermocouple is at the temperature of the thing you want to measure; you may need to arrange for a pressure contact, and perhaps put a dab of thermal material or adhesive.

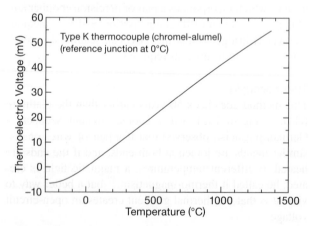

Figure 9x.107. Thermoelectric voltage for the ever-popular Type-K thermocouple. (Data from NIST Monograph 175, as found on NIST ITS-90 Thermocouple Database.)

A useful temperature-sensor factoid: the most serious sensor errors are offset (rather than gain). So "zeroing" the sensor error by calibrating at a known temperature near the important region yields a more accurate sensor (with

Table 9x.5: Thermometry Methods

Method	Advantages	Disadvantages
Finger test	always available zero cost (unless $T>120°C$)	poor accuracy limited temperature range: −20°C to +150°C
Crickets	charming	limited availability limited temperature range: 0°C to 60°C poor accuracy
Change-of-state indicators	quick visual indication inexpensive wide temp range: −30°C to +2000°C passive, non-metallic, no electronics assurance surface was not overheated	slow poor accuracy irreversible (except liquid xtal)
Forward-looking IR (FLIR)	non-contacting wide temperature range: −40°C to +2000°C locate hot spots visually	limited accuracy (depends on emissivity) very expensive
IR Thermometer (pyrometer)	non-contacting wide temperature range: −50°C to +2500°C	limited accuracy (depends on emissivity) not well localized
RTD	accurate, repeatable, highly stable wide temperature range: −250°C to +900°C highly linear	low tempco: requires precision electronics some are delicate expensive self-heating
Silicon (V_{BE})	inexpensive easy to use (e.g., 10mV/°C)	limited temperature range: −55°C to +175°C (can be extended to −270°C)
Thermistor	inexpensive small (to 0.005″) large tempco good to excellent stability good accuracy (~1°C)	temperature range: −100°C to +450°C nonlinear: circuit or computation complexity (many tempco curves) self-heating
Thermocouple	very small (to 0.001″) wide temperature range: −270°C to +1800°C rugged inexpensive self-powered	some nonlinearity – requires computation requires reference requires stable amplifier (μV levels) moderate linearity, requires correction

gain errors accumulating gradually at the range extremes). Many digital sensors include a unique serial number.[90] A batch of these can be clamped to a copper block, and calibrated as a group before assembly, with the measured offsets added to the code.

Wrapup

Need advice? Check out our handy summary in Table 9x.5.

---

[90] We've heard of folks adding Maxim's DS1822 sensor to their PCB, primarily to get its readable serial number in their product; but Maxim also makes the DS2401 for just that purpose.

9x.25 Transient Voltage Protection and Transient Thermal Response

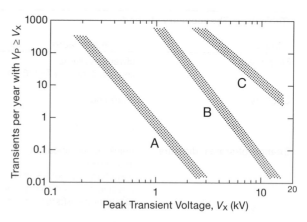

Figure 9x.108. Annual occurrence of powerline transients. A. "quiet" location (underground power feed). B. US composite. C. rural (overhead lines).

9x.25.1 The problem

Take an example: Current-carrying inductors have the potential to blow out semiconductors, because they do not permit a discontinuous change in current. Indeed, the collapsing magnetic field in an inductor generates a voltage to fight that change in current, according to $V = L\,dI/dt$. If you try to stop abruptly the current flowing through an inductor, it will defeat you by generating a voltage large enough to cause breakdown in your switch, or in the connecting wiring, or within the inductor itself; this can easily reach kilovolts, even with harmless-looking inductors. The inductor stores an amount of energy $E = \frac{1}{2}LI^2$, and is prepared to use it all to accomplish its task. That's often enough to destroy even a husky semiconductor.

Power engineers are quite aware of this, and have evolved systems with layers of protective devices, including cascades of gas-filled discharge tubes, varistors, and zeners, combined with current-limiting series impedances ("shunt-mode" protection); alternatively, you sometimes see powerline protection systems that use substantial low-pass series inductors and shunt capacitors, followed by a further stage of power-diode clamping to larger capacitors to limit the peak "let-through" voltage ("series-mode" protection). But designers of plain ordinary circuits need to understand how to prevent damage in circuitry that drives relays, solenoids, or transformers. Destructive impulses can come from outside, as well: the wallplug ac socket routinely has voltage spikes of a few hundred volts (largely from switching of inductive loads such as motors), occasionally reaching a kilovolt or more; a nearby lightning strike will induce momentary transients of 1–10 kV.[91] These can pass through the dc supply's transformer and wind up on the low-voltage dc rails.

9x.25.2 The solution

In the simple case of unipolar drive of an inductive load, the simplest protection is a reverse diode across the inductor, as we showed in Fig. 1.84. However, this won't work where the drive can assume both polarities. In addition it has the disadvantage of slowing the collapse of current (which is proportional to the voltage across the inductor: $dI/dt = V/L$), because it allows only a diode drop (plus the drop across the inductor's internal series resistance).[92] You can use a resistor instead of a diode (or in series with the diode) to speed the decay of current, choosing its resistance such that the peak voltage $V_{pk} = I_0R$ keeps the voltage applied to the semiconductor devices within their ratings; in this case the current decays more quickly, in fact exponentially with a time constant $\tau = L/(R + R_{int})$, i.e., $I(t) = I_0 e^{-[L/(R+R_{int})]t}$, where R_{int} is the inductor's internal series resistance.

For inductive loads powered by ac, you cannot use a shunt diode, because it would conduct during normal powering; instead you can use a resistor, or a series RC. However, these methods contribute additional operating current; this can be avoided by using a bidirectional voltage-limiting clamp device, either a bidirectional zener (a pair of back-to-back zeners in series), or a varistor. These devices are also known as *transient voltage suppressors* (TVS). They are particularly effective in this application, because they clamp the voltage during inductor turn-off, thereby minimizing the decay time of the inductive current for a given allowable voltage. TVSs are also widely used as protection devices for externally produced transients. Let's look at them.

---

[91] Typically you can expect a kilovolt spike 100–1000 times per year, and a 5 kV spike once per year; see Fig. 9x.108.

[92] The voltage drop across the inductor's internal series resistance R_{int} will usually dominate, producing an exponential decay of current with time constant $\tau = L/R_{int}$.

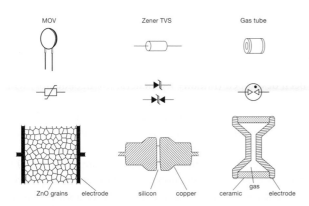

Figure 9x.109. Transient voltage suppressors.

9x.25.3 TVS devices

There are several varieties of TVS, most commonly gas tubes, metal-oxide varistors (MOV), and zener-type clamps. Each has its niche – for example, gas tube devices have very low capacitance, so they're good for signal or RF lines – and you sometimes see them used in combination (for example to protect incoming ac lines at the service entry box). Here is a compact summary (see also Fig. 9x.109) of these three types of surge arrestors:

A. Gas surge arrestors

These devices, also known as gas plasma arrestors or gas discharge tubes (GDTs), consist of a closely spaced electrode pair in a small sealed tube (typically 8–10 mm diameter and 6 mm long) filled with some inert gas. They are triggered at some critical "firing voltage," usually in the range of a few hundred volts, whereupon they go into conduction, holding the voltage at a much lower "arc voltage" of roughly ten volts until the triggering voltage source drops below the "holdover voltage," typically 65% of the firing voltage. They can withstand peak currents of thousands of amperes for an "8/20" test pulse ($8\mu s$ rise-time, exponential decay with $20\mu s$ width; more on this later), and hundreds of amperes for a "10/1000" test pulse ($10\mu s$ rise-time, exponential decay with 1 ms width).

Gas tube devices excel in low capacitance (1–2 pF) and high off-state impedance ($\sim 10^{10}\,\Omega$); but they are limited by relatively high minimum firing voltage (typically 75 V), and by sluggish response, particularly for a slowly rising voltage. They also degrade with use (more precisely, with triggering and consequent pulse energy absorption), a phenomenon known as "wear-out." Gas tube TVSs are widely used to protect telecommunication lines from destructive lightning-induced spikes.[93]

Figure 9x.110. Large transients caused the gas surge arrestor (white cylindrical object on right side) to "fail short"; the lower of the two power resistors then exploded. Lots of smoke!

When gas surge arrestors fail, they can become a short circuit. Figure 9x.110 shows a portion of a commercial 6-outlet surge protector in which the failure of a gas arrestor (induced by transients from the *load* end) caused the series protective device (a $10\,\Omega$, 2 W resistor) to explode, with the release of plenty of black smoke.

B. Metal oxide varistors

Varistors, also called MOVs, consist of a sintered pill of material made primarily of granular zinc oxide, with electrodes coated onto the two surfaces. The individual grains form junctions at their boundaries, each acting like a low-voltage junction that breaks down at 2–3 V. The overall device is effectively a series-parallel assembly of many such junctions; you can think of it as a "statistical bidirectional zener," whose breakdown voltage is set by the thickness of the pill, and whose peak current capability is set by the diameter of the pill (common sizes are 7 mm, 14 mm, and 20 mm). Voltage ratings go from about 10 V to a kilovolt, with peak current ratings of a few thousand amperes for an 8/20 pulse.

MOVs are inexpensive, inherently bidirectional, and fast (<1 ns, effectively limited by lead inductance), with a steep knee, reasonably well controlled breakdown voltage (typically $\pm10\%$), and very high peak current ratings. They are limited by high capacitance, and by wear-out (with consequent reduction of breakdown voltage, as grain boundaries

---

[93] When looking for low-capacitance clamps, don't neglect the alterna-

tive of simple diode clamps to a pair of reference voltages, suitably bypassed.

become shorted); it is worth noting that the "peak pulse current" rating shown on MOV datasheets is often the value for which the unit survives only 1 or 2 such pulses. MOVs are widely used to clamp energetic transients (from large inductive transients and from lightning) on the incoming ac line.[94]

Figure 9x.111. Transient-induced varistor failure. The right-hand varistor "failed short," belching black smoke from two edges.

When a MOV fails, it usually becomes a short circuit. The result can be dramatic – we've seen these things disassemble into blackened pieces, or just get blown to bits. Figure 9x.111 shows the innards of a commercial 2-outlet surge protector after failure of one of its varistors (130 Vac, 10 mm size). It split open along both sides, depositing some black stuff on the circuit board; the series protective fuse opened up before more damage occurred.[95]

Figures 9x.112 and 9x.113 show the circuits of two commercial surge protection outlet strips. They connect MOVs across all three wire pairs (line-neutral, line-ground, and neutral-ground), with series fusing that opens in the event of MOV failure; they include also some series inductance and shunt capacitance, for passive filtering of spikes and high-frequency interference.

---

[94] Though there are good reasons to favor the more expensive approach of series-reactance slew-rate and current limiting, followed by power-diode clamping. Such "series-mode" transient protectors, offered for example by ZeroSurge, SurgeX, and BrickWall, provide enhanced protection. Furthermore, they do not use (sacrificial) MOVs, and offer superior transient suppression in terms of endurance ("Grade A") and let-through ("Class 1"), according to the ANSI C62.41 specification. But these puppies aren't cheap! See §9x.25.5.

[95] The regulatory specifications for surge suppressors were revised in 1998 (UL 1449 Second Edition) to deal with the potential for fire, following some unfortunate incidents.

C. Zener TVSs

Ordinary zener diodes are voltage clamps, though generally not intended for large peak currents or absorption of large pulse energies. However, a variety of zeners have been developed (initially by General Semiconductor) with a large junction area bonded to substantial heat-spreading copper electrodes, to enhance the peak current and pulse energy capabilities of these devices. Popular types are the P6KE and 1.5KE series, in both axial-lead and surface-mount packaging, with voltage ratings from 6.8 V to 540 V, and with both unidirectional and bidirectional polarity types in each voltage (the latter consist of a series pair of zeners). These series are rated at 600 W and 1500 W peak pulse power (10/1000 pulse), respectively.

Zener-type TVSs (also knows as silicon avalanche devices, or SADs) are inexpensive, fast (<1 ns, effectively limited by lead inductance), with a steep knee and very tightly controlled breakdown voltage; they have no inherent wear-out mechanism. They are limited by high capacitance (particularly for low-voltage types), and relatively low peak pulse current (particularly for high-voltage types). To make the latter quantitative, a 20 mm MOV for 115 Vac line protection (type V130LA20C) can tolerate a hundred isolated 8/20 pulses of 1000 A peak current before failure; the analogous zener TVS (type 1.5KE200A) can tolerate only 25 A pulses, but in unlimited number (no wear-out). Zener TVS devices are widely used within electronic circuitry to clamp reactive and switching spikes, overvoltage and reverse-polarity mishaps, and powerline transients of modest energy content.

If you want to use zener TVSs to protect *signal* lines, you've got to do something about the large shunt capacitance. For example, a 1.5K-series SAD to stand off 10 V (1.5KE12A) looks like 3300 pF at 0 V, and 2200 pF at 10 V. One trick is to put a forward rectifier in series with the (unbiased) zener TVS, so that you see the rectifier's (smaller) capacitance; this combination is available commercially as the LCE series, which has properties similar to the 1.5KE series, but with zero-bias capacitance of 100 pF for all member of the series (which go from 6.5 V to 28 V). Another approach is to use (reverse-biased) diode clamps to the bounding voltages; the latter could be formed from forward-biased zeners, or (for transient protection only) from storage capacitors suitably biased. See Figure 9x.114, which shows clamping devices to handle relatively high currents (but which are afflicted with high capacitance).

Recently some nice low-capacitance clamps have been introduced, for protection of signal lines such as USB. Check out parts like the CD143A-series from Bourns; their

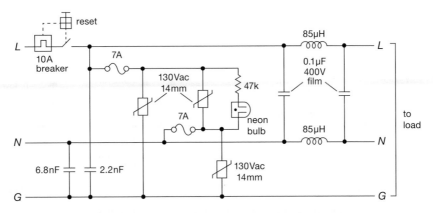

Figure 9x.112. A low-cost 2-outlet surge suppressor using MOVs.

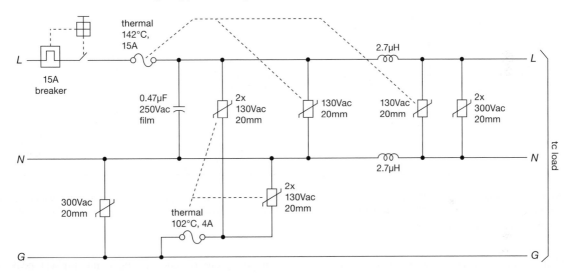

Figure 9x.113. This 10-outlet surge suppressor uses 9 MOVs.

CD143-SR05LC, for example, can clamp up to 10 A or more peak currents in each of its 2 signal lines, while presenting a shunt capacitance of just 3 pF. There's more discussion on low-capacitance protective zeners in Chapter 1x (§1x.7.1G). The very useful CD143 is seen in Figures 1x.136 and 1x.139, and the similar D1213A from Diodes, Inc., in Figures 3x.62 and 4x.147; these primarily protect against human-induced sparks.

As with MOVs, zener TVSs are likely to "fail short" when driven beyond their ratings. We recently decommissioned a hilltop observatory, collecting a half dozen high-quality 6-outlet surge suppressors of commercial manufacture (see Figure 9x.115); they used a combination of MOVs, TVSs, and gas surge arrestors, along with LC filtering. Half of them rattled when shaken! That's because the

TVSs had failed over time, causing the 10 Ω series resistor to explode into pieces.

9x.25.4 MOV versus zener TVS

The manufacturers of MOVs and silicon TVSs are enthusiastic advocates of their respective products, with good datasheets and application notes from companies like Harris (Littelfuse) and Panasonic (for MOVs); and On Semiconductor (formerly Motorola), Vishay (the General Semiconductor line), and Microsemi (for silicon zener TVSs). Both varieties have their strong and weak points. We generally favor zener-type TVSs for most circuit protection applications, particularly at lower voltages; that is because of their tighter tolerances, steeper conduction, and absence of wear-out. MOVs are still preferred, however, in situa-

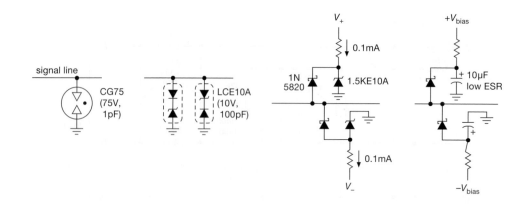

Figure 9x.114. Low-capacitance high-current clamps.

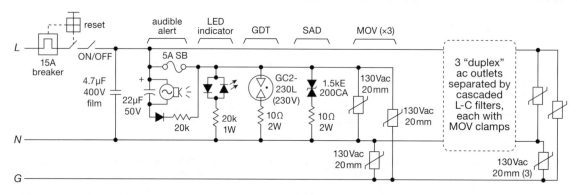

Figure 9x.115. A highly rated 6-outlet transient suppressor. It combines MOVs, zener TVS, and gas surge arrestor, with extravagant downstream *LC* filtering, and a triple-MOV finale. While they were at it, they added an audible "loss of protection" alert, and put the whole thing in a robust extruded aluminum case.

tions where extremely high energy impulses are expected to occur, but only rarely – for example for line voltage protection of occasional large transients that have not been removed by upstream devices. It's perfectly reasonable to combine the advantages of both, as in Figure 9x.115 – "belt and suspenders."

9x.25.5 "Series-mode" transient protection

Transient suppression with shunt MOVs, zener TVSs, or gas tubes has a certain brute-force flavor – when the incoming voltage exceeds a threshold, clamp it with a robust avalanche device, and hope for the best. In practice the "let-through" voltage during an energetic transient can reach 500 V; and, as the photographs attest, they can produce smoky devastation.

One problem is that the avalanche device is being asked to dissipate enormous transient power – several hundred kilowatts, for example, during a 1000 A transient – which invites failure. Another problem is that the fixed avalanche

voltage has to be chosen above the highest peak voltages that are normally encountered; when combined with production tolerances, this dictates a minimum rating of ~ 200 V for use on a 120 V (nominal) ac powerline. The clamping voltage of a standard 20 mm MOV with this voltage rating, for example, will reach 420 V during a 1000 A transient.

An alternative approach is the use of power rectifier clamping to a large capacitor that is charged to the line's peak voltage, combined with an upstream inductor large enough to limit the transient peak currents and slew rates. Figure 9x.116 shows a commercial implementation.[96] Here the relatively large storage capacitor C_1 is charged and held at the peak line voltage (about 170 V for a 120 Vac powerline); brief-duration input transients beyond ± 170 V are clamped by the capacitor. Of course, during the transient the capacitor's voltage is rising, according to

---

[96] See "Power Line Surge Suppressor" patents by J.R. Harford, US 4,870,528 and 4,870,534.

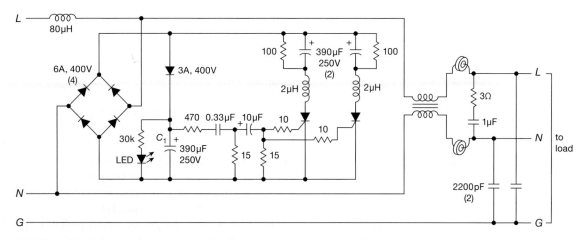

Figure 9x.116. Industrial-strength "series-mode" transient suppressor. Normal-mode transients are diode-clamped to a large capacitor (held at the peak line voltage), with additional uncharged capacitors waiting in the bullpen.

$I = C\,dV/dt$, so you want to use a pretty large capacitor. For example, a current pulse of 500 amperes would cause C_1 to charge at $dV/dt = 1.28\,\mathrm{V}/\mu s$; so a "$8\times20\,\mu s$" transient of this amplitude (see §9x.25.7A) would raise the capacitor's voltage by about 25 volts during the pulse, still an admirable degree of suppression.[97]

To handle larger transients, the circuit of Fig. 9x.116 has another trick up its sleeve: it keeps another pair of capacitors, in an uncharged state, ready to be switched across the downstream line by a pair of SCRs (see §9.1.1C) according to C_1's slew rate (note the *RC* differentiator). Yet another trick, available in some series-mode surge suppressor models, is the use of *feedforward cancellation*: A secondary winding on the input inductor takes a sample of the transient input voltage, which is then connected in series with the output, reversed in phase to cancel the transient.

These series-mode surge suppressors use no sacrificial devices, such as MOVs, and in fact they do not use avalanche devices of any sort. They deliver better suppression and robustness (we had absolutely no success destroying any of our samples). The manufacturers deliberately do not provide clamping to the ground terminal, arguing that conducting large transient currents back through the protective ground line induces large voltage spikes on the ground; such ground transients invite damage to the inputs of interconnected equipment that are not powered from the same "protected" outlet. As a consequence, common-mode

transients are not clamped (but are filtered somewhat in the downstream filter of Fig. 9x.116).

9x.25.6 TVS circuit example

The transients that can arrive, uninvited, on the incoming ac line are poorly characterized, and unpredictable. General practice, as seen in the preceding figures, is to use a combination of an incoming ac line filter, followed by a line-voltage TVS (MOV or zener type), along with fusing to protect against fire, and just hope for the best. Enhanced protection (traded off against size, weight, and cost) comes with the use of series-mode suppressors. Using a commercial surge suppressor device on the incoming power adds some security. To take some of the guesswork out of this, you can design and test for survival against several "standard transient" waveforms, described later; but in this business there are no guarantees.

On the other hand, transients generated by understood phenomena in your own circuitry can be well characterized. Let's look at a real example from an instrument we designed.

A. Fast-switching magnet

For a certain sodium ion trapping experiment we needed to energize a robust copper coil with 875 A of dc current (stable to 10 ppm), then bring it rapidly to zero. The coil's inductance was $2.3\,\mu H$, and the inductance of the connecting cables and circuit wiring brought the total inductance to $10\,\mu H$; the combined resistance was less than $10\,\mathrm{m}\Omega$. The idea is to use paralleled n-channel power MOSFETs to pull the low side of the coil to ground, with the high

---

[97] We've ignored the effect of the capacitor's parasitic series reactance – ESR and ESL (see §§1x.3.2 and 1x.3.3) – which permit an additional voltage drop of $\Delta V = R_s I + L_s dI/dt$; these figure importantly in the capacitor selection.

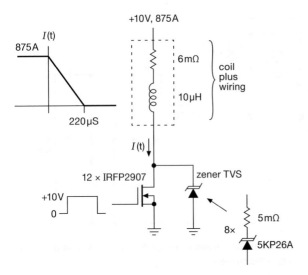

Figure 9x.117. TVS clamp for high-current electromagnet. Not shown are the long (and inductive!) wiring to the remote magnet coil (see for example Fig. 3x.125).

side powered from a highly stable 875 A current source whose output was limited to +10 V. Upon switching off the MOSFETs, of course, the inductance of the coil and wiring strives to maintain this very high current by bringing its low side to a large positive voltage, destroying the MOSFETs (and probably everything nearby!); we satisfy its craving for a current path by connecting a TVS to ground.

Figure 9x.117 shows the switches and zener TVS transient clamp. For the switch we used 24 pieces of IRCP054 (60 V, 70 A, 14 mΩ) MOSFETs, but if doing the design now we'd choose the inexpensive IRFP2907, an "automotive MOSFET" with a pleasant combination of maximum voltage, current, and power dissipation, and low R_{ON}, namely 75 V, 90 A, 400 W, and 4.5 mΩ. A dozen of these in parallel, operating as common-source switches, will handle the current, and the power dissipation in each is a very modest 25 W. When the switches are opened the TVS will clamp the drain voltage to approximately 50 V, and the inductor current will decay approximately[98] as a ramp, given by $V = L\,dI/dt$.

Let's figure out what's needed. From L and V_{clamp} we find that the current decays at $dI/dt = 4\,\text{A}/\mu\text{s}$, so it goes from 875 A to zero in 220 μs. And so the energy stored in the inductor, $E = LI^2/2 = 3.8\,\text{J}$, is delivered to the clamp over that period, as shown.

Now we look at the datasheets for zener TVS devices.

---

[98] The zener clamp voltage depends somewhat on current, so the ramp isn't perfect.

We might start with the 1.5KE36A, from the popular series of 1.5 kW peak power devices (for the standard 1 ms exponential decaying pulse), which specifies a maximum clamp voltage of 50 V at 30 A. Because our pulse is shorter than the 1 ms reference pulse, we read from the datasheet's curves that the peak power (for the half-width of our pulse) is 4 kW, which corresponds to 80 A peak current. Owing to the linear (rather than exponential) current profile, we gain an additional factor of 1.44, for a rated peak current of 115 A. To handle the 875 A peak current, then, we need to use eight of these in parallel (with small "ballasting" resistors to equalize the current; something like 25 mΩ, producing a 3 volt drop, will do nicely).

For conservative design we would probably specify a dozen TVSs in parallel. That's a lot of parts, and so a better route is to choose a larger TVS. The 5KP or 15KP series (5 kW and 15 kW peak power, 1 ms) would be good choices: with no safety margin we would need to parallel three 5KP30As, or we could use a single 15KP30A. In practice we wound up using eight 5KP26As in parallel, each with a Dale 0.05 Ω LVR-3 non-inductive ballast resistor. An additional 0.47 μF with 3.3 Ω in parallel was required to suppress ringing at the end of the fast current shutoff.[99]

9x.25.7 Transient test circuit

When designing circuits to deal with such prodigious peak power, it is often worth building an instrument to generate well-characterized inductive current transients. Then you can measure the properties of the transient suppressors, and the performance of the overall transient suppression circuit. For example, you can measure the actual clamp voltage and current waveforms, useful for determining appropriate ballasting resistor values to accommodate the production spread of clamping voltages at the peak operating current. You can use the tester also to "burn in" the circuit, with a series of transients larger than will be encountered in the actual application, to ensure reliability. It's far better to torture your device to failure on the test bench, rather than find out from disasters in the field that you overstressed your semiconductors or skimped on protection devices.

Fig. 9x.118 shows an example. A large bulk storage electrolytic capacitor of low ESL and ESR delivers current through an inductor L that should match that in the intended application circuit. Look for changes in the clamp

---

[99] For more information, ask about the RIS-254 instrument (email to winfieldhill@yahoo.com), which included two independent 900 A switches, and is rated at 1800 A in parallel operation.

voltage as the test device is repeatedly pulsed, preferably under conditions somewhat more severe than in the target circuit.

As seen in Figure 9x.119, when testing at very high voltages, it's better to use an IGBT switch. In that case it's a good idea to include some circuitry to protect the IGBT if the TVS "fails open" – for example, a power diode from the high side (the collector) to a bulk capacitor that is precharged to a voltage higher than V_{DD}, but less than the IGBT's maximum collector voltage. (You could do the same for MOSFETs, but in many cases you can probably rely on its avalanche rating.)

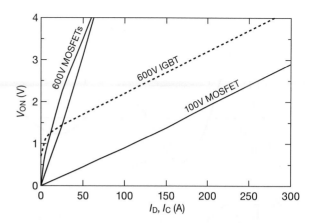

Figure 9x.119. For testing TVS devices at low voltages (say <200 V), a MOSFET is a satisfactory switch for Fig. 9x.118's tester. But at higher voltages an IGBT is far better, because it saturates well, compared with the anemic performance of a MOSFET of comparable voltage ratings.

ANSI have enshrined these in a set of "standard impulses" (Fig. 9x.120):

0.5 μs–100 kHz ring wave This is meant to represent the *voltage* waveform seen on indoor ac wall outlets, caused by distant lightning or distribution line switching. The standard pulse[100] rises to its peak voltage (often specified as 6 kV) in 0.5 μs, then decays as a low-Q damped 100 kHz oscillation.

1.2×50 μs and 8×20 μs pulses These are unipolarity pulses, intended to simulate lightning induced transients closer to the source, for example near the ac "service entrance," or outdoors.[101] The 1.2×50 μs pulse is an open-circuit *voltage* pulse, rising in 1.2 μs and then decaying exponentially to half amplitude in 50 μs; the 8×20 μs pulse is an analogous *current* pulse into a low-impedance load. In both cases you specify the amplitude, e.g., a certain 20 mm MOV is specified to survive an isolated "6500A peak current 8×20 μs pulse."

10×1000 μs pulse This is a unipolarity pulse meant to simulate inductive transients from motors and transformers. It is a *current* waveform seen by a low-impedance load, risetime of 10 μs and decay time (to 50% amplitude) of 1000 μs; this standard waveshape is specified by its peak current.

human body model This models, crudely, the discharge of an electrostatically charged person – a serious business in ensuring longevity of delicate integrated circuits.

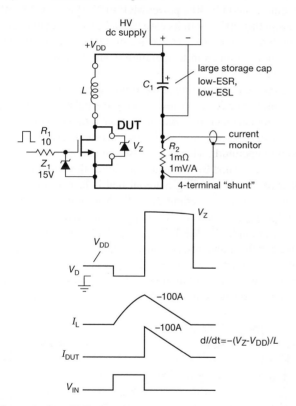

Figure 9x.118. TVS tester. The heavy lines show the high-current path through the MOSFET (or IGBT) switch and the TVS "device under torture" (DUT). For voltages above 200 V or so, substitute an IGBT for the MOSFET shown here.

A. Standard test pulses

In contrast to the predictable transients you get when shutting off an inductor, you don't have much control over what comes in the ac power line. Over time the power industry has gained experience with the sort of stuff you can expect from lightning, from inductive transients (motors for example), and the like. Standards groups like IEEE, IEC, and

---

[100] Officially known as IEEE std C62.41 "category A."
[101] IEEE C62.41 categories B and C, respectively.

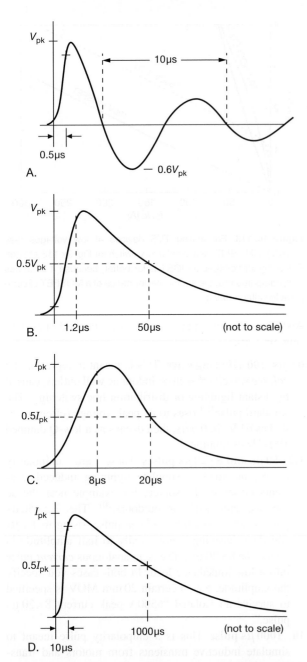

Figure 9x.120. Standard Pulses. A. 0.5 μs–100 kHz ring wave. B. 1.2×50 μs pulse. C. 8×20 μs pulse. D. 10×1000 μs pulse.

We saw this in §§3.5.4H and 12.1.5: it consists of a 100 pF capacitor in series with a 1.5k resistor, initially charged to one of several choices of voltage (e.g., 2 kV, 8 kV, or 15 kV). When discharged into a (noninductive) victim that brings it close to ground, the resulting current waveform is a step (to $I = V_{peak}/1.5k$), with an expo-

nential discharge of time constant $\tau = RC = 150$ ns (not shown in Fig. 9x.120).

9x.25.8 Transient thermal response

Power semiconductors operating with some steady-state power dissipation require heatsinking to remove the heat by thermal conduction and convection. As explained in the power chapter (Chapter 9), power devices have a *thermal resistance* from junction to case, $R_{\Theta JC}$ (sometimes called Θ_{JC}), with units of °C/W, so that the temperature rise of the junction relative to the case is just $\Delta T = P_{diss}R_{\Theta JC}$. Thus, for example, an IRF540N power MOSFET ($R_{\Theta JC} = 1.2$°C/W) dissipating 50 W, with its case at 50°C, has a junction temperature of $T_J = T_C + P_{diss}R_{\Theta JC}$, or 110°C; at 104 W dissipation the junction temperature would be the maximum allowed, 175°C.

Here's the interesting news: if you apply a short pulse of power, the peak power level during the pulse can be quite a bit higher than the maximum steady-state power whose removal by conduction results in maximum specified junction temperature. That is because the energy of the pulse can be absorbed by the heat capacity of the junction, plus that of the surrounding material that is heated during the pulse, without exceeding T_{Jmax}.[102] You often see a graph of "peak pulse power" versus pulse duration; Fig. 9x.121 is an example, for the 1.5KE zener TVS series. Engineers like familiar concepts, so you sometimes see them converting this into an effective R_Θ, which they call "transient thermal impedance," or "thermal response"; Fig. 9x.122 is an example, for the same 1.5KE device.

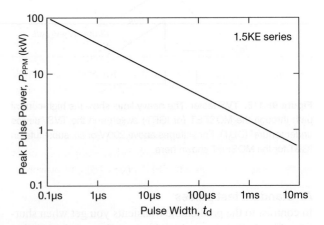

Figure 9x.121. Peak pulse power P_{PPM} *vs* 1/*e* pulse width t_d for 1.5KE-type TVS.

---

[102] Recall §1x.2.6, the analogous situation with power resistors.

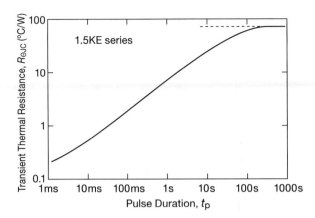

Figure 9x.122. Transient thermal resistance $R_{\Theta JC}$ *vs* width of rectangular pulse t_p for 1.5KE-type TVS. For long pulses $R_{\Theta JC}$ is asymptotic to the dc value of 75°C/W.

Some interesting things to note: First, the effective thermal capacity increases roughly as the square root of pulse duration (Fig. 9x.122), so the maximum allowable pulse energy also rises as $\sqrt{\tau}$; this means that the peak pulse *power* decreases as $1/\sqrt{\tau}$, as shown in Fig. 9x.121.

Second, in a series of MOVs of given disk diameter, for which the disk thickness increases proportional to clamping voltage rating, the short-pulse *current* rating is independent of the clamping voltage rating; for example, a family of 20 mm diameter MOVs is rated at 6500A (for an $8\times20\,\mu$s pulse). In other words, within a family of a given diameter, the peak *power* (or total pulse energy) increases proportional to the clamping voltage rating; this makes sense, since the volume of MOV material is proportional to the clamping voltage rating. In contrast, for a device like the zener TVS, with a standard silicon die size that does not depend upon clamp voltage rating, it is the short-pulse *power* (or total pulse energy) rating that is independent of clamping voltage rating – for example 1500 W (for a $10\times1000\,\mu$s pulse) for the 1.5KE series. This makes sense, also, because a fixed silicon die size implies a fixed pulse energy rating. A consequence is that peak pulse *current* rating decreases proportional to $1/V_{\text{clamp}}$.

Similar thermal pulse considerations apply to other power devices, such as MOSFETs (§3x.13) and power resistors (§1x.2.6).

PARTS INDEX

KEY

suffixes	ff	"and pages following"
	n	footnote
	g	graph
	r	review section
	p	photograph
	s	screen shot ('scope or spectrum analyzer)
	t	table
	T	tabular
	boldfaceff	main entry, folios following
	(box)	in a box

1N4937, 86g

SUBJECT INDEX